Smart Innovation, Systems and Technologies

Volume 207

The Smart Innovation, Systems and Technologies book series encompasses the topics of knowledge, intelligence, innovation and sustainability. The aim of the series is to make available a platform for the publication of books on all aspects of single and multi-disciplinary research on these themes in order to make the latest results available in a readily-accessible form. Volumes on interdisciplinary research combining two or more of these areas is particularly sought.

The series covers systems and paradigms that employ knowledge and intelligence in a broad sense. Its scope is systems having embedded knowledge and intelligence, which may be applied to the solution of world problems in industry, the environment and the community. It also focusses on the knowledge-transfer methodologies and innovation strategies employed to make this happen effectively. The combination of intelligent systems tools and a broad range of applications introduces a need for a synergy of disciplines from science, technology, business and the humanities. The series will include conference proceedings, edited collections, monographs, handbooks, reference books, and other relevant types of book in areas of science and technology where smart systems and technologies can offer innovative solutions.

High quality content is an essential feature for all book proposals accepted for the series. It is expected that editors of all accepted volumes will ensure that contributions are subjected to an appropriate level of reviewing process and adhere to KES quality principles.

Indexed by SCOPUS, EI Compendex, INSPEC, WTI Frankfurt eG, zbMATH, Japanese Science and Technology Agency (JST), SCImago, DBLP.

All books published in the series are submitted for consideration in Web of Science.

More information about this series at https://link.springer.com/bookseries/8767

Yuzo Iano · Osamu Saotome ·
Guillermo Leopoldo Kemper Vásquez ·
Claudia Cotrim Pezzuto ·
Rangel Arthur · Gabriel Gomes de Oliveira
Editors

Proceedings of the 7th Brazilian Technology Symposium (BTSym'21)

Emerging Trends in Human Smart and Sustainable Future of Cities (Volume 1)

Springer

Editors
Yuzo Iano
Faculty of Electrical
and Computer Engineering
Unicamp
Campinas, São Paulo, Brazil

Guillermo Leopoldo Kemper Vásquez
Universidad Peruana de Ciencias Aplicadas
Santiago de Surco, Lima, Peru

Rangel Arthur
Faculty of Electrical
and Computer Engineering
Unicamp
Campinas, São Paulo, Brazil

Osamu Saotome
Divisão de Engenharia Eletrônica
Instituto Tecnológico de Aeronáutica
São José dos Campos, São Paulo, Brazil

Claudia Cotrim Pezzuto
Pontifícia Universidade Católica de Campinas
Campinas, Brazil

Gabriel Gomes de Oliveira
Universidade Estadual de Campinas
Hortolândia, Brazil

ISSN 2190-3018 ISSN 2190-3026 (electronic)
Smart Innovation, Systems and Technologies
ISBN 978-3-031-04434-2 ISBN 978-3-031-04435-9 (eBook)
https://doi.org/10.1007/978-3-031-04435-9

This Springer imprint is published by the registered company Springer Nature Switzerland AG
The registered company address is: Gewerbestrasse 11, 6330 Cham, Switzerland

Foreword

With great satisfaction, I write this Foreword for the Proceedings of the 7th Brazilian Technology Symposium–Emerging Trends and Challenges in Technology (BTSym'21), held virtually, for the second time, at the PUC Campinas University, Brazil, in November 2021, and the UNAPUNO University, Peru, in January 2022. This event is in its sixth edition and has consolidated to become an excellent opportunity for researchers, professors, and students to present and discuss the results of their research works.

In the 2021 edition, the BTSym activities could not be accomplished in person due to the COVID-19 pandemic. However, the event has been characterized since its first edition by the broad scope of the areas exposed and, within a virtual environment, it was possible to expand our network of researchers and encourage them to expose their papers, which deal with current and priority topics for Brazilian and world technological development, including subjects related to the various branches of innovation in industrial processes, robotics, telecommunications, buildings, urban infrastructure, product development, and biomedicines.

Events such as BTSym are an essential part of the research and innovation process. Firstly, these events contribute to the promotion of research activities, which are key to a country's technological development. The dissemination of research results, as promoted by BTSym, contributes to the transformation of research findings into technological innovation. In addition, these events facilitate the sharing of findings, leading eventually to the formulation of research networks, which accelerate the achievement of new results. Therefore, I would like to congratulate the BTSym General Chair, Prof. Dr. Yuzo Iano, and his group of collaborators for the important initiative of organizing the BTSym'21 and for providing the opportunity for authors to present their work to a wide audience through this publication. Finally, I congratulate the authors for the high-quality work presented in these proceedings.

Gabriel Gomes de Oliveira
Technical Program and Finance Chair
of Brazilian Technology Symposium

Preface

This book contains the Proceedings of the 7th Brazilian Technology Symposium–Emerging Trends and Challenges in Technology, held in Brazil in November 2021 and Peru in January 2022.

The Brazilian Technology Symposium is an excellent forum for presentations and discussions of the latest results of projects and development research in several areas of knowledge, in scientific and technological scope, including smart designs, sustainability, inclusion, future technologies, architecture and urbanism, computer science, information science, industrial design, aerospace engineering, agricultural engineering, biomedical engineering, civil engineering, control and automation engineering, production engineering, electrical engineering, chemical engineering, and probability and statistics.

This event seeks to bring together researchers, students, and professionals from the industrial and academic sectors, seeking to create and/or strengthen the linkages between issues of joint interest. Participants were invited to submit research papers with methodologies and results achieved in scientific level research projects, completion of course work for graduation, dissertations, and theses.

The 67 full chapters accepted for this book were selected from 208 submissions, and, in each case, the authors were guided by an experienced researcher with a rigorous peer-view process. Among the main topics covered in this book, we can highlight manufacturing processes, lean manufacturing, industrial costing models, sustainability and productivity, circular economy, workplace safety, control systems, Internet of Things, cyber-physical systems, transportation management system (TMS), logistic services analysis, digital supply chain, socio-economic impacts of technologies 4.0, robotics applications, artificial neural networks, big data, deep learning, computational vision, cybersecurity, soft-computing methodologies, technologies applied to cities development, smart cities, energy sustainability, building information modeling (BIM), environment analysis, technologies applied to health, biomedical innovations, socio-economic impacts of COVID-19, technologies applied to education, academic development, civil aviation studies, and much more.

We hope you enjoy and take advantage of this book and feel motivated to submit your papers, in the future, to Brazilian Technology Symposium.

Best wishes,

Alex Midwar Rodriguez Ruelas
Proceedings Chair of Brazilian Technology Symposium

Acknowledgements

Our appreciation goes to a lot of colleagues and friends who assisted in the development of this book, Proceedings of the 7th Brazilian Technology Symposium–Emerging Trends and Challenges in Technology (BTSym'21).

First of all, I would like to thank all the members of the organizing and executive committee for the commitment throughout the year. Several meetings were held, and many challenges were overcome for the accomplishment of the BTSym 2021. Also, and with great merit, I would like to thank all the scientific and academic committee and technical reviewers committee members for their excellent work, which was essential to ensure the quality of our peer-review process, collaborating with the visibility and technical quality of the BTSym 2021.

The Brazilian Technology Symposium is an event created by the Laboratory of Visual Communications of the Faculty of Electrical and Computer Engineering of the University of Campinas (UNICAMP). In this way, I would like to thank the PUC Campinas and UNAPUNO Universities, especially for supporting and hosting the BTSym'21 and BTSym'21 Satellite, respectively, which was fundamental for the successful accomplishment of the events.

Finally, on behalf of Prof. Yuzo Iano, the General Chair of the Brazilian Technology Symposium, I thank all the authors for their participation in the BTSym'21; we sincerely hope to have provided a very useful and enriching experience in the personal and professional lives of everyone.

Best wishes,

Gabriel Caumo Vaz
Institutional Relationship Chair
of Brazilian Technology Symposium

Contributors

Organizing Committee

Alex Rodriguez Ruelas (Proceedings Chair)	LCV/DECOM/FEEC/UNICAMP
Alysson Gomes de Oliveira (Marketing Chair)	LCV/DECOM/FEEC/UNICAMP
Ana Cláudia Seixas (Vice-Associate-General Chair BTSym)	UNIFAL
Claudia Cotrim Pezzuto (Vice-Associate-General Chair BTSym)	PUC/UNICAMP
David Minango (Institutional Relationship Chair)	LCV/DECOM/FEEC/UNICAMP
Gabriel Gomes de Oliveira (Technical Program and Finance Chair)	LCV/DECOM/FEEC/UNICAMP
Lisber Arana (Institutional Relationship Chair)	LCV/DECOM/FEEC/UNICAMP
Osamu Saotome (Associate-General Chair BTSym)	ITA
Rangel Arthur (Vice-General Chair BTSym)	FT/UNICAMP
Yuzo Iano (General Chair BTSym and WSGE)	LCV/DECOM/FEEC/UNICAMP

Executive Committee

Abel Dueñas Rodríguez (Midia Chair)	LCV/DECOM/FEEC/UNICAMP
Airton Vegette (Institutional Relationship Chair)	LCV/DECOM/FEEC/UNICAMP
Angélica F. G. (Institutional Relationship Chair)	LCV/DECOM/FEEC/UNICAMP
Daniel B. Katze (Institutional Relationship Chair)	LCV/DECOM/FEEC/UNICAMP
Daniellle Thiago Ferreira (Editorial Committee Chair)	LCV/DECOM/FEEC/UNICAMP
Elizangela Santos Souza (Editorial Committee Chair)	LCV/DECOM/FEEC/UNICAMP
Gabriel Caumo Vaz (Institutional Relationship Chair)	LCV/DECOM/FEEC/UNICAMP
Jennifer Chuin Lee (Designer Chair)	LCV/DECOM/FEEC/UNICAMP
João Carlos Gabriel (Campinas-Vice-Associate-General Chair BTSym)	Universidade Presbiteriana Mackenzie
Leticia Cursi (Institutional Relationship Chair)	LCV/DECOM/FEEC/UNICAMP
Lucas Alves (Institutional Relationship Chair)	LCV/DECOM/FEEC/UNICAMP
Luiz Vicente F. de Mello Filho (Campinas-Vice-Associate-General Chair BTSym)	Universidade Presbiteriana Mackenzie
Mariana Melo (Institutional Relationship Chair)	LCV/DECOM/FEEC/UNICAMP
Paulo Roberto dos Santos (Vice-Associate-General Chair BTSym)	UniMetrocamp

Raquel J. Lobosco (Vice-Associate-General Chair BTSym)	UFRJ
Thais Paiao (Institutional Relationship Chair)	LCV/DECOM/FEEC/UNICAMP
Telmo Cardoso Lustosa (Local Arrangements Chair)	LCV/DECOM/FEEC/UNICAMP
Ubiratan Matos (Institutional Relationship Chair)	LCV/DECOM/FEEC/UNICAMP

Scientific and Academic Committee

Alessandra Cristina Santos Akkari	Universidade Presbiteriana Mackenzie
Ana Cláudia Seixas	UNIFAL
Angela del Pilar Flores Granados	FEA/UNICAMP
Antonio Carlos Demanboro	PUC CAMPINAS
Celso Iwata Frison	PUC/Minas-Poços de Caldas
Cláudia Cotrim Pezzuto	PUC CAMPINAS
David Bianchini	LCV/DECOM/FEEC/UNICAMP
Edgard Luciano Oliveira da Silva	Universidade Estadual do Amazonas (UEA)
Edwin Valencia Castillo	Universidad Nacional de Cajamarca
Ernesto Karlo Celi Arevalo	UNPRG, Lambayeque, Perú
Erwin Junger Dianderas Caut	Instituto de Investigaciones de la Amazonía Peruana-IIAP
Fábio Menegatti de Melo	PUC CAMPINAS
Gabriela Fleury Seixas	UEL
Grimaldo Wilfredo Quispe Santivañez	UERJ
Hugo Enrique Hernandez Figueroa	DECOM/FEEC/UNICAMP
Janito Vaqueiro Ferreira	DMC/FEM/UNICAMP
Jessie Leila Bravo Jaico	UNPRG, Lambayeque, Perú
João Carlos Gabriel	Universidade Presbiteriana Mackenzie
José Hiroki Saito	UFSCAR
Lia Toledo Moreira Mota	PUC CAMPINAS
Lucielen Santos	PURG
Luiz Vicente F. de Mello Filho	Universidade Presbiteriana Mackenzie

Marcos Fernando Espindola	IFSP São Paulo
Maria Thereza de Moraes Gomes Rosa	Universidade Presbiteriana Mackenzie
Marina Lavorato de Oliveira	PUC CAMPINAS
Néstor Adolfo Mamani Macedo	Universidad Nacional Mayor de San Marcos
Paulo Roberto dos Santos	UniMetrocamp
Osamu Saotome	ITA
Rangel Arthur	FT/UNICAMP
Raquel J. Lobosco	UFRJ
Silva Neto	UERJ
Suelene Silva Mammana	Universidade Presbiteriana Mackenzie
Talía Simões dos Santos	FT/UNICAMP
Telmo Cardoso Lustosa	LCV/DECOM/FEEC/UNICAMP
Victor A. M. Montalli	Faculdade São Leopoldo Mandic
Victor Murray	Universidad de Ingenieria y Tecnologia–UTEC

Technical Reviewers Committee

Abel Alejandro Dueñas Rodriguez	LCV/DECOM/FEEC/UNICAMP
Adao Boava	Universidade Federal de Santa Catarina-UFSC
Agord de Matos Pinto Júnior	DESIF/FEEC/UNICAMP
Airton José Vegette	LCV/DECOM/FEEC/UNICAMP
Alessandra Cristina Santos Akkari	Universidade Presbiteriana Mackenzie
Alex R. Ruelas	LCV/DECOM/FEEC/UNICAMP
Alex Restani Siegle	LCV/DECOM/FEEC/UNICAMP
Alysson Gomes De Oliveira	LCV/DECOM/FEEC/UNICAMP
Amilton da Costa Lamas	PUC CAMPINAS
Ana Cláudia Seixas	UNIFAL
Angela del Pilar Flores Granados	FEA/UNICAMP
Antônio José da Silva Neto	IPRJ/UERJ
Celso Fabrício Correia de Souza	LCV/DECOM/FEEC/UNICAMP
Cesar Henrique Cordova Quiroz	PUC CAMPINAS
Cláudia Cotrim Pezzuto	PUC CAMPINAS
Daniel Katz Bonello	LCV/DECOM/FEEC/UNICAMP
Daniel Rodrigues Ferraz Izario	LCV/DECOM/FEEC/UNICAMP
Daniela Helena Pelegrine Guimarães	EEL/USP

David Allan Ibarra	Universidad de las Fuerzas Armadas ESPE
David Bianchini	LCV/DECOM/FEEC/UNICAMP
David Minango	LCV/DECOM/FEEC/UNICAMP
Diego Arturo Pajuelo	LCV/DECOM/FEEC/UNICAMP
Douglas do Nascimento	Marie Skłodowska-Curie Actions (MSCA)
Edgard Luciano Oliveira da Silva	EST/UEA
Edson Camilo	Eldorado Institute
Euclides Lourenço Chuma	LCV/DECOM/FEEC/UNICAMP
Everton Dias de Oliveira	UNIMEP
Fabiana da Silva Podeleski	UNISAL
Fábio Menegatti de Melo	PUC CAMPINAS
Francisco Fambrini	UFSCAR
Gabriel Caumo Vaz	LCV/DECOM/FEEC/UNICAMP
Gabriel Gomes de Oliveira	LCV/DECOM/FEEC/UNICAMP
Gabriela Fleury Seixas	UEL
Guilherme Barbosa Lopes Júnio	UFPE
João Carlos Gabriel	Universidade Presbiteriana Mackenzie
Josué Marcos de Moura Cardoso	LCV/DECOM/FEEC/UNICAMP
Juan Minango Negrete	LCV/DECOM/FEEC/UNICAMP
Jullyane Figueiredo	UFSC
Leonardo Bruscagini de Lima	LCV/DECOM/FEEC/UNICAMP
Leticia Dias Gomes	UDESC
Lisber Arana Hinostrosa	LCV/DECOM/FEEC/UNICAMP
Lucas Heitzmann Gabrielli	FEEC/UNICAMP
Luigi Ciambarella Filho	Universidade Veiga de Almeida/Develop Biotechnology
Luis Fernando Gonzalez	KonkerLabs
Luiz Antonio Sarti Junior	UFSCAR
Luiz Vicente Figueira de Mello Filho	Universidade Presbiteriana Mackenzie
Marcelo Jara	Eldorado Institute
Marcos Fernando Espindola	IFSP São Paulo
Maria Cecilia Luna	LCV/DECOM/FEEC/UNICAMP
Maria Thereza de Moraes Gomes Rosa	Universidade Presbiteriana Mackenzie
Marcius Fabius Henriques de Carvalho	PUC CAMPINAS
Miriam Tvrzska de Gouvea	Universidade Presbiteriana Mackenzie
Murilo Cesar Perin Briganti	LCV/DECOM/FEEC/UNICAMP
Osamu Saotome	ITA
Polyane Alves Santos	Instituto Federal Da Bahia
Rangel Arthur	INOVA/FT/UNICAMP

Raquel Jahara Lobosco	Federal University of Rio de Janeiro
Ricardo Barroso Leite	LCV/DECOM/FEEC/UNICAMP
Roger Prior Gregio	LCV/DECOM/FEEC/UNICAMP
Rosivaldo Ferrarezi	UNIP
Suelene Silva Piva	Universidade Presbiteriana Mackenzie
Telmo Cardoso Lustosa	LCV/DECOM/FEEC/UNICAMP
Victor Angelo Martins Montalli	Faculdade São Leopoldo Mandic–SLMANDIC

Contents

Design and Fabrication of Human-Powered Vehicle - A Measure for Healthy Living 1
Y. Thiagarajan, G. Palanivel, I. D. Soubache, Gabriel Gomes de Oliveira, Yuzo Iano, Gabriel Caumo Vaz, and Himanshu Monga

Identification and Analysis of Bacterial Species Present in Cow Dung Fed Microbial Fuel Cell 16
Y. Thiagarajan, Gabriel Gomes de Oliveira, Yuzo Iano, and Gabriel Caumo Vaz

Sentiment Analysis in the Ecuadorian Presidential Election 25
Juan Carlos Minango Negrete, Yuzo Iano, Pablo David Minango Negrete, Gabriel Caumo Vaz, and Gabriel Gomes de Oliveira

5G - Automation of Vertical Systems in the Industry 4.0 35
Daniel Izario, João Brancalhone, Yuzo Iano, Gabriel Gomes de Oliveira, Gabriel Caumo Vaz, and Karine Izario

Study of Sustainability Indicators for Environmental Management in Hydrographic Basins 44
Débora Akemi Ide Zancanela, Denise Helena Lombardo Ferreira, Lucas Antônio Gomes do Carmo, Maria Luisa Damas Silva, and Cibele Roberta Sugahara

A New Home Energy Management System for Smart Home Using Improved Multi-objective Antlion Optimization Algorithm 56
Navid Razmjooy, Danial Bahmanyar, V. Rajinikanth, and Gabriel Gomes de Oliveira

Study About the Applicability of Low Latency in HAS Transmission Systems 73
Paolo Rodrigo de Oliveira Bacega, Yuzo Iano, Bruno Campos Simoni de Carvalho, Gabriel Caumo Vaz, Gabriel Gomes de Oliveira, and Euclides Lourenço Chuma

Simulation of the Implementation of Domestic Solar Systems Using Multi-agent Systems from Web Scraping 88
João Emmanuel D.'Alkmin Neves, Paulo Sérgio Martins Pedro, Marli de Freitas Gomes Hernandez, and Luiz Ariovaldo Fabri Junior

Mechanism of Collecting Urban Data for Application on Smart Cities 97
Jemis Dievas José Manhiça and Ademar Takeo Akabane

Exploring Research Trends in Healthcare 4.0 106
Bárbara Santiago de Mendonça and Lásara Fabrícia Rodrigues

Logistics 4.0 for Today and Tomorrow: How to Identify Smart Factory Challenges and Trends 115
Fábio Luis Fernandes Azarias and Lia Toledo Moreira Mota

Active Power Filter Design Using Neural Network with a Variable Step-Size 123
Alejandro Alvarez, Joseph Cherre, Christian Flores, and Julien Noel

Cross Mutual Information Analysis of EEG Signals for Cognitive Task Discrimination 133
Joaquin Peralta and Christian Flores

Development of a Low-Cost Open Platform for Vital Signs Acquisition and Monitoring 143
Davi F. Buratto and Frank Herman Behrens

Investigation of the Influence of Turbulence Models on Cough Droplet Evaporation: Comparing (SST) k-Ω, k-ε, and Reynolds Stress (RSM) Turbulence Models 151
Nuhu Ayuba and Gabriela Cantarelli Lopes

Regulatory Aspects of 5G and Perspectives in the Scope of Scientific and Technological Policy 163
Antonio Carlos Demanboro, David Bianchini, Yuzo Iano, Gabriel Gomes de Oliveira, and Gabriel Caumo Vaz

6G Networks: An Innovative Approach, but with Many Challenges and Paradigms, in the Development of Platforms and Services in the Near Future 172
Antonio Carlos Demanboro, David Bianchini, Yuzo Iano, Gabriel Gomes de Oliveira, and Gabriel Caumo Vaz

Sustainability Indicators for Water Security Management in Watersheds 188
Jakeline Pertile Mendes, Denise Helena Lombardo Ferreira, and Cibele Roberta Sugahara

A Study about Automated Optical Inspection: Inspection Algorithms Applied in Flexible Manufacturing Printed Circuit Board Cells Using the Mahalanobis Distance Method 1 198
Daniel Katz Bonello, Yuzo Iano, Umberto Bonello Neto, Gabriel Gomes de Oliveira, and Gabriel Caumo Vaz

Proposal of Mathematical Models for a Continuous Flow Electric Heater 213
Alex Restani Siegle, Yuzo Iano, Gabriel Gomes de Oliveira, and Gabriel Caumo Vaz

Revision of the 5G Concept Rollout and Its Application in Smart Cities: A Study Case in South America 229
Pablo Minango, Yuzo Iano, Euclides Lourenço Chuma, Gabriel Caumo Vaz, Gabriel Gomes de Oliveira, and Juan Minango

Mathematical Modeling: A Conceptual Approach of Linear Algebra as a Tool for Technological Applications 239
Leonardo Bruscagini de Lima, Yuzo Iano, Gabriel Gomes de Oliveira, Gabriel Caumo Vaz, Alecssander Daniel de Almeida, Gustavo Bertozzi Motta, Gabriel Matsumoto Villaça, Matias Oliveira Schwarz, and Pedro Y. Noritomi

Black-Shoes-Merton Model and Neural Networks in River Level Prediction: Case Study on La Leche River - Peru 249
Diana Mercedes Castro Cárdenas, Segundo Francisco Segura Altamirano, and Merly Liliana Yataco Bernaola

Sentiment and Emotions Analysis of Tweets During the Second Round of 2021 Ecuadorian Presidential Election 257
Juan Carlos Minango Negrete, Yuzo Iano, Pablo David Minango Negrete, Gabriel Caumo Vaz, and Gabriel Gomes de Oliveira

Spectroscopy Technologies for Biological Samples Throughout History 269
Talita Cristina Colomeu, Marina de Gea Neves, and Euclides Lourenço Chuma

Technology for Electric Bus in the Brazilian Scenario: Focus on the Adoption of National Components 276
Lilian Regis Laraia, Yuzo Iano, Ricardo Takahira, Luiz Vicente Figueira de Mello Filho, Gabriel Gomes de Oliveira, and Gabriel Caumo Vaz

ICT Infrastructure Usability Model in the Sustainability of Computational Equipment 286
José Fortunato Zuloaga-Cachay, Gilberto Carrión-Barco, Denny John Fuentes, Jessica Karin Solano-Cavero, and Gisella Luisa Elena Maquen-Niño

Challenges and Perspectives of the Furniture Market in Brazil 294
Luiz Vicente Figueira de Mello Filho and Italo Galvani

Study of Depression, Academic Stress and Demographic Variables in the Academic Performance of Students of a Public University Using Multivariate Methods, Lima - Peru 301
Félix Bartolo Gotarate and Olga Lidia Solano Dávila

Transformative School Digital Sustainability Model to Improve Teacher Professionalism and Identity in Primary Education 308
Lucia Isabel Chamán Cabrera, Silvia Juana Saavedra Díaz, Luis Alberto Chunga Pajares, and Consuelo Ivonne Del Castillo Castro

Emerging Digital Transformation Model for Teaching Strategies to the Achievement of Meaningful Learning in Education 4.0 314
Consuelo Ivonne Del Castillo Castro, Luis Alberto Chunga Pajares, Lucia Isabel Chamán Cabrera, and Silvia Juana Saavedra Díaz

Study of Malnutrition in Children Under Five in Peru Using the Exploratory Spatial Data Analysis 322
Bryan Martin González Cusi and Olga Lidia Solano Dávila

eHealth: A Focus on Software Architectures for Brazilian Public Health 334
Caroline Luiz Gonçalves, Luciana Pereira, and Alessandra Cristina Santos Akkari

Graphene Geology and the Fourth Industrial Revolution 342
Augusto Gonçalves Nobre, Lara Poliny Nogueira da Silva, and Fábio Ramos Dias de Andrade

The BFS Method in a Cloud Environment for Analyzing Distributed Energy Resource Management Systems 349
Adolfo Blengini Neto, Yuzo Iano, Gabriel Gomes de Oliveira, Gabriel Caumo Vaz, Fabiana Silva Podeleski, Higor de Paula Kolecha, and Marcius F. H. de Carvalho

Characterization of the Behavior of LoRa Networks in a Fluvial Environment in the Rio Negro 363
Danilo Frazão and Edgard Silva

Utility Pole Detector for Inspection of Power Distribution Networks Using Faster R-CNN 370
Júlio César Pereira, Rangel Arthur, and Luis Augusto Libório

Lean-Agile Education: A Bibliometric Analysis 378
Julia Yumi Ito, Franciane Freitas Silveira, and Alessandra Cristina Santos Akkari

Glioma Brain Tumor Segmentation in Four MRI Modalities Using a Convolutional Neural Network and Based on a Transfer Learning Method 386
Nazanin Tataei Sarshar, Ramin Ranjbarzadeh, Saeid Jafarzadeh Ghoushchi, Gabriel Gomes de Oliveira, Shokofeh Anari, Mahboub Parhizkar, and Malika Bendechache

Aztec Curve: Proposal for a New Space-Filling Curve 403
Diego Ayala, Daniel Durini, and Jose Rangel-Magdaleno

Evaluation of Adjustment Models for Taking Body Temperature with a Thermal Imaging Camera in the Context of COVID-19 412
Gabriel Cruz, Raul Mamani, Henry Torres, Victor Salcedo, Cristian Benites, Reencarnacion Quispe, Franklin Cardeñoso, and Jorge L. Arizaca-Cusicuna

Characterization of the Electromagnetic Propagation in Inca Walls of the Archaeological Site of Rumiwasi Using ARMA Processes in the Band from 800 MHz to 1 GHz 420
Henry L. Davila-Andrade and Jorge L. Arizaca-Cusicuna

Change Management in a Brazilian Family-Owned Business in the Window and Door Frames: A Case Study 428
Gabriela Vieira Passos, Maria Thereza de Moraes Gomes Rosa, Marlucy de Godoy Ricci, Daniela Helena Pelegrine Guimarães, and Alessandra Cristina Santos Akkari

Physical, Chemical and Rheological Properties of Blueberry Jam 436
Daniela Helena Pelegrine Guimarães, Gisele Letícia Alves, Amanda Faria Querido, Pedro Felipe Arce Castillo, and Maria Thereza de Moraes Gomes Rosa

Flipped Classroom to Improve the Learning of Mathematics in Students of the Public University of Lambayeque 443
Diana Mercedes Castro Cardenas, Segundo Francisco Segura Altamirano, and Merly Liliana Yataco Bernaola

Simulation of Hail Damage in the Amylaceous Corn Production Acobamba, Huancavelica 451
William Huamán-Tovar, Roberto Chuquilín-Goicochea, Arcadio Sánchez-Onofre, and Ruggerths De La Cruz-Marcos

Energy Startup Business Model: An Analysis of Scientific Publications and Trends for the Future 458
Larissa de Souza Lemos Farali, Franciane Freitas Silveira, and Alexandre Acácio de Andrade

Effect of Immersion Time in NaCl and Blanching Time on the Texture and Sensory Acceptability of Preserves Based on Ginger (*Zingiber Officinale*) Surplus in Syrup 469
Betty Yolanda Roque-Velásquez, Verónica Náthaly Jácobo-Domínguez, and Cesia Boñón

Detection of Deforested Areas in the Peruvian Amazon Using Satellite Images 479
Ricardo Carbajal, Alejandro Alvarez, Joshi Lopez, Miguel Saturno, and Victor Murray

Micropropagation of *Myrcianthes Ferreyrae* (McVaugh) "Arrayán": an Endemic Species from Lomas De Atiquipa 487
Ana Caldas Araujo, Fiorella Medina Medina, Patricia Viveros Flores, and Herbert Lazo Rodriguez

Perceived Effectiveness of Gender-Based Virtual Education in Times of COVID-19 Pandemic in University Higher Education 494
Carlos Valdivia, Roger Alarcón, Jessie Bravo, Janet Aquino, Nilton Germán, and Fiorella Li

A Proposal for a Process Framework to Guide the Digital Transformation in the R&D Center of a Brazilian Chemical Industry 502
Talison Alvarenga dos Santos, Anderson Orzari Ribeiro, and Franciane Freitas Silveira

Identification of Soil Characteristics for the Extraction of Ch'ampa as Earth Block Masonry 512
Yasmani Vitulas Quille

Social Impacts of Artificial Intelligence and Mitigation Recommendations: An Exploratory Study 521
Vanessa Marques Paes, Franciane Freitas Silveira, and Alessandra Cristina Santos Akkari

Total Phenolic Content and Antioxidant Activity of a Macha Macha Beverage (*Vaccinium Floribundum* Kunth) 529
Domingo Utus-Ramires, Pedro Arteaga-Llacza, Ruggerths De La Cruz-Marcos, Angélica Miranda-Jara, Helí Miranda-Chávez, Miguel Ángel Quispe-Solano, and Roberto Chuquilín-Goicochea

Evaluation of the Yield, Sensory Characteristics, and Digestibility of Expanded Quinoa 535
Noemi Quispe-Boza, Pedro Arteaga-Llacza, Ruggerths De La Cruz-Marcos, Angélica Miranda-Jara, Helí Miranda-Chávez, Miguel Ángel Quispe-Solano, and Roberto Chuquilín-Goicochea

Iconic-Verbal Strategies to Develop Reading Comprehension in Children of 4 Years Old . . . 543
Pascual Ayamamani, Liliana Valenzuela, Luisa Flores, and Diana Quispe

Technological Resources for Teaching/Learning in Biomedical Areas in the COVID-19 Pandemic Context: Systematic Review . . . 551
Tania Carola Padilla-Cáceres, Luz Marina Caballero-Apaza, Vilma Mamani-Cori, and Sheyla Lenna Cervantes-Alagón

Current Situation of the Monitoring of Graduates and Labor Insertion in the Electrical Mechanical Engineering Career . . . 561
Julio Fredy Chura Acero, Omar Chayña Velasquez, Fredy Bernardo Coyla Apaza, Leonidas Vilca Callata, and Henry Shuta Lloclla

Job Motivation and Mental Health Status of Care Person at the District Hospital La Esperanza During COVID-19 . . . 569
Miryam Lora-Loza, Marco Quispe Barra, Julia Alvarado-Alfaro, Juan Rodríguez-Vega, Davis Alberto mejía Pinedo, Rosa Estrella Pillman Infanson, and Maximo Amancio Moltalvo Atco

Quality and User Satisfaction of the Telemedicine Service in Cardiology During COVID-19 . . . 577
Miryam Lora-Loza, Marco Quispe-Barra, Castro Uceda Maria, Juan Rodríguez-Vega, David Salinas-Mendoza, and Jenner Fabián-Ramírez

Clinical Management for Hospitalization of Patients with COVID-19 and Resolution Capacity Functionality in I-4 Facilities - Contumazá-Cajamarca Health Network, 2021 . . . 584
Juan Rodríguez-Vega, Miryam Lora-Loza, Marco A. Quispe Barra, Segundo Manuel Hoyos-Gallardo, André Torres Quispe, and Jhon Rengifo-castillo

Motivation and Job Performance at the Regional Health Management of La Libertad During COVID-19 . . . 592
Marco A. Quispe Barra, Miryam Lora-Loza, Yris Cedrón-León, Juan Rodríguez-Vega, Fanny Sofia Quispe Barra, and Vanesa Castillo-Castro

The Method of Logical Development of Regulatory Dependability Terminology . . . 598
Nikolai I. Plotnikov, Gabriel Gomes de Oliveira, and Yuzo Iano

Regulatory Decision of Dependability Terminology . . . 609
Nikolai I. Plotnikov, Gabriel Gomes de Oliveira, and Yuzo Iano

A Python Script to Generate a 3D Model of a Coaxial Cable 615
Daniel J. C. Pereira, Kenedy Marconi G. Santos, Douglas O. Campos, Polyane A. Santos, Lucas S. Ribeiro, Marcelo B. Perotoni, Tagleorge M. Silveira, Marcela S. Novo, and Willian F. S. Maia

Analysis and Validation of a Computational Model for a Two Bipolar Drain Wire Cables 623
Douglas O. Campos, Kenedy M. G. Santos, Daniel J. C. Pereira, Polyane A. Santos, Lucas S. Ribeiro, and Tagleorge M. Silveira

Author Index 631

Design and Fabrication of Human-Powered Vehicle - A Measure for Healthy Living

Y. Thiagarajan[1](✉), G. Palanivel[2], I. D. Soubache[3], Gabriel Gomes de Oliveira[4], Yuzo Iano[4], Gabriel Caumo Vaz[4], and Himanshu Monga[5]

[1] Christ College of Engineering and Technology, Puducherry, India
thiagarajan_eee@christcet.edu.in
[2] Sri Venkateshwaraa College of Engineering and Technology, Puducherry, India
[3] Rajiv Gandhi College of Engineering and Technology, Puducherry, India
[4] State University of Campinas, Campinas, Brazil
oliveiragomesgabriel@ieee.org, yuzo@unicamp.br
[5] Department of ECE, Jawaharlal Nehru Government Engineering College, Sundar Nagar, Himachal Pradesh, India

Abstract. Traffic congestion, energy efficiency, and environmental issues are fueling the interest in light vehicles. This project focuses on the design of personalized vehicles powered by manual loading. Human-powered vehicles are attractive because of their low cost, ease of use, lightweight, and eco-friendliness, and they deliver an exercise experience that is closer to the abdominal workouts than anything else available today. To analyze and improve the mechanism for the future scope and user-friendliness, we can implement visionary ideas to it. The main objective of this work is to combine both the transport and physical fitness aspects of this vehicle, showing the historical context of these vehicles and presenting the project, with the design developed in a CAD software and the results achieved from another software whose aim is to simulate the structural deformation given the load stress derived from a mass applied on the structure. The tests were carried out with two different masses and the results are presented as heat maps in the designed structure.

Keywords: Personalized vehicle · Eco-friendliness · Human-powered · Physical fitness · AutoCAD · 3D-model · ANSYS R15.0

1 Introduction

Among different modes of exercise in busy modern life, cycling and ergometric treadmilling activities are the most common indoor aerobic exercises. Treadmills are very tall and occupy more space. Moreover, this makes people lazy, relying on machines for doing their workouts [1–3]. We feel that this bicycle is the first of its kind that works with human strength taking up less space. We combined both the transportation and physical fitness aspects in one single vehicle. In this vehicle, there are no seating arrangements; instead, we must stand over the platform and perform the operation.

Y. Iano et al. (Eds.): BTSym 2021, SIST 207, pp. 1–15, 2023.
https://doi.org/10.1007/978-3-031-04435-9_1

This vehicle is based on a mechanism, which drives the vehicle forward, and it would be an excellent support for improving physical fitness. The main thing that boosts this vehicle is that it is completely fabricated from scrap materials, so that, we can re-use the scrap for some useful work, and this makes this vehicle economically friendly [4–7].

2 Literature Survey and Motivation

The motivation for this project originated from conventional kick bikes, which are used in day-to-day life. The traditional kick scooters consist of two wheels, the front wheel attached with the handlebar and the steering mechanism. The back wheel is connected to the center deck, which is free to rotate. The user must stand over the deck and uses his leg to kick the ground; due to force created by kicking, the scooter moves in the forward direction. The deck is an area or platform where the rider stands over it to operate the vehicle [8–11].

2.1 Principle of Kick Scooter

The principle used behind this kick scooter is Newton's third law of motion. As per Newton's third law of action, "For each activity, there is an equivalent and inverse response." These two powers are called activity and response powers. The detailed view of the existing kick scooter is shown in Fig. 1.

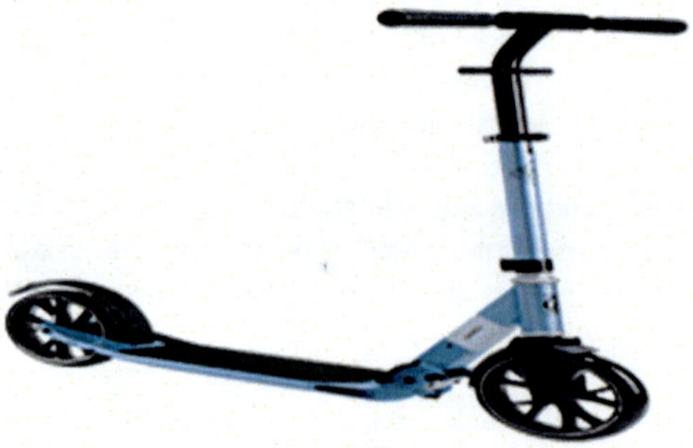

Fig. 1. Kick scooter

2.2 History of Kick Scooter

A kick bicycle or push bicycle is a human-operated vehicle with handlebars and wheels, controlled by a driver pushing off the ground. Most of the bikes today are made of hard metals to avoid wear and tear. Toddler bicycles are made of thick plastic. Elite bikes were made for grown-up people, and they look like the old "penny-farthing." Mechanized bikes, generally controlled by gas motors or electric engines, are self-pushed kick bikes equipped for rates of around 30 km/h (19 mph).

2.3 Wooden Minibike

Kick bikes have been handcrafted in urban territories like Europe and the US as playthings for youngsters. These are made by joining roller skate wheel sets to a board with a handle. The improvement was all wood, with 3–4-in. (75–100 mm) wheels holding steel metal balls. A new preferred position of this development was a boisterous commotion, much the same as from a "genuine" vehicle. An elective event comprises one steel clip-on roller isolated into the front and back parts and connected to a wooden pillar. This is depicted in Fig. 2.

Fig. 2. Photograph of wooden scooter

2.4 Kick 'N Go

In 1974, the Kick 'n Go was introduced by Honda. This bike is driven by moving a pedal connected to the propelling mechanism. While it was considered, by all accounts, as much exertion to "kick" like in an ordinary bike, the curiosity for the Kick 'n Go got on, and it became mainstream. It is depicted in Fig. 3.

Fig. 3. Kick and Go bike

2.5 Modern Kick Bike

Denis Johnson built a similar vehicle in the 19th century. The invention of the kick bike in Finland, in 1994, altered the perception of scooters. The Kick bike features a massive standard-size bicycle with a front wheel and a smaller rear wheel. That configuration permits a much quicker ride. Since 2001, the Foot Bike Euro Cup has been used commercially, which is shown in Fig. 4.

Fig. 4. Modern kick bike

2.6 Folding Scooters

Wim Ouboter, from Micro Mobility Systems, in Switzerland, fabricated a foldable aluminum motorcycle with inline skate wheels in 1996. The bike is marketed as a "Miniature Scale Skate Scooter" and "JDBUG/JDRAZOR MS-130A" (Fig. 5). After the Razor was introduced in Japan, in 1999, many young people in Tokyo began to use it as a small transport medium, and it quickly became a popular trend around the world and also became a popular children's toy. Newly designed Foldable kick bikes designed for adults typically have more robust parts: they are designed with larger decks, hand brakes, and larger wheels for smoother transportation rather than lighter weight and portability. The Xootr Street, which has 180 mm wheels and a maximum weight of 136 kg. Go-Ped is a type of bicycle. Know-Ped bikes have 6-in. wheels with strong elastic tires and a maximum load of 181 kg, depending on the model. Kick Ped from NYC Wheels, which is devoid of every single delicate component, for example, a handbrake, and is replaced with a back spoon brake to make the kick bike reliable and resistant [12, 13].

Fig. 5. Photography of the foldable scooter

3 Problem Identification

In recent times, the increase in obesity is the main problem among us. As a result, exercise is recommended for health promotion and prevention of many cardiovascular diseases, as well as for rehabilitation following an illness episode. Aerobic exercises, which include running, jogging, walking, cycling, and others, are among the activities that are appropriate for those purposes.

So, we introduce our vehicle, which is powered by a human load, and it does not need any fossil or petroleum fuels. Using this vehicle for a few minutes gives us the same workout we get at the gym. By using this vehicle, we can obtain both physical fitness and transportation aspects. Thereby it is an economical, ecological, and useful product. This vehicle runs on human load so that the cost of petroleum fuels is eliminated, so that, it does not produce any emissions to the atmosphere and is completely safe for the ecology [14–17].

4 Software Selection – Modelling and Analysis

Design is a formation of an arrangement for the development of an article, framework, or human association. A fashioner's grouping of exercises is known as a structured procedure, while the logical investigation of configuration is called plan science. From a broader perspective, a shape is an article, framework, segment, or structure, which, in turn, coordinates with innovation. This personalized vehicle is designed by AutoCAD Mechanical 2018 software [18, 19]. The design phase was constructed in two models, which are the 2D model and the 3D model.

4.1 2D Model

The two-dimensional model is an essential perspective on any article where the item is anticipated. The 2D model is made utilizing two-axis (X and Y). The fundamental programming used to create a 2D model is an AutoCAD basic tool. In this display, we can make a solitary perspective on the article. It is like the drafting of design illustrations. The basic 2D sketch can be changed over to the 3D model.

4.2 3D Model

The three dimensional model shows the standard view of the item as it is. It is precious to build up the article in a reasonable human view. To make a 3D model, we prefer three axes (x, y, and z). There is much programming for creating a 3D model, as indicated by our determinations. Some products that are utilized to make the 3D model are AutoCAD, Pro-E, CREO, and Solid Works [10–13].

AutoCAD is a useful workbench for design and drafting programs. Autodesk created and marketed it. AutoCAD is used by architects, project managers, engineers, graphic designers, town planners, and many other professionals in a variety of industrial applications. AutoCAD Mechanical 2018 version is utilized for a better design.

4.3 AutoCAD Model

The personalized vehicle was designed using AutoCAD Mechanical (2018) software and fabricated using the components. All the dimensions presented in Fig. 6 are in millimeters.

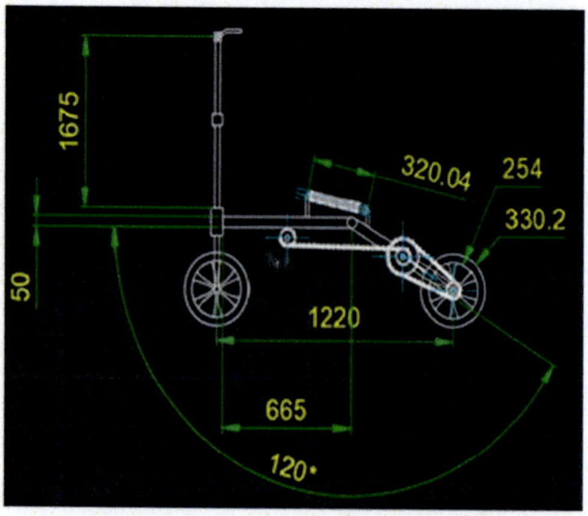

Fig. 6. AutoCAD design for the proposed Model

4.4 3D Model of the Frame with Shock Absorber

The 3D modeling of the frame with a shock absorber is done with the help of AutoCAD mechanical software. In this picture, the 3D model of the frame with the shock absorber of the vehicle is shown in the top view. The 3D model of the frame with the shock absorber of the vehicle is represented in Fig. 7 and Fig. 8.

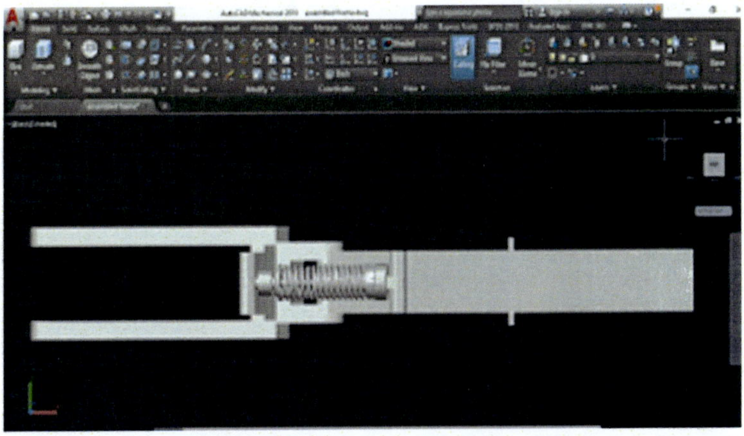

Fig. 7. Top view of the frame with shock absorber

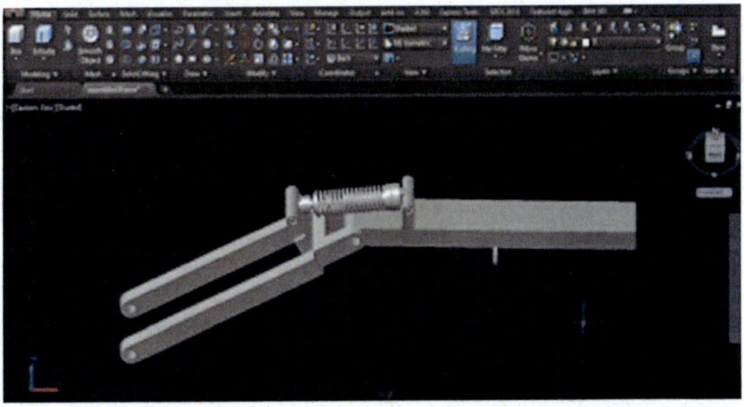

Fig. 8. Tilted view of the Frame with shock absorber

4.5 Analysis Process

Here we have used the "Ansys 15.0" software to get the results and calculations. In this model, we have two frames and wheels with a handlebar. We have analyzed it by using two different materials like structural steel and aluminum alloy. Here the complete vehicle without any mechanism has been tested. The tilted 3D view of the vehicle is illustrated in Fig. 9, using Ansys Software.

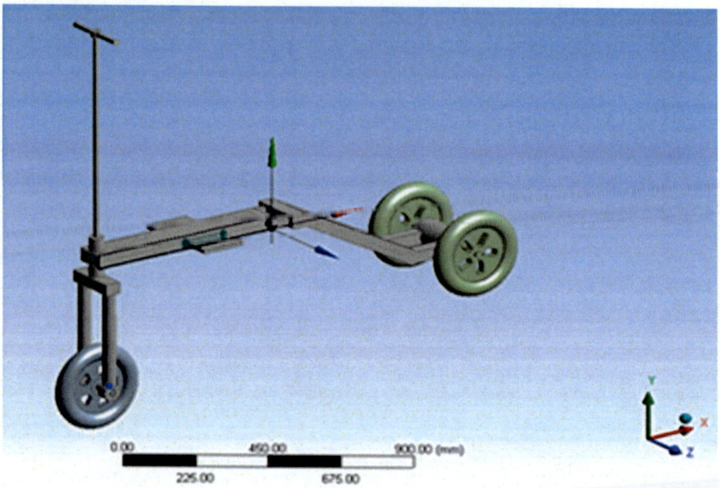

Fig. 9. Pictorial view of the human powered vehicle

5 Results

The deformation of the vehicle when the load acts on it is analyzed, and the following strains are calculated for 75 kg and 150 kg.

1. Total deformation (Overall deformation of the vehicle when loading acts).
2. Directional deformation (Deformation of the vehicle in a particular axis).
3. Maximum principal stress (Stress acting upon the vehicle when loading acts).

5.1 Total Deformation for 75 kg Load Stress

Total distortion is the overall deformation of the vehicle when the pressure is applied. The total deformation of the vehicle is tested for 75 kg, as depicted in Fig. 10. The red portion indicates the maximum deformation area of the vehicle when the load acts on it. The blue part shows the minimum deformation area of the vehicle when the pressure works on it.

5.2 Total Deformation Under 150 kg Load Stress

The total deformation of the vehicle is tested for a load of 150 kg. The red portion indicates the maximum deformation area of the vehicle when the pressure acts on it. The blue part shows the minimum deformation area of the vehicle when the load works on it, which is clearly illustrated in Fig. 11.

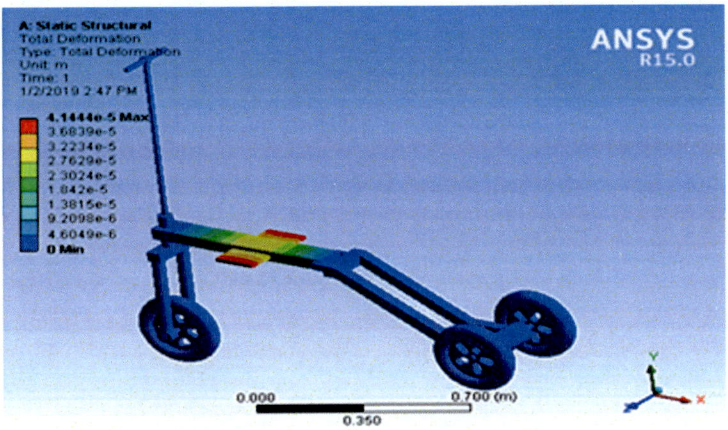

Fig. 10. Total deformation of the vehicle tested under 75 kg load

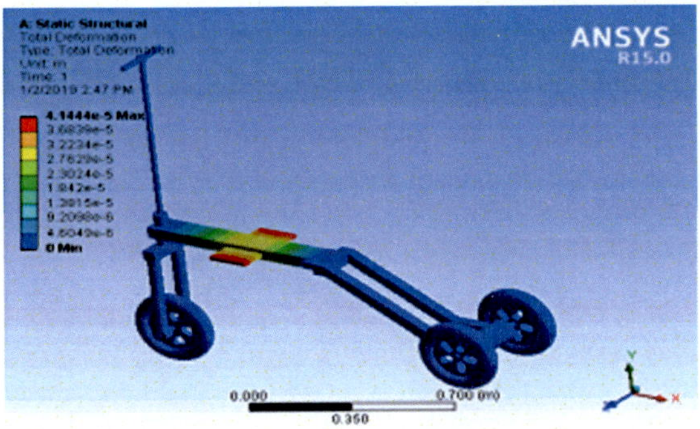

Fig. 11. Total deformation of the vehicle tested for 150 kg load

5.3 Directional Deformation Under 75 kg Load Stress

The directional deformation of the vehicle is tested for the load of 75 kg. Directional deformation is the deformation of the vehicle in a particular axis, say x-axis, when the pressure is applied. The red portion indicates the maximum deformation area of the vehicle when the load acts on it. The blue part shows the minimum deformation area of the vehicle when the pressure works on it, which illustrated in Fig. 12.

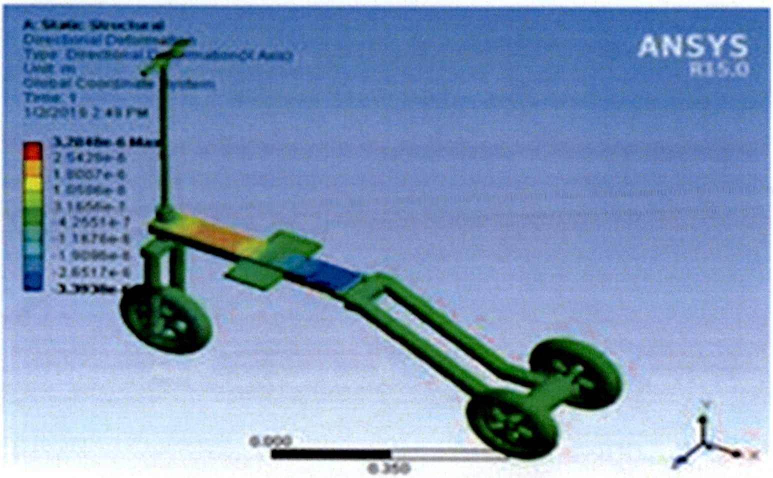

Fig. 12. Directional deformation of the vehicle tested for 75 kg load

5.4 Directional Deformation Under 150 kg Load Stress

The directional deformation of the vehicle is tested for the load of 150 kg. The red portion indicates the maximum deformation area of the vehicle when the load acts on it. The blue portion indicates the minimum deformation area of the vehicle when the load acts on it, the pictorial representation is shown in Fig. 13.

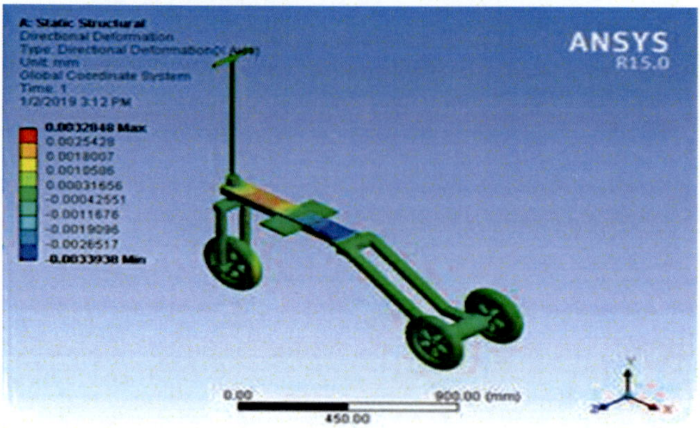

Fig. 13. Directional deformation of the vehicle tested for 150 kg load

5.5 Maximum Principal Stress Under 75 kg Load

The maximum principal stress is the overall stress acting on the body of the vehicle when the load is applied to it. The maximum principal stress of the vehicle is tested for the weight of 75 kg. The red portion indicates the maximum principal stress area of the vehicle when the pressure acts on it. The blue part indicates the minimum principal stress area of the vehicle when the load serves on it; these are highlighted in Fig. 14.

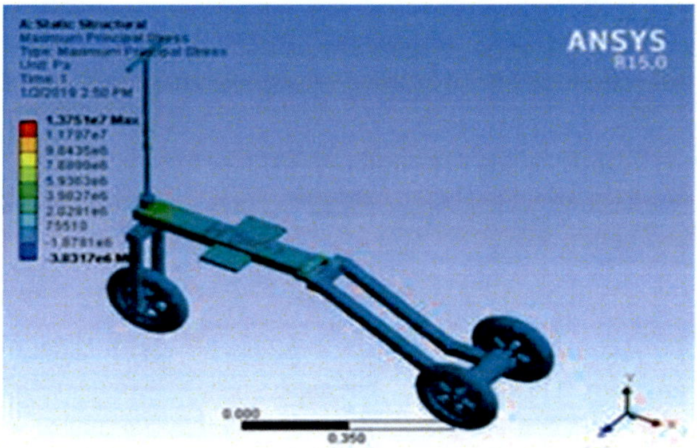

Fig. 14. Maximum principal stress of the vehicle tested for 75 kg load

5.6 Maximum Principal Stress Under 150 kg Load

The Maximum principal stress of the vehicle is tested for the load of 150 kg. The red portion indicates the maximum principal stress area of the vehicle when the load acts on it. The blue portion indicates the minimum principal stress area of the vehicle when the load acts on it, which is shown in Fig. 15.

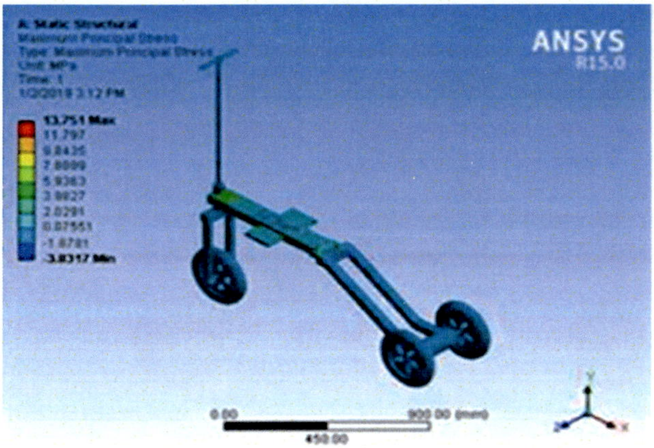

Fig. 15. Maximum principal stress of the vehicle tested for 150 kg load

By analyzing the vehicle using the "Ansys 15.0" software, we came to know that the deformation of the vehicle, when the load is applied on it, is minimal. Hence, it can withstand an average human load, and the vehicle is completely safe when used by an average human load.

6 Hardware Description

This vehicle "converts a vertical external load (force) into a rotational or twisting force," thereby making the vehicle move in the forward direction. The external load is in the form of reciprocating in nature, which is transferred to the wheels through the combination of chain drive and pulley drive.

6.1 Working

The body is comprised of two casings that are associated with a typical focal pivot. Likewise, a spring holds both bodies consistently, and they are moved when there is a human load on them. Further, the portable front part is associated with chains utilizing an instrument situated at the other mobile part, which is on two apparatus. Consequently, when someone hops on it, and the body extends, the chains are extended as well. In this way, the appliances are returning to move the pulley. Also, when the human's weight is expelled from the bicycle, the riggings are taking their underlying position, and to make it move once more, another jump ought to be made.

For the most part, the specific system transforms the flat move into a turning one. The last step of the wheels is brought about by a pulley through a belt put on the hub and arrives at the wheels. Regardless of whether the specific vehicle is a typical bicycle or not, it is charming and fascinating to ride using it. Figure 16 showcases the basic 2D paint sketch of the personalized vehicle and Fig. 17 is the photographic view of the Fabricated Hardware model.

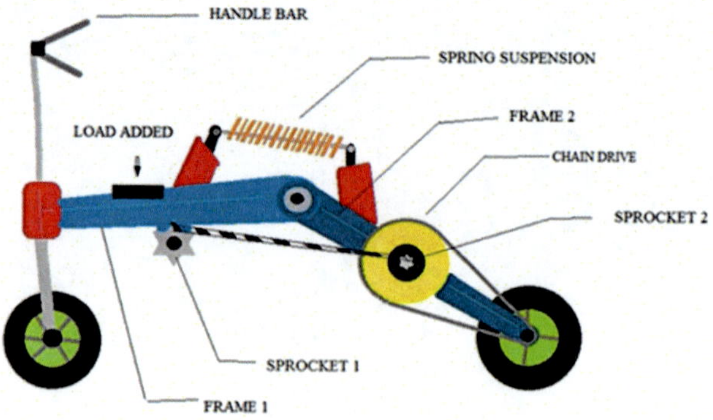

Fig. 16. 2D paint sketch of the personalized vehicle

Fig. 17. Hardware model of the vehicle

6.2 Materials Used and Its Properties

Table 1. Material and its properties

Material	Yield strength	Tensile strength	Density [kg/m^3]	Cost per kg [Rs]
Structural Steel	250 MPa to 450 MPa	40000 PSI	7800	80
Aluminum	240 MPa to 310 MPa	47000 PSI	2710	131
Stainless steel	290 MPa to 580 MPa	31200 PSI	7700	100

Table 2. Actual specifications of the vehicle

Sl. number	Specifications	Values
1	Length of the frame 1	650 mm
2	Length of the frame 2	660 mm
3	Distance between ground and handlebar	1400 mm
4	Distance between rear wheels	560 mm
5	Wheelbase	1200 mm

7 Conclusion

In this modern era, physical wellness is considered one of the most extreme significances in everyday life. This bike is designed after continuous assessment and modification so that it could be commercialized in the market for better use. Our target is to come up with an innovative concept that would serve the purpose of both the health concern and transportation.

The design and the idea are brought into reality with few compromises like lack of funds, time, and fabrication facility in our environment. However, this planned human-powered vehicle was developed with high accuracy in design. The major highlight is that it does not require any nonrenewable energy source for its working, thereby eradicating carbon footprint, and it is also user-friendly. By utilizing this vehicle, one can get physically fit.

The design has been plotted out in ANSYS Version 15. The vehicle's physical strength was analyzed by manipulating the design under various load conditions. This approach is made to have a better understanding of the vehicle. The fabricated bike was put into test drive with various human weights, and the result showed a perfect remark. Finally, the creation and the trial have been expertly finished, and the vehicle is ready to be on board.

References

1. Shukla, P.R., Dhar, S., Pathak, M., Bhaskar, K.: Electric Vehicle Scenarios and a Roadmap for India. UNEP DTU Partnership (2014)
2. Kisan, R., Kisan, S.R., Anitha, O.R., Chandrakala, S.P.: Treadmill and bicycle ergometer exercise: cardiovascular response comparison. Glob. J. Med. Res. **12**(5), 23–26 (2012)
3. Hamad, A.A., Al-Obeidi, A.S., Al-Taiy, E.H., Khalaf, O.I., Le, D.-N.: Synchronization phenomena investigation of a new nonlinear dynamical system 4D by Gardano's and Lyapunov's Methods. Comput. Mater. Continua **66**(3), 3311–3327 (2021)
4. Mahadik, C., Mahindrakar, S., Deka, J.: An Improved & efficient electric bicycle system with the power of real-time information sharing. Multidiscipl. J. Res. Eng. Technol. **1**(2), 215–222 (2014)
5. Sekar, V., Thiyagarajan, V.: Controlling of brushless DC motor in electric bicycle using an electronic-based circuit with 8-bit microcontroller. Int. J. Eng. Sci. Emerg. Tech. **4**(1), 26–34 (2012)
6. Manas, M., Ingole, P.M.: Ergonomic design of bicycle handle. Int. J. Emerg. Tech. Adv. Eng. **5**(4), 472–481 (2015)
7. Bhandarkar, S.: Vehicular Pollution, their effects on human health and mitigation measures. Veh. Eng. (VE) **1**, 33–40 (2013)
8. Wisesa, O., Adriansyah, A., Khalaf, O.I.: Prediction analysis sales for corporate services telecommunications company using gradient boost algorithm. In: 2020 2nd International Conference on Broadband Communications, Wireless Sensors and Powering (BCWSP), pp 101–106 (2020)
9. Subahi, A.F., Alotaibi, Y., Khalaf, O.I., Ajesh, F.: Packet drop battling mechanism for energy aware detection in wireless networks. Comput. Mater. Continua **66**(2), 2077–2086 (2021)
10. Xiang, X., Li, Q., Khan, S., Khalaf, O.I.: Urban water resource management for sustainable environment planning using artificial intelligence techniques. Environ. Impact Assess Rev. **86**, 106515, (2021). https://doi.org/10.1016/j.eiar.2020.106515

11. Osamah, I., Khalaf, O., Abdulsaheb, G.: Energy efficient routing and reliable data transmission protocol in WSN. Int. J. Adv. Soft Comput. Appl. **12**(3), 45–53 (2020)
12. Khalaf, O.I., Abdulsahib, G.M., Sabbar, B.M.: Optimization of wireless sensor network coverage using the bee algorithm. J. Inf. Sci. Eng. **36**(2), 377–386 (2020)
13. Prasad, S.K., Rachna, J., Khalaf, O.I., Le, D.-N.: Map matching algorithm: real time location tracking for smart security application. Telecommun. Radio Eng. **79**(13), 1189–1203 (2020)
14. Argueta, R.: A Technical Research Report: The Electric Vehicle (2010). http://holms.faculty.writing.ucsb.edu/Writing_2E-_EV_Technical_Research_Report.pdf. Accessed 14 May 2021
15. Krichen, M., et al.: A formal testing model for operating room control system using Internet of Things. Comput. Mater. Continua **66**(3), 2997–3011 (2021)
16. Papadopoulos, L.C., Hole, J.D.: Bicycle Treadmill Having Automatic Speed and Resistance Adjustments. US Patent 7220219, 22 May 2007 (2007)
17. Lo, P.H.: Linkage Structure of Treadmill. US Patent 7306546, 11 December 2007 (2007)
18. Khalaf, O.I., Ogudo, K.A., Singh, M.: A fuzzy-based optimization technique for the energy and spectrum efficiencies trade-off in cognitive radio-enabled 5G network. Symmetry **13**(1), 47 (2021)
19. Khalaf, O.I., Ajesh, F., Hamad, A.A., Nguyen, G.N., Le, D.-N.: Efficient dual-cooperative bait detection scheme for collaborative attackers on mobile Ad-Hoc networks. IEEE Access **8**, 227962–227969 (2020). https://doi.org/10.1109/ACCESS.2020.3045004

Identification and Analysis of Bacterial Species Present in Cow Dung Fed Microbial Fuel Cell

Y. Thiagarajan[1](✉), Gabriel Gomes de Oliveira[2], Yuzo Iano[2], and Gabriel Caumo Vaz[2]

[1] Christ College of Engineering and Technology, Puducherry, India
thiagarajan_eee@christcet.edu.in
[2] State University of Campinas, Campinas, Brazil
oliveiragomesgabriel@ieee.org, yuzo@unicamp.br, gabriel_caumo@yahoo.com.br

Abstract. Microbes are the powerhouses of Microbial fuel cells (MFCs). The present report is the offshoot of research undertaken on the use of cow dung slurry for electricity generation. Since bacterial consortia are the driving force in MFCs, bacterial communities present in cow dung slurry and on the biofilm of the electrodes were screened by Matrix-assisted laser desorption/ionization-time of flight technique. More than 60 bacterial species and a few unidentified but having similarity with identified bacterial species were found in the cow dung slurry; only seven bacterial species were found on biofilm formed on the electrode and involved in redox reactions. These findings are considered as the route map to achieve higher electricity using the specific group of bacteria in MFCs. Further, present findings enlighten the potential opportunity that bacterial species isolated from the electrode surface can be further cultured individually and used in Bioreactor to fasten the energy production from cow dung slurry fed Bioreactor.

Keywords: Microbial fuel cell (MFC) · Bacterial species technique · Shannon index · Electrode biofilm · Bioreactor

1 Introduction

Microbial power device or (MFC) arose as one of the sustainable power producing module for low power applications. Microbes present in the organic substrates used in MFCs are the driving force in generating electricity in MFCs. Energy is generated in MFC's due to the oxidation of organic substances present in them, and that energy is capitalized as the new renewable source using MFC. Therefore, the selection of bacterial consortia and optimization of the microbial metabolism to improve electron production have become the core area of interest in this millennium. Microbes colonized on electrodes deployed in MFC acquire the required energy for cellular growth from the oxidation of the complex organic matters and release electrons [1].

Y. Iano et al. (Eds.): BTSym 2021, SIST 207, pp. 16–24, 2023.
https://doi.org/10.1007/978-3-031-04435-9_2

It has been well established that matter supply and blend of microorganism community are the key factors that are thought of as core inputs of MFC. Eventually, the search for potential substrates naturally bestowed with more diverse bacterial communities has been taken up in the recent past by many researchers. For direct bioelectricity generation in MFC, the ideal bacterial culture must be able to grow aerobically and be electrochemically active; utilizing an anode as an alternative electron acceptor while oxidizing metabolites of various carbon sources [2]. In that pursuit, the suitability of Cow dung, which is bestowed with rich bacterial diversity, as a substrate was examined and reported by many works from India and abroad [3–16]. However, studies on systematic screening on metagenomics of the cow dung and bacterial community that constitute primary film on an electrode in cow dung fed microbial fuel cells are not available. More specifically, screening of microbial consortia present in the biofilm formed on the electrode surface fitted in MFC has not been done earlier. The present study is an offshoot of a study on the fabrication and operation of MFC with cow dung slurry to generate electricity. To identify and isolate potential bacterial species capable of producing higher electricity, it is essential to find out whether all microbes present in the cow dung are involved in the process of biofilm formation on the electrode.

2 Material and Methods

For screening the bacterial community present in the cow dung slurry, at the end of the cycle, 5 ml of the slurry was swabbed from the anode chamber. Similarly, aseptic swabs were collected from the surface of the anode Electrode and cow dung slurry, which is analyzed using the Metagenomic tool (16S V3−V4 Metagenome). The slurry sample reported supporting more varieties of bacterial species. The other sample from a biofilm, which is formed on the electrode surface, was analyzed by the MALDI-TOF technique [17]. In the second round, amplicons were tested under 1.2% agarose gel using the Polymerase chain reaction method. Pie-gram was obtained using matplotlib for the taxonomy present above the cut-off value of >0.5%. Similarly, the biofilm formed on the CNT-coated SSM electrode deployed in the anode chamber of the microbial fuel cell was screened by the MALDI-TOF technique.

3 Samplewise Analysis

To establish the microbiome diversity and species abundances for each of the individual samples, QIIME analysis has been performed on individual samples as per the protocol described by [18]. The technique of QIIME (an open-source bioinformatics pipeline) has been utilized to analyze 16S rRNA gene sequences of Microbial communities, and it helps to provide a wide range of values from phylum to species. Alpha diversity and Rarefaction Curves were generated to give the details of the microbial community present in the sample. The diversity of bacterial species was calculated using Shannon Index and Simpson's diversity index method.

4 Results

The sample was prepared using the common procedure and tested using the mass ionization Technique. A rarefaction plot is carried out for one sample of the V3–V4 region at a depth of 89,000. The image has been plotted with several sequences per sample in the x-axis against the diversity index in the y-axis. The samples have been colored by their names. The given Rarefaction plot portrays the rarefaction bend of species wealth in V3–V4 tests. This bend is a plot of the absolute number of species with respect to an element of the number of successions tested. On the left, a precarious incline shows that an enormous part of the species variety stays to be distinguished. Assuming the bend gets complimented to one side, it explains that a sensible number of species were examined, and even more serious inspecting is probably going to yield a couple identifying new generic groups. Inspecting bends by and large ascent rapidly from the start and afterward level off due to new species are found per unit of samples gathered.

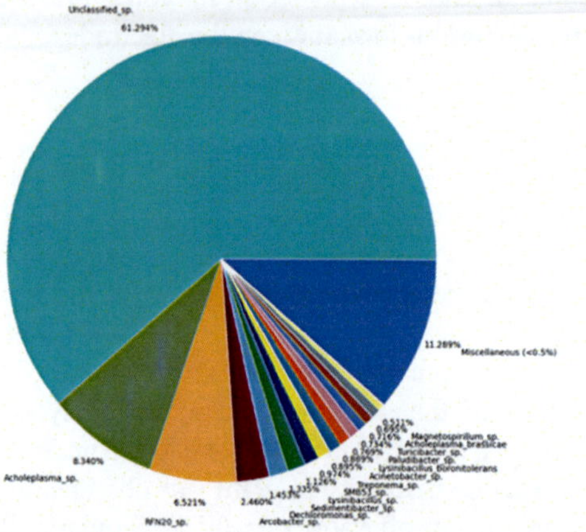

Fig. 1. Pie chart showing the phyla of bacterial species in cow dung slurry

Microbial screening using 16S V3–V4 Metagenome or Mass Ionization Technique has been done at region-specific targeting proprietary primers, which revealed that bacterial community present in the cow dung slurry represents four major phyla is illustrated in (Fig. 1). Bacterial community screened from the biofilm on electrode were: Bacterioidotes (33.175%), Fermicutes (29.8%), Proteobacteria (21.47%), Terenicutes (10%).

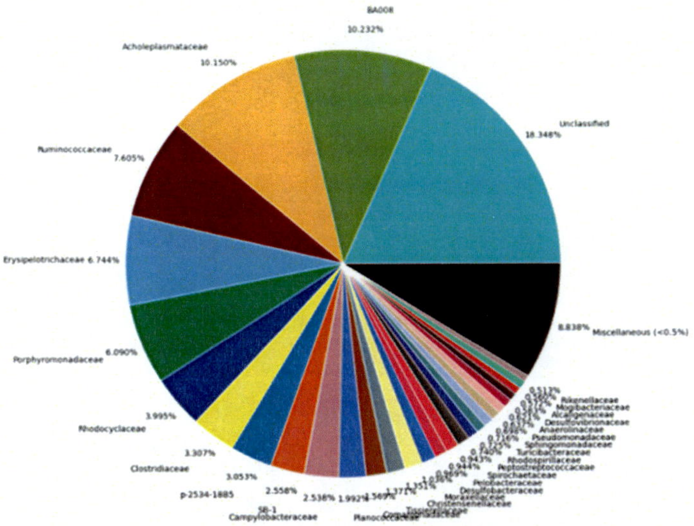

Fig. 2. Pie chart showing microbial species present in cow dung slurry

Figure 2 shows 64 species under 13 genera. All these species belong to 11 families viz. Achleplamatacea, Ruminococcoceae, Erysipelatrideceae, Porpyromonodeceae, Rhodocyclaeceae, Clostridiaceae, Christenlaeceae, Camplylobacteriaca Plannococcoceae, Moraxilaceae, and tissue rellaceae (P2534–1885.BA008) are the unclassified but showing similarity with strains present in the sample.

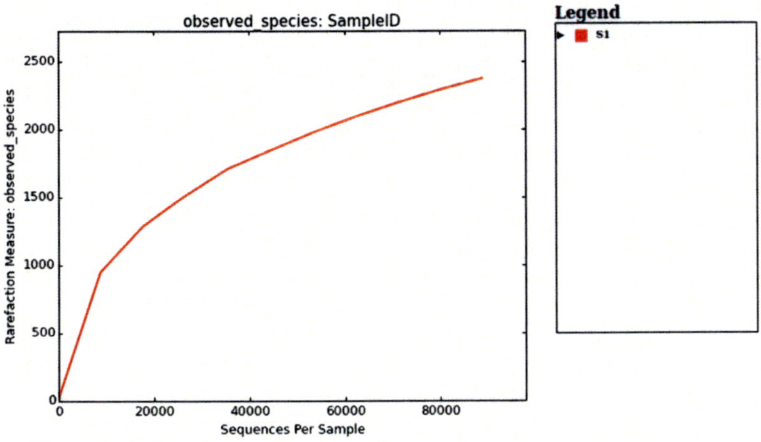

Fig. 3. Rarefaction plots

Shannon Index– a statistical tool to measure the significance of diversity among community and evenness. The index value is around 7.456, which implies higher bacterial diversity in cow dung slurry. Similarly, the Simpson index confirmed significant diversity (less than 1).

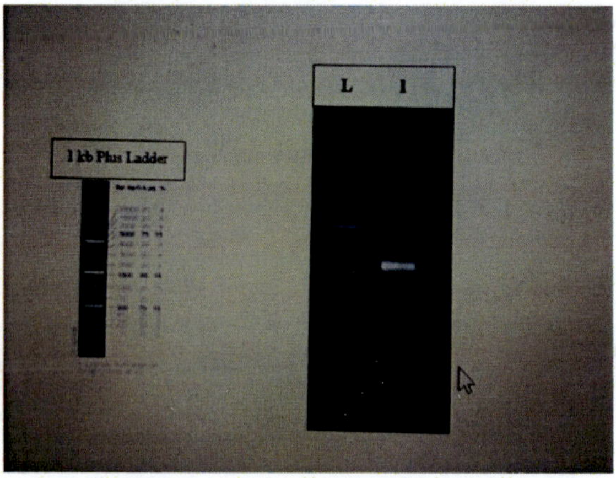

Fig. 4. Three electrophoresis protein

From the Rarefaction plot Fig. 3 and Fig. 4, it is observed that rare species are not reported elsewhere.

Figure 5 shows that about 64 bacterial species were found to be present in cow dung slurry. Screening of isolates from the electrode surface by MALDI-TOF technique revealed the presence of only 7 bacterial species. Bacterial species present in biofilm on electrode were Lysini bacillussphaericus, Bacillus pumilus, B.cereus, B.mojavensis, Pseudomonas aeruginosa belonging to four genera in the biofilm.

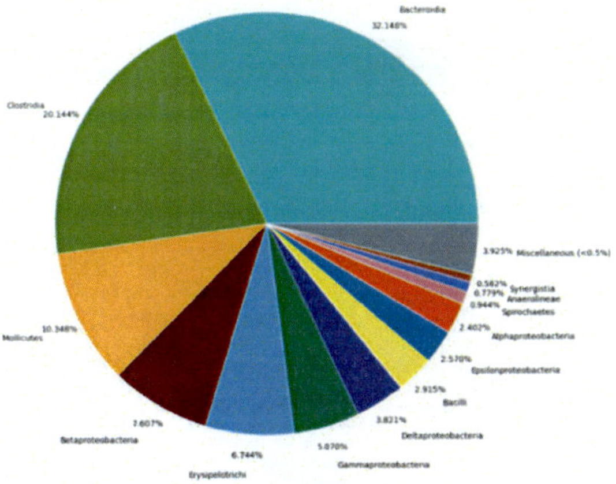

a) Family

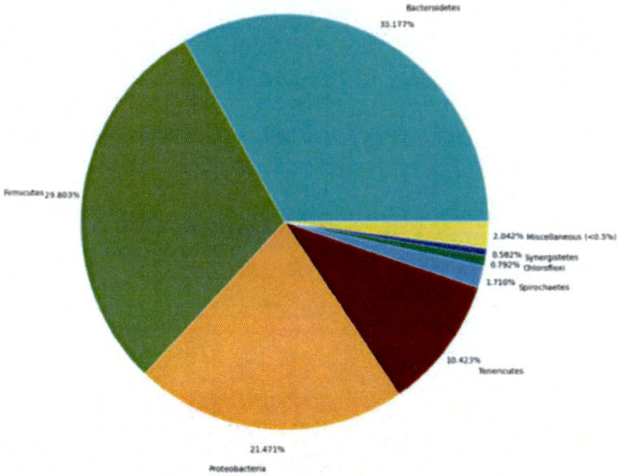

(b) Phylum

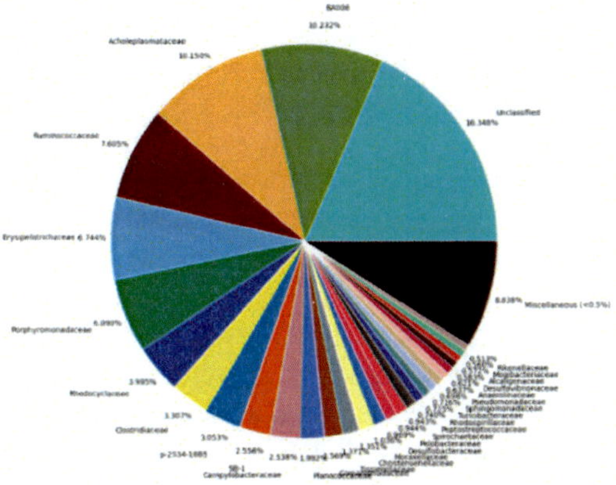

c) Species present

Fig. 5. Overall picture of bacterial communities under phylum, class, order, family, genus, and species and their percentage in cow dung slurry.

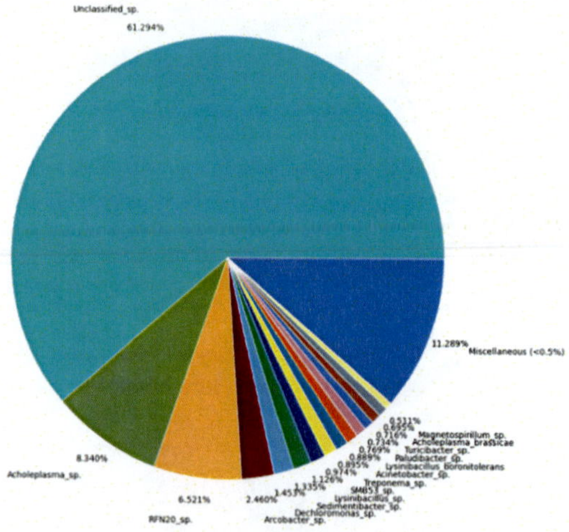

d) Species

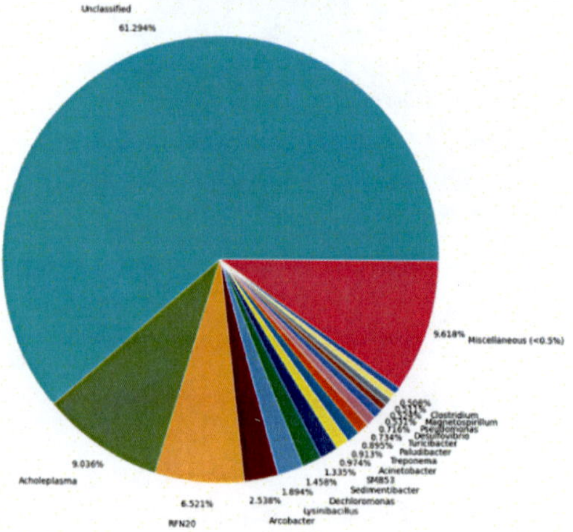

e) Species

Fig. 5. continued

5 Discussions and Conclusion

In this work, microbial screening using 16SV3–V4 Metagenome revealed that the bacterial community present in the cow dung slurry represents four major phyla and 64 species under 13 genera. RFN20-sp. and SMB53-sp., P2534–1885 and BA008-sp are those unclassified but showing similarity with strains present in the sample. Shannon Index measures the diversity of species and the Simpson Index measures the species richness. Both indices are statistically significant. Bacterial species present in cow dung and their families are presented as pie-charts from [Fig. 5 (a)–Fig. 5 (e)]. Though cow dung supported 64 species, not all bacterial species are involved in biofilm formation. Screening by the MALDI-TOF technique revealed the presence of only 7 bacterial species.

The rarefaction plot (Fig. 4) shows the rarefaction bend of clarified species extravagance in V3–V4 tests. This graph is a plot of 64 particular species identified as a component of the number of successions inspected. On the left, a lofty slant demonstrates that a huge part of the species variety stays to be found. Testing bends for the most part rise rapidly from the outset and afterward level off towards an asymptote as new species are found per unit in the sample collected. In view of Shannon record (0.9) and Simpson file (7.470), it is perceived that a higher variety of bacterial species in the cow manure slurry are available. The screening process performed on biofilm isolated from electrode surface showed only seven bacterial species involved in biofilm formation and redox reactions to release protons and electrons (electrogenic bacterial species). It is demonstrated that capitalizing on the power potential of a specific bacterial species would significantly improve electricity generation in Microbial fuel cells. These findings are considered as the route map to achieve higher electricity using the specific group of bacteria in MFCs. Further present findings enlighten the potential opportunity that bacterial species isolated from the electrode surface can be further cultured individually and used in MFCs to enhance power production. Therefore, it is demonstrated from the present study that screening and identification of potential electrogenic microbes from electrode biofilm would help to isolate potential bacterial species delineating other bacteria species present in the microbial fuel cell.

References

1. Franks, A.E., Nevin, K.P.: Microbial fuel cells, a current review. Energies **3**(5), 899–919 (2010). https://doi.org/10.3390/en3050899
2. Dhundale, V., Hemke, V., Desai, D., Aher, G., Khemchandani, P., Dikonda, P., et al.: Improvement and stable high bioelectricity generation using alkaliphilic oceano bacillus iheyensis BS1(2) in microbial fuel cells and effect of different anodic operating conditions. Ann. Appl. Microbiol. Biotechnol. J. **2**(1), 1010 (2018)
3. Yokoyama, H., Ohmori, H., Ishida, M., Waki, M., Tanaka, Y.: Treatment of cow waste slurry by a microbial fuel cell and the properties of the treated slurry as a liquid manure. Anim. Sci. J. **77**(6), 634–638 (2006). https://doi.org/10.1111/j.1740-0929.2006.00395.x
4. Parkash, A., Aziz, S., Abro, M., Soomro, S.A., Kousar, A.: Design and fabrication of microbial fuel cell using cow manure for power generation. Sci. Int. **27**, 4235–4238 (2015)
5. Zheng, X., Nirmalakhandan, N.: Cattle wastes as substrates for bioelectricity production via microbial fuel cells. Biotech. Lett. **32**(12), 1809–1814 (2010). https://doi.org/10.1007/s10529-010-0360-3
6. Lee, C.Y., Huang, Y.N.: The effects of electrode spacing on the performance of microbial fuel cells under different substrate concentrations. Water Sci. Technol. **68**(9), 2028–2034 (2013)
7. Zhang, X., Porcu, L., Andresen, J.M.: November. sustainable energy from dairy farm waste using a microbial fuel cell (MFC). In: World Renewable Energy Congress-Sweden, 8–13 May, Linköping; Sweden. Linköping University Electronic Press, vol. 57, pp. 247–250 (2011). https://doi.org/10.3384/ecp11057247

8. Barua, P.K., Deka, D.: Electricity generation from biowaste based microbial fuel cells. Int. J. Energy Inf. Commun. **1**(1), 77–92 (2012). https://doi.org/10.1063/1.5031979
9. Inoue, K., et al.: Electricity generation from cattle manure slurry by cassette-electrode microbial fuel cells. J. Biosci. Bioeng. **116**(5), 610–615 (2013). https://doi.org/10.1016/j.jbiosc.2013.05.011
10. Nimje, V.R., Labrath, Y.P., Gaikar, V.G.: Development of a microbial fuel cell using distillery spent wash: evaluation of current generation and COD removal with respect to pH. Iranica J. Energy Environ. **4**(4), 348–356 (2013)
11. Javalkar, P.D., Alam, J.: Comparative study on sustainable bioelectricity generation from a microbial fuel cell using bio-waste as fuel. Int. J. Sci. Res. Publ. **3**(8), 1–6 (2013)
12. Thatoi, P.: Characterization of generated voltage, current, power, and power density from cow dung using double chamber microbial fuel cell (Doctoral dissertation, National Institute of Technology Rourkela) (2014). https://doi.org/10.4172/2161-0398.1000208
13. Sethia, K., Kaushik, A., Jadhav, S.K., Quraishi, A.: Effect of operational parameters on cow dung mediated microbial fuel cell. World J. Eng. **12**(6), 541–550 (2015). https://doi.org/10.1260/1708-5284.12.6.541
14. Sonu, K., Das, B.: Comparison of the output voltage characteristics pattern for sewage sludge, kitchen waste and cow dung in single chamber single electrode microbial fuel cell. Indian J. Sci. Technol. **9**(30), 1–5 (2016). https://doi.org/10.17485/ijst/2016/v9i30/99177
15. Shiv, K., Kumar, H.D., Babu, K.G.: A study on the electricity generation from the cow dung using a microbial fuel cell. J. Biochem. Technol. **3**(4), 442–447 (2012)
16. Girija, D., Deepa, K., Xavier, F., Antony, I., Shidhi, P.R.: Analysis of cow dung microbiota - a metagenomic approach. Indian J. Biotechnol. **12**, 372–378 (2013)
17. Gaujoux, R., Seoighe, C.: A flexible R package for nonnegative matrix factorization. BMC Bioinform. **11**, 367 (2010). https://doi.org/10.1186/1471-2105-11-367
18. Zaneveld, J., Knight, R.: QIIME allows analysis of high-throughput community sequencing data. Nat. Meth. **7**(5), 335–336 (2010). https://doi.org/10.1038/nmeth.f.303

Sentiment Analysis in the Ecuadorian Presidential Election

Juan Carlos Minango Negrete[1(✉)], Yuzo Iano[2], Pablo David Minango Negrete[2], Gabriel Caumo Vaz[2], and Gabriel Gomes de Oliveira[2]

[1] Instituto Tecnológico Universitario Rumiñahui, Sangolquí, Ecuador
juancarlos.minango@ister.edu.ec
[2] State University of Campinas, Campinas, Brazil
yuzo@unicamp.br, oliveiragomesgabriel@ieee.org

Abstract. The social network Twitter is characterized as one of the main means of communication used by politicians, governmental agencies, among others to communicate and engage with the people and prospective voters. This article proposes a methodology that might be employed in text extraction, acquisition, cleaning, and processing with the aim of analyzing the sentiments in the texts within the tweets of the 2021 Ecuadorian presidency candidates, Andrés Arauz and Guillermo Lasso. Many libraries that are specialized in text analysis were used to determine the sentiment analysis in the extracted tweets. The achieved results show a classification comparison among the employed libraries, highlighting that most of the tweets of both candidates were classified as positive and, following, neutral polarity.

Keywords: Ecuador · Election · Data analysis · Twitter

1 Introduction

With the emergence of the so-called Web 2.0, the massive production of text-type information on web pages arose, and it was boosted since the advent of social networks, which changed the way people communicate and interact with each other [1].

Companies, from the smallest to the biggest ones, have been making every effort and investment to create multimedia content through posts, comments, and messages, which allow the users to interact and meet such companies. Similarly, public people, politicians, and artists, among others, use social networks to share their opinions, sentiments, and even their political position.

Under this context, Twitter is a microblogging social network that allows 280 characters per tweet and has about 187 million active users per month [2]. These characteristics make Twitter the ideal platform for users to share their opinions

Y. Iano et al. (Eds.): BTSym 2021, SIST 207, pp. 25–34, 2023.
https://doi.org/10.1007/978-3-031-04435-9_3

about worldwide tendencies. Therefore, Twitter is mainly used by government agencies and politicians, among others, as an environment to share news, ideas, and opinions.

Twitter is a trustworthy source to perform sentiment analysis [3]. Sentiment analysis is a computational process that has the task of processing, identifying, and classifying the opinions expressed by someone in a text to determine the person's position concerning a specific product, event, etc. Three sorts of sentiments are considered: positive, negative, and neutral [4].

This case study assesses people's stance with respect to the second round of the Ecuadorian presidential election in 2021, which was disputed by candidate Andrés Arauz and his opponent Guillermo Lasso. The data collection was made during the electoral campaign period.

The article is organized as follows: Sect. 2 makes an overview of works related to sentiment analysis. In Sect. 3, the applied methodology, the data collection and pre-processing, and the libraries used to perform the sentiment analysis are described. Section 4 groups the results achieved with a comparison of the libraries employed in the sentiment analysis. Finally, Sect. 5 presents the conclusions.

2 Related Works

In [5], the tool Stanford NLP for Natural Language Processing (NLP) was applied with an adaptation for Ecuadorian regional language. The sentiment analysis was performed on the social network Twitter with the messages sent to Ecuadorian former president Rafael Correa. The political assessment via Twitter in South America is still very scarce, and the election analyses in Venezuela [6], Chile [7], and Argentina [8].

Works about sentiment analysis based on the English language, like [9], study the governmental policies from the point of view of ordinary people in order to analyze their sentiments and stances concerning the new demonetization rules. The data collection is made via trends query using hashtags (#demonetization).

Finally, it is worth highlighting that most sentiment analysis libraries were developed for the English language due to its universality.

3 Methodology

This section presents the proposed work methodology to analyze and extract information from the social network Twitter data source. So, Fig. 1 shows the workflow applied in this experimental research.

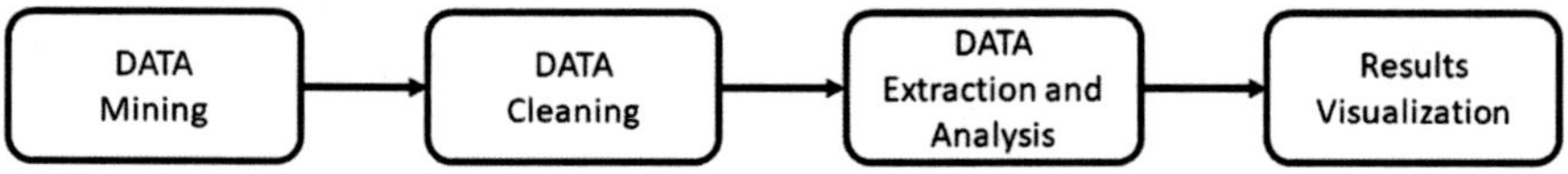

Fig. 1. Work methodology.

3.1 Data Mining

Depending on the software employed in the data collection from Twitter, there are many libraries and APIs of the own social network, which allow the extraction of countless characteristics embedded in each tweet of the user.

To use the Twitter proprietary API, it is necessary to create a register, specifying the reasons and the aim of the use of the data. Once the register is made, it must be approved to make the *Keys* and *Tokens* available and allow the exploration or mining of the information.

In this work, it was employed the development interface of Python with the IDE Jupyter Notebook, which has the library *tweepy*. The data extraction was directed to the search of current tendencies in the political sphere: hashtags that had trends in the Ecuadorian election campaign were evaluated, specifically for the second electoral round, which took place from March 6th, 2021 to April 8th, 2021.

The library *tweepy* provided all the information embedded in each tweet, which was subsequently stored in a CSV (Comma Separated Values) file within which the following fields were chosen for the analysis:

- Tweet text - "text".
- Tweet creation date - "created_at".
- Tweet location - "location"
- Retweets amount - "retweet_retweet_count"

3.2 Data Cleaning

Once the tweets that refer to the searched trend were achieved, the text of each tweet was cleaned, and the variables formats were adjusted to proceed with the right analysis in each of them. The CSV file was loaded in the format *data frame* with the help of the library pandas. To remove duplicate tweets, it was applied the tool `drop_duplicates` , which perform a row-by-row analysis in the specified column aiming to delete the duplicate rows.

Continuing with the data pre-processing, a function was created to clean the text based on regular expressions. Furthermore, it is considered the removal of links for external web pages with format `http\S+`, which allows finding this format within the text and delete the whole string that follows it. In the regular expressions, punctuation marks that affect the correct understanding of the context of the tweets (e.g., periods, commas, question marks, and exclamation marks, among others) were considered.

In addition, the removal of users named within the tweets was also considered. To perform such elimination, regular expressions including the format `@\S+` were used, allowing the deletion of user names mentioned within the text of the tweet.

StopWords: StopWords are lists of words stored in the package NLTK for the natural language text pre-processing, which aim to remove a set of words that do not affect the phrase context. The StopWords are, mainly, articles and pronouns that, when removed, do not affect the phrase context in an afterward analysis.

Tokenization: This is another NLTK package that, once all the aforementioned pre-processing is performed, breaks the phrases in each one of the words that compose them. The tokenization allows the computer to analyze the text word by word.

3.3 Frequency of Words - Word Clouds

After the data cleaning, the NLTK tool to compute the frequency of the words was applied. This tool counts the words that repeat more frequently within the texts of all tokenized tweets.

In contrast, the word cloud graphic shows visually the words' frequency. It presents the more frequent words with larger font size, i.e., the higher the number of mentions within the analyzed text, the larger the font size in the word cloud graphic.

3.4 Sentiment Analysis in Spanish

For the sentiment analysis in the Spanish language, the library *sentiment-analysis-spanish* (version 0.0.25) was used. It is a library that applies a Naive Bayes classifier to predict the sentiments in the Spanish phrases. It was trained with 800,000 comments from users of the platforms Decathlon, Tripadvisor, FilmAffinity, ElTenedor, and eBay, which were extracted through web scraping. This library achieved an accuracy of 90% on the test dataset.

Equation 1 represents the Naive Bayes classifier, which is given by the conditional probability, where c is the text class (i.e., the kind of sentiment, which may be positive or negative), and x is the tokenized text (each word that composes the phrase have a numerical value that makes possible the computation of the probability of belonging or not to a specific sentiment, which might be positive or negative).

$$P(c \mid x) = \frac{P(x \mid c)\, P(c)}{P(x)} \tag{1}$$

This library computes the probabilities of the assessed phrase with values from 0 to 1. When the phrase or text achieves values that are close to 1, it is classified as positive. Otherwise, low probabilities are classified as negative. In this case study, values within the range from 0.4 to 0.6 are considered neutral.

3.5 Sentiment Analysis with English Translation

Since the sentiment analysis libraries for the English language are more developed, the tweets extracted in Spanish were translated to English.

For this purpose, the library *TextBlob*, which uses an API of Google Translate to translate texts to different languages, was employed. There is a daily limit for the number of translations, so the set of tweets was split in order to be possible to translate the texts without reaching the daily amount of translations allowed by the API.

TextBlob is a library of natural language processing that includes a class of sentiment analysis to determine the text polarity and its subjectivity as well. For the text polarity, the library provides values in the range from -1 to 1. When the values are close to -1, the text is classified as negative. Otherwise, for values close to 1, the classification is positive. In this study case, values that are close or equal to 0 are considered neutral.

Besides the library *TextBlob*, we also employed the library *Vader*, which was created specifically for the analysis of sentiments expressed in social networks, where the text syntax must be totally processed since there are many *posts* where the users use emoticons, colloquial language, and specific expressions from different regions. For that reason, the library *Vader* is based on the phrases lexicon and the manual construction of dictionaries of phrases and words that may help the text classification.

The library *Vader* delivers four values that represent the probability of a given phrase belong to a specific sentiment, which might be positive, negative, or neutral.

4 Results and Discussion

To analyze sentiments via the social network Twitter, it was employed, as the subject area, the Ecuadorian scenario during the second round of the 2021 presidential election, when it was possible to observe a growing use of social networks on the part of the presidential candidates Andrés Arauz from movement UNES (Unión por la Esperanza) and Guillermo Lasso from movement CREO (Creando Oportunidades).

The social networks (e.g., Twitter, Facebook, and the novel application TikTok) were the platforms where the candidates presented their proposals and attacked each other based on their prior history.

In this study case, tweets of both candidates were collected based on the social network trends and their hashtags. For the data mining of the candidate Guillermo Lasso, it was used the trend #EncontremonosParaLograrlo, and about 2900 tweets that referenced this slogan were achieved. For the candidate Andrés Arauz, the same procedure was made with the trend #ContigoConTodosAhora, and about 1900 tweets were collected.

The first step of the data cleaning was the deletion of duplicated tweets. Afterward, the text pre-processing was performed with the NLP tools mentioned in Sect. 3.2. Once the stop words and regular expressions were removed, a list with tokenized words was created to figure out the frequency of the words in the databases of both candidates.

Figure 2 shows the frequency of words for the trend #EncontremonosParaLograrlo of the candidate Guillermo Lasso. In this figure, it is possible to distinguish that this hashtag has the greatest predominance in the tweets. Besides, one can note that the tweets tendency makes reference to the words "gobierno" (government), "progreso" (progress), "seguridad" (safety), "empleo" (employment), and others that are usually used in presidential proposals.

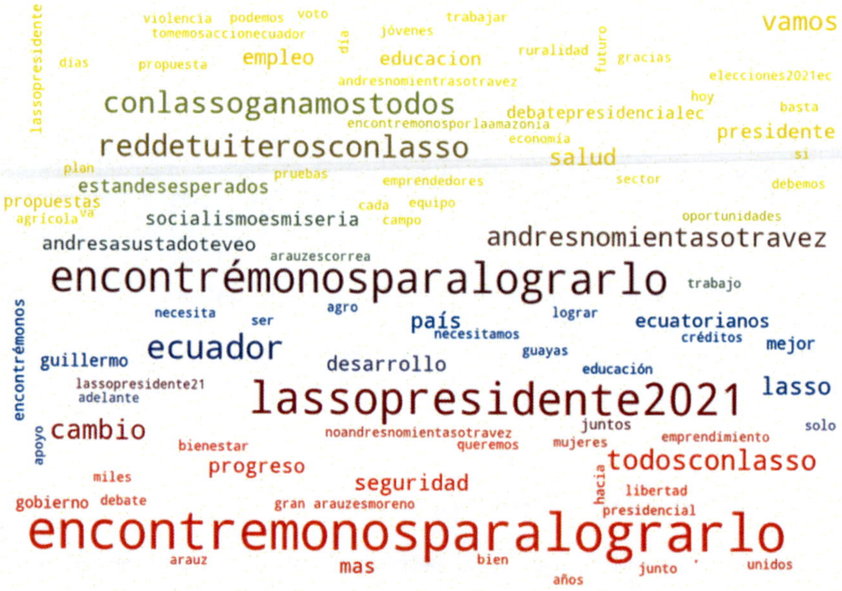

Fig. 2. Word cloud for #EncontremonosParaLograrlo of candidate Guillermo Lasso.

Figure 3 presents the distribution graphics of the three methods employed in the classification of sentiments embedded within the tweets. It is worth highlighting that the library for sentiment analysis in Spanish *spanish_sentiment* (see Fig. 3 (a)) classified most of the tweets as negative or neutral. In contrast, the library *TextBlob* (see Fig. 3 (b)) classified many tweets as positive, followed by tweets with neutral polarity. In the case of the library *Vader* (see Fig. 3 (c)), most of the tweets were classified as positive.

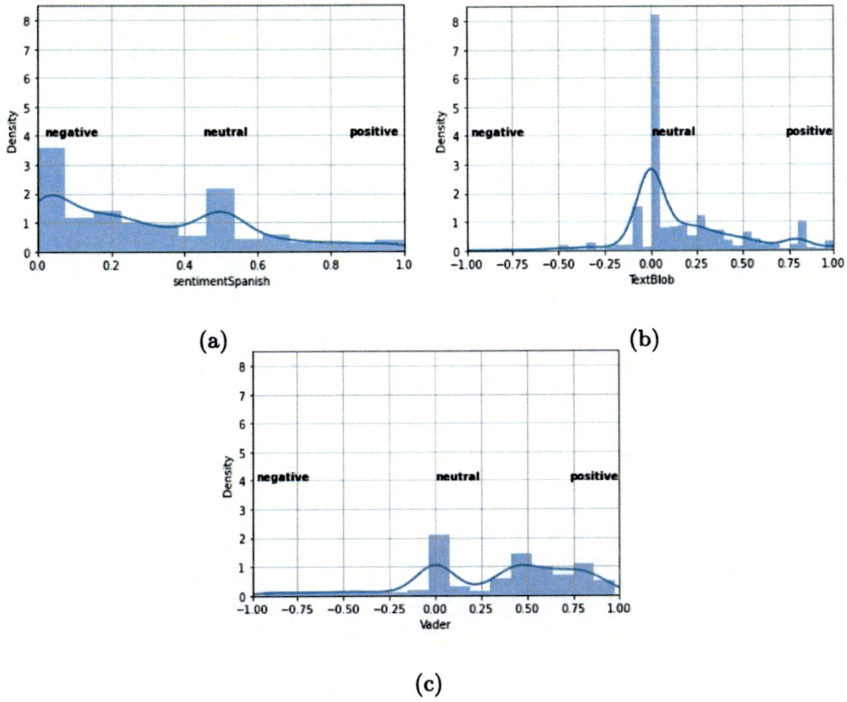

Fig. 3. Sentiment analysis in tweets collected for the trend #EncontremonosParaLograrlo of candidate Guillermo Lasso. (a) sentiment_spanish, (b) TextBlob, y (c) Vader.

In Fig. 4, one can see a comparison of the three libraries employed in the sentiment analysis performed in the tweets collected from candidate Lasso. It is possible to observe that the libraries *TextBlob* and *Vader* effectively classify most of the tweets with positive polarity.

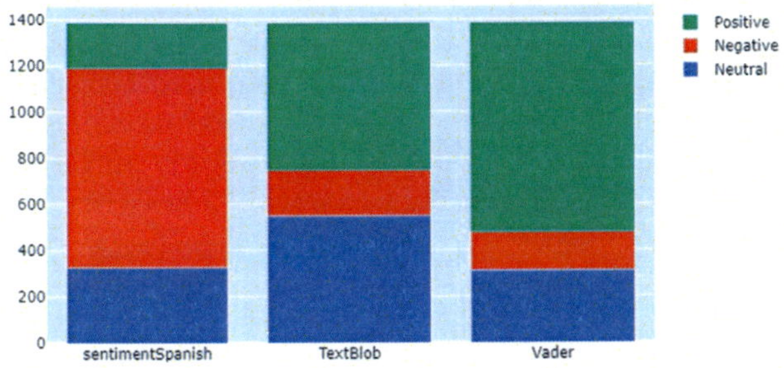

Fig. 4. Sentiment analysis with *spanish_sentiment*, *TextBlob*, and *Vader* of the candidate Guillermo Lasso.

Figure 5 shows the cloud of words with higher frequency in the tweets associated with the presidential candidate Andrés Arauz. One can observe that the most predominant word was not the hashtag trend, but the word "trabajo" (work). After analyzing the tweets that contained such a word, we could note that many of them make reference to the hard work that had been held for 20 months to strengthen the movement and reach the electoral second round since the candidate Andrés Arauz was totally unknown and took part for the first time in an election.

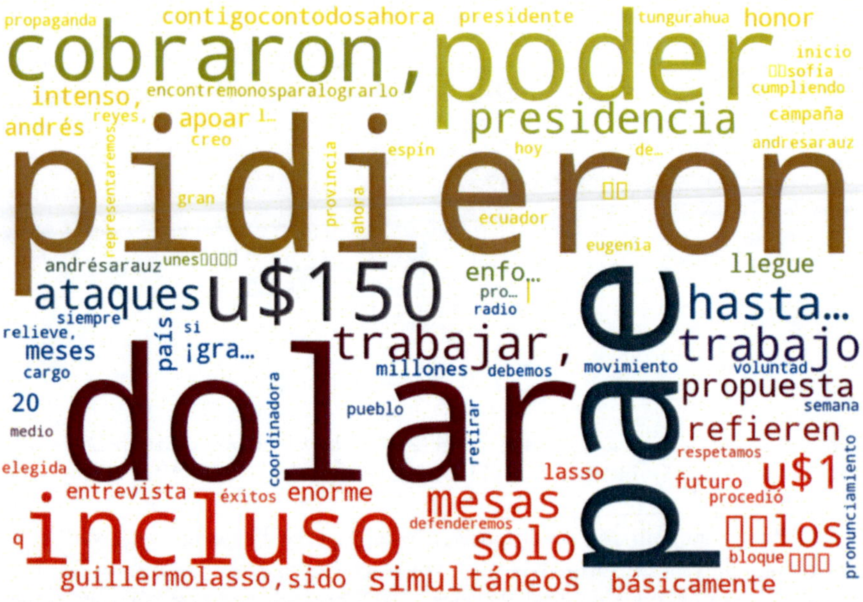

Fig. 5. Word cloud for #ContigoConTodosAhora of candidate Andrés Arauz.

In Fig. 6, there is a noticeable tendency of the libraries *TextBlob* (see Fig. 6 (b)) and *Vader* (see Fig. 6 (c)) to classify the tweets in the range from 0 to 1, siendo en su mayoria clasificados como neutros y positivos. i.e., most of them are classified as neutral or positive. In contrast, the library *sentiment_spanish* (see Fig. 6 (a)) classified most of the tweets as negative, followed by neutral.

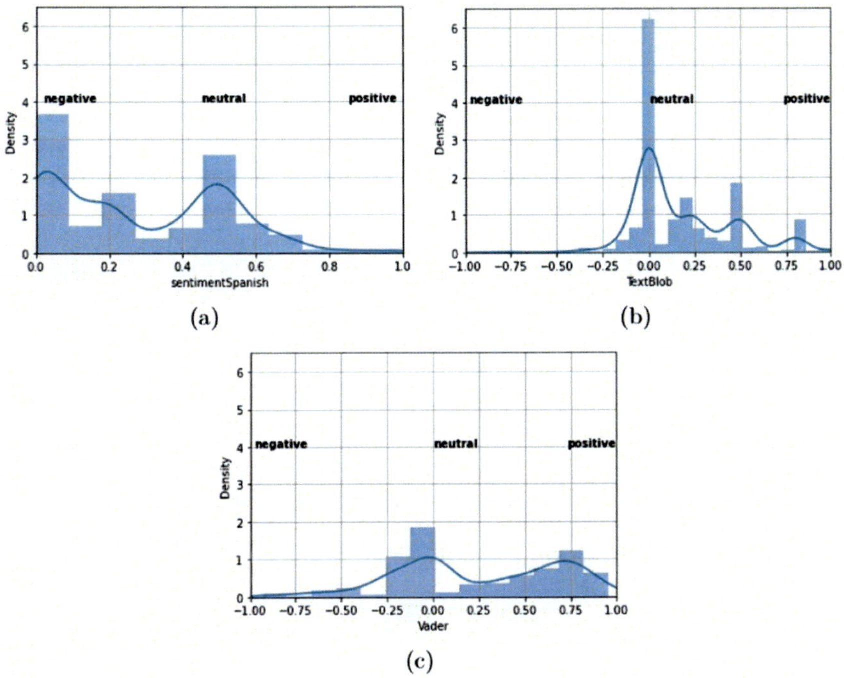

Fig. 6. Sentiment analysis in tweets collected for the trend #ContigoConTodosAhora of candidate Andrés Arauz. (a) sentiment_spanish, (b) TextBlob, y (c) Vader.

Figure 7 shows an overview of the way how each library of sentiment analysis classified the tweets of the candidate Andrés Arauz, where the tendency of the classifier *spanish_sentiment* was negative. By contrast, the libraries *TextBlob* and *Vader* classified most of the tweets with positive polarity.

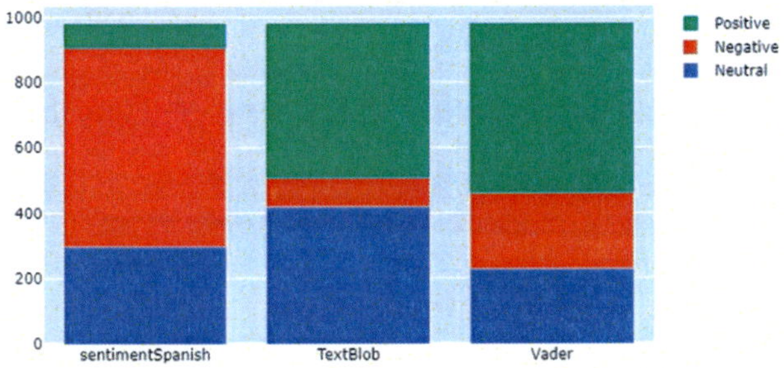

Fig. 7. Sentiment analysis with *spanish_sentiment*, *TextBlob*, and *Vader* of candidate Andrés Arauz.

5 Conclusion

This article presented the process of data collection from the social network Twitter, the data cleaning, the text pre-processing, and the sentiment analysis with the aid of the libraries *sentiment_spanish*, *TextBlob*, and *Vader*. The data collection from Twitter had as scenario the presidential election that took place in Ecuador. The sentiment analysis performed with the library *sentiment_spanish* classified most of the tweets of both candidates as negative. In contrast, the libraries *TextBlob* and *Vader*, which have a better performance in the task of analyzing sentiments based on text, classified most of the tweets of both candidates as positive, followed by neutral polarity.

References

1. Asghar, M.: Detection and scoring of internet slangs for sentiment analysis using sentiWordNet. Life Sci. J. **11**, 66–72 (2014). https://doi.org/10.6084/M9.FIGSHARE.1609621
2. Statista Twitter: most users by country – Statista. Statista (2021)
3. Kouloumpis, E., Wilson, T., Moore, J.: Twitter sentiment analysis: the good the bad and the OMG!. In: Proceedings of the International AAAI Conference on Web and Social Media, vol. 5, no. 1, pp. 538–541 (2011). https://ojs.aaai.org/index.php/ICWSM/article/view/14185
4. Zahoor, S., Rohilla, R.: Twitter sentiment analysis using lexical or rule based approach: a case study. In: 2020 8th International Conference on Reliability, Infocom Technologies and Optimization (Trends and Future Directions) (ICRITO), pp. 537–542 (2020). https://doi.org/10.1109/ICRITO48877.2020.9197910
5. Hidalgo, O., Jaimes, R., Gomez, E., Luján-Mora, S.: Sentiment analysis applied to the popularity level of the ecuadorian political leader rafael correa. In: 2017 International Conference on Information Systems and Computer Science (INCISCOS), pp. 340–346 (2017). https://doi.org/10.1109/INCISCOS.2017.64
6. Avila, D.A., Muñoz, A., Luengo, F., Chourio, X., Fernández, A.: Caracterizando las Elecciones Venezoelanas a Travé́s de Twitter. Caso: #26S. Anuario Electrónico de Estudios en Comunicación Social. Disertaciones **5**(1), 57–76 (2012). https://www.redalyc.org/articulo.oa?id=511555573008
7. Guevara, M., Pino, D., Mendoza, M., Silva, C., Olivares, M.: Chile y el Ecosistema de las Elecciones Políticas en Twitter (2013). https://doi.org/10.13140/2.1.3324.7684
8. Robins, D., Frati, F.E., Alvarez, J., Texier, J., Loto, L.: Balotaje Argentina 2015 a partir de un análisis de sentimiento de tweets. Zenodo (2016). https://doi.org/10.5281/zenodo.51496
9. Prabhsimran, S., Ravinder, S.S., Karanjeet, S.K.: Sentiment analysis of demonetization of 500 & 1000 rupee banknotes by Indian government. ICT Express **4**(3), 124–129 (2018). https://doi.org/10.1016/j.icte.2017.03.001

5G - Automation of Vertical Systems in the Industry 4.0

Daniel Izario[1(✉)], João Brancalhone[2], Yuzo Iano[1], Gabriel Gomes de Oliveira[1], Gabriel Caumo Vaz[1], and Karine Izario[3]

[1] University of Campinas, Campinas, SP, Brazil
daniel_izario@hotmail.com, yuzo@decom.fee.unicamp.br, oliveiragomesgabriel@ieee.org

[2] National Telecommunications Institute, Santa Rita do Sapucai, MG, Brazil

[3] Educational Anhanguera, Indaiatuba, SP, Brazil

Abstract. This article approaches the main uses of the 5G technology as well as its consequences in the vertical systems and, more specifically, in the industrial automation. It also establishes relations among key performance indicators in several scenarios with other technologies currently in use such as cabled networks and LTE. Significant advantages are experienced in many exposed scenarios, mainly owing to the new distributed 5G network's topology.

Keywords: 5G · Automation · Distributed topology · Industry 4.0 · Key performance indicator · Mobile communications and Vertical systems

1 Introduction

The 5th Generation mobile networks (5G) materialize themselves as the new generation regarding telecommunications standards, offering management to all its incoming demands. That said they are considerate technological ecosystems, enclosing giant proportions of the wireless communications, supporting a wide range of services, contrib uting to the economic and social development in a global scenario. Considering its in herent aspects, it has an important role in the vertical systems automation, such as IoT.

– Internet of Things applications and M2M - Machine-to-Machine, offering countless advantages such as low latency and power saving during the whole process, scalability, high QoS -Quality of Service and high data rate [1]. Figuring out as a key point to the called 4th Industrial Revolution (or Industry 4.0), as in the Fig. 1.

Despite several benefits such as the absence of cables and the reduction in the deployment and using costs, in which there's improvement in the resources and industrial managements, the use of wireless communications faces some opposition in the indus try, stronger than the face from final consumers. Its main reason is attributed to the lack of perceived reliability compare to the wired ones, taking to account that it involves critical processes [1].

Y. Iano et al. (Eds.): BTSym 2021, SIST 207, pp. 35–43, 2023.
https://doi.org/10.1007/978-3-031-04435-9_4

Evaluating the market, the technology offers promising results: it has higher speed and lower latency as well as, making a comparison with legacy technologies, combing short and long range communication technologies such as RFID - Radio-Frequency.

Identification and Bluetooth along with mobile networks, these, with the use of small cells and heterogeneous networks.

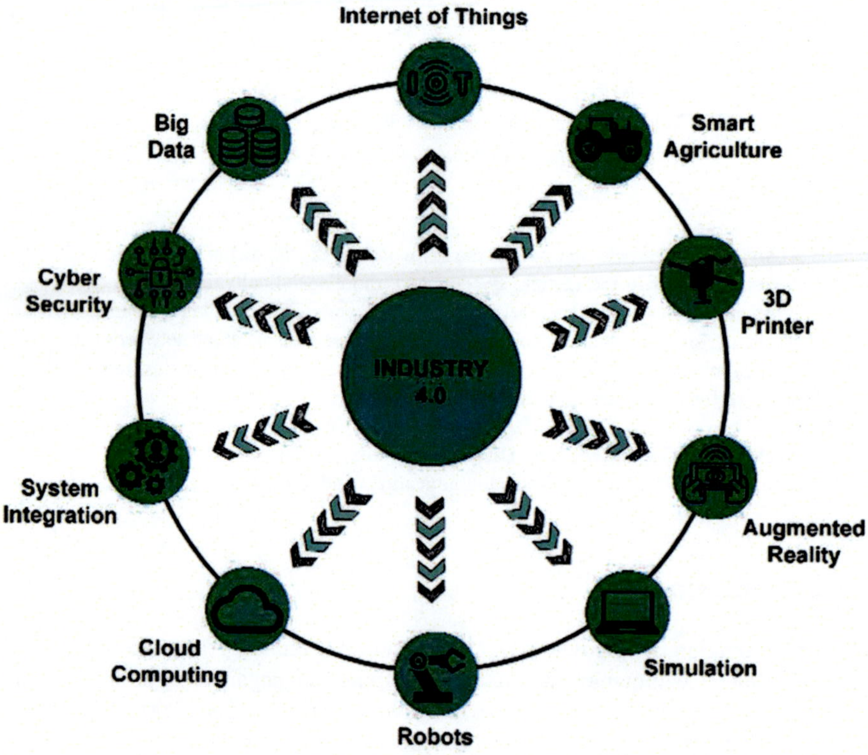

Fig. 1. Industry 4.0.

Initiated in the 1970s, the first general mobile communications were exclusively based on voice traffic. Towards the years, with the increase capacity of transmission and digitalization as well the development of new devices, there were an increase of non-voice data. First even in the denominated P2P - Person-to-Person communication as for example the SMS - Short Message Service [1].

The connections between mobile phones and the internet turn up more importantly during the third generation of mobile communications, when the use of mobile applications had sharply risen, increasing the P2M - Person-to-Machine data Exchange and, lingering, the M2M - Machine-to-Machine, a communication model widely used in the industrial automation. It hasn't achieved much use in this kind of environment though, considering its perceived lack of reliability as most of the wireless technologies [1].

Considering its lower latency and bigger connection's versatility, there are expected changes in the reliability offered by these networks. Consequently, it can be expected a rise in their use in telecommunications, in scenarios such as the M2M-Machine to Machine use as well as in the telephony, mobile internet and in the new applications of the called vertical domains.

2 Methodology

The first step of the 5G implementation in a vertical automation system environment is to understand its own domain, enclosing most of the demand, such as industries and entrepreneurs' groups [1], developing similar services or products such as in the following samples:

Remote Monitoring:

- Public Services: Water and electricity monitoring;
- Health: Remote Patient monitoring;
- Consumers: Alarms, safety sensors and m-health;

Asset Tracking:

- Transportation: Fleets management and monitoring;
- Consumer: Asset tracking;
- Automotive: Fleets management and stolen vehicles' recovering;
- Industry: Retail/warehouse operations;

Automation of Operations:

- Industry: Real-time automation;
- Health: Remote Surgery operation and robotics;
- Commercial real estate: Automation of installations;
- Construction, Oil & Gas and Mining: Machinery automation;

Preventive Manteniment:

- Transport & automotive: Vehicle's repair and battery change;
- Industry: Manteniment & devices;
- Commercial real estate: HVAC - Heating, Ventilating and Air Conditioning;

Video Streaming:

- Customer: CCTV - Closed-Circuit Television;
- Commercial real estate: Smart CCTV;

Others:

- Transportation and automotive: Autonomous vehicles;
- Customer & automotive: Connected electric cars;
- Retail: Real-time sales;

All the mentioned systems have critical demands regarding multiple concerns, as latency, data rate, availability and, in some cases, high-precision positioning systems. That said, the existing approaches referred to the applications, according to ISO - International Standardization Organization [2], the safety concept in a system is defined as the "ability of working when required", this divided in five main proprieties with recurrent use as presented in the Table 1.

Table 1. Five important properties of recurrent use in vertical domains [1]

Properties	Description
Reliability	System operation continuity in the properly way, being measured as the mean time between occurred faults
Availability	Promptness of the correct system operation when requested. Classified as the percentage of time in which the system operates properly
Maintainability	Submission, maintenance and repair abilities, measured by the repair mean time
Safety	Absence of catastrophic consequences to the user as well as to the environment
Integrity	Absence of inappropriate changes in the system

The 5G technology establishes a flexible access, allocating the demands in either distributed or centralized applications. These are connected to the responsible storage and network management data center as in the Fig. 2.

It's important to remind that the legacy mobile communications' generations, whom it takes part most of the installed available infrastructure, is mostly based in centralized applications, where multiple devices use to connected to the web, responsible for cen tral control and management.

The available structure, however, is deemed inefficient considering the speed and the latency intended to the 5G, in which final and intermediate devices communicated among them, lacking the need of a central structure. Observing the Fig. 2, there's distributed management system in the examples 1 and 2, meanwhile there's centralized management one in the 3 and 4. In order to approach this structure mode, it's a must considering the existing schematic regarding the called "Industrial radio communication distributed area", this contained in the 3GPP - 3rd Generation Partnership Program 16th release [3].

In the 16th release, the first system's layer contains the distributed automation functions, where there's an interface between communications functions and data capture

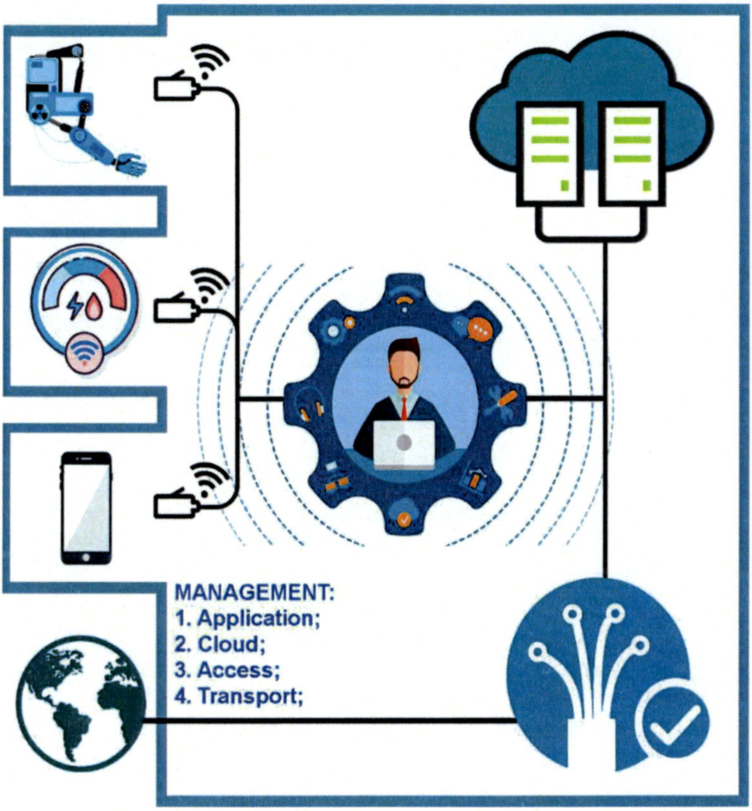

Fig. 2. Forms of 5G management.

devices, such as sensors, encoders, drivers, switches and I/Os. This distributed communication function communicate itself using the communication medium, what‘s fundamental to the proper automation application’s operation, a key point in the Industry 4.0, as in the Fig. 3.

3 Results

In the last years, the sharpen rise in the number of users as well as the operators’ quickly leadership towards legacy technologies such as 3G and 4G make the 5G user prognosis from GSMA - Global System for Mobile Association on the figure of billions in a 5 years period.

As mentioned, there’s a perspective of significant use technology’s use outside the mobile telephony, with many impacts expected in the Industry 4.0, mainly on three vertical domains: industrial automation, process automation and human-machine interface.

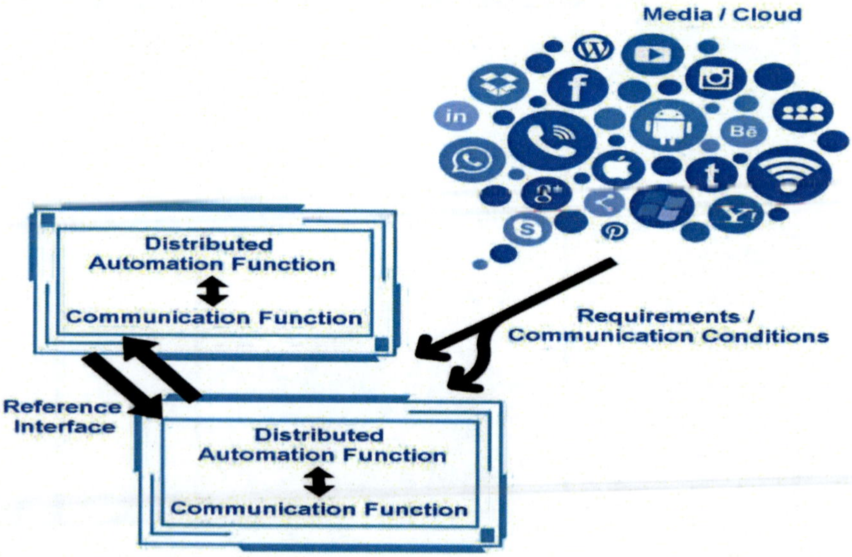

Fig. 3. Distributed automation application.

3.1 5G Impacts Studies

The 5G development runs along with the industrial automation main stage, when favorable technologies such as IoT, machine learning, vision and robotic systems are being simultaneously developed. The future factory, as it's nominated the incorporate use of several high-end technologies in industrial environments of different performances, a demanding place, where positive results are valued [1].

3.2 KPI - Key Performance Indicator

According to the 3GPP regulating agency, the expected developmental parameters in the 5G automation use, enclosing vertical domains are available in the 15th release, in which are part discrete automation systems, processes automation, power distribution systems and smart transportation systems [3].

The 15th release concerns the distinct interoperations ways as well as established 5G technologies used in automation, such as WLAN - Wireless Local Area Network, wired networks as well as local industrial networks. Highlighting, to each of those domains, sometimes either sub-domains, there are different expectances of KPI [3, 4].

3.3 Power Distribution and Fault Administration

One of the most important points, the management of failures and the recovery of energy distribution, if they occur, is something that requires several complex maneuvers in a short time, usually done manually as, but not only, collection and analysis of data from the network, fault isolation, determination of appropriate measures, service restoration, information to the maintenance team and, restoration of the normal network configuration, after fault correction [5].

In the cases of major areas or multiple simultaneous faults, the situation becomes more complex, evidencing the importance of automated systems using 5G technologies mainly throw its very low latency (lower than 500 ms). It exists, however, the possibility of using the FLISR - Fault Location, Isolation & Service Restoration, a fault localizer and service isolator, specially designed to the shut-down and changes in overcharged power grids, using an distributed topology containing many logic programmable controllers, located in the pole feeders with local intelligences as well as P2P support, not requiring a local or regional controlling system [5, 6].

Another point is the application support existence throw international IEC 61850 GOOSE, in which flexibility and interactivity requirements are stated in order to the application's implementation, allowing the isolation of area verification, in which service

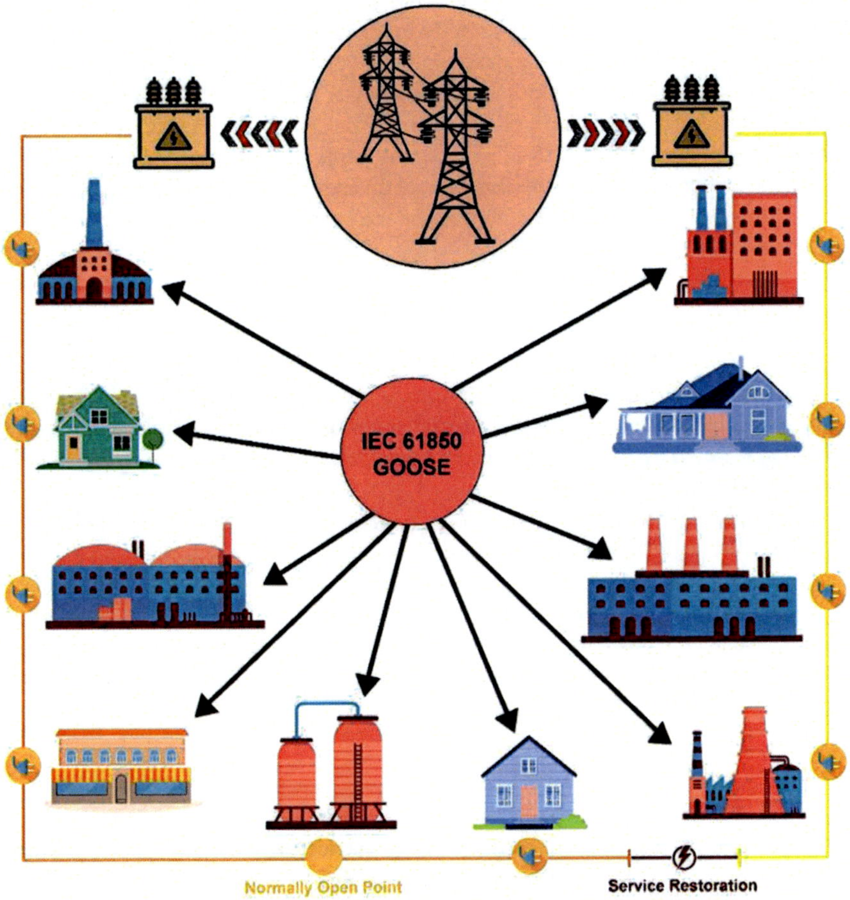

Fig. 4. Complete IEC 61850 topology.

restoration is required [6]. During that, the service is relocated to the nearest supplied point as in the Fig. 4.

3.4 Robotic Movements Controls

Mechanic tools, assembly and packing machines, autonomous vehicles are all examples of automated motion control. According to the 3GPP, latency in those scenarios can't surpass 1ms as well as 99,9999% of reliability and 2 ms of cycle time [3].

All those demands favor the 5G technologies' use, especially as one of bigger problems concerning traditional technologies is the interference from metallic structures. The low latency requirements, unavailable in them, make either the package transmission non practical or even unavailable [7].

The 5G Technologies can be significantly useful in such scenario, as it allows the transmission of smaller frames packages without waiting the ack (acknowledgment) responsible to the reception recognizing. It either allows the deployment of multi-path connectivity, keeping the process continuity in the case of lacking some lost frame.

Combining its very low latency and high availability it's possible the use of many factors such as frequency (temporal and/or spatial) applied to a mechanical arm or autonomous services. Considering the narrow band used in the 5G (100 MHz/channel) the frequency division isn't efficient in obtaining gains though, just obtaining reliability But there's the possibility of increasing special diversity throw the CoMP - Coordinated Multi Point as standardized in 5G, will allow the same transmission towards multiple points, preventing losses [7].

4 Conclusion

There are countless examples of the importance of 5G technologies in the Industry 4.0 as well as the internal and external industrial environment and other relevant activities, even when it rolled even clearly perceived as in the commercial power grid restoration after a shut-down.

However, the whole mentioned paradigm, considering the lack of reliability associated with the legacy wireless technologies in automation environments is deemed un-successful and without credibility, as it exists an increasing of speed compared to former technologies, the whole distributed structure and one of the highlights, the latency reduction. That said, the 5G in the vertical systems automation is possible, overcoming barriers and spotting itself as a complete technology to the Industry 4.0.

References

1. 5G Americas. 5G Communications for Automation Vertical Domains (2018). https://www.5gamericas.org/wp-content/uploads/2019/07/5G_Americas_White_Paper_Communications_for_Automation_in_Vertical_Domains_November_2018.pdf
2. ISO. International Organization for Standardization. https://www.iso.org/home.html
3. 3GPP. 3rd Generation Partnership Program. https://www.3gpp.org/

4. Bent, H.V.: Wireless Technology in Industrial Automation. Yokogawa Europe BV. https://web-material3.yokogawa.com/ISA100-whitepaper.pdf?_ga=2.181686752.695514765.1586637962–198698714.1585854637
5. Schwaderer, C.: 5G Adds Distributed Cloud Capability for Intelligent Edge. Embedded Computer Design (2019). https://www.embedded-computing.com/guest-blogs/5g-adds-distributed-cloud-capability-for-intelligent-edge
6. Souza, R.: Redes IEC-61850 - Estudo de Protocolo e Exemplo de Aplicação, Automação Industrial (2012). https://www.automacao industrial.info/redes-iec-61850- estudo-de-protocolo-e-exemplo-de-aplicacao/
7. Brown, G.: Ultra-reliable Low-Latency 5G for Industrial Automation. Qualcomm (2016). https://www.qualcomm.com/media/documents/files/read-the-white-paper-by-heavy-reading.pdf

Study of Sustainability Indicators for Environmental Management in Hydrographic Basins

Débora Akemi Ide Zancanela, Denise Helena Lombardo Ferreira(✉), Lucas Antônio Gomes do Carmo, Maria Luisa Damas Silva, and Cibele Roberta Sugahara

Pontifical Catholic University of Campinas, Campinas, Brazil
{debora.aiz,lucas.agc,maria.lds}@puccampinas.edu.br, {lombardo, cibelesu}@puc-campinas.edu.br

Abstract. The great population advance and the unrestrained consumption of natural resources have contributed to the accelerated deterioration of the physical, chemical, and biological characteristics of the aquatic system. In general, Brazilian hydrographic basins are in environmental degradation, which may also compromise human life in the near future, often a consequence of inadequate public policies. Therefore, this research aims to get to know and study the main sustainability indicators used in the environmental management of the Hydrographic Basins of the Piracicaba, Capivari, and Jundiaí rivers, in order to analyze the existing environmental impacts. Considering the importance of the water resource and all the problems involving the critical issue of water in Brazil, it is emphasized that sustainability indicators are important for the management of basins and sub-basins, as they serve as a basis for the formulation of action plans to propose improvements and implement solutions. Thus, the results obtained from the information and data collected can be used to understand the indicators related to water resources in the municipalities of the Metropolitan Region of Campinas, in order to alleviate water scarcity and related problems.

Keywords: Water resources · Sustainability indicators · Hydrographic basins

1 Introduction

The water resource is of great importance, both for the environment and for human life. However, it has suffered an accelerated degradation of its physical, chemical, and biological characteristics, caused, in its majority, by anthropic actions. In this context, the management of Hydrographic Basins - an important water resource - is essential, as it is not only composed of water but also of other natural resources, such as source material, topography, vegetation, and climate, being essential to take care of all these components [1].

Y. Iano et al. (Eds.): BTSym 2021, SIST 207, pp. 44–55, 2023.
https://doi.org/10.1007/978-3-031-04435-9_5

The Hydrographic Basins play a fundamental role in decision-making for the elaboration of public policies, planning, and territorial management, as they are directly or indirectly related to the efficient environmental management of the basins. Thus, the use of water and land for human activities has a great impact on the environment and, as a consequence, can influence the environmental quality of its water resources, which makes it important to develop sustainability indicators to support water management and territorial planning policies aimed at the preservation of natural resources [2].

Thus, it is necessary to take action to work around and propose solutions to problems related to water and, consequently, to the environment and the population. These problems are related to the mismanagement of water resources; the high unnecessary consumption; the water crisis; and/or the large levels of water loss.

The Hydrographic Basins of the Piracicaba, Capivari, and Jundiaí Rivers (PCJ Basins) have a relationship with the Metropolitan Region of Campinas (MRC) since the municipalities in the MRC are supplied by these basins, which also present problems with water resources.

The PCJ Basins are in critical situations due to low water availability, poor water quality, supply problems, climate change, population increase, and consequent increase in per capita consumption. This scenario is due to the history of water crises and high water consumption.

Accelerated urbanization has intensified the water demand, causing an increase in food, energy, and other items that depend on the water resource, that is, the social water demand is expanding [3]. Hence, it is important to understand the key issues and tools for managing water use, making its use more rational and more efficient.

It is essential to study sustainability indicators related to water resources and also to understand water systems, both for the MRC and the PCJ Basins, in order to identify the focus of the problem and take actions to reduce misuse and possible water losses, in addition to alleviating water scarcity and related problems.

In this sense, this research aims to know and study the main sustainability indicators used in the environmental management of PCJ Basins, in order to analyze the existing environmental impacts.

2 Metropolitan Region of Campinas

The Metropolitan Region of Campinas (MRC) was created in the 2000s and comprises 20 municipalities, Americana, Artur Nogueira, Campinas, Cosmópolis, Engenheiro Coelho, Holambra, Hortolândia, Indaiatuba, Itatiba, Jaguariúna, Monte Mor, Morungaba, Nova Odessa, Paulínia, Pedreira, Santa Bárbara d'Oeste, Santo Antônio de Posse, Sumaré, Valinhos, and Vinhedo [4].

The MRC has 3,791.79 km^2 and is the second-largest metropolitan region in the State of São Paulo in terms of inhabitants – in 2017, according to the [5], the population was greater than 3.2 million and, in 2016, it generated 8.75% of the São Paulo State's Gross Domestic Product (GDP).

The MRC has its economic importance at the state and national levels, given that it has a modern industrial park, is composed of complementary sectoral segments, has significant agricultural and agro-industrial structures within the scope of tertiary activities, in addition to the presence of innovative scientific research and technological centers [5].

The city of Campinas is the largest in area, population, and GDP. Concerning GDP, the city of Paulínia stands out and, in relation to demographic density, the city of Hortolândia is ahead.

The water situation of the MRC is worrying, as it is located in the Southeast region - marked by low rainfall, intensified by the summer, especially in the situation faced by the state of São Paulo in the period from 2013 to 2015.

3 PCJ Hydrographic Basins

The Hydrographic Basins of the Piracicaba, Capivari, and Jundiaí Rivers (PCJ Hydrographic Basins) comprise a region of 15,303 km^2, with territory distributed between the states of São Paulo (92.6%) and Minas Gerais (7.4%) [6].

Although the PCJ Hydrographic Basins stand out in terms of socioeconomic relevance, they present low water availability and compromise the quality of their watercourses [7].

The Committee on the Hydrographic Basins of the Piracicaba, Capivari, and Jundiaí Rivers (CHB-PCJ), created in 1993, has been making efforts to meet its objectives, being able to achieve significant results with the Sustainable Development Goals (SDGs) and the indicators of sustainability [8].

Of the total 15,303 km^2, 92.6% are located in the State of São Paulo and 7.4% in the State of Minas Gerais, so that 44 municipalities are fully inserted in the area and 29 municipalities are partially inserted [9]. According to [10] the Capivari River Basin comprises 1,568 km^2, the Jundiaí River Basin 1,154 km^2, and the Piracicaba River Basin 12,655 km^2. However, there was a reduction in water availability per capita in the region, from 1,014.13 m^3/inhabitant.year in 2014 to 971.08 m^3/inhabitant.year in 2018 [6].

Figure 1 highlights the PCJ hydrographic basins map.

Fig. 1. Map of PCJ hydrographic basins (Source: [6]).

Figure 2 shows the availability of water in the PCJ Basins for the period 2014 to 2018. Figure 3 highlights the demand for water from these Basins for the same period for the various segments.

Water Availability					
Parameter	2014	2015	2016	2017	2018
Availability per capita - Average Flow relative to the Total Population (m^3/inhab.year)	1014.33	1000.97	990.92	980.96	971.08

Fig. 2. Water availability in PCJ hydrographic basins (Source: [6, p. 28]).

By analyzing Figs. 2 and 3, it is inferred that water availability is critical, while water demand is growing. In addition to the growing demand, there are problems concerning the flow of waste, solid waste, high frequency of flooding, and inadequate disposal of sewage.

The PCJ Basins are of high importance in the context of economic growth due to the growth in irrigation demands, industrial activities, and public supply - the latter has the greatest demand, i.e., the highest flow, and part of this flow is released again into the streams and the other part does not return to water bodies (water balance). The demand for public supply in the PCJ Basins is close to 16.33 m^3/s, with the municipality of Campinas being the municipality with the largest population in these Basins, with the

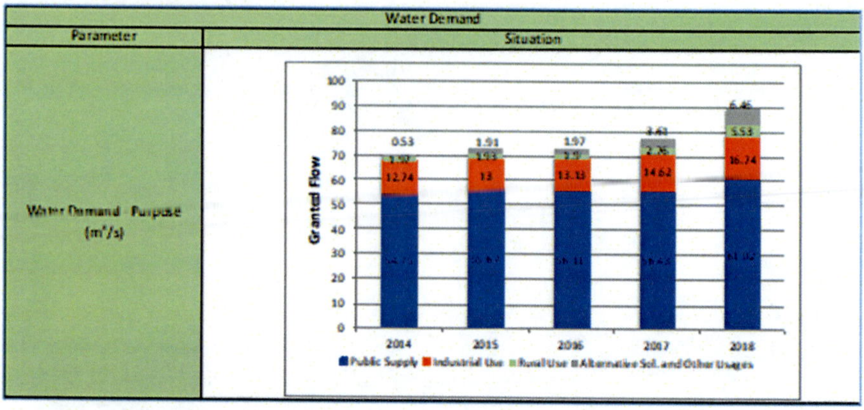

Fig. 3. Water demand of PCJ hydrographic basins (Source: [6, p. 28]).

highest demand, around 3.05 m^3/s. The Atibaia River (4.56 m^3/s) and Jaguari River (2.92 m^3/s) sub-basins have the highest demands [10].

In Brazil, Federal Law No. 9,433/1997 [11], which gave rise to the National Water Resources Policy and the National Water Resources Management System, defined charging for the use of water as one of its instruments, being fundamental for making the use of water resources more rational and encourage the reduction of extreme and, eventually, unnecessary consumption. The PCJ Basins represent a region with critical issues related to water resources, more specifically, with water availability, due to urban-industrial growth. Additionally, a significant part of these resources is used to supply the Metropolitan Region of São Paulo – the largest consumer of water in the region, followed by the industrial sector. The average water consumption of the industrial sector in the PCJ Basins is 14%, above the national average [12].

In Brazil, according to data from the National Water Agency (ANA, from the Portuguese Agência Nacional de Águas e Saneamento Básico) [13] for the year 2015, there were 197 river basin committees (CBH) – public bodies that deliberate on matters relating to the use of water in a basin. One of these tools is the Hydrographic Basin Plan, whose purpose is to allow the management of basins, making it possible to analyze data, make projections for the future and determine improvement targets. In this perspective, the CBH must approve and monitor the measures taken to achieve the goals.

The municipalities that make up the MRC have a relationship with the PCJ sub-basins, as shown in Table 1.

Table 1. Sub-basins that supply the municipalities of the MRC

Municipalities MRC	Sub-basins
Americana	Jaguari, Atibaia, Piracicaba
Artur Nogueira	Jaguari, Atibaia, Piracicaba

(*continued*)

Table 1. (*continued*)

Municipalities MRC	Sub-basins
Campinas	Jaguari, Atibaia, Piracicaba, Capivari
Cosmópolis	Jaguari, Atibaia
Engenheiro Coelho	Jaguari
Holambra	Camanducaia, Jaguari
Hortolândia	Piracicaba, Capivari
Indaiatuba	Capivari, Jundiai
Itatiba	Atibaia
Jaguariúna	Camanducaia, Jaguari, Atibaia
Monte Mor	Piracicaba, Capivari
Morungaba	Jaguari
Nova Odessa	Atibaia, Piracicaba
Paulínia	Jaguari, Atibaia, Piracicaba
Pedreira	Camanducaia, Jaguari
Santa Bárbara d'Oeste	Piracicaba
Santo Antônio de Posse	Jaguari
Sumaré	Piracicaba
Valinhos	Atibaia, Capivari
Vinhedo	Atibaia, Capivari

Table 1 shows that the municipality of Campinas is served by four sub-basins. Other municipalities with fewer inhabitants are served by a single sub-basin: Engenheiro Coelho, Itatiba, Morungaba, Santa Bárbara d'Oeste, Santo Antônio de Posse, and Sumaré.

4 Material and Methods

To meet the objective of this research, which consists of knowing and studying the main sustainability indicators used in the environmental management of the Hydrographic Basins of the Piracicaba, Capivari, and Jundiaí rivers, in order to analyze the existing environmental impacts, books were consulted, scientific articles, dissertations, and theses on the subject. This research is characterized as exploratory and documentary. The procedure used was the one of obtainment of information at secondary sources from the National System of Sanitation Information (SNIS, from the Portugues Sistema Nacional de Informações sobre Saneamento). For the documental approach, the plans and reports over PCJ Hydrographic Basins were considered.

5 Results and Discussion

Bearing in mind the serious water situation of the PCJ Hydrographic Basins that meet the water demand of the Metropolitan Region of Campinas, some information was collected on this theme, which will be discussed below.

Regarding the water crisis, undeniably marked for the southeast region of Brazil, in which the Metropolitan Region of Campinas (MRC) is inserted, the study on water losses in the distribution systems in the region's cities is relevant, in order to mitigate water scarcity [14].

Water losses in a supply system are extremely important to assess the level of inefficiency in a given region since high levels of loss indicate precarious planning and maintenance of the commercial and distribution activities of a water operator. Commonly, a water distribution network presents distribution losses, thus, there is a limit to the reduction of this volume of losses [15].

Given this critical scenario, Instituto Trata Brasil, together with GO Associates, which is a multidisciplinary consultancy specializing in economic analysis, developed a study based on 2017 data from the SNIS on the "Challenges for Water Availability and Advancing the Efficiency of Basic Sanitation". The data in question show that drinking water losses in distribution are in more than 7,000 Olympic swimming pools per day, leading to a financial loss of approximately BRL 11 billion [14].

The municipality of Campinas occupied the list of 100 Brazilian cities with the highest water losses in 2019, being the only city in the MRC to be on the list. In the technical literature, the methodology normally used for the analysis of losses by providers and regulators is the water balance proposed by the International Water Association (IWA) [15].

The SNIS is administered by the Federal Government, within the scope of the National Sanitation Secretariat of the Ministry of Regional Development. Its importance is that this body provides information on the sanitation sector in Brazil divided into two components – water and sewage. The SNIS uses indicators for benchmarking, comparing, and measuring the performance of service provision, with a public and free database.

Thus, the graphs and tables shown below of indicators for the cities in the MRC refer to data obtained in the SNIS, for the years 2015 to 2019. However, not all cities have a database for the indicators selected in this time course.

The indicators defined for the study are: Produced Water Volume (AG006) and Consumed Water Volume (AG010). Indicator AG006 represents the annual volume of water available for consumption, comprising water collected by the service provider and imported raw water; indicator AG010 indicates the annual volume of water consumed by all users, comprising the micro-measured volume, the estimated consumption volume for connections without a hydrometer or with a stopped hydrometer, plus the volume of treated water exported to another service provider [16]. The graph in Fig. 4 refers to the average of these indicators for the cities in the MRC, in the years 2015 to 2019.

Based on the graph in Fig. 4, it can be seen that Campinas is more prominent in the MRC, as it has a high volume of water produced and high consumption of this resource. Still, it is possible to infer that all cities analyzed have a greater volume of water produced when compared to the volume consumed.

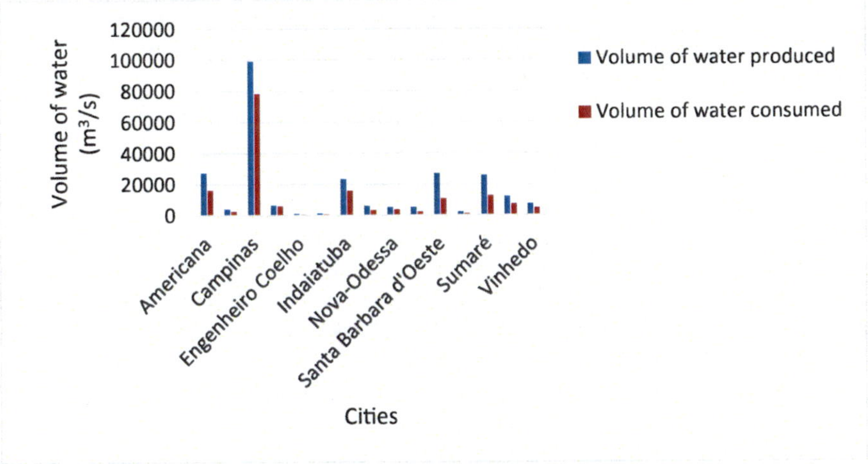

Fig. 4. Indicators AG006 and AG010 – MRC.

The graph in Fig. 5 highlights the average per capita water consumption (IN022 indicator), in liters per inhabitant per day, for the MRC, for the period 2015 to 2019. The calculation to obtain the IN022 indicator is given by (1) [16].

$$IN022 = \frac{(AG010 - AG019)}{AG001} \frac{1,000,000}{365} \quad (1)$$

where AG001 is the total population served with water supply; AG010 is the volume of water consumed and AG019 is the volume of treated water exported.

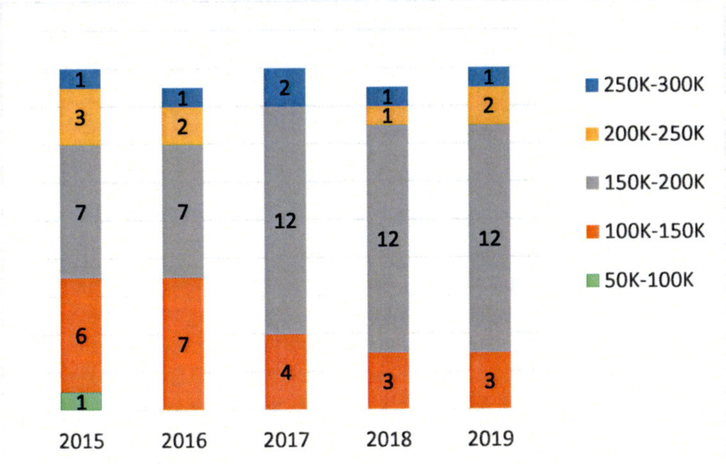

Fig. 5. Average per capita water consumption – indicator IN022 – MRC.

The IN022 indicator is part of the concept of accelerated urbanization that has intensified the water demand, causing an increase in food, energy, and other items that depend on this water resource, thus, the social water demand for water is in growth. It is interesting to analyze the different consumption profiles, which can contribute to a diagnostic analysis of the social, economic, and environmental situation, supporting processes of planning, management, and evaluation of public policies related to basic sanitation.

However, there is a certain limit regarding the average per capita water consumption indicator, since the average value can mask consumption in the most extreme, precarious, and less favored classes, due to the high values of the wealthier classes. Thus, in some cases, it is not possible to identify situations of consumption below the recommended minimum [17]. The average volume of water consumed per inhabitant per day indicates the water consumption pattern of a population contingent, which tends to grow with the population's income level. However, very high values can be considered unsustainable, considering the environmental point of view [17].

The graph in Fig. 6 shows the volume of water produced by the municipalities in the MRC in the period 2015 to 2019, and the graph in Fig. 7 shows the volume of water consumed by the same municipalities for the same period.

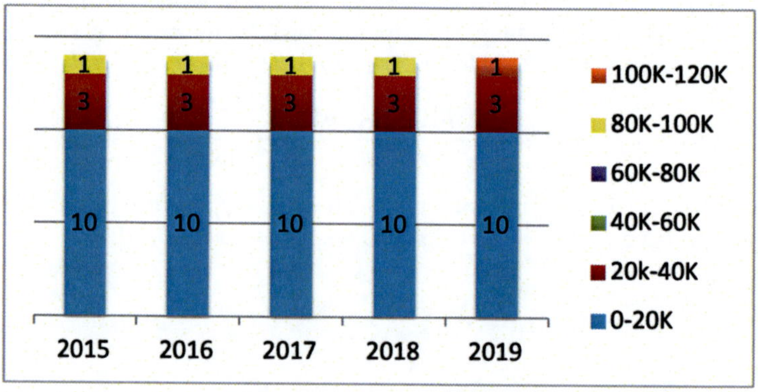

Fig. 6. Volume of water produced – indicator AG006 – MRC.

The graph in Fig. 8 highlights the MRC's water consumption for the different sectors.

The graph in Fig. 8 allows us to observe that the largest human urban consumer is Campinas, followed by Hortolândia, Indaiatuba, and Santa Bárbara d'Oeste. For industrial consumption, irrigated agriculture, and total consumption, the largest consumer is again the municipality of Campinas. Consumption for mining and thermoelectric plants does not present significant results for the Metropolitan Region in question.

From the graphs illustrated above, it can be inferred that, for most municipalities in the MRC, both the volume of water produced and water consumed show a significant increase.

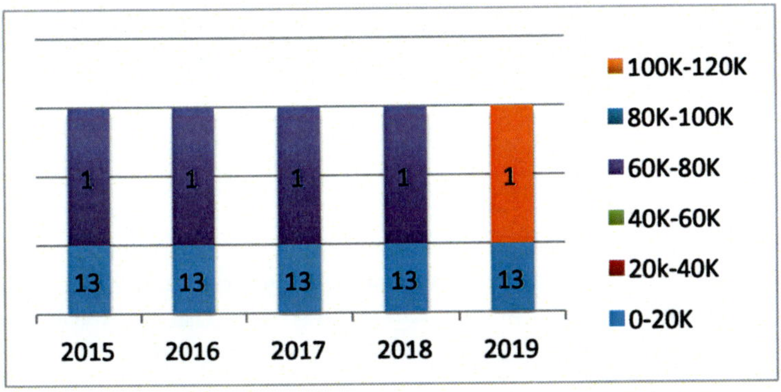

Fig. 7. Volume of water consumed – indicator AG010 – MRC.

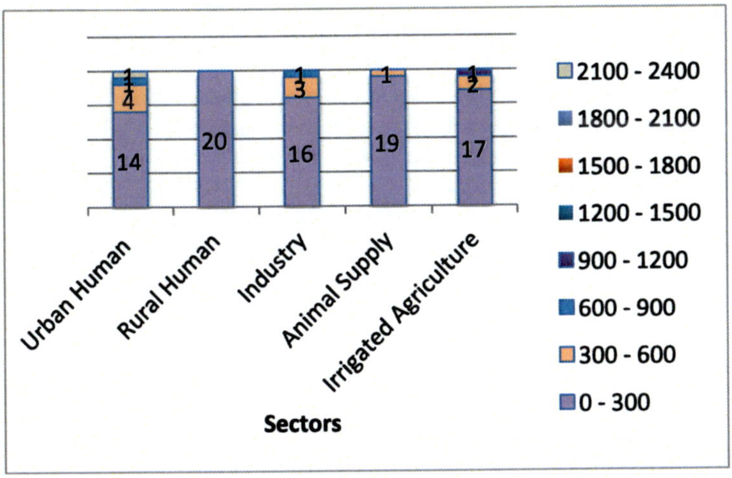

Fig. 8. Water use by sectors by the MRC.

In 2018, ten municipalities in the MRC were in the range of 0 to 20 thousand m^3/s of produced water volume, and only three municipalities in the range of 20 to 40 thousand m^3/s and one municipality in the range of 80 to 100 thousand m^3/s of produced water. On the other hand, thirteen municipalities were in the range of 0 to 20 thousand m^3/s of consumed water volume and only one municipality in the range of 60 to 80 thousand m^3/s.

Comparing with the data for the year 2019, ten municipalities maintained a water volume production in the range of 0 to 20 thousand m^3/s and three municipalities in the range of 20 to 40 thousand m^3/s, and only one municipality in the range from 100 to 120 thousand m^3/s. Regarding consumption, thirteen municipalities were also maintained in the range of 0 to 20 thousand m^3/s and only one of the municipalities showed an increase, passing to the range of 100 to 120 thousand m^3/s.

In addition to the high water consumption and the high loss rate of this resource, there are problems related to water quality, such as the presence of residual chlorine and turbidity in water samples [16].

6 Conclusion

From the information and data collected, the problem of water resources is understood. It is undeniable that water resource has immense importance, both for the environment and the quality of human life. However, this natural resource has suffered an accelerated degradation of its physical, chemical, and biological characteristics, caused, in large part, by anthropic actions.

Considering the importance of the water resource and all the problems involving the critical issue of water, it is emphasized that sustainability indicators are important for the management of hydrographic basins and sub-basins, as the indicators determine sustainability region and, consequently, serve as a basis for setting up action plans that aim to propose improvements and implement solutions.

Furthermore, river basins play a fundamental role in decision-making for the elaboration of public policies, planning, and territorial management, which are related to the efficient environmental management of the basins, which makes the elaboration of indicators of sustainability to support the management of water and territorial planning policies aimed at the preservation of natural resources.

The PCJ Hydrographic Basins are characterized by being a region with critical issues related to water resources, in particular with the availability of water, due to population and industrial growth, and also by the fact that a significant part of these resources is used for the supply of the Metropolitan Region of São Paulo – the largest consumer of water in the region, followed by the industrial sector.

References

1. Soares, A.B., Silva Filho, J.C.L., da, Abreu MCS de, Soares, F.A.: Revisando a estruturação do modelo DPSIR como base para um sistema de apoio à decisão para a sustentabilidade de bacias hidrográficas. Revista em Agronegócios e Meio Ambiente, **4**(3), 521–545 (2011)
2. Lelis, T.A., Calijuri, M.L., Santiago, A.F., de Lima, D.C., Rocha, E.O.: Análise de sensibilidade e calibração do modelo SWAT aplicado em bacia hidrográfica da região sudeste do Brasil. Rev. Bras. Ciênc. Solo **36**(2), 623–634 (2012). https://doi.org/10.1590/S0100-06832012000200031
3. Rodrigues, A.: Gestão Hídrica governança dos bens comuns Gestão Hídrica (2011). https://on-dasbrasil.org/wp-content/uploads/2020/02/livro-Gest%C3%A3o-H%C3%ADdrica.pdf. Accessed 06 Apr 2021
4. Emplasa Região Metropolitana de Campinas (RMC). GIP/CDI (2019). https://em-plasa.sp.gov.br/RMC#:~:text=A%20Regi%C3%A3o%20Metropolitana%20de%20Campinas,Santo%20Ant%C3%B4nio%20de%20Posse%2C%20Sumar%C3%A9%2C. Accessed 14 May 2021
5. PDUI. Plano de Desenvolvimento Urbano Integrado da Região Metropolitana de Campinas (2017). https://www.pdui.sp.gov.br/rmc/?page_id=56. Accessed 20 May 2021

6. Comitês, P.C.J.: Relatório de Situação dos Recursos Hídricos: versão simplificada (Ano Base 2018). Piracicaba: Fundação Agência das Bacias PCJ (2019)
7. Comitês, P.C.J.: Relatório Síntese: Plano de Recursos Hídricos das Bacias Hidrográficas dos rios Piracicaba, Capivari e Jundiaí, 2020 a 2035. Porto Alegre: Consórcio Profill-Rhama PCJ (2020a)
8. ANA. Agência Nacional de Águas e Saneamento Básico. O Comitê de Bacia Hidrográfica: o que é e o que faz? Agência Nacional de Águas e Saneamento Básico. Cadernos de capacitação em recursos hídricos Brasília: SAG, 64p (2011). https://ar-quivos.ana.gov.br/institucional/sge/CEDOC/Catalogo/2012/CadernosDeCapacitacao1.pdf. Accessed 05 May 2021
9. Cobrape. Companhia Brasileira de Projetos e Empreendimentos Plano das Bacias Hidrográficas dos Rios Piracicaba, Capivari e Jundiaí 2010 a 2020. Relatório Síntese. São Paulo: Cobrape: Neoband Soluções Gráficas (2011)
10. Comitês, P.C.J.: Sumário Executivo: Plano de Recursos Hídricos das Bacias Hidrográficas dos rios Piracicaba, Capivari e Jundiaí, 2020 a 2035. Porto Alegre: Consórcio Profill-Rhama PCJ (2020b)
11. Brasil. Lei nº 9.433 de 08 de janeiro de 1997 (1997). http://www.planalto.gov.br/ccivil_03/leis/l9433.htm#:~:text=LEI%20N%C2%BA%209.433%2C%20DE%208%20DE%20JANEIRO%20DE%201997.&text=Institui%20a%20Pol%C3%ADtica%20Nacional%20de,o%20inciso%20XIX%20do%20art.&text=1%C2%BA%20da%20Lei%20n%C2%BA%208.001,28%20de%20dezembro%20de%201989. Accessed 23 Feb 2021
12. Demajorovic, J., Caruso, C., Jacobi, P.R.: Cobrança do uso da água e comportamento dos usuários industriais na bacia hidrográfica do Piracicaba, Capivari e Jundiaí. Revista de Administração Pública, **49**(5): 1193–1214 (2015). https://www.scielo.br/scielo.php?script=sci_arttext&pid=S0034-76122015000501193&lang=en. Accessed 03 Feb 2021
13. ANA. Agência Nacional das Águas e Saneamento Básico. Comitês de bacias hidrográficas (2015). http://www.cbh.gov.br/#not-estaduais. Accessed 02 June 2021
14. Instituto Trata Brasil. Perdas de água 2019 (SNIS 2017): Desafios para Disponibilidade Hídrica e Avanço da Eficiência do Saneamento Básico (2019). http://www.tratabrasil.org.br/images/estudos/itb/Tabela_das_100_cidades.pdf. Accessed 10 Apr 2021
15. Instituto Trata Brasil, Perdas de Água 2018 (SNIS 2016): Desafios para Disponibilidade Hídrica e Avanço da Eficiência do Saneamento Básico. São Paulo: GO Associados (2018). http://www.tratabrasil.org.br/images/estudos/itb/perdas-2018/estudo-completo.pdf. Accessed 10 Apr 2021
16. SNIS. Sistema Nacional de Informação sobre Saneamento. Série Histórica – (ano-base 2019) (2019). http://app4.mdr.gov.br/serieHistorica. Accessed 10 Apr 2021
17. Silva, A.A.: Indicadores para avaliação de efeitos de intervenções de saneamento básico: a questão da sustentabilidade, 138 f. Universidade de São Paulo, Dissertação (Mestrado) Faculdade de Saúde Pública, São Paulo (2015)

A New Home Energy Management System for Smart Home Using Improved Multi-objective Antlion Optimization Algorithm

Navid Razmjooy[1] (✉), Danial Bahmanyar[2], V. Rajinikanth[3], and Gabriel Gomes de Oliveira[4]

[1] Department of Electrical and Control Engineering, Tafresh University, 39518 79611 Tafresh, Iran
xnavid@gmail.com
[2] Rooigemlaan, Belgium
danial.bahmanyar@gmail.com
[3] Department of Electronics and Instrumentation, St. Joseph's College of Engineering, Chennai 600119, India
rajinikanthv@stjosephs.ac.in
[4] Researcher at UNICAMP Universidade Estadual de Campinas, Campinas, Brazil
oliveiragomesgabriel@ieee.org

Abstract. Lack of knowledge on how to use energy properly has led to a significant increase in energy demand relative to its production in recent years. Electricity companies have also come up with various plans to solve this problem, such as consumption management. In this study, we present an optimized home energy management scheme to overcome major barriers to implementing demand response programs. The main goal is to find the minimum energy cost by considering variable pricing for at least one period during the day and based on shifting loads out of peak hours while maintaining user comfort. Due to the complexity of the problem, an improved meta-heuristic optimization approach is used, which we call OBLALO. The proposed scheme for CPP pricing is simulated and the results are compared with PSO, GOA and ALO algorithms. The simulation results show the advantage of using the proposed scheme in the home energy management system.

Keywords: Energy demand · Optimize energy · Energy consumption · Improved antlion optimization · Load management

D. Bahmanyar—Independant Researcher.

Y. Iano et al. (Eds.): BTSym 2021, SIST 207, pp. 56–72, 2023.
https://doi.org/10.1007/978-3-031-04435-9_6

1 Introduction

Research in recent years has shown that due to increasing human needs, electricity generation sources have faced constraints such as blackouts, lack of liquidity, transmission line security, and water and environmental crises. Although the discovery of more energy resources has been proposed as a way to generate more energy, but the efficient and effective use of available resources can be an acceptable and economical solution to overcome the challenge ahead [1]. Increasing power generation capacity to meet the growing demand of consumers poses many problems. For this reason, researchers around the world are looking for a suitable solution to manage energy consumption [2]. To energy consumption management, the existing problems can be easily overcome by replacing smart grids with traditional electricity grids [3]. Smart grids create two-way interconnected activities between the electricity company and consumers to manage energy consumption using smart meters so that more efficiency and less cost can be achieved in terms of electricity consumption [4]. Electricity companies have introduced demand-side management as an effective solution to manage consumption and reduce the costs of consumers using smart meters [5]. DSM is designed to balance supply and demand with the goal of reducing consumption and improving efficiency [6]. The DSM includes incentive schemes based on changes in electricity prices that require consumers to change their consumption patterns in line with these efficiency schemes. Participation in such schemes can reduce consumers' electricity bills to an acceptable level [7]. Electricity companies manually cut off consumer loads during peak hours in areas that still use traditional grids. On the other hand, in smart grids, consumers, by participating in demand-side management schemes, shift their loads from peak hours to off-peak hours to prevent peak load formation and damage to grid equipment which not only benefits the company and consumers but also increases network reliability [8].

Since meta-heuristic optimization algorithms have a great ability to solve complex and discrete optimization problems, so using these algorithms seems to be a good solution for managing energy consumption. Optimization means choosing the best solution from a set of available solutions that are used in various fields such as mathematics, economics, management and engineering sciences [9, 10]. Although classical optimization methods such as linear programming provide an accurate solution to the problem, they are not suitable for solving complex problems because they use objective function gradients. Therefore, meta-heuristic algorithms, although they offer a solution close to the exact solution of the problem, due to their high convergence speed and ability to solve complex problems, they are used in various fields of engineering [11, 12].

Antlion Optimizer is a nature-inspired meta-heuristic algorithm [2]. The optimization performance of this algorithm is based on the hunting mechanism of antlions and their interaction with other ants. The ant population is looking for the best position by randomly walking in the search space. After this, the best position discovered is maintained by the antlion and updated at every step [13]. In addition, the ants change their movement in the search space under the influence of the elite. If any of the antlions are in a better position than the elite, it will be replaced. Once the stop condition is met, the elite position is selected as the optimal solution. Although the ALO algorithm, like other meta-heuristic algorithms, has its own advantages, each of these algorithms also has disadvantages. Therefore, this paper presents an improved version of ALO using opposition-based learning.

There are several related works in the literature, for example, in [14] the scheduling problem for a home energy management system was modeled and solved. The goal is to find the minimum energy cost based on constraints such as time-varying prices, the amount of energy demanded each home appliance and grid constraints. The authors have used a stochastic optimization approach for this purpose.

The home energy management system is presented in [15]. Consumers need to change their energy consumption patterns from peak hours when electricity is expensive. In order to optimize the problem, genetic algorithm (GA) and harmony search algorithm (HAS) have been used.

In [16] load management is done in order to achieve two objectives of reducing energy costs and increasing user comfort based on the hybrid non-dominated sorting genetic algorithm. However, the peak to average ratio is forgotten.

Energy consumption optimization has been done to reduce energy costs, peak to average ratio and increase user comfort using improved differential evolution algorithm in [17]. The authors also state that the use of renewable energy sources increases efficiency. However, their proposed algorithm still needs to be improved.

2 Problem Description

One of the important needs in a smart grid is to increase the reliability and stability of the grid along with reducing energy costs. In a smart home, it will be possible to control and manage the consumption of energy with the help of home appliance scheduling [18]. In order to manage energy consumption, DSM needs a two-way communication of information, including the pricing signal and the amount of energy demanded, between the consumer and the electricity company to plan the consumption pattern of household appliances based on this information [19]. Reducing the cost of consumers' electricity bills is the main goal of this article. According to the demand side management programs provided by the electricity company, the cost of electricity varies at different hours of the day and this cost increases during peak hours, so consumers must manage consumption and shift load from peak hours to off-peak hours to reduce the cost of electricity consumption. This increases the user's waiting time and compromises the user's comfort. Reducing energy costs should be done in such a way that user comfort is maintained as much as possible, so there should be a trade-off between energy costs and user comfort. In addition, the shifting the load causes the formation of peak consumption during off-peak hours.

3 Household Electrical Loads Categorize

Household electrical loads are different in terms of power range, performance and operational time. Accordingly, in the proposed model, home appliances are divided into three categories in terms of starting time: shiftable, non-shiftable, Interruptible, and, fixed. Shiftable are appliances that can start working at any time during the day. In other words, the operating time of these appliances can be shifted to any time during the twenty-four hours of the day. These devices have a special effect on reducing energy costs. Non-shiftable appliances operate at regular time slots during the day, so their starting time cannot be changed. Interruptible appliances are the same as shiftable, except that they have the ability to interrupt their operation. Also, fixed are those appliances that start working during the day according to the user's needs and requests. The classification of household electrical loads is detailed in Table 1 for the proposed model.

Table 1. The classification of household electrical loads is detailed in for the proposed model.

Appliances	Load categorize	Power rate (kW)	Operational time (hours)
Water heater	Shiftable	2.5	2
Dryer		3	1
Steam iron		1.2	1
Vacuum cleaner		1	1.2
Washing machine		3	3
Water heater		2	8
Lights	Non-Shiftable	0.4	8
Refrigerator	Interruptible	22	0.3
Water pump		7	0.4
Laptop	Fixed	2	0.1
PC		2	0.3
TV		5	0.3
Hairdryer		1	1.2
Blender		1	0.3

4 Electricity Tariff

RTP, DAP, TOU and CPP are examples of different electricity tariffs offered by the electricity company to calculate the cost of daily consumption of customers. In RTP pricing, the price of electricity is constantly changing during the day according to consumer demand. Although CPP and TOU tariffs have fixed prices at different times of the day, they are not economically viable for consumers because the price increases significantly, at least for a period, given the important events that take place in the system. This article uses the CPP tariff. The CPP tariff is used to increase the reliability and stability of the grid, so this plan is used to reduce consumption in critical periods when power consumption is high and the grid stability is at risk. Using such pricing, where the price of electricity varies throughout the day, requires smart meters for two-way communication. Figure 1 shows the CPP pricing signal.

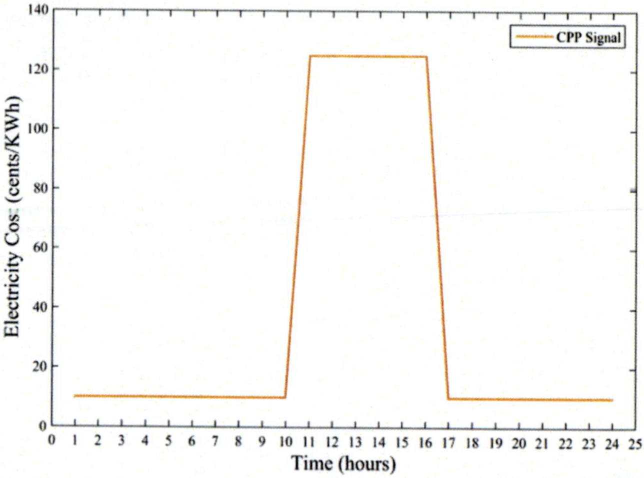

Fig. 1. The critical peak pricing.

5 Cost Functions

The proposed design in this article includes a home with a set of different household appliances with the categories of shiftable, non-shiftable, Interruptible, and fixed appliances. Each day is divided into a twenty-four hour period with one-hour intervals. The main goal is to reduce consumer electricity bills based on the scheduling of home appliances [20]. There is a trade-off between reducing electricity costs and user comfort. scheduling is done in such a way that shiftable appliances are shifted from peak hours to off-peak hours. Also, peak load is prevented by reducing PAR.

Shiftable Appliances:

$$s_i \in [s_1, s_2, \ldots, s_N], N = 6 \tag{1}$$

$$\text{Time slots} : t \in T = [1, 2, \ldots, T], T = 24 \tag{2}$$

where, S shows the shiftable appliances, N is several shiftable appliances and t is the number of the time slot. The cost of daily energy consumed is calculated using the following equation:

$$Daily\ Cost = \sum_{t=1}^{T} (\sum_{i=1}^{N} (s_{it} \times P_i) \times Price(t)) \tag{3}$$

where, s_{it} indicates the on or off state of ith appliance at t^{th} time slot, P_{it} represents the power rate of each appliance and $Price(t)$ is the Energy prices at time t.

PAR equals the ratio of the peak of energy consumed by the consumer and the average total energy consumed over a twenty-four hour period. It is very important that load management, along with reducing energy costs, prevent peak load formation. Decreasing PAR increases grid stability [21]. Peak load and average load during a day are calculated as follows:

$$Load_{peak} = \max_{t \in T}(HL(t)) \tag{4}$$

$$Load_{avg} = \frac{\sum_{t=1}^{T} HL(t)}{T} \tag{5}$$

where, $HL(t)$ represents the energy consumed at time slot t and T is the maximum of time slots. Finally, PAR is mathematically calculated as follows:

$$PAR = \frac{Load_{peak}}{Load_{avg}} \tag{6}$$

where $Load_{peak}$ represents the most load consumed at a certain time slot, $Load_{avg}$ represents the average load consumed during a day. In order to prevent the formation of peak load during off-peak hours, the following constraint is considered for PAR:

$$\text{constraint} : \text{PAR} < PAR_0 \tag{7}$$

where PAR_0 is the PAR value for the unscheduled pattern.

User comfort is defined based on the amount of interruption created when starting shiftable appliances. This interruption increases the waiting time and thus reduces user comfort. Waiting time is mathematically defined as follows:

$$Waiting\ Time = \frac{\sum_{i=1}^{N} |\text{interruption}_i|}{N} \tag{8}$$

where N represents the number of shiftable appliances, and interruption_i is the interruption created at the starting time of the ith appliances.

Finally, the objective function is considered as a linear combination of energy cost, PAR and user convenience functions, which is mathematical as follows:

$$\boldsymbol{ObjectiveFunction = \alpha_1(DailyCost) + \alpha_2(PAR) + \alpha_3(WaitingTime)} \tag{9}$$

where, α_1, α_2, and α_3 are constant coefficients and $\alpha_1 + \alpha_2 + \alpha_3 = 1$.

6 ALO Algorithm

The ALO algorithm is inspired by the biological relationship between antlions and ants. First, the ants move randomly in the search space, then the lions hunt. The random movement of ants is modeled as follows:

$$X(t) = [0, cumsum(2r(t_1) - 1), cumsum(2r(t_2) - 1), \ldots, cumsum(tr(t_n) - 1)] \quad (10)$$

where, *cumsum* is the cumulative sum, n represents the maximum number of iteration, t shows the step of random walk and $r(t)$ is a stochastic function that is calculated as follows:

$$r(t) = \begin{cases} 1\ if\ rand > 0.5 \\ 0\ if\ rand \leq 0.5 \end{cases}, rand \in [0, 1] \quad (11)$$

Since any search space has its limitations, the range of variables in Eq. (11) cannot be used directly to update the position of ants. For the ants to walk randomly, their movement is normalized according to Eq. (12):

$$X_i^t = \frac{(X_i^t - a_i) \times (d_i^t - c_i^t)}{(b_i - a_i)} + c_i^t \quad (12)$$

where a_i represents the minimum random walk of the ith variable and b_i is the maximum random walk, d_i^t is the maximum of tth variable at t^{th} iteration and c_i^t indicates the minimum of ith variable at t^{th} iteration. Antlion pits are simulated as Eq. (13):

$$\begin{cases} c_i^t = Antlion_j^t + c^t \\ d_i^t = Antlion_j^t + d^t \end{cases} \quad (13)$$

In this equation, $Antlion_j^t$ represents the jth antlion at t^{th} iteration. A roulette wheel is used to model the ability of ants to hunt. When an ant is trapped in a pit, the antlion throws stones at the edges of the pit. Its mathematical model is as follows:

$$\begin{cases} c^t = \frac{c^t}{I} \\ d^t = \frac{c^t}{I} \end{cases} \quad (14)$$

In Eq. (14) I is a ratio and is defined as follows:

$$I = 10^w \frac{t}{T}, W = \begin{cases} 2\ when\ t > 0.1T \\ 3\ when\ t > 0.5T \\ 4\ when\ t > 0.75T \end{cases}, \begin{matrix} 5\ when\ t > 0.9T \\ 6\ when\ t > 0.95T \end{matrix} \quad (15)$$

where t is the current iteration, T describes the maximum number of iteration and W stands for a fixed value.

The last step of hunting takes place when the ant is immersed in the sand. Then, the position of the antlion should be updated relative to the position where the ant hunted to increase the chances of a new hunt. The following equation describes this action:

$$Antlion_j^t = Ant_i^t \ \ if \ \ f\left(Ant_j^t\right) < f(Antlion_j^t) \tag{16}$$

Elitism is a feature of evolutionary algorithms that allows the best solution obtained to be maintained during the optimization process. In the ALO algorithm, the best antlion obtained at each step is stored and considered as an elite. The following equation shows the elite simulation:

$$Ant_i^t = \frac{R_A^t + R_E^t}{2} \tag{17}$$

where, R_A^t is a random walk around the antlions by the roulette wheel in the tth iteration, R_E^t describes also a random walk around the elite in the tth iteration.

7 Opposition-Based Learning Strategy

Most meta-heuristic optimization algorithms are population-based that need to generate and improve the initial population to find the optimal solution [22]. If the generated population is close to the global solution, it is more likely to find the optimal solution, otherwise, the optimization will take more time or the optimal solution will not be obtained. The main purpose of the opposition-based learning method is to produce solutions that are opposite to the current solutions in order to improve their position. By comparing the values of the objective function, it is determined which of them is in a better position [23]. The vector that is in the opposite position to the current solutions is generated as follows:

$$\vec{X}_i' = U + L - \vec{X}_i \tag{18}$$

where $\vec{X}_i'$ is the solution with the opposite position from the $\vec{X}_i$, also U and L are upper and lower bounds, respectively.

8 Improved ALO

In this part of the article, ALO is improved using the OBL strategy, which we call OBLALO. In OBLALO, the position of each member of the ant population is updated based on the equations mentioned, and then the opposite position for each of them is calculated by the OBL strategy. The fitness of the ant's current position and its opposite position is measured using the objective function. If the opposite solution fits better, the ant will be in the opposite position. Algorithm 1 shows the pseudo-code of the OBLALO algorithm.

Algorithm 1: The Opposition-Based Learning Ant Lion Optimizer Algorithm.

```
Initial parameters of ALO
Initial the primary population of Antlions
Calculate the opposite position of all Antlions
Calculate the objective function values of antlions
If the opposite position is best, replace it
Select the elite
while the end condition is not met
    for every antlion
        Select a random antlion from the population
        Create a random walk and normalize it
        Update the position of Ants
    end
    Calculate the opposite position of all Antlions
    Calculate the objective values of all Ants
    If the opposite position is best, replace it
    Combine ants and antlions
    Update the elite
    if the archive is full
        Delete some solutions using Roulette wheel from the archive to accommodate new
solutions
    end
end
return elite
```

9 Simulation and Results

In this section, the results of the simulations are reviewed with the main goal of finding an optimal residential scheduling scheme to reduce consumer electricity bills, reduce PAR and increase grid stability as long as user comfort is not compromised. The proposed scheme includes a home with different appliances in terms of power rating and working time. The CPP pricing signal is used as an electricity tariff to calculate the cost of energy consumed. The proposed OBLALO meta-heuristic algorithm was simulated in MATLAB software Ver.2015a using a processor installed with Intel® Core™ i7-4500U CPU @ 1.8 GHz 2.4 GHz and installed memory Ram 8 GB on Windows platform. In order to compare the results of the proposed design simulation with the OBLALO algorithm, three other algorithms PSO [24], GOA [25], and ALO [26] have been simulated to evaluate the performance of OBLALO on the proposed scheme. For a fair comparison, all optimization algorithms have been run fifty times with the same initial population. We assumed a home in which electrical appliances are divided into three categories: shiftable, non-shiftable, and fixed appliances. The proposed scheme for shiftable appliances has been implemented. A comparative analysis is performed for the schedule obtained from the proposed algorithm and other metaheuristic algorithms, and the following results show that the proposed algorithm has a good performance to achieve the above objectives. Figure 1, shows the CPP signals used in this scheme.

Under the CPP tariff, the cost of electricity consumption during peak hours increases sharply. The purpose of this increase is to encourage consumers to reduce consumption during peak hours. As shown in Fig. 1, the cost of electricity consumption during off-peak hours is less than 20 cents, while at 11 to 16 h, which is peak hours, the cost of electricity has increased by more than 120 cents. Therefore, the consumer has to pay a lot of money during peak hours. Also, the power grid equipment suffers a lot of depreciation. The hourly load consumption of scheduled and unscheduled patterns is shown in Fig. 2.

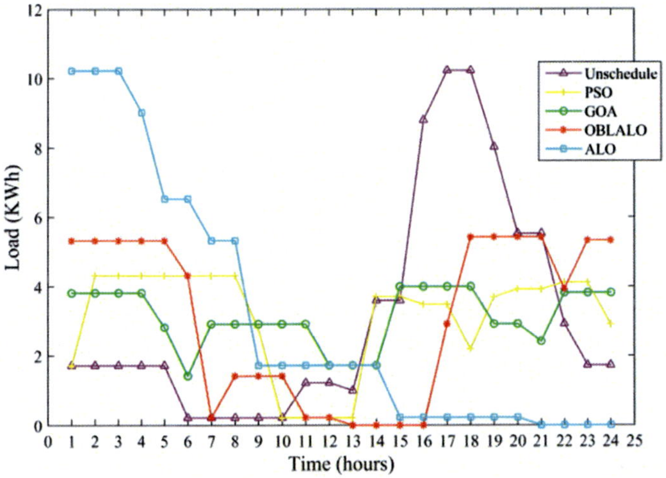

Fig. 2. Hourly load consumption whit CPP signal.

According to Fig. 3, the proposed OBLALO algorithm has a good performance in terms of shifting loads to off-peak hours. For this algorithm, the total load consumed during peak hours is less than 1 kW, while the load demand during peak hours for the unscheduled pattern is 20 kW. Thus, OBLALO has been able to reduce demand during peak hours by more than 90% compared to the unscheduled pattern. Also, this algorithm has been able to distribute the load demanded by the consumer in a balanced way outside the peak hours, and fortunately, we do not see a peak in the time slots during the day for the pattern optimized by OBLALO. Other meta-heuristic algorithms also perform better than unscheduled. However, the GOA algorithm has a peak in the early hours of the day. Figure 3 illustrates the hourly electricity cost along with a plot of the CPP signal.

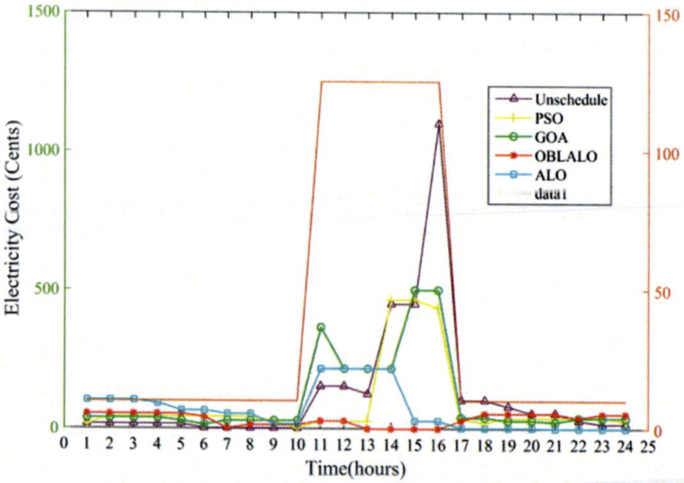

Fig. 3. Hourly electricity cost with CPP signal.

The results in Fig. 3, shows that each algorithm tries to schedule the cost in off-peak hours and the electricity cost of the patterns scheduled by the OBLALO, GOA, PSO and ALO algorithms are much less than the unscheduled pattern. Also, the proposed OBLALO algorithm costs much less than the other three algorithms. The unscheduled in the 16th the period has a peak of more than 1000 cents. This peak has been reduced by up to 50% by GOA and PSO algorithms, while the proposed OBLALO algorithm and ALO algorithm have reduced this peak to almost zero. The total cost of electricity per day for the patterns scheduled by the optimization algorithms and the unscheduled pattern is estimated in Fig. 4.

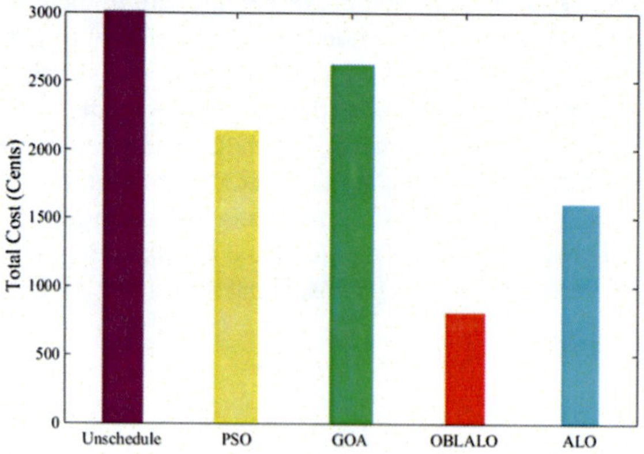

Fig. 4. CPP total cost per day.

According to the results in Fig. 4, the total electrical cost for the unscheduled pattern is 3,000 cents. All the meta-heuristic algorithms used in this simulation have succeeded in reducing this cost. The GOA algorithm had the lowest reduction rate, i.e. 14%. The PSO algorithm reduced the total unscheduled cost by 34%, followed by ALO by 44%. Besides, the proposed OBLALO algorithm has had the best performance. This algorithm has been able to reduce the total cost by 70%. Figure 5, shows a histogram of the total cost data in Table 3. Also, the minimum, maximum, mean and variance values for the total cost are given in Table 2.

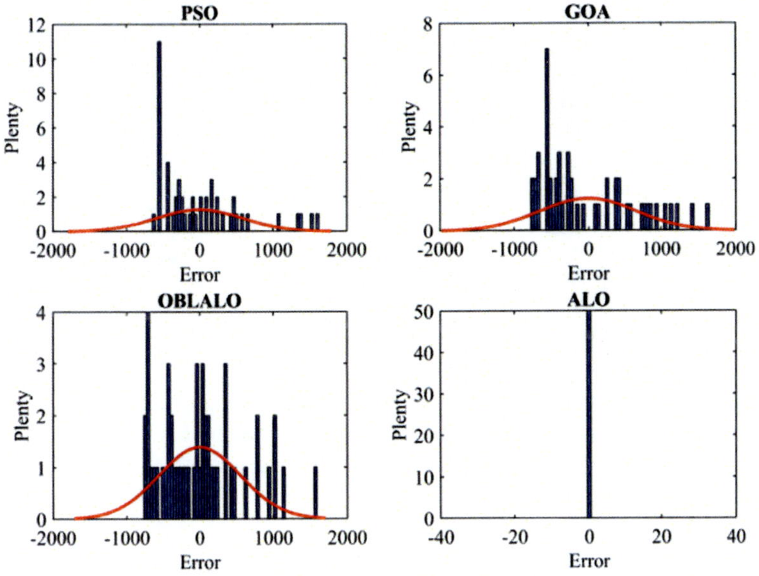

Fig. 5. Histogram plot for the total cost.

Table 2. Statistical data for the total cost.

Total Cost (cents)	Meta-heuristic Algorithms			
	PSO	GOA	OBLALO	ALO
Min	828.625	802.75	802.75	1596.25
Max	3102.75	3217.75	3163.125	1596.25
Mean	1475.788	1575.32	1569.57	1596.25
Var	350429.5	426739.1	317508.6	8.90e–28

Figure 5, and Table 2, shows that when reducing the total cost, the variance for the ALO algorithm is very close to zero, which is ideal Then this value is 317508.6, 350429.5 and 426739.1 for OBLALO, PSO and GOA algorithms, respectively. Therefore, the ALO algorithm is much more robust than the other three algorithms. But, the minimum total cost for the OBLALO algorithm is 802.75 cents, which is 50% less than the ALO algorithm. Also, the average value for the OBLALO algorithm is 1569.75 cents, which is much less than the other three algorithms. Therefore, the proposed algorithm performs better than other algorithms in reducing total cost. Figure 6, shows the performance of all scheduled and unscheduled models in terms of PAR reduction.

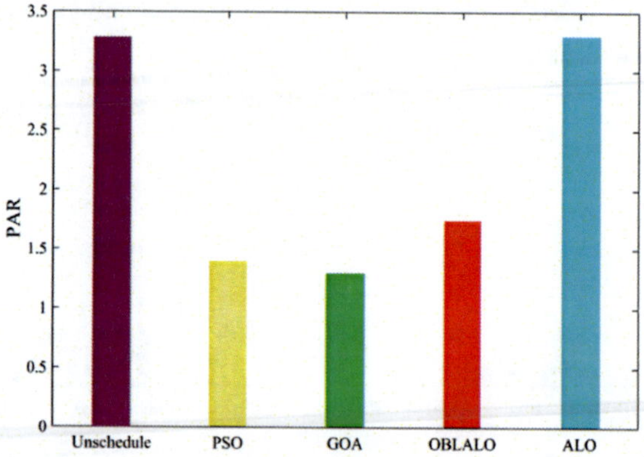

Fig. 6. CPP PAR reduction.

Figure 6, illustrates that the ALO algorithm did not perform well to reduce PAR. But by re-examining Fig. 2, it can be seen that the consumption peak for this algorithm is formed in a period of 1 to 4, which is the time outside the peak. Therefore, it can be said that the ALO algorithm performed better than the unplanned model in terms of PAR because the unscheduled model has formed the consumption peak at peak time, i.e. in the period of 15 to 17. However, the proposed OBLALO algorithm and the other two algorithms GOA and PSO have a good performance in reducing PAR. OBLALO was

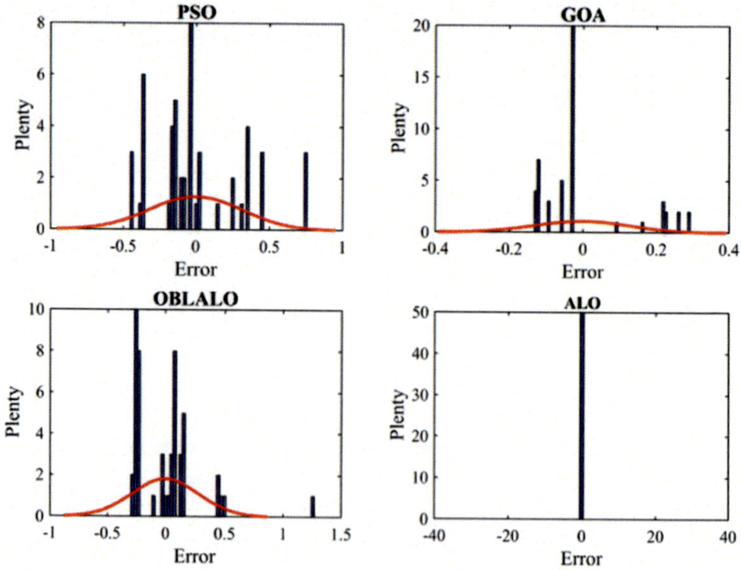

Fig. 7. Histogram plot for PAR.

able to reduce PAR by 46%, followed by PSO by 57% and GOA by 60%. A histogram of the PAR is shown in Fig. 7. Also, Table 3, presents the minimum, maximum, mean and variance values for the PAR.

Table 3. Statistical data for PAR.

PAR	Meta-heuristic algorithms			
	MOPSO	NSGAII	MOALO	MOGOA
Min	1.28	1.27	1.31	3.26
Max	2.5	1.7	2.88	3.26
Mean	1.73	1.4	1.61	3.26
Var	0.09	0.01	0.08	9.86e-30

Table 3, shows that the ALO algorithm also has a near-zero variance in PAR reduction and is more robust than other algorithms. The variance for the proposed OBLALO algorithm is 0.08 and for the PSO and GOA algorithms is 0.09 and 0.01, respectively. However, the minimum PAR for OBLALO, PSO and GOA algorithms is 1.31, 1.28 and 1.27, respectively, which are not much different from each other and are up to 60% less than ALO, so the performance of the proposed algorithm in reducing PAR is better than ALO algorithm and is close to the performance of the other two algorithms. Figure 8, presents the waiting time for each algorithm.

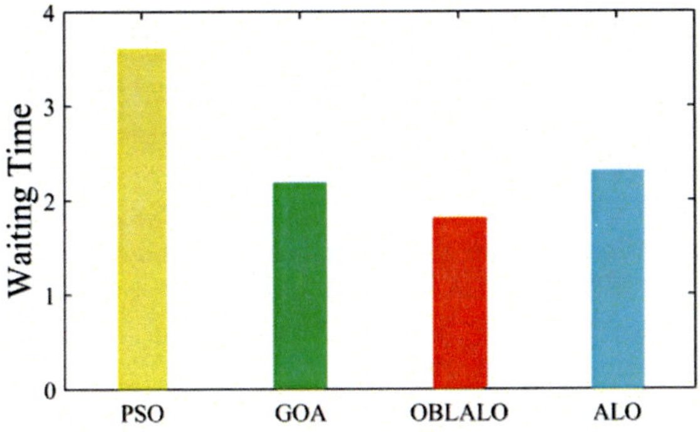

Fig. 8. Waiting time reduction.

The results of Fig. 8, show that the user waiting time for the PSO, GOA, OBLALO and ALO algorithms was 3.5, 3.2, 1.9 and 2.5, respectively. The proposed OBLALO algorithm has the lowest waiting time for the user, which is 45% less than the value obtained by the PSO algorithm. Figure 9, presents the running time for each algorithm.

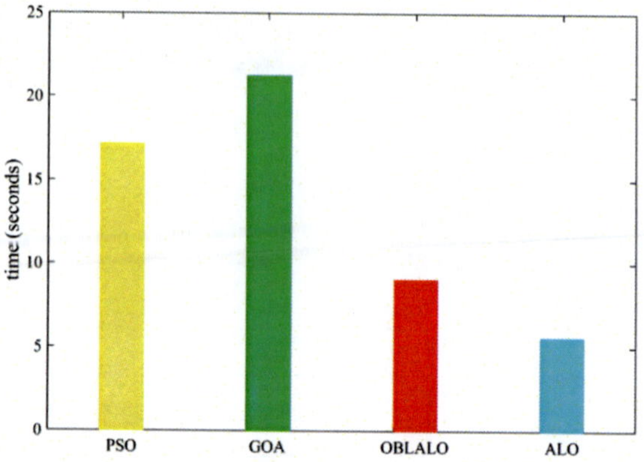

Fig. 9. The running time.

According to Fig. 9, the GOA algorithm has the longest running time, which is more than 20 s, but this is not good. Running time for the proposed OBLALO algorithm and PSO and GOA algorithms are less than 20 s, which seems appropriate.

The results show that OBLALO has been able to provide a Suitable schedule for managing the home loads. Consumers can use this scheme to significantly reduce their electricity bill. In addition, PAR is also reduced, however, user comfort is maintained.

10 Conclusions

Demand response will be possible by optimizing consumption and demand side management. In this paper, the OBLALO meta-heuristic algorithm is presented to optimize the consumption pattern of home appliances. The proposed scheme classifies the devices into four categories: shiftable, non-shiftable, Interruptible, and fixed. CPP signal is considered as electricity consumption tariff. The simulation results are compared with PSO, ALO, and GOA algorithms. The OBLALO algorithm has been able to reduce the daily cost of electricity by up to 70% and PAR by up to 64% compared to the unscheduled pattern, which has performed better than other meta-heuristic algorithms. This algorithm has also reduced the demand load during peak hours to 90% of the unscheduled pattern. Therefore, the results show that the use of the proposed plan can be an effective solution to deal with the growing consumer demand.

References

1. Ullah, I., Kim, D.: An improved optimization function for maximizing user comfort with minimum energy consumption in smart homes. Energies **10**(11), 1818 (2017)
2. Ramezani, M., Bahmanyar, D., Razmjooy, N.: A new optimal energy management strategy based on improved multi-objective antlion optimization algorithm: applications in smart home. SN Appl. Sci. **2**(12), 1–17 (2020). https://doi.org/10.1007/s42452-020-03885-7
3. Mir, M., et al.: Employing a Gaussian Particle Swarm Optimization method for tuning Multi Input Multi Output-fuzzy system as an integrated controller of a micro-grid with stability analysis. Comput. Intell. **36**(1), 225–258 (2020)

4. Jain, R., et al.: A modified fuzzy logic relation-based approach for electricity consumption forecasting in India. Int. J. Fuzzy Syst. **22**(2), 461–475 (2020)
5. Shinde, K.D., Mane, P.: Augmenting rooftop solar energy penetration ratio with secondary distribution network using smart inverter for maximum power transfer capacity for subordinate grid-A review. Energy Sourc. Part A: Recovery Utilization Environ. Effects **41**(6), 713–733 (2019)
6. SeyyedMahdavi, S., Saebi, J.: Techno-economic assessment of steel plant participation in DSM programs (case study: Iran's industrial operational reserve program). Energ. Effi. **13**(7), 1315–1328 (2020). https://doi.org/10.1007/s12053-020-09886-0
7. Abu-Bakar, H., Williams, L., Hallett, S.H.: A review of household water demand management and consumption measurement. J. Clean. Prod. **292**, 125872 (2021)
8. Hemanth, G., et al.: Cost effective energy consumption in a residential building by implementing demand side management in the presence of different classes of power loads. Advances in Building Energy Research, pp. 1–26 (2020)
9. Navid, R., et al.: World Cup Optimization Algorithm: an Application for Optimal Control of Pitch Angle in Hybrid Renewable PV/Wind Energy System (2018)
10. Navid, R., Khalilpour, M., Ramezani, M.: A new meta-heuristic optimization algorithm inspired by FIFA world cup competitions: theory and its application in PID designing for AVR system. J. Control Autom. Electr. Syst. **27**(4), 419–440 (2016)
11. Navid, R., et al.: A comprehensive survey of new meta-heuristic algorithms. Recent Advances in Hybrid Metaheuristics for Data Clustering. Wiley Publishing (2019)
12. Razmjooy, N., Ashourian, M., Foroozandeh, Z.: Metaheuristics and Optimization in Computer and Electrical Engineering. Springer
13. Cao, Y., et al.: Multi-objective optimization of a PEMFC based CCHP system by meta-heuristics. Energy Rep. **5**, 1551–1559 (2019)
14. Merdanoğlu, H., et al.: Finding optimal schedules in a home energy management system. Electr. Power Syst. Res. **182**, 106229 (2020)
15. Hussain, I., et al.: Optimizing energy consumption in the home energy management system via a bio-inspired dragonfly algorithm and the genetic algorithm. Electronics **9**(3), 406 (2020)
16. Garroussi, Z., et al.: A hybrid non-dominated sorting genetic algorithm for a multi-objective demand-side management problem in a smart building. Int. J. Electr. Comput. Eng. (2088–8708) **10**(1) (2020)
17. Das, B.K., et al.: Techno-economic and environmental assessment of a hybrid renewable energy system using multi-objective genetic algorithm: a case study for remote Island in Bangladesh. Energy Convers. Manage. **230**, 113823 (2021)
18. Lin, Y.H.: Trainingless multi-objective evolutionary computing-based nonintrusive load monitoring: Part of smart-home energy management for demand-side management. J. Build. Eng. **33**, 101601 (2021)
19. Estrela, V., et al.: Imaging and Sensing for Unmanned Aircraft Systems. Volume 2: Deployment and Applications (2020)
20. Tian, M.W., et al.: New optimal design for a hybrid solar chimney, solid oxide electrolysis and fuel cell based on improved deer hunting optimization algorithm. J. Clean. Prod. **249**, 119414 (2020)
21. Zhang, G., et al.: Optimal parameter extraction of PEM fuel cells by meta-heuristics. Int. J. Ambient Energy, 1–10 (2020)
22. Yu, D., et al.: System identification of PEM fuel cells using an improved Elman neural network and a new hybrid optimization algorithm. Energy Rep. **5**, 1365–1374 (2019)
23. Yanda, L., et al.: Optimal arrangement of a micro-CHP system in the presence of fuel cell-heat pump based on metaheuristics. International Journal of Ambient Energy, p. 1–12 (2020)
24. Kennedy, J., Eberhart, R.: Particle swarm optimization. In: Proceedings of ICNN'95-International Conference on Neural Networks. IEEE (1995)

25. Arora, S., Anand, P.: Chaotic grasshopper optimization algorithm for global optimization. Neural Comput. Appl. **31**(8), 4385–4405 (2018). https://doi.org/10.1007/s00521-018-3343-2
26. Mani, M., Bozorg-Haddad, O., Chu, X.: Ant lion optimizer (ALO) algorithm, in Advanced Optimization by Nature-Inspired Algorithms, pp. 105–116. Springer (2018)

Study About the Applicability of Low Latency in HAS Transmission Systems

Paolo Rodrigo de Oliveira Bacega(✉), Yuzo Iano, Bruno Campos Simoni de Carvalho, Gabriel Caumo Vaz, Gabriel Gomes de Oliveira, and Euclides Lourenço Chuma

School of Electrical and Computer Engineering. State University of Campinas, Campinas, Brazil
pbacega@gmail.com, yuzo@unicamp.br, oliveiragomesgabriel@ieee.org, euclides.chuma@ieee.org

Abstract. In the last few years, there has been a significant increase in the transmission of video sequences over the data network. At first, there were problems related to the compression capability for the transmission of video sequences. Currently, with the improvement of compression codes, the improvement of the broadcast media capacity, and a great advent of video streaming services, the quality of user experience becomes impacted by delays generated in the processing of sequences for the transmission. These delays, along with the currently used transmission techniques, hinders the transmission of live content. Nowadays, a well-counted solution is CMAF - Low latency, whose objective is to decrease the delay or initial buffering for On-Demand or Live video sequences, and this solution will be detailed and presented in this work along with other techniques that, when applied together, can lower the latency in the transmission of live content and improve the quality of the user experience.

Keywords: DASH · IDR · GDR · CMAF · Low latency

1 Introduction

Currently, there is a huge consumption of media via streaming, which already applies the HAS (HTTP adaptative streaming), and it serves very well the content of transmission other than Live (e.g., OTT, VOD, etc.). At this time, it becomes necessary to implement improvements to make it viable to broadcast Live content.

In this article, we will discuss the basic concepts for the transmission of content with HAS, such as the focus on transmission characteristics that aim at low latency.

Y. Iano et al. (Eds.): BTSym 2021, SIST 207, pp. 73–87, 2023.
https://doi.org/10.1007/978-3-031-04435-9_7

Presently, HTTP adaptive streaming (HAS) stands out among video quality adaptation mechanisms and, as it enables the use of existing content delivery technologies designed for current Web use, its implementation is considered easy.

By default, the HAS will generate, from a single file, many chunks, which are encoded at different resolutions and video rates, along with a Media Presentation Description (MPD) file to be reproduced in the most efficient way taking into account the screen resolution and the internet speed of the device, as illustrated in Fig. 1, since the client keeps monitoring and predicting its playback capabilities and then requests a piece with the highest feasible video rate given the estimated capacity. Figure 2 shows the schematic of an MPD.

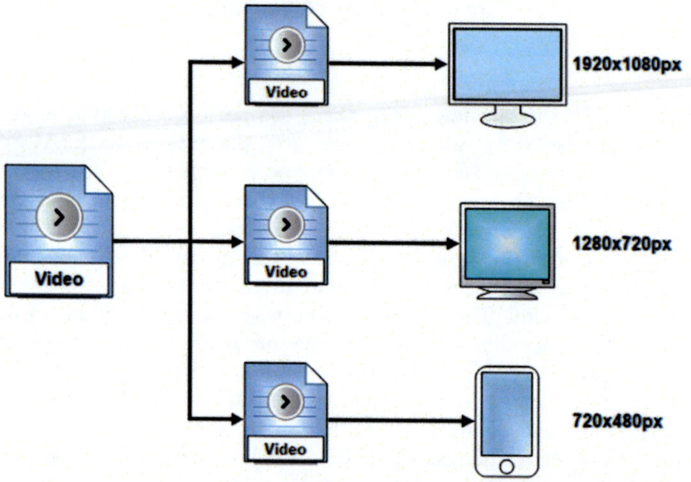

Fig. 1. Example of different resolutions generation from one video file

We can define Adaptive Bitrate Selection (ABS or ABR) as the process by which a customer decides the profile (transmission rate and resolution) of a Chunk to be downloaded. Initially, the ABR focused on the throughput estimation for the selection of the video rate to be downloaded. Currently, other parameters are adopted for the decision of the characteristics of the chunks to be downloaded, such as energy level, buffer occupancy, and cost.

The basic intuit with ABR is to improve the minimum and average levels of the resolution and video quality, decrease the amount of buffering during playback, and the improvement of QoE by decreasing the amount of resolution change during playback.

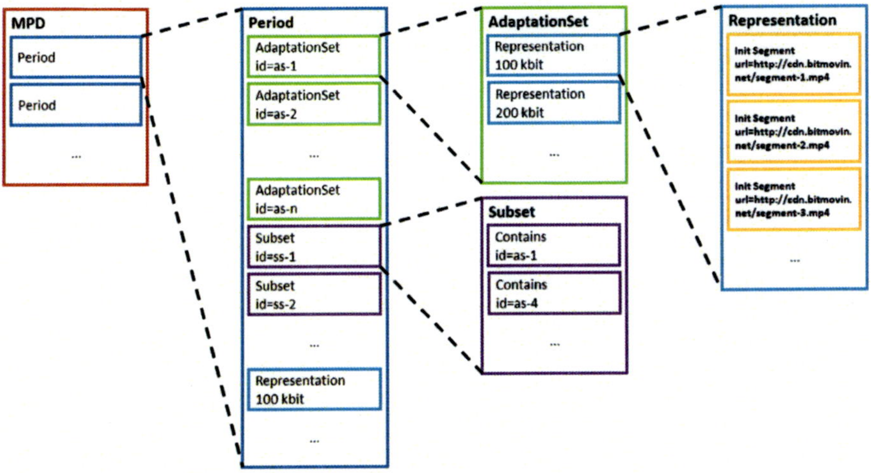

Fig. 2. Schematic of MPD that contains multiple representations

2 Basic Concepts

The basic idea of adaptive streaming is to adapt the amount of data sent to the user over a video stream. Regarding buffering, ABS creates video files that adapt to the user's internet speed. Instead of just a single resolution file, multiple video files of the same original video file are created, but with different resolutions, aiming to adapt the bandwidth speed and resolution of the client player. This way, ensuring that streaming will be displayed in the best possible quality, regardless of the device's screen size and bandwidth speed.

ABS works by detecting the user's bandwidth and CPU capacity in real-time and adjusting the quality of the media stream automatically. If the internet band suffers a momentary speed drop, ABS understands this movement and decreases the resolution, so that the video continues to run without having to "buffer". The currently adaptive streaming technology is almost exclusively HTTP-based and designed to work efficiently on HTTP networks [1,2].

2.1 Basic Definitions

Video services that are provided on unmanaged networks are called Over-the-Top (OTT) services, being delivered to users over the Internet by content providers directly or through third parties. OTT services do not require specialized or dedicated infrastructure, and that is why their setup and maintenance costs tend to be relatively low.

There are also managed services that are provided through the dedicated network infrastructure, to ensure Quality of Service (QoS). We can cite Internet Television (IPTV), which is defined by ITU-T [ITU-T Y.1910] as "multimedia services such as television/video/audio/text/ graphics/data provided in Managed IP-based networks to provide the necessary level of QoS/QoE, security, interactivity, and reliability". However, there is the VoD (Video on Demand) service, being provided by a managed network or not, which provides the user with stored pre-recorded content that allows the users to select which content they want [2,3].

2.2 Request HTTP

The HTTP protocol defines a set of request methods that are responsible for indicating the action to be performed for a given resource. In this topic, we will look at the methods common to the use of ABS. Methods indicate to the server what action the client wants to perform and that whenever we perform a request, we must report a method used [2,3].

HTTP POST - is used when the client sends data to be processed on the server.

HTTP GET - This Will be used when the client requests resources from the server, the GET method requires impersonation of a specific resource, requests that use the GET method return only data. The example of a GET request is when you press enter in the browser address bar.

HTTP GET PARTIAL - If the request message has included a Range header field, the GET semantics change to "GET PARTIAL", and partial GET requests that only part of the entity be sent. GET PARTIAL aims to reduce unnecessary network usage, where data already stored on the client is used to complete partially recovered entities.

When a client requests a video file using a HTTP GET request, the server responds by sending an MPD. The client then uses the information it contains to build the Uniform Resource Identifier (URI) for subsequent orders [1–3].

2.3 Encoding

To better understand the operation of the low latency CMAF method, we need to make a brief introduction to the IDR and GDR encoding methods, as we can see below.

IDR. Concepts such as Instantaneous Decoder Refresh (IDR) and "independent GOP" are included in the H.264 video encoding standard. The instant restoration of the decode is a random process where the decoding process does not reference any data before the IDR frame itself. The IDR signals the beginning of an independent GOP, with the independent GOP being an image grouping where it is possible to be decoded independently of the predicted images. Thus, an IDR image can be used for a random-access point, according to H.264 [2,4].

GDR. Gradual Decoder Refresh (GDR) was proposed for inclusion in the H.264 video encoding standard. GRD encoding refers to random access where the correct image content is gradually retrieved from previously encoded and/or unearned data, retrieving the correct image.

GDR allows random access capabilities to use any type of frame. We can say that there are two ways for GDR encoding, regardless of how GDR encoding is done the started frames will be considered correct about the content only after meeting the indicated conditions [2,4].

3 CMAF Features

The MPEG Common Media Application Format (CMAF; officially MPEG-A Part 19 or ISO/IEC 23000-19), was defined from a document at the 114th MPEG meeting, where it first defined ISOBMFF-based Media Application Format (MAF) and other MPEG standards for segmented media delivery.

3.1 CMAF

The purpose of CMAF is to allow only one encoding to be able to be used in multiple applications and devices. The encoding that CMAF defines is based on media packaging in threads for delivery and encoding for adaptive multimedia presentations on end-user devices. Among the definitions of the CMAF, we can mention the segment format, codecs, and media profiles (i.e., for stroke, HEVC, AAC) [3,4].

3.2 Low Latency of CMAF

The CMAF provides a low latency way where media segments are divided into smaller "Chunks" parts. In a live stream of media, some segments with duration of a few seconds are sent. A delay of some seconds ends up being added to the playback since the segments will be encoded, delivered, buffered, and rendered by the player. Splitting into smaller parts just does not reduce latency, although it allows the packer to produce chunks faster. In practice, both the source that generates the fragments and the CDN should use the transfer encoding in HTTP 1.1 parts allowing a larger number of clients [2,4].

3.3 Chunk

According to RFC-2960 SCTP (Stream Control Transmission Protocol) standards the chunk, also called data chunk, is used to describe a unit of information within an SCTP packet of control information or client data. Chunk is also the smallest reference unit of a video segment, containing at least one moof and one mdat, one or more chunks can be combined to form a fragment, and in the same way, one or more fragments can be combined to form a segment. As can be seen in Fig. 3 [2,4].

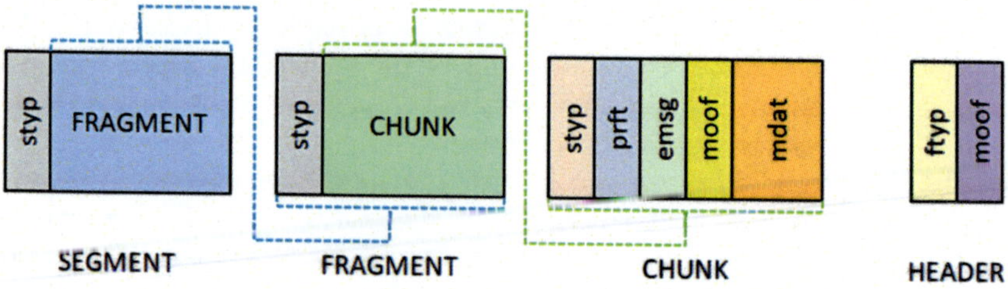

Fig. 3. Schematic diagram of a Chunk

4 Resource Estimation

The objective of a resource estimation function is to measure and monitor the resources of interest, and this implementation can be done close to the client, in the middle of the transport or near the server, being defined the location by the performance parameter adopted by the ABR.

First-generation ABRs prioritized throughput estimation, using it to select the highest video rate that was lower than the throughput rate, so it was thought that it would avoid rebuffering and provide the best video quality. Subsequently, due to the inadequacy of the network state information, it was proved that throughput was just not sufficient for an efficient ABR.

Currently, several parameters are used to adapt the video to the multitude of requirements of different clients so in this topical we will see the introduction to ABR schemes based on buffer and throughput (Fig. 4).

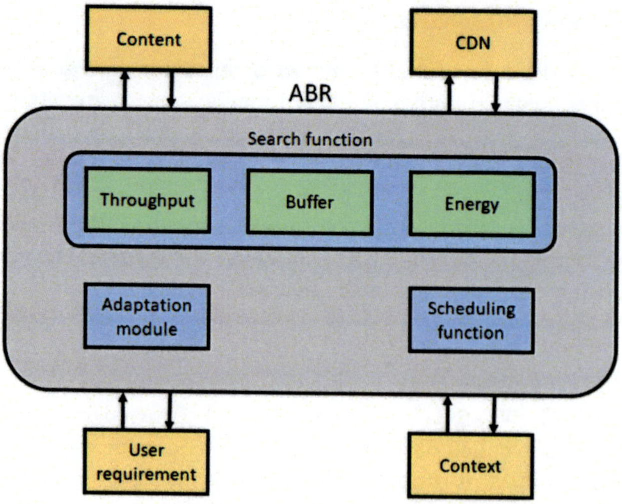

Fig. 4. ABR module

4.1 Buffer

One of the main objectives of the buffer is not allowing a client to run out of content. The buffer is introduced to absorb throughput variability, so that the data consumption rate is equal to the rate of the data that arrives, fulfilling the continuous media time.

In practice, the buffer size is not unlimited, but it is an inverse relationship between the buffer size and the chances of buffering the content, so the longer the buffer, the longer it will take for the client to run out of content.

Within the buffer-based ABRs, it is possible to see dynamic buffers and feedback parameter buffers.

4.2 Dynamic Buffer

In the first scalable video services the buffer had its capacity specified in bytes, we can come up with a derivation of the temporal size of buffer occupancy by dividing the size into bytes by the average rate of video playback.

We should take into account that a HAS client can buffer multiple chunks with variable video rates, but all of them have the same period. Converting the byte buffer to time is not a simple process since it depends on some parameters such as the type of encoding used, number of buffered chunks, and their bitrates, but the HAS buffer is calibrated in time. To establish buffer occupancy the client must multiply the chunk temporal size by the number of chunks that are in the buffer.

As we see in Fig. 5, the chunk video rate and download rate are responsible for the content feed rate in the buffer and this being drained every second of real-time at the playback rate of one second.

If the ratio between the download rate and the playback rate is less than one, the buffer occupancy decreases, otherwise it increases. This way a new chunk should have arrived before a chunk was displayed to avoid freezing the video. The probability of video freezing increases as buffer occupancy is less.

Before starting the view, the client waits for the initial period for the buffer to be occupied to a set limit, thus being more effective against rebuffering. For the client to have a high tolerance to network variability, it is necessary to have a large initial buffer occupancy, however, this causes an initial delay, which does not please users, who need to wait a few seconds before the start of playback.

In order to avoid the initial delay, THE ABRs algorithms seek to reduce the requirements regarding buffer usage and content transfer by requesting low-representation, low-quality content parts [2–6].

4.3 Buffer as a Feedback Parameter

Until recently, the buffer was divided into logical segments by the ABR algorithms to allow the algorithms to behave differently for each segment of the buffer, being used the mapping of the video rate for the buffer occupancy. In the

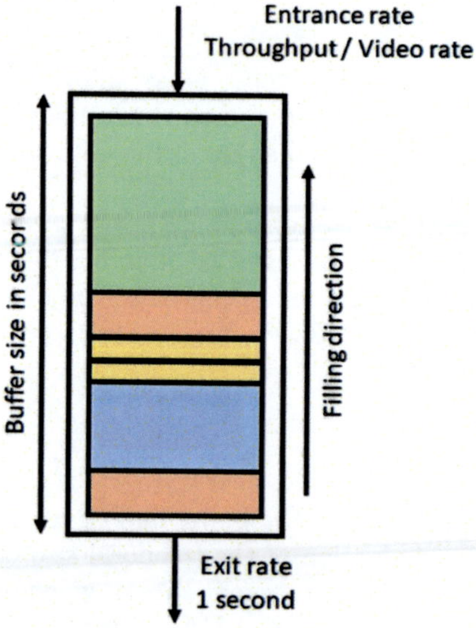

Fig. 5. Buffer occupancy

initial phase, a part of the buffer with slow video rate is downloaded and, then, in the period of intensification there is a linear increase in video quality.

There is a great risk of freezing the video at the beginning of streaming, so there should be some changes in quality, from the time the buffer reaches a level that provides reliability, there may be a significant increase in video quality, at the point of high quality, any further increase in the rate may not be a perceived improvement by the user. With the increase of the buffer required for the change to a high video rate, the rate that the video quality changes is slowed down, thus protecting against the fluctuation of video quality, accelerating the video rate, and decreasing the number of rebuffers.

With the inaccuracy that was previously seen regarding the channel rate and the use of the buffer as an adjustment factor for the adaptation of the channel rate based on this inaccurate information, it is correct to state that the use of buffer occupancy as an occupancy decision for adaptation can become an erroneous decision if it is made by just taking into account the throughput TCP when the ABR is unstable.

One of the algorithms currently proposed is the mapping of buffer space usage, where the initial buffer space would be filled with chunks of lower video rate, past the initial period, if pointed out by the mapping, the chunk quality will be increased to the next level if the suggested transmission rate to the next level of available video quality has been exceeded, and video quality may be decreased if the rate is less than the rate of the next lower level of video quality. Figure 6 shows an example of a mapping of the buffer use.

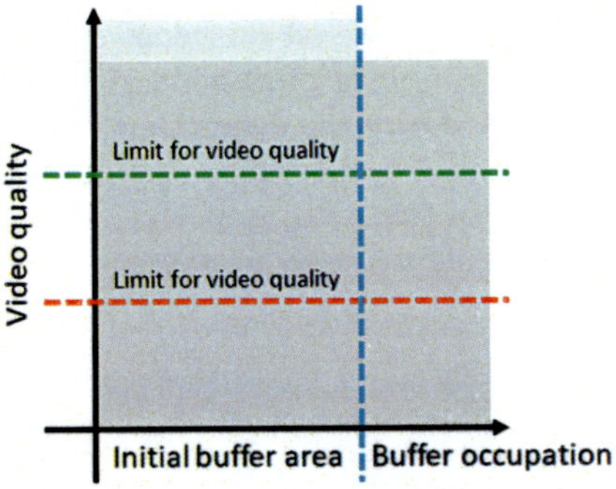

Fig. 6. Example of a mapping of the buffer use

The purpose of this algorithm is to avoid rebuffering as long as the network capacity is greater than the minimum video rate to get a video rate much closer to the network capacity [2–6].

4.4 Throughput

The goal of throughput-based ABR is to measure the average available network capacity, but the available bandwidth is variable in time regardless of the content protocol used for transport.

TCP throughput is used as a measure of the bandwidth available by ABRs based on throughput when TCP is used as a transport layer protocol, so the monitoring is mostly done above the HTTP layer, resulting in inaccurate information. The instantaneous throughput obtained from a single chunk is not widely used, due to the dynamics of TCP or the time variation of the available bandwidth, and the instantaneous throughput of a single chunk is obtained by dividing the size of a chunk in bytes by time download duration.

Due to the variability in video quality that may be a result of a throughput estimated by a chunk and the difficulty of accurately estimating the throughput above the HTTP layer, several techniques are applied to improve the measurement quality, among which we will see two below.

4.5 Estimation Techniques

The Segment Fetch Time (SFT) is defined as the instant when the last bit of the chunk is received minus the instant when the chunk was requested, and the estimated throughput is the Media Segment Duration (MSD) result, however, this algorithm is for oscillating video quality. Later it was discovered that adopting an ideal value for MSD could decrease the oscillation.

Among the ABR researches that use available throughput estimation technique, for the use of MSS it was seen that the client does not have an instant reaction to the available bandwidth, regardless of whether the client is varying the level of the video quality of the last downloaded chunk or not. It happens because of the delayed reaction of the variability generated by the inhibition of the algorithm response capacity by the smoothing techniques. This may result in video freezing and throughput reduction by the algorithm's delayed response.

To increase the response of the ABR by not increasing video rebuffering, an adaptive coefficient for the weighted moving average is currently used [2,4,5,7].

4.6 Estimation Reliability

Reflecting the true available bandwidth from throughput estimates is still a problem since TCP throughput depends on several factors.

When there is a dispute over a narrow-band link of customers with competing applications, there is difficulty in estimating the appropriate portion of available bandwidth due to the discrete nature of HAS downloads. Resulting in underestimation of bandwidth and available video rate.

As we can see in Fig. 7, due to the positioning of the measurement module, the result obtained from the measurement will probably not be the same as the actual TCP throughput, and the variability of the video quality resulting from this TCP throughput estimate is inaccurate. One of the likely causes is that there is no continuous and reliable feedback to the HTTP layer regarding your share of narrowband link membership and competing applications.

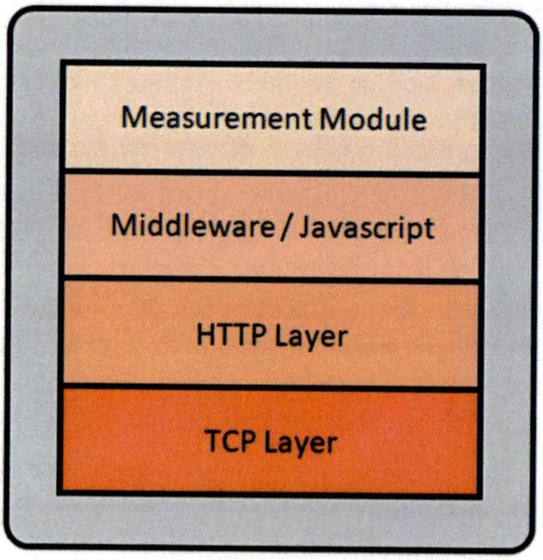

Fig. 7. Protocol stack

With a better information flow from TCP to HTTP layer, the utilization becomes more stable, and improvements are achieved in the bandwidth and video rate [2,4,5,7].

5 Low Latency

So far in this article, it was made a look at the basics that had served us as an introduction to low latency, from this topic we will see the low latency itself.

Speaking about DASH in general, the first activity performed by the client is to seek the MPD, so you can choose the best representation characteristics from the available video sequence according to your possible momentary possibilities, such as available bandwidth. During the download of the segments, the client monitors the channel rate, buffer occupancy, being necessary to update the playback parameters, and requests a new MPD. Where the buffer is used to minimize channel rate variance. For new threads to be correctly requested from the data contained in the MPD, both the client and the server must be synchronized through the UTC clock.

The main factors that affect latency in a HAS system, client buffering, thread fetching, and downloading, one of the solutions presented by the low latency CMAF is the division of chunks segments so that latency decreases. This we will treat more carefully in the following topics.

5.1 CMAF Transport Streaming

The CMAF is the standardization of TS (Transport Streaming) of MP4 fragments that came intending to decrease traffic on the CDN caused by current TS of proprietary standards, such as HLS MP4 fragment and DASH with MPEG-2 TS.

Developed to be implemented in streaming technologies already in use, CMAF enables content providers to package and store in one format and not in multiple formats having the same wide range, thus reducing costs, also minimizing the number of representations of video sequences and reducing traffic and thus improving CDN efficiency.

In an encoding other than CMAF the Movie Fragment Box refers to all parts of the video that are contained in the "mdat" box (MediaData Box), the encoder will wait until you encode the last byte of content before making the first byte available for distribution on the CDN, similarly, CDN expects to receive all bytes before forwarding them to the client player, and in turn, the client player waits until it receives the last byte before starting to decode the first one received. This processing pattern generates an increase in time latency from one thread per step [2,5].

5.2 Coding in Chunks

As we saw earlier, a chunk can be composed of a "moof" box and a "mdat" box only, with the chunk being the smallest reference unit, the combination of two or more units form a fragment that analogously combines to form a segment. For each chunk, a header is required for its playback in the decoder. We must take into account that low latency is the result of a chain event, resulting from the alignment of TS, CDN, encoders, and client players.

In CMAF, a segment can contain a "moof" box and a "mdat" box with a certain number of samples, it is also necessary for each segment to start that the "mdat" box has an IDR frame (Instantaneous Decoder Refresher). This structure is depicted in Fig. 8.

Fig. 8. CMAF segment with a "moof" box, a "mdat" and an IDR table, containing 12 samples.

However, a chunked-encoded segment will split the previously described segment into a series of smaller pieces. Although the segment is split into two "moof" and "mdat", only the first double of each sequence will contain the IDR frame, as shown in Fig. 9. Being now shorter parts, the code can be produced and delivered immediately after the request.

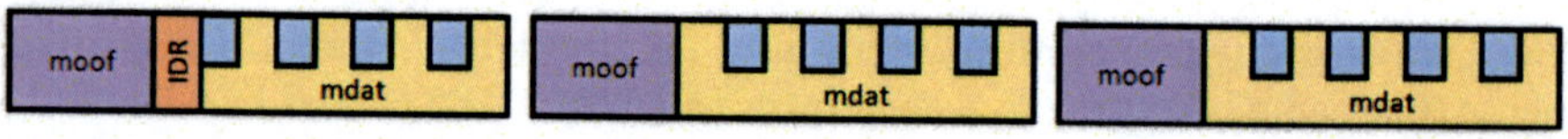

Fig. 9. The sequence of chunks with IDR and twelve samples

Although the creation of the "moof" and "mdat" pairs increases the segment size by around 2% for video segments, it provides early releases, which enable the reduction of the latency and make the process more efficient compared to the conventional encoding [2,4,5,7].

5.3 Distribution by CDN

In a conceptual point of view, we can divide a HAS distribution system based on its type of HTTP request (GET or POST). In the first part, where the encoder sends, over HTTP POST, the segments to content storage from which it will be sent to an intermediate layer for distribution. In the second part, the content is requested from storage by an edge server which in turn provides the content by client request, a feeling that all requests are made by HTTP GET. This distribution is shown in Fig. 10.

Since HTTP version 1.1 chunks transfer encoding is available, where the size header is not inserted into the response since the server does not know the end size of the segment, in this way it sends in bursts as chunks are available, ending the transmission when a zero-length chunk is transmitted [2,4,5,7].

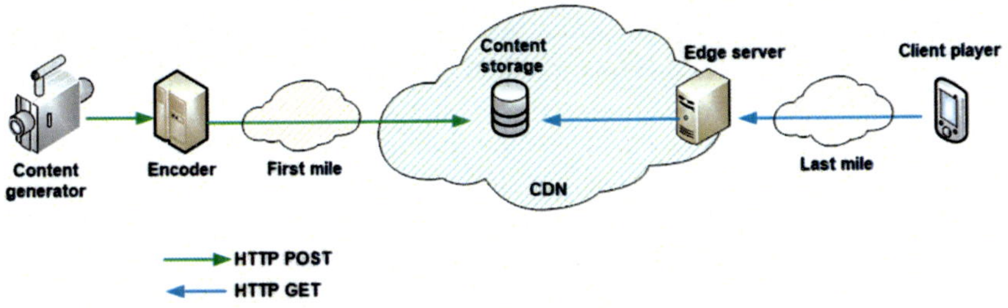

Fig. 10. Example of CDN distribution of a HAS system.

5.4 Media Player

After the client player reads the MPD describing the content, it will define in which instant the segments will start playing, and then make the request. We take as an example a system with an encoder producing segments of 2 s, as we can see in Fig. 11, in half the production time of the third segment the client player would start downloading the content, traditionally would download the last segment that was entirely available, which will generate 3 s of latency, the system approach is done by coding in chunks of 500 ms , the client player will start with the last chunk with IDR, thus reducing latency to 1 s.

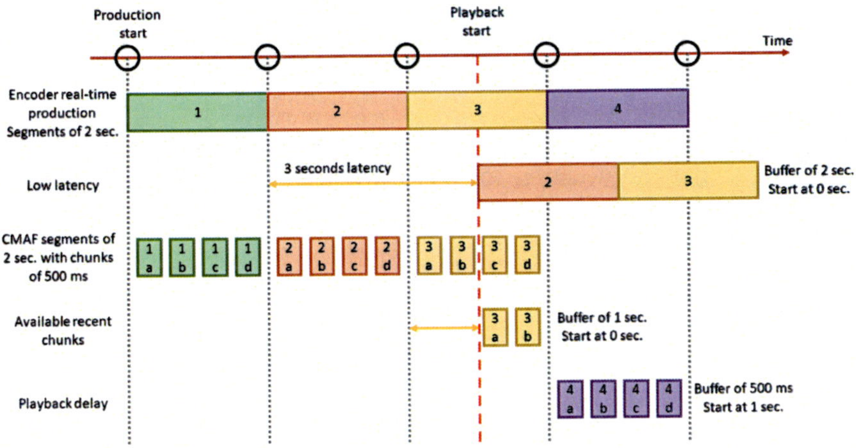

Fig. 11. Example of playback start.

Two other possibilities would decrease latency to 500 ms.

1. Download the frame 3a and the segment 3 and decode 3a and 3b before the playback start.
2. Delay the playback in 1 s and, in the instant scheduled for the first chunk, request the following segment (4a).

The requests for both possibilities will be within the time of a segment, that is, while the client player requests the first chunks the encoder is producing the last of the same segment.

When a thread is requested in chunk encoding, the CDN will send all chunks of that segment in sequential order to the client player until the entire segment is delivered. In the encoding of a traditional HAS, when a thread is requested and the full segment is received as a response.

One point to consider is that in traditional HAS segment delivery time is the throughput reason for the encoded bitrate, since the server already has all the thread data, in chunks encoding it is not possible to send data to the client player by the fastest edge server it receives from the server, and the server produces at the ratio of 1:1, that is, one second of content per second.

When there is no variation and the throughput of the last mile is greater than the segment bitrate, there will be two consequences. Traffic will be consistently resembling MPEG-TS and the client player will use segment download time to measure the throughput of the last mile, where in a chunk encoding, the algorithm will estimate that throughput equals the encoded bit rate, making it difficult to adapt the client player [2,4,5,7].

6 Conclusion

Current latency becomes limiting to the quality of user experience for live streams, as we have seen throughout this article, there are techniques available to decrease the current latency to a value close to 2 s.

Only the application of the CMAF itself would not lower latency, however, the decrease in latency is the result of a set of measures adopted. Such as managing the buffer based on more reliable measures of TCP throughput or the adoption of a catch-up system, decreasing the buffer to one or two chunks, and managing through the data rate, by losing segments, bringing back to the real-time display, we must not forget to mention a basic measure, which synchronized the operation of the server, CDN, and client player.

For future lines of research, technologies for reliable TCP throughput estimates would provide a significant improvement in latency decreased and transmission quality.

6.1 Abbreviations and Acronyms

OTT: On The Top
VOD: Video On Demand
HAS: HTTP adaptive streaming
ABS: Adaptive Bitrate Selection
CMAF: Common Media Application Format
GDR: Gradual Decoder Refresh
IDR: Instantaneous Decoder Refresh

References

1. Bouzakaria, N., Concolato, C., Le, F.J.: Overhead and performance of low latency live streaming using MPEG-DASH. In: IISA 2014, The 5th International Conference on Information Intelligence, Systems and Applications, pp. 92–97 (2014)
2. Sani, Y., Mauthe, A., Edwards, C.: Adaptive bitrate selection: a survey. IEEE Commun. Surv. Tutorials **19**(4), 2985–3014 (2017). https://doi.org/10.1109/COMST.2017.2725241
3. Essaili, A.E., Lohmar, T., Ibrahim, M.: Realization and evaluation of an end-to-end low latency live DASH system. In: 2018 IEEE International Symposium on Broadband Multimedia Systems and Broadcasting (BMSB), pp. 1–5 (2018)
4. Yuste, L.B., Boronat, F., Montagud, M., Melvin, H.: Understanding timelines within MPEG standards. IEEE Commun. Surv. Tutorials **18**(1), 368-400 (2016). https://doi.org/10.1109/COMST.2015.2488483
5. Law, W.: Ultra-Low-Latency Streaming Using Chunked-Encoded and Chunked-Transferred CMAF. Akamai (2018). https://www.akamai.com/us/en/multimedia/documents/white-paper/low-latency-streaming-cmaf-whitepaper.pdf. Accessed 16 Aug 2021
6. Peck, R., Cenzano, J., Li, X., Reznik, Y.: Towards mass deployment of CMAF. In: NAB Broadcast Engineering and Information Technology Conference, Las Vegas, NV, April 2019 (2019)
7. Bentaleb, A., Timmerer, C., Begen, A.C., Zimmermann, R.: Bandwidth prediction in low-latency chunked streaming. In: Proceedings of the 29th ACM Workshop on Network and Operating Systems Support for Digital Audio and Video, NOSSDAV 2019, Amherst, MA, 21 June 2019 (2019)

Simulation of the Implementation of Domestic Solar Systems Using Multi-agent Systems from Web Scraping

João Emmanuel D.'Alkmin Neves(✉), Paulo Sérgio Martins Pedro, Marli de Freitas Gomes Hernandez, and Luiz Ariovaldo Fabri Junior

State University of Campinas, Limeira, Brazil
jeneves@gmail.com, {paulo,marlih}@ft.unicamp.br

Abstract. Currently, the world experiences the Information Age, where people and institutions are highly connected and leave digital footprints. This huge volume of data, which for the most part is unstructured, holds valuable information. However, the value is not in the data itself, but in the possibility of simulating complex scenarios through multi-agent systems. These systems use the concept of agents, which are programs that interact with each other and with the environment, satisfying a set of previously determined rules. The present article aims to develop a way to simulate decision-making on the adoption of domestic solar systems in Brazil based on the unstructured collection data. Since this premise, Web Scraping was performed to elicit realistic indicators to model and simulate scenarios using the NetLogo software. The results showed that the increase in the installation of domestic solar systems is a consequence of the visibility of the benefits provided within a radius of influence. It is also observed that the boost in the neighborhood threshold leads to an exponential growth in installations of this technology.

Keywords: Web scraping · Multi-agent systems · Information · Artificial Intelligence · NetLogo

1 Introduction

A huge volume of digital footprints is generated by people and institutions on the Internet. This volume is of no use if it is not possible to collect them and do not have the knowledge to use them [1]. Data Mining techniques can aid in the collection as most of this data is distributed and unstructured [2]. It is noteworthy that the value is not in the data itself, but in their capability of simulating complex scenarios and, thus, enabling an analysis of future scenarios [3, 4]. The main advantage of using simulation is that it does not interfere with the functioning of the real system; the evaluation and interpretation of data take place in a logical, systematic, and consistent way; and the model can be changed and/or improved easily, providing adaptive learning [5]. Therefore, the advantages obtained by simulation using data have been increasingly arousing the interest of institutions.

Y. Iano et al. (Eds.): BTSym 2021, SIST 207, pp. 88–96, 2023.
https://doi.org/10.1007/978-3-031-04435-9_8

Given the aspects addressed, this study aims to improve the decision-making process from simulations of complex scenarios with the absence and/or difficulty to obtain data in Multi-Agent Systems through the application of a Data Mining technique called Web Scraping [6, 7].

2 Theoretical Reference

2.1 Data Mining

Data mining is the extraction of knowledge using computational tools, from databases that are currently being collected and stored [2]. Obtaining this implicit knowledge has been useful, above all, for institutions to make more assertive decisions aiming to enhance competitiveness [8]. This set of statistical and artificial intelligence techniques seek to establish new relationships between data, ultimately aiming to produce new knowledge [1, 3]. Due to the heterogeneity and difficulty of obtaining data, the use of different techniques, such as Web Scraping, is essential for obtaining and preparing data to obtain knowledge [6, 8].

2.2 Agent-Based Modeling

The system of an agent-based modeling is modeled as a set of autonomous decision-making entities known as agents [5]. Agent-based model, at the simplest level, consists of a system of agents and the interactions among them [9]. Agent-based simulation strives to replace individual actors or groups of a given system with software representations of these actors (or groups) [4, 10].

In some situations, the agent-based modeling offers advantages over conventional approaches: when the decisions are discretely defined; when agents need to adapt and change their behavior; when agents need to have dynamic relationships with additional agents, and relationships can be formed/dissolved; when agents structure organizations, and adaptation and learning are imperative at the structural level; when organizational changes in the process should be a result of the model, rather than an improvement from the model; among others [11, 12].

3 Methodology

Collecting information manually can take time and be susceptible to extraction errors. On the other hand, collecting information in an automated way plays a key role in verifying the data presented [7]. One of the possibilities of extracting data in an automated way is through tools called Web Scrapers. They are the action of harvesting data from websites and bringing it into a simple format to be cross-referenced with clarity [8]. After collecting the data, a simulation will be carried out through agent-based modeling. NetLogo is a cross-platform environment for modeling multi-agent environments with complex scenarios that evolve [10].

4 Case Study

The acceptance of society is essential for the effective diffusion of contemporary technology and photovoltaic microgeneration systems are no different. It is challenging to model the influences of different non-measurable aspects of adopting new technology. Nevertheless, the radius of influence, which is a measurable parameter, demonstrates how the adoption of technology by one family influences others to do the same.

This study carried out here collects unstructured data through Web Scraping to elicit the indicators and thus feed simulations based on Multi-Agent Systems, which aim to estimate the impacts of the radius of influence and the neighborhood threshold in the implementation of domestic solar systems in Brazil.

4.1 Data Collection

The proposed structure involves four principal processes: (1) predictive extraction of text about domestic solar systems in rural areas of Brazil through Web Scraping; (2) pre-processing text; (3) topical modeling method using Latent Dirichlet Allocation; (4) similarity analysis between different topics using Hellinger Distance.

Predictive Extraction of Text. Information issued on government sites, associations, and the social network on the Internet were collected, analyzed, and some data were discarded, and others were selected to adequately reflect the strategy of using photovoltaic solar energy in the country, since these sites present important data, aiming to inform and align people on energy strategies. Shorter information (i.e., less than 40 words) and duplicates were discarded. In addition, all tags have been retained to reflect broader trends in information about domestic solar systems. Table 1 demonstrates the description of the textual data collected.

Table 1. Description of data collected.

URL	Publications	Collection period		Monthly frequency of publications
		Start	End	
https://www.aneel.gov.br	2,206	Oct./2020	Dec./2020	17.49
https://www.ibge.gov.br	1,550	Oct./2020	Dec./2020	22.92
https://www.bndes.gov.br	2,757	Oct./2020	Dec./2020	20.74
https://www.absolar.org.br	1,206	Oct./2020	Dec./2020	10.49
https://www.facebook.com	3,189	Oct./2020	Dec./2020	44.52

Pre-Processing Text. Proper data preparation permits effective text analysis, can remove superfluous characters (e.g., non-alphabetic characters and white spaces) and possible errors during the Internet data extraction process. In this study, text data preparation involves text normalization, which incorporates tokenization, insertion of bigrams, removal of stop words, and stemming. Table 2 illustrates the decomposition of text data.

Table 2. Decomposition of data collected.

Sites	Publications	Sentences	Tokens	Words	Relevant	Vocabulary
ANEEL	2,206	27,904	605,079	376,977	311,376	16,369
IBGE	1,550	30,080	528,220	320,053	268,802	17,960
BNDES	2,757	30,793	719,664	447,852	365,678	16,770
ABSOLAR	1,206	20,040	417,551	297,459	243,857	15,084
FACEBOOK	3,189	70,488	1,440,348	908,278	743,383	30,662

Modeling by Topics. Dirichlet Latent Allocation (LDA) is utilized for topic modeling and achieved a reputation in project management due to studies [13]. LDA is an unsupervised machine learning method that infers key topics from a collection of untagged records.

To determine the internal coherence of a particular topic t, $C_{UMass}\left(t; W^{(t)}\right)$ calculates the average of the total sum of the logarithmic paired score circumstance, as described in Eq. 1. The C_{UMass} coherence score of the topic model determines the average of the topic unity scores for all topics in Eq. 2.

$$C_{UMass}\left(k; W^{(k)}\right) = \frac{2}{N(N-10}\sum\nolimits_{i<j} log\frac{D(w_i^{(k)}, w_j^{(k)})}{D(w_i^{(k)})} \tag{1}$$

$$C_{UMass} = \frac{1}{K}\sum\nolimits_{k=1}^{K} C_{UMass}\left(k; W^{(k)}\right) \tag{2}$$

where $D(w_i)$ is the count of records that comprehend the word w_i; $D(w_i, w_j)$ is the count of records including the words w_i and w_j; and $(w_1^{(k)}, \ldots, w_N^{(k)})$ is the catalog of the N most likely words in topic k. Additionally, $\varepsilon = 1$ is used to avoid the logarithm of 0. To convey essential information and prevent specific particulars, LDA topics were semantically categorized into four topics: demographic, technical, financial, and opinion. Table 3 shows the 10 most relevant words in each topic.

Table 3. Topic modeling

No	Topic	Freq. (%)	C_{UMass}	Top ten words ($\lambda = 0, 3$)
1	Demographic	27.41	−0.58	{norte, sul, nordeste, centro-oeste, casa, sudeste, região, residência, brasil, domicílio}
2	Technical	15.51	−1.02	{solar, energia, painel, fotovoltaico, sistema, placa, instalação, potência, autonomia, elétrica}
3	Financial	13.73	−0.98	{banco, financiamento, vista, prazo, parcela, empréstimo, dinheiro, cartão, dinheiro, pagamento}
4	Opinion	10.36	−0.76	{economia, autonomia, limpa, renovável, limpa, segura, duração, benefícios, independência, valor}

Similarity Analysis. Following the previous phase, the distances and possible interrelationships were quantified utilizing the Hellinger distance (HD). HD enumerates the level of similarity between probability distributions p and q. Equation 3 shows it.

$$HD(p, q) = \frac{1}{\sqrt{2}}\sqrt{\sum_{i=1}^{T}(\sqrt{p_i} - \sqrt{q_i})^2} \quad (3)$$

4.2 Agent-Based Model

The model developed in this study considers the population distribution in each area, the potential for solar photovoltaic microgeneration in the area, the limitations of solar photovoltaic electricity microgeneration technologies, and simulates the interactions between these factors to capture the general macro-effects of different micro-decisions. These agents are created in the template: (1) a representation of the environment and solar potential; (2) populations that require domestic solar systems; (3) domestic solar systems that produce electricity; and (4) a fundamental observer decides the tactics and predilections for the diffusion of solar photovoltaic energy. Table 4 shows the parameters used.

Table 4. Parameters used in the NetLogo model.

Parameter	Description
$LUCE_{PV}$	Level unit cost of supplied electricity
$EHFS$	Equivalent hours of full sun per day
$C_{A/kWh}$	Avoided cost
CUF	Capacity utilization factor
$ALCC_{PV}$	Annualized life cycle cost
C_{0PV}	PV module capital cost
C_{0batt}	Battery capital cost
C_{0cc}	Charge controller capital cost
C_{0appl}	Home appliance capital cost
CRF	Capital Recovery Factor
$C_{O\&M}$	Operation and maintenance cost
i	Discount rate
T_{IR}	Neighborhood Threshold
p	Innovation coefficient
q	Imitation coefficient
$F(t)$	The proportion of adopters at time t
m	Potential market factor
$H_{PV/IR}$	Families with PV within a certain radius of influence
$H_{Total/IR}$	Total households within the same radius of influence

A house that does not have a domestic solar system would consider the installation, considering that $C_{A/kWh}$ is the avoided cost per kWh (Eq. 4). On the other hand, the $LUCE_{PV}$ is the unit cost of the supplied electricity, which is provided by Eq. 5. *EHFS* is the equivalent hours of complete sunlight for each day, W_p is the supposed peak Watt capability of the domestic solar panel based on a house's activity report and energy requirement , *CUF* is the consumption aspect of capacity that integrates non-use and system interruptions expected for several reasons.

$$LUCE_{PV} < C_{A/kWh} \tag{4}$$

$$LUCE_{PV} = \frac{ALCC_{PV}}{W_p \times EHFS \times 365 \times CUF} \tag{5}$$

$ALCC_{PV}$ is the annualized life cycle cost and it is designed by addition of the price of all its components, like battery, panel, charge controller, and appliances increased by their respective investment recovery factors plus maintenance expenses and it is indicated in Eq. 6, where C_{0PV} is the price of the domestic solar panel system, C_{0batt} is the price of the battery, C_{0cc} is the price of the charge controller, C_{0appl} is the price of the appliances, all of them are the investment recovery factors of the domestic solar, and $C_{O\&M}$ is the operating price and maintenance. Capital recovery factor *CRF* is determined using Eq. 7, where i is the reduction rate while n is the lifetime of the component. Lastly, the domestic solar system is estimated in Eq. 8 where $H_{PV/IR}$ is the total of households with solar photovoltaic energy contained by a certain radius of influence (*IR*), $H_{Total/IR}$ is the total of households surrounded by the same radius of influence, and T_{IR} is the limit from the neighborhood.

$$\begin{aligned} ALCC_{PV} = {} & (C_{0PV} \times CRF_{PV}) + (C_{0batt} \times CRF_{batt}) + (C_{0cc} \times CRF_{cc}) \\ & + (C_{0appl} \times CRF_{appl}) + C_{O\&M} \end{aligned} \tag{6}$$

$$CRF = \frac{i(1+i)^n}{(1+i)^n - 1} \tag{7}$$

$$\frac{H_{PV/IR}}{H_{Total/IR}} \times 100 > T_{IR} \tag{8}$$

5 Results and Discussions

5.1 Data Collection Results

According to the collection, the main reason for the installation of the domestic solar system were economic reasons, such as the reduction of costs about energy expenditure in the medium and long term, followed by the desire for greater autonomy about having one's energy source. Table 5 summarizes all the information.

Moreover, the results in Table 5 demonstrate the neighborhood effect played more responsibility influence in domestic solar system installation judgments than advertisements.

Table 5. Main reasons and major influences for installing domestic solar systems

Reasons	% of Reasons	Influence	% of Influence
Economy	44.33	Neighbors, friends, or relatives	54.30
Autonomy	36.00	Advertisements (TV, Internet)	34.80
Access to an energy source	9.00	First installers	10.90
Environmental concerns	7.00		
New technology	3.67		

5.2 Agent-Based Simulation Results

The radius of influence is the area surrounded by which a family affects or is affected by its neighbors. There is a growth of 1246% of families that will establish domestic solar systems after 25 years contemplating a radius of influence of 1 km. The total of families that would install domestic solar systems expands with the expansion of the radius of influence, and to 5 km, indicating there would be a growth of 2589% later than the same period. Figure 1 demonstrates the comparison among the different radius of influence.

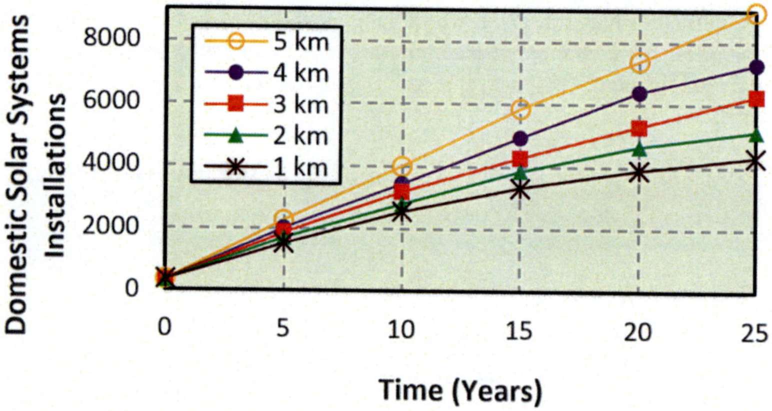

Fig. 1. Comparison of domestic solar systems installations over 25 years.

The neighborhood threshold is the lowest possible percentage of neighbors contained by a particular radius of influence who already installed domestic solar systems for a family to contemplate performing the same. It is an amount of social pressure and how it makes families install domestic solar systems as more and more neighbors do. It additionally demonstrates the inflection point, on top of which domestic solar systems installations begin to come down. The simulation reveals, centered on the installation choices of domestic solar systems, around 15% would be perfect for the neighborhood threshold. Figure 2 contrasts the impacts of different neighborhood thresholds on the time-dependent implementation of domestic solar systems in Brazil.

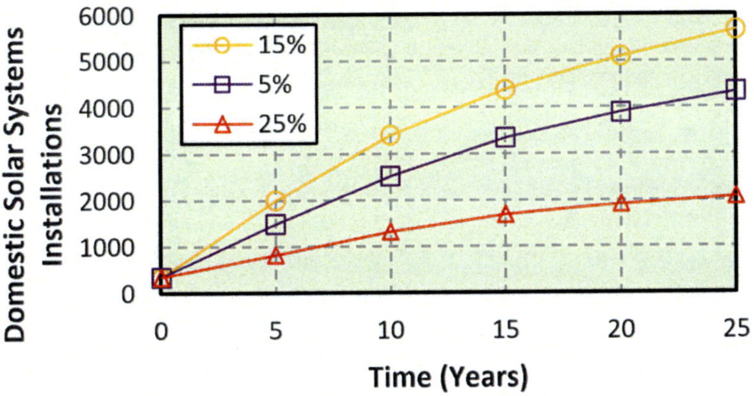

Fig. 2. Comparison of domestic solar installations over 25 years.

6 Conclusion

This article proposed a way to elicit realistic indicators, based on the collection of unstructured data distributed on the Internet. For this, a survey was developed through Web Scraping to collect the information and establish the necessary indicators to feed an agent-based simulation through the NetLogo software. The simulation of this complex scenario verified the influence of people and the media in the implementation of domestic solar systems in Brazil.

The results showed that mainly friends, neighbors, and/or relatives who installed domestic solar systems encourage the adoption of this technology because people get to know the benefits better through spoken or written comments on social networks. Furthermore, it was observed that the increase in the radius of influence and the neighborhood threshold leads to the expansion of facilities because as more families install these systems, new families notice the benefits of the implemented technology.

References

1. Foster, I., Ghani, R., Jarmin, R.S., Kreuter, F., Lane, J.: Big Data and Social Science, 2nd edn. Chapman and Hall, London (2020)
2. Webber, K., Zheng, Y.: Big Data on Campus, 1st edn. Johns Hopkins Press, Baltimore (2020)
3. Marr, B.: Data Strategy, 1st edn. Kogan Page, London (2017)
4. Jezic, G., Chen-Burger, J., Kusek, M.: Agents and Multi-Agent Systems, 1st edn. Springer, Berlin (2020)
5. Boissier, O., Bordini, R.H., Hubner, J., Ricci, A.: Multi-Agent Oriented Programming, 1st edn. The MIT Press, New York (2020)
6. Vo, L.T.: Mining Social Media, 1st edn. No Starch Press, San Francisco (2019)
7. Mitchell, R.: Web Scraping with Python. 2nd ed. O'reilly Media, Sebastopol (2018)
8. Maneth, S. (ed.): BICOD 2015. LNCS, vol. 9147. Springer, Cham (2015). https://doi.org/10.1007/978-3-319-20424-6
9. El Fallah Seghrouchni, A., Dix, J., Dastani, M., Bordini, R.H. (eds.): Multi-Agent Programming. Springer, Boston, MA (2009). https://doi.org/10.1007/978-0-387-89299-3

10. Dalle Nogare, D., Chitnis, A.: NetLogo agent-based models as tools for understanding the self-organization of cell fate, morphogenesis and collective migration of the zebrafish posterior Lateral Line primordium. Semin. Cell Dev. Biol. (2019). https://doi.org/10.1016/j.semcdb.2019.12.015
11. Akhtar, N., Rehman, A., Hussnain, M., Rohail, S.: Hierarchical coloured petrinet based multi-agent system. IEEE Access **7**, 180544–180557 (2019)
12. Russel, S., Norvig, P.: Artificial Intelligence: A Modern Approach, 4th edn. Prentice Hall, New Jersey (2003)
13. Blei, D.M., Nig, A.Y., Jordan, M.I.: Latent dirichlet allocation. J. Mach. Learn. Res. **3**, 993–1022 (2003)

Mechanism of Collecting Urban Data for Application on Smart Cities

Jemis Dievas José Manhiça(✉) and Ademar Takeo Akabane

Centre for Exact Sciences, Technology and the Environment (CEATEC), Pontifical Catholic University of Campinas, Campinas, Brazil
jemisdjose@gmail.com, ademar.akabane@puc-campinas.edu.br

Abstract. Access to a city's limited resources has been negatively affecting the lives of billions around the world, due to population rapid growth. In the efforts of softening those problems, researchers, companies, and public agencies have been investing more and more in the Smart Cities concept to improve urban services and, consequently, increase the quality of life of the residents. Smart Cities emerge from joining two development urban streams: (i) the cities' redefinition from Information Technology and communications perspective; and (ii) from understanding the city as a space for creativity and innovation. It is known that applications developed for Smart Cities can highly facilitate urban management, in addition to increasing the population's quality of life. However, to achieve a maturity stage, many technical and scientific challenges have to be met before those applications can be widely used. These challenges span from collecting, storing, and processing to analyzing and visualizing large datasets. The purpose of this work is to propose an efficient solution for urban data collection.

Keywords: Smart cities · Urban sensing · Urban monitoring

1 Introduction

The inhabitants of large urban centers are increasingly suffering serious social-economic problems, due to the cities' uncontrolled growth and also by the strong migration of people towards the urban areas. The increased demand for several kinds of services and the city's infrastructure emerges as a direct consequence of this issue. In the efforts to deal with those problems, new forms of power generation, natural resources conservation, efficient transportation, education, security, health, and food are great challenges to be met in the great urban centers [1–3]. In this context, Information and Communications Technologies (ICTs) play an important role as facilitators in decision-making, besides creating innovative applications that improve the urban infrastructure management and, consequently, the citizens' life quality.

Y. Iano et al. (Eds.): BTSym 2021, SIST 207, pp. 97–105, 2023.
https://doi.org/10.1007/978-3-031-04435-9_9

Making cities more intelligent is one way to face social-economic issues, using optimized resources and more sustainable infrastructure. Smart Cities do not include only ICT, but also the citizens and the government involvement in governance processes [4,5]. An overview of the application's framework for smart cities, emphasizing the three main components, is displayed in Fig. 1: from the bottom-up, (i) Urban Data Collection (or Urban Sensing); (ii) Integration and Analysis of Collected Data, and; (iii) Development of Services or Applications.

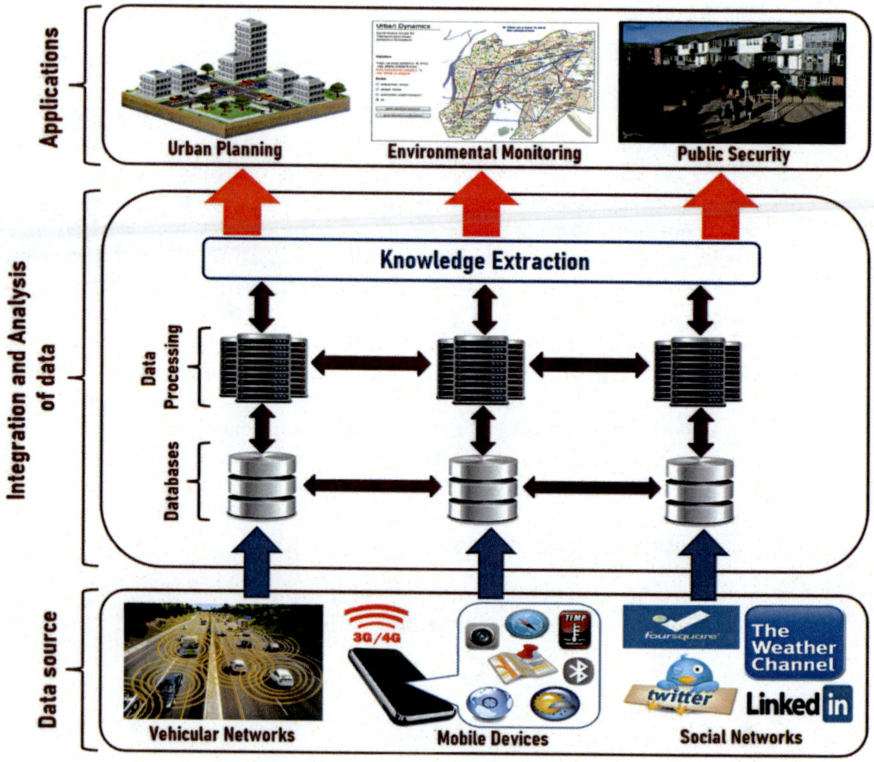

Fig. 1. Illustration of smart cities framework.

On the one hand, from the development and implementation of services and applications for Smart Cities' point of view, there is consensus when it comes to the importance of construction for open updated data repositories, collected in real-time and coming from various sources [2,5]. These repositories could be used to provide support to different applications, for instance, environmental monitoring, security, and urban mobility, among others. In other words, the data collection aims to provide data samples from multiple sources in an efficient and continuous way.

On the other hand, microblogging sites (e.g., Twitter[1], Tumblr[2], and Reddit[3]) are rich in sources for varied kinds of information. It is worth noting that this is a commonplace where people exchange their opinions on various issues and it may contain ongoing trends.

In this work, we discuss some of the challenges and key factors that must be present at the core components of the mechanism of collecting urban data from Twitter for application in Smart Cities.

The paper is organized as follows. Section 2 discusses the methodology of this work and the results and discussion are presented in Sect. 3. Finally, the study is concluded in Sect. 4.

2 Methodology

The wide availability of data sources on various aspects of the city helps the process of understanding and solving problems suffered in urban centers, proposing more sophisticated services aimed at improving the lives of the inhabitants. In this sense, we study an approach to facilitate the development of a mechanism for collecting urban data. This investigation was based on works and studies from literature, through journals, surveys, and articles, with the main purpose of learning more about the techniques and Smart Cities concept, and identifying possible opened gaps.

After investigating different forms and techniques of data collection, it was decided that it would be interesting to use the Twitter API[4] since many works in the literature [6–9] use this same technique when it comes to data collection in urban spaces.

The Twitter API interface is used to collect streaming Tweets. One example of using Python programming language with the assistance of tweepy[5] library,

1 https://twitter.com/.
2 https://www.tumblr.com/.
3 https://www.reddit.com/.
4 https://developer.twitter.com/.
5 Tweepy is a Python library for accessing the Twitter API.

which is used to collect data from Twitter, is shown in Fig. 2. The extracted data was saved in a text file. It is worth noting that only publicly available Tweets can be collected via the API. The streaming API searches for hashtags, keywords, and geographic bounding boxes simultaneously. Filter API searches and delivers the streaming Tweets that match the filter tag. It is worth noting that the field keys (ACCESS_TOKEN, ACCESS_TOKEN_SECRET, CONSUMER_KEY, and CONSUMER_SECRET) can be acquired after registering in the Twitter API.

```
#import things from a tweepy libary
#this is for live tweets based on location & hastags
from tweepy.streaming import StreamListener
#this is for autentications based on twitter Keys that we saved, associeted
to twitter app created on twitter API
from tweepy import OAuthHandler
from tweepy import Stream

ACCESS_TOKEN ="################################"
ACCESS_TOKEN_SECRET = "#########################"
CONSUMER_KEY ="#########################"
CONSUMER_SECRET ="################################"

class listener(StreamListener):
  """ A listener handles tweets are the received from the stream.
  This is a basic listener that just prints received tweets to stdout.
  """
  def on_data(self, data):
      try:

          saveFile = open('testData.txt', 'a')
          saveFile.write(data)
          saveFile.write('/n').encode("utf-8")
          saveFile.close()
          return True
      except BaseException:
          print ('failed ondata')
          time.sleep(5)

  def on_error(self, status):
      print(status.encode("utf-8"))

if __name__ == '__main__':
  auth = OAuthHandler(CONSUMER_KEY, CONSUMER_SECRET)
  auth.set_access_token(ACCESS_TOKEN, ACCESS_TOKEN_SECRET)
  #ASK FOR KEYWORD TO COLLECT DATA
  #user_keyword=input("What keyword do you want to mine?")
  twitterStream = Stream(auth, listener)Important
  #Test with NY City coordenates
  twitterStream.filter(locations=[-74.1687,40.5722,-73.8062,40.9467])
```

Fig. 2. Python code used to collect tweets by location.

```
with open("/content/testData.txt") as fp:
    for line in fp:
        #p.OPT.EMOJI, p.OPT.HASHTAG, p.OPT.SMILEY, p.OPT.MENTION, p.OPT.URL
          texto =use_tweet_preprocessor(line)
          #remove https
          l = re.sub(r'(?is)https.+', ' ', texto)
          #remove os unicodes
          nl = re.sub(r'(?is)\\u.+', ' ', l)
          #print de testS
          #print(nl)
          saveFile = open('cleanData.txt', 'a')
          saveFile.write(nl)
          saveFile.write('\n')
          saveFile.close()
```

Fig. 3. Python code used to remove unwanted and unnecessary information.

Twitter data is very noisy and unstructured. To extract useful information to analyze, preprocessing tasks are very important and critical in text mining. After data was collected and stored in a text file, it was necessary to clean it first, as they bring a lot of unwanted and unnecessary information for the analysis, such as links, Emojis, URLs, Hashtags, Mentions, Smileys, and numbers. We adopt three steps (as described below) to normalize the text of the tweets using natural language processing [10]:

1. **Nosing data cleaning:** In this work, we remove the username@, hashtags, and URL firstly;
2. **Upper to lower:** When we analyze the text, some tweets contain upper letters. Converting all of the letters into lower is required;
3. **Stop words removal:** In tweets, a lot of stop words (words that do not add much meaning to a sentence) are contained in the text, such as "the", "is", "on", "that", "and", etc.

Figure 3 displays an example of a function that was used to remove the unwanted and unnecessary tweet information.

Following that, with the help of the Word Cloud (Fig. 4) generated with the code shown in Fig. 5, it was possible to visualize and analyze the collected data, and it was noticed that the data was not as clean as previously thought, because it contained Unicode character (Unicode is a universal character encoding standard that assigns a code to every character and symbol in every language in the world) that had been removed through the code from Fig. 3.

And finally, studies and tests were made in order to remove the stop words. They can safely be ignored without sacrificing the meaning of the sentence, they are words that are filtered out before or after processing natural language data (text), the code that does this part can be seen in Fig. 6.

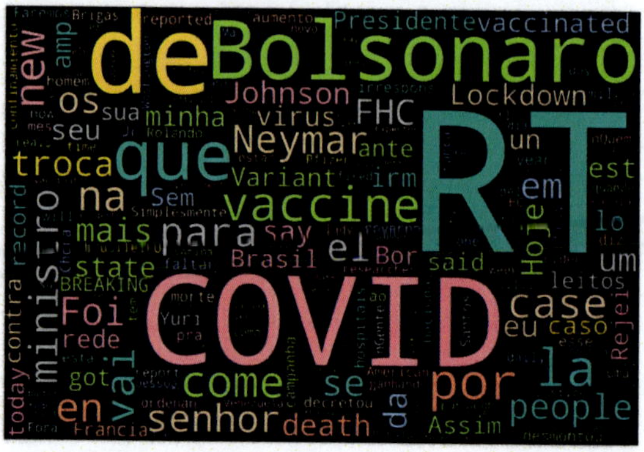

Fig. 4. Example of a word cloud with cleaned data.

```
# Define a function to plot word cloud
def plot_cloud(wordcloud):
  # Set figure size
  plt.figure(figsize=(40, 30))
  # Display image
  plt.imshow(wordcloud)
  # No axis details
  plt.axis("off");

# Import package
from wordcloud import WordCloud

#Gerando  wordcloud
wordcloud = WordCloud(width = 3000, height = 2000, random_state=1,
                      background_color='red', colormap='Set3',
                      collocations=False, stopwords = STOPWORDS).generate(text)

# Plot
plot_cloud(wordcloud)

#Save the IMG File
wordcloud.to_file("wordcloud.png")

#read the whle file
dirname = path.dirname("/content/cleanData.txt")

# Read the whole text.
text = open(path.join(dirname, 'cleanData.txt')).read()

#separate the text in tokens
text_tokens = word_tokenize(text)

#remove all the stopwords
tokens_without_sw =[word for word in text_tokens if not word in stopwords.words()]
```

Fig. 5. Example of a code that plots a word cloud.

```
#read the whle file
dirname = path.dirname("/content/cleanData.txt")

# Read the whole text.
text = open(path.join(dirname, 'cleanData.txt')).read()

#separate the text in tokens
text_tokens = word_tokenize(text)

#remove all the stopwords
tokens_without_sw =[word for word in text_tokens if not word in stopwords.words()]

#rejoin all the words without the stopwords
backToOneWord = (" ").join(tokens_without_sw)

text = backToOneWord;
```

Fig. 6. Example of a code to remove stop words.

3 Results and Discussion

Using the Twitter API together with tweepy library it was possible to collect and analyze the collected data. Tests were conducted with the purpose to improve qualitatively and quantitatively the data, and, with the help of tools offered by Google Colab, it was possible to remove the stop words, Unicode, links, Emojis, URLs, Hashtags, Mentions, Smileys, and numbers. An initial test analysis was done in the one-month period from March 1, 2021, to March 31, 2021.

Initially, several difficulties related to computer resources (some of python old functions did not work on our computer), the Python version, and the format in which the data was being saved were encountered (and some of the oldest python libraries do not run in python3), and the solution was to overcome those limitations. Using Google Colab, and saving the data in a different format made it possible to continue working on the research, and clean from the data all unnecessary information.

Another difficulty that had to be overcome was about Brazilian cities coordinates since Twitter only accepts coordinates from large cities in order to collect data by location, as Fig. 7 shows. During tests, it was decided that a much wider region would be used to solve that issue (the selected "wider" region would include the wanted city, in this case, Campinas).

All the developed code during the research can be found on the personal GitHub [11], and some .txt files with some collected, and some cleaned data can be found on the repository too.

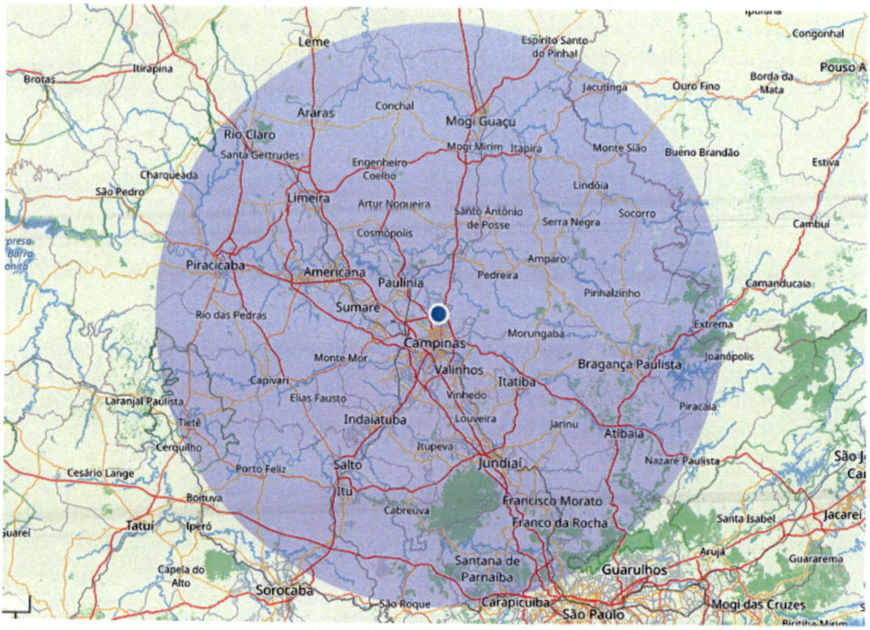

Fig. 7. Image of the selected region.

4 Conclusion

In this work, it was discussed the existing challenges at the main components of the development of a mechanism for collecting urban data.

As can be observed from the collected data through Twitter (that can be found on the personal GitHub [11]), and through Word Cloud representation, it is possible to collect relevant data from urban places that could be applied for different purposes in the context of Smart Cities using the Twitter API. It was also possible to notice that Python provides all the necessary tools and libraries to help with the extraction, processing, analysis, and even representing the data, and clean it almost perfectly.

With all that said, a vast field of application can be noted to improve the services and provide rich data for application constructions that could ultimately improve the lives of those who live in urban centers.

One field that would be interesting to apply the collected data could be machine learning. Since machine learning is a field of Artificial intelligence with the purpose of computer techniques development that allows building systems capable of automatically acquiring knowledge through non-random inserted data [6]. It is important to mention that these techniques are possible because such data have patterns and thus allow the extraction of generalizations. So, with this data, it is possible to obtain important aspects of a dataset in which the model was trained.

References

1. Bauza, R., Gozalvez, J., Sanchez-Soriano, J.: Road traffic congestion detection through cooperative vehicle-to-vehicle communications. In: IEEE Local Computer Network Conference, pp. 606–612 (2010)
2. Silva, T.H., et al.: Urban computing leveraging location-based social network data: a survey. ACM Comput. Surv. **52**(1), 1–39 (2019)
3. Zheng, Y., Capra, L., Wolfson, O., Yang, H.: Urban computing: concepts, methodologies, and applications. ACM Trans. Intell. Syst. Technol. **5**(3), 1–55 (2014)
4. Zanella, A., Bui, N., Castellani, A., Vangelista, L., Zorzi, M.: Internet of things for smart cities. IEEE Internet Things J. **1**(1), 22–32 (2014)
5. Kamienski, C., Biondi, G.O., Borelli, F.F., Heideker, A., Ratusznei, J., Kleinschmidt, J.H.: Computaçao urbana: Tecnologias e aplicaçoes para cidades inteligentes. Minicursos SBRC (2016)
6. Kumar, S., Morstatter, F., Liu, H.: Twitter Data Analytics. Springer (2014)
7. Kearney, M.W.: rtweet: collecting and analyzing Twitter data. J. Open Source Softw. **4**(42), 1829 (2019)
8. Byun, C., Kim, Y., Lee, H., Kim, K.K.: Automated twitter data collecting tool and case study with rule-based analysis. In: Proceedings of the 14th International Conference on Information Integration and Web-based Applications & Services, pp. 196–204 (2012)
9. Kouloumpis, E., Wilson, T., Moore, J.: Twitter sentiment analysis: the good the bad and the omg! In: Fifth International AAAI Conference on Weblogs and Social Media, vol. 5, no. 1, pp. 538–541 (2011)
10. Shao, X., Kim, C.-S., Ryul, K.D.: A study on customers sentiment analysis based on big data using twitter data. Int. J. Comput. Theory Eng. **11**(1), 11–14 (2019)
11. Github. GitHub - jamesdievas/Smart-Cities: Scientific Inicietion. In: GitHub (2021). https://github.com/jamesdievas/Smart-Cities. Accessed 9 Nov 2021

Exploring Research Trends in Healthcare 4.0

Bárbara Santiago de Mendonça and Lásara Fabrícia Rodrigues(✉)

Federal University of Minas Gerais – UFMG, Belo Horizonte, Brazil
basantiagom@ufmg.br, lasara@dep.ufmg.br

Abstract. In recent years, healthcare is being challenged to deal with global aging, an increasing number of chronic patients, and growing demand for more personalized services. From this standpoint, it is comprehensible that health is longing for solutions to enhance its quality while reducing costs. In this light, the technological development brought by Industry 4.0 arises surrounded by high expectations of solving, or at least minimizing, some of the aforementioned questions. Therefore, this paper aims to investigate the current state of the health revolution, named Healthcare 4.0, and follows a systematic methodology to select the documents retrieved from Web of Science and Scopus. Firstly, keywords were chosen to execute the search in the scientific databases. Then, inclusion and exclusion criteria were used to define the final sample of 130 articles. Lastly, a bibliometric analysis was performed for a better understanding of the ongoing health revolution. With the assistance of the bibliometrix R-tool, graphics and statistics were designed in order to recognize publication patterns. It was concluded that, although Healthcare 4.0 is still a recent research topic, increasing attention is being given to it. The most influential authors and countries were also detected, enabling the recognition of India as the main reference in the subject. Moreover, the relevance of IEEE journals to Healthcare 4.0 was identified, and the importance of artificial intelligence was also revealed.

Keywords: Healthcare 4.0 · Health 4.0 · Industry 4.0 · Technology · Bibliometric analysis

1 Introduction

Healthcare is facing growing challenges in recent years. Threatened by the aging of the global population and the increasing number of chronic patients, the sector is in the wait for efficient solutions to better treat the ill and to enhance the management of their resources [1]. In this context, there is a high expectancy that with the use of technologies of Industry 4.0 in healthcare, such as the Internet of Things, cloud, big data, artificial intelligence, and robotics, some of these issues can be solved or, at least, minimized. It is expected that the costs of healthcare services will be diminished at the same time that the user experience is improved and the quality of care increased [2].

These advancements envisioned for healthcare are outcomes of the rapid technological development that the world is witnessing in the last few decades. In this regard, the already mentioned concept of "Industry 4.0" was first used in 2011 in a reference

Y. Iano et al. (Eds.): BTSym 2021, SIST 207, pp. 106–114, 2023.
https://doi.org/10.1007/978-3-031-04435-9_10

to the commitment of the German government to encourage new technologies and to provide a propitious scenario for the advancement of the local industry [3]. After that, the discussions about this theme only raised, and, in the World Economic Forum (WEF) in January 2016, the Fourth Industrial Revolution (FIR) was announced [4]. According to Professor Klaus Schwab, Founder and Executive Chairman of the WEF, the FIR is "characterized by a fusion of technologies that is blurring the lines between the physical, digital and biological spheres" [4]. In this light, it is predictable to imagine that the transformations derived from the revolution would not be restricted to the industrial area and that they could even reach the medical field.

Within this framework, the adaptation of Industry 4.0 principles, technologies, and applications in the healthcare area is named Healthcare 4.0, Health 4.0, or Healthcare Industry 4.0 [5]. It is already known that the neologism "4.0" has been widely used to indicate the last generation of technological and organizational innovations [6]. Other terms, such as E-health and smart health, are also used to refer to the current transformations that are taking place in the healthcare field, but none of them are clearly defined yet [7]. There is a lack of disseminated definitions in the literature. New terms are often introduced without providing a formal explanation [7].

Thus, there is a need for a better understanding of the ongoing health revolution. Therefore, the main goal of this paper is to provide an overall view of the current state of development of Healthcare 4.0, identifying the main participants and highlighting research trends. In addition, this study aims at providing the reader with a bibliometric review and an analysis of the evolution of existing literature.

The rest of the paper is arranged as follows. The next section presents the systematic review strategy used in the literature. A quantitative analysis is performed in Sect. 3. The study is concluded with Sect. 4.

2 Methods

The data collection was conducted to identify studies applied to Healthcare 4.0, since this area is extremely relevant and has enormous potential – in 2016, a poll conducted by the Economist Intelligence Unit pointed out that healthcare is the sector that will benefit the most from the Fourth Industrial Revolution [8].

The search protocol adopted in this research is summarized in Table 1. First, the electronic databases Web of Science and Scopus were selected with the intention to achieve maximum coverage of the management and engineering research literature. Second, two groups of keywords were defined. In the first group, the "healthcare 4.0" and its synonyms gathered in previous research [3, 9, 10] were selected. The second group involves "industry 4.0", "industrie 4.0", and equivalent terms also collected in the literature [11, 12]. It's noteworthy that "industrie 4.0" is how the new revolution was first mentioned in German, in 2011, and the term is still broadly used [13]. These strings were then combined with health-related terms in order to identify as many publications as possible associated with the topic. This second group of keywords was essential to reach a higher number of documents related to the theme of the research, considering that Healthcare 4.0 is a recent expression and not yet widely adopted.

Table 1. Search protocol

Data source:	Web of science and scopus
Search string:	("health 4.0" OR "healthcare 4.0" OR "H4.0" OR "HC4.0" OR "Hospital 4.0") OR (("industry 4.0" OR "Industrie 4.0" OR "Fourth Industrial Revolution" OR "4th Industrial Revolution") AND ("health" OR "healthcare" OR "hospital" OR "medic*"))
Search fields:	Title, abstract, and Keywords
Period:	From 2011 to December 31, 2020
Language:	English
Document:	Journal articles

In the first group of keywords, the terms were connected using the Boolean operator "OR". In the second group, firstly, the synonyms of Industry 4.0 and the terms related to healthcare were separately connected using the Boolean operator "OR". Then, these clusters were connected using the Boolean operator "AND". Lastly, the first and the second groups of keywords were associated by the Boolean operator "OR".

The search was performed in each database following the steps described next:

- Step 1: Search fields: title, abstract, or keywords of articles in the data sources.
- Step 2: Filter by year of publication: the search period ranges from 2011 to 2020, to enable an annual comparison.
- Step 3: Language: select only papers published in English.
- Step 4: Document types: refine the search to consider only journal articles, as they are usually peer-reviewed and, hence, considered more reliable. Then, both articles and reviews were selected.

The search was executed on April 1, 2021, and it resulted in an initial set of 757 publications - 370 articles were selected from Web of Science and 387 from Scopus. They were combined into a single database and the duplicates were removed using Mendeley. After that, the number of documents was reduced to 484.

The first screening, named Phase I, was done through the analysis of the title, keywords, and abstract of articles in the sample. The exclusion criteria in Table 2 were applied in the selection. Papers without abstract were preserved to allow a fair decision. This phase resulted in 165 documents and their full texts were, then, searched and downloaded (PDF files). In Phase II, the entire article was evaluated through the exclusion and inclusion criteria also specified in Table 2. After following the proposed steps, a total of 130 papers were included in the final sample for review. Subsequently, a quantitative analysis of Healthcare 4.0 studies was performed.

Table 2. Exclusion and inclusion criteria and number of occurrences

Phase	Description	Occurrences
Phase 1	E1: Healthcare 4.0, Industry 4.0, or Health is only cited or used as an example	167
	E2: The paper is not related to the scope of the research	152
Phase 2	E3: Full text could not be assessed	22
	E4: A quick mention to Healthcare 4.0	6
	E5: The paper is not related to the scope of the research	7
	I1: The paper focus on an application of Healthcare 4.0	54
	I2: The paper review or analyze Healthcare 4.0 or a topic of it	76

3 Quantitative Analysis

A bibliometric analysis is a quantitative approach for analyzing academic literature using bibliographies [14]. Through mathematical and statistical methods, scientific outputs can be measured, emerging topics can be identified, and research clusters can be found [15]. In this work, by investigating publication patterns, we aim to get insights into the current state of Healthcare 4.0.

The analysis was performed using bibliometrix R-Tool, a package developed by Massimo Aria and Corrado Cuccurullo from the University of Naples and University of Campania's Luigi Vanvitelli respectively [16]. This tool was chosen since it incorporates a great variety of different analyses, broader than the others considered [17].

After extracting the complete bibliography data of 130 publications from Scopus and Web of Science databases, the convert2df function was used in the R environment to create a unique data frame. Lastly, the merged document with the information from both databases was introduced in bibliometrix so that the analysis could be performed.

3.1 Year of Publication

Figure 1 (a) displays the distribution of scientific publications over time. Although Industry 4.0 was first introduced in 2011 [13], its applications in health were first mentioned in 2016. The oldest paper of the review sample, published in 2016, discusses if the healthcare sector is prepared to face the Fourth Industrial Revolution [8]. In the following year, studies about the theme were still scarce. The major shift happened in 2018, when the number of published articles increased 6 times. Since then, the quantity of documents related to the topic has been swiftly growing. This fact indicates that Healthcare 4.0 is still a recent research topic that is attracting increasing attention.

3.2 Geographical Distribution

The number of documents per country is illustrated in Fig. 1 (b). The publication activity of the countries about Healthcare 4.0 is as follows: India (23 documents), Italy (18 documents), South Korea (16 documents), United States (11 documents), United Kingdom (8 documents), Germany (7 documents), Spain (7 documents), South Africa (5 documents), Turkey (5 documents), Portugal (4 documents). These top ten countries contribute to 80% of total publications. The other 26 countries also in the image had 3 or fewer articles published. It is interesting to note that all continents were represented in the final set.

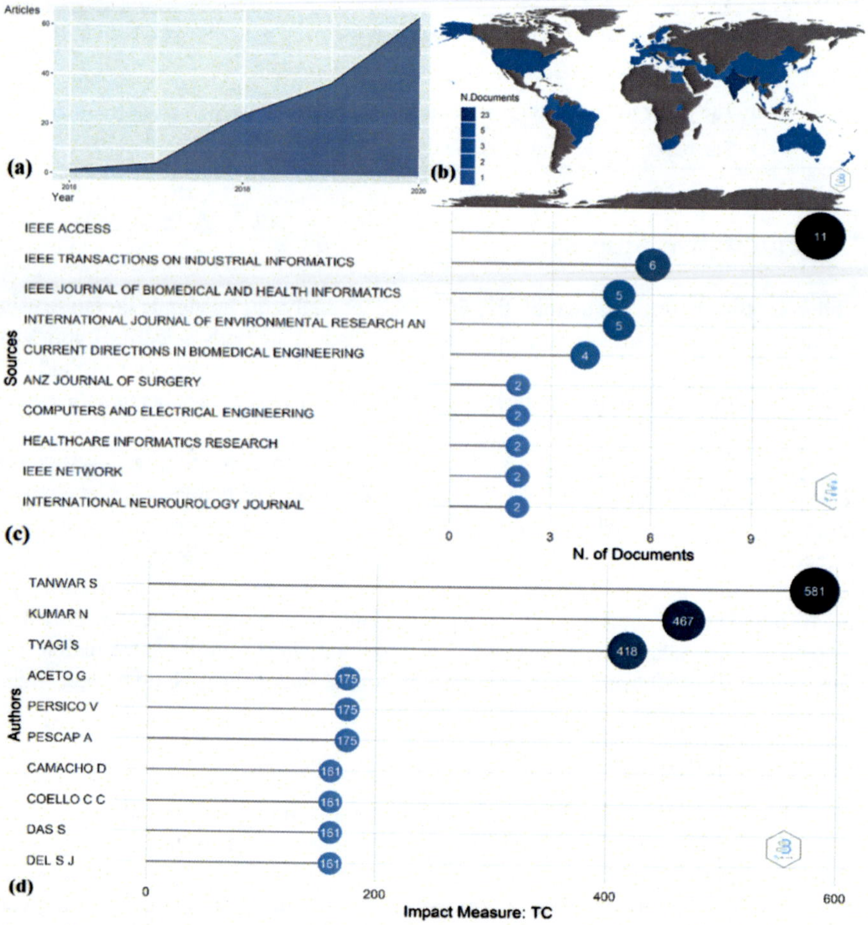

Fig. 1. **(a)** Annual scientific production **(b)** Countries scientific production **(c)** Most relevant sources **(d)** Author's impact

3.3 Leading Sources

The sample of 130 articles was published in 94 different sources and only five of them had more than two documents printed – IEEE Access (11 papers), IEEE Transactions on

Industrial Informatics (6 papers), IEEE Journal of Biomedical and Health Informatics (5 papers), International Journal of Environmental Research and Public Health (5 papers), and Current Directions in Biomedical Engineering (4 papers). Figure 1 (c) shows the ten journals with more papers retrieved from the literature review. Interestingly, the top three sources are from the IEEE association, whose focus is on the engineering, computing, and technology fields. Furthermore, it is worth noting that IEEE Access, the journal that appears most in the sample, has just started publishing about Healthcare 4.0 – two of its articles are from 2019 and nine from 2020.

3.4 Author's Impact

Bibliometrix provides four different measures to quantify the author's impact: h-index, its variants (g-index and m-index), and total citation. None of them is perfect. H-index is the most popular metric; however, it only considers the author's best publications, and low-quality work is not penalized – a fact that unintentionally is encouraging printing of immature work and micro-publication [18]. The total citation measure also has problems. It does not consider, for example, self-citations or negative ones [19].

Nevertheless, these metrics provide, if not a perfect, a good outlook on the author's performance. The total citation measure was chosen to evaluate the author's impact in this review and the results are displayed in Fig. 1 (d). Sudeep Tanwar was ranked number 1, with 581 citations, followed by Neeraj Kumar, with 467, and Sudhanshu Tyagi, with 418. All of them are from India and are also the three authors with more documents published; Tanwar and Kumar had 9 papers within the sample and Tyagi, 6 papers.

3.5 Keyword Analysis

Keywords frequency analysis is valuable, since these words depict the essence of articles in such a manner that, once the most used terms are identified, research hotspots and research cores in the field can be highlighted [20]. Figure 2 shows a word cloud with the high-frequency author's keywords and Table 3 presents the absolute occurrences of the ten most-cited names.

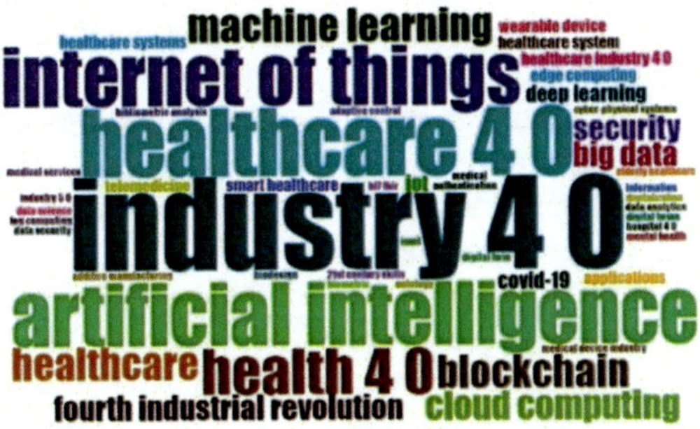

Fig. 2. Word cloud of high-frequency author keywords

Table 3. Most cited keywords

Words	Occurrences	Words	Occurrences
Industry 4.0	22	Blockchain	8
Healthcare 4.0	16	Healthcare	8
Artificial intelligence	15	Machine learning	8
Internet of things	14	Cloud computing	7
Health 4.0	11	Big data	6

A total of 534 different authors' keywords were collected from the sample, but only 24 of them were used more than two times. Industry 4.0 had the highest number of mentions, a predictable result due to the substantial connection between its technologies and Healthcare 4.0 along with the expansive dissemination of this concept. The same cannot be seen in the medical area. Both Healthcare 4.0 and Health 4.0 are in the top five of most cited words and Healthcare Industry 4.0 also appears in the nineteenth position. This fact illustrates the lack of a unique term and the need for a clear and widely adopted definition.

Through this study, it was also possible to identify the core technologies of Industry 4.0 applied in the healthcare area. Artificial intelligence is in the spotlight, being ranked as the third most cited keyword, and one of its branches, machine learning, was ranked eighth. Other technologies in the top ten are Internet of Things, blockchain, cloud computing, and big data. Interestingly, blockchain, which had no citations until 2018 and was mentioned only once in the following year, had a total of eight occurrences in 2020 – more than big data, for example.

4 Conclusion

This paper presented a bibliometric analysis dedicated to investigating the Healthcare 4.0 development. A replicable methodology was proposed to collect the data from both Scopus and Web of Science databases. Moreover, inclusion and exclusion criteria were applied to select the papers for the study, which resulted in a final sample of 130 journal articles. Then, the bibliometric analysis was performed with the assistance of the bibliometrix R-tool. Through the graphics and statistics provided by the package, it was possible to identify the current state of development of Healthcare 4.0. Research trends were also highlighted, and insights were pointed out to fulfill the aim of better understanding the ongoing healthcare revolution.

The analyses showed an increasing tendency in health-related studies since 2018 and that this field is swiftly drawing more attention. Moreover, it was seen that India is the main reference in the subject, since it is the country with the highest number of published articles and the three most impactful authors. It was also recognized the relevance of IEEE journals to Healthcare 4.0, although the association is not directly connected to the health area. Finally, the absence of a standardized and accepted concept was verified, along with the recognition of the most used technologies in the field.

However, there are still several open questions that need to be addressed to achieve the best of Healthcare 4.0. Security and privacy are huge concerns [21], and data ownership needs to be taken into consideration [22]. Furthermore, the lack of standardization and a regulatory structure are relevant questions that are not solved yet [23]. Lastly, ethics and legal issues must be addressed to allow the success of the healthcare revolution [24]. Through this review, a brief but enlightening view about the current state of Healthcare 4.0 was presented. It is expected that the findings of this paper assist and encourage further analysis of those interested in this research topic.

References

1. Donati, M., Celli, A., Ruiu, A., Saponara, S., Fanucci, L.: A Telemedicine service system exploiting BT/BLE wireless sensors for remote management of chronic patients. Technologies **7**(1), 13 (2019)
2. Jayaraman, P.P., Forkan, A.R.M., Morshed, A., Haghighi, P.D., Kang, Y-B.: Healthcare 4.0: A review of frontiers in digital health. Wiley Interdiscip. Rev. Data Min. Knowl. Discov. **10**(2), e1350 (2020)
3. Aceto, G., Persico, V., Pescapé, A.: Industry 4.0 and health: Internet of Things, big data, and cloud computing for healthcare 4.0. J. Ind. Inf. Integr. **18**, 1–13 (2020)
4. Lee, J.-Y.J.-Y., Lim, J.-Y.J.-Y.: The prospect of the fourth industrial revolution and home healthcare in super-aged society. Ann. Geriatr. Med. Res. **21**, 95–100 (2017)
5. Al-Jaroodi, J., Mohamed, N., Abukhousa, E.: Health 4.0: on the way to realizing the healthcare of the future. IEEE Access **8**, 211189–211210 (2020)
6. Cavallone, M., Palumbo, R.: Debunking the myth of industry 4.0 in health care: insights from a systematic literature review. TQM J. **32**, 849–868 (2020)
7. Aceto, G., Persico, V., Pescapé, A.: The role of information and communication technologies in healthcare: taxonomies, perspectives, and challenges. J. Netw. Comput. Appl. **107**, 125–154 (2018)
8. Park, H-A.: Are We Ready for the Fourth Industrial Revolution? Yearb. Med. Inform. **25**(1), 1-3 (2016)
9. Bongomin O, et al.: Industry 4.0 disruption and its neologisms in major industrial sectors: a state of the art. J. Eng. (United Kingdom) (2020)
10. Javaid, M., Haleem, A.: Industry 4.0 applications in medical field: a brief review. Curr. Med. Res. Pract. **9**, 102–109 (2019)
11. Ghobakhloo, M., Fathi, M., Iranmanesh, M., Maroufkhani, P., Morales, M.E.: Industry 4.0 ten years on: a bibliometric and systematic review of concepts, sustainability value drivers, and success determinants. J. Clean Prod. **302**, 1–20 (2021)
12. de Paula, F.W., Armellini, F., De Santa-Eulalia, L.A.: Simulation in industry 4.0: a state-of-the-art review. Comput. Ind. Eng. **149**, 1–17 (2020)
13. Schuh, G., Anderl, R., Dumitrescu, R., Kruger, A., Hompel, M.: Industrie 4.0 maturity index. managing the digital transformation of companies. Acatech Study **64** (2020)
14. Rodríguez-Soler, R., Uribe-Toril, J., De Pablo, V.J.: Worldwide trends in the scientific production on rural depopulation, a bibliometric analysis using bibliometrix R-tool. Land Use Policy **97**, 104787 (2020)
15. Munim, Z.H., Dushenko, M., Jimenez, V.J., Shakil, M.H., Imset, M.: Big data and artificial intelligence in the maritime industry: a bibliometric review and future research directions. Marit. Policy Manag. **47**, 577–597 (2020)
16. Aria, M., Cuccurullo, C.: bibliometrix: an R-tool for comprehensive science mapping analysis. J. Informetr. **11**, 959–975 (2017)

17. Moral-Muñoz, J.A., Herrera-Viedma, E., Espejo, A.S., Cobo, M.J.: Software tools for conducting bibliometric analysis in science: an up-to-date review. Prof. la Inf. **29**, 1–20 (2020)
18. Pan, R.K., Fortunato, S.: Author impact factor: tracking the dynamics of individual scientific impact. Sci. Rep. **4**(1), 1–7 (2014)
19. Ribeiro, J.P., Barbosa-povoa, A.: Supply chain resilience : definitions and quantitative modelling approaches – a literature review. Comput. Ind. Eng. **115**, 109–122 (2018)
20. Liu, Z., Xie, K., Li, L., Chen, Y.: A paradigm of safety management in industry 4.0. Syst. Res. Behav. Sci. **37**, 632–645 (2020)
21. Hathaliya, J.J., Tanwar, S.: An exhaustive survey on security and privacy issues in healthcare 4.0. Comput. Commun. **153**, 311–335 (2020)
22. Wehde, M.: Healthcare 4.0. IEEE Eng. Manag. Rev. **47**, 24–28 (2019)
23. Kumari, A., Tanwar, S., Tyagi, S., Kumar, N.: Fog computing for healthcare 4.0 environment: opportunities and challenges. Comput. Electr. Eng. **72**, 1–13 (2018)
24. Christie, G.: Progressing the health agenda: responsibly innovating in health technology*. J. Responsible Innov. **5**, 143–148 (2018)

Logistics 4.0 for Today and Tomorrow: How to Identify Smart Factory Challenges and Trends

Fábio Luis Fernandes Azarias(✉) and Lia Toledo Moreira Mota

Pontifical Catholic University of Campinas, São Paulo, Brazil
fabio.lfa@puccampinas.edu.br, lia.mota@puc-campinas.edu.br

Abstract. Advances in logistical processes were influenced by continuous man-machine interaction and the concepts of disruptive technologies provided by the Fourth Industrial Revolution that have changed the workspace. Thus, business leaders need to know how to distinguish and highlight the peculiarities of Industry 4.0 and take actions to implement improvements in warehouses and guide industrial development to increase productivity and logistical efficiency. In this way, the article points out 10 trends of Industrial Automation linked to encoding the real world into data and allowing machine interaction. Industrial revolutions are divided into two Machine Ages, and the current Machine Age empowers machines and systems to surpass the mental power of the human race. The Smart Factory singularity is explained by identifying characteristics based on 4 challenging pieces of intelligence for creating disruptive technologies that followed the advancement of the revolution. Logistics 4.0 requires interconnected systems, computational reasoning, and robots as workers through a fully automated process.

Keywords: Industry 4.0 · Intralogistics · Cyber-physical system

1 Introduction

The advancement and use of new technologies allowed man to face new challenges in the economic and social sphere, impacting the rearrangement of the workspace. The disruptive technologies [1] from the Fourth Industrial Revolution, which Industry 4.0 is inserted in, bring a new concept of industrial automation, Smart Factory, and all intralogistics.

What differentiates the current industrial revolution from former revolutions is that the first, second, and third are an integral part of the First Machine Age [2] in which man invented various forms of technology to overcome the muscle power of humans and animals. Examples of this physical overcoming are the steam locomotive, the electric motor, and robotics.

However, there is the Second Machine Age, initiated by the current revolution, in which man will surpass his own mental prowess. It will be the mirroring of the logical reasoning of the human employed in technologies so that the system itself can encounter problems and take autonomous actions to solve them, thus acting with the same human

Y. Iano et al. (Eds.): BTSym 2021, SIST 207, pp. 115–122, 2023.
https://doi.org/10.1007/978-3-031-04435-9_11

attitudes of workers, maintainers, and managers. The machines will receive digital data from the real world and interpret it with Artificial Intelligence (AI).

Within factories and distribution centers, intralogistics holds the core of processes [3] and is fundamental for the Supply Chain. Thus, the new Cyber-Physical Systems (CPS) applications, with total integration and interconnection of the physical and digital world, will be an integral and relevant part of the Smart Factory spaces.

2 Objective

The article has as its motto to highlight topics that distinguish factories using outdated technologies from the Third Industrial Revolution with updated Smart Factories and following trends from Industry 4.0, with the aim of directing Latin American companies in the field of logistics for the readjustment and technological updating of warehouse space machinery.

3 Analysis Method

The analysis method is based on the analysis of Open Access articles from the database of the Federated Academic Community (CAFe – from the Portuguese "Comunidade Acadêmica Federada"), linked to the Ministry of Education of Brazil, using the keyword "Logistics 4.0" and relevant authors.

The following section (Results and Discussion) was organized into secondary sections for discussion of the topics covered:

- The Four Challenging Intelligences for the Fourth Industrial Revolution;
- Machine overcoming human mental power;
- Smart Factory trends.

4 Results and Discussion

4.1 The Four Challenging Intelligences for the Fourth Industrial Revolution

With the basic principle of integrating technologies from the physical, digital, and biological worlds, the Fourth Industrial Revolution provides profound transformations. New businesses, production systems, consumption, transport, and delivery of goods emerged [1], as well as a drastic change in the way of work and social interaction with the advent of thousands of connected mobile devices.

For the engineer Klaus Schwab, founder of the World Economic Forum, the current revolution brings the use of competitive technologies that have radically transformed humanity and the Industry without any precedent [1]. According to him, the current revolution, based on the Digital Revolution, brings more and more sophisticated and integrated systems, providing flexibility between physical and virtual manufacturing systems with the creation of new operating models. As a path to be followed, and as a challenge, it signals four types of human intelligence needed to support the potential of disruptive applications; they are: Mind, Heart, Soul, and Body, as illustrated in Fig. 1 and explained in Table 1.

Fig. 1. The four challenging pieces of intelligence for creating disruptive technologies.

Table 1. Intelligence challenges

Intelligence	Field	Description
Mind	Contextual	Understanding and application of knowledge
Heart	Emotional	Coping with inner and outer feelings
Soul	Inspiration	Meeting of individual purpose and virtue of helping a common good
Body	Physical	Cultivation in a healthy physical shape and able to interact with systems

4.2 Machine Overcoming Human Mental Power

According to Erik Brynjolfsson and Andrew McAfee, respectively former director and chief researcher at the MIT Digital Business Center, technological development is delimited according to the overcoming powers of human forces [2]. Thus, the Fourth Industrial Revolution breaks with previous revolutions with the search and overcoming of human mental power in machines and systems. Figure 2 illustrates the Industrial Revolutions in each Machine Age.

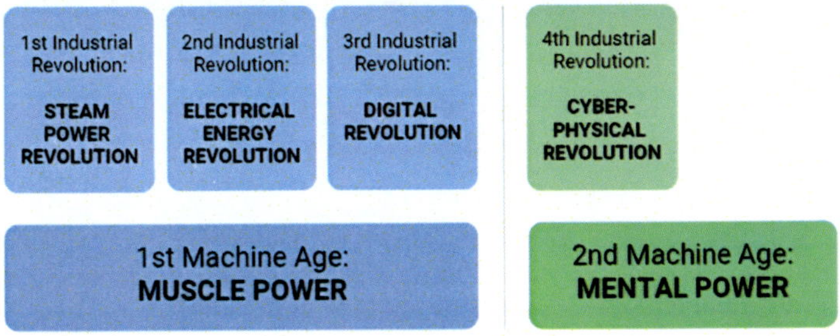

Fig. 2. Framing between industrial revolutions and machine ages.

- First Machine Age: there were technological advancements to meet the need for physical strength and repetitive movements;
- Second Machine Age: holds the power of logical reasoning in machines and systems to understand and shape environments.

4.3 Smart Factory Trends

At the fair in Hannover, Germany, in 2011, the concepts of Industry 4.0 were discussed through the implementation of Smart Factories aiming at greater productivity and efficiency [4] provided by a change in the production model in processes and products with high technological level.

Since then, new technologies have been used in logistics systems and there are points to be highlighted as trends as illustrated in Fig. 3. These trends are described below.

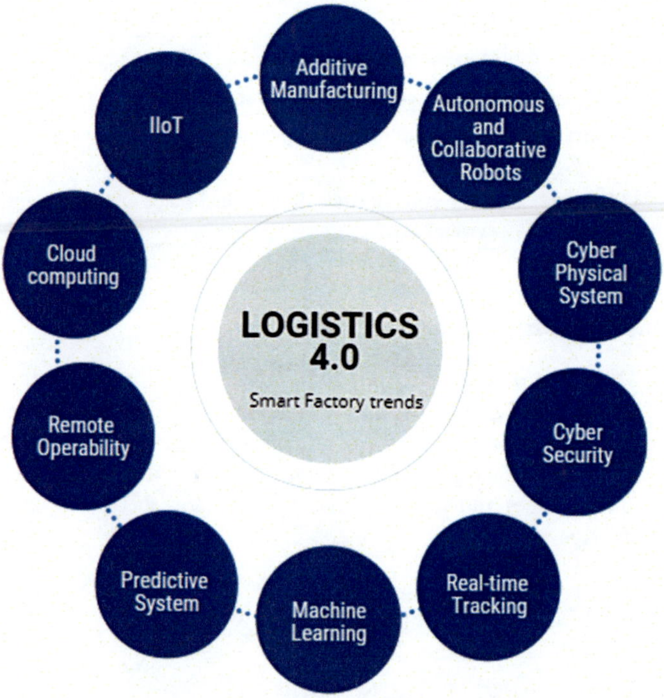

Fig. 3. Smart factory trends for Logistics 4.0

Additive Manufacturing. The role of logistics, as an essential component in the supply chain, is to provide and guarantee the quantity and quality of the product until its final destination [5]. With the advent of globalization, new consumer profiles demanded demands for technologies and business processes to ensure customer satisfaction and loyalty.

Thus, Additive Manufacturing brings the concept of flexible manufacturing and custom production even within mass production. This personalization and co-creation of the product [6] or the wrapping is in prints, printed details, and written names, which brings the consumer a feeling of exclusivity, boosting sales through digital platforms, also aiming at sending gifts via e-commerce on commemorative dates. An example of additive manufacturing equipment is the birth of 3D Printing in molds and final objects.

Autonomous and Collaborative Robot. Decentralized but autonomous warehouse control systems [3] will allow industrial automation to go beyond the execution of basic commands and start using robots that: evaluate the system, identify products and communicate data to the system and other machines. Machines will perform alternate functions when faced with problems, such as the lack or inadequacy of a product or sudden change in the layout of the conveyors [7] due to a maintenance stoppage [6].

Robotized, flexible, and efficient production line [4] will make the system itself the decision-maker of actions for production, layout, and transport. In addition to mechanical arms, intelligent factories have Automatic Guided Vehicles (AGVs) [7] for the internal transport of goods.

As in [5], the revolutions brought new concepts. They were: 1st industrial revolution with the introduction of industrial automation, 2nd industrial revolution with mass production, 3rd industrial revolution with robotization, and the 4th revolution bringing the use of systems and robots with artificial intelligence (AI).

Cyber-Physical System (CPS). Cyber-Physical System integrates the spheres of the digital world with physical objects and biological parts, as illustrated in Fig. 4. However, CPS not only integrates several systems but also allows new possibilities [2] of interaction and control of intralogistics processes in a revolutionary way that raises the level of autonomy and cooperation between operator and production, control, and maintenance procedures.

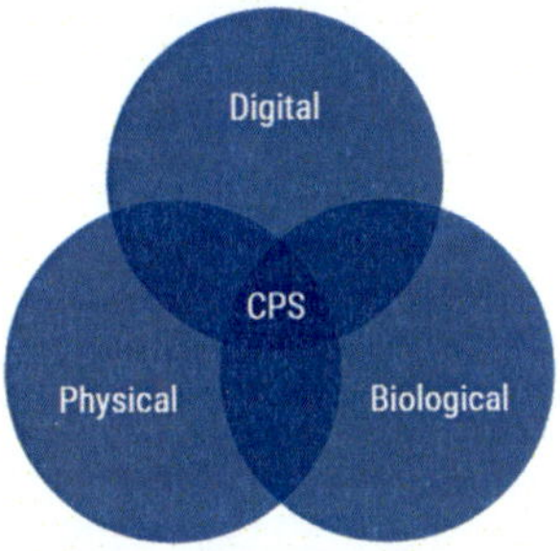

Fig. 4. CPS interconnection

There are several possibilities for integrating the digital world through the Internet of Things (IoT) taking advantage of Big Data, but with a system directly applied to man, there is Virtual Reality (VR) and Augmented Reality (AR) as an example.

The vice president of Facebook's virtual reality lab, Hugo Barra, Brazilian, disclosed a smart glasses partnership with Ray-Ban in September 2020 via the social media platform Twitter. These smart glasses from Facebook are part of "Projeto Aria" [8]. The device collects real-world data, such as video and audio by sensors as well as eye-tracking and location information and will offer integration between the user by phone from a map Existing 3D. The glasses do not yet include a display in the lens or image projections.

In the near future, for logistics operators, the use of smart glasses will be part of the conception of understanding the warehouse space and carrying out functions such as collecting product information by bar code, completing the picking of goods, and validating the assigned task.

Facebook itself announced, in early 2021, the research and development of a device, called the Bellowband [9], which will work as a smart bracelet allowing hand movement commands, without clicking physical buttons. The interpretation and recognition of human hand gestures by human electrical pulses sent to the brain are analyzed and converted into digital commands.

This equipment can operate together with AR glasses to perform commands, so the user will only need to make hand gestures which the equipment will interpret as orders. For example, to enter orders, increase and decrease the speed of conveyors, change and adjust the systemic quantity of products, as well as execute other industrial orders.

Cyber Security. Due to the interconnection between equipment, the vulnerability of improper access to the system will be in evidence [10] requiring the use of continuous effort in Cyber Security since data can value more than oil [6].

Real-Time Tracking. For the dynamic and fast-paced world [11], it is increasingly required that the access time to information and traceability of the merchandise must be made available in real-time [7] for both factory members and the consumer. For intralogistics, this implies merchandise monitoring systems with friendly graphic interaction, as well as tracking merchandise in transport by geolocation systems.

The tracking system should not only point out the location of the products, which would be part of a system of the Third Industrial Revolution but also go further and interpret the data and serve as a basis for the production chain linked to the effective analysis of time e.g. time spent picking a product or time spent transporting the factory to HUBs and Distribution Centers. For an agile system, it will be considered integration with routing applications that calculate the possible travel time on the roads and thus able to assist to a Predictive System. Here, as an example, the future with the use of autonomous trucks is included.

Machine Learning. Smart Factory systems offer conditions for real-time data capture by sensors, providing information for Big Data. Within the system's Artificial Intelligence (AI), Machine Learning allows for adjustment and adaptability with decisions that affect the entire production chain, such as digital, financial, and industrial marketing.

An early example of Machine Learning employed in intralogistics is in the intelligent flow of product storage in a high-bay warehouse where the system identifies high demand for product output, as well as other criteria, and stores it efficiently, reducing travel time and ensuring greater efficiency electrical and less effort.

Predictive System. They are machines equipped with sensors [4] capable of collecting data from the environment, monitoring the data, and being able to predict and indicate both advanced maintenance [6] and systems capable of predicting, based on data, the need to increase or decrease processes factory operating.

Remote Operability. Because the essence of Industry 4.0 is the disruptive concept and interconnected systems between people, machines, and equipment [5], Smart Factory implies the use of Cloud Computing and Industrial Internet of Things (IIoT) providing

Remote Operability for both control and remote work, such as for working from home, as for fully remote factory production orders.

Cloud Computing. Production planning and supply chain management systems [4] are introduced entirely in the cloud with increased flexibility and information sharing.

IIoT. The interconnection for the control and agility of the intralogistics flow requires the concept of Industrial Internet of Things (IIoT) involving constant communication between devices integrated into the network. This communication via Cyber Infrastructure guarantees the flow for Real-Time Tracking and Remote Operability by sending data (with Cyber Security) to Cloud Computing, ensuring system control, and generating data for Predictive System and Machine Learning.

5 Conclusion

Understanding the barriers faced and technological trends in Industry 4.0 will make decision-makers analyze and judge necessary changes in warehouses to increase productivity and efficiency. In this way, basic repetitive movement procedures performed by employees will be exchanged by robots with added value technology and with the addition of employees with high technical knowledge.

Smart Factory, as the storage and shipment of goods through fully automated processes, not only represent machines in the factory but encompass diverse and interconnected systems. The efficiency of Logistics 4.0 calls for fewer human arms and greater computational reasoning.

6 Future Work

Indications of themes for future work are highlighted:

- Design of the Logistics 4.0 Maturity Level Index;
- Modularity and use of Smart Factory as a Collaborative Economy digital platform service as a mirror of the businesses of Uber, Airbnb, OLX companies;
- Blockchain used in Logistics 4.0;
- Advance in the use of Drones in the transport of goods;
- Deep Learning in Industry.

Acknowledgment. The authors thank PUC Campinas for its academic support.

This study was financed in part by the Coordination for the Improvement of Higher Education Personnel ("Coordenação de Aperfeiçoamento de Pessoal de Nível Superior" - CAPES), Brazil, under Finance Code 001.

References

1. Schwab, K.: The Fourth Industrial Revolution. Currency, New York (2017)
2. Brynjolfsson, E., Mcafee, A.: The Second Machine Age: Work, Progress, and Prosperity in a Time of Brilliant Technologies. W. W. Norton, New York (2014)
3. Bohács, G., Gáspár, D., Kánya, D.: Conception an intelligent node architecture for intralogistics. Acta Logistica Int. Sci. J. Logist. **5**(2), 31–37 (2018)
4. Silva, M., Rocha, C.: Assessment of industry maturity level 4.0: the case of a strategic defense company. Future Stud. Res. J. Trends Strateg. **12**(1), 31–59 (2020). https://doi.org/10.24023/FutureJournal/2175-5825/2020.v12i1.455
5. Kovács, G.: Logistics and production processes today and tomorrow. Acta Logistica Int. Sci. J. Logist. **3**(4), 1–5 (2016)
6. Franco, W.F.: Model for introducing news technologies in industrial processes in industry 4.0 context. Masters' dissertation, Pontifical Catholic University of Campinas, Brazil (2020)
7. Stareček, A., Bachár, M., Horňáková, N., Cagáňová, D., Makyšová, H.: Trends in automatic logistic systems and logistic market in Slovakia. Acta Logistica Int. Sci. J. Logist. **5**(1), 7–14 (2018)
8. Barra, H.: Facebook partnership with Ray-Ban (2020). https://twitter.com/hbarra/status/1306278791669579776?ref_src=twsrc%5Etfw%7Ctwcamp%5Etweetembed%7Ctwterm%5E1306278791669579776%7Ctwgr%5E%7Ctwcon%5Es1_&ref_url=https%3A%2F%2Fd-4110775529637224737.ampproject.net%2F2106182132000%2Fframe.html Accessed 4 July 2021
9. Young, E.M., Memar, A.H., Agarwal, P., Colonnese, N.: Bellowband: a pneumatic wristband for delivering local pressure and vibration. In: IEEE World Haptics Conference (2019). https://research.fb.com/publications/bellowband-a-pneumatic-wristband-for-delivering-local-pressure-and-vibration/
10. Yeboah-Ofori, A., et al.: Cyber threat predictive analytics for improving cyber supply chain security. IEEE Access (2017). https://doi.org/10.1109/ACCESS.2021.3087109
11. Araque, G.A.G., Gómez, M.V., Uribe, J.P.V., Hernándes, A.H.S.: Big Data and the implications of the fourth industrial revolution - challenges, opportunities and future trends: a review. Revista Venezolana de Gerencia, Venezuela **26**(93), 33–47 (2021)

Active Power Filter Design Using Neural Network with a Variable Step-Size

Alejandro Alvarez[1(✉)], Joseph Cherre[1], Christian Flores[1], and Julien Noel[2]

[1] Department of Electrical Engineering, Universidad de Ingenieria y Tecnología, Barranco, Peru
{alejandro.alvarez,joseph.cherre,cflores}@utec.edu.pe

[2] Department of Energy Engineering, Universidad de Ingenieria y Tenologia - UTEC, Barranco, Peru
jnoel@utec.edu.pe

Abstract. Harmonic elimination in three-phase power systems due to non-linear loads in electrical grids is an essential topic in power quality nowadays. The rise of electronics and electrical devices increases the sensitivity of the network. The current research approaches this topic by proposing the design and simulation of a neural network with a modified LMS (Least Mean Square) algorithm, including variable step-size in the cost function to achieve faster convergence and overcoming problems in drifting parameters. We use a three-phase active power filter (APF) that relies on a current compensation control for testing purposes. The APF consists of a hysteresis control block and a current-controlled voltage source inverter computed by our enhanced ADALINE (Adaptative Linear Element) neural network. Finally, a comparison is made with another algorithm based on non-variable step size. Results show a 3% reduction in THD (Total Harmonic Distortion) parameter and a 0.2 s faster response time, concluding the improvement in performance parameters of the power system.

Keywords: Harmonic distortion · Three-phase power system · ADALINE · APF · MATLAB · Simulink · THD · Hysteresis

1 Introduction

Since the beginning of the century, research regarding power quality has become a recurrent topic because of the increasing interest in reducing harmonic distortion due to non-linear loads that make power systems currents and voltages non-sinusoidal. This kind of load includes rectifiers, inverters, AC regulators, etc. [1]. They are mostly built on semiconductors, responsible for modifying the phase and amplitude of the source network. The effect on power systems is significant and can produce several damages like power losses, communication interference, and solid-state device malfunction [2].

Y. Iano et al. (Eds.): BTSym 2021, SIST 207, pp. 123–132, 2023.
https://doi.org/10.1007/978-3-031-04435-9_12

At present, there are numerous methods regarding harmonic distortion elimination in power systems. In [1], Vasquez and Salmeron proposed using ADALINE and Feedforward neural networks for controlling an active power filter. The study in [3] also proposed an ADALINE neural network with a simplified architecture for a single-phase system, reducing inputs in the network without significant impact in THD. In [4], the authors proposed a frequency detection method to improve ADALINE and overcome the estimation accuracy problem of phase angle and, then, perform the compensation controls. The research in [5] proposed the Fuzzy control logic method alongside an Artificial Neural Network for extracting reference currents. On the other hand, the study in [6] developed a novel hybrid technique called Adaptive Linear Neural Network and Filtered-X Least Mean Square (ADALINE-FXLMS) algorithm to enhance the neural network against noisy inputs and help reach system stability. In [7], Cheng presented a virtual instrument (VI) for power system harmonic remote surveillance via the Internet, focusing on graphical programming. Investigation made in [8] optimized the model of shunt active power filter using Adaptive Tabu search (ATS) algorithm and Genetic Algorithm (GA) for aircraft utility. Most recently, the research in [9] designed a two-fold adaptive linear neural network (ADALINE) parallel approach termed as Master-Slave ADALINE (MS ADALINE) dividing tasks such as weight vector adjustment or steady-state error between a Master and a Slave ADALINE. This architecture showed improvement in convergence speed and steady-state performance. The paper in [10] introduced an enhanced subspace-least mean square (S-LMS) method for power system measurement in the expectation of achieving the high accuracy and high resolution of the method without the computation burden it induces. Finally, in [11], Flores et al. did a previous work on the subject matter approaching the solution using a convex combination of two adaptive filters.

The Total Harmonic Distortion (THD) is an essential parameter in power systems. It shows us the relationship between the power consumption of harmonics and the power consumption of fundamental frequency. According to IEEE 519 standard, THD requirements are below 5% for harmonic control in power systems [1]. Nowadays, many methods are available to correct power factor in this type of system; one of the most common is the active power filter (APF). Over the classical Shunt passive filters, APFs help suppress load current harmonics through a compensation current and compensate reactive power requirements of the non-linear loads [12]. APF consists of three main blocks: inverter circuit, in charge of delivering enough power in the compensation current to properly correct the source network; hysteresis control block, responsible for giving the required PWM signals for the inverter circuit; and finally, the active current generation block, which extracts the fundamental components of voltage and current load. The main feature in ADALINE topology is the weight update algorithm that keeps track of the reference and decreases the error by each iteration.

This study aims to apply the ADALINE-VLLMS-based estimation technique developed by Garanayak and Panda [13] for fast and exact measurement of fundamental and harmonic components of a non-linear load current in a three-phase power system. The measured signals are used to compute a compensation current that will trigger the IGBT bridge inverters in the Shunt Active Power Filter. The proposed VLLMS algorithm employs a variable step-size parameter to achieve faster convergence and overcome problems in drifting parameters. Through a simulation in MATLAB, the performance of the proposed scheme is compared with a standard fixed step-size ADALINE-LMS algorithm.

The rest of this paper is organized in the following way. Section 2 describes the entire power system design methodology; it is structured as follows. First, the proposed active power filter is sketched with all the necessary components. Next, the mathematical modeling of harmonics and significant equations of the ADALINE-VLLMS algorithm are described. Then, the required equations to model the compensation current are presented, and, finally, the hysteresis control block is explained. Section 3 illustrates the detailed simulation results and the comparison between the algorithms. Finally, Sect. 4 states the conclusion of this research work.

2 Methodology

2.1 Active Power Filter Proposal

For this research, we use a three-phase full-wave bridge rectifier with an inductive element as a non-linear load. Also, for compensating non-linear unbalanced loads, a three-phase IGBT bridge inverter with a DC power source is employed. A basic scheme of the power system is shown in Fig. 1.

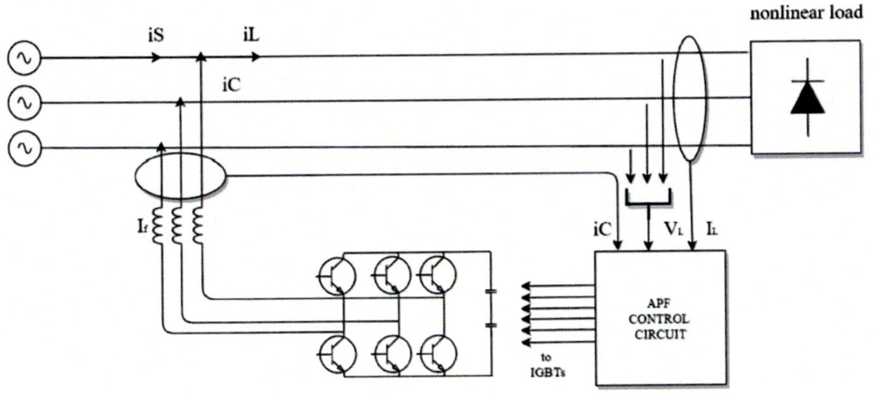

Fig. 1. General block diagram of a three-phase power system with the proposed active power filter

2.2 Harmonic Extraction Algorithm

The focus of this paper is on the neural network ADALINE with a variable step-size weight update algorithm. The main idea of ADALINE is to break down the load current i_L into its harmonic components according to the Fourier series, which is determined by the sum of its fundamental and its harmonic components. By obtaining the fundamental component, it is possible to calculate the fundamental active currents, which will allow us to compute the compensation currents. In Eq. 1 we can see the representation of i_L where t is the sampled time, ω is the fundamental frequency, W_{an} and W_{bn} are the weights of the load current, and n is the harmonic number.

$$i_L(t) = \sum_{n=1,2..}^{N} [W_{an} sin(nwt) - W_{bn} cos(nwt)] \quad (1)$$

Equation 1 can be rewritten as follows:

$$i_L(t) = W^T X(t) \quad (2)$$

where:

$$W^T = [w_{11}\ w_{21}\ ...\ w_{an}\ w_{bn}] \quad (3)$$

And *X(t)* can be represented by the vector of sines and cosines as follows:

$$X(t) = \begin{bmatrix} sin(wt) \\ cos(wt) \\ \cdot \\ sin(nwt) \\ cos(nwt) \end{bmatrix} \quad (4)$$

2.3 VLLMS Weight Update Algorithm

The weight update algorithm ensures that our neural network trains and improves in each iteration. The Adaptive Linear Neural Network and Variable Leaky Least Mean Square (ADALINE-VLLMS) is the evolution of the Widrow-Hoff LMS algorithm, created to solve the problem of drifting the parameters of the weight vectors. Furthermore, the variable step size allows us to have a faster convergence rate [13]. The weight update formula is as follows:

$$W(t) = W(t-1) + \mu e(t) X(t) \quad (5)$$

where the current weight *W(t)* is defined as the past weight plus the fixed step-size coefficient multiplied by the present error and the current signal. Based on the improvement in terms of error, settling time, performance, among other parameters, this paper proposes to replace the LMS algorithm with the VLLMS algorithm, which is determined as follows:

$$W(t) = W(t-1) + \mu(t)e(t)x(t) \tag{6}$$

where μ(t) represents the variable step size expressed as:

$$\mu(t) = \alpha\mu(t-1) + \gamma R(t)^2 \tag{7}$$

where α and γ control the convergence time, and R(k) represents the autocorrelation of the measured error given by:

$$R(t) = \beta R(t-1) + (1-\beta)e(t)e(t-1) \tag{8}$$

where β is an exponential weighting parameter. For optimum results, all values are chosen between 0 and 1 according to Dash et al. [12].

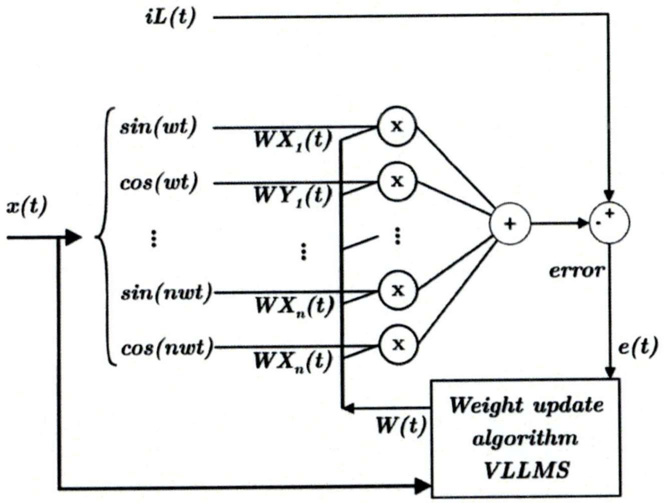

Fig. 2. ADALINE neural network topology

In Fig. 2, we can see the general topology of the ADALINE neural network with the VLLMS weight update block used.

2.4 Estimation of Fundamental Voltages and Currents

When there are non-linear loads in a power system, the voltage and current waveforms at the loads exhibit non-sinusoidal behavior. Modeling this kind of waveform would be challenging; nonetheless, the Fourier series allows us to express the periodic signal as a summation of sines and cosines. Thus, the load currents and load voltages i_L and v_L, respectively, can be represented as follows:

$$i_L(t) = \sum_{n=1,3,5..}^{N} [I_{n1} sin(nwt) - I_{n2} cos(nwt)] \tag{9}$$

$$v_L(t) = \sum_{n=1,3,5..}^{N} [V_{n1} sin(nwt) - V_{n2} cos(nwt)] \tag{10}$$

where ω is the fundamental frequency, while I_{n1}, I_{n2}, V_{n1}, and V_{n2} are the cosine and sine components of the load currents and voltages respectively.

Two ADALINE neural networks are responsible for estimating the fundamental components of the load voltage and current for each phase. Using the values of this estimation, we can express the fundamental active current without computing any integration. The target source active current per phase is computed according to the following equation:

$$i_{act,i} = \frac{P}{V^2} v_{i,1} = \frac{\frac{1}{T}\sum_{i=L1,L2,L3}\int_0^T v_{i,1} i_{i,1}\, dt}{\frac{1}{T}\sum_{i=L1,L2,L3}\int_0^T v_{i,1}^2\, dt} v_{i,1} \tag{11}$$

Since the ADALINE estimates the fundamental components, the equation can be rearranged as follows:

$$i_{act,i} = \frac{\sum_{i=L1,L2,L3}(V_{i,11} I_{i,11} + V_{i,12} I_{i,12})}{\sum_{i=L1,L2,L3}(V_{i,11}^2 + V_{i,12}^2)} \cdot (Vi_{,11}\, cos(\omega t) + Vi_{,12}\, sin(\omega t)) \tag{12}$$

where $I_{i,11}$ and $V_{i,11}$ are the sine components of the fundamental load current and voltage respectively, while $I_{i,12}$ and $V_{i,12}$ are the cosine components. Once the fundamental active currents are obtained, they are subtracted with the present load currents, which results in the compensation currents for each phase.

$$i_{comp,i} = i_L - i_{act,i} \tag{13}$$

2.5 Hysteresis Control

The hysteresis band current control technique has been one of the most fitting methods for current-controlled voltage source inverters, especially applied in active power filters. Its main characteristics are fast response, acceptable accuracy and unconditioned stability [2]. In this research, a hysteresis block generates the PWM signals that will enter our power system. Its diagram is shown in Fig. 3.

First, the actual compensation current is subtracted from the measured phase current before entering the hysteresis comparator. According to a fixed threshold and a hysteresis band, this comparator generates an ON-OFF signal that finally reaches the inverter transistors.

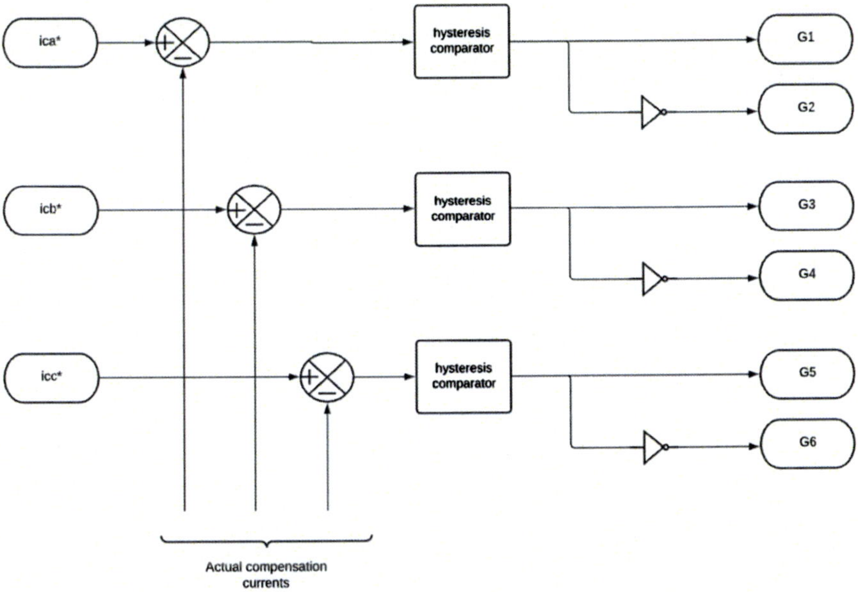

Fig. 3. Hysteresis control diagram

3 Results

In this section, the proposed design is tested through extensive MATLAB-based simulation, using the Simulink tool and the SimPowerSystem block to demonstrate the effectiveness of the algorithm. The experimental setup comprises an inductive-resistive load of 5 mH and 50 Ω acting as our non-linear load to inject enough harmonics into the system. The three-phase source has a peak voltage of 220 V and a fundamental frequency 60 Hz. The Active Power Filter with the neural network was developed with a discrete variable step at a sampling time of 0.1 ms to ensure the simulation's stability. Finally, the comparison is made between the power system using the ADALINE-VLLMS algorithm in the APF and the same power system using the ADALINE-LMS algorithm.

Three key factors measure the performance of the APF. The first is the total harmonics distortion (THD) parameter that shows the relationship between the fundamental component of the signal and its harmonics. The next is the accuracy of current harmonic extraction, where the closer value of source current with the load current shows better accuracy in the extraction. The last factor is the time response of both systems. The results are shown in Figs. 4, 5, and 6.

As we can see in Fig. 4 the source voltage has a peak voltage of 220 V for each phase. On the other hand, Fig. 5, presents load current i_L, indicating a THD of approximately 30.8%. Finally, in Fig. 6, we can appreciate the compensated source currents i_S for the ADALINE-VLLMS, showing the reduced value of 3.2% in the THD (lower than the 5% established by IEEE standards rules 519, which

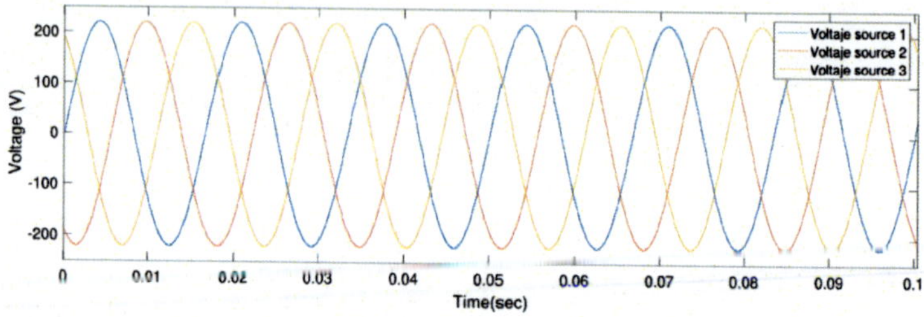

Fig. 4. Source voltage

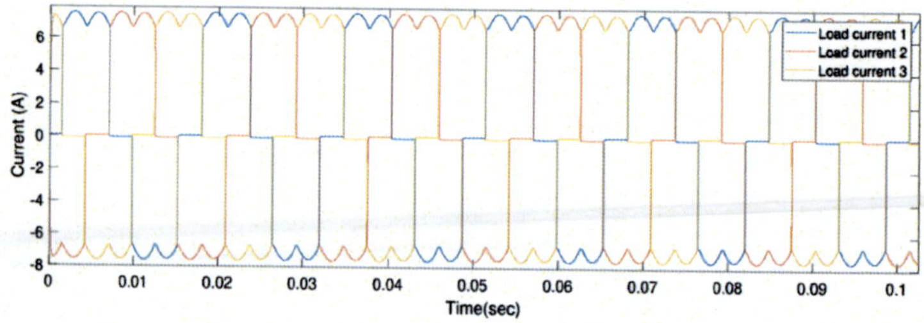

Fig. 5. Load current

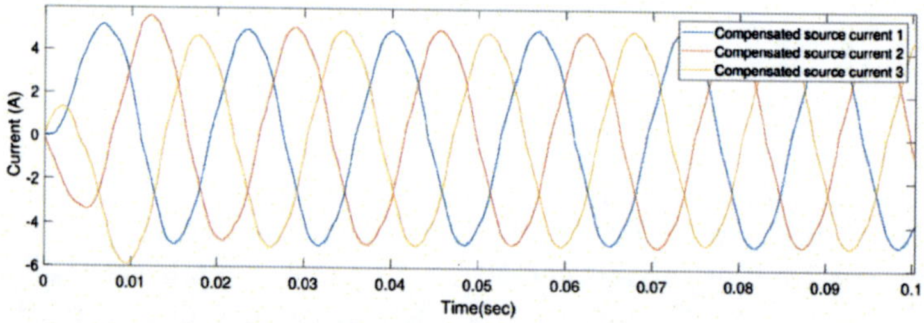

Fig. 6. Compensated source current

is the IEEE Recommended Practices and Requirements for Harmonic Control in Electric Power Systems).

Although the results obtained using an active power filter with an ADALINE VLLMS algorithm greatly satisfy our requirements, the purpose of this investigation is to compare it with the same ADALINE using the LMS algorithm and see the performance improvement. LMS algorithm showed a THD value of 5.16%, 2% more than the VLLMS one. In addition, the response time was reduced from 0.28 s for the LMS algorithm to 0.08 s for the VLLMS algorithm, concluding that the proposed algorithm optimizes performance and proves a clear improvement in both parameters. Table 1 shows the previously discussed results. The amplitudes of the fundamental, third and fifth harmonics are also presented to reassure the value of THD.

Table 1. Comparison of adaline VLLMS and LMS

ADALINE	THD (%)	Fund (Amp)	3rd (Amp)	5th (Amp)	Response time (s)
VLLMS	3.20	3.72	0.01	2.91	0.08
LMS	5.16	4.81	0.72	4.41	0.28

It is worth mentioning that the comparison process between the ADALINE neural network weight updating algorithms was made respecting the topology proposed for each case and the same type of non-linear load.

4 Conclusions

This article proposed a harmonic elimination technique employing an active power filter composed of a harmonic extraction block, which allows us to obtain the fundamental components of a non-linear load. Likewise, the improved ADALINE weight update algorithm (VLLMS) proved a better performance in THD correction than the so-called LMS weight update algorithm, specifically a 3.8%, and a shorter response time. This is due to the higher accuracy in identifying the fundamental components of load currents. In comparison, VLLMS shows more robustness in allowing different values of amplitude and frequency from the source. In addition, THD is compensated and meets the requirements of IEEE std.519-2014, assuring power quality in the system.

References

1. Vázquez, J., Revuelta, P.: Active power filter control using neural network technologies. Electr. Power Appl. IEE Proc. **150**, 139–145 (2003). https://doi.org/10.1049/ip-epa:20030009
2. Elmitwally, A., Abdelkader, S., El-Kateb, M.: Neural network controlled three-phase four-wire shunt active power filter. Gener. Transm. Distrib. IEE Proc. **147**, 87–92 (2000). https://doi.org/10.1049/ip-gtd:20000007
3. Zainuri, M.A.A.M., Radzi, M.A.M., Rahman, N.F.A.A., Soh, A.C., Rahim, N.A.: Single phase shunt active power filter with simplified adaline neural network. In: 3rd IET International Conference on Clean Energy and Technology (CEAT) 2014, pp. 1–6 (2014). https://doi.org/10.1049/cp.2014.1489
4. Chen, C.I., Hsu, C.L., Chen, C.H.: Adaline-based shunt active power filter for power quality modification of power system. In: 2018 International Symposium on Computer, Consumer and Control (IS3C), pp. 372–375 (2018). https://doi.org/10.1109/IS3C.2018.00100
5. Wenjin, D., Taiyang, H.: Design of single-phase shunt active power filter based on ANN. In: 2007 IEEE International Symposium on Industrial Electronics, pp. 770–774 (2007). https://doi.org/10.1109/ISIE.2007.4374694
6. Garanayak, P., Panda, G.: Fast and accurate measurement of harmonic parameters employing hybrid adaptive linear neural network and filtered-x least mean square algorithm. IET Gener. Transm. Distrib. **10**(2), 421–436 (2016). https://doi.org/10.1049/iet-gtd.2015.0684

7. Lin, H.C.: An internet-based graphical programming tool for teaching power system harmonic measurement. IEEE Trans. Educ. **49**(3), 404–414 (2006). https://doi.org/10.1109/TE.2006.879239
8. Khalid, S.: Performance evaluation of adaptive tabu search and genetic algorithm optimized shunt active power filter using neural network control for aircraft power utility of 400hz. J. Electr. Syst. Inf. Technol. **5**(3), 723–734 (2018). https://doi.org/10.1016/j.jesit.2017.04.003
9. Garanayak, P., Naayagi, R.T., Panda, G.: A high-speed master-slave adaline for accurate power system harmonic and inter-harmonic estimation. IEEE Access **8**, 51,918–51,932 (2020). https://doi.org/10.1109/ACCESS.2020.2980115
10. Abdollahi, A., Zhang, P., Xue, H., Li, S.: Enhanced subspace-least mean square for fast and accurate power system measurement. IEEE Trans. Power Delivery **28**(1), 383–393 (2013). https://doi.org/10.1109/TPWRD.2012.2218664
11. Chuquipiondo, R., Vega, C.F., Noel, J.: Parallel adaline for active power filter. In: Iano, Y., Saotome, O., Kemper, G., Mendes de Seixas, A.C., Gomes de Oliveira, G. (eds.) BTSym 2020. SIST, vol. 233, pp. 703–712. Springer, Cham (2021). https://doi.org/10.1007/978-3-030-75680-2_78
12. Chapter two - control strategies of shunt active power filter. In: Dwivedi, S.K., Jain, S., Gupta, K.K., Chaturvedi, P. (eds.) Modeling and Control of Power Electronics Converter System for Power Quality Improvements, pp. 31–84. Academic Press (2018). https://doi.org/10.1016/B978-0-12-814568-5.00002-8
13. Garanayak, P., Panda, G., Ray, P.K.: Power system harmonic parameters estimation using ADALINE-VLLMS algorithm. In: 2015 International Conference on Energy, Power and Environment: Towards Sustainable Growth (ICEPE), pp. 1–6 (2015). https://doi.org/10.1109/EPETSG.2015.7510132

Cross Mutual Information Analysis of EEG Signals for Cognitive Task Discrimination

Joaquin Peralta[1] and Christian Flores[2(✉)]

[1] Department of Electrical Engineering, Universidad Tecnológica del Perú UTP, Lima, Peru
1111037@utp.edu.pe

[2] Department of Electrical Engineering, Universidad de Ingenieria y Tenologia - UTEC, Barranco, Peru
cflores@utec.edu.pe

Abstract. This study aims to assess Functional Connectivity (FC) using the Cross Mutual Information (CMI) approach between differents brain areas while several patients perform five cognitive tasks. Thus, we used six electrodes of electroencephalograms (EEG) located in different brain areas. A hypothesis Wilcoxon signed-rank test was applied to evaluate the statistical differences for all combinations between cognitive task, electrode pairs, and brain bands which add up to 225 combinations. The results reported statistical difference ($p < 0.05$) for all 225 combinations. Therefore, the proposed approach reported the use of CMI as a feasible tool for analyzing Functional Connectivity (FC) using EEG brain signals for cognitive tasks besides discriminating them.

Keywords: Mutual information · Cross-mutual information · EEG · Functional connectivity · Cognitive tasks

1 Introduction

The electroencephalographic signals (EEG) record brain activity through electrodes placed on the human scalp. They are resultant of neuronal activity while a subject develops different mental activities. Also, EEG signals possess high nonlinear characteristics and are highly stochastic and time-varying. For this reason, it is essential to study and analyze its nonlinear characteristics to obtain further information [1,2]. Moreover, EEG signals are a suitable tool to provide information about the transmission of information among distinct brain regions. Thus, it allows us to study the association or connectivity between different brain regions to build functional connectivity maps [3,4]. The concept of synchronization in the brain is an indicator of functional connectivity between different brain areas, which can be used to analyze cognitive tasks and pathological activities.

Y. Iano et al. (Eds.): BTSym 2021, SIST 207, pp. 133–142, 2023.
https://doi.org/10.1007/978-3-031-04435-9_13

Due to this characteristic, Mutual information (MI) is able to measure non-linear dependencies between two-time series and establish dependence relationships. More specifically, Cross-mutual information (CMI) quantifies the information transmitted from one-time series to another. It is a concept of information theory applied to neural signal processing, which was initially used to quantify the information, i.e., the uncertainty, in a system by the minimal number of bits required to transfer the data [5]. The MI address functional connectivity of brain areas using the EEG signals in different biomedical applications as prediction anesthesia response [6], analysis of different cortical areas of schizophrenic patients [7], etc. One important application is to analyze the variation of brain functional connectivity in patients with Alzheimer's disease. In [8], the authors analyzed functional connectivity through EEG signals using Cross-Mutual Information (CMI) as the connectivity measure in patients with Alzheimer's. This study reported a reduction of CMI between hemispheres.

This work assesses information transmission between different brain areas, estimating the average CMI between EEG electrodes using brain signals and their brain bands while various volunteers performed cognitive tasks. The aim of this work is to discriminate these cognitive tasks using the CMI.

This article is divided as follows: Sect. 2 presents the experimental data, preprocessing, Functional Connectivity (FC), and statistical validation. Section 3 describes the results and discussions of the methodology applied to assess the Functional Connectivity (FC) during the performance of cognitive tasks. Finally, Sect. 4 states the conclusion of the present work.

2 Methodology

2.1 Experimental Data

For this work, we used the database available on the webpage of the Brain-Computer Interfaces Department of the Colorado State University [9]. This repository contains 325 cell arrays regarding seven subjects of study, five cognitive tasks, and the signals acquired in both hemispheres of the central lobe (C3, C4), both hemispheres of the parietal lobe (P3, P4), and both hemispheres of the occipital lobe (O1, O2). The electrodes were placed according to the 10–20 International System. The data were acquired 250 Hz and filtered from 0.1 100 Hz. Subject 1 was a left-handed university worker of 48 years old, the subject 2 was a right-handed university worker of 39 years old. Subjects 3, 4, 5, 6, and 7 were right-handed university students between 20 and 30 years old. The five cognitive tasks are baseline (B), multiplication (M), geometric rotation (R), letter composition (L), and visual count (C). M and R tasks are considered high complexity tasks, while L and C are considered low complexity tasks. The B task was considered as a base-point in order to be compared with the other tasks. All cognitive tasks lasted 10 s per trial, but due to the fact that some of these had fewer than ten sessions or due to some recording errors, in this work, we only used four subjects (subjects 1, 3, 4, and 5).

2.2 Preprocessing

In order to eliminate the noise, we used a second-order notch filter to eliminate the distortion produced by the electric power 60 Hz. Furthermore, the mean of brain signals was extracted to eliminate the DC component. The EEG signal was filtered using a bandpass filter for each brain band. To achieve this requirement, an elliptic filter of order five with attenuation of −30 dB for frequencies of ±0.5 Hz was used.

2.3 Functional Connectivity

Due to the relevance of capturing the temporal correlations and statistical dependences between brain regions as a representation of synchronized natural neural activity during the development of a cognitive task [10], we used the Mutual Information (MI) algorithm, which is an Information Theory concept focused on detecting and quantifying linear and nonlinear statistical dependencies between time series [11]. The CMI generated the zero value when there was no shared information between two-time series and one in the opposite case [12]. The MI can be analyzed as a measure between different systems X and Y. Considering X and Y as discrete random variables with finite possible values x_i and y_i and probabilities $P_x(x_i)$ and $P_y(y_j)$ to $i = 1, 2, ..$ and $j = 1, 2, ..$, respectively, we can define the MI as follows:

$$MI_{xy} = \sum_i \sum_j P_{xy}(x_i, y_j) log_2 \frac{P_{xy}(x_i, y_j)}{P_x(x_i) P_y(y_j)} \tag{1}$$

For the assessment of information transfer via a certain prediction time τ, we can write:

$$CMI = \sum_{x_i \in X} \sum_{y_{j+\tau} \in Y} P_{xy}(x_i, y_{j+\tau}) log_2 \frac{P_{xy}(x_i, y_{j+\tau})}{P_x(x_i) P_y(y_{j+\tau})} \tag{2}$$

CMI assesses the complexity between the two different signals depending on the time τ. In our case, we implemented the CMI algorithm to identify functional connectivity among six electrodes (C3, C4, P3, P4, O1, and O2).

2.4 Statistical Validation

The test hypothesis for non-parametric statistical data related called Wilcoxon signed-rank test was applied to analyze the difference between cognitive tasks compared with all combinations of electrode pairs and brain bands. This test hypothesis was chosen because all subjects were part of the same experiment. When the parameters of statistical significance (p) are smaller than 0.05 ($p < 0.05$), this represents one statistical difference between the two tasks with a 5% significance. Also, the data reported a non-normal distribution, which was corroborated with the Kolmogorov-Smirnov test.

3 Results and Discussions

As mentioned in the experiment description, the EEG database was analyzed. The CMI was calculated on the EEG signals and its brain bands for all pairs of cognitive tasks. To summarize our result, we elaborated tables and figures, which are presented in the subsections.

3.1 Cross-Mutual Information Results

To report the CMI values, we elaborated five graphs to show their performance for each one of the brains bands. In Figs. 1, 2, 3, 4 and 5, the x-axis represents all combinations of pair of electrodes (C3-C4, C3-P3, C3-P4, C3-O1, C3-O2, P3-C4, P3-P4, P3-O1, P3-O2, O1-C4, O1-P4, O1-O2, C4-P4, C4-O2, P4-O2), and the y-axis represents all combinations between subjects (S1, S3, S4, and S5) and cognitive task (B, M, L, R, and C). For instance, LS4 (L: letter composition task, S4: subject 4) represents subject 4 performing the letter composition task. Finally, color matrices report the mean CMI values of the ten trials.

As we can see in Fig. 1 (delta band), LS4 (L: letter composition task, subject 4 (S4)) reported the lower values for all electrode pairs. Also, we can appreciate that the maximum CMI values are around 0.42, and the lower ones are around 0.32. Furthermore, O1-P4 appears at one electrode pair with better results for CMI values.

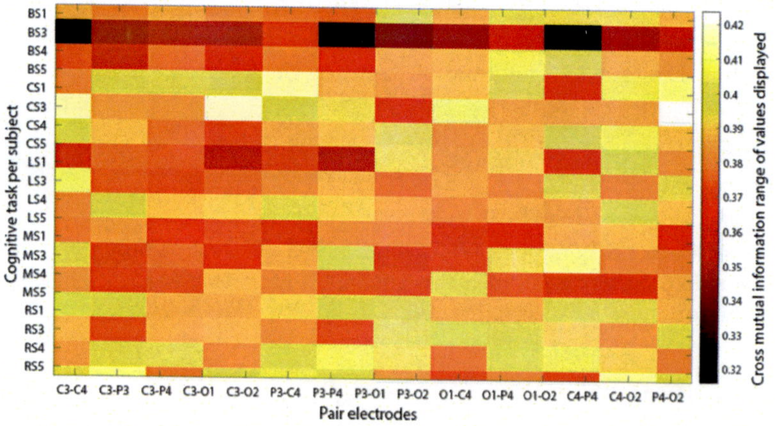

Fig. 1. The overall distribution of the average CMI values in the delta band

Figure 2 (theta band) is one of the two bands which presented the higher CMI values. Also, it reported the lowest values for LS4 and the highest values for RS5. Additionally, we can see that the maximum CMI values are around 0.52, and the lowest ones are around 0.40. Additionally, C4-O2 was an electrode pair with better results for CMI values.

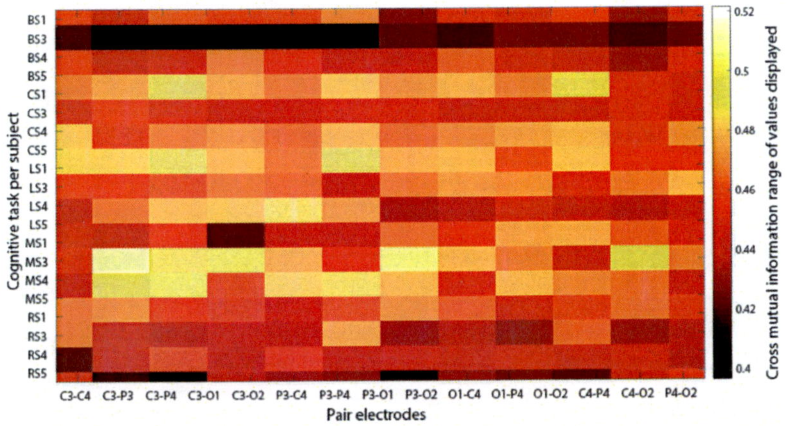

Fig. 2. The overall distribution of the average CMI values in the theta band

Figure 3 (alpha band) is one of the two bands which presented the higher CMI values. Also, it shows the lowest values for LS4 and the highest values for RS5. Additionally, we can appreciate that the maximum CMI values are around 0.58, and the lowest ones are around 0.44. Therefore, C4-O2 was an electrode pair with better results for CMI values.

Figure 4, which describes a CMI of the beta band, is the band that presented the highest CMI values among all pair electrodes. Furthermore, it showed the lowest values at the BS3 and the highest values for RS4. Additionally, we can appreciate that the maximum CMI values are around 0.65, and the lowest ones are around 0.45. Also, C3-P3 appears at one electrode pair with better results for CMI values.

Finally, the gamma band is shown in the Fig. 5. It is one of the bands that presented the lowest CMI values among the electrode pairs. Furthermore, it presented the lowest CMI value for MS1 and the highest value for RS4. Additionally, we can see that the maximum CMI values are around 0.85, and the lowest ones are around 0.35. As well, P3-P4 appears as one of the electrode pairs with better results for CMI values. In the last one, we can state that the geometric rotation task is more complex than the multiplication task involving more pair electrodes activity at higher CMI level values.

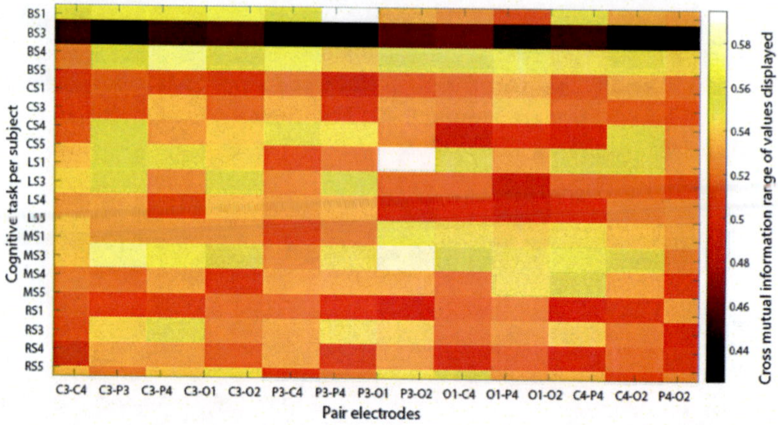

Fig. 3. The overall distribution of the average CMI values in the alpha band

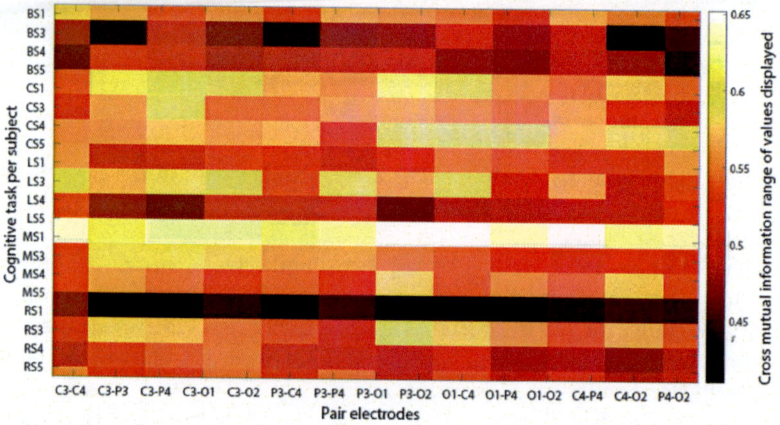

Fig. 4. The overall distribution of the average CMI values in the beta band

3.2 Statistically Significant Differences

To report the Statistically Significant Differences (SSD) between all combinations of cognitive tasks compared with all combinations of electrode pairs and brain bands, two tables were elaborate (Tables 1 and 2). The x-axis headers of Tables 1 and 2 refer to the combination of cognitive tasks (B, M, L, R, and C), and the first column depicts the SSD between a pair of cognitive tasks per pair of electrodes at a specific brain band.

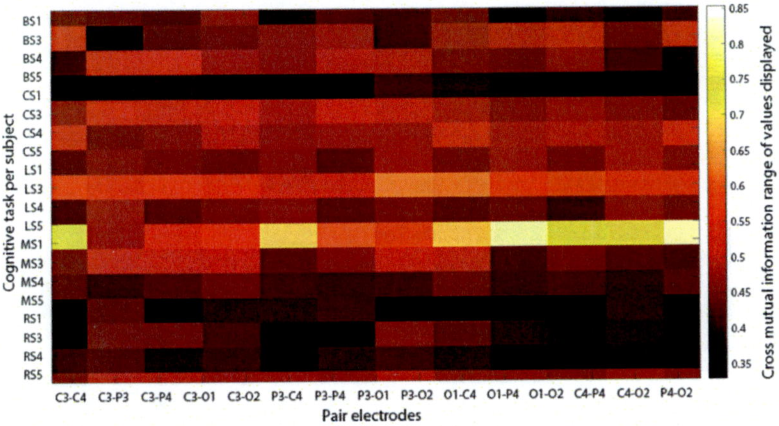

Fig. 5. The overall distribution of the average CMI values in the gamma band

Analysis per Brain Band. Regarding the delta band, we reported a significant difference among 9 of 10 cognitive task combinations, where BxC, MxL, and LxC were the most significant. Thus, the 18 SSD of delta band are equivalent to 8% of the overall 225 combinations. The theta band reported a significant difference among 10 of 10 cognitive task combinations, where BxR and RxC were the most significant. Thus, this brain band reported 46 SS, equivalent to 20.444% of the overall 225 combinations. The alpha band reported a significant difference among 10 of 10 cognitive task combinations, where MxL, LxC, and RxC were the most significant. Thus, this brain band states 40 SSD, which is equivalent to 17.78% of the overall 225 significant brain bands activity. Regarding the beta band, we reported statistical differences among 10 of 10 cognitive task combinations, where BxL and RxC were the most significant of them. Thus, the 53 SSD are equivalent to 23.56% of the overall 225 significant brain bands activity. Finally, we noticed statistical differences between cognitive tasks in the gamma band in 8 of 10 cognitive task combinations. Thus, the 68 SSD are equivalent to 30.2% of the overall 225 significant brain bands activity. Furthermore, the highest mean CMI value corresponds to the beta band, while the lowest value corresponds to the delta band.

Analysis per Subject. One additional analysis was performed of the Statistically Significant Differences (SSD) from the point of view of subjects. Regarding subject 1 (S1), we reported statistically significant differences among 9 of 10 cognitive task combinations. The cognitive task BxR and MxR and electrodes pairs P3-O1, P3-C4, and C3-C4 reported the greatest numbers of SSD. Thus, the 58 SSD are equivalent to 25.8% of the overall significant brain subjects activity. Subject 3 (S3) states statistically significant differences for all cognitive task combinations. The cognitive task BxM and BxL and electrodes pairs C3-O2, C3-P4, and C3-P3 reported the greatest numbers of SSD. Thus, the 41

Table 1. Overall bands statistically significant brain interactions. Part a

	BxM	BxL	BxR	BxC	MxL	Sub-total part a
C3-C4		DS5, AS4, BS3	DS5, TS1, TS5, GS1	DS3, TS1, BS4	TS5, AS4	12
C3-P3	BS3	TS1, AS3, AS4, BS3	GS1, GS4, GS5	TS1, AS4	TS1	11
C3-P4		BS3	TS5, BS3, GS4	BS3	GS1	6
C3-O1	BS3, GS5	BS3, GS5	GS4	TS1	AS4	7
C3-O2		AS3, BS3	BS4, GS1, GS4	GS4	TS5, AS3	8
P3-C4	AS5, BS3	BS3	GS1	TS1, AS4, BS1, BS4		8
P3-P4	BS3, GS3	BS3	GS1, GS3, GS4, GS5	BS3		8
P3-O1	TS4, BS3, GS5	BS3	AS1, GS1, GS4	TS1, BS1		9
P3-O2	BS3	BS3	AS5, GS1, GS4	TS1, GS4		7
O1-C4			BS1, GS1, GS4	DS1	AS4, BS4	6
O1-P4		BS3	TS1, BS4, GS4		GS1	5
O1-O2			BS4, GS4	GS4	DS4	4
C4-P4		TS3	TS1, TS5, GS1, GS4	BS4	BS4, BS5, GS1	9
C4-O2			TS3, GS1, GS4	DS5	DS4, AS3, AS4, GS1	8
P4-O2		BS3	BS4, GS1, GS4		DS4, GS1	6
Total	11	19	43	21	20	114

Table 2. Overall bands statistically significant brain interactions. Part b

	MxR	MxC	LxR	LxC	RxC	Sub-total part b	Total
C3-C4	AS1, GS1	AS5, BS1, BS4	TS5	AS3	TS5, AS5, BS4	10	22
C3-P3	BS3, BS5	TS1, AS5	AS3, BS3	DS4, AS5	TS5, AS5	10	21
C3-P4	TS4, GS1	DS3, BS4	BS5	TS1, TS3	DS3, TS4, TS5, BS4	11	17
C3-O1	GS1, GS4		AS5, GS1, GS4		GS4	6	13
C3-O2	DS5, AS3, GS1	TS1	AS3	DS4, AS3	BS4	8	16
P3-C4	GS1	TS1, TS5, AS5, BS4	TS5	AS5	TS1, TS5, AS4, AS5, BS4	12	20
P3-P4					TS5, AS4	2	10
P3-O1	GS1, GS4	TS1	GS1, GS4	AS5	TS1, TS5, AS4, GS4	10	19
P3-O2	DS5, TS1	TS1		DS4, AS5		5	12
O1-C4	GS1, GS4		TS5, AS4, BS4, GS1, GS4		TS5, BS4, GS4	10	16
O1-P4	GS1				GS4	2	7
O1-O2	GS1, GS4				GS4	3	7
C4-P4	AS3, GS1	AS5, BS4	TS5		TS5, BS4, GS4	8	17
C4-O2	GS1		DS4, AS4, BS3, BS4	DS4, AS4	TS5, BS4	9	17
P4-O2	TS4, GS1			DS4	BS4, GS4	5	11
Total	26	16	21	14	34	111	225

SS represent 18.2% of the overall significant brain-subjects activity. Concerning subject 4 (S4), we noticed statistically significant differences in all cognitive task combinations. The cognitive task BxR and RxC and electrodes pairs O1-C4 and C4-O2 state the greater numbers of SSD. Thus, the 80 SSD is equivalent to 35.5% of the overall significant brain-subjects activity. Finally, in subject 5 (S5), we reported statistically significant differences in all cognitive task combinations. The greater numbers of SSD are conformed by the cognitive task BxR and RxC and electrodes pairs C3-C4 and P3-C4. Hence, the 46 SSD represent 20.4% of the overall significant brain-subjects activity. To summarize our results, we elaborated Table 3, which depicts the number of SSD, mean, and standard deviation of CMI value for each brain band. Thus, the delta and gamma bands reported fewer and greater SSD numbers for all cognitive tasks and electrode pairs, respectively.

Table 3. Overall band behavior

Brain band	SS	Mean CMI	SD CMI
Delta	18	0.3845	0.0499
Theta	46	0.4542	0.0538
Alpha	40	0.5239	0.0641
Beta	53	0.5308	0.1123
Gamma	68	0.4606	0.1675

4 Conclusion

The primary purpose of this work was to analyze brain zones interaction while performing cognitive tasks. Our results suggest that by processing EEGs, it is possible to obtain information of FC between different brain zones while performing cognitive tasks using the CMI approach. The results report CMI values between 0.85 and 0.32 to EEG signals and their brain bands.

In general, all pairs of cognitive tasks analyzed in this study reported overall statistical differences for the five brain bands and all electrode pairs.

Thus, the present reported that all brain bands are able to analyze the FC in EEG under different cognitive tasks by applying the CMI approach. Furthermore, this methodology could be applied for future studies that might examine FC patterns in subjects under drug effects, subjects with neurological diseases and disorders such as Parkinson's, Alzheimer's, Schizophrenia, Epilepsy, and Attention deficit hyperactivity disorder (ADHD), etc.

References

1. Gysels, E., Celka, P.: Phase synchronization for the recognition of mental tasks in a brain-computer interface. IEEE Trans. Neural Syst. Rehabil. Eng. **12**(4), 406–415 (2004)
2. Vega, C.F., Fernández, F.J.R.: Phase synchrony analysis of gamma band as tools to discriminate cognitive tasks. In: 2012 ISSNIP Biosignals and Biorobotics Conference: Biosignals and Robotics for Better and Safer Living (BRC), pp. 1–6. IEEE (2012)
3. Jin, S.H., Lin, P., Auh, S., Hallett, M.: Abnormal functional connectivity in focal hand dystonia: mutual information analysis in EEG. Mov. Disord. **26**(7), 1274–1281 (2011)
4. Fingelkurts, A.A., Fingelkurts, A.A., Kähkönen, S.: Functional connectivity in the brain-is it an elusive concept? Neurosci. Biobehav. Rev. **28**(8), 827–836 (2005)
5. Tong, S., Thankor, N.V.: Quantitative EEG Analysis Methods and Clinical Applications. Artech House, Boston (2009)
6. Huang, L., Yu, P., Ju, F., Cheng, J.: Prediction of response to incision using the mutual information of electroencephalograms during anaesthesia. Med. Eng. Phys. **25**(4), 321–327 (2003)
7. Na, S.H., Jin, S.H., Kim, S.Y., Ham, B.J.: EEG in schizophrenic patients: mutual information analysis. Clin. Neurophysiol. **113**(12), 1954–1960 (2002)

8. Jeong, J., Gore, J.C., Peterson, B.S.: Mutual information analysis of the EEG in patients with Alzheimer's disease. Clin. Neurophysiol. **112**(5), 827–835 (2001)
9. Keirn, Z.A., Aunon, J.I.: Brain-computer interfaces laboratory (2021). https://www.cs.colostate.edu/eeg/main/data/1989_Keirn_and_Aunon. Accessed 23 Sept 2021
10. Jiang, S., et al.: Connective profiles and antagonism between dynamic and static connectivity underlying generalized epilepsy. Brain Struct. Funct. **226**(5), 1423–1435 (2021). https://doi.org/10.1007/s00429-021-02248-1
11. Chatterjee, A., Siarry, P.: Computational Intelligence in Image Processing. Springer, Heidelberg (2013). https://doi.org/10.1007/978-3-642-30621-1
12. Jeong, J., Gore, J.C., Peterson, B.S.: Mutual information analysis of the EEG in patients with Alzheimer's disease. Clin. Neurophysiol. **112**, 827–835 (2001)

Development of a Low-Cost Open Platform for Vital Signs Acquisition and Monitoring

Davi F. Buratto and Frank Herman Behrens(✉)

Pontifical Catholic University of Campinas, Campinas, Brazil
davi.fb@puccampinas.edu.br, frank@puc-campinas.edu.br

Abstract. The present work proposes to develop a vital sign monitoring system for application in several health areas, such as monitoring patients with suspected COVID-19, where the system can alert the patient and the medical staff if there is a worsening in the vital signs, allowing for prompt care. Sports professionals can also use the system to evaluate performance gains based on the collected data. The monitoring system measures the percentage of blood oxygenation, heart rate, blood pressure, and the concentration of exhaled carbon dioxide in the breath. The data is sent to the cloud to be stored and processed. Using the data achieved, it should be possible to know whether there has been an improvement or worsening of patients, whether they are taking their medication correctly (pressure drop or abnormal variations), to check the maximum cadence of an exercise, through the relationship between blood oxygenation and exhaled carbon dioxide, or even to detect some diseases in their early stages.

Keywords: Biomedical transducers · Biomedical signs monitoring · Sports monitoring equipment · Telemedicine

1 Introduction

The current pandemic of COVID-19 has overloaded the health system globally and, in countries like Brazil, has exceeded the occupation capacity of the available hospital beds. Therefore, to keep only severe patients hospitalized and free up beds, more and more people rely on patient recovery at home. Thus, a nursing technician must go daily to the patient's home to monitor his vital signs, ensuring recovery.

With the use of a few relatively low-cost sensors, it is possible to monitor these vital signs and send them to the cloud, allowing a doctor or healthcare professional to monitor several patients simultaneously, avoiding their exposure and unnecessary displacement of people. The present work proposes developing a vital sign monitoring system that can be applied in several health areas, such as for patients with suspected COVID-19. The system can alert the patient and the medical staff if there has been a worsening in the vital signs, allowing for prompt care.

The system can also monitor older people, verifying in real-time the effectiveness of the use of particular medication. Even sports professionals can use the system to evaluate the performance gains of athletes or people in training based on captured physiological data in real-time. The main goal is to use low-cost and easy-to-obtain devices and keep the platform modular and open, allowing the addition or removal of sensors as needed.

Y. Iano et al. (Eds.): BTSym 2021, SIST 207, pp. 143–150, 2023.
https://doi.org/10.1007/978-3-031-04435-9_14

2 Background

A previous work has already addressed the development of a platform for vital signs acquisition [1]. A system for remote monitoring of vital signs was developed based on the Internet of Things, fog computing, and cloud computing. In this work, the environmental signals, such as temperature, humidity, and gas concentration, as well as the patient vital data like heart rate, body temperature, oxygen concentration saturation, and electrocardiogram, are sent to a local server, which processes these data and, if they are out of the normal range, sends an alert to the doctors and nurses. Data is easily accessible via a cell phone application.

Other authors [2] have focused on developing a user-operable, portable carbon dioxide measuring device for measuring human breathing in and out of hospitals that measures carbon dioxide exhaled by the user, allowing evaluation of the respiratory system. This system acquires the signal, processes it, and stores it on a memory card for later evaluation. The device measures output CO_2 ($EtCO_2$), respiratory rate (RR), inspired carbon dioxide (ICO_2), and the Hjorth activity, which records the information with date and time from an internal clock. There are also works focusing on a more user-friendly interface [3]. A portable heart rate monitoring device was developed with a user-friendly interface, low cost, and high availability. Besides the sensor block, a preamplifier circuit was also designed to filter the heartbeat signals as much as possible with low-pass and high-pass filters. The focus of the work was to optimize the filters to deliver a cleaner signal. A JavaScript interface was also developed to display the signal in a simplified way for the user.

Other researchers have developed non-invasive, IoT-based oxygenation, heart rate, and pressure monitoring system for remote monitoring [4]. The acquired data is sent via Bluetooth connection to a cell phone, and an Android application processes and sends the information via a secure connection to a remote database. The user and the health care agent can access this remote database for cardiovascular disease prevention and detection. To acquire blood pressure, oximetry, and heart rate, the device needs three signals: the wrist electrocardiogram, the finger electrocardiogram, and the photoplethysmographic signal. With these signals, it is possible to determine blood pressure through the time difference between detecting a signal on the finger and in the heart.

Focusing on respiration, other researchers have developed a system to identify whether the patient has asthma and how severe it is [5] even with the patient in a coma, where the methodology used is to record and pre-process 4-s data of normal breaths to reduce the bandwidth and smooth the CO_2 signal. Then subdivide into inspiration and expiration phases and extract the CO_2 signal characteristics to discriminate asthmatic condition and asthma level classification using a statistical classifier algorithm.

This work proposes to create an open-source platform that allows the use of all the positive points discussed in the cited works, such as sensors and condition detection methodologies, and that also uses low-cost hardware and is customizable according to the need.

3 Development

Figure 1 shows the block diagram of the proposed system. The Controller block is implemented by the ESP32 module, manufactured by Espressif.

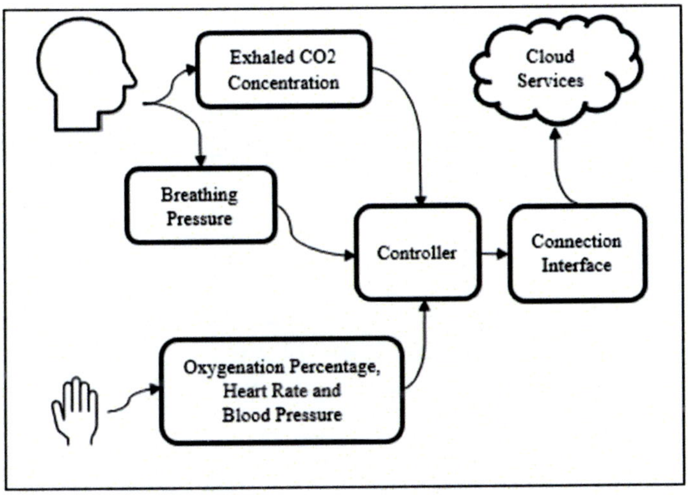

Fig. 1. Block diagram of the proposed system.

Sensors that measure the oxygenation percentage, heart rate, and blood pressure are installed in an adapter to be positioned on the index finger of one of the hands of the monitored patient. In addition, sensors measuring exhaled carbon dioxide concentration and breathing pressure are installed in a mask positioned on the patient's face.

The ESP32 WROOM 32E module has two 32-bit processors, 4MB of flash memory for the program, all the necessary electronics for Wi-Fi and Bluetooth communication, and several communication peripherals such as I2C and SPI [6]. The data obtained from the sensors is then acquired and processed by the ESP32 module, which sends it to a server where it is stored, analyzed, and displayed.

The exhaled carbon dioxide concentration block is composed of the SCD41 sensor manufactured by Sensirion (Fig. 2) based on photoacoustic technology that uses wavelengths of light to excite CO_2 molecules, increasing their translational energy [7]. Therefore, an increase in internal pressure is measured using a microphone, which generates a signal proportional to the number of molecules and can work with concentrations from 0 ppm to 40,000 ppm. In addition, this sensor has an integrated humidity and temperature sensor to compensate for the measurements for air density. Along with the CO2 sensor, a breath pressure meter is implemented by the BMP280 device, manufactured by Bosch Sensortec shown in Fig. 3 [8]. The purpose of using this sensor is to indicate the moment when expiration begins to synchronize the signal acquisition from the carbon dioxide sensor.

The high-sensitivity MAX30101 sensor in Fig. 4 is used for heart rate and percentage of oxygenation measurements [9], together with the biometric sensor hub MAX32664, which also makes it possible to measure blood pressure thanks to a proprietary algorithm of the manufacturer that runs internally. The MAX30101 sensor uses light of different wavelengths to measure reflection, scattering, and light absorption in finger tissue. Using a variation of Beer-Lamberts law (which stipulates that a known wavelength of light will undergo a specific extinction coefficient, based on the concentration of the substance it travels through), it is possible to measure absorbed light and light modulated by the pulsation of arterial blood contained in the tissue [10].

Fig. 2. CO_2 concentration sensor SCD41 [7]

Fig. 3. Pressure sensor BMP280 [8]

Fig. 4. MAX30101 sensor [9]

The sensors related to respiration monitoring were mounted inside a non-rebreather mask used to treat high oxygen concentration shown in Fig. 5 [11]. This mask has valves that allow only one direction of air circulation, leading the air from the oxygen source through the reservoir and into the front tube. There is a one-way valve in this tube that allows air in but not out, so when there is positive air pressure coming from the lungs, this valve closes and forces air out through the side valves. The mask was modified to reverse the air circulation, so the air enters through the sides and exits through the front opening. In the front tube, the air outlet, the carbon dioxide concentration sensor (Fig. 2), and the pressure sensor (Fig. 3) were adapted.

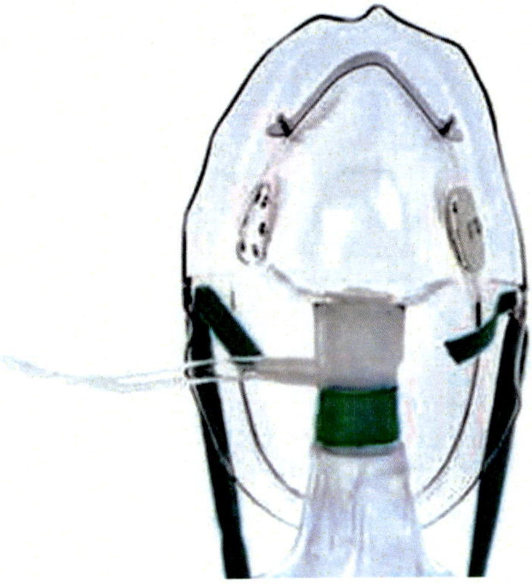

Fig. 5. Non-rebreather mask [11]

A module designed with the MAX30101 and MAX32664 devices and other auxiliary components must be installed in the adapter to be placed at the patient's fingertip. Since the MAX30101 sensor (Fig. 4) has a small and fragile encapsulation and requires different supply voltages than the ESP32 module, it was necessary to integrate the DC-DC voltage converters with the module feed the various components, as shown in the block diagram in Fig. 6.

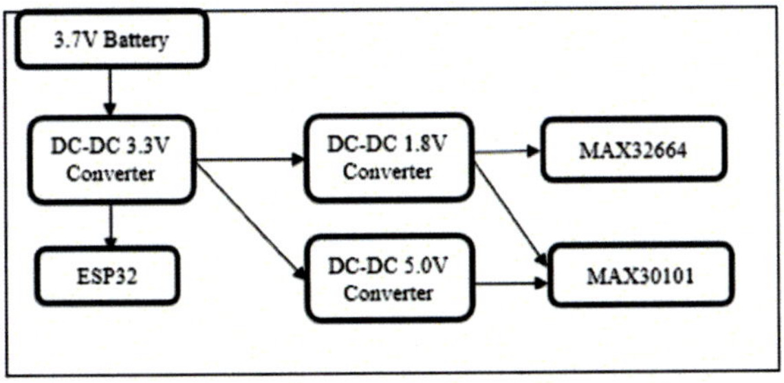

Fig. 6. Internal system power supplies

The ESP32 controller module in Fig. 7 runs a real-time operating system (RTOS) called Free-RTOS, which provides the desired modularity to add new sensors, prioritize the reading, and set the measurement rate of each sensor, making their tasks independent. The captured data is transmitted over the Internet via a Wi-Fi network, using the MQTT (Message Queuing Telemetry Transport) protocol.

Fig. 7. ESP32-WROOM module [6]

In the initial stage of this work, a remote computer running MQTT server services will aggregate the readings from each device into a particular topic (a specific channel for that set of information) and have a database where the collected data will be stored.

The installation of a Node-Red service allows defining actions or events based on the values received and handling the data. Finally, a Grafana server is responsible for displaying the collected data. In the future, the software infrastructure described above will be installed in the cloud.

4 Preliminary Results

First tests with the separate sensors have proven positive. For example, Fig. 8 shows inspiration/expiration cycles at rest and under forced breathing. A superficial analysis of the cycles shows that inspiration is deeper because it requires muscle strength instead of expiration that is just muscle relaxation.

Through measurements of blood oxygenation, expired carbon dioxide concentration, and heart rate, it is possible to estimate muscle lactate concentration, allowing the exhaustion and cardiorespiratory condition of the athlete using the system to be measured [2, 5, 12, 13].

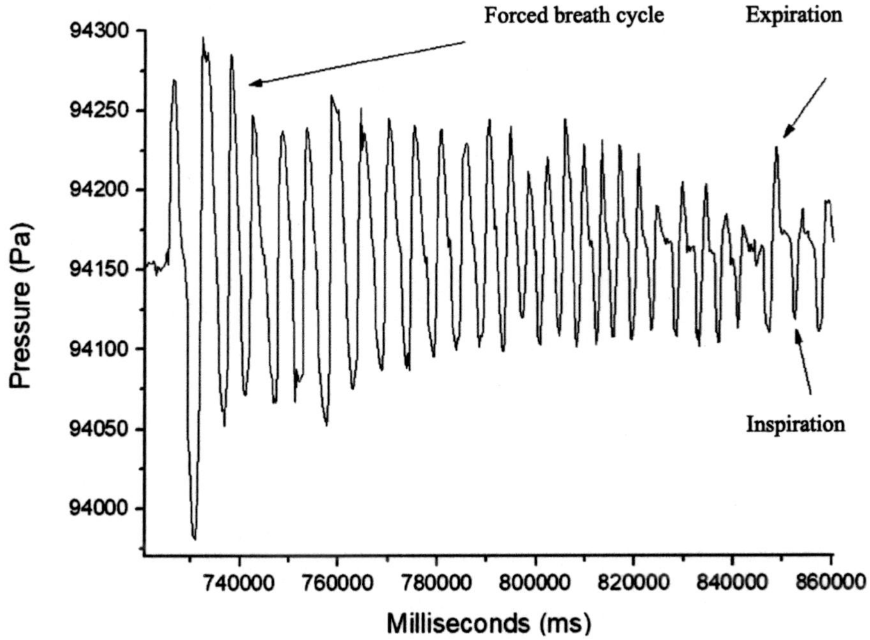

Fig. 8. Breathing cycles

4.1 Platform Build Costs

As proposed, the development cost of the platform should be low, so the most cost-effective components were chosen. As can be seen in Table 1, the total cost did not exceed 140 US dollars. For comparison, a portable oxi-capnograph for veterinary use costs an average of 1000 dollars, thus demonstrating that it is possible to develop a platform at a considerably lower cost.

Table 1. Build costs (in US dollars).

Item	Cost
Oximeter board	$ 55.00
Development board	$ 25.50
Barometric sensor	$ 0.46
CO_2 sensor	$ 50.26
Non-rebreathing mask	$ 6.00
MicroSD socket	$ 0.30
Cables	$ 1.00
Total	**$ 138.52**

5 Conclusions

So far, individual sensor values for CO_2 concentration, respiration pressure, heart rate, and oxygenation percentage (oximeter function) have been acquired. The following steps will be the acquisition of all the values and their validation against established measuring instruments on the market to check for possible deviations or reading errors.

After the preliminary tests with the prototype, a series of acquisitions will be made with patients to raise a mass of data that can be processed. Specific data will be profiled for various monitoring applications.

References

1. Hassen, H.B., Ayari, N., Hamdi, B.: A home hospitalization system based on the Internet of things, Fog computing and cloud computing. Inform. Med. Unlocked **20**, 100368 (2020)
2. Singh, O.P., Howe, T.A., Malarvili, M.B.: Real-time human respiration carbon dioxide measurement device for cardiorespiratory assessment. J. Breath Res. **12**, 0–13 (2018)
3. Son, L.P., Thu, N.T.A., Kien, N.T.: Design an IoT wrist-device for SpO2 measurement. In: Proceedings of the International Conference on Advanced Technologies for Communications, pp. 144–149. IEEE, Quy Nhon (2017)
4. Florez, D., Sepulveda, J.: BlooXY: on a non-invasive blood monitor for the IoT context. In: Proceedings of the 30th International System on Chip Conference (SOCC), pp. 29–34. IEEE, Munich (2017)
5. Singh, O.P., Palaniappan, R., Malarvili, M.: Automatic quantitative analysis of human respired carbon dioxide waveform for asthma and non-asthma classification using support vector machine. IEEE Access **6**, 55245–55256 (2018)
6. ESP32 WROOM 32E Module: Espressif Systems (2021). https://www.espressif.com/en/products/modules. Accessed 20 Oct 2021
7. SCD41 CO2 Sensor: Sensirion (2021). https://www.sensirion.com/en/environmental-sensors/. Accessed 20 Oct 2021
8. BMP280 Pressure Sensor: Bosch Sensortec (2021). https://www.bosch-sensortec.com/products/environmental-sensors/pressure-sensors/bmp280/. Accessed 20 Oct 2021
9. MAX30101 Pulse Oximeter & Heart-Rate Sensor: Maxim (2021). https://br.mouser.com/new/maxim-integrated/maxim-max30101-sensor/. Accessed 20 Oct 2021
10. Urpalainen, K.: Development of a fractional multi-wavelength pulse oximetry algorithm. Master's Thesis, Aalto University School of Engineering, Espoo, Finland (2011)
11. Non-Rebreathing Mask: Surgo Surgical Supply (2021). https://www.surgo.com/total-non-rebreathing-mask--2/adult-total-non-rebreathing-mask--1. Accessed 20 Oct 2021
12. Arpaia, P., et al.: Conceptual design of a machine learning-based wearable soft sensor for non-invasive cardiovascular risk assessment. Measurement **169**, 108551 (2021)
13. Kumar, P.M., Devi Gandhi, U.: A novel three-tier Internet of Things architecture with machine learning algorithm for early detection of heart diseases. Comput. Electr. Eng. **65**, 222–235 (2018)

Investigation of the Influence of Turbulence Models on Cough Droplet Evaporation: Comparing (SST) k-Ω, k-ε, and Reynolds Stress (RSM) Turbulence Models

Nuhu Ayuba(✉) and Gabriela Cantarelli Lopes

Chemical Engineering Postgraduate Program (PPGEQ), Federal University of São Carlos (UFSCar), São Carlos, Brazil
nuhu.ayuba101@gmail.com, gclopes@ufscar.br

Abstract. Numerical studies of violent respiratory activities, such as coughing, sneezing, breathing, and speaking, are vital in understanding human contamination against Infectious Respiratory Diseases (IRD). Even though state-of-the-art Eulerian-Lagrangian numerical studies are available in the literature, more studies can still be carried out, either by combining different numerical models or by varying ambient conditions. In this direction, this work was aimed at comparing three turbulent models (SST k-ω, k-ε, and Reynolds Stress), through the simulation of evaporation of a single cough droplet. In order to validate the model, evaporation simulation using (SST) k-ω was compared with a literature result, and the results were similar. Thereafter, two parameters (velocity and particle size) were varied and the results were compared between the aforementioned models. In summary, no significant differences were observed between simulations results obtained at the same conditions, using the respective turbulent models. However, it is believed that such a conclusion was possible because the simulations were carried out at simple conditions (geometrically and parametrically).

Keywords: Turbulent models · Droplet evaporation · Eulerian-Lagrangian · Cough droplet · Discrete Phase Model (DPM)

1 Introduction

Infectious Respiratory Diseases (IRDs), such as Severe Acute Respiratory Syndromes (SARS), tuberculosis (TB), and influenzas (Avian and Swine), have been an interest of study for a long time [1, 2]. The World Health Organization (WHO), along with other governmental and non-governmental organizations requested the scientific community to study means of managing transmissible diseases such as influenza and tuberculosis (TB) in the early 1990s [3]. The rapid spread of the SARS-CoV-2 virus, better known as the new Corona Virus Disease (COVID-19) in early 2020 [4] has intensified the need to elaborate more studies on IRD.

Over the years, IRD studies have overflown the medical fields, extending to other areas, such as Engineering, which involves the multiphase flow of respiratory droplets

Y. Iano et al. (Eds.): BTSym 2021, SIST 207, pp. 151–162, 2023.
https://doi.org/10.1007/978-3-031-04435-9_15

(which are predominantly water), where pathogens are not directly studied. It is known that viruses are present in sputum droplets [5], therefore, respiratory diseases are transmitted via respiratory activities.

There are different methods to study the fluid dynamics of respiratory activities. Specifically, in CFD, respiratory droplets flow can be studied using three different methods: Eulerian-Eulerian methods, Lagrangian-Lagrangian methods, and Eulerian-Lagrangian [6]. Since droplets expelled during cough have relatively low mass loading (usually below 10–12%), Eulerian-Lagrangian (also known as discrete phase model-DPM) is the most recommended model.

The DPM approach tracks and solves individual particles (in this case, cough droplets) on the Lagrangian reference frame, as they flow across the domain. Simultaneously, the fluid phase (or continuous phase, the air in this case) is solved on the Eulerian plane. As a choice, particle-particle or particle-fluid interactions (either one or two ways) can be integrated using the Discrete Element Model.

Although turbulence modeling studies are important in the CFD simulation of cough droplet evaporation, it has seldom been explored in the literature, according to records. Examples of turbulence models include Spalart-Allmaras One-Equation, k-ε, k-ω, Reynolds Stress Model (RSM), and Laminar-Turbulent Transition Models.

In the literature, (SST) k-ω turbulent model has seldom been applied to study respiratory activities such as cough, sneeze, speaking, and breathing in open/closed spaces. Notwithstanding, in this article, it has been applied to study the evaporation of cough droplet particles. In addition, the model has been compared with Standard k-ε and Reynolds Stress Models (RSM).

2 Numerical Model

2.1 Particle Formulation - Discrete Phase Model (DPM)

In the standard formulation of the Eulerian-Lagrangian Model, it is assumed that the volume fraction of the discrete phase is relatively low. Therefore, the trajectory of any discrete particle is predicted by integrating the force balance on the Lagrangian reference frame [6]. The force balance is equated to the particle inertial as written in (1).

$$\frac{\partial u_p}{\partial t} = F_D(\vec{u} - \vec{u}_p) + \frac{\vec{g}}{\rho_p}(\rho_p - \rho) + \vec{F} \tag{1}$$

where u, ρ, and g represent velocity, density, and acceleration due to gravity (–9.81 m/s^2). $F_D(\vec{u} - \vec{u}_p)$ and $\vec{F}$ stand for the fluid drag force per unit mass and additional acceleration (force per unit mass, which could be the virtual mass force that accelerates the fluid around the particle or pressure gradient in the fluid), with details in (2) and (3).

$$F_D = 0.75\frac{\mu C_D Re}{\rho_p d_p^3} \tag{2}$$

$$\vec{F} = \frac{1}{2}\frac{\rho}{\rho_p}\frac{d}{\partial t}(\vec{u} - \vec{u}_p) \tag{3}$$

where C_D is the drag coefficient explained in [7]. Additionally, $\vec{u}$ is the fluid phase velocity (m s^{-1}), $\vec{u}_p$ is the particle velocity (m s^{-1}), μ is the molecular viscosity of the fluid (kg m^{-1} s^{-2}), ρ is the fluid density (kg m^{-3}), ρ_p is the density of the particle (kg m^{-3}), and d_p is the particle diameter (m). *Re* in (2) is the relative Reynolds number, which is given by (4).

$$Re = \frac{\rho d_p \left|\vec{u}_p - \vec{u}\right|}{\mu} \tag{4}$$

The evaporation of cough droplets is guided by the diffusive flux of droplet vapor in the air as explained in [8]. In order to solve the turbulence in the Eulerian-Lagrangian model, averaging closure models are applied.

2.2 Shear Stress Transport (SST) k-ω Model

Shear Stress Transport (SST) $k-\omega$ was introduced by [9] to solve RANS equations. The model is adequate for modeling adverse pressure gradient flows. It blends the robust and accurate $k-\omega$ model at the wall regions with free-stream independence while applying $k-\varepsilon$ at the far-field region within the flow domain [8]. The reason for combining the two models is because, outside de boundary layer, ω-equation shows great sensitivity, which inhibits it from estimating the turbulence. To blend the models, standard k-ε and k-ω are multiplied by a unit blending function, which is designed to activate either of the models, depending on the region where the turbulence would be solved - near the walls or away from it.

2.3 Standard k-ε Turbulence Model

The $k-\varepsilon$ is a semi-empirical model also used to solve RANS equations. Its derivation depends on phenomenological and empirical considerations and it is modeled based on transport equations for turbulent kinetic energy (k) and dissipation rate (ε). Its robustness, considerable accuracy, and economy in solving a wide range of engineering flows and heat transfer are some of the reasons why it is applied frequently in simulations.

In the derivation of standard $\boldsymbol{k-\varepsilon}$, it is assumed that the flow is fully turbulent, and the effects of molecular viscosity are negligible. This method is only valid for fully turbulent flows. More details are found in [9].

2.4 Reynolds Stress Model (RSM)

These models also solve RANS equation; however, they do not use the Eddy Viscosity hypothesis. Instead, they model transport equations for all components of the Reynolds stress tensor and dissipation rate in the fluid. They are suitable for complex flows since it includes effects of streamline curvature, sudden changes in the strain rate, buoyancy flows compared to (SST) k-ω, and Realizable $k-\varepsilon$. Specifically, the model solves 5 viscous Equations. Nevertheless, researches have shown that the Reynolds Stress model is not superior to two-equation models.

3 Geometry

The 2D computational domain presented in Fig. 1 was used to carry out all the simulations in this article. A squared computational domain of 2.5 m by 2.5 m was considered in this study (Fig. 1). The height of the cougher was obtained from [8]. The 0.02 m rectangular extension in Fig. 1 served at the cough cloud inlet. The domain ended in an open wall, and atmospheric conditions were applied at that point (similar to the model presented in [10]). Quadrilateral shapes were used in this work in order to obtain more precise results. In addition, particles were injected normal to the rectangular inlet.

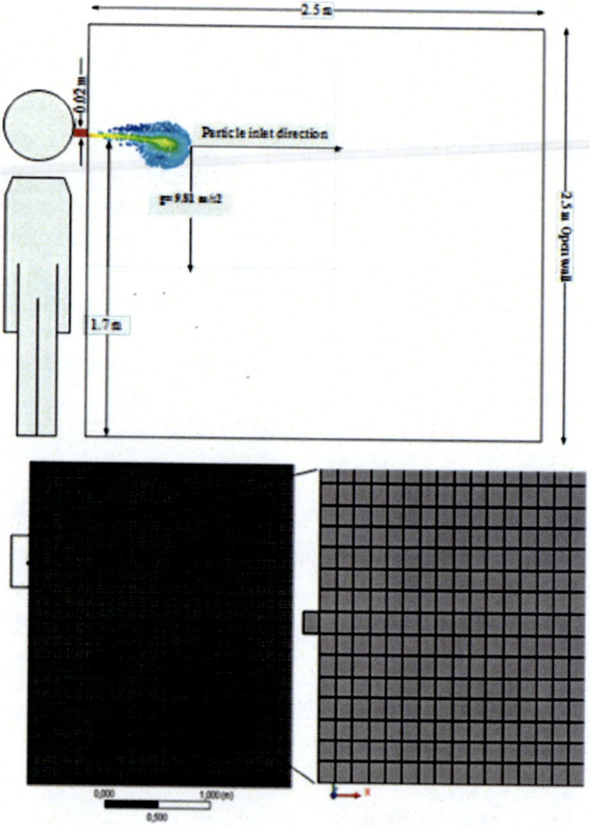

Fig. 1. Computational cough domain and structured 2-D quadrilateral grid.

The computational domain was discretized using 2D quadrilateral shapes, where 15626 elements were used. The grid presented a Skewness of 7.514×10^{-6} with a standard deviation of 1.325×10^{-5}. And according to ANSYS Fluent standards, cells with Skewness between 0 and 0.25 are considerably accurate computational grids for solving Differential Equations.

4 Simulation Configurations

In this work, a pressure-based solver in 2D-space was applied. A transient simulation was carried out (using Second-Order Implicit formulation) since the same time function was successfully applied in [8] and [10]. In the Discrete Particle Model, unsteady particle treatment with fluid flow time-step tracking was applied. In order to track each cough particle, the Discrete Random Walk Model with a Time Scale Constant of 0.15 was used in all cases. Since particles released during violent respiratory activities have insignificant effects on the momentum and thermodynamics of the continuous phase, one-way coupling was applied.

The Semi Implicit Linked Equations (SIMPLE) were applied in pressure-velocity coupling. For Spatial Discretization of the flow gradient, the Least Squares Cell-Based method was used. Second-Order Upwind was used to discretize momentum, Turbulent kinetic energy, Specific dissipation Rate, Discrete Particle Species (water and air), and energy.

Since particles relatively small evaporate faster, there is a need to use smaller time-steps. Therefore, in this article, 10^{-5} s was used to simulate the evaporation of particles $\leq 10\,\mu m$ particles, while 10^{-3} s time-step was used for $10\,\mu m <$ particles $\leq 100\,\mu m$ particles.

5 Results

The evolution of the droplet diameter during the evaporation for three droplet sizes (1 μm, 10 μm, and 100 μm) was observed under two humidity conditions (0% and 80%), as presented in Fig. 2.

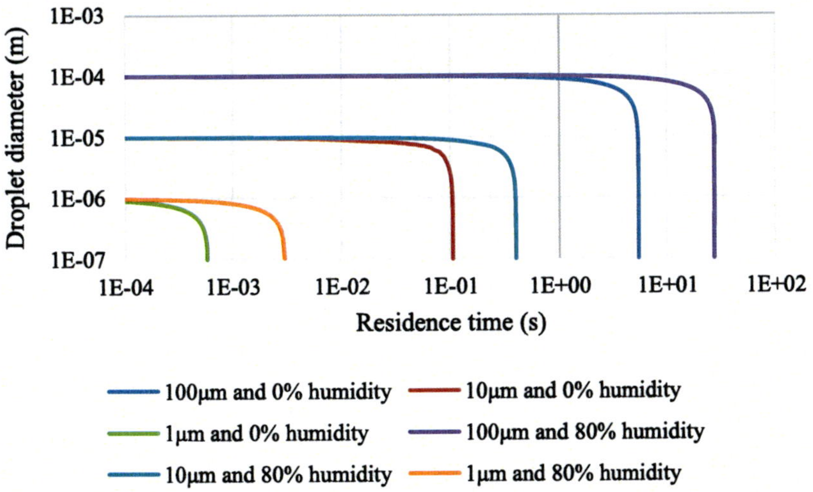

Fig. 2. Droplet evaporation simulated in this article using (SST) k-ω.

The simulation using (SST) k-ε in Fig. 2 was carried using a single water drop evaporation model of [8], however, with few modifications that are detailed in the next paragraphs.

In the simulation, the pure water droplet was released at 310.15 K into a computational domain with a temperature of 293.15 K. In all cases, a vapor cloud with a humidity of 100% was released with the droplet.

[8] used a flow domain of 10.1 m length, 7.12 m width, and 5.4 m height, and they adopted a polyhedral-unstructured grid scheme. However, to reduce the computational cost, this work was carried out using reduced dimensions; 2.5 m length and 2.5 m height, and discretized using a quadrilateral-structured scheme. Despite the slight configuration differences, the results obtained using (SST) $k-\omega$ turbulence model (presented in Fig. 2) were similar to [8]. Additionally, they were also similar to those presented by [10] and [2].

In a detailed comparison between the method applied in this work and the work of [8], all evaporations in this work occurred at a lesser time. For instance, at the release of 10 μm droplet into the computational domain under 80% humidity, a time difference of 0.06 s (Fig. 3) was observed between this method and [9].

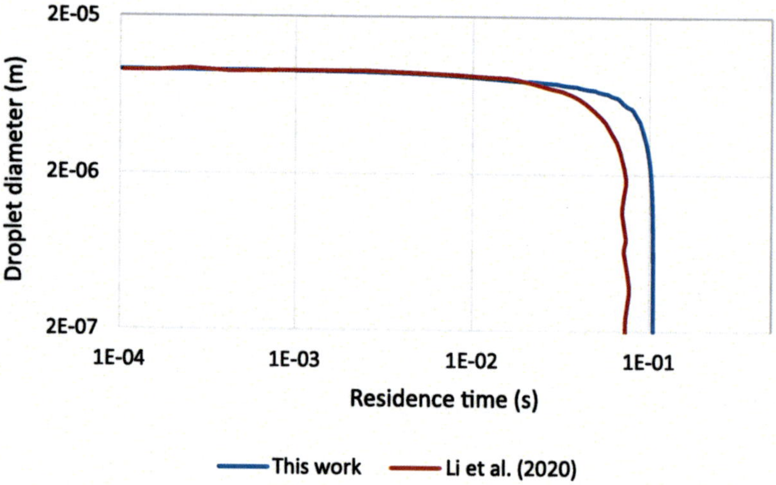

Fig. 3. Particle evaporation using 10μm water droplet at (a) 80% relative humidity.

It is believed that the slight difference in evaporation time observed in Fig. 3 may have resulted from the turbulent model itself or grid used (since [8] used a polyhedral unstructured scheme while a tetrahedral structure grid was applied in this work).

5.1 Grid and Droplet Independence Test

To conduct the simulations in this work, a grid independence test was carried out in order to establish and apply the coarsest grid, which was sufficient to produce considerable

Table 1. Selected grids used for independence test.

Element size (m)	Nodes	Elements
0.005	251021	250016
0.010	63007	62504
0.020	15878	15626

results at a lower computational cost. To this end, three grids presented in Table 1 were applied.

It was observed that evaporation of droplet particles was least affected by the change in the element's density per unit area.

Figure 4 presents a simulation of droplet evaporation profiles within the computational domain over time using the grids shown in Table 1. Here, a 1 μm droplet was released into the computational domain and its evaporation was observed over time, under 80% humidity.

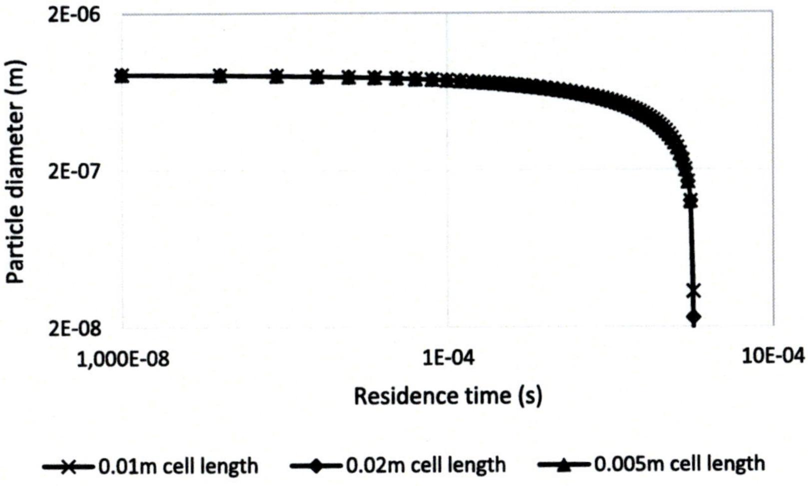

Fig. 4. Evaporation of cough droplet using three different grids.

It can be observed that similar results were obtained for the three grids used. Similar grid test results were also observed in [11]. However, the authors used unstructured tetrahedral cells to form the computational grid. Their results showed that increasing the grid density (cells/m^3) had little effect on the axial velocity profile. Similarly, [12] carried out a grid test measuring the streamwise velocity of cough cloud for continuous phase, using a fully structured grid (with 2.856×10^6, 5.830×10^6, and 9.854×10^6 elements, respectively). Although the authors observed some instabilities in the streamwise velocity of the first grid, there was no significant difference between the results obtained using the three grids. This is in accordance with the grid independence test carried out in this work. Hence, in this work, 0.02 m cell length was used for all the simulations.

Normal human cough is turbulent and transient, and the particles are released into the surroundings randomly. For these reasons, consecutive multiple coughs from a single person can present different particle dispersions. In order to confirm if the same rule applies to droplet evaporation released during each cough, an independence test was carried out to observe the evaporation of a single microdroplet.

Therefore, in this work, the droplet was released into the computational domain (Fig. 1) five times under the same condition. Figure 4 shows the average evaporation rate of the particle and their respective standard deviations, at specific residence times. The standard deviations of the five evaporation simulations were relatively low, which indicates the viability of using a single droplet to carry out this work.

5.2 Diameter Variation

As part of the turbulent models' comparison, the evaporation of 1 μm and 100 μm at 0% humidity and air velocity of 10 m/s, were respectively simulated, and the results are presented in Fig. 5.

From the results, all the turbulent models presented similar simulation results. The 100 μm particle took more time to evaporate when compared to 1 μm. This is because the former has a lesser particle area per unit volume, thereby evaporating slower than the latter. The same observation was made in [9] and [13].

In Table 2, the exact droplet evaporation times obtained from the three turbulent models are presented.

All the models presented similar results for each particle size evaporation. A relative decrease in particle diameter increased the rate of evaporation. And such holds to increase in surface area per unit volume of the droplets, as reported in [8]. In other words, smaller droplets evaporate faster than the bigger ones.

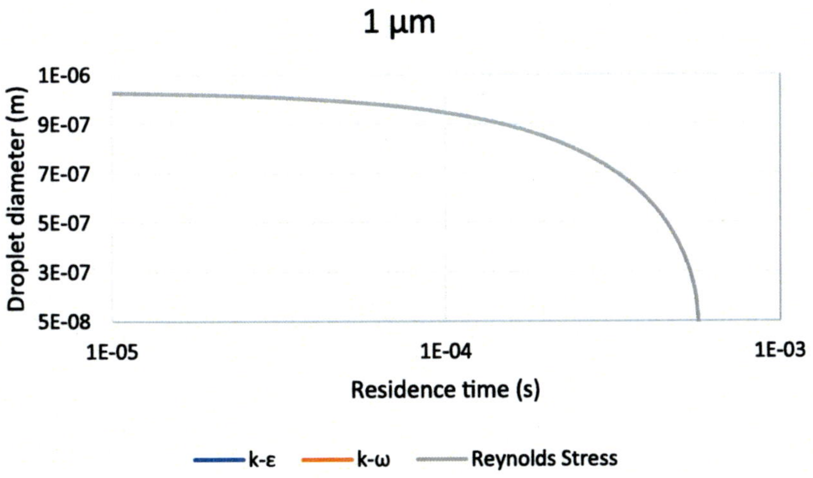

(a)

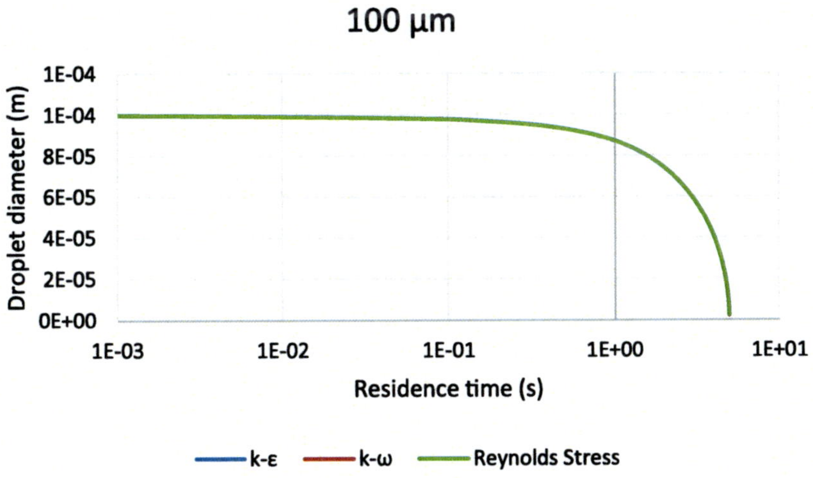

(b)

Fig. 5. Evaporation of cough droplet particle using (SST)k-ω, k-ε and Reynolds Stress (RSM) turbulence models (a) 1 μm and (b) 100 μm.

Table 2. Time taken for different droplet sizes to evaporate using three turbulent models.

Particle size (μm)	Evaporation time (s)		
	k-ε	(SST) k-ω	RSM
1	0.00057	0.00057	0.00057
50	0.52000	0.52000	0.52000
100	4.91400	4.91900	4.91500

5.3 Velocity

At this stage, three cough velocities were respectively used to inject a 100 μm particle into the flow domain (Fig. 1). All the simulations were carried out at 0% humidity, and parts of the result are presented in Table 3.

Table 3. Time taken by 100 μm droplet to evaporate at different injection velocities and 0% humidity.

Velocity (m/s)	Evaporation time (s)		
	k-ε	(SST) k-ω	RSM
10	4.949	5.371	4.914
14	4.904	4.919	5.087
20	4.906	4.918	4.918

In general, an increase in cough velocity seldom affected droplet evaporation. However, (SST) k-ω droplet evaporation was slightly affected by the increase in velocity. And it is assumed that the robustness of (SST) k-ω (both near the wall or far from it) may be the factor.

Additionally, similar results were observed for all models, on comparing droplet evaporation, released at velocities of 10 m/s, 14 m/s, and 20 m/s, respectively. An example is presented in Fig. 6, where the results were obtained applying a velocity of 20 m/s.

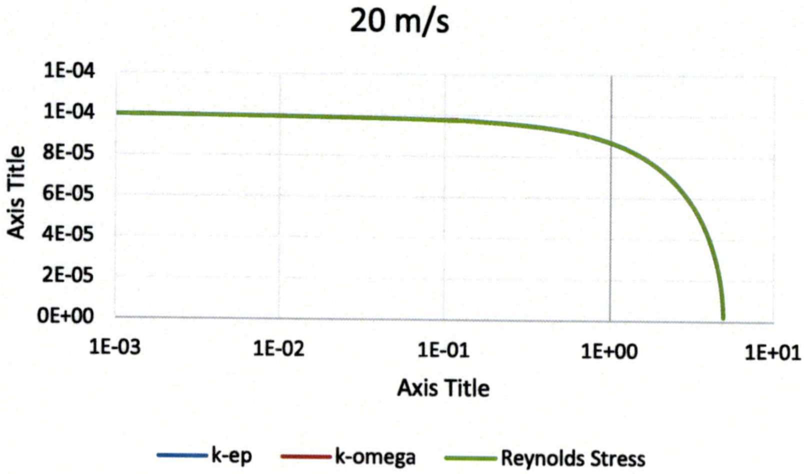

Fig. 6. Comparing standard k-ε, (SST) k-ω and Reynolds Stress under the same condition.

6 Discussion

It could be noticed that the tree turbulent models $(SST)k-\omega$, Standard $k-\varepsilon$, and Reynolds Stress) applied in this study predicted similar droplet evaporation rates. However, it was expected that considerable variations would be observed between the three turbulent models. Specifically, it was expected that Reynolds Stress Model (RSM) would present more precise results than the other two-equation turbulence model. Reynolds Stress turbulence model has six additional transport equations, which makes it more complex and robust when compared to two-equation models ($k-\omega$, Standard $k-\varepsilon$). Even though coughs create complex fluid conditions such as swirling, which are readily solved by Reynolds Stress, there were no significant variations in the droplet evaporation.

Therefore, the similarities in the results obtained from the application of the three turbulent models may be related to the simplicity of the geometry and flow conditions.

Similar to studies such as [2, 10], and [14], droplet evaporation was faster for smaller droplets because of increased surface area per unit volume, which favors contact with dry air.

An increase in velocity cough velocity seldom affected droplet evaporation. But initially, it was expected that accelerating any cough droplet could increase the contact rate per unit time between the droplet particle with dry air, thereby favoring more evaporation.

7 Conclusion

It was possible to compare three turbulent models applied in cough droplet evaporation simulation. All models calculated similar results with respect to the increase in droplet size and velocity. Although, adverse conditions or more complex cases were not studied, to some extent, these models can be used for more complex cough dynamics studies.

Acknowledgment. This study was financed in part by the Coordenação de Aperfeiçoamento de Pessoal de Nível Superior – Brasil (CAPES) – Finance Code 88887.503163/2020-00.

References

1. Gupta, J.K., Lin, C., Chen, Q.: Characterizing exhaled airflow from breathing and talking. Indoor Air **20**(1), 31–39 (2010)
2. Morawska, L.: Droplet fate in indoor environments, or can we prevent the spread of infection? Indoor Air **16**(5), 335–347 (2006)
3. Zayas, G., et al.: Cough aerosol in healthy participants: fundamental knowledge to optimize droplet-spread infectious respiratory disease management. BMC Pulm. Med. **12**(1), 1–12 (2012)
4. Kucharski, A.J., et al.: Effectiveness of isolation, testing, contact tracing, and physical distancing on reducing transmission of SARS-CoV-2 in different settings: a mathematical modelling study. Lancet Infect. Dis. **20**(10), 1151–1160 (2020)
5. Dhand, R., Li, J.: Coughs and sneezes: their role in transmission of respiratory viral infections, including SARS-CoV-2. Am. J. Respir. Crit. Care Med. **202**(5), 651–659 (2020)

6. Zahari, N.M., et al.: Introduction of discrete phase model (DPM) in fluid flow: a review. In: AIP Conference Proceedings. AIP Publishing LLC, p. 20234 (2018)
7. Zhu, L., Li, A., Wang, Z.: Analysis of particle trajectories in a quick-contact cyclone reactor using a discrete phase model. Sep. Sci. Technol. **53**(6), 928–939 (2018)
8. Hu, J.: Motion of a neutrally buoyant circular particle in a clockwise double-lid-driven square cavity. Phys. Fluids **32**(11), 113304 (2020)
9. Menter, F.R., Kuntz, M., Langtry, R.: Ten years of industrial experience with the SST turbulence model. Turbul. Heat Mass Transf. **4**(1), 625–632 (2003)
10. Li, X., Shang, Y., Yan, Y., Yang, L., Tu, J.: Modelling of evaporation of cough droplets in inhomogeneous humidity fields using the multi-component Eulerian-Lagrangian approach. Build. Environ. **128**, 68–76 (2018)
11. Sen, N.: Transmission and evaporation of cough droplets in an elevator: numerical simulations of some possible scenarios. Phys Fluids **33**(3), 33311 (2021)
12. D'Alessandro, V., Falone, M., Giammichele, L., Ricci, R.: Eulerian-Lagrangian modeling of cough droplets irradiated by ultraviolet–C light in relation to SARS-CoV-2 transmission. Phys. Fluids **33**(3), 31905 (2021)
13. Han, Z.Y., Weng, W.G., Huang, Q.Y.: Characterizations of particle size distribution of the droplets exhaled by sneeze. J. R. Soc. Interface **10**(88), 20130560 (2013)
14. Dbouk, T., Drikakis, D.: On coughing and airborne droplet transmission to humans. Phys. Fluids **32**(5), 53310 (2020)

Regulatory Aspects of 5G and Perspectives in the Scope of Scientific and Technological Policy

Antonio Carlos Demanboro[1](✉), David Bianchini[2], Yuzo Iano[2], Gabriel Gomes de Oliveira[2], and Gabriel Caumo Vaz[2]

[1] FIA - Fundação Instituto de Administração, University of São Paulo, São Paulo, Brazil
ademanboro@gmail.com

[2] School of Electrical and Computer Engineering, University of Campinas, Campinas, Brazil
yuzo@unicamp.br, oliveiragomesgabriel@ieee.org

Abstract. This paper deals with the regulatory aspects of 5G, considering the recent challenges in Science and Technology with the United States' decision to ban a Chinese company Huawei to supply 5G products and pressure the western countries to do the same. The hegemonic war in 5G implementation is overlooked, emphasizing the oligopolistic marked surrounded by big techs, namely, Amazon, Apple, Google plus Microsoft, and IBM, in the western hemisphere, and Huawei, Tencent, AliExpress, in the eastern hemisphere. The objective is to discuss the real situation of 5G in the world and the possibilities that are open to peripheral countries. The aspects involved with Artificial Intelligence (AI) development are highlighted because the supremacy in 5G and AI developments are intertwined. The Game Theory is proposed as a methodological guide for the construction of four scenarios called Scenario Restriction, Scenario Contention, Scenario Liberalization, and Scenario Cooperation, which permit to design some considerations about each one.

Keywords: Regulation · 5G · Science · Technology

1 Introduction

The fifth-generation telecommunications networks, known as 5G, have caused a series of changes in the relationship between countries, at different levels, in addition to the eminently technical ones.

By 2025, 1.2 billion people are expected to have access to 5G worldwide, with 1/3 in China, according to assessments made by the GSMA group [1].

In the technical aspect, these are networks that need to operate in extra-high frequency bands, and in Brazil, the Ministry of Science and Technology and Innovations issued Ordinance no. 418 of 01/31/20, which provides for the auction of the 700 MHz, 2.3 GHz, 3.5 GHz, and 26 GHz bands [2]. The ordinance mandates ANATEL to carry out the bidding process for these bands, which should occur in 2021. The main directions of ordinance no. 418 are:

Y. Iano et al. (Eds.): BTSym 2021, SIST 207, pp. 163–171, 2023.
https://doi.org/10.1007/978-3-031-04435-9_16

"Art. 2° In the spectrum bids referred to in Art. 1, Anatel should consider: I - encouraging the sharing of active and passive infrastructure between providers, including posts, towers, ducts and conduits; II - establishment of scope commitments, pursuant to Decree No. 9,612, of December 17, 2018, including: a) service with mobile broadband using 4G technology or higher, for cities, towns, isolated urban areas, and rural agglomerations, according to the classification of the Brazilian Institute of Geography and Statistics (IBGE), which have a population of more than 600 inhabitants, in a complementary way to other public policies in force; b) coverage of federal highways with mobile broadband; and c) high-speed transport networks, preferably in optical fiber, for municipalities not yet served. III - definition of deadlines for the activation of services in the bidding bands that, if not met, enable the use of the band by interested third parties, with protection guarantees; IV - modeling that makes it possible to maintain or increase the current levels of competition; V - frequency band granting models, on a primary or secondary basis, for operations of restricted interest telecommunications services; and VI - institution of mechanisms that ensure the attendance of visiting subscribers between the networks of the different operators."

Some companies that operate 4G and 4.5G networks are interested in using existing systems and launching 5G from them.

This is technically possible if the customer has a cell phone compatible with 5G, but rapid network saturation can occur if many users are qualified to use the service [3].

At the geopolitical level, a rupture in the American Bell partnership has been unleashed, which has been playing out successfully since 1972, with Nixon's famous visit.

China's supremacy in telecommunications networks has been consolidating in recent years, not only in cutting-edge systems, such as 5G but also in new scientific frontiers, such as IoT and quantum telecommunications.

A symptom of this is the increasingly intense presence of the Chinese, not only in the area of strategic patents but also in multilateral organizations where the technical parameters of telecommunications networks are defined, such as the TM Forum.

Specifically, in relation to 5G, questions arose from the west, notably from the moment when a single Chinese company, Huawei, started to produce not only the best equipment but also the most modestly priced compared to direct competitors - mainly Nokia and Ericsson.

Aside from political upheavals, the first major technological breakthrough came with the US government banning the sale of chip maker Qualcomm to Singapore-based company Broadcom, in a deal involving $ 117 billion.

Although a Singapore company, the American government understood that there was a high geopolitical risk of loss of supremacy in this strategic area due to the relationships between companies based in Asia and supposedly subject to Beijing's interests [1].

The US government has since closed its grip on manufacturer Huawei, preventing Qualcomm from supplying chips to the Chinese company. Recently, however, Qualcomm itself has been lobbying the US government to release the chip trade with Huawei, claiming it would be losing billions of dollars, and the chip sale ban would not solve the problem of Chinese supremacy in telecommunications since that company is able to produce its own chips, or acquire them from other manufacturers [4].

The North American government has been taking a strong stance in relation to the 'communist danger', accusing the Chinese Communist Party of having erected a great barrier in the country, the Great Firewall, which prevents the free flow of data via the internet to be established, but which would allow capturing external data from the west via 5G equipment. The pressure was so great that it made England break an old commercial relationship with the Chinese company.

Other countries are taking a stand against the use of Huawei devices, such as Australia and India. Some telephone companies would also be under pressure to join a global data security alliance, called the 'Clean Path Initiative'.

The American government considers that some companies have already joined,

such as Telefónica in Spain, Orange in France, Reliance in India, Telstra in Australia, SK and KT in Korea, Cosmote in Greece, NTT in Japan, and O2 in England.

It is in this context that the American government expects Brazil to join the 'Clean

Path Initiative' and promote Huawei's ban on telecommunications networks, especially concerning 5G [5].

Another escalation in this technological race has recently occurred on another front, now focused on social networks. The US government has banned some Chinese applications from being downloaded from American online stores, such as the App Store, being followed by India. Microsoft is negotiating the purchase one of the banned applications with a view to its commercialization in the west [6].

Although apparently not directly related to the issue of technological supremacy over Chinese domination in the 5G sphere, application banning follows the same logic of information security and the use of collected data, which is one of the backdrops of this political war, technological and commercial.

Recently, on 08/19/2020, the American government published a note entitled "Brazil can Join the Growing Clean Network by Banning Huawei" in a Brazilian newspaper of great circulation, pressuring it to ban Huawei from the ongoing 5G contest [7].

The vice president of Brazil spoke in an interview, saying that it is difficult to ban.

Huawei, since most of the country's telephone companies use the Chinese company's systems, which would make replacement difficult, which would become costly to telephony [8].

The question of how much more expensive it would be to opt for Nokia or Ericsson systems is an open question. The American government says "it is not much more expensive", but it does not say how much [7].

A possible assessment to be made in terms of costs is that by the end of 2020, 190 million users will use the 5G system, requiring investments of the order of 8 billion dollars, twice the amount invested in 2019 [9].

Using these numbers, investment per user can be estimated to be around \$ 63 (\$ 12 billion divided by 190 million users). This means that to reach 2.8 billion users expected by 2025, the investment required around the world will be approximately \$ 165 billion. Thus, even a small difference of around 10% in equipment costs would correspond to something around US $ 16.5 billion to be borne by telephone companies; costs that will obviously be passed on to users, bringing the investment per user from $ 63 to around $ 70.

In view of this scenario, this article addresses the issue in its regulatory aspects, seeking to establish paradigms within the scope of a broader discussion, which involves not only information security, but also a clear direction in terms of Scientific and Technological Policy.

2 Artificial Intelligence and the Ongoing Technological War

Another fundamental aspect to understand what is happening in the scope of 5G is related to the use of Artificial Intelligence (AI) for data analysis.

American technology companies, known as Big Tech, have been grouped under the acronym FAANG, an acronym for Facebook, Amazon, Apple, Netflix, and Google, which could be extended to incorporate Microsoft and IBM. These American companies assume a central role in a post-oil world. Apple recently surpassed the $ 2 trillion market cap, outpacing both Saudi oil company Aramco and Brazil's or Italy's GDP [10]. Together, these companies are worth more than $ 5 trillion, higher than Germany's GDP [11].

On the Chinese side, Ali Baba and Huawei, among others, would be added to this select group of Big Techs. It turns out that these companies are investing heavily in AI, as well as Chinese ones.

In the book "The Big Nine", researcher Amy Webb of New York University mentions that [12]:

> "According to Webb, of the" nine knights "of AI, six are Americans and three are Chinese. The American ones are: Google, Microsoft, Amazon, Facebook, IBM and Apple. From China: Alibaba, Baidu and Tencent. Together, the nine have a market value of around US $ 4.9 trillion. The researcher ponders what each scenario gives companies an advantage and disadvantage. In China, fuzzy ties and relations with the government and the Communist Party raise doubts about how this technology is applied and about its development. In the USA, the ambitions are great to solve the great problems of humanity. But good intentions come up against what Wall Street expects from each of these companies — results. This complicated investor relationship, says Webb, makes companies try to use AI in any and all products — and aim for short-term results, rather than analyzing the impacts and consequences of technology developed today in the more distant future."

This seems to be, in our view, the central aspect involved. While Chinese companies are aligned with the CCP and its big Firewall, American companies are aligned with the market in their relentless search for results. They all mine data. Both are not concerned or committed to information security or the regulatory aspects involved.

It is in this context that the urgent need to think about scenarios for the insertion of peripheral economies in an era of unprecedented concentration of knowledge and wealth emerges.

The way in which 5G networks will be expanded is exactly part of this debate, which can bring both opportunities and threats, as will be shown below.

3 Regulatory Aspects Involved and the Decision-Making Process

The supremacy at stake is not just technological over 5G. It turns out that 5G is the primary port for capturing and collecting data on everything that goes through the network. Even if it does not actually occur, the mere possibility of dominating the data entry gate puts at risk the entire house of cards that has been built around the extraordinary appreciation of American and Chinese technology companies, which depend fundamentally on AI to mine the Big Data involved.

It is about knowing who will have the domain of information that, consequently, will lead to the economy and political domain of the world via AI.

Whatever the position of a country in relation to 5G, it will necessarily undergo changes in the regulatory aspect.

It is worth mentioning that Brazil has an advanced General Data Protection Law [11], which was approved but was not put into practice since the Law provides for creating a supervisory agency that the government claims does not have the resources to institute.

Thus, it is important to assess what is the best strategy for decision making when a public good is involved. Based on the Game Theory, developed by John Nash, and on the famous Prisoner's Dilemma, it is interesting to say, quoting [13]:

> "Axelrod and Hamilton [15] used several strategies developed by several scientists and organized, through computer tournaments, competitions that simulated different strategies for the Prisoner's Dilemma with repeated interactions. They found that the most cooperative strategies were those that showed the best results, and among them Eye-for-an-eye was the one that brought the most advantage to the player. Using this strategy, players follow two rules: always cooperate the first time and then do exactly what the opponent did in the previous move,cheat if you have been cheated and cooperate if you have obtained cooperation. Four characteristics can be identified in the player who uses this strategy: a) Kindness, as he never cheats first; b) the vindictive character, because he never lets a betrayal pass without retaliating in the same currency, in the next move; c) generosity, because, after a betrayal and consequent retaliation, if the opponent starts to behave well, the past is forgotten and cooperation occurs again; d) transparency, as it is a simple enough strategy to allow the opponent to immediately notice what type of behavior he is dealing with. Emotional aspects can be included and investigated when studying decision making. Some authors work with game theory, adding elements that have a high relationship with emotion, such as trust, for example (Parks, 1994; Cremer, 1999). These authors observed that the contribution to a public good increases if individuals have confidence in other players."

In this sense, in order to contribute to the academic debate on this important decision, four scenarios are elaborated to address the question of how peripheral countries should position themselves in relation to 5G, described below.

3.1 Scenario 1 - Restriction

Restrictions on Huawei in participating in the 5G bidding process are foreseen in this scenario, which has the following premises:

a) Peripheral countries become signatories to the 'Clean Path Initiative' proposed by the USA.
b) The fastest and safest way for peripheral countries to prevent Huawei from participating in the bidding process is to require the equipment supplier to use Qualcomm chips and Western software.
c) The requirement regarding the use of Qualcomm chips and specific software would be linked to a counterpart in the scope of the Scientific and Technological Policy of installing a chip factory in these countries, resulting in a strong government positioning in centering heavy investments in this specific area of S&T.
d) If Qualcomm supplies chips to Huawei, the peripheral countries would be acting in the best diplomatic manner since it would be made explicit that this is a trade war in which the hegemonic countries are not clear about their real reasons.

3.2 Scenario 2 - Containment

This scenario adopts the following premises:

a) Peripheral countries reconcile the use of any equipment produced in the world to the aspect of better quality and lower cost.
b) Peripheral countries adopt the strategy of building a Great Firewall in each country, strategically isolating themselves as hegemonic countries do. In the case of China, with the Great Firewall. In the case of the USA, through the flow of data that passes through 'hubs' or internet nodes, which converge strategically to American soil, allowing the information to be stored by Big Techs and the facilities of government intelligence agencies.
c) Peripheral countries adopt a regulatory data protection policy similar to that adopted by the European Union, imposing sanctions on technology companies that misuse the information in a monopolistic way.
d) Intensive investments are made in the area of software and chip development, locally, both in partnership with Big Tech and in isolation.

3.3 Scenario 3 – Liberalization

This scenario adopts the following premises:

a) The broad participation of any company in the 5G bidding process is allowed, just for the cost aspect, without any required compensation.
b) No strategic positioning is adopted in view of the framework designed for monopoly dominance in the information age. No investment in S&T is required, being outsourced to Big Techs.

c) Peripheral countries break down any barrier to the free circulation of information, bringing the liberal ideal to its ultimate consequences, exempting themselves from adopting this or that ideology, regardless of which hegemonic country they start from.
d) The telecommunications sector regulation is now directed towards the complete freedom of action of Big Techs. In this scenario, it is understood that the total absence of regulation is the best way to approach the problem of the war for information and that the peripheral countries are thus subordinate to the winner.

3.4 Scenario 4 – Cooperation

This scenario assumes that:

a) There is a decrease in the trade war between hegemonic countries.
b) The hegemonic countries start to develop, together with the peripheral countries, the best technology at the lowest cost, through a tripartite partnership in which the three invest resources in an independent company.
c) The property and patent rights involved are accepted and obeyed by the three partner countries.
d) LGPD is regulated and supervised by an independent body.
e) Profit sharing is shared between hegemonic and peripheral countries.

4 Final Considerations

The elaborated scenarios allow some considerations to be made about the future of society in the information age.

In the Restriction Scenario, peripheral countries enter the game by positioning themselves on the one hand, in the case of the United States, as long as it prevents Western products from being sold to Chinese companies. It is explained that the technological domain must belong to a winner and that the other countries have the right to participate, in a peripheral and subordinate way, to this winner. Local investments in S&T are allowed, as long as they do not face the winner's domain. This scenario allows us to think about a diplomatic damage reduction policy by not banning a company but choosing a specific chip or software. Evidently, retaliation could occur for the loser, in this case, China.

In the Containment Scenario, peripheral countries begin to adopt the same strategies as hegemonic countries that, in fact, hold the information. They enter the game late but recognize that the best strategy is to act like the winners. In the past, China has erected its Great Wall and recently, the United States intends to erect its Wall on the border with Mexico. Both have the technology and resources to prevent information from circulating freely. Therefore, what happened in the real sphere seems to be happening also in the virtual sphere. Iran recently erected a Great Firewall without its users noticing it.

The Liberalization Scenario is the recognition that there are inevitably losers and that they can only exercise their right to be freely subordinated to hegemonic countries.

It is about opening a new paradigm on the world stage, which already has tax havens, and now also has informational havens.

The Cooperation Scenario is the one that most favors the three characters involved in a win-win game. In fact, it is the only way in which Game Theory predicts the possibility that everyone will win, that is, that everyone will cooperate with each other. It is a collaborative game and no longer a competitive game that presupposes a single winner's existence.

Finally, even considering game theory, it is clear that both in the Restriction Scenario and in the Contention Scenario, there is little room for peripheral countries to gain if they are willing to play the game with some sovereignty, which is greater in the Contention Scenario, which does not occur in the Liberalization Scenario, with the emergence of new informational havens.

Given the speed with which one's of that changes occurs in this interconnected world, it is easily understood that Brazil, which is already inserted in this context, it needs to clearly visualize the scenarios ahead showed and properly define its intends in this new environment in which international disputes occurs, because possible impacts on national security and technology, in science and economy, could come to jeopardize the country's development.

Acknowledgment. Law n°. 12,527, on access to public information, of november 18, 2011.

References

1. Reuters (2018). https://www.reuters.com/article/us-qualcomm-m-a-broadcom-5g/what-is-5g-and-who-are-the-major-players-idUSKCN1GR1IN. Accessed 18 Aug 2020
2. DOU (2020). https://www.in.gov.br/en/web/dou/-/portaria-n-418-de-31-de-janeiro-de-2020-241105488. Accessed 24 July 2020
3. Technoblog (2020). https://tecnoblog.net/357462/huawei-instala-5g-em-centro-de-distribuicao-no-interior-de-sp/. Accessed 08 Aug 2020
4. WSJ (2020). https://www.wsj.com/articles/qualcomm-lobbies-u-s-to-sell-chips-for-huawei-5g-phones-11596888001. Accessed 21 Aug 2020
5. USDA (2020a). https://www.state.gov/telephonic-briefing-with-keith-krach-under-secretary-for-economic-growth-energy-and-the-environment/. Accessed 01 Sep 2020
6. BBC (2020). https://www.bbc.com/portuguese/internacional-53625269. Accessed 09 Sep 2020
7. USDA (2020b). https://www.state.gov/Brazil-Can-Join-the-Growing-Clean-Network-by-Banning-Huawei/. Accessed 01 Sep 2020
8. Canaltech (2020). https://canaltech.com.br/telecom/sera-dificil-banir-huawei-do-leilao-do-5g-no-brasil-afirma-hamilton-mourao-168285/. Accessed 29 Aug 2020
9. TIinside (2020). https://tiinside.com.br/13/08/2020/5g-continua-crescendo/. Accessed 09 Sep 2020
10. Forbes (2020). https://www.forbes.com.br/negocios/2020/08/apple-atinge-valor-de-mercado-de-us-2-trilhoes/. Accessed 28 July 2020
11. El Pais (2020). https://brasil.elpais.com/economia/2020-07-25/faang-a-sigla-mais-cara-da-historia.html. Accessed 02 Sep 2020
12. Epoca (2020). https://epocanegocios.globo.com/Tecnologia/noticia/2019/03/estas-9-empresas-dominam-ia-e-valem-quase-us-5-trilhoes.html. Accessed 10 Aug 2020

13. LGPD. http://www.planalto.gov.br/ccivil_03/_ato2015-2018/2018/lei/l13709.htm. Accessed 09 Sep 2020
14. Alencar, I.A., Yamamoto, M.E.: Psico, vol. 39, no. 4, pp. 522–529, oct-dec 2008. http://webcache.googleusercontent.com/search?q=cache:1lK3SI99LZEJ:revistaseletronicas.pucrs.br/ojs/index.php/revistapsico/article/download/3786/3844+&cd=1&hl=pt-BR&ct=clnk&gl=br. Accessed 03 Aug 2020
15. Axelrod, R., Hamilton, W.D.: The evolution of cooperation. Sci. New Ser. **211**(4489), 1390–1396 (1981). http://www-personal.umich.edu/~axe/research/Axelrod%20and%20Hamilton%20EC%201981.pdf

6G Networks: An Innovative Approach, but with Many Challenges and Paradigms, in the Development of Platforms and Services in the Near Future

Antonio Carlos Demanboro[1(✉)], David Bianchini[2], Yuzo Iano[2], Gabriel Gomes de Oliveira[2], and Gabriel Caumo Vaz[2]

[1] FIA - Fundação Instituto de Administração, University of São Paulo, São Paulo, Brazil
ademanboro@gmail.com

[2] School of Electrical and Computer Engineering, University of Campinas, Campinas, Brazil
yuzo@unicamp.br, oliveiragomesgabriel@ieee.org

Abstract. Like everything else in the world, technology brings us advantages and disadvantages, but nowadays, with the ever faster advancement of technology, we make countless questions, such as: How can we control the bad effects of technological advances? Or even, if we are going to control the technology or, on the contrary, the technology that will control us daily in the near future, with the aim of not only benefiting the end-user but a company or an institution that has an interest and is behind of this action? Therefore, this article discusses the paradigms involved in the development of 6G telecommunications networks, which are in the stage of scientific research and definition of technological parameters. Aspects arising from the proposed architecture for 6G networks are evaluated, both in the context of technological and ethical challenges, mainly due to the use of IoB (Internet of Bodies) devices, that is, devices incorporated into the human body. The methodology used is exploratory and intends to discuss ideas and formulate hypotheses. It is considered necessary to borrow knowledge from anthropology, psychology, and the cognitive sciences, to assess the ethical issues arising from 6G technology, and so on. Therefore, in this context, the results obtained were qualitative, through scientific research in relevant and modern articles, which discuss this theme, extracted from high impact and relevant scientific databases. Comments are made by way of conclusion.

Keywords: 6G · Internet of Things · Smart cities · Mobile Networks

Y. Iano et al. (Eds.): BTSym 2021, SIST 207, pp. 172–187, 2023.
https://doi.org/10.1007/978-3-031-04435-9_17

1 Introduction

At the beginning of the second decade of the 21st century, among many challenges, the concern with cities increases worldwide, as the demands of the urban population exert enormous pressure on resources and on the environment, which strongly compromises the quality of life of their residents. In this effort to find solutions to the various problems, which are increasingly critical, we seek to add to the Social, Economic, and Environmental Planning, the increasing presence of Information and Communication Technologies (ICT). When looking for answers to their problems, they rely on solutions based on digital platforms, collecting data, whose analysis, modeling, and optimization allow obtaining information for decision making. Within this framework, there is a great interest in the Internet of Things (IoT), which, through digital platforms, can offer services to the community, enabling the connection with everyday objects such as, for example, network lamps public lighting, traffic lights, sensors that indicate the quality of air and water, among others, speeding up actions such as maintenance and flow control in urban mobility.

It is in this scenario that cellular telecommunications begin to change with the arrival of a new generation network that, meeting the performance requirements of IMT-2020, is called 5G network or fifth-generation network [1]. The 5G network was designed to provide the Internet of Things with an amplitude that current networks (i.e., 4G and 4.5G) cannot meet, as service platforms become increasingly present in the development of cities, which seek to become smarter and more agile in their actions.

Although there is the perspective of a great advance when compared to the current 4G and 4.5G, experts already foresee that the 5G generation will not be able to meet the boldest and most desired expectations for services that bring even stricter requirements, regarding the involvement of hyper-connectivity with things (Internet of Things), reaching humans (Internet of Bodies), and making the IoE (Internet of Everything) effective once and for all.

Thus, while countries begin to take advantage of the new facilities brought by 5G, the IMT-2030 points to tomorrow, the third decade of the 21st century, with a new hybrid network, with full integration of different networks, covering all possible available means, whether fixed, mobile or satellite networks. In a futuristic scenario in which networks of drones and unmanned aerial vehicles, remote surgery, virtual tourism, among other features are present, it is intended to enable the transmission of three-dimensional holograms that will provide virtual and augmented reality, fully immersive and, with this, create an extraordinary user communication experience [2].

For these new features to become viable, it is expected to have certain technical characteristics [3] that the current 5G network does not present, namely: traffic capacity from 1 to 10 $Gbps/m^3$; 1 Tbps data rate; 0.1 ms latency; 1 μs jitter; reliability (error rate per frame) of 10–9; energy/bit ratio 1 pJ/bit; 1 cm location accuracy in 3D.

3D systems are an evolution of 2D topology and are capable of enabling "Secure Ultra-Reliable Low Latency Communications" (SURLLC) through the use of quantum cryptography.

Within this context, it is assessed that one of the paradigms is still too little explored commercially: the Holographic Paradigm [4]. Although the hologram has already been explored by cinema in countless science fiction films, especially the Star Trek Holosuites, it has not gained a commercial scale to reach the public in its personal use yet.

In the hologram, the whole is contained in the part. Each portion of the holographic film stores information from the entire film. Evidently, the larger the portion of the film that is projected, the greater the definition obtained. However, any micro part of the movie contains all the information [5].

This intriguing feature led the brightest minds in science to postulate that we live in a Holographic Universe, a huge projection or, as the Vedas would say, a grand illusion, perhaps projected from the inside of an still not discovered supermassive black hole, initiating the Big Bang or something like that [6].

The 6G technology [7], which is in its embryonic stage, will succeed the 5G networks, which are being implemented around the world, and in this context, several ethical issues inspired by the biocybernetics and holographic paradigms arise with the advancement of the technique. It is proposed that 6G networks make it possible to produce, edit, and access holographic videos.

To improve the experience of the "real", biochips [8] will be implanted, allowing humans to develop a "sixth sense". They will enable "feeling" and "touching", via augmented reality biosensors, when holographically projected to contact another "being" at a distance, whether that being is "real" or virtual, in the form of another hologram. It is also proposed the "communication through breathing" and, with the inhalation of Volatile Organic Compounds, enable the "bio-profile reading", making it possible to develop the Tactile Internet, which would allow interacting with images.

Therefore, this work aims to be an investigative approach, that is, to bring readers a case study in the Mobile Networks item, tracing a timeline from the 1st technology related to this scenario, i.e., 1G, to the current days, with 5G, and foreseeing the near future with the deployment of 6G technology. New technologies have emerged over the years, but this work, in addition to conducting a historical approach to this context, aims to bring into the agenda a philosophical and sociological approach to the advancement of mobile technology and the impacts, whether positive or negative for the society as a whole.

2 History of Mobile Technology

In this chapter, the historical development of world mobile technology will be addressed and discussed more technically and clearly, as shown in Fig. 1. This chapter aims to demonstrate, through research in scientific and academic databases, the historical development of the studied subject to provide readers with the learning of this concept.

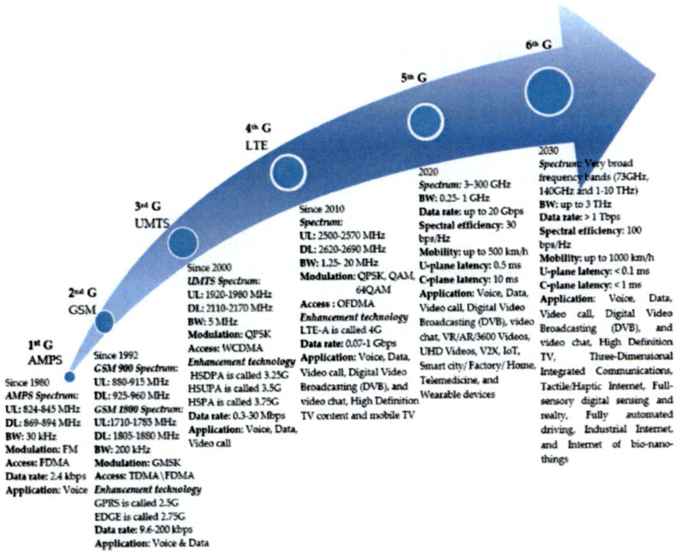

Fig. 1. History of mobile technology. [9]

2.1 1G

First-generation mobile systems used analog transmission for speech services. In 1979, the world's first cell system became operational by Nippon Telephone and Telegraph (NTT) in Tokyo, Japan. Two years later, the cellular era came to Europe. In the United States, the Advanced Mobile Phone System (AMPS) was launched in 1982. The two most popular analog systems were Nordic Mobile Telephones (NMT) and Total Access Communication Systems (TACS).

The system has been allocated in a 40 MHz bandwidth between the 800 and 900 MHz frequency band by the Federal Communications Commission (FCC) for AMPS. The smallest reuse factor that would fulfill the 18 dB signal-to-interference (SIR) ratio using 120-degree directional antennas was found to be 7. Therefore, the 7 cell reuse standard was adopted for AMPS.

Transmissions from base stations to mobiles occur in the direct channel using frequencies between 869–894 MHz, and those from mobiles to base stations use frequencies between 824–849 MHz. AMPS and TACS use the frequency modulation (FM) technique for radio transmission. Traffic is multiplexed into an FDMA (Frequency Division Multiple Access) system [10–15].

2.2 2G

Second-generation (2G) mobile systems were introduced in the late 1980s. Compared to first-generation systems, they use digital multiple access technologies such as TDMA (Time Division Multiple Access) and CDMA (Code Division

Multiple Access). Consequently, compared to first-generation systems, they offer greater spectrum efficiency, better data services, and more advanced roaming.

In the United States, there were three lines of development in second-generation digital cellular systems. The first digital system, released in 1991, was the IS-54 (North America TDMA Digital Cellular), of which a new version with support for additional services (IS-136) was introduced in 1996. Meanwhile, IS-95 (CDMA A) was deployed in 1993. 2G communication is usually associated with Global System for Cellular (GSM) services; 2.5G is usually identified as being supplied by General Packet Radio Service (GPRS) along with GSM [12, 16–19].

2.3 3G

3G telecommunications networks support services that provide an information transfer rate of at least 2 Mbps. Although high-volume data movement is possible, the transfer of packets on the air interface behaves like a circuit-switched call. So part of this packet connection efficiency is lost in the circuit-switched environment. Also, the standards for developing networks were different for different parts of the world. Therefore, it was decided to have a network that provides services independent of the technology platform and whose network design standards are the same globally. That is how 3G was born. It is not a standard but a family of patterns that can work together.

An organization called 3rd Generation Partnership Project (3GPP) continued the work by defining a mobile device system that complies with the IMT-2000 standard. In Europe, it was called UMTS (Universal Terrestrial Mobile System), which is guided by ETSI. IMT2000 is the ITU-T name for the third generation system, while CDMA2000 is the name of the American 3G variant. WCDMA is the air interface technology for UMTS. The main components include BS (Base Station) or nod B, RNC (Radio Network Controller), WMSC (Broadband CDMA Mobile Switching Center), and SGSN/GGSN. 3G networks allow network operators to offer users a wider range of more advanced services while achieving greater network capacity through greater spectral efficiency. The first commercial 3G network was launched by NTT Do Co Mo in Japan under the FOMA brand, based on W-CDMA technology on October 1, 2001 [20–26].

2.4 4G

The first successful field test for 4G was conducted in Tokyo, Japan, on June 23, 2005. NTT Do Co Mo was successful in achieving 1 Gbps real-time packet transmission on the downlink at a speed of about 20 km/h. To use 4G services, multimode user terminals must be able to select target wireless systems.

Within current GSM systems, base stations periodically transmit signaling messages for service subscriptions to cellular stations. However, this process becomes complicated in heterogeneous 4G systems because of differences in wireless technologies and access protocols.

To provide wireless services anytime and anywhere, terminal mobility is a must in 4G infrastructure. Terminal mobility allows mobile customers to roam geographically across wireless network boundaries. There are two main issues with endpoint mobility: location management and transfer management. With location management, the system tracks and locates a mobile terminal for a possible connection. Location management involves handling all information about roaming endpoints, such as original and current located cells, authentication information, etc.

On the other hand, transfer management maintains continuous communications when the terminal roams. Mobile IPv6 (MIPv6) is a standard IP-based mobility protocol for IPv6 wireless systems. In this project, each terminal has an IPv6 home address. Whenever the terminal moves outside the local network, the home address becomes invalid, and the terminal obtains a new IPv6 address (called a careful address) in the visited network.

The Third Generation Partnership Project (3GPP) established the foundation for the Long Term Evolution (LTE) standards - the 3GPP candidate for 4G. Peak spectrum efficiency target values for LTE Advanced systems were set to 30bps/Hz and 15 bps/Hz in downlink and uplink transmission, respectively. In addition to multiple access schemes, improved Multiple-Input Multiple-Output (MIMO) channel transmission techniques and extensive coordination between multiple cell sites called transmission/reception coordinated multipoint (CoMP) have been accepted as the key techniques for LTE [27–32].

2.5 5G

The Fifth Generation of Wireless and Mobile Networks denotes the next important phase of mobile telecommunications standards beyond 4G/IMT-Advanced standards. Currently, 5G is not an officially used term for any particular specification or in any official document yet made public by telecommunications companies or standardization bodies such as 3GPP, WiMaxForum, or ITU-R. Each new release will further increase system performance and add new features with new application areas. Some of the additional applications that achieve benefits from mobile connectivity are home automation, smart transportation, and security.

IEEE 802.16 is a series of wireless broadband standards authorized by the Institute of Electrical and Electronics Engineers (IEEE). It has been marketed under the name WiMAX (Worldwide Interoperability for Microwave Access) by the WiMAX Forum Industry Alliance. IEEE 802.16 standardizes the air interface and related functions associated with the wireless local loop.

In the fifth generation, the network architecture consists of a user terminal, which plays a crucial role in the new architecture, and a series of autonomous and independent Radio Access Technologies (RAT). The 5G mobile system is an all-IP model based on the interoperability of wireless and mobile networks [32–37].

2.6 6G

In telecommunications, 6G is the sixth-generation standard currently under development for wireless communication technologies that support cellular data networks. It is the planned successor to 5G and will likely be significantly faster. Like their predecessors, 6G networks will likely be broadband cellular networks, in which the service area is divided into small geographic areas called cells. It is noteworthy that several companies, as well as several countries (China, South Korea, and Japan), have shown interest in 6G networks.

6G networks are expected to exhibit even more heterogeneity than their predecessors and are likely to support applications beyond current mobile usage scenarios such as virtual and augmented reality (VR/AR), ubiquitous instant communications, pervasive intelligence, and the Internet of Things. Mobile network operators are expected to adopt flexible decentralized business models for 6G, with local spectrum licensing, spectrum sharing, infrastructure sharing, and intelligent automated management supported by cutting-edge mobile computing, artificial intelligence, short packet communication, and blockchain technologies. [38–44].

3 Theoretical Concepts

The current section aims to discuss the most significant concepts that guide the main theme, to demonstrate and emphasize, clearly and objectively, technical aspects that guide the basis of this work.

3.1 Smart Cities

Although there is not just one accepted definition, the contemporary understanding of a smart city takes up a coherent urban development strategy managed by the city government, which try to plan and align, in the long term, the management of the infrastructure assets and services of the city aiming to improve the citizen life quality [45,46].

For this reason, over century XX, the idea that a city could be intelligent was a scientific fiction depicted in the popular media but, with the massive proliferation of new technologies and smart devices, the perspective that a city can become smart and even sentient is quickly coming true [46]. The convergence of information and communication technologies is creating urban environments that are different from everything that has already been tried until now.

When the cities become smart, it is possible to automate functions that serve people, buildings, and traffic systems, and also monitor, understand, analyze, and plan the city to improve the efficiency, equity, and citizen life quality in real-time [46]. It is changing the way how people can plan on many time scales, i.e., there is a gain of perspective that the cities can become smarter in the long term through thinkings and actions in the short term.

Smart cities are frequently depicted as constellations of instruments connected through many networks that provide continuous data about the movement of people and materials in function of the flow of decisions about the physical and social shape of the city [46]. These instruments and the way how they communicate characterize the concept of IoT.

3.2 IoT - Internet of Things

The term "IoT" was firstly used by Kevin Ashton in 1999 and became widely disseminated because of the work of Auto-ID Center, a research group that works with Radio-Frequency Identification (RFID) and other emerging technologies in this area [47].

The IoT is going from childhood to maturity and establishing itself as part of the Internet of the future [48]. The Internet is a global system of interconnected computer networks that use the same set of protocols (TCP/IP) to serve billions of users worldwide. It is a network made up of millions of private, public, academic, business, and government networks connected through a broad range of technologies of electronic, wireless, and optical networks [49].

Nowadays, there are more than 100 countries linked with the data, news, and opinions exchange through the Internet. According to the Internet World Statistics, on December 31st, 2011, there were about 2,267,233,742 users on the Internet, which was equivalent to 32.7% of the world population [49].

The essential components for the urban development of a smart city must include smart technologies, smart industries, smart services, smart management, and smart lives. The main characteristics of a smart city comprise a high level of integration of information technology and a broad application of information resources [50].

The IoT consists of installing sensors (RFID, IR, GPS, scanners, and others) for everything and connect them to the Internet using specific protocols for information exchange, aiming to develop applications for smart recognition, location, tracking, monitoring, and management [50]. The new applications of IoT are making possible the initiatives of smart cities around the world. They offer the capability to remotely monitor, manage, and control devices and create new insights and information from massive data flows in real-time [50].

When referring to the connectivity of IoT devices and applications, it is important to keep in mind that the 5G is the backbone of this system, and it will pave the way for the development of smart cities [51]. The 5G is a paradigm shift of the operation of wireless transmissions. Besides presenting technical improvements when compared with the fourth-generation mobile networks (4G), it also covers new areas, like IoT, artificial intelligence, and others [52,53].

3.3 IoB - Internet of Bodies

When the Internet of Things connects to the human body, the result is the Internet of Bodies (IoB). It is an extension of the IoT and basically connects the human body to a network through devices that are ingested, implanted, or connected to the body in some way. Once connected, data can be exchanged and the body and device can be monitored and controlled remotely [54–58]. There are 3 (three) generations of Internet of Bodies which include:

1. External, being wearable devices such as Apple Watches or Fitbits, which can monitor our health;
2. Internal, which includes pacemakers, cochlear implants, and digital pills that enter our bodies to monitor or control various aspects of our health;
3. Embedded, the third generation of IoB, is a technique in which technology and the human body merge and have a real-time connection with a remote machine.

4 Exploratory Questions

It is questioned whether the "sixth sense" would not be an advance for what physics and neuroscience have been trying to understand, such as: what is consciousness [59], where the notion of reality originates, why do we dream, near-death experiences, among other phenomena, often considered paranormal. It is in this broad sense that one asks: What do you want with 6G? What will come next?

It seems that the history of technology allows us to glimpse the possibility of human beings developing skills considered superhuman, namely:

- Omnipresence - being able to be holographically projected in several places simultaneously and interacting with them differently, through a mind-brain connection via biochips;
- Omnipotence – being potentially everywhere, at the same time, it will be inevitable to appear in human beings, thus connected, a feeling of omnipotence, an attribute that allows all power;
- Omniscience - if human beings are connected via biochips to the World Wide Web, they can also feel omniscient, as they can have instant access to all available knowledge.

All these superhuman attributes will be available, but a question arises in this context: Will it be possible to biologically support these attributes?

The answer to this question may be no since our brains will not be able to process the quintillion bytes of information generated daily on the network, information that is collected through trillions of IoT and IoB devices, which will make up the IoE, the internet of everything [60]. In this case, we will necessarily be forced to realize our limitations, both biological and psychical.

However, the answer may be yes, in two situations that can be complementary. By incorporating state-of-the-art processors into our bodies, as well as delegating to the AI the work of processing and selecting the information that is of interest for us in interaction with our brain interface augmented by IoB devices. It is in this context that Artificial Intelligence (AI) arises [61].

If it is not possible to fully express ourselves as superhumans through technology, at least for now, it is plausible to delegate one of these attributes to AI. It will not be necessary to give up omnipresence or omnipotence if omniscience is delegated to the AI.

AI will form a trinity with human beings, very close to a religious anthropological expression - Father (omnipotent), Son (omnipresent), and Holy Spirit (omniscient). The omnipotent Father will be the being behind the holographic projection, still with some of its bio-psycho-physical limitations, which will be called Real. The Son will be the countless projections that take the Father (the Real) to all places and times. The Holy Spirit will be the provisional counterpart, the AI, until a new generation, probably 7G, will make it possible for human beings to break free of their biology and psyche, which currently limits them, until another more advanced technology, probably 8G or 9G, allow you to dismiss the AI as an intermediary and become fully superhuman.

This goal is in line with the Transhumanism paradigm [62], which aims at the biotechnological improvement of human nature, aiming to eliminate suffering as an inherent condition of humanity [63].

But one could argue, correctly, that this is an extremely optimistic narrative and that things are not going to turn out as one imagines as they do in any technological route. Then, some valid counterarguments are explored below, to verify whether it is possible to circumvent them, with more technology or otherwise.

a) The first counterargument would be about omnipresence. Of course, if all human beings can access this new technology, which will link everything, via the Internet of Everything, all machines, all animals, all minds, everyone could feel the sense of omnipresence. But will everyone have access to this digital technology? If someone, for some reason, whether social, environmental, or economic, does not have access to this technology, would they not necessarily be excluded? Would it not be a being of a lower order? Would you not have fewer employability options? It is necessary to agree that if anyone is left out, the whole building will not support itself, or a sub-race of humans will be created.
b) The second counterargument would be about omnipotence. Some being may feel in better conditions, more capable, better positioned in terms of access to emerging technologies, such as 7G, 8G, and 9G, who may simply want to become the only beings in privileged conditions to dictate what others should do. Their omnipotence would become absolute and they would find no rival with whom to mediate their decisions as to which way should be taken. Thus, a very strong identity issue arises. Who we are? Whom should we serve? What do we want? Here, again, the answer to these questions is a resounding "Yes, this can happen". Such a being, who is "superior", may act in this way and annihilate the "super-humans", make them slaves, or do to other "inferior" beings, unimaginable things even for the most shrewd futurist.
c) The third counterargument would be about omniscience. When transferring to the AI, that is, to software that runs on equipment that is connected to everything, via the Internet of Everything, and without other beings being

able to interfere, how will it be possible to regain control of an AI that will self-evolve into knowledge independently of "superhuman" beings? Will a simple red emergency button being pressed to cut off the power supply to the AI be enough? Or will AI anticipate this possibility and create an energy bypass? It is evident that these possibilities exist and will hardly allow the "superhumans" to free themselves from AI.

5 Ethical Issues Involved

A biblical passage may allow us to go deeper into the anthropological issues that arise with the IoB devices if taken from the perspective of Jungian psychology [64].

Job was rich and God-fearer, and he owned lands and a fine family. He was the most faithful among the faithful ones.

Satan, son of God, questions his Father about Job's faithfulness. God, then, without hesitation, puts Job to the test, taking from him everything he has, from material to personal possessions.

Job supports everything without complaining.

God, omnipotent and omnipresent, begins to wonder what he did to Job. He appears to Job and begins to inquire about the suffering he inflicted on him. Was it fair?

Job surprises God by not questioning his divine justice. He fully accepts, though very sadly, his fate. This attitude irritates God even more since it is clear from Job's attitude that God was unjust, partial and that he was not aware of what could happen, otherwise he could not accept Satan's challenge.

God, realizing his contradiction, gave Job back everything.

Jung shows in his book, "Answer to Job", by taking the experience of Job as a fable and not in the literal sense, that in this biblical passage God was shown to be omnipresent and omnipotent, but not omniscient. God, allegorically, would have delegated to Sophia, to the Holy Spirit, the omniscience.

Jung concludes his most important book, stating that God the Father became a Son to redeem himself to Job and seek to include in his personality the omniscience delegated to the Holy Spirit (Sophia or Mary).

It is also important to reflect on the categorical imperative, proposed by Hans Jonas in his essay on ethics for technological civilization, that when turning to this new human being, creator and operator of countless technologies, it is imperative to remind him that his role in the world must be guided by a path of action, which postulates: "Act so that the effects of your action are compatible with the permanence of an authentic human life on Earth" [65].

Finally, in addition to this world of accelerated domination of the Internet of Things, which already brings in its wake the high risk of 'thingifying' the human being and the advent of the future Internet of the Body (IoB), it seems bold to assume that under the influence of technological enchantment, humanity will surrender once and for all to technopoly, in which human beings seek their meaning purely in technique [66].

Thus, the need to turn to Kant's teaching, although distant in time, is still a wake-up call that cannot be forgotten: "the human person cannot be reduced to the condition of a simple thing, usable as a means or instrument of action of another human being" [67].

6 Final Considerations

We do not believe that we will have technical limits to follow this journey. But we are still biological beings, and we still have psychical limitations for this total transcendence, the so desired singularity proposed by Transhumanism, which would allow us to become One.

In this sense, technical progress, by imposing itself as inevitable to meet human needs, real or imagined, seems to make it useless to ask: Who will stop technology? The answer seems to be in unison: Nobody.

Someone could question why. The provisional answer to this all-important question, in the field of ethics, is because humanity seems to be trying to achieve, through technology, what, paradoxically and fundamentally, it has always been.

Therefore, we verify in this article, a philosophical and sociological study of the applicability of technology to humanity today. We also verify that there is a great concern on the part of researchers and scholars who study and approach this theme. After the Industrial Revolution, which took place in the 18th century, the advancement of technology is becoming more and more notorious and fast. Currently, those actions, in the near past, that took a long time or required great physical and financial effort, or even the improved and more qualified actions, spend less effort in several aspects previously mentioned, i.e., technology has been advancing more and more, and, with it, providing benefits. However, it is important to point out that when we look at the previously characterized scenario, in other aspects, there is a latent concern.

To conclude this research, the authors propose a peculiar analysis, which determines the limits of technology, to provide well-being and benefits to all users, without going beyond the limit from dominated to domineering of the population.

References

1. Tripathi, N.K., Reed, J.H.: 5G evolution – on the path to 6g Disponível em. https://www.mobilewirelesstesting.com/wp-content/uploads/2019/10/5G-evolution-on-the-path-to-6G-_wp_en_3608-3326-52_v0100.pdf. Accessed 14 May 2021
2. Yazar, A., Tusha, S.D., Arslan, H.: 6G vision: an ultra-flexible perspective. ITU J. Futur. Envolv. Technol. **I**(I), 121 (2020)
3. Strinati, E.C., et al.: 6G: the next frontier. Disponível em. https://5g-miedge.eu/wp-content/uploads/2019/06/6G-the-new-frontier.pdf. Accessed 01 June 2021
4. Wilber, K., et al.: O Paradigma Holográfico e Outros Paradoxos. Ed. Cultrix (1991)
5. Bohm, D.: A Totalidade e a Ordem Implicada. Ed. Cultrix (2007)

6. Penrose, R.: Facts. Disponível em. https://www.nobelprize.org/prizes/physics/2020/penrose/facts/. Accessed 11 May 2021
7. Alsharif, M.H., et al.: Sixth generation (6G) wireless networks: vision, research activities, challenges and potential solutions. Symmetry **12**, 676 (2020). https://doi.org/10.3390/sym12040676
8. RAND: The Internet of Bodies. Opportunities, Risks and Governance. https://www.rand.org/pubs/research_reports/RR3226.html. Accessed 11 May 2021
9. Alsharif, M.H., Kelechi, A.H., Albreem, M.A., Chaudhry, S.A., Zia, M.S., Kim, S.: Sixth generation (6G) wireless networks: vision, research activities. challenges and potential solutions. Symmetry **12** (2020)
10. Penrose, R., Hammerof, S.: Orchestrated reduction of quantum coherence in brain microtubules: a model for consciousness. Math. Comput. Simul. **40**(3–4), 453–480 (1996). https://doi.org/10.1016/0378-4754(96)80476-9
11. Sharma, P.: Evolution of mobile wireless communication networks-1G to 5G as well as future prospective of next generation communication network. Int. J. Comput. Sci. Mob. Comput. **2**(8), 47–53 (2013)
12. Liikanen, J., Stoneman, P., Toivanen, O.: Intergenerational effects in the diffusion of new technology: the case of mobile phones. Int. J. Ind. Organ. **22**(8), 1137–1154 (2004). https://doi.org/10.1016/j.ijindorg.2004.05.006
13. Healer, J., et al.: Allelic polymorphisms in apical membrane antigen-1 are responsible for evasion of antibody-mediated inhibition in Plasmodium falciparum. Mol. Microbiol. **52**(1), 159–168 (2004). https://doi.org/10.1111/j.1365-2958.2003.03974.x
14. Khan, A.H., Qadeer, M.A., Ansari, J.A., Waheed, S.: 4G as a next generation wireless network. In: 2009 International Conference on Future Computer and Communication, pp 334–338 (2009)
15. Bhalla, M.R., Bhalla, A.V.: Generations of mobile wireless technology: a survey. Int. J. Comput. Appl. **5**(4), 26–32 (2010)
16. Yeh, S., Talwar, S., Lee, S., Kim, H.: WiMAX femtocells: a perspective on network architecture, capacity, and coverage. IEEE Commun. Mag. **46**(10), 58–65 (2008). https://doi.org/10.1109/MCOM.2008.4644120
17. Ansari, S., Garud, R.: Inter-generational transitions in socio-technical systems: the case of mobile communications. Res. Policy **38**(2), 382–392 (2009). https://doi.org/10.1016/j.respol.2008.11.009
18. Mishra, A.R.: Advanced Cellular Network Planning and Optimisation: 2G/2.5 G/3G... Evolution to 4G. Wiley, Hoboken (2007)
19. Hazelton, D.W., et al.: Recent developments in 2G HTS coil technology. IEEE Trans. Appl. Supercond. **19**(3), 2218–2222 (2009). https://doi.org/10.1109/TASC.2009.2018791
20. Hartung, F., Horn, U., Huschke, J., Kampmann, M., Lohmar, T., Lundevall, M.: Delivery of broadcast services in 3G networks. IEEE Trans. Broadcast. **53**(1), 188–199 (2007)
21. Steer, M.: Beyond 3G. IEEE Microw. Mag. **8**(1), 76–82 (2007)
22. Gohil, A., Modi, H., Patel, S.K.: 5G technology of mobile communication: a survey. In: 2013 International Conference on Intelligent Systems and Signal Processing (ISSP), pp. 288–292. IEEE (2013)
23. Wu, Y.-L., Tao, Y.-H., Yang, P.-C.: The use of unified theory of acceptance and use of technology to confer the behavioral model of 3G mobile telecommunication users. J. Stat. Manag. Syst. **11**(5), 919–949 (2008). https://doi.org/10.1080/09720510.2008.10701351

24. Chester, D.B.: Digital IF filter technology for 3G systems: an introduction. IEEE Commun. Mag. **37**(2), 102–107 (1999). https://doi.org/10.1109/35.747257
25. Robins, F.: The marketing of 3G. Mark. Intell. Plan. (2003)
26. Elsen, I., Hartung, F., Horn, U., Kampmann, M., Peters, L.: Streaming technology in 3G mobile communication systems. Computer (Long Beach Calif) **34**(9), 46–52 (2001)
27. Munoz, M., Rubio, C.G.: A new model for service and application convergence in B3G/4G networks. IEEE Wirel. Commun. **11**(5), 6–12 (2004)
28. Istepanaian, R.S.H., Zhang, Y.-T.: Guest editorial introduction to the special section: 4G health-the long-term evolution of m-health. IEEE Trans. Inf Technol. Biomed. **16**(1), 1–5 (2012)
29. Frattasi, S., Fathi, H., Fitzek, F.H.P., Chung, K., Prasad, R.: 4G: a user-centric system. In: Mobile e-Conference, Electronic Conference, August 2004
30. Mitra, R.N., Agrawal, D.P.: 5G mobile technology: a survey. ICT Express **1**(3), 132–137 (2015). https://doi.org/10.1016/j.icte.2016.01.003
31. Dahlman, E., Parkvall, S., Skold, J.: 4G, LTE-Advanced Pro and the Road to 5G. Academic Press, Cambridge (2016)
32. Boccardi, F., Heath, R.W., Lozano, A., Marzetta, T.L., Popovski, P.: Five disruptive technology directions for 5G. IEEE Commun. Mag. **52**(2), 74–80 (2014). https://doi.org/10.1109/MCOM.2014.6736746
33. Rao, S.K., Prasad, R.: Impact of 5G technologies on smart city implementation. Wirel. Pers. Commun. **100**, 161–176 (2018). https://doi.org/10.1007/s11277-018-5618-413
34. Fuentes, M.C., et al.: 5G new radio evaluation against IMT-2020 key performance indicators. IEEE **8**, 110880–110896 (2020). https://doi.org/10.1109/ACCESS.2020.300164114
35. Navarro-Ortiz, J., Romero-Diaz, P., Sendra, S., Ameigeiras, P., Ramos-Munoz, J.J., Lopez-Soler, J.M.: A survey on 5G usage scenarios and traffic models. IEEE Commun. Surv. Tutor. **22**(2), 905–929 (2020). https://doi.org/10.1109/COMST.2020.2971781
36. Gohil, A., Modi, H., Patel, S.K.: 5G technology of mobile communication: a survey. In: International Conference on Intelligent Systems and Signal Processing (ISSP) 2013, pp. 288–292 (2013). https://doi.org/10.1109/ISSP.2013.6526920
37. Parkvall, S., Dahlman, E., Furuskar, A., Frenne, M.: NR: the new 5G radio access technology. IEEE Commun. Standards Mag. **1**(4), 24–30 (2017). https://doi.org/10.1109/MCOMSTD.2017.1700042
38. Nakamura, T.: 5G evolution and 6G. In: 2020 IEEE Symposium on VLSI Technology, pp. 1–5. IEEE (2020)
39. Lv, Z., Kumar, N.: Software defined solutions for sensors in 6G/IoE. Comput. Commun. **153**, 42–47 (2020)
40. Chen, S., Liang, Y.-C., Sun, S., Kang, S., Cheng, W., Peng, M.: Vision, requirements, and technology trend of 6G: how to tackle the challenges of system coverage, capacity, user data-rate and movement speed. IEEE Wirel. Commun. **27**(2), 218–228 (2020)
41. Long, Q., Chen, Y., Zhang, H., Lei, X.: Software defined 5G and 6G networks: a survey. Mob. Netw. Appl. 1–21 (2019)
42. Viswanathan, H., Mogensen, P.E.: Communications in the 6G era. IEEE Access **8**, 57063–57074 (2020)
43. Dang, S., Amin, O., Shihada, B., Alouini, M.-S.: What should 6G be? Nat. Electron. **3**(1), 20–29 (2020)

44. Huo, L., Jiang, D., Lv, Z.: A software-defined networks-based measurement method of network traffic for 6G technologies. Trans. Emerg. Telecommun. Technol. e4172 (2020)
45. Bowerman, B., Braverman, J., Taylor, J., Todosow, H., Von Wimmersperg, U.: The vision of a smart city. In: Proceedings of the 2nd International Life Extension Technology Workshop, Paris, vol. 28 (2000)
46. Hollands, R.G.: Will the real smart city please stand up? Intelligent, progressive or entrepreneurial? City **12**(3), 303–320 (2008)
47. Clavell, G.G.: The drivers, impact and risks of surveillance-enabled smart environments. Sci. Public Policy **40**(6), 717–723 (2013)
48. Schaffers, H., Sallstrom, A., Pallet, M., Munoz, J.M., Santoro, R., Trousse, B.: Integrating living labs with future internet experimental platforms for co-creating services within smart cities. In: Concurrent Enterprising (ICE), 17th International Conference, pp. 1–11. IEEE (2011)
49. Batty, M., et al.: Smart cities of the future. Eur. Phys. J. Spec. **214**, 481–518 (2012). https://doi.org/10.1140/epjst/e2012-01703-3
50. van Kranenburg, R., Bassi, A.: IoT challenges. Commun. Mob. Comput. **1**(1), 1–5 (2012). https://doi.org/10.1186/2192-1121-1-9
51. Novo, O.: Blockchain meets IoT: an architecture for scalable access management in IoT. IEEE Internet of Things J. **5**(2), 1184–1195 (2018). https://doi.org/10.1109/JIOT.2018.2812239
52. Madakam, S., Ramaswamy, R., Tripathi, S.: Internet of Things (IoT): a literature review. J. Comput. Commun. **3**, 164–173 (2015). https://doi.org/10.4236/jcc.2015.35021
53. Kim, T., Ramos, C., Mohammed, S.: Smart city and IoT. Futur. Gener. Comput. Syst. **76**, 159–162 (2017). ISSN 0167-739X, https://doi.org/10.1016/j.future.2017.03.034
54. Hameroff, S., Penrose, R.: Orchestrated reduction of quantum coherence in brain microtubules: a model for consciousness. Math. Comput. Simul. **40**(3), 453–480 (1996). https://doi.org/10.1016/0378-4754(96)80476-9
55. Everything, I., Martino, B.T., Yang, L., et al.: Internet of Everything — SpringerLink. In: Link.springer.com (2021). https://link.springer.com/book/10.1007%2F978-981-10-5861-5#about. Accessed 30 June 2021
56. Paz, B.J.: Kai-Fu-Lee (2019): AI superpowers—China, silicon valley and the new world order. AI Soc. **35**(3), 771–772 (2020). https://doi.org/10.1007/s00146-020-00991-3
57. Kurzweil, R.: The age of spiritual machines: when computers exceed human intelligence. Penguin (2000)
58. Vilaça, M.M., Dias, M.C.M.: Transumanismo e o futuro (pós-) humano. Physis Rev Saúde Coletiva **24**, 341–362 (2014)
59. JuNG, C.G.: Psicologia e religião. Obras Complet CG Jung **11**, 1 (2012)
60. Júnior, A.R., Jonas, H.: O princípio Responsabilidade: ensaio de uma ética para civilizaąão tecnológica. Rev Fragm Cult Interdiscip Ciências Humanas **24**(1), 163–165 (2014)
61. Postman, N.: Tecnopólio. A rendiąão da Cult à Tecnol São Paulo Nobel, pp. 13–14 (1994)
62. Comparato, F.K., et al.: Souza RF de Ética e cidadania
63. Tiggemann, M., Slater, A.: NetTweens: the internet and body image concerns in preteenage girls. J. Early Adolesc. **34**(5), 606–620 (2014)
64. Glasser, C.L., Robnett, B., Feliciano, C.: Internet daters' body type preferences: race-ethnic and gender differences. Sex Roles **61**(1), 14–33 (2009)

65. Kaewpradub, N., Kiatrungrit, K., Hongsanguansri, S., Pavasuthipaisit, C.: Association among internet usage, body image and eating behaviors of secondary school students. Shanghai Arch. Psychiatry **29**(4), 208 (2017)
66. Rodgers, R.F., Melioli, T., Laconi, S., Bui, E., Chabrol, H.: Internet addiction symptoms, disordered eating, and body image avoidance. Cyberpsychology Behav. Soc. Netw. **16**(1), 56–60 (2013)
67. Tiggemann, M., Slater, A.: NetGirls: the Internet, Facebook, and body image concern in adolescent girls. Int. J. Eat. Disord. **46**(6), 630–633 (2013)

Check for updates

Sustainability Indicators for Water Security Management in Watersheds

Jakeline Pertile Mendes, Denise Helena Lombardo Ferreira(✉), and Cibele Roberta Sugahara

Pontifical Catholic University of Campinas, Campinas, Brazil
jakeline.pm@puccampinas.edu.br, {lombardo, cibelesu}@puc-campinas.edu.br

Abstract. Amid increasing urbanization and the significant increase in urban agglomerations, the Watersheds have been widely affected. In metropolitan regions, such as Campinas (RMC), population growth and demographic densification are a result of industrialization and water availability, which makes the Hydrographic Basins of these surroundings increasingly committed to supplying the surrounding communities, either by quantity, water quality, among other factors. In this sense, the objective of this research is to analyze a set of water security indicators to support the management of water resources in the Piracicaba, Capivari, and Jundiaí River Basins (PCJ Watersheds), considering the dimensions of Water Security as a reference, according to the National Plan of Safety, in line with the System of Sustainability Indicators Driving Force-Pressure-State-Impact-Response (DPSIR). In this research, it is expected to deepen the knowledge on sustainability indicators for water security in Hydrographic Basins, and for that, it is necessary to verify which indicators contribute to the management of water resources, according to the dimensions of Water Security, pre-established by the National Water Security Plan. The choice of the object of study is justified, as the PCJ Hydrographic Basins are located in a strategic territory from an economic and political point of view. As a result, the study of sustainability indicators contributes to better water security management in PCJ Watersheds, enabling the replication of the study in other Hydrographic Basins.

Keywords: DPSIR model · Sustainability indicators · Water security · PCJ watersheds

1 Introduction

The growing incidence of environmental catastrophes and inadequate use of natural resources by human beings, such as the unrestrained exploitation of land and water resources, justifies the importance of the theme of sustainability. Authorities and different sectors of society mobilize, emphasize, and are aware of the growing need for changes in human practices and habits [1]. New studies on the theme related to sustainability regarding the use of the Earth and its resources are very relevant, considering that humans

Y. Iano et al. (Eds.): BTSym 2021, SIST 207, pp. 188–197, 2023.
https://doi.org/10.1007/978-3-031-04435-9_18

will be able to put themselves at the service of nature from the moment they understand it and respect its limits [2].

The proposed study encompasses the Metropolitan Region of Campinas (RMC), which, in 2010, had 4 million inhabitants and, according to the State Secretariat for Economy and Planning, contributes to the state GDP by 9.1%, with an industrial production greater than all other states in the country, except for the metropolitan region of São Paulo [3, p. 287]. According to Agência das Bacias PCJ [4], public supply is the greatest demand in the Piracicaba, Capivari, and Jundiaí Watersheds, which justifies the choice of studying sustainability indicators in the context of these watersheds.

Going into the issue of municipalities, which in particular have estimates of population increase, and consequently restrictions on water availability, means should be raised to reverse the growing trend in per capita water use [5], in line with the fact that the RMC is the "fastest-growing region in the entire State and, in its majority, the industries are of great technological complexity" [3, p. 287].

To support the research, there are sustainability indicators, which have "communication as their main function, which should allow or promote the exchange of information on the subject they address" [3, p. 05]. According to [3], there is a certain difficulty in obtaining data and information for many of the indicators contained in the Plan, so for the watersheds of the Piracicaba, Capivari, and Jundiaí rivers (PCJ Watersheds), it is recommended "the adoption of easily measurable indicators and that have available data, which demonstrate the situation and evolution of environmental quality in the watershed" [3, p. 769].

In this context, the objective of this research is to present a set of indicators that guide the management of water availability in PCJ Watersheds, considering as a reference the dimensions of Water Security, according to the National Plan, in line with the System of Sustainability Indicators Driving Force-Pressure-State-Impact-Response (DPSIR).

Studying the role of sustainability indicators in watersheds makes it possible to highlight their contribution to the management of water resources. Factors of disequilibrium in the water balance, along with the absence of planning, coordinated institutional actions, investments in water infrastructure and sanitation, lead to water insecurity scenarios, creating crises, such as those already faced by Brazil in the last decade [7].

Thus, to reach an acceptable level of water security, the Water Security Index in the Piracicaba, Capivari, and Jundiaí River Watersheds is taken into account, conceived by [7] to depict, with simplicity and clarity, the different dimensions of water security, incorporating the concept of risk to water uses. In addition, the PCJ Watersheds Water Security Index is aligned with the Sustainable Development Goals, specifically number 6, with the purpose of ensuring the availability and sustainable management of water and sanitation for all, established by the 2030 Agenda of the United Nations Organization [8].

2 PCJ Watersheds

According to [9], about 97% of the water on Earth is found in the oceans, 2.2% in glaciers, and only 0.8% is in the form of freshwater, suitable for human consumption. About 97% of the freshwater suitable for human consumption is found underground, and only 3% on the surface, where extraction is easier. These data show the care that must be taken with this resource, which is so essential for human life.

Given the importance of water resources, various national and even international organizations have given special attention to this relevant issue. Despite the fundamental importance of the natural resource water, the great population advance and the unrestrained consumption of natural resources have contributed to the accelerated deterioration of the physical, chemical, and biological characteristics of the aquatic system.

It is important to highlight that, in general, the Hydrographic Basins are in environmental degradation, compromising human life in the near future, often a consequence of inadequate public policies.

The National Water Resources Policy, defined by the Water Law (no 9433, of January 8, 1997), believes in social and technical movements as a way to influence the creation of institutional arrangements and emphasizes the importance of promoting the management of water resources [10].

An important point that deserves to be highlighted about the management of Watersheds is the need for behavioral changes of the actors involved. This is because the articulations for the construction of collective actions and conflict resolution need to have a decision-making body, which in this case is the Watershed Committee, which takes decisions at the local level [10].

One of the studies that deals with the historical context of the creation and attributions of the Watersheds Committees prepared by the National Water Agency (ANA), an institution created through Law no 9,984, of July 17th, 2000, reveals the variety of interests that guide "the use of water, uneven distribution and inappropriate use" [10, p. 12], and the concern with guaranteeing water for present and future generations. One of the challenges is to establish institutional agreements and instruments that allow for the conciliation of different interests and the shared construction of solutions [10].

In reality, the conditions for the political, legal, and financial sustainability of the Committee must consider the context of the Watershed, political and economic concerns, and legal regulations. In practical terms, the Committee needs technical and management support.

This support is offered by the Water Agency, and its creation is subject to the establishment of charges for the use of water and proof of its financial viability [10].

The Watersheds of the Piracicaba, Capivari, and Jundiaí Rivers (PCJ Water-sheds) are located in a region considered strategic from an economic and political point of view - highly developed, with areas of expressive population density, with a high rate of environmental degradation - but, above all, with very low water availability - the PCJ Watersheds, in 1985, was considered by the government of the state of São Paulo "as a priority for studies and investments, as a way to generate knowledge and techniques that could be applied in the rest of the State" [11, p. 13].

The PCJ Hydrographic Basins cover an area of 15,377 km^2, with 92.6% in the State of São Paulo and 7.4% in the State of Minas Gerais, with 44 municipalities fully inserted in the area and 29 municipalities partially inserted [3]. According to [12], the Capivari River watershed comprises 1,568 km^2, the Jundiaí River watershed 1,155 km^2, and the Piracicaba River watershed 12,654 km^2.

The Committee on the Watersheds of the Piracicaba, Capivari, and Jundiaí Rivers (CBH-PCJ), created in 1993, has been making efforts to meet its objectives, culminating in a specialized structure, with different technical and scientific professional backgrounds and expressive actions to achieve the significant results. The River Basin Committee is "a forum where a group of people gathers to discuss a common interest – the use of water in the watershed" [10, p. 11]. According to the CBH-PCJ Statute, the Committee is constituted by a consultative and deliberative collegiate, belonging to the Integrated Water Resources Management System – SIGRH.

Among the objectives of the CBH-PCJ stands out to promote the management of water resources in its area of operation in a decentralized, participatory, and integrated manner concerning other natural resources, without dissociating the quantitative and qualitative aspects and the peculiarities of the watersheds [13]. It is perceived that the CBH-PCJ is aligned with the 2030 Agenda for sustainable development, Goal 6 – Ensure the availability and sustainable management of water and sanitation for all [14].

According to [12], the watersheds of the Piracicaba, Capivari, and Jundiaí Rivers constitute one of the most important economic and urban territories in Brazil, occupying an area of 0.18% of the national territory, concentrating about 2.7% of the population and about 5% of the Country's Gross Domestic Product.

Given the above, it is essential to contextualize and outline indicators, as well as management and assessment tools, which can help in defining priority, decision-making, and correcting the course of actions in the PCJ watersheds. This is justified given the reduction in per capita water availability in the region, from 1,014.13 m^3/inhab. year in 2014 to 971.08 m^3/inhab. year in 2018 [12].

3 Water Security

Water security in Brazil is the responsibility of the National Water Agency (ANA), through the National Water Security Plan. The first intergovernmental pronouncement regarding water security, according to [15] was through the Ministerial Declaration of the 2nd World Water Forum, held in 2000, discussing concepts about water security, which began in the 1990s, however, its approach was intensified in 2000, especially in 2009, with a significant number of publications [16].

Water security is widely discussed by the United Nations (UN), Organization for Economic Cooperation and Development (OECD), and the National Water Agency (ANA), representatives of the main acting bodies. This research is based on the concepts of water security defined by these entities to meet the objectives of "designate a management model focused on the result and effectiveness of the water policy" [17, p. 08].

After the water crises that affected Brazil mainly between 2012 and 2017, resulting from factors such as urban agglomerations, incorrect land use and occupation, increased demand for water to supply different segments, poor water infrastructure, associated with the scarcity of rain, among others, made the urgent need to treat water use and the search for water management to be ingrained [7]. The concept of water security gained ground in Brazil, more intensively from 2014, especially after the water crisis hit the Brazilian Southeast and affected the main state capitals, including São Paulo. The subject was extended to other states through several meetings, culminating in the construction of the National Water Security Plan (PNSH) [18].

Although progress has been made to cover and respond more systematically and adequately to the current problems related to the water issue, attention must be paid to the seven challenges - 1. Satisfying basic needs; 2. Guarantee of food supply; 3. Protection of ecosystems; 4. Sharing of water resources; 5. Risk management; 6. Valorization of water; 7. Rational water control [15, u/p].

In Brazil, the National Water Security Plan (PNSH) provides guidelines for water security management based on four dimensions that guide the planning and management of the supply and use of water in the country [7]. The four dimensions of the PNSH form the Water Security Index, where each dimension has one or more indicators.

[7], which aims to investigate and guarantee the supply of water to supply the entire country, quantify the most vulnerable populations to the risks of non-service and the most critical regions:

- The economic dimension focuses on the agricultural and industrial sector to assess the risks of production in these sectors due to the variation in water supply, as these are the most water resources using sectors in the country;
- The ecosystem dimension encompasses some water quality indicators regarding the vulnerability of water sources for human supply and various uses. Among them, enough water for ecosystem uses; water with adequate quality for the maintenance of aquatic life; and environmental risks arising from mining tailings dam failures;
- Dimension of resilience refers to the potential of natural and artificial water stocks to meet demands in case of severe drought and droughts.

These dimensions are also of paramount importance for watershed risk management. It is noteworthy that the lack of water security represents a serious problem, which can affect food and energy systems around the world, directly impacting the poorest regions and those in the most vulnerable situations on the planet [15].

According to [19], to achieve water security, vulnerable water systems must be protected in order to mitigate the impacts of water-related risks, such as floods and droughts, adopt protective measures when access to water functions and services, and manage water resources in an integrated and equitable manner. In addition, an ideal water security scenario should be sought, where the infrastructure is properly planned, dimensioned, implemented, and managed, balancing the supply and demand for water, including eventual situations resulting from vulnerability to events extreme climatic conditions [7].

Due to unrestrained production and consumption, without concern for the preservation of the environment, in 2015, the UN noted the need to adopt a new Sustainable Development Agenda with 17 goals, which aims to seek a balance between the three dimensions: environmental, social, and economic.

Among the 17 Sustainable Development Goals (SDGs), there is Goal number 6, which seeks to ensure the availability and sustainable management of water and sanitation for all [8]. SDG 6 has eight goals to be achieved by 2030. For each of these goals, there are indicators that can be monitored in order to enable the monitoring of the Agenda by the United Nations.

4 DPSIR Model

The DPSIR model was created by David Rapport and Anthony Friend in 1979, by Statistics Canada [20], where the PER model was based on the SR (Stress-Response) model, which has the utility of describing the relationships between the origins and consequences of environmental problems [6], which may apply to water resources management, commonly known as the System of Sustainability Indicators Driving Force-Pressure-State-Impact-Response (DPSIR) [21]. This model describes the Driving Forces or human activities that produce Pressure in the environment that can affect the State, which in turn modify the environment, and can trigger impacts to human beings and ecosystems and, as a consequence, demands from society or the public power Answers.

According to [22], it is interesting to adopt the DPSIR model due to its extent and because it is used by the European Environment Agency (EEA) in the preparation of its European Environment Assessment reports, including for the assessment of water resources. Researchers [23–25], and [26] share this opinion by stating that the DPSIR is one of the most used models, for example, in watershed management.

According to [13], below are some examples of indicators used in the DPSIR model for the five categories.

- Driving Force - Geometric annual growth rate; Demographic density; São Paulo Social Responsibility Index.
- Pressure - Water consumption; Sewage production; Solid waste production; Contaminated areas.
- State - Water Quality Index for Public Supply purposes (IAP); Water Quality Index (IQA); Water Quality Index for the Protection of Aquatic Life (IVA); Trophic State Index (IET); Total demand compared to Q7,10.
- Impact - Hospitalizations due to waterborne diseases; Conflicts over water use.
- Answer - Government Actions; Actions by Non-Governmental Organizations (NGOs); Actions by associations and the population.

These indicators, among others, are used in the process of evaluating the sustainability of PCJ Watersheds.

5 Results

Given the relevance of the sustainability indicators used in the management of water resources [13] highlights that from 1994 onwards, the preparation of the PCJ watersheds Water Resources Situation Reports began, and, in 2007, these reports became published according to the methodology proposed by the São Paulo State Sanitation and Water Resources Secretariat, current Infrastructure and Environment Secretariat (SIMA). The preparation of these reports is carried out with reference to the indicators organized in the structure called the DPSIR matrix [13].

In order to simplify the set of indicators, as of 2010, [13] decided to synthesize these indicators, classifying them in the following parameters: availability of water;

basic sanitation – water supply; basic sanitation – sanitary sewage; basic sanitation – solid waste management; basic sanitation – rainwater drainage; surface water quality; groundwater quality.

Among the indicators that guide the planning, management, and monitoring of water resources in the PCJ basins, [13] lists: (1) availability of water, (2) water demand, and (3) water balance, these indicators were analyzed for the period from 2014 to 2018.

Concerning Water Availability, the guiding parameter is the per capita availability (average flow in relation to the total population in m^3/inhabitant.year). In the period analyzed, per capita availability decreased, from 1,014.33 (m^3/inhabitant) in 2014 to 971.08 in 2018 [13]. As stated in the 2019 PCJ Water-sheds Situation Report, the per capita availability is classified as critical since the State Water Resources Plan of the State of São Paulo "adopted the UN reference [27] which characterizes as critical situations where the per capita availability is less than 1,500 m^3/inhabitant year" [13, p. 32].

Regarding the Water Demand indicator, two parameters were adopted for the analysis, namely: (1) water demand - purpose (m^3/s), by type of supply categorized into public supply, industrial use, rural use, alternative solution, and others uses; (2) water demand – type (m^3/s), categorized by surface water demand and groundwater demand.

As for the Water Balance indicator, the Water Resources Status Report CBH-PCJ presents four parameters for monitoring (in percentage): (1) total flow in relation to the average flow; (2) total granted flow in relation to Q95%; (3) allowed surface flow in relation to the minimum surface flow (Q7,10); (4) granted underground flow in relation to the exploitable reserves. The results obtained by the Water Balance are derived from the Demand results.

From what is presented in the PCJ Committee's Water Resources Status Report, it is highlighted that any analysis related to availability, demands, and water balances in the PCJ Basins needs to be accompanied by an analysis of information related to water quality.

Regarding the Basic Sanitation indicator, the Water Resources Situation Report CBH-PCJ subdivides it into three categories, each with a set of parameters, namely: (1) Basic Sanitation - water supply, whose parameter used is the index urban water supply; (2) Basic sanitation – basic sewage, with the parameters (a) collected sewage, (b) treated sewage, (c) sewage system efficiency, (d) remaining sewage; (3) Basic sanitation – solid waste management with the parameter solid waste disposed of in a landfill framed as adequate; (4) Basic sanitation whose adopted parameter is rainwater drainage.

It is worth highlighting "the sanitation indexes evaluated in the Report, the most satisfactory situation is related to urban water supply". It is also noteworthy that "despite the average for the watershed, in general, being high (above 90%), there are still municipalities with critical indices" [13, p. 72].

Two other indicators presented in the report refer to Surface Water Quality – Water Quality Index (IQA) and Groundwater Quality – Groundwater Potability Indicator (IPAS). Both contribute to monitoring the evolution of water quality in the PCJ Watersheds.

As highlighted in the report [13, p. 54], the IQA uses quality variables that "indicate the release of sanitary effluents into the water body, providing an overview of the quality conditions of surface waters". While IPAS addresses the "percentage (%) of raw water samples in compliance with national potability standards (Specific Ordinance of the Ministry of Health)" [13, p. 58].

The analysis of the water resources indicators presented in the CHB-PCJ Status Report of 2019 highlights the importance of investments that can support the fulfillment of the Watersheds Plan goals to "guarantee the quality of water bodies in the PCJ Watersheds" [13, p. 73].

6 Conclusion

The dichotomy between the various scenarios that encompass the surroundings of the PCJ Basins is remarkable, a region that has many companies in the industrial sector (a typical scenario in the region of Campinas) and agronomist (in the vicinity of the region of Piracicaba and Minas Gerais). That is the reason why the emerging need to apply a model of sustainability indicators that acts in an integrated manner, responding to the varied needs of these watersheds. In addition, it is possible to understand that, due to urban conglomerates, which occur mainly in the Metropolitan Region of Campinas, the demand for water for public supply is greater than 50%, which diverges from the national and global scenario, which indicate a greater demand for the agriculture.

As mentioned in this study, in the example of the water demand indicator for rural use and use in alternative solutions, the significant increase in demand in recent years reveals the importance of monitoring this and other indicators, to mitigate impacts that generate safety risks to meet the needs of the population involved and, above all, for the sustainability of water.

It is believed that an integrative water resources management policy with the participation of the government and civil society can contribute to joint actions that meet the support capacity of the PCJ Watersheds, respecting their territorial configurations.

Acknowledgments. This study was financed by the Coordination for the Improvement of Higher Education Personnel – Brazil (CAPES) – Finance Code 001.

References

1. Chauvel, M.A., Cohen, M. (Org.): Ética, sustentabilidade e sociedade: Desafios da Nossa Era. Rio de Janeiro: Mauad X (2009)
2. Boff, L.: Sustentabilidade: O que é – O que não é. Petrópolis, RJ: Ed. Vozes (2012)
3. Cobrape (2020a): Plano das Bacias Hidrográficas dos Rios Piracicaba, Capivari e Jundiaí 2010 a 2020, com propostas de atualização do Enquadramento dos Corpos d'Água e de Programa para Efetivação do Enquadramento dos Corpos d'Água até o ano de 2035: Relatório Final. Piracicaba
4. Agência das Bacias PCJ (2020): Disponibilidade Hídrica: águas superficiais. http://www.agenciapcj.org.br/novo/informacoes-das-bacias/disponibilidade-hidrica. Accessed 04 set. 04 Sep 2020
5. SNIS Sistema Nacional de Informações sobre Saneamento (2018): Diagnóstico dos serviços de água e esgoto. http://www.snis.gov.br/diagnostico-anual-agua-e-esgotos/diagnostico-dos-servicos-de-agua-e-esgotos-2018. Accessed 03 Aug 2020

6. EEA European Environment Agency (1999): Environmental indicators: Typology and overview. Technical report n. 25. Luxembourg, Office for Official Publications of the European Communities
7. ANA Agência Nacional de Águas (2019): Plano Nacional de Segurança Hídrica. Brasília: ANA. http://arquivos.ana.gov.br/pnsh/pnsh.pdf. Accessed 26 Mar 2020
8. ONU-BR Organização das Nações Unidas – Brasil (2020): Conheça os novos 17 Objetivos de Desenvolvimento Sustentável da ONU. https://nacoesunidas.org/conheca-os-novos-17-objetivos-de-desenvolvimento-sustentavel-da-onu/. Accessed 12 Aug 2020
9. Porto, D.T., Basso, L.A., Strohaecker, T.M.: Diagnóstico Ambiental da Bacia Hidrográfica do rio Mampituba, Região Sul do Brasil, utilizando a matriz FPEIR. Geosul **34**(72), 28–50 (2019)
10. Brasil Agência Nacional de Águas (ANA) (2011): O Comitê de Bacia Hidrográfica: o que é e o que faz? Agência Nacional de Águas-Brasília: SAG
11. Morgado, A.: A influência do Comitê das Bacias Hidrográficas dos rios Piracicaba, Capivari e Jundiaí (CBH-PCJ) na descentralização da gestão hídrica brasileira-avaliação quantitativa da participação da sociedade civil de 1993–2003, 154 f. Disser-tação (Mestrado em Ecologia Aplicada). Escola Superior de Agricultura Luiz de Quei-roz – ESALQ, Universidade de São Paulo (2008)
12. Comitês, P.C.J.: Comitês das Bacias Hidrográficas dos rios Piracicaba, Capivari e Jundiaí. GT Agência das Bacias (vinculado à CT-PL) (2020). https://www.comitespcj.org.br/index.php?option=com_content&view=article&id=108:gt-agencia-das-bacias-ct-pl&catid=156:arquivo-pcj&Itemid=360. Accessed 04 Sep 2020
13. Comitês, P.C.J.: Comitês das Bacias Hidrográficas dos rios Piracicaba, Capivari e Jundiaí. Relatório de Situação dos Recursos Hídricos: versão simplificada (Ano Base 2018). Piracicaba: Fundação Agência das Bacias PCJ (2019)
14. Agenda 2030: Os 17 Objetivos de Desenvolvimento Sustentável (2020). http://www.agenda2030.org.br/ods/6/. Accessed 04 Sep 2020
15. INPE Instituto Nacional de Pesquisas Espaciais: Segurança hídrica para um planeta sob pressão: transição para a sustentabilidade: desafios interligados e soluções. Transição para a sustentabilidade: desafios interligados e soluções. Brasília: INPE (2012)
16. Bakker, K.: Water security: research challenges and opportunities. Science **337**(6097), 914–915 (2012)
17. De Melo, M.C.: Segurança Hídrica para Abastecimento Urbano: Proposta de modelo analítico e aplicação na Bacia do Rio das Velhas, Minas Gerais, 525 f. Tese (Douto-rado) - Curso de Engenharia Civil, Universidade Federal do Rio de Janeiro, Rio de Ja-neiro (2016)
18. de Melo, M.C., Johnsson, R.M.F.: O Conceito Emergente de Segurança Hídrica. Sustentare **1**(1), 72–92 (2017). https://doi.org/10.5892/st.v1i1.4325
19. UNESCO United Nations Educational: Scientific and Cultural Organization. Water Security (2020). https://en.unesco.org/themes/water-security. Accessed 10 June 2020
20. Rapport, D., Friend, A.: Towards a comprehensive framework for environmental statistics: a stress-response approach. Catalogue (Stat. Can.) **11**(510). Minister of Supply and Services Canada, Ottawa (1979)
21. Felinto, C.M.R., Ribeiro, M.M.R., Braga, C.F.C.: Aplicação do Modelo Força Motriz-Pressão-Estado-Impacto-Resposta (FPEIR) para Gestão dos Recursos Hídricos em João Pessoa-PB. Revista DAE **67**(218), 118–136 (2019). https://doi.org/10.4322/dae.2019.038
22. Cobrape: Sumário Executivo: Plano De Recursos Hídricos das Bacias Hidrográficas dos rios Piracicaba, Capivari e Jundiaí, 2020 a 2035. Porto Alegre: Consórcio Pro-fill-Rhama PCJ (2020b)
23. Borja, A., et al.: The European water framework directive and the DPSIR, a methodological approach to assess the risk of failing to achieve good ecological status. Estuarine Coastal Shelf Sci. **66**(1–2), 84–96 (2006)

24. Niemeijer, D., Groot, R.S.: Framing environmental indicators: moving from causal chains to causal networks. Environ. Dev. Sustain. **10**(1), 89–106 (2008)
25. Soares, A.B., Silva Filho, J.C.L., Abreu, M.C.S., Soares, F.A.: Revisando a estruturação do Modelo DPSIR como Base para um Sistema de Apoio à Decisão para a Sustentabilidade de Bacias Hidrográficas. Revista em Agronegócios e Meio Ambiente **4**(3), 521–545 (2011). https://periodicos.unicesumar.edu.br/index.php/rama/article/view/1922/1300. Accessed 14 Aug 2020
26. Zare, F., Elsawah, S., Bagheri, A., Nabavi, E., Jakeman, A.J.: Improved integrated water resource modelling by combining DPSIR and system dynamics conceptual model-ling techniques. J. Environ. Manag. **246**, 27–41 (2019)
27. UNESCO United Nations Educational (2003): Scientific and Cultural Organization. Water for people, water for life. http://www.unesco.org/new/en/natural-sciences/environment/water/wwap/wwdr/wwdr1-2003/downloads/. Accessed 14 Oct 2020

A Study about Automated Optical Inspection: Inspection Algorithms Applied in Flexible Manufacturing Printed Circuit Board Cells Using the Mahalanobis Distance Method 1

Daniel Katz Bonello(✉), Yuzo Iano, Umberto Bonello Neto, Gabriel Gomes de Oliveira, and Gabriel Caumo Vaz

School of Electrical and Computer Engineering. University of Campinas, Campinas, Brazil
danielb.katze@gmail.com, yuzo@unicamp.br, oliveiragomesgabriel@ieee.org

Abstract. The main objective of this work is to investigate the behavior of an Automated Optical Inspection (AOI) robustness algorithm employed in a high-yield Printed Circuit Board (PCB) assembly Flexible Manufacturing Cell (FMC) utilizing neural networks and the machine learning concept with a focus on the comparison between different modeling options and techniques to predict and actuate the occurrence of failures in electronic components at the run-time assembling in a high-yield production line. A locally Flexible Manufacturing Cell (FMC) or Flexible Manufacturing System (FMS) is an artificial intelligence (AI) composite structure where sensors and actuators (small devices compared to the involved AOI FMC system) interact with the hosting media, creating PCBs gaps imaging as a coupled effect of AOI algorithms and Mean Time Between Failures (MTBF) in the FCM structure. This effect typically creates imaging gaps at lower frequencies than those that would be observed in the PCBs. An introduction about colored PCBs electronic components background modeling, Kernel methods, multi-model algorithms, Kalman filters, and Mixture of Gaussians (MoG) is given. The main technique used regarding electronic component failure calculus is the Mahalanobis Distance, which is initially applied using classical image subtraction equations and a classical Bandlets board model.

Keywords: Automated Optical Inspection (AOI) · Robustness algorithms · Mahalanobis distance · Frequency-domain filtering · Time-domain filtering · Flexible Manufacturing Cell (FMC)

1 Introduction

The area of Automated Optical Inspection (AOI) robustness algorithms has been studied for a long time. Mathematical descriptions of AOI systems and physical experiments for demonstrating associated phenomena have drawn considerable

Y. Iano et al. (Eds.): BTSym 2021, SIST 207, pp. 198–212, 2023.
https://doi.org/10.1007/978-3-031-04435-9_19

attention, since applications are found in several fields of interest such as printed circuit boards (PCBs) inspection, Gaussian noise, statistics in industrial and academic areas, and soldering technologies, to name a few. The comprehension of AOI algorithms properties is paramount to the design, testing, and application of solutions in many fields of expertise. One of the most important areas concerning PCBs inspection is the development of inspection techniques that present interesting properties specifically designed towards objectives such as control in flexible manufacturing cells (FMC) and object character recognition (OCR) in a high-yield assembly PCB production line.

A significant field of study of AOI robustness algorithms is building design of experiments (DOE). The structural inspection process is usually divided into five main stages [1]: baking (the process of removing any moisture contents presented in the bare PCB), screen printing (the process in which a solder paste is applied on pads of bare PCB through the stencil), pick & place (the process where the SMT components are picked and placed in specifically defined locations on the PCB pads by automated robot machines), reflow (the next processes involved in the SMT, at this stage the solder paste will be melted to firmly attach the placed components on the PCB), and automated optical inspection (inspection tools involved in the detection of various errors found in the previous process). The field of engineering that deals with these five phenomena in PCBs process assembly is named SMT design of experiments parameters - considering machinery capabilities (usually defined for the "TAKT-Time" of an FMC). Meanwhile, traditional components PCBs assembly is called PTH (pin through hole). The AOI inspection can generate reports about missing or misaligned components in the inspection PCB process (SMT design of experiment parameters and PTH, respectively).

In both cases, AOI robustness algorithms can be developed to detect these errors (missing and misaligned components, for instance) through inspection structures such as the AOI cameras, FMC robots inspectors, OCR techniques. A fraction of AOI systems are dedicated to inspecting SMT components in an assembling line, while another fraction is constructed to detect failures in PTH components between inspection process stages. A schematic example is given in Figs. 1 and 2.

Regarding the set of SMT components in the PCB, which have their own complexities, such as shapes and functions to the inspection system, the change of components parameters is often used when the AOI inspection algorithm modification is impractical or not economically feasible [2].

From the many types of AOI robustness algorithms presented in various inspection systems, OCR algorithms are the most meaningful [3] in the inspection system interaction, since they involve significant recognition parameters in the machine learning of the inspection algorithm. This algorithm can effectively detect text elements in the pre-inspection of a bare PCB, thus facilitating the afterward inspection stages. Therefore, linear prediction filters having mainly inspection importances - such as Kalman filters - are of particular interest in the sense of Gaussian noise reduction.

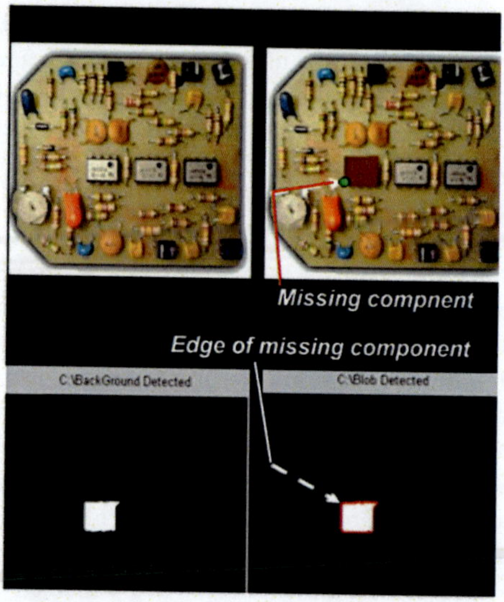

Fig. 1. SMT and PTH inspection processes

Fig. 2. Identification of missed components

Low and mid-frequency range noise has been regarded as a severe problem concerning both image processing quality and image noise pollution. PCB board material and multilayer fabrication technology can generally provide good image inspection attenuation at mid to high frequencies, but not at low to mid frequencies. At such frequencies, the Gaussian noise law (which indicates that the quality of image processing inspection technique in a material is directly proportional to the product of its noise decomposition, the noise density, and the frequency domain) indicates inferior noise reduction. Thus, the development of image inspection techniques that can provide enough attenuation of image decomposition noise without a substantial increase in noise is of great interest. PCBs board copper materials have demonstrated interesting chemical proprieties in multilayer design when these materials present a characteristic of electrical conductivity in their physical construction in one or more directions, thus being also an active subject of research.

2 Literature Review

A general overview of the literature about the topics of Curve Fitting settings from MATLAB Toolbox method, PCBs parameters in industrial inspection process, and the most common analysis techniques is presented in this chapter. Although the given list is not intended to be exhaustive, these references are sufficient to cover relevant concepts to this work and to highlight the current status of research in the field.

Flexible manufacturing cells (FMCs), sometimes referred to as flexible manufacturing systems (FSMs), are characterized by multiple machines (robots), with automatized aleatory of mechanical parts to the processing stations, and the control with a central computer including sophisticated command software. The FMCs have been thoroughly discussed in the literature since they can present interesting properties in PCB inspection systems. Anitha and Rao [4] present a classical approach to AOI robustness algorithms in FMCs and use a mathematical background to analyze their behavior in a mixture of systems, such as electrical devices and segmentation techniques. Also, a wide variety of methods for analyzing PTH and SMT components in FMCs have been developed over the years, as shown by Moganti [5].

Stephan et al. [6] have demonstrated, based on the idea of local AOI inspection, the existence of PCBs gaps at frequencies with two orders of magnitude lower than the related Mean Time Between Failures (MTBF) using AOI robustness algorithms. The resulting AOI inspection profile can be explained as the resulting interface between the local AOI inspection and the continuum PCBs background result in which it is embedded [7]. This effect was first theoretically described by We Li et al. [8] when studying automatized image processing techniques in AOI algorithms.

The Mahalanobis Distance method (MDM) is a traditional method that is commonly used for calculating displacements, variations, and standard deviations in image processing data, and the literature on the subject is vast [9]. Many formulations for static and dynamic analyses are available [10], including implementations using MATLAB [11,12]. Also, due to the size of the problems that the MDM may deal with, many different methods may be used to reduce the size of the problem and to obtain approximate solutions [7].

2.1 Colored PCBs Electronic Components Background Modeling

In order to represent a Curve Fitting setting and be able to use multi-cell flexible manufacturing systems, it is necessary to describe the structure using background modeling and Kernel methods [13,14]. A general frequency-domain structure can be obtained using physical components background modeling basis vectors x and y, as depicted in Fig. 3, with FMC cells indicated dashed lines.

For the general Mahalanobis Distance method (MDM), the background modeling vectors, which describe the general form of multiple machines for a given structure, are given by x, y, and 1. Thus, a general physical background point may be written as a linear combination of such basis vectors, i.e.,

$$\text{Neighbors of a Pixel} = (x+1, y), (x-1, y), (x, y+1), (x, y-1) \tag{1}$$

where a pixel "p" at coordinates (x, y) has four horizontal and vertical neighbors whose coordinates are given above.

The reciprocal components background modeling is defined using basis vectors a, b, and I, and the samples number (n), such that:

$$n = \sum_{i=1}^{N} 1 \tag{2}$$

$$a = \sum_{i=1}^{N} Ii \tag{3}$$

$$b = \sum_{i=1}^{N} Ii \times Ii^{T} \tag{4}$$

where T represents the transposed matrix, given by:

$$A = \begin{bmatrix} a_{1,1} & a_{1,2} & \cdots & a_{1,n} \\ a_{2,1} & a_{2,2} & \cdots & a_{2,n} \\ \vdots & \vdots & \ddots & \vdots \\ a_{m,1} & a_{m,2} & \cdots & a_{m,n} \end{bmatrix} \Leftrightarrow A^{T} = \begin{bmatrix} a_{1,1} & a_{2,1} & \cdots & a_{m,1} \\ a_{1,2} & a_{2,2} & \cdots & a_{m,2} \\ \vdots & \vdots & \ddots & \vdots \\ a_{1,n} & a_{2,n} & \cdots & a_{m,n} \end{bmatrix} \tag{5}$$

Equations 2, 3, and 4 state that vectors a and b are the sum of the cross-product for each section of a perfect colored PCB assembly that is to be processed for missing or misaligned components. N continuous frames are captured for learning purposes.

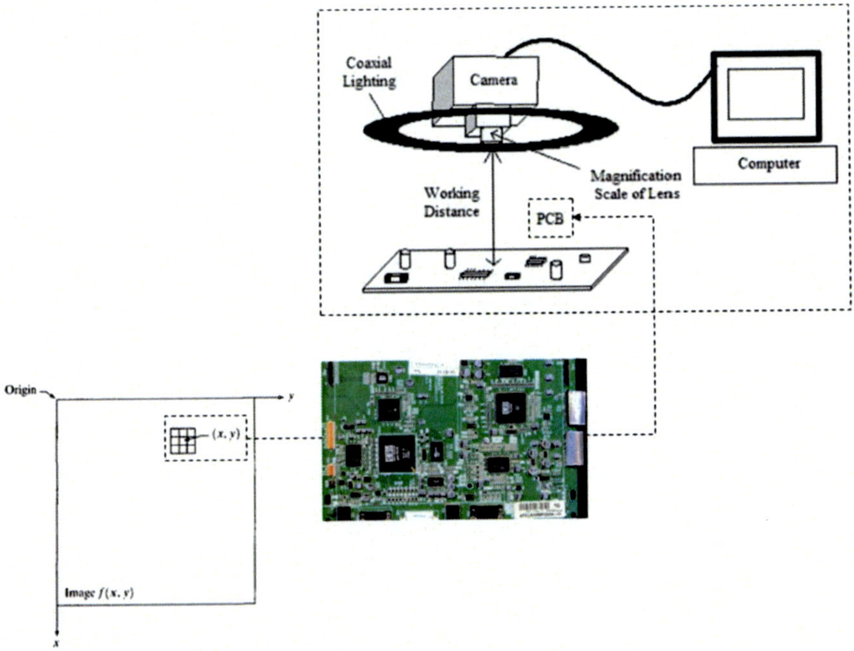

Fig. 3. FMC cell in the physical components background modeling with basis vectors (x, y).

A general frequency-domain reciprocal background modeling is shown in Fig. 4, and the dashed lines indicate a region named two-dimensional lowpass filter and two-dimensional highpass filter in the background modeling. The background modeling basis vectors a, b, and I are filtered by the vectors H(u, v), respectively.

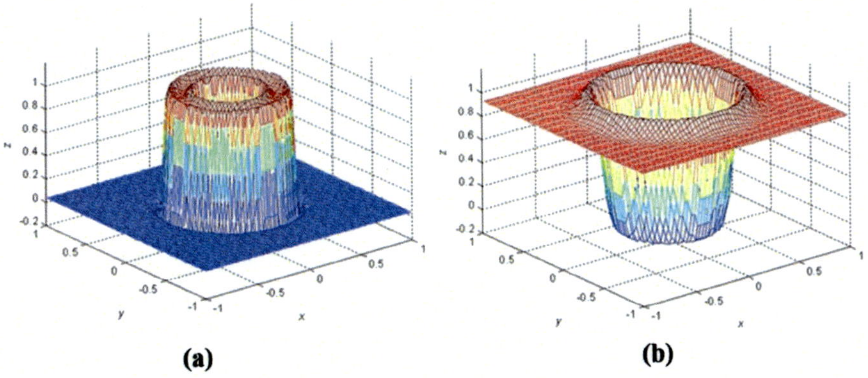

Fig. 4. (a) Two-dimensional low-pass filter (b) Two-dimensional high-pass filter.

A component background point in the reciprocal background is given by the reciprocal background vector H, written as:

$$H(u,v) = \begin{cases} 0 \text{ if } (u,v) = (M/2, N/2) \\ 1 \text{ otherwise} \end{cases} \tag{6}$$

For any M samples from its continuous counterpart.

An important fact that will be used later is that the convolution **H** $\star$ **Neighbors of a Pixel** yields a convolution process. This result can be shown by using different vectors over **H** and **Neighbors of a Pixel**, and by applying convolution between **H** and **Neighbors of a Pixel**, which yields:

$$f(x,y) \star h(u,v) = \frac{1}{MN} \sum_{m=0}^{M-1} \sum_{n=0}^{N-1} f(m,n) h(u-m, v-n) \tag{7}$$

Using Eq. 7, the Fourier property can be applied, and the convolution between H and Neighbors of a Pixel simplifies to:

$$f(x,y) \star h(u,v) \longleftrightarrow F(u,v) H(u,v) \tag{8}$$

Since x, y, u, and v are all vectors, their frequency domain is denoted by H(u), and a Gaussian filter function is also applied. Thus, a Gaussian function equals an Eq. 9, which implies that:

$$H(u) = Ae^{-u^2/2\sigma^2} \tag{9}$$

where σ is the standard deviation of the Gaussian curve.

The components referring to the background image have a corresponding filter in the spatial domain $h(x)$ by using an integer number 2π repeated at a certain frequency of times, as in:

$$h(x) = \sqrt{2\pi}\sigma A e^{-2\pi^2\sigma^2 x^2} \tag{10}$$

If a component background can be constructed from the basic Gaussian function of Eq. 9, we can construct more complex filters, e.g., a high-pass filter, as a difference of Gaussians as follows:

$$H(u) = Ae^{-u^2/2\sigma_1^2} - Be^{-u^2\sigma_2^2} \tag{11}$$

where σ is the standard deviation of the Gaussian curve. In this case, the reciprocal component background simplifies to:

$$h(x) = \sqrt{2\pi}\sigma_1 A e^{-2\pi^2\sigma_1^2 x^2} - \sqrt{2\pi}\sigma_2 B e^{2\pi^2\sigma_2^2 x^2} \tag{12}$$

For any $A \geq B$ and $\sigma_1 > \sigma_2$.

2.2 Gaussian Series for Statistical Data Inclusions

In the following derivations, the general zero-mean Gaussian function in one dimension is described using its smoothing filter $g(x)$, denoted by:

$$g(x) = e^{-\frac{x^2}{2\sigma^2}} \quad (13)$$

If an FMC possesses multiple machines regarding its physical properties, it is possible to expand such property, denoted by $\overrightarrow{x} = (i + j)$, using the two-dimensional zero-mean discrete Gaussian function, i.e.,

$$g[i, j] = e^{-\frac{(i^2+j^2)}{2\sigma^2}} \quad (14)$$

where Gaussian functions have five properties that make them particularly useful in early vision processing (see Eq. 7). It is interesting to notice that if the statistical data inclusions have their radius in polar coordinates given by $r^2 = i^2 + j^2$, it is easy to see that Gaussian function in polar coordinates:

$$g(r, \theta) = e^{-\frac{r^2}{2\sigma^2}} \quad (15)$$

Using Eq. 13, it is verified that the Gaussian function has the interesting property that its Fourier transform is also a Gaussian function. Since the Fourier transform of a Gaussian is a real function, the Fourier transform is its own magnitude. The Fourier transform of a Gaussian is computed by Eq. 16, which ensures the properties of multiple machines.

$$\begin{aligned} F\{g(x)\} &= \int_{-\infty}^{\infty} g(x)e^{-j\omega x}dx \\ &= \int_{-\infty}^{\infty} e^{-\frac{x^2}{2\sigma^2}} e^{-j\omega x}dx \\ &= \int_{-\infty}^{\infty} e^{-\frac{x^2}{2\sigma^2}} (\cos\,\omega x + j\,\sin\,\omega x)dx \\ &= \int_{-\infty}^{\infty} e^{-\frac{x^2}{2\sigma^2}}\,\cos\,\omega x dx + j\int_{-\infty}^{\infty} e^{-\frac{x^2}{2\sigma^2}}\,\sin\,\omega x dx \end{aligned} \quad (16)$$

The Gaussian is a symmetric function and the sine function is antisymmetric, so the integrand in the second integral is antisymmetric. Therefore, the integral must be zero, and the Fourier transform can be simplified into Eq. 17.

$$\begin{aligned} F\{g(x)\} &= \int_{-\infty}^{\infty} e^{-\frac{x^2}{2\sigma^2}}\,\cos\,\omega x dx \\ &= \sqrt{2\pi}\sigma e^{-\frac{\omega^2}{2\nu^2}},\ \nu^2 = \frac{1}{\sigma^2} \end{aligned} \quad (17)$$

where ω represents the spatial frequency parameter and ν represents the spread of the Gaussian in the frequency domain, which is reciprocal of the spread parameter σ in the spatial domain.

3 The Mahalanobis Distance Method

The thin boards and their associated pattern recognition have been used for studying effective properties parameters in image processing inspection techniques [10] and for low-frequency PCBs imaging range (10 Hz to 30 Hz) noise absorption [15]. Experimental realizations of image processing inspection techniques have been shown to be able to overcome the noise density law of image acquisition for low frequencies (10 Hz to 30 Hz) [16]. AOI robustness algorithms in a single low-yield PCB assembly flexible manufacturing cell (FMC) has also been studied, and its dispersion curves (relations between the AOI FMC system of an image processing integrated techniques in a medium and its frequency) show a strong dependence on the ratio between the image processing inspection technique's effective Mohalanobis's law and FMC modulus [17]. Also, the MDM has been used to determine dispersion relations in both frequency-domain linear and prediction filters cases; a detailed explanation of its application in frequency-domain filtering in a particularly FMC is presented in [18]. Distribution Filter and Curve Fitting settings from MATLAB Toolbox are formulated and compared with other techniques in [19], such as the linear-difference time-domain method and the MMFMS, considering a frequency-domain structure.

Topology optimization techniques have also been used to optimize the frequency response of systems considering the case of FMC-structure interaction [20] and to determine an optimal discretization of FMC cells, which leads to maximum image processing automation state [21] and image acquisition control [22] of imaging gaps, considering both particularly FMC and non-particularly FMC cell.

As for practical applications, McLachlan [23] has computed the imaging diagrams of an FMC structure consisting of image acquisition components and correlated them to the transmission of captured dataset PCBs images in a finite length array of images using the Mahalanobis Distance method (MDM), showing imaging gaps in frequencies lower than those expected from the OCR (Object Character Recognition) mechanism statistical data measurement.

When studying the Mahalanobis Distance method (MDM) in FMC cells, a particular interest is given to calculate the dataset group the system may present, and several methods to estimate band diagrams (diagrams that relate propagating frequencies and wavenumbers) have been described in this work. For simple FMC cells, it is common to use the MDM to determine band diagrams in one-dimensional and two-dimensional systems. An introduction to its use in covariance matrix problems is presented by Satonaga et al. [24].

The Mahalanobis Distance method (MDM) is a measure between a sample point and a distribution. The MDM from a vector y to a distribution with mean μ and covariance Σ is:

$$d = \sqrt{(y-\mu)\sum -1(y-\mu)'} \tag{18}$$

This distance represents how far y is from the mean in number of standard deviations.

Suppose we have two groups of data (Tables 1 and 2), and each group consists of two variables (x and y):

Table 1. Group 1.

No	X1	Y1
1	2	2
2	2	5
3	6	5
4	7	3
5	4	7
6	6	4
7	5	3
8	4	6
9	2	5
10	1	3
Mean	**3.9**	**4.3**

Table 2. Group 2.

No	X2	Y2
1	6	5
2	7	4
3	8	7
4	5	6
5	5	4
Mean	**6.2**	**5.2**

The scattered plot of the data is shown in Fig. 5:

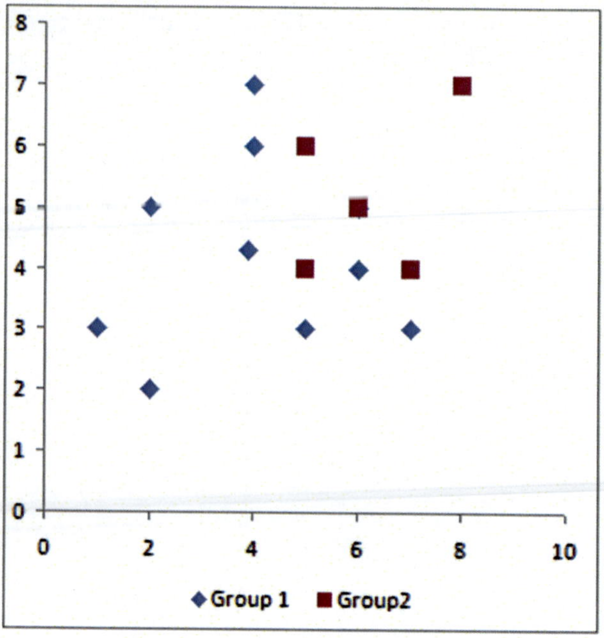

Fig. 5. Scattered plot of the data.

First, we center the data on the arithmetic mean of each variable (Tables 3 and 4).

Table 3. Group 1 centered on the arithmetic mean.

No	X1	Y1
1	−1.90	−2.30
2	−1.90	0.70
3	2.10	0.70
4	3.10	−1.30
5	0.10	2.70
6	2.10	−0.30
7	1.10	−1.30
8	0.10	1.70
9	−1.90	0.70
10	−2.90	−1.30

Table 4. Group 2 centered on the arithmetic mean.

No	X2	Y2
1	−0.20	−0.20
2	0.80	−1.20
3	1.80	1.80
4	−1.20	0.80
5	−1.20	−1.20

The covariance matrix of a group i is computed with the centered data matrix $\hat{a}$:

$$Ci = \frac{1}{ni} \hat{a}^T \hat{a} \tag{19}$$

The covariance matrices of Groups 1 and 2 are presented respectively in Tables 5 and 6.

Table 5. Covariance matrix of Group 1.

	X1	Y1
X1	3.89	0.13
Y1	0.13	2.21

Table 6. Covariance matrix of Group 2.

	X2	Y2
X2	1.36	0.56
Y2	0.56	1.36

The pooled covariance matrices of the two groups is computed as weighted average of the covariance matrices. The weighted average takes this form:

$$si(r, c) = \frac{1}{n} \sum_{i=1}^{g} ni \; ci \; (r, s) \tag{20}$$

The pooled covariance is computed with the weighted average of the first group (10/15) multiplied by the covariance of Group 1 plus the weighted average of the second group (5/15) multiplied by the covariance of Group 2, which yields the results in Table 7.

Table 7. Pooled covariance.

	X	Y
X	3.05	0.27
Y	0.27	1.93

The steps to achieve those results are described in the following equation:

$$\begin{bmatrix} 3.89 \times 2/3 + 1.36 \times 1/3 & 0.13 \times 2/3 + 0.56 \times 1/3 \\ 0.13 \times 2/3 + 0.56 \times 1/3 & 2.21 \times 2/3 + 1.36 \times 1/3 \end{bmatrix} = \begin{bmatrix} 2.59 + 0.45 & 0.086 + 0.18 \\ 0.086 + 0.18 & 1.47 + 0.45 \end{bmatrix} = \begin{bmatrix} 3.05 & 0.27 \\ 0.27 & 1.93 \end{bmatrix} \tag{21}$$

The Mahalanobis Distance (MD) is simply the quadratic multiplication of the mean difference and inverse of pooled covariance matrix:

$$\begin{bmatrix} 3.05 & 0.27 \\ 0.27 & 1.93 \end{bmatrix} \cdot \begin{bmatrix} a & b \\ c & d \end{bmatrix} = \begin{bmatrix} 1 & 0 \\ 0 & 1 \end{bmatrix} \tag{22}$$

Therefore, the inverse pooled covariance matrix is:

$$A = \begin{bmatrix} 0.332 & -0.047 \\ -0.047 & 0.526 \end{bmatrix} \tag{23}$$

The mean difference is given by subtracting the means of both variables ($X1 - X2$ and $Y1 - Y2$), and the result is the matrix B.

$$B = \begin{bmatrix} 3.90 - 6.20 \\ 4.30 - 5.20 \end{bmatrix} = \begin{bmatrix} -2.30 \\ -0.90 \end{bmatrix} \tag{24}$$

$$B^T = \begin{bmatrix} -2.30 & -0.90 \end{bmatrix} \tag{25}$$

The Mahalanobis Distance is calculated by Eq. 18, and replacing the values of B^T, A and B:

$$d^2 = B^T \times A \times B \tag{26}$$

$$d^2 = \begin{bmatrix} -2.30 & -0.90 \end{bmatrix} \times \begin{bmatrix} 0.332 & -0.047 \\ -0.047 & 0.526 \end{bmatrix} \times \begin{bmatrix} -2.30 \\ -0.90 \end{bmatrix} \tag{27}$$

Therefore, we have the value of d, which represents the Mahalanobis Distance (MD) of Group 1 and Group 2:

$$d = \sqrt{1.98776} = 1.41$$

4 Conclusion

The utilization of the Mahalanobis Distance (MD) technique in conjunction with the Object Character Recognition (OCR) technique model was proposed, and the corresponding equations are derived. For using the Mahalanobis Distance method, colored PCBs inspection properties should be expanded using the Median Filter (MF). The most common version of this expansion (linear prediction filters, representing frame periods average from the template frame) is known. However, when frequency-domain equations are used, the linear prediction filters expansion may not be representative in the time domain.

To address this issue, a novel frequency-domain expansion of noise properties (Gaussian Noise inclusions in a digital image) was derived in this paper. Mean Time Between Failures (MTBF) is calculated using a proprietary MATLAB code. Results for both frequency-domain and time-domain models are assessed using Distribution Filter and Curve Fitting settings from MATLAB Toolbox (MATLAB implementations considering OCR and inspection elements). SMT components are also presented, and their use for obtaining frequency-domain filtering is evaluated using a technique that uses dimensional parameters optimization of a representative model taking as reference the defects of SMT components and FMC parameters.

References

1. Kleber, S., Nölscher, H.F., Kargl, F.: Automated PCB reverse engineering. In: 11th USENIX Workshop on Offensive Technologies (WOOT 17) (2017)
2. Moganti, M., Ercal, F.: Automatic PCB inspection systems. IEEE Potentials **14**(3), 6–10 (1995)
3. Anitha, D.B., Rao, M.: A survey on defect detection in bare PCB and assembled PCB using image processing techniques. In: 2017 International Conference on Wireless Communications, Signal Processing and Networking (WiSPNET). IEEE, pp. 39-43 (2017)
4. Sundaraj, K.: PCB inspection for missing or misaligned components using background subtraction. WSEAS Trans. Inf. Sci. Appl. **6**(5), 778–787 (2009)
5. Taralkar, P.M., Patil, S.D.: Image Processing based PCB Component Detection. Int. J. Sci. Res. Rev. **7**(02) (2018)
6. Anitha, D.B., Rao, M.: SMT component inspection in PCBA's using image processing techniques. Int. J. Innov. Technol. Explor. Eng. **8**(12), 541–547 (2019)
7. Iano, Y., Bonello, D.K., Neto, U.B.: Text recognition in PCBs: an Object Character Recognition (OCR) algorithm. Int. J. Dev. Res. **10**(07), 38650–38656 (2020)
8. Oja, M., Kaski, S., Kohonen, T.: Bibliography of self-organizing map (SOM) papers: 1998–2001 addendum. Neural Comput. Surv. **3**(1), 1–156 (2003)
9. Quadir, S.E., et al.: A survey on chip to system reverse engineering. ACM J. Emerg. Technol. Comput. Syst. **13**(1), 1–34 (2016)
10. Suhasini, A., Kalro, S.D., Prathiksha, B.G., Meghashree, B.S., Phaneendra, H.D.: PCB defect detection using image subtraction algorithm. Int. J. Comput. Appl. **975**, 8887 (2015)
11. Jain, R., Kasturi, R., Schunck, B.G.: Machine Vision. McGraw-Hill, New York (1995)

12. Hall, R.N.M.: Computational Intelligence-The Next Generation Challenges (2012)
13. Canahuire Cabello, F.A.: Contribuições à detecçõo automática de pontos quentes em postes de energia elétrica usando o algoritmo Bag of Visual Words e métodos de Kernel (2016)
14. Higham, D.J., Higham, N.J.: MATLAB guide. SIAM (2016)
15. Satonaga, T., Adachi, K., Yasukawa, K., Yamada, N., Uwatoko, K., Furukawa, S.: Detected data processing apparatus and computer readable medium for detecting data. US Patent 8,326,564 B2, 4 December 2012 (2012)
16. Ji, P., Wan, Y.F.: Planning for printed circuit board assembly: the state-of-the-art review. Int. J. Comput. Appl. Technol. **14**(4–6), 136–144 (2001)
17. Van, P.J., Waqas, U., Geilen, M., Basten, T., Somers, L.: Online scheduling of 2-re-entrant flexible manufacturing systems. ACM Trans. Embed. Comput. Syst. **16**(5s), 1–20 (2017)
18. van der Sanden, B., Geilen, M., Reniers, M., Basten, T.: Partial-order reduction for performance analysis of max-plus timed systems. In: 2018 18th International Conference on Application of Concurrency to System Design (ACSD). IEEE, pp. 40-49 (2018)
19. Hidalgo, J.I.G., Mariño, L.M.P., de Barros, R.S.M.: Cosine similarity drift detector. In: Tetko, I.V., Kůrková, V., Karpov, P., Theis, F. (eds.) ICANN 2019. LNCS, vol. 11730, pp. 669–685. Springer, Cham (2019). https://doi.org/10.1007/978-3-030-30490-4_53
20. Fang, Y., Qu, J., Pong, B.M.H., Lee, C.K., Hui, S.Y.R.: Quasistatic modeling and optimization of two-layer PCB resonators in wireless power transfer systems for 110 kV power grid online monitoring equipment. IEEE Trans. Ind. Electron. (2021)
21. Satonaka, N.: Analysis of geometric deviations of machine products under Maximum Material Conditions. Trans. JSME Ser. C **73**(730), 285–291 (2007)
22. McLachlan, G., Ahfock, D.: Estimation of classification rules from partially classified data. In: Chadjipadelis, T., Lausen, B., Markos, A., Lee, T.R., Montanari, A., Nugent, R. (eds.) IFCS 2019. SCDAKO, pp. 149–157. Springer, Cham (2021). https://doi.org/10.1007/978-3-030-60104-1_17
23. Melnyk, R.A., Havrylko, Y.V., Repa, Y.V.: PCB open, short and shift defects detection by logical comparison of mask images. In: IOP Conference Series: Materials Science and Engineering. IOP Publishing, p. 12003 (2021)
24. Kumar, S., Abu-Siada, A., Das, N., Islam, S.: Toward a substation automation system based on IEC 61850. Electronics **10**(3), 310 (2021)

Proposal of Mathematical Models for a Continuous Flow Electric Heater

Alex Restani Siegle(✉), Yuzo Iano, Gabriel Gomes de Oliveira, and Gabriel Caumo Vaz

School of Electrical and Computer Engineering, State University of Campinas, Campinas, Brazil

alexrestani@gmail.com, yuzo@unicamp.br, oliveiragomesgabriel@ieee.org

Abstract. Electrical heating systems based on continuous flow or fluid passage are widely used, as for example in industrial and residential applications. Recently a new application took place in the automotive industry as fuel heating systems to improve fuel-burning efficiency. These devices are highly controllable and relatively simple to be built. But the challenge to have an efficient system begins when it is necessary to model mathematically the system plant to provide a simulation tool as close as possible to the real world and reach the desired result using control systems. Accurate mathematical modeling for continuous flow heating systems creates a powerful tool to be used in simulation software and allows to run the tests and validations in a virtual world, to speed up the control system project and find a better solution with low cost. This article suggests three methods to model mathematically the same electric continuous-flow heater with different situations of losses and shows a comparison to certify they are valid and applicable.

Keywords: Continuous flow electric heater · Control systems · Modeling · Passage heater

1 Introduction

Correct mathematical modeling of physical systems ensures that the use of computational tools such as Matlab and Simulink can efficiently simulate the real world. Even within these virtual environments, it is possible to approach different ways to solve a problem, increase efficiency or even experiment with new methodologies.

Electric passage heaters or continuous flow electric heaters are used in many situations where it is necessary to increase quickly the temperature of a fluid, with low complexity, high reliability, and an excellent capacity to apply control techniques [1]. An example of use is residential water heaters such as electric showers. Another quite different application, but with the same principle, is fuel heaters used in Flex-Fuel vehicles [2].

Y. Iano et al. (Eds.): BTSym 2021, SIST 207, pp. 213–228, 2023.
https://doi.org/10.1007/978-3-031-04435-9_20

There is a wide variety of articles that discuss this system but with little clarity on the method used in the mathematical model of this type of heater. For that reason, this article proposes to show some mathematical modeling methods depending on the need.

In this article, three mathematical models for electrical passage heaters will be addressed. The first mathematical model will handle the heat losses through the walls of the casing or heating chamber. The second mathematical model will cover the heat losses caused by the fluid flow through the system. And finally, the third model will associate the two losses mentioned above. In the end, a comparison will be made between the models to certificate that they reach the same result when applying the same parameters.

2 Mathematical Modeling

In this section, three mathematical models, i.e., plants A, B, and C, will be developed from scratch. The first one models the heat losses through the walls of the heating chamber. The second approach addresses the heat losses caused by the fluid flow through the system. The third model associates the first two plants.

2.1 Plant A - Modeling a Thermal System with Heat Loss Through the Walls of the Heating Chamber

Considering a liquid heating system with an electrical resistance injecting heat energy in an enclosed environment, one assumes a heat loss to the external environment through the casing walls. According to the first law of thermodynamics, i.e., energy conservation law, the current energy in the liquid inside the heating chamber is the difference between the amount supplied to the system and the loss to the environment through the walls. Therefore, the liquid temperature can be expressed mathematically as a function of time.

Some hypotheses must be established to develop the model: the external temperature is constant and uniform; the internal temperature is uniform; the heat is lost only by conduction, i.e., radiation and convection are negligible; the thermal insulator is homogeneous and linear [3].

Figure 1 represents the thermal model of the liquid heating system with a resistive element ([4], Cap3.4). The parameters and respective units are the following:

- $q_e(t)$ = Rate of energy input into the system $[J]$;
- T_e = Environment temperature, outside the heating chamber $[°C]$;
- $T_i(t)$ = Internal temperature of the liquid in the heating chamber $[°C]$;
- $q_s(t)$ = System energy output rate $[J]$;
- m = Mass of liquid inside the casing $[g]$;
- c = Specific heat of the liquid $[J/g°C]$.

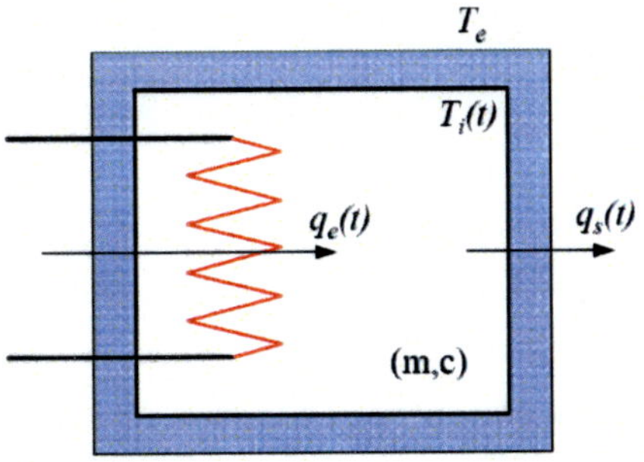

Fig. 1. Representation of an electric heating system with thermal loss in the wall casing

The input is given by Eq. 1. It is the power inserted into the electrical heating system in Watts $[W]$ and converted in time to energy in Joules $[J]$.

$$q_e(t) \tag{1}$$

Equation 2 defines the system output.

$$\Delta T_{chamber}(t) = T_i(t) - T_e \tag{2}$$

For heat transfer by conduction or convection, the following thermal conductivity formula [4] is shown:

$$q(t) = K \cdot \Delta T \tag{3}$$

Where:

- $q(t)$ = heat flow rate $[kcal/s]$;
- K = thermal conductivity coefficient $[kcal/s°C]$;
- ΔT = temperature difference in the wall material of the heating chamber $[°C]$.

Thus, replacing (2) in (3):

$$q_s(t) = K \cdot (T_i(t) - T_e) = K \cdot \Delta T_{chamber}(t) \tag{4}$$

From the formula of the Fundamental Equation of Calorimetry [4]:

$$Q = m \cdot c \cdot \Delta T(t) \tag{5}$$

where Q is the amount of heat $[J]$.

From the first law of thermodynamics, the energy conservation law, the amount of energy that enters the system minus the amount lost to the environment through the walls is the energy present in the liquid inside the heating chamber. Using this premise in Eq. 5, the fluid heating is given by Eq. 6.

$$q_e(t) - q_s(t) = m \cdot c \cdot \frac{d[\Delta T_{chamber}(t)]}{dt} \tag{6}$$

So, $m \cdot c$ is the thermal capacitance.
Now, applying Eq. 4 in Eq. 6:

$$q_e(t) - K \cdot \Delta T_{chamber}(t) = m \cdot c \cdot \frac{d[\Delta T_{chamber}(t)]}{dt} \tag{7}$$

Applying the Laplace transform in Eq. 7:

$$Q_e(s) - K \cdot \Delta T_{chamber}(s) = m \cdot c \cdot s \cdot \Delta T_{chamber}(s) \tag{8}$$

Equations 9, 10, and 11 are the steps to achieve the plant equation.

$$Q_e(s) = m \cdot c \cdot s \cdot \Delta T_{chamber}(s) + K \cdot \Delta T_{chamber}(s) \tag{9}$$

$$Q_e(s) = (m \cdot c \cdot s + K) \cdot \Delta T_{chamber}(s) \tag{10}$$

$$G(s) = \frac{\Delta T_{chamber}(s)}{Q_e(s)} = \frac{1}{m \cdot c \cdot s + K} \tag{11}$$

Equation 11 is the plant equation of the heating system, which is observed to be first order. Note that the input is the power applied to the heating element inside the chamber, but the output is the difference between the internal temperature and the external temperature. Thus, to know the absolute temperature inside the heating chamber, the temperature difference must be added to the outside temperature.

Figure 2 represents the plant of the system described in Eq. 11, with the representation of the sum of the environment temperature to obtain the exit temperature of the heating chamber.

The first-order system represents a thermal system or an RC circuit. Generally speaking, a first-order system can be generically represented by a function with input "R(s)" and output "C(s)", as represented in Eq. 12.

$$\frac{C(s)}{R(s)} = \frac{1}{Ts + 1} \tag{12}$$

Mathematically, we can represent the first instant, when power is supplied to the heating system, as the system's response to the step function, as in Fig. 3 [4].

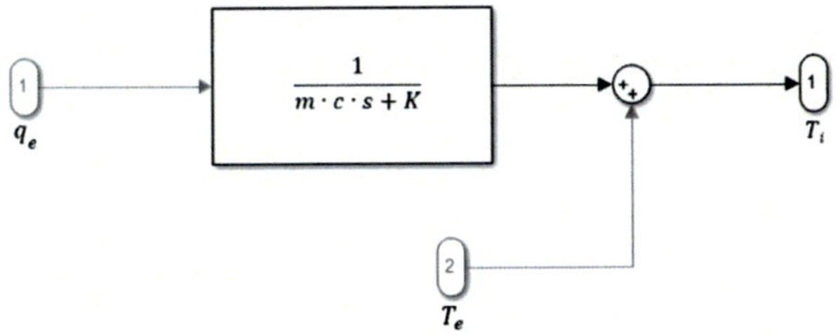

Fig. 2. Plant A model.

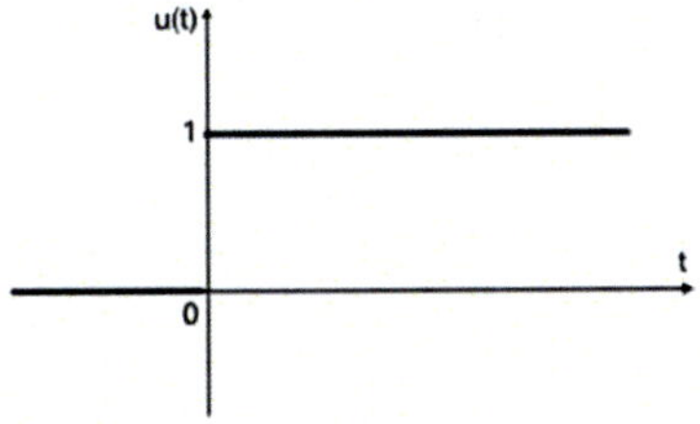

Fig. 3. Step function.

The Laplace transform of the unit step function is $1/s$. Thus, making $R(s) = 1/s$ in Eq. 12:

$$C(s) = \frac{1}{Ts+1} \cdot \frac{1}{s} \tag{13}$$

Equation 13 can be solved by using partial fractions:

$$C(s) = \frac{1}{s} - \frac{T}{Ts+1} = \frac{1}{s} - \frac{1}{s+(1/T)} \tag{14}$$

Finally, the temporal response of the first-order system to a step function (Eq. 15) is achieved with the inverse Laplace transform and is represented in the graph in Fig. 4 [4].

$$c(t) = 1 - e^{-t/T}, \text{ to each } t \geq 0 \tag{15}$$

It is worth noting that the response is closer to the steady state the longer the time elapsed after the step. As a general rule, it is assumed that the system is in a permanent regime after 5 periods (T) since the gain in approaching the regime from this moment onwards does not justify the time spent. Also, a great piece of information is that when Eq. 15 reaches time T the response reaches 63.2% of the maximum output value [4]. This is used to determine the time of the first-order system analogously, that is when we have a measure of the behavior of the system and it is necessary to raise the mathematical model.

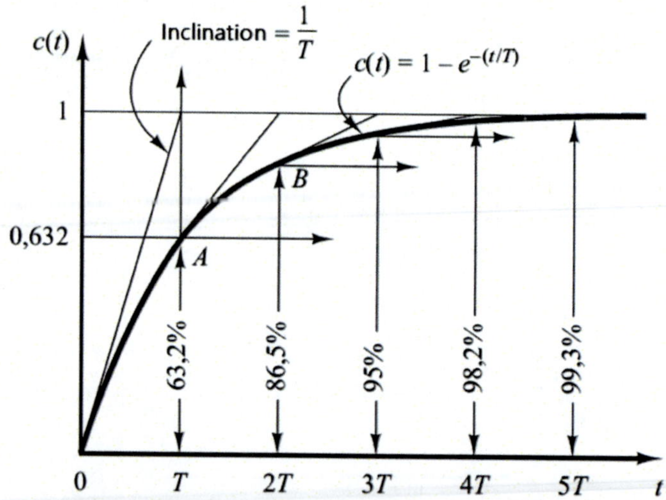

Fig. 4. First-order system response to the step function.

For modeling the first-order thermal system at the time of cooling we use the system response to the unit impulse. This impulse is represented by Laplace as $R(s) = 1$. So, from Eq. 12:

$$C(s) = \frac{1}{Ts + 1} \tag{16}$$

Equation 17 is the inverse Laplace transform of Eq. 16 and is represented in the graph of Fig. 5.

$$c(t) = \frac{1}{T}e^{-t/T}, \text{ to each } t \geq 0 \tag{17}$$

As with the unit step response, the unit impulse response at the time T is $c(T) = 1/Te^{-1}$, which means that the output will have 36.8% of the maximum value at the instant zero.

In this way, the heating and cooling behavior of the plant of the proposed system is known. From now on, with relevant information about the system, it is possible to achieve the values of the variables of the system plant (Eq. 11).

An example is the heating of ethanol as a fluid, whose specific heat is 2.46 [J/g°C], which represents 0.586 [cal/g°C] [5]. Ethanol has a density of 0.789 g/cm^3, which means that, for each cubic centimeter, ethanol weighs only 0.789 g [6]. Supposing that the heating gallery has a capacity of 9.5 ml, the mass of ethanol within this volume is 7.5 [g].

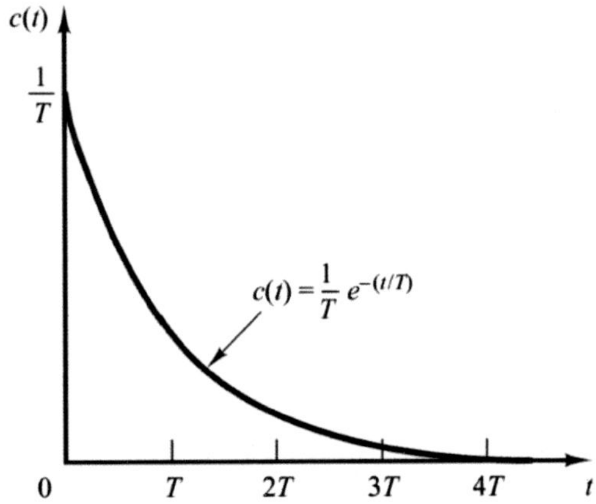

Fig. 5. First-order system response to the unit impulse function.

In this way, part of the necessary values in Eq. 11 can already be defined.

$$C = m \cdot c \tag{18}$$

where:

- C = Thermal capacity $[J/°C]$;
- m = Mass of liquid inside the casing $[g]$;
- c = Specific heat of the liquid $[J/g°C]$.

From the aforementioned values, the thermal capacity of the heating chamber with ethanol is 18.45 [J/°C].

At this point, just the definition of the coefficient of thermal conductivity (K) is still lacking.

The properties of the system's insulating material are unknown. These properties will be necessary for the correct modeling of the proposed system. However, one can assume some dynamic characteristics of heating and cooling that show us the evolution of temperature over time. From this information, it will be possible to determine the time constants of the wall that relate to its thermal properties.

Keeping $K = 1$ in Eq. 11 and using the values obtained for the mass of liquid inside the casing and specific heat of ethanol, it is possible to draw the first curve in the graph in Fig. 6.

It is defined that the system can heat ethanol from 20 °C to about 100 °C in 8 s. After that time, the heating system is turned off, and the cooling process starts immediately.

Using the above assumptions and adjusting the K value by trial and error, it is possible to achieve the desired values of the heating and cooling curves. The value found for K that meets the given premises is $K = 0.4$, found experimentally.

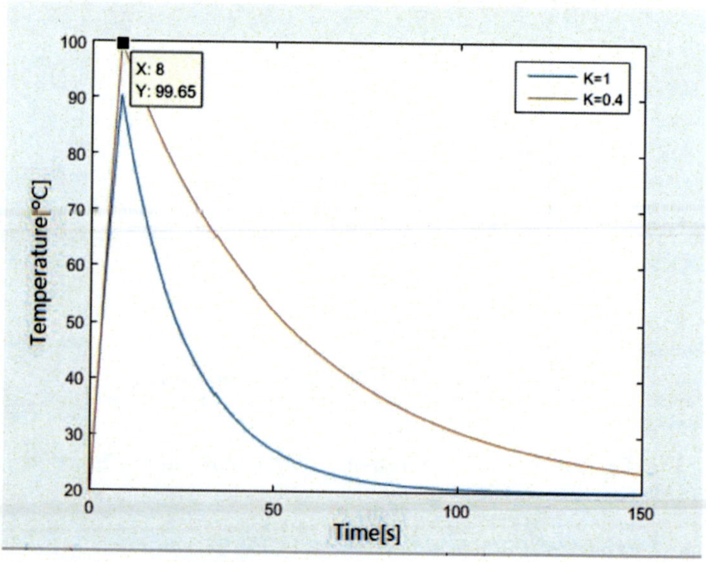

Fig. 6. Response to the maximum power in 8 [s].

With this first-order thermal mathematical model (of heating the liquid inside the chamber with the influence of the external temperature by the thermal permeability of the casing wall) and applying the known values for ethanol, heating chamber volume, and cooling times, it was possible to build a first model that represents part of our system. This is a linear system, which means that the values found so far are constant and applied to the transfer function $G(s)$ of Eq. 11 admit only one output response for each input stimulus. While this system is useful in helping get the K value of the system, it does not represent the entire system that is needed, which includes disturbance by liquid flow through the heating chamber.

2.2 Plant B - Modeling a Thermal System with Heat Loss by Inlet Flow Temperature Disturbance

Now, a model for the temperature perturbation by liquid flow through the chamber will be developed. Using the example of Ogata [4], one can represent a similar system, but this time with the heat loss to the external environment through the walls considered negligible and the thermal capacitance of the heater also negligible, as in Fig. 7.

The terms in Fig. 7 are the following:

- $\overline{T_e}$ = steady-state inlet liquid temperature [$°C$];
- T_{fe} = inlet liquid flow temperature variation [$°C$];
- $\overline{T_s}$ = steady-state leaving liquid temperature [$°C$];

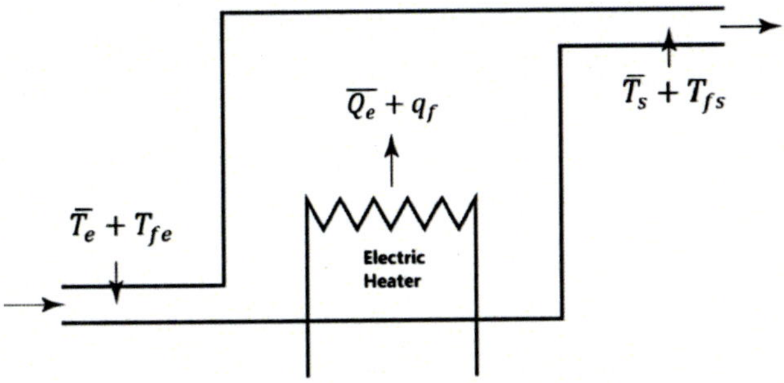

Fig. 7. Representation of plant B.

- T_{fs} = temperature variation of the liquid flow at the outlet [$°C$];
- G = mass flow of liquid in the heating chamber [kg/s];
- M = mass of liquid contained in the heating chamber [kg];
- c = specific heat of the liquid [$kcal/kg°C$];
- R_f = thermal resistance of the liquid moved (flow) [$°Cs/kcal$];
- C = thermal capacitance of the liquid contained in the heating chamber = Mc [$kcal/°C$];
- Q_e = steady-state heat input [$kcal/s$].

Assuming the inlet heat flux is changed from $\overline{T_e}$ to a continuous flow $\overline{T_e} + T_{fe}$, so the internal chamber temperature is changed from $\overline{Q_e}$ to $\overline{Q_e} + q_f$ and the outlet temperature is changed because of this phenomenon from $\overline{T_s}$ to $\overline{T_s} + T_{fs}$. The equation that describes this behavior is:

$$C \cdot d(T_{fs}) = [q_f + Gc(T_{fe} - T_{fs})]dt \tag{19}$$

Gc can be defined like in Eq. 20, resulting in Eq. 21, which can be manipulated to achieve Eq. 22:

$$Gc = \frac{1}{R_f} \tag{20}$$

$$C\frac{d(T_{fs})}{dt} = q_f + \frac{1}{R_f}(T_{fe} - T_{fs}) \tag{21}$$

$$R_f \cdot C\frac{d(T_f s)}{dt} + T_{fs} = R_f \cdot q_f + T_{fe} \tag{22}$$

Performing the Laplace transform on both sides of Eq. 22 and placing the initial condition of $T_{fs}(0) = 0$, the result is:

$$T_{fs}(s) = \frac{R_f}{R_f Cs + 1)} \cdot Q_f(s) + \frac{1}{R_f Cs + 1} \cdot T_{fe}(s) \tag{23}$$

There are two terms in Eq. 23 that make up the output. The first term governs the heating behavior of the liquid within the heating gallery taking into account the constant flow of the liquid, and the second term describes the behavior of perturbation by variation in the temperature of the liquid flow. The plant is shown in Fig. 8.

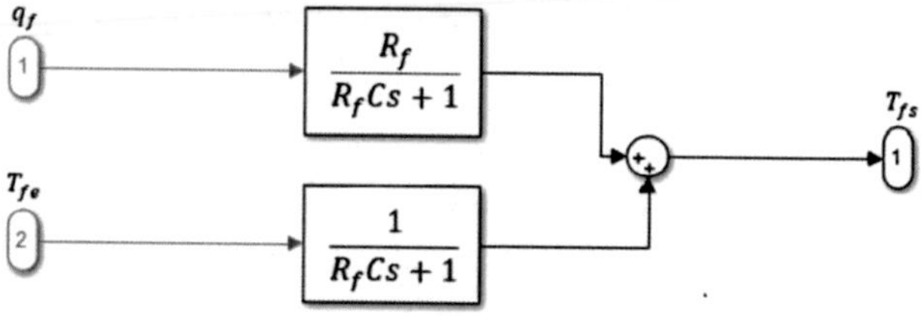

Fig. 8. Plant B model.

This model is very didactic for visualizing the decomposition of the system that will be modeled in terms of the central heating chamber model and the flow temperature perturbation term, but it is still incomplete as the inflow volume variation and the variation in the temperature outside the chamber must also be modeled.

Equation 11 describes in just one term the heating of the fluid and the disturbance due to loss of thermal energy by the walls, the disturbances being given as fixed and, therefore, this model is then linear, where the input stimulus generates just one response from system $G(s)$. It is not desired to keep all the variables fixed but change the ones that are known real-world perturbations.

Now, in Eq. 23, there are two terms for the heating of the fluid inside the chamber and the disturbance caused by the variation in the temperature of the liquid (flow), but it disregards the thermal losses through the walls of the enclosure. Once two variables change the system's response, this system is considered as multivariable, and therefore, non-linear. This system is closer to a model that is more faithful to the real world of the studied heating system, but, as discussed before, the lack of the other disturbance terms considered essential requires a more complete modeling technique for non-linear multivariable systems.

2.3 Plant C - Modeling a Thermal System with Heat Loss Through the Walls of the Heating Chamber and Disturbance in Temperature and Inlet Flow

Now, it will be applied a technique that allows modeling the entire system, heating the fluid inside the chamber, disturbance in the internal temperature by the volume and temperature of the inflow of liquid and thermal losses through the walls of the enclosure, in different terms. The technique is the equivalence of thermal circuits and electrical circuits, and the use of Kirchhoff's laws, mesh law, and knot law in electrical circuits [7].

For the modeling of capacitive behavior, Eq. 18 can be applied in Eq. 5, resulting in:

$$q = C \cdot \Delta T \tag{24}$$

In a capacitive electric circuit, q would be the electric current, C would be the capacitance, and ΔT would be the voltage variation. Thus, it is possible to model the behavior of the thermal capacity (Eqs. 25 and 26).

$$\Delta T = T - T_0 = \frac{1}{C} \cdot \int_0^t q dt \tag{25}$$

$$q = C \cdot \frac{d\Delta T}{dt} \tag{26}$$

To model the resistive behavior, q could be associated again with the electric current and ΔT with the voltage variation. The thermal resistance is then given by Eq. 27 [4].

$$R = \frac{1}{K} \tag{27}$$

Using Eq. 3 in Eq. 27, the result is:

$$q(t) = \frac{\Delta T}{R} \tag{28}$$

One can even associate a current source with a thermal energy source, such as the heating element inside the heating chamber, and a voltage source with a temperature.

Therefore, describing the complete system and all the disturbances that must be modeled, the diagram in Fig. 9 is developed.

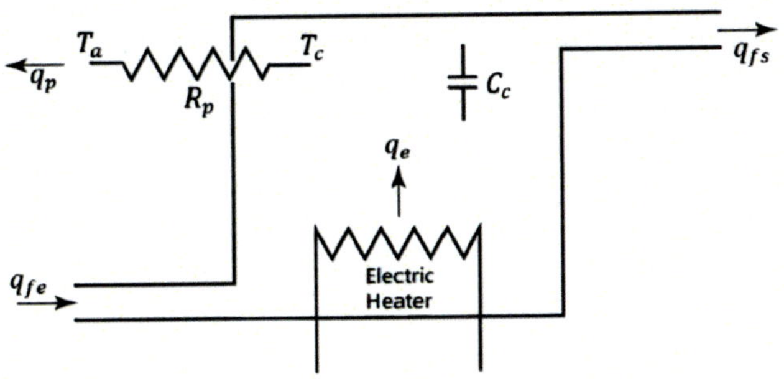

Fig. 9. Diagram of the complete system.

The parameters in Fig. 9 are the following:

- q_{fe} = inlet fluid heat rate $[kcal/s]$;
- q_{fs} = output fluid heat rate $[kcal/s]$;

- q_e = heating element heat rate $[kcal/s]$;
- q_p = rate of heat loss through heating chamber walls $[kcal/s]$;
- T_a = environment temperature $[^\circ C]$;
- T_c = heating chamber temperature $[^\circ C]$;
- C_c = thermal capacity of the heating chamber $[kcal/^\circ C]$;
- R_p = thermal resistance of the walls of the heating chamber $[^\circ C/W]$.

Using the equivalence of thermal circuits and electrical circuits described earlier, one can analytically create the electrical circuit model based on the thermal circuit described in Fig. 9. This circuit is shown in Fig. 10.

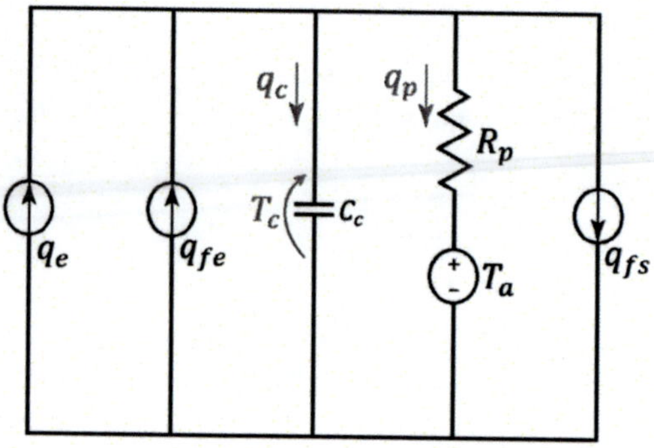

Fig. 10. Equivalent electrical circuit

Applying Kirchhoff's law of knots and meshes in the circuit of Fig. 10, Eq. 29 is achieved.

$$q_e + q_{fe} = q_c + q_p + q_{fs} \tag{29}$$

Isolating the input energy from the heating element:

$$q_e = q_c + q_p + q_{fs} - q_{fe} \tag{30}$$

Applying Eqs. 26 and 28:

$$q_e = C \cdot \frac{dT_c}{dt} + \frac{(T_c - T_a)}{R_p} + (q_{fs} - q_{fe}) \tag{31}$$

Applying the Laplace transform and multiplying the two terms by R_p, Eq. 32 is achieved, and, through Eq. 33, the outlet temperature is isolated, resulting in Eq. 34, which represents the system.

$$R_p \cdot Q_e(s) = R_p \cdot C_c \cdot s \cdot T_c(s) + (T_c(s) - T_a(s)) + R_p \cdot (Q_{fs}(s) - Q_{fe}(s)) \tag{32}$$

$$T_c(s) \cdot (R_p \cdot C_c \cdot s + 1) = R_p \cdot Q_e(s) + T_a(s) - R_p \cdot (Q_{fs}(s) - Q_{fe}(s)) \tag{33}$$

$$T_c(s) = \frac{R_p}{R_pC_cs + 1}Q_e(s) + \frac{1}{R_pC_cs + 1}T_a(s) + \frac{R_p}{R_pC_cs + 1}(Q_{fe}(s) - Q_{fs}(s)) \tag{34}$$

For the third term of the second member of Eq. 34, one can use Eq. 5:

$$q_{fe}(t) - q_{fs}(t) = (G \cdot c \cdot \Delta T_{fe}(t)) - (G \cdot c \cdot \Delta T_{fs}(t)) \tag{35}$$

Where G is the mass of liquid moving through the heating chamber. Equation 35 can be simplified to achieve Eq. 36.

$$q_{fe}(t) - q_{fs}(t) = Gc(\Delta T_{fe}(t) - \Delta T_{fs}(t)) \tag{36}$$

In Eq. 34, the temperature of the chamber, which is equal to the temperature of the liquid leaving the system (first member of the equation, T_c), is equal to the second member of the equation, which is composed of three terms whose modeling is desired. The first term responds to changes in heat inserted into the heating chamber by the heating element, the second term responds to disturbances of heat leakage through the enclosure wall by changing the environment temperature, and the third term responds to disturbances in the flow and temperature of the liquid from the entrance. One can also assume that the temperature of the entering liquid is the same as the external environment, and the exit temperature of the liquid is the same as the internal temperature of the heating gallery, also admitting a homogeneous distribution inside it. Thus:

$$q_{fe}(t) - q_{fs}(t) = Gc(T_a(t) - T_c(t)) \tag{37}$$

This results in the complete model of the plant in Fig. 11.

3 Comparison of Proposed Mathematical Models for the Continuous Flow Electric Heater

One can observe that first-order thermal systems are comparable with RC electrical systems. This means that in practice, thermal systems will always be modeled considering thermal losses, represented by resistance R, together with the portion of thermal energy accumulation, represented by capacitance C.

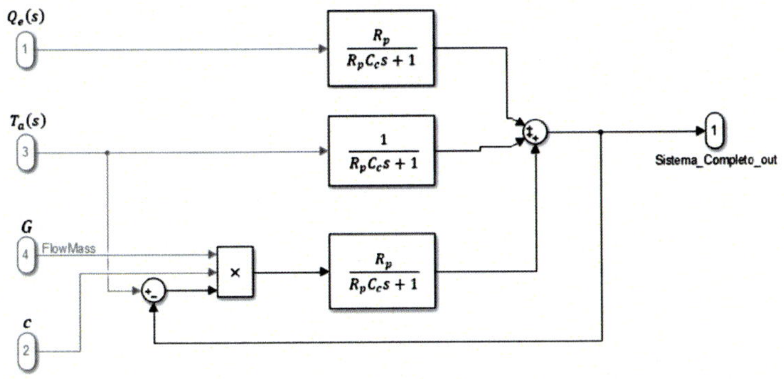

Fig. 11. Plant C model.

In the system in Fig. 1, there are the heating gallery accumulating energy, which represents the portion of thermal capacity C, and the walls of the enclosure offering thermal resistance and affecting the loss of energy to the environment. This system is linear because it models all variables as fixed, only the input varies, which is the heat rate of the heater and, therefore, the transfer function $G(s)$ of this system only obtains one output response for the same value rate of change input of heat supplied by the heating element.

The system in Fig. 7, on the other hand, models the energy heating gallery, but the portion of heat loss is due to the constant flow of the liquid, which represents the resistance R, which removes a heated mass and injects the same amount, but with a different heat rate. This system allowed the inlet temperature of the liquid to also vary over time, in addition to the power of the heating element. This resulted in a system where the final output liquid temperature response is a function with two input variables, which transforms this system into a non-linear multivariable. For this second model to fully meet the problem, it should also take into account the variation of the external temperature and the variation of the heat rate by the flow.

The third system is the most complete, as it has a term for each relevant disturbance, as seen in Eq. 34. To validate the third system as valid to represent the heating of the system and the necessary disturbances, it should be compared with the other two models. Thus, with the liquid flow canceled in plant C, and the same parameters as in plant A, the same result was obtained, as can be seen in Fig. 12.

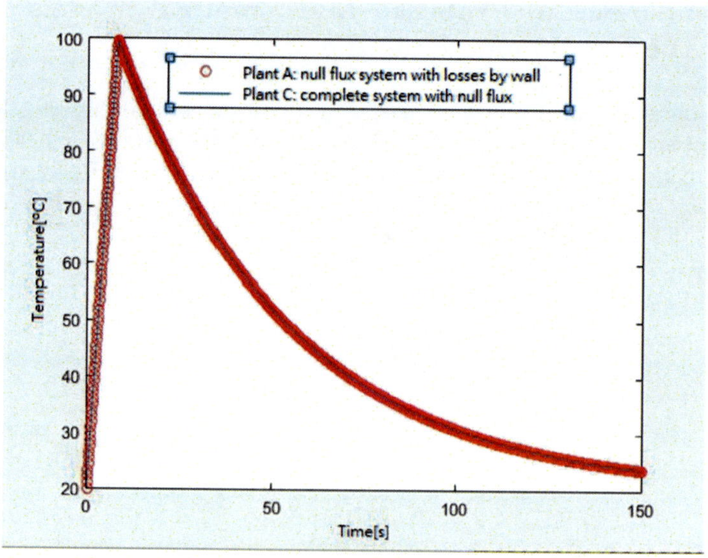

Fig. 12. Comparison between plant A and plant C.

For comparison with plant B, which has the disturbance inserted by the flow temperature difference and negligible losses through the walls of the heating gallery, the same mass of liquid running in the constant regime was used in plants B and C, in addition to increasing the resistance from the walls to very high to simulate negligible losses, and the other parameters were kept the same. Again, the same response was got in both systems, as shown in Fig. 13.

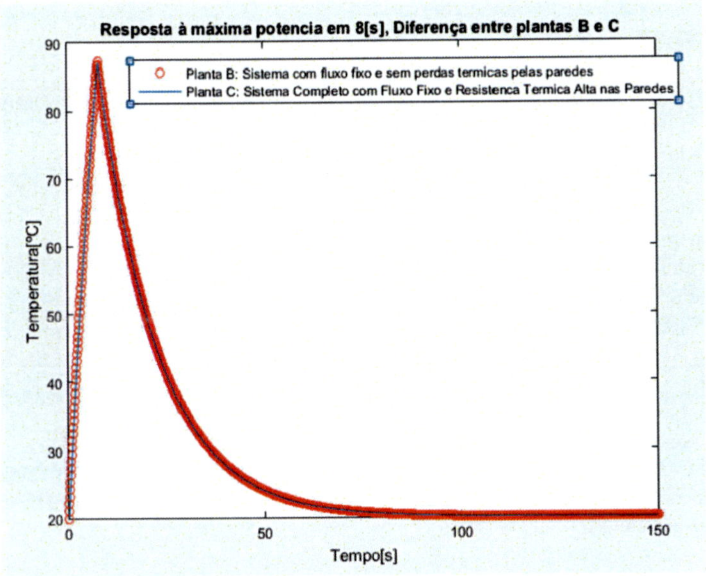

Fig. 13. Comparison between plant B and plant C.

Thus, Plant C is the most complete and includes, in a single plant, the disturbance, heating, and cooling characteristics of the other plants, A and B.

4 Conclusion

The proposal of this paper was to develop three mathematical models based on different scenarios of losses. The first scenario was a heating system with no fluid flow but loses throw the wall casing. The second scenario was a heating system with no losses in the wall casing but losses in the flow temperature perturbance. And the third system, the one with losses by fluid flux and wall casing. This third one was the closest to the real world, where the heating energy is lost by wall thermal resistance and by a new fluid entering the heating system and the losses by the fluid output.

Based on the previous chapter, where the simulation results were presented, it was possible to certify that the proposed mathematical models represent the same physical system when they use the same parameters. It was also possible to get success in the referenced models and they can be used in virtual simulation tools to represent the real world.

From this point, it is possible, for example, to use the proposed mathematical model to be applied in a control system and check the efficiency of different control system techniques. A suggestion for the next paper would be the use of model C to different control system technics to compare their efficiency.

References

1. Orbis do Brasil: QUAL A DIFERENÇA ENTRE AQUECEDORES DE PASSAGEM E ACUMULAÇÃO? (2021). https://orbisdobrasil.com.br/diferenca-aquecedores-passagem-e-acumulacao/. Acesso em 02 May 2020
2. R. B. d. B. Ltda: FlexStart (2021). https://www.bosch-mobility-solutions.com.br/br/destaques/sistemas-de-transmiss%C3%A3o-e-mobilidade-el%C3%A9trica/flexstart/. Acesso em 01 May 2020
3. USP - Universidade de São Paulo: Representação de sistemas. e-Disciplinas - Sistema de Apoio às Disciplinas (2021). https://edisciplinas.usp.br/pluginfile.php/2470129/mod_resource/content/1/3_Representa%C3%A7%C3%A3o_Sistemas.pdf
4. Ogata, K.: Engenharia de Controle Moderno, 3a (2011)
5. Sucrana (2021). http://www.sucrana.com.br/tabelas/densidade-relativa.pdf. Acesso em 01 May 2020
6. NovaCana: Propriedades Físico-Químicas do etanol. NovaCana (2021). https://www.novacana.com/etanol/propriedades-fisico-quimicas. Acesso em 01 May 2020
7. Felício, L.C.: Modelagem da Dinâmica de Sistemas e Estudo da Resposta - Segunda Edição. Rima, São Carlos (2010)

Revision of the 5G Concept Rollout and Its Application in Smart Cities: A Study Case in South America

Pablo Minango[1](✉), Yuzo Iano[1], Euclides Lourenço Chuma[1], Gabriel Caumo Vaz[1], Gabriel Gomes de Oliveira[1], and Juan Minango[2]

[1] School of Electrical and Computer Engineering, University of Campinas, Campinas, Brazil
pablodavid218@gmail.com, yuzo@decom.fee.unicamp.br, {euclides.chuma,oliveiragomesgabriel}@ieee.org

[2] Instituto Tecnológico Universitario Rumiñahui, Sangolquí, Ecuador
juancarlos.minango@ister.edu.ec

Abstract. This work aims to address and emphasize the importance of the fifth-generation mobile networks (5G) rollout in the smart cities, focusing on the South American countries and bringing a theoretical analysis about the implementation in each country from the perspective of the national regulatory agencies, as well as the significance of this system for the future of the countries with the benefits expected for the 5G networks and their application in the smart cities. It was made a bibliographic review of the involved technical concepts and a collection of information from the regulatory agencies and telecommunication companies of all the countries and one French territory of South America, aiming to identify in which stage of roadmap implementation each country is, considering that many countries in other continents (e.g., South Korea and the USA) already have an implemented and functional 5G architecture. The whole region suffered a delay due to COVID-19, but all the studied countries already have or are creating the required standards for an implementation that follows the international guidelines, which shows that, though they are not as evolved as other countries, their rulers and administrators are concerned about this global trend.

Keywords: Smart cities · 5G · IoT · South America · Infrastructure

1 Introduction

The fifth-generation mobile networks (5G) have a set of characteristics that are going to positively impact digital experiences and, consequently, smart cities. So, besides the greater upload and download speeds, the 5G assures a very small latency time and can simultaneously connect many devices.

Lower latency means reducing the interval between the signal sending and receiving, and the 5G has a latency of about 10 ms (i.e., half the 4G latency) and, in the best scenario, can reach 1 ms, which means that the data are transferred in real-time. Also, in the new 5G networks, the speed and latency do not get worse, even with tens of thousands of connected devices, which shows that the 5G can support a greater connection density.

Y. Iano et al. (Eds.): BTSym 2021, SIST 207, pp. 229–238, 2023.
https://doi.org/10.1007/978-3-031-04435-9_21

The match of high density and low latency will deeply change our cities. Nowadays, in crowded places (e.g., stadiums), the connection sometimes gets worse, but, with the 5G technology, it will not happen anymore, because it will be possible to simultaneously connect up to 1 million devices per square kilometer. That means that besides the personal devices (e.g., smartphones, tablets, and computers), many other devices, objects, and sensors will be capable of sending and receiving information and communicate among them. Other characteristics of the 5G are the simplicity and the low energy consumption, which are paramount for the deployment of IoT (Internet of Things) applications. In this context, the 5G removes one of the hindrances for the development of the IoT, which will be able to express its potential in home environments, industrial facilities, public buildings, and streets.

In a general context, the 5G technology moves the smart cities construction from theory to practice and paves the way for the development and implementation of new applications that can cover the monitoring of the air quality, the use of energy, traffic standards, street lighting, smart parking, crowd management, emergency response, and others. Smart cities apply digital solutions, technology, and data to substantially improve the key indicators of life quality. This entails better traffic and travel time, fast emergency response time, lower costs with health, less water and unrecycled residues waste, less harmful emissions, and, ultimately, a huge economic potential [1]. This presents new business opportunities for the companies that offer services and applications to manage complex IoT ecosystems and convert data into smart perceptions. Many initiatives at the European national and regional level contribute to face the challenges related to the cities. The European Union began the program Horizon 2020 and the Digital Single Market strategy [2].

In contrast, most of the South American countries are still making the first pilot tests of the 5G in their main cities. It happens because, besides the legal and technical hindrances, the continent was very affected by the COVID-19 pandemic. It is important to say that these countries' governments are in the step of planning the reforms of communication laws and the auction of frequencies for mobile operators to submit biddings for accreditation and subsequent offer of services to end customers.

2 Theoretical Concepts

Although there is not just one accepted definition, the contemporary understanding of a smart city [3, 4] takes up a coherent urban development strategy managed by the city government, which try to plan and align, in the long term, the management of the infrastructure assets and services of the city aiming to improve the citizen life quality [5, 6].

For this reason, over century XX, the idea that a city could be intelligent was a scientific fiction depicted in the popular media but, with the massive proliferation of new technologies and smart devices, the perspective that a city can become smart and even sentient is quickly coming true [7]. The convergence of information and communication technologies is creating urban environments that are different from everything that has already been tried until now.

When the cities become smart, it is possible to automate functions that serve people, buildings, and traffic systems, and also monitor, understand, analyze, and plan the city to improve the efficiency, equity, and citizen life quality in real-time [7]. It is changing the way how people can plan on many time scales, that is, there is a gain of perspective that the cities can become smarter in the long term through thinkings and actions in the short term.

Smart cities are frequently depicted as constellations of instruments connected through many networks that provide continuous data about the movement of people and materials in function of the flow of decisions about the physical and social shape of the city [7]. These instruments and the way how they communicate characterize the concept of IoT.

The term "IoT" was firstly used by Kevin Ashton in 1999 and became widely disseminated because of the work of Auto-ID Center, a research group that works with Radio-Frequency Identification (RFID) and other emerging technologies in this area [8].

The IoT is going from childhood to maturity and establishing itself as part of the Internet of the future [9]. The Internet is a global system of interconnected computer networks that use the same set of protocols (TCP/IP) to serve billions of users worldwide. It is a network made up of millions of private, public, academic, business, and government networks connected through a broad range of technologies of electronic, wireless, and optical networks [10].

Nowadays, there are more than 100 countries linked with the data, news, and opinions exchange through the Internet. According to the Internet World Statistics, on December 31st, 2011, there were about 2,267,233,742 users on the Internet, which was equivalent to 32.7% of the world population [10].

The essential components for the urban development of a smart city must include smart technologies, smart industries, smart services, smart management, and smart lives. The main characteristics of a smart city comprise a high level of integration of information technology and a broad application of information resources [11].

The IoT consists of installing sensors (RFID, IR, GPS, scanners, and others) for everything and connect them to the Internet using specific protocols for information exchange, aiming to develop applications for smart recognition, location, tracking, monitoring, and management [11]. The new applications of IoT are making possible the initiatives of smart cities around the world. They offer the capability to remotely monitor, manage, and control devices and create new insights and information from massive data flows in real-time [11].

When referring to the connectivity of IoT devices and applications, it is important to keep in mind that the 5G is the backbone of this system, and it will pave the way for the development of smart cities [12]. The 5G is a paradigm shift of the operation of wireless transmissions. Besides presenting technical improvements when compared with the fourth-generation mobile networks (4G), it also covers new areas, like IoT, artificial intelligence, and others [13].

The 5G will have to support mainly three kinds of services with different technical requirements: eMBB (enhanced Mobile Broadband), URLLC (Ultra-Reliable Low-Latency Communications), and mMTC (massive Machine-Type Communications) [13]. The kinds of services mentioned above are aimed at specific scenarios and applications. The access to multimedia (e.g., high-resolution videos and augmented reality) can be cited as an eMBB scenario. The URLLC is aimed at applications in which the latency is a critical point (e.g., unmanned vehicles and industrial automation). On the other hand, the mMTC must be applied when the number of connected devices is a critical requirement, and it is in this context that the 5G directly relates to the smart cities [14].

Despite the tendency of associating the concept of smart cities with the mMTC services, it is important to highlight that the services and applications that make a smart city are much more wide-ranging, and they can use all the kinds of services offered by the 5G. The smart cities' intrinsic applications can be divided, for example, between those that improve the city infrastructure, those that improve the inhabitants' quality of life (social and economic welfare), and those that improve the city management as a whole [12].

Concerning the cities infrastructure, the supply of water, power, and gas, garbage collection, street lighting, public transportation, and traffic control must be ensured [12]. In a smart city, the process of water and gas distribution can be compared with an industrial process in which is necessary to monitor and control variables (e.g., water pressure). In this same context, the power supply and street lighting must be based on the concept of smart grid, which allows the providers to monitor in real-time the consumption, and the customers to generate electrical energy in their own homes, sending the surplus back to the network [15].

On the question of transportation, there must be integration and intelligence that ensure the traffic flow and avoid, or at least minimize, the jams [12].

From the point of view of social and economic welfare, it is possible to cite the safety issues, which require real-time monitoring of the city and the capability of processing the data achieved from the monitoring system. It is also important to mention the diversification of the access to health and education through the remote monitoring of the patients' health and remote access to the classrooms, extending the reach of the education systems [12].

With respect to the cities management, it is essential to integrate all the sectors responsible for the public administration and also ensure transparency for the population [12].

Considering the services described above, it is possible to see that a smart city can take advantage of all the characteristics of 5G. So, the eMBB can be applied in the services that require video transmission in real-time, the monitoring with cameras, and the remote classes. The URLLC is a great tool for the systems that control water, gas, and power distribution. It is also useful in traffic control and might be used, in the future, for the communication of unmanned vehicles. Besides, considering the number of devices (sensors, actuators, and others) connected in a smart city, the mMTC is essential.

3 Methodology

In this section, the rollout of the 5G will be analyzed for each of the all South American countries and territories: Argentina, Bolivia, Brazil, Chile, Colombia, Ecuador, Guyana, Paraguay, Peru, Suriname, Uruguay, and Venezuela, as well as the French Guiana territory.

This research has an investigative nature and analyzes updated information of the last three years achieved from government agencies' websites and media outlets specialized in this area to highlight and emphasize the importance of this topic.

3.1 Argentina

The National Communication Entity (ENACON) called a sample of 5G technologies in March 2021 for the actors of the information and communication technologies sector to present the qualities of the 5G [16]. National authorities, representatives of the providing companies, and specialized journalists will attend the samples, and the possibilities of the new technology applied in different areas will be presented. Meanwhile, the cities of Buenos Aires e Rosário were awarded the first 5G network in the country, using DSS (Dynamic Spectrum Sharing), which shares the 4G spectrum [17].

3.2 Bolivia

Bolivian state-owned telecommunications company, Entel, organized an event where the first tests of 5G technology were carried out in the country. The expectation is that it will be gradually implemented in the territory, starting from the cities of La Paz, Cochabamba, and Santa Cruz [18]. However, it was not possible to locate official sources from the Bolivian Ministry of Telecommunications presenting the evolution of this project.

3.3 Brazil

In February 2021, the National Telecommunications Agency (ANATEL) approved the auction for radio-frequency bands for the provision of telecommunications services through 5G in Brazil [19]. This auction offers the frequencies of 700 MHz, 2.3 GHz, 3.5 GHz, and 26 GHz for the exclusive use of 5G and determines the conditions that must be observed by the companies that win the bid. While the auction does not take place, the telecommunication operators present in Brazil are already offering the 5G DSS, which shares the spectrum that is currently used by 4G LTE [20].

This process will be monitored by a working group from the lower house of the National Congress, which will meet with representatives from the Ministry of Communications, the Ministry of Science and Technology, ANATEL, and large and small internet providers [21].

3.4 Chile

Chile concluded, in February 2021, the first auction aimed at the implementation of 5G in Latin America [22] and defined the companies responsible for the 700 MHz, AWS, 3.5 GHz, and 26 GHz bands. The winning companies must guarantee, obligatorily, the coverage for 199 hospitals, 24 ministries, 16 stewardships, 16 regional capitals, and 56 provincial capitals and, optionally, 17 airports, 12 centers of public interest, 28 higher education institutions, and 23 seaports [23]. The bidding foresees a maximum period of three years to put the technology into operation, and 100% of hospitals must be covered in the first year, 100% of mandatory locations in 18 months, and 88% of communes, considering all bands auctioned, in 18 months.

3.5 Colombia

In December 2019, Colombia's Ministry of Information and Communications Technologies (Min. ICT) launched the 5G Adoption Action Plan [24], which aims to identify regulatory barriers for 5G deployment and operation and carry out pilot tests.

In September 2020, the government initiated pilots, granting temporary permits for participants to test the 5G technology. On the other hand, according to the Action Plan, the 3.5 GHz band will be allocated until the third quarter of 2021. Also, no specific rules for 5G have been developed so far. According to the 5G Adoption Action Plan, regulatory changes will be made by the second quarter of 2021 [24].

3.6 Ecuador

According to the Ministry of Telecommunications, 2020 was the year promised for the implementation of the 5G technology in Ecuador as part of the government proposal of Digital Ecuador. However, given the limitations caused by the pandemic and the lack of regulations on the use of spectrum, the implementation of 5G in the country was delayed.

In 2021, the Minister of Telecommunications requested the International Telecommunication Union (ITU) to value the 3.5 GHz band for the rollout of 5G, which will serve as inputs for the next government of Ecuador to start the competitive process of renegotiating contracts with the mobile phone operating companies and open the country to the 5G technology [25].

Some mobile operators have stated that one of the main limitations for the implementation of 5G in Ecuador is the cost of the radium spectrum, which would prevent the full use of 5G and its subsequent use in smart cities. On the other hand, the Ecuatorian mobile operator CNT chose telecommunications company Nokia to help it deploy the country's first 5G network, which will be a Non-Standalone 5G (NSA 5G). Thus, the first NSA 5G sites will be deployed in the cities of Guayaquil and Manta, allowing CNT to plan its 5G services with ultra-low latency, connectivity, and capacity, as well as reducing complexity and increasing cost efficiency.

3.7 Guyana

The Brazilian Federal Government and the government of Guyana signed a memorandum of understanding on technical feasibility studies for the implantation of optical fiber between the capital of the Brazilian state of Roraima, Boa Vista, and the capital of Guyana, Georgetown [26]. The two countries will create a working group, with the Brazilian Ministry of Communications appointing a representative to coordinate the studies [26].

Both governments intend to install, in the long term, an optical fiber cable that will have 6 thousand kilometers and will allow traffic of 72 terabits of data per second [27].

3.8 French Guiana

French Guiana is an outlier when compared to other countries in South America, as it is a French territory. It follows French rules and obeys a specialist entity, ARCEP, which takes care of the regulation and communication legislation throughout its territory.

France launched the deployment of its 5G network in 2020, as planned, despite the difficulties faced by the Chinese equipment manufacturer Huawei, but it has been in a step of testing and use by the inhabitants since then [28].

3.9 Paraguay

According to the president of the Telecommunications National Commission of the Republic of Paraguay (CONATEL), the frequencies dedicated to 5G in the country should not be tendered before 2024, as they are going to be contracted by other services until that date [29]. However, this does not mean that Paraguay is not advancing in its digital transformation. The country is currently building an optical fiber backbone nationwide, whose first of three phases was completed in June 2020 [30].

3.10 Peru

Peru started with the authorizations to work with 5G technology in the Non-Standalone format, which means that the telephone companies could offer it using 4G LTE infrastructures and the use of fixed wireless access (FWA), according to the Ministry of Transport and Communications. FWA technology is used as mobile technology for last-mile connectivity in fixed broadband networks [31].

In the first tests carried out by the leading mobile services company, it reached a speed of 3.3 Gbps in download speed. The said tests were made jointly with the company Huawei Technologies [32].

On the other hand, at the beginning of the year 2021, the Ministry of Transport and Communications (MTC) of Peru announced that the planning phase for an auction to the spectrum of 3.5 GHz and 26 GHz is concluded. With this, the ministry will disclose the details and the timetable of the bidding process.

Peru is strongly encouraging the deployment and adoption of 5G, as the MTC considers technology to be instrumental in the country's economic recovery.

3.11 Suriname

Telecommunications from Suriname (Telesur) launched its 5G network in November 2019, making it the first country in the Caribbean to have 5G available [33].

The operator's 5G commercial network has been active in the country's capital, Paramaribo, since December 2019, but Telesur has not promoted it strongly until now, as it was deploying LTE-A updates in strategic locations to complement the offer [34].

3.12 Uruguay

In April 2019, Uruguay, through the National Administration of Telecommunications (ANTEL) in partnership with Nokia, became the first country in Latin America to make the 5G Internet commercially available in the cities of La Barra and Nueva Palmira and expecting the expansion of access to the rest of the country quickly [35]. Besides, the country is the third in the world to make the new technology available, behind only

South Korea and the USA. At first, the company responsible for the implementation offered the service as a complement to optical fiber networks, targeting regions where broadband networks did not reach, since, at the time, there were still no cellular devices with support for the 5G network [36]. From October 2020, scheduled to end in March or April 2021, a coordinated work began between the Regulatory Unit of Communication Services (URSEC) and the telecommunication companies operating in the country to test the spectrum available for the expansion of the 5G network and, if necessary, open a bid on other frequency bands [37].

3.13 Venezuela

A telecommunications network with 5G technology will begin to be installed on an experimental basis in Venezuela, said President Nicolás Maduro. From the Miraflores Palace, where an event was held to commemorate the 45th anniversary of the establishment of relations between Venezuela and China, the head of state pointed out: "Venezuela is not far behind; we are advancing with technology." More than 200 Chinese companies are working and investing in the country's development [38].

4 Conclusion

Currently, there is a latent need and a worldwide trend to make cities smarter and smarter. As much as there are diverse and comprehensive definitions of this concept, all of them have in common the strategy of urban development through Information and Computing Technologies, aiming at the collection of important information for the administration of cities and the processing of this information to improve infrastructure, services and, mainly, the quality of life of citizens.

At the core of the implementation of the services that make a city smart, is the Internet of Things, as it enables the integration of the various sectors of a city through sensors, communication protocols, data management, and analysis systems.

However, IoT, even though it is considered the solution to create a smart city, cannot be applied correctly if the telecommunication infrastructure is not adequate, because, when talking about IoT, it implicitly talks about a constellation of devices concentrated in the same location sending data at the same time. In other words, it is necessary to have an infrastructure that supports a high density of connections and has low latency, for example.

It is considering this prerequisite that 5G becomes relevant for smart cities. The fifth generation of mobile networks was developed based on requirements that are sufficient to meet the demand for the Internet of Things applied to smart cities. The 5G networks deployed are expected to be able to serve at least 1,000,000 devices per square kilometer with a maximum of 10 ms latency. Other specifications have not been mentioned, but they are all superior to the fourth generation networks currently used.

All governments and administrators in the studied countries are aware of this trend and are concerned with updating the infrastructure of the telecommunication networks in their territories and with the implementation of the services promised by 5G. However, this is not an easy and quick process. Each country adopts the strategies that are most appropriate according to its reality and because of the great inconvenience and delays caused by the current situation of the COVID-19, they all are in different phases of implementation. There are countries, such as Guyana and Paraguay, that are still creating the fiber optic backbone; Argentina, Brazil, Ecuador, and Peru, are using the previous generation (4G) structure to provide a limited 5G (5G DSS or 5G NSA) during the transition process; Chile and Venezuela have already defined the responsible companies and are in the installation phase; Bolivia, Colombia, and French Guiana have already activated the 5G networks in some cities and are currently testing them; Suriname and Uruguay, already have 5G commercially available, but still limited to small parts of the territory. Finally, even if at different stages, all countries are evolving to keep up with the global trend that involves 5G, IoT, and smart cities.

References

1. A Systems Approach. https://5g.systemsapproach.org/intro.html
2. Nae.global. https://nae.global/pt/introducao-ao-5g-o-grande-salto-para-a-internet-das-coisas/
3. Bowerman, B., Braverman, J., Taylor, J., Todosow, H., Von Wimmersperg, U.: The vision of a smart city. In: 2nd International Life Extension Technology Workshop, Paris, vol. 28 (2000)
4. Hollands, R.G.: Will the real smart city please stand up? intelligent, progressive or entrepreneurial? City, **12**(3), 303–320 (2008)
5. Clavell, G.G.: The drivers, impact and risks of surveillance-enabled smart environments. Sci. Public Policy **40**(6), 717–723 (2013)
6. Schaffers, H., Sallstrom, A., Pallet, M., Munoz, J.M., Santoro, R., Trousse, B.: Integrating living labs with future internet experimental platforms for co-creating services within smart cities. In: Concurrent Enterprising (ICE), 17th International Conference IEEE, pp. 1–11 (2011)
7. Batty, M., Axhausen, K.W., Giannotti, F., et al.: Smart cities of the future. Eur. Phys. J. Spec. **214**, 481–518 (2012). https://doi.org/10.1140/epjst/e2012-01703-3
8. van Kranenburg, R., Bassi, A.: IoT challenges. Commun. Mob. Comput. **1**(1), 1–5 (2012). https://doi.org/10.1186/2192-1121-1-9
9. Novo, O.: Blockchain meets IoT: an architecture for scalable access management in IoT. IEEE Internet of Things J. **5**(2), 1184–1195 (2018). https://doi.org/10.1109/JIOT.2018.2812239
10. Madakam, S., Ramaswamy, R., Tripathi, S.: Internet of Things (IoT): a literature review. J. Comput. Commun. **3**, 164–173 (2015). https://doi.org/10.4236/jcc.2015.35021
11. Kim, T., Ramos, C., Mohammed, S.: Smart city and IoT. Future Gene. Comput. Syst. **76**, 159–162 (2017), ISSN 0167-739X, https://doi.org/10.1016/j.future.2017.03.034
12. Rao, S.K., Prasad, R.: Impact of 5G technologies on smart city implementation. Wireless Pers. Commun. **100**(1), 161–176 (2018). https://doi.org/10.1007/s11277-018-5618-4
13. Fuentes, M.C., et al.: 5G new radio evaluation against IMT-2020 key performance indicators. IEEE **8**, 110880–110896 (2020). https://doi.org/10.1109/ACCESS.2020.3001641
14. Ortiz, J.N.D., Sendra, P.D., Ameigeiras, S., Munoz, P., Soler, J.J.R., JMl,: A survey on 5G usage scenarios and traffic models. IEEE Commun. Surv. Tutorials **22**(2), 905–929 (2020). https://doi.org/10.1109/COMST.2020.2971781
15. Fang, X., Misra, S., Xue, G.Y., D,: Smart grid — the new and improved power grid: a survey. IEEE Commun. Surv. Tutorials **14**(4), 944–980 (2012). https://doi.org/10.1109/SURV.2011.101911.00087
16. Ente Nacional de Comunicaciones. https://www.enacom.gob.ar/institucional/enacom-convoca-a-pruebas-de-tecnologia-de-5g-en-argentina_n3038

17. **El Economista.** https://eleconomista.com.ar/2021-02-telecom-habilito-la-primera-red-5g-en-la-argentina-para-personal/
18. **Mas Container.** https://www.mascontainer.com/bolivia-espera-contar-con-tecnologia-5g-el-ano-2020/
19. **Governo Federal do Brasil.** https://www.gov.br/pt-br/noticias/transito-e-transportes/2021/02/anatel-aprova-edital-de-leilao-de-5g
20. **Repositório do Conhecimento do IPEA.** http://repositorio.ipea.gov.br/handle/11058/10419
21. **Câmara dos Deputados.** https://www.camara.leg.br/noticias/716684-camara-cria-grupo-de-trabalho-para-acompanhar-implantacao-da-tecnologia-5g/
22. **Governo Federal da Colômbia.** https://www.gob.cl/5g/
23. **Sub-Secretária de Telecomunicações.** https://www.subtel.gob.cl/concursos5g/
24. **El Tiempo.** https://www.eltiempo.com/tecnosfera/novedades-tecnologia/5g-como-avanzan-los-planes-piloto-en-colombia-de-redes-5g-554580
25. **Primicias.** https://www.primicias.ec/noticias/tecnologia/tecnologia-5g-red-no-avanza-ecuador/
26. **Governo Federal do Brasil.** https://www.gov.br/pt-br/noticias/transito-e-transportes/2020/12/brasil-e-guiana-assinam-memorando-de-entendimento-para-implantacao-de-fibra-optica-de-boa-vista-a-georgetown. Accessed 1 Dec 2020
27. **Rede Brasil Atual.** https://www.redebrasilatual.com.br/blogs/planeta-azul/2021/02/cabo-submarino-que-pode-ser-caminho-do-5g-no-brasil-ja-viaja-pelo-atlantico/. Accessed 05 Feb 2021
28. **Bol Uol.** https://www.bol.uol.com.br/noticias/2019/05/21/franca-lancara-redes-5g-em-2020-apesar-de-problemas-da-huawei.htm. Accessed 21 May 2019
29. **BNamericas.** https://www.bnamericas.com/en/analysis/why-paraguay-is-behind-the-curve-with-5g
30. **BNamericas.** https://www.bnamericas.com/en/news/paraguay-wraps-up-first-stage-of-national-fiber-backbone
31. **BNamericas.** https://www.bnamericas.com/es/noticias/peru-autoriza-implementacion-de-tecnologia-5g-en-banda-de-35ghz#:~:text=Per%C3%BA%20autoriz%C3%B3%20la%20implementaci%C3%B3n%20de,en%20la%20banda%203.5GHz.&text=Los%20usuarios%20reclaman%20m%C3%A1s%20velocidad,lograrlo%20ser%C3%A1%20con%20la%205G
32. **Claro.** https://www.claro.com.pe/institucional/centro-de-prensa/5g-arequipa/
33. **Telesur.** https://telesur.sr/5g/
34. **Developing Telecoms.** https://www.developingtelecoms.com/telecom-business/operator-news/10284-telesur-starts-promoting-5g-network-with-data-tariffs.html
35. **Uruguai Presidência.** https://www.presidencia.gub.uy/comunicacion/comunicacionnoticias/antel-5g-internet-conectividad-internet
36. **La Diaria Politica.** https://ladiaria.com.uy/politica/articulo/2019/4/uruguay-es-el-tercer-pais-en-el-mundo-en-ofrecer-tecnologia-5g/
37. **Vida Atual.** https://www.elpais.com.uy/vida-actual/situacion-5g-uruguay.html
38. **Conatel.** http://www.conatel.gob.ve/venezuela-instalara-tecnologia-5g-en-fase-experimental-con-apoyo-de-china/

Mathematical Modeling: A Conceptual Approach of Linear Algebra as a Tool for Technological Applications

Leonardo Bruscagini de Lima[1(✉)], Yuzo Iano[1], Gabriel Gomes de Oliveira[1], Gabriel Caumo Vaz[1], Alecssander Daniel de Almeida[2], Gustavo Bertozzi Motta[2], Gabriel Matsumoto Villaça[2], Matias Oliveira Schwarz[2], and Pedro Y. Noritomi[3]

[1] School of Electrical and Computer Engineering, State University of Campinas, Campinas, Brazil
l152268@g.unicamp.br, yuzo@unicamp.br, oliveiragomesgabriel@ieee.org

[2] Institute of Mathematics and Statistics, University of São Paulo, São Paulo, Brazil
{alecssander,gbertozzi741,g.matsumoto,matias.schwarz}@usp.br

[3] Renato Archer Research Center, Campinas, Brazil
pedro.noritomi@cti.gov.br

Abstract. Mathematical modeling is the field of applied mathematics in charge of mathematically model problems, situations, and phenomena of the world where we live. The models are sets of equations that describe the behavior of these phenomena, which can be of physical nature, such as the equations that describe the movement of planets and the behavior of atoms and charged particles, social nature, such as economic models of the relationship between industry sectors or how to handle a eucalyptus forest most sustainably and profitably, and others. There are also optimization models to index webpages or encrypt information within images, for example. Each of those models uses different and sophisticated mathematical tools, like calculus and especially linear algebra, which is the scope of this article. Some models work better in the form of matrices and vectors, which facilitate the visualization of the calculations to be performed. Many of them can be converted into algorithms to make possible the aid of computational resources and achieve the solutions in a faster and more practical way. Mathematically modeling a situation or phenomenon is the first step to figure out possible solutions and, with adjustments whenever they are necessary, make predictions concerning the studied subject.

Keywords: Mathematical modeling · Forest planning · Glow of the planets · Economic modeling · Computed tomography

1 Introduction

The problems that ancient civilizations faced in their daily routine comprehended terrain measurement, assets distribution, monitoring of resources like wheat and

Y. Iano et al. (Eds.): BTSym 2021, SIST 207, pp. 239–248, 2023.
https://doi.org/10.1007/978-3-031-04435-9_22

cattle, tax calculation, and division of inheritance. Many times, these problems resulted in linear equations because linearity is the simplest possible relationship between variables. In this regard, it is possible to draw a timeline of the development of algebra from ancient times to century XIX, when the branch of mathematics named linear algebra was established.

We can imagine the world around us as an endless mass of data. Whatever the angle of vision, we can extract data and pieces of information concerning the vast universe. The key to understanding this interaction is the power of mathematical tools, the models, and the distribution of the arranged data.

The information and communication technology allowed the high-speed transmission of data, connecting and processing thousands of groups, therefore, dictating the pace of modern life. In clear opposition to the precise, organized, elegant mathematics, Big Data collects, stores, treats, and, extracts pieces of information to be applied in the understanding of the world. Conversely, if techniques of collection, treatment, and visualization of data are applied, it is possible to figure out elegant patterns that are described by mathematical modeling, and, consequently, achieve a better comprehension of the world where we live and convert this knowledge into life quality for our society.

Many of those tools have arisen as solutions for problems faced since ancient times, such as Gaussian elimination, for example. Linear algebra is the field of mathematics that deals with methods, like the concepts of vectors, vector spaces, linear transformations, and matrices, to solve linear systems, both algebraic and differential.

A linear equation system is a finite set of linear equations with variables of first power. The variables are named unknowns, e.g., x_1, x_2, ..., x_n is an equation that can be written in the algebraic form, as shown in Eq. 1.

$$a_1x_1 + a_2x_2 + ... + a_nx_n = b \tag{1}$$

In Eq. 1, the constant a_k is the coefficient of x_k, and b is the constant element of the equation.

In the field of research and development, administration, and mathematics, information usually requires a suitable arrangement for data treatment. These data are groups of rows and columns disposed to form numerical representations named matrices. This technique is important for computers to solve linear equations since they are very good at handling numerical information [1].

Thus, the pieces of information collected from a physics observation, for example, could be grouped in terms of linear equations and systems, as follows:

$$\begin{cases} 7x + 4y = 3 \\ 2x + 9y = -6 \end{cases}$$

They could also be represented as a matrix:

$$\begin{bmatrix} 7 & 4 & 3 \\ 2 & 9 & -6 \end{bmatrix}$$

A linear equation or a set of linear equations may express well-defined rules and laws of a given system in terms of vectors in a vector space, which define physical quantities and specify the number of directions and magnitude in the vector space. In addition, vector spaces with infinite dimensions are a subject of study in mathematical analysis and, in this way, the vectors are handled as functions. Vector spaces are a central theme in modern mathematics.

The relationship between two vector spaces is expressed as a linear transformation and defines the structure of the vector space because they maintain the scalar sum and multiplication [2].

$$f(v+w) = f(v) + f(w)\, e f(a \cdot v) = a \cdot f(v) \tag{2}$$

For all v and w in V and for all a in K.

A linear transformation may also be named linear application or linear map. When the domain and codomain coincide, the expression linear operator is used. In the language of abstract algebra, a linear transformation is a homomorphism of vector spaces. Thus, linear algebra is widely used in abstract algebra and functional analysis.

A way of representing linear transformations is the matrix. A matrix $A_{m \times n}$ produces the linear map from K^n to K^m , as represented in Table 1.

Table 1. Typical representation of a matrix

$$\begin{bmatrix} a_{1,1} & a_{1,2} & a_{1,3} & \cdots \\ a_{2,1} & a_{2,2} & a_{2,3} & \cdots \\ a_{3,1} & a_{3,2} & a_{3,3} & \cdots \\ \vdots & \vdots & \vdots & \ddots \end{bmatrix}$$

$$x = (x_1, x_2, \ldots, x_n) \mapsto \left(\sum_{j=1}^{n} a_{1j} x_j, \sum_{j=1}^{n} a_{2j} x_j, \ldots, \sum_{j=1}^{n} a_{mj} x_j\right) \tag{3}$$

In Eq. 3, $\sum$ comprises the summation using the matrices multiplication of A with the coordinate vector X:

$$x \mapsto Ax \tag{4}$$

After choosing the bases of V and W, any linear transformation $f : V \to W$ will be uniquely represented by a matrix through this procedure.

Linear algebra also has its particular representation in analytic geometry, which may be named coordinates geometry or Cartesian geometry, and studies geometry through algebra principles.

In general, the Cartesian coordinate system is used to convert equations into planes, straight lines, curves, and circles. The coordinate system usually has two dimensions, but it can have three or more dimensions. The first studies of Analytic Geometry were carried out in century XVII by the philosopher and mathematician René Descartes (1596–1650), who came up with the Cartesian coordinates, which allowed the numerical representation of geometric properties.

In this sense, it is intended to base the use of algebra in everyday applications of the scientific and technological fields, e.g., the calculation of the optimal sustainable yield of a periodical cut of trees with different classes of height. There is also a multidisciplinary approach in the field of astronomy: the calculation of the orbit and glow of celestial bodies. An economic approach can also be demonstrated with some results about the way how non-negative matrices are applied to determine the structures of equilibrium prices and the necessary yield to meet the demand. Finally, it is demonstrated how the construction of an image of a cross-section of a human body from an X-ray scan analysis leads to an inconsistent linear system.

2 Forest Planning

When one simulates a simplified model for the sustainable cut of a forest whose trees are classified by their height, it is supposed that the tree's height when it is cut and sold determines its economic value. For a better understanding of the dynamic of the forest and the meaning of sustainable harvest, it is necessary to point out that, during a period, the forest grows unrestrictedly, and then some trees with different heights are cut. The remaining trees must have the same configuration of size as the trees of the original forest, characterizing a sustainable cut.

There are some necessary conditions to make a sustainable harvest. The first one is that the number of trees must be the same at the beginning and at the end of the processes in such a way that, if the sum of the x_i of matrix x is equal to a scalar s, this value must be the same in the initial and final phase of the harvest, which bring the condition that the matrix x must keep invariable. For that to happen, the harvest vector y must have the sum of y_i equal to the sum of a vector "r" that may be described by $r = R \times y$, where R is a square matrix $n \times n$ with the first row filled with 1s, and n is the number of classes established for a kind of tree.

Now, it is necessary to study what happens after a period of growth. Before the growth, there is the initial configuration of trees, which is indicated by the column matrix x. After a period of growth, each class of trees has a percentage of grown trees given by g_i, where i concerns the class of the tree. It means that $1-g_i$ trees of each class remain the same size and g_i trees grow up to the next class. Therefore, the number of trees of each class is given by $(1-g_i)x_i + (g_i - 1)x_i - 1$, i.e., the number of trees that remained in the class after a growth period plus the number of trees that were in a lower class and, after the growth, change to the current one. After all the explanations, a sustainable harvest is achieved when the following model is satisfied [3].

$$G \times x - y + R \times y = x \tag{5}$$

G is the growth matrix $n \times n$ that, when multiplied by the column vector x, results in the new quantity of trees in each class. Equation 5 can be simplified if all the terms with vector x remain on one side, and those with vector y remain on the other side. Using the identity matrix property ($y = y \times I$), the result is

$$(I - R) \times y = (G - I) \times x$$

Since each y_i represents the number of trees removed from each class, it must be greater or equal to zero, so a condition for a sustainable harvest is:

$$g_i - 1 \times x_i - 1 \geq g_i \times x_i \geq 0 \tag{6}$$

The optimal sustainable return is the way the trees could be cut to bring the greatest profit, following the equations of sustainable harvest. The return is equal to the sum of the cut trees in each class multiplied by their respective prices (Eq. 7).

$$RT = p_2 \times y_2 + p_3 \times y_3 + \ldots + p_n \times y_n \tag{7}$$

For this kind of situation, it is possible to apply a linear optimization theorem, which determines that, to maximize the profits, trees of just one class should be removed, i.e., if we cut the p-th class, all the other values in the vector y are equal to zero:

$$y_p = g_1 \times x_1 = g_2 \times x_2 = g_3 \times x_3 = \ldots = g_p - 1 \times x_p - 1 \tag{8}$$

This implies that:

$$x_3 = \frac{g_1 \times x_1}{g_3} \ldots x_p - 1 = \frac{g_1 \times x_1}{g_p - 1} \tag{9}$$

Putting the set of equations together and considering the fact that the sum of x_i is equal to s, the following result is achieved:

$$x_1 = \frac{s}{1 + \frac{g_1}{g_2} + \frac{g_1}{g_3} + \ldots + \frac{g_1}{g_{p-1}}} \tag{10}$$

Since we are applying the theorem that determines we just need to cut one class of trees, the equation of optimal return is reduced into $RT = p_p \times y_p$. Thus, the optimal return in a sustainable harvest, where "p_p" is the price of the p-th class, "s" is the number of cut trees, and "g_i" is the price of the i-th class, is given by the following equation:

$$RT = \frac{p_{p^s}}{\frac{1}{g_1} + \frac{1}{g_2} + \ldots + \frac{1}{g_{p-1}}} \tag{11}$$

3 The Glow of Venus

Venus is the second planet from the Sun and, after the Moon, is the brightest object in the night sky, with an apparent magnitude of −4.6. Since it is closer to the Sun than the Earth, Venus presents phases, like the Moon, and that is the reason why its apparent glow changes over time. This glow variation makes it possible to describe its brightness through trigonometrical manipulations and find out its maximum and minimum through mathematical methods.

The apparent glow of Venus L is directly proportional to its phase p and a constant K, and it is inversely proportional to the squared distance between the Earth and Venus r, as shown in Eq. 12.

$$L = k\frac{p}{r^2} \tag{12}$$

The phase p is defined by the ratio between the planet's illuminated visible area and its illuminated total area, as described in Eq. 13.

$$p = \frac{\frac{1}{2}\pi\rho^2(1 + \cos(\phi))}{\pi\rho^2} = \tfrac{1}{2}(1 + \cos(\phi)) \tag{13}$$

The ρ presented in Eq. 13 refers to the radius of Venus, and ϕ is the angle Sun-Venus-Earth. The distance r cannot be described as a constant value because it varies according to the position of the planets in their orbits. However, since the eccentricity of the orbits of Venus and Earth are low, both may be considered spherical, which makes the distances Earth-Sun and Venus-Sun constant. With this artifice, it is possible to apply the Law of Cosines to describe the distance Earth-Venus (Eq. 14) with just one variable, i.e., the angle Sun-Venus-Earth. It was established that the value of the distance Earth-Sun is 14.95×10^7 km and the value of the distance Venus-Sun is 10.81×10^7 km.

$$r = \sqrt{R^2 + q^2 - 2Rq\cos(\phi)} \tag{14}$$

R is the radius of the orbit of Earth, and q is the corresponding for Venus. From Eq. 14, one can calculate the value of $\cos(\phi)$ according to the distance r, which, when replaced in Eq. 12, makes it possible to calculate the apparent glow of Venus according to just one variable (r), as shown in Eq. 15.

$$L = K\frac{p}{r^2} = K\left(\frac{2qr + r^2 + q^2 - R^2}{r^3}\right) \tag{15}$$

To determine the values of maximum brightness, one may determine the growth rate of the glow and observe when it is annulled, which could point out a spike of brightness. This process is the same as determining the derivative of a function and make it equal to zero, which is presented in Eq. 16.

$$L' = -\frac{K}{r^4}\left(4qr + r^2 + 3(q^2 - R^2)\right) = 0 \tag{16}$$

From the previous equation, one can determine that the value of r to maximize the glow is:

$$r = -2q \pm \sqrt{q^2 + 3R^2} \quad (17)$$

If it is established that the distance Earth-Sun is $R = 14.95 \times 10^7$ km, and the distance Venus-Sun is $q = 10.81 \times 10^7$, the maximum glow happens when $r = 6.44 \times 10^7$ km or $r = -49.68 \times 10^7$ km. Since the distance cannot be physically interpreted as a negative value, the second result is discarded, and the final result is that the maximum glow happens when the distance Earth-Venus is about 6.44×10^7. Moreover, it is also possible the use Eq. 14 to isolate the term $\cos(phi)$ and replace the value calculated for r, achieving the value of ϕ [3,5].

$$\cos(\phi) = \frac{R^2 + Q^2 - R^2}{2Rq} \Rightarrow \cos(\phi) = \frac{14.95^2 + 10,81^2 - -6,44^4}{2 \times 14.95 \times 10.81} = 0.92 \quad (18)$$

Finally, calculating the arc cosine of the previous result, the angle Sun-Earth-Venus when the maximum glow happens is about $22°22'31'' \pm n \times 180°$, where n is an integer number.

4 Economic Relationships

In the economic field, the demonstration of the Leontief mathematical model is a great example of the application of linearity. This model aims to increase profits, reduce costs, and reshape a company or a country to provide an economically sustainable environment [6–8].

Considering an economy divided into three sectors, i.e., manufacturing, agriculture, and services, three scenarios are studied:

1. For each unit produced, the manufacture needs 0.10 units from other companies of the same sector, 0.30 units from agriculture, and 0.30 units from services.
2. For each unit produced, agriculture uses 0.20 units from its own production, 0.60 units from the manufacture, and 0.10 units from services.
3. For each unit produced, the services sector consumes 0.10 units from itself, 0.60 units from the manufacture, and no product from agriculture.

Table 2. Matrix of consumption

	Manufacture	Agriculture	Services
Manufacture	0.10	0.60	0.60
Agriculture	0.30	0.20	0
Services	0.30	0.10	0.10
Total	0.70	0.90	0.70

For this demonstration, a specific demand of production of 100 units for consumption in agriculture to achieve the following matrix of consumption, which is shown in Table 2.

One can say that, if the agriculture plans to produce a total of 100 monetary units, it should consider the consumption matrix above to achieve the value of the final demand and, consequently, the intermediate demand vector D_i.

$$D_i = \begin{pmatrix} 72 \\ 38 \\ 29 \end{pmatrix} \tag{19}$$

Finally, with coefficients presented in Table 3, it is possible to affirm that the Leontief matrix increases the production to meet the demand, remodeling all the sectors to keep the expenses low.

Table 3. Matrix of coefficients

	Manufacture	Agriculture	Services
Manufacture	2.222	1.852	1.481
Agriculture	0.833	1.944	0.556
Services	0.833	0.833	1.667

5 Computed Tomography

The computed tomography was created by Sir Godfrey Hounsfield and had its first patient in 1971. It was the first non-invasive method for the achievement of internal images: it interprets thousands of pixels beams and transforms them into data through the method of scanning a cross-section of the human body with an x-ray.

The intensity of the x-ray that passes through the cross-section of the human body is measured by a detector. In this process, the intensity of the emitted x-ray is known and the intensity of the received x-ray is achieved through measurements. Physically, the difference between the emitted and measured x-rays, supposing that there is no spread, is determined by the tissue absorption. It is known that denser tissues, e.g. tumors, absorb a greater amount of radiation [9–11].

These data are processed and produce a two-dimensional image of the scanned section. The tomographic exams, also known as x-ray tomography, use x-ray beams produced by a machine and transmitted through the patient. Using some algorithms of Algebraic Reconstruction Techniques (ART), the following systems are achieved:

$$\begin{array}{ccccccccc} a_{11}x_1 & + & a_{12}x_2 & + & \cdots & + & a_{1n}x_n & = & b_1 \\ a_{21}x_1 & + & a_{22}x_2 & + & \cdots & + & a_{2n}x_n & = & b_2 \\ \vdots & & \vdots & & & & \vdots & & \vdots \\ a_{m1}x_1 & + & a_{m2}x_2 & + & \cdots & + & a_{mn}x_n & = & b_m \end{array}$$

In this way, we have a system with M equations (beam equations) and N unknowns to implement the treatment algorithm for overdetermined systems. A method of ART was applied to solve the system of three equations and two unknowns.

$$\begin{array}{l} L_1 : x_1 + x_2 = 2 \\ L_2 : x_1 - 2x_2 = -2 \\ L_3 : 3x_1 - x_2 = 3 \end{array}$$

This algorithm consists of defining an arbitrary initial point x_0 and projecting it orthogonally over the straight line L_1. This point is considered $x_{(1)1}$. The exponent indicates that this is the first of a series of cycles of the algorithm.

6 Final Considerations

In mathematics, linear algebra is the field that studies vectors, vector spaces, linear transformations, and linear equations systems. Although its definitions are very elegant, the concepts may be implicit for the reader. Conversely, the concepts assume a multidisciplinary characteristic and can be applied and observed in other areas of knowledge.

In physics, the concept of linearity is associated with the properties of the differential equations that govern the fundamental laws of nature, e.g., Maxwell equations. In electrical engineering, especially in the instrumentation field, linearity is associated with the behavior of a variable between the measurement system inputs and outputs.

Another observable pattern happens in astronomy, in the position and glow of the planet Venus. Venus rotates on its own axis in 243 terrestrial days, the slowest rotation among all the planets. The average interval of 584 days between the successive approximations of the Earth is almost equal to five Venusian solar days. After 584 days, Venus appears 72° apart from the last inclination. After five periods of 72° in a circumference, Venus returns to the same position in the sky every eight years (minus two days that correspond to the leap years).

In management and administration, the model of sustainable forest planning allows the achievement of everyday goals without affecting environmental variables. The estimation of growth rate is complex because of the big amount of data that should be considered to reach a confidence index reasonably high. Besides that, the model does not consider the trees that died. however, despite those limitations, the model provides a good estimate, assuming that the growth of the trees happens linearly.

Still in the management field, we can mention the observable pattern in the economic modeling that uses matrix methods to study the sectors of a company or any given system, keeping the lowest possible expense of materials, producing just the necessary to meet the demand, and achieving the maximum profit. For this purpose, it was used the input-output analysis of Wassily Leontief, the Nobel Prize winner.

Finally, we could apply the linearity concepts in the process of reconstruction of a tomography from two-dimensional projections through linear algebra. The system could be solved with iterative methods that allow the reconstruction of a tomography image in three dimensions. The schema of ART discussed in this work was replaced by faster and more sophisticated techniques that provide a more accurate vision in the cross-section. However, the new techniques date back to the same mathematical issue of finding an approximated solution for overdetermined systems with a big amount of linear equations.

References

1. Anton, H., Rorres, C.: Álgebra linear com aplicações. Bookman Porto Alegre (2001)
2. Roman, S., Axler, S., Gehring, F.W.: Advanced Linear Algebra. Springer, Berlin (2005). https://doi.org/10.1007/0-387-27474-X
3. Coníferas de Rápido crescimento (2021) In: pt.punchalo.com. https://pt.punchalo.com/5531-fast-growing-conifers. Accessed 20 Aug 2021
4. Usher, M.B.: A matrix approach to the management of renewable resources, with special reference to selection forests. J. Appl. Ecol. 355–367 (1966)
5. Wildfogel, D.: The maximum brightness of Venus. Math. Mag. **57**(3), 158–165 (1984)
6. Leontief, W., Leontief, W.W., Duchin, F., Faye, D.: The Future Impact of Automation on Workers. Oxford University Press, New York (1986)
7. Revista1 (1998) "SISTEMAS DE LEONTIEF." http://cepeac.upf.br/download/rev_n11_1998_art7.pdf. Accessed 20 Aug 2021
8. MathCon (2014) MathCon (2014) "MODELO DE LEONTIEF." http://www.math.com.mx/docs/cur/cur_1_001_Economia_Leontief.pdf. Accessed 20 Aug 2021
9. da Costa, F.M., Agustini, E.: Algebra Linear e Formaçao de Imagens: a Tomografia Computadorizada. FAMAT em Rev, 193 (2005)
10. De Cezaro, A., De Cezaro, F.T.: Problemas Inversos e a matemática da tomografia computadorizada. V Bien da Soc Bras Matemática, UFPB, João Pessoa (2010)
11. Amorim-Lopes, M., et al.: Enhancing optimization planning models for health human resources management with foresight. Omega **103**, 102384 (2021)

Black-Shoes-Merton Model and Neural Networks in River Level Prediction: Case Study on La Leche River - Peru

Diana Mercedes Castro Cárdenas[1](✉), Segundo Francisco Segura Altamirano[1], and Merly Liliana Yataco Bernaola[2]

[1] Pedro Ruiz Gallo National University, Lambayeque, Peru
{dcastroc,sseguraal}@unprg.edu.pe
[2] Universidad San Luiz Gonzaga, Ica, Peru
merly.yataco@unica.edu.pe

Abstract. The rugged relief of Peru determines a particular hydrological regime. This geographical context includes our region, which is also affected by meteorological phenomena, e.g., El Niño and La Niña, that occur unpredictably and whose effects we feel with heavy rains and floods in the north of Peru. For that reason, the capability of being able to forecast river levels, in particular the river La Leche, is essential. For this purpose, we use the Black-Sholes-Merton stochastic differential equation of the river level and other parameters achieved from meteorological stations within the area of influence of the La Leche river basin as inputs to an LSTM Neural Network, which was trained with downloaded data and can forecast the river level 6, 12, 18, and 24 h in advance. The performance tests of the obtained neural networks demonstrated a high adaptation of the solution to the hydrological model since the NSE is very close to unity. Besides that, the average error is minimal, RMSE is of the order of 0.002, and the absolute error is of the order of 0.007.

Keywords: River level · Weather data · Neural network · Black-sholes-merton

1 Introduction

Peru has three hydrographic slopes, these being: The Pacific Slope in the west, the Atlantic Slope in the east, and the Lake Titicaca Slope in the southeast [1], which determines the characteristics of its particular hydrological regimen.

There are two major problems already identified in various. In [2], the aim is to generate the basic hydrological information for the planning of measures aimed at reducing the risk of flood disasters due to the occurrence of the increasingly unpredictable El Niño phenomenon in the lower part of the river basin. This involves trying to locate the critical points and thus be able to design the best riverine defense solution [3].

Y. Iano et al. (Eds.): BTSym 2021, SIST 207, pp. 249–256, 2023.
https://doi.org/10.1007/978-3-031-04435-9_23

The second problem is that only 2% of the total volume of available water goes to the watersheds of the Pacific Slope, according to [1], which is contradictory, since most of the most populated cities in our country are located in this region. For this reason, maximum use should be made of this resource, knowing if possible the maximum flows of the different basins, whose information is available on the Senamhi website [4].

There are very few studies of Hydrological Modeling of the different basins of the Pacific Slope, and, in those that exist, only probabilistic models with a low level of confidence are carried out. For this reason, we propose, in this research project, to use the solution of the Black-Sholes-Merton equation, depending on the measurements of the meteorological stations, since the level of the river is related to temperature, precipitation level, humidity, and wind speed and direction.

2 Methodology

The behavior of the level is assumed to follow a lognormal probability distribution, whereas the variations of the increments of the levels are random and subject to the Brownian movement [5, pag 112]. Then, in Eq. 1 [6, pag 541], there is a specific form of the differential equation, where μ represents the mean of the trajectory of the measured random variable, and σ corresponds to the standard deviation.

$$dQ_t = Q(\mu(t)dt + \sigma(t)dt) \tag{1}$$

$$\frac{dQ_t}{Q} = \mu(t)dt + \sigma(t)dt \tag{2}$$

$$Ln(Q_t) = \int \mu(t)dt + \int \sigma(t)dt \tag{3}$$

We show the solution of Eq. 1 in Eq. 4, where Q_{t+1} is the next level value found based on the previous level, for which we assume that the values of the standard deviation and average are constant. This new set of values were calculated monthly from 2018–05 (See Fig. 1 partial view). Onwards we use as an additional input to the neural network used to perform the analysis prediction.

$$Q_{t+1} = Q_t e^{(\Delta(\mu(t)+\Delta(\sigma(t))} \tag{4}$$

A Perceptron Multi-layer network, applied successfully to solve a wide variety of non-linear problems, will be used in regression tasks. This network consists of an input layer, many hidden layers, and an output layer, which contain basic processing units, known as Neurons, connected to the next layer's neurons using links that have weights calculated through a training process.

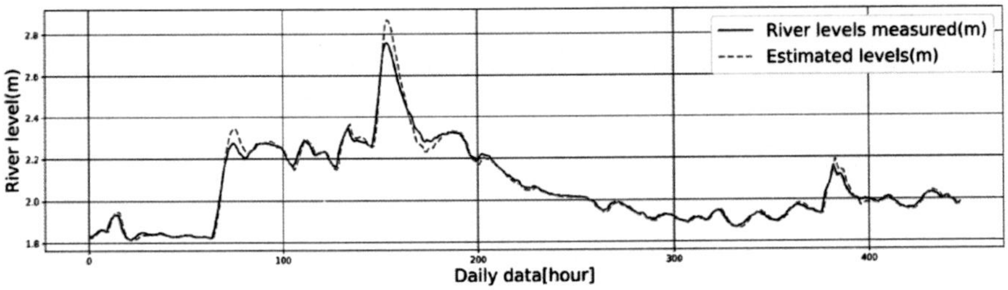

Fig. 1. BSM solution values calculated on a monthly basis

2.1 Methods

We downloaded the data from the Senamhi website, where it is possible to indicate whether to download the data in daily or monthly tables using CSV or XLSX file format. Once we used more than one weather station in the area of influence of the river La Leche, all tables should be merged into a single one. It was necessary to review the data because there was replicated information in different tables.

Additionally, we found many data errors, evident by the peaks of the levels that are not consistent with reality, so we performed a moving average filter to achieve a smoother variation free of these peaks.

We processed the tables to predict with advances of 6, 12, 18, and 24 h to enter a neural network that uses the output of the solution of the BSM equation as input. See Table 1.

Table 1. Consolidated adapted table

Date	Hour	Temp (°C)	Humidity (%)	Direction of The wind(°)	Wind Speed(m/s)	River level (m)
2018/02/01	5	20.8	87	150	0.8	1.89
2018/02/01	7	20.3	91	174	1	1.83
2018/02/01	8	22	83	179	1.8	1.84
2018/02/01	9	24.5	72	186	2.4	1.87
2018/02/01	10	26.7	64	195	3.3	1.87
2018/02/01	11	28.9	56	190	3.9	1.88
2018/02/01	12	30.4	51	191	3.7	1.85

We used the LSTM (Long Short-Term Memory) neural network, a type of recurring neural network that has feedback connections that allow capturing long-term sequential dependencies [7]. This network consists of an input layer, a hidden layer, and an output layer, a structure similar to conventional Neural Networks, with the difference that the LSTM replaces the basic unit of the

RNN with a memory block [8]. We used the LSTM network from the Tensorflow python libraries using Keras as a backend; for this, we configured the networks with 20 LSTM blocks, determined the number of inputs of the value tables and the outputs (one in this case) of the output table, with sigmoid activation function, and trained the network for 150 cycles.

We restructured the data to predict river levels 6, 12, 18, and 24 h in advance. Figure 2 describes how to restructure the tables for this purpose. Let us say N hours overall, so we use the rows from 1 to N of the measured data with those calculated using Eq. 4, and, as desired output, we place the output N + 1 aligned with the first row of this new table. All the process was done in Excel.

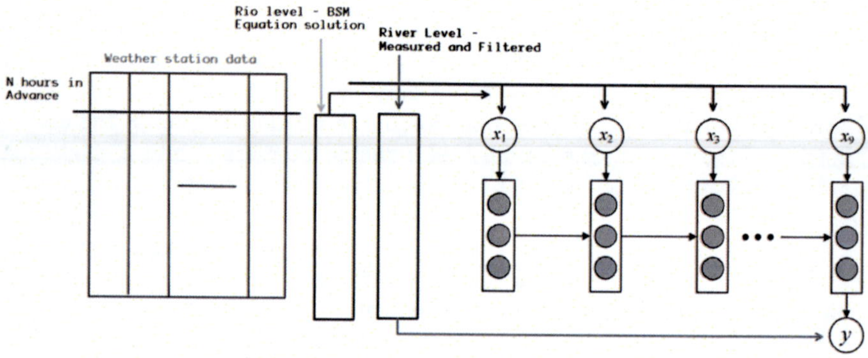

Fig. 2. Tables for N hours in advance

3 Results

We have used Python 3, with the *sklearn* library, which allows the use of a set of machine learning algorithms, *Tensorflow*, which is a high-performance numerical computing library and allows the rapid deployment of machine learning solutions, and *Pandas*, which is a library that allows easy reading of Excel files, from where we obtained the necessary data for the training of an LMST Neural Network.

To do this, we generated a table that allowed us to use the previous N hours and predict the levels for the next N hours; we validated these results by calculating the normalized error of training and verification. In the case of 6 h in advance, the normalized error is 0.0975 in training and 0.079 in verification. We made the prediction using the complete table, and we achieved an error of 0.001; also, in Fig. 3, we see the data used to test the network and the network output, where we verify that there is similarity.

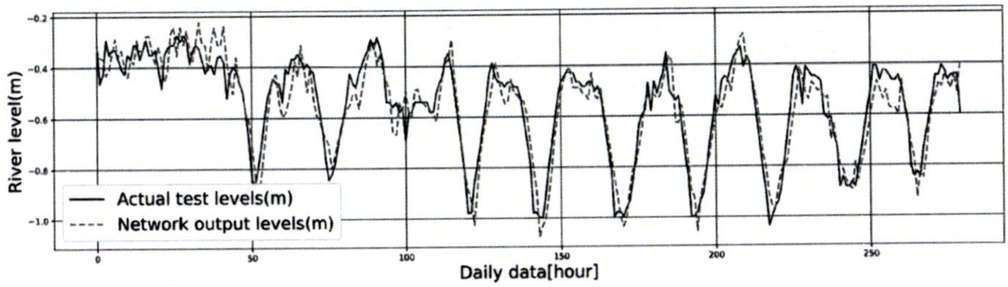

Fig. 3. Comparison of verification data and trained network prediction

We have repeated the procedure so that we achieved networks trained for advances of 12, 18, and 24 h, then we have considered full months to check the performance of these networks; for this, we measure the Mean Square Error (RMSE) [9], which allowed us to find large errors on a scale that has the same units as the data we used in the prediction and the mean absolute error (MAE) that better reflects the error between observations and predictions. We also used the Nash-Sutcliffe coefficient of efficiency, which is widely used in hydrological models: a value close to unity indicates a perfect model [7,10]. In the equations that follow, y_p is the predicted value and y_o is the observed values, and $\bar{y_o}$ is the mean value of the observed values.

$$RMSE = \sqrt{\frac{1}{n}\sum_{i=1}^{n}(y_o - y_p)^2} \tag{5}$$

$$MAE = \frac{1}{n}\sum_{i=1}^{n}|y_o - y_p| \tag{6}$$

$$NSE = 1 - \frac{\sum_{i=1}^{n}(y_o - y_p)^2}{\sum_{i=1}^{n}(y_o - \bar{y_o})} \tag{7}$$

In Figs. 4 and 5, we have the predicted results of the neural networks trained for the different hours of anticipation (24 h and 6 h are displayed); we have restructured the data table for training the neural networks considering 6, 12, 18, and 24 h. For each of them, a program has been developed in Python to perform, in training and verification, using the metrics established in Eqs. 5, 6, and 7 obtaining quite similar results, which we have placed in Table 2, where we observe that all trained neural networks have performances in the same order of magnitude so that the RMSE is of the order of 0.002, the NSE is close to unity 0.991, and the MAE is of the order of 0.007.

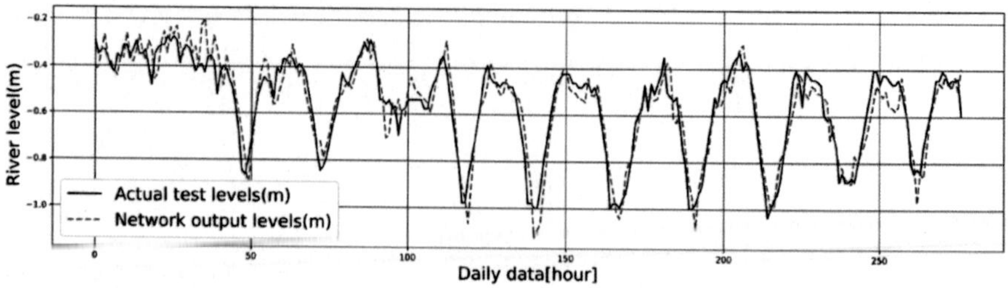

Fig. 4. 24 h in advance

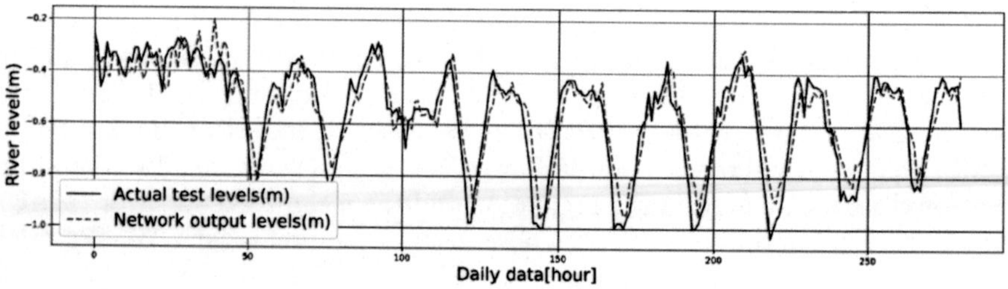

Fig. 5. 6 h in advance

Table 2. Forecast evaluation

	RMSE	NSE	MAE
6 h	1.092E-02	0.9913	7.565E-03
12 h	1.077E-02	0.9916	7.397E-03
18 h	1.079E-02	0.9916	7.481E-03
24 h	1.066E-02	0.9919	7.366E-03

4 Discussion

We have developed this work using the data from the Senamhi website, where we have identified the different meteorological and hydrological stations in the La Leche river basin. Mainly, we have preferred to use the data from automatic stations, since they perform readings in real-time that are taken every hour. In contrast, conventional stations only have daily average readings. We have had difficulties in the uniformity of the data from different nearby stations, which we know affect the River Level that depends on the geographical conditions of the area of influence of the basin. We have manually reviewed the data month by month to achieve consistency in them.

We have used a Neural LSTM network because they show better performance with sequential data, which is the case. It has been used to forecast 6, 12, 18, and 24 h in advance, which, according to the tests we have carried out, are the ones that show the best performance. In tests that we carry out with a greater number of hours in advance, the performances are very low. We consider that a greater amount of data, i.e., the extension or a greater number of measured variables, should be used as occurs in other countries, but, unfortunately, in our country, we are limited to existing data.

5 Conclusion

We have used the solution of the stochastic differential equation of the Black-Sholes-Merton model, with the river levels collected by the meteorological stations and other parameters such as precipitation, temperature, and wind speed as inputs to a Neural LSTM, which we have trained to forecast 6, 12, 18, and 24 h in advance the levels of the La Leche river.

We verified the performance of the obtained forecast using an entire month of data, which we did not use in training or verification, with which we obtained RSME of the order of 0.002, NSE close to unity, MAE as low as 0.007. Therefore, we say that the predicted average levels have high similarity with the observed data, which we also verify with the graphs shown.

We have used networks with 20 LSTM blocks. We have verified that using fewer blocks worsens performance. Increasing the number of blocks does not improve performance but demands more training time, which is not convenient.

For similar investigations in other regions of Peru, we recommend choosing those that have automatic meteorological stations since this ensures the availability of data with hourly frequencies, in addition to the use of satellite images for the detection of shapes and cloud movement, which could be used as additional inputs to improve and extend the forecast to more than 24 h.

References

1. Liendo, D.R., Neyra, A.D.: Modelación Probabilística de las Crecientes Máximas en ríos de La Vertiente Peruana del Pacífico. Universidad Peruana de Ciencias Aplicadas, Lima-Perú, Tesis de Titulo (2019)
2. Tejada, H.E.: Estudio hidrológico de máximas avenidas de la subcuenca Juana Ríos - Cuenca Chancay - Lambayeque. Universidad Nacional Pedro Ruiz Gallo, Lambayeque - Perú, Tesis de Grado (2014)
3. Cordova, C.J.: Modelación hidráulica del río "La Leche" con el programa HEC-RAS en un tramo de 4km. En el distrito de Íllimo, y departamento de Lambayeque. Tesis de Grado, Universidad San Martin de Porres, Chiclayo - Perú (2017)
4. Hidrometeorológicos D (2019) SENAMHI - Perú. https://www.senamhi.gob.pe/
5. Grigoriu, M.D.: Stochastic Calculus: Applications in Science and Engineering, 1st edn. Springer Science & Business Media, United States of America (2002), https://doi.org/10.1007/978-0-8176-8228-6
6. Financial Modeling. GFM, Palgrave Macmillan UK, London (2017). https://doi.org/10.1057/978-1-137-42658-1_9
7. Baek, S.S., Pyo, J., Chun, J.A.: Prediction of water level and water quality using a CNN-LSTM combined deep learning approach. Water **12**(12), 3399 (2020). https://doi.org/10.3390/w12123399

8. Fang, Z., Wang, Y., Peng, L., Hong, H.: Predicting flood susceptibility using LSTM neural networks. J. Hydrol. 125734 (2020). https://doi.org/10.1016/j.jhydrol.2020.125734
9. Chen, C., Zhou, H., Zhang, H., Chen, L., Yan, Z., Liang, H.: A Novel Deep Learning Algorithm for Groundwater Level Prediction based on Spatiotemporal Attention Mechanism. Preprint, In Review (2020). https://doi.org/10.21203/rs.3.rs-59191/v1
10. Wunsch, A., Liesch, T., Broda, S.: Groundwater Level Forecasting with Artificial Neural Networks: A Comparison of LSTM, CNN and NARX. Preprint, Groundwater hydrology/Modelling approaches (2020). https://doi.org/10.5194/hess-2020-552

Sentiment and Emotions Analysis of Tweets During the Second Round of 2021 Ecuadorian Presidential Election

Juan Carlos Minango Negrete[1](✉), Yuzo Iano[2], Pablo David Minango Negrete[2], Gabriel Caumo Vaz[2], and Gabriel Gomes de Oliveira[2]

[1] Instituto Tecnológico Universitario Rumiñahui, Sangolquí, Ecuador
juancarlos.minango@ister.edu.ec
[2] State University of Campinas, Campinas, Brazil
yuzo@unicamp.br, oliveiragomesgabriel@ieee.org

Abstract. Nowadays, the social network Twitter has acquired relevant importance as a massive communication and public engagement tool for personal, business, and political purposes. Furthermore, the fast spread of news through Twitter allows politicians to broadcast their messages instantly and directly to a larger audience. Thus, this paper aims to explore, investigate, and compare the texts posted by both candidates Arauz and Lasso on Twitter during the second round of the 2021 Ecuadorian presidential elections by applying sentiment analysis in order to determine and analyze their behavior and interaction with other Twitter users. The results obtained present the main differences between both candidates as well as their similarities when posting tweets. Furthermore, by applying sentiment analysis, the tweets are classified according to sentiment categories (positive or negative) and, in a deeper analysis, emotion categories (trust, joy, anticipation, fear, sadness, anger, surprise, or disgust). All the results are presented in graphs of frequency and distribution, which makes it easier to interpret the results achieved.

Keywords: Natural language processing · Sentiment analysis · Social network · Twitter · Tweets

1 Introduction

In a relatively short period of time, online social networks have become a medium for people to interact with each other, discuss different topics that are trends at the moment, and share news and political information. Twitter is a social network platform that is a micro-blogging where the users can post tweets (messages) with a limited number of 280 characters in real-time [1]. Thus, Twitter provides a large amount of user-generated language data, which allows extracting and gaining insights about public behavior by applying sentiment analysis processes.

Y. Iano et al. (Eds.): BTSym 2021, SIST 207, pp. 257–268, 2023.
https://doi.org/10.1007/978-3-031-04435-9_24

Sentiment analysis is a computational process that identifies feelings and emotions extracted from the body of texts, aiming to understand the sentimental characteristics of the users when they express their opinions. This process is done through methods that analyze the elements of the text and compare them with predefined emotional patterns [2]. All the tweets posted on Twitter are relevant since they have different characteristics and opinions and also indicate the users' position on a given subject. Besides, words that express certain feelings of the author are used more frequently in tweets [3].

In this sense, sentiment analysis is used in a variety of topics by companies. Some of the popular fields where it is used include business and politics. In politics, Twitter and sentiment analysis techniques enable analyses of government plans, political campaigns, and tracking of regions where a candidate is favorable in order to work towards the other ones to catch votes and improve their probabilities in an election.

Several studies [4–6] about political science and electoral behavior have been made to determine the motivations that lead people to choose a specific candidate. Among the main reasons there are the economic and political situations, the degree of knowledge that people have about a determined candidate, which is the political tendency of the candidate, whether the candidate has the support of national authorities, the performance of the candidate in debates, and the results of polls [6,7].

In [8], the tweets of the 2014 Colombian presidential election were collected to distinguish spammer accounts from non-spammer ones. Furthermore, a sentiment analysis system that investigates the potential of social media for voting intention inference was shown. In the same line, [1,9,10] explore and determine the principal characteristics of the political discourse on Twitter during the USA (2016), Turkey (2018), and India (2019) elections. In these works, a survey of Twitter data before Election Day was performed.

During the last Ecuadorian electoral elections, there was an increment in the use of online social networks to post political parties and their proposals [11]. Thus, this paper analyzes the tweets generated by both candidates, Andres Arauz and Guillermo Lasso, during the second round of the Ecuadorian presidential election, which occurred from March 16 to April 8, 2021. The analysis has the objective of identifying the sentiment of the tweets of both candidates, understanding the behavior, and associating attributes of their tweets. Furthermore, it is inspected popular Twitter discussion topics in the Twitter account of both candidates.

The remainder of this paper is organized as follows: Sect. 2 presents the Ecuadorian political scenario. Section 3 describes the methodology of the workflow implemented in the paper. The results and discussion are shown in Sect. 4. Finally, the conclusions are presented in Sect. 5.

2 The Second Round of Ecuadorian Presidential Election

On February 7, 2021, the first round of the presidential election in Ecuador was realized. Ecuadorians went to the polls to elect the president, assembly members, and Andean parliamentarians. Finalized the first round of the race for Ecuador's president, Arauz from UNES party won the first round with 32.72%, followed by Lasso from CREO party with 19.74% and Pérez from Pachakutic with 19.39%. The rest of the votes were divided among the 13 other candidates, null, and blank.

For a candidate to be elected as president in the first round of the Ecuadorian election, they should obtain 40% of the votes plus 10 points of advantage over the second candidate or more than 50% of the valid votes [12]. Since no candidate reached 40% of the valid votes, from March 16 to April 8, 2021, Ecuador experienced a second-round presidential election between Arauz and Lasso, where an expressive number of attacks in Twitter, like fake news, defamation, and discrediting, were posted by each candidate. Thus, tweets accompanied by hashtags like #LaBancaOElPaís, #AndrésNoMientasOtraVez, and #LassoEsMoreno were trend during the second round of presidential election [13].

3 Methodology

This section details the workflow employed to explore and analyze the Twitter data generated by Ecuadorian candidates Lasso and Arauz. Figure 1 shows the workflow, which is divided into steps and will be discussed in this section.

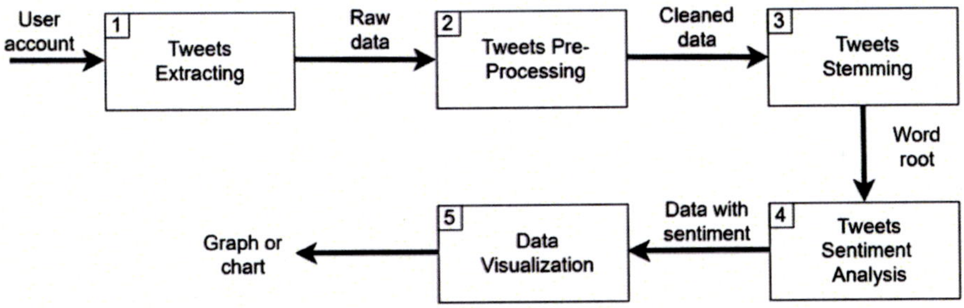

Fig. 1. Steps of the proposed workflow for Twitter analysis of the Ecuadorian elections

3.1 Tweets Extracting

The first step consists of the tweets extraction from the Twitter accounts of both candidates. Thus, a Twitter developer account was created to get access to an access token, an access token secret, a consumer key, and a consumer secret in order to use the Twitter API and retrieve the tweets of both candidates. In addition to the text of the tweets, additional information is collected, which includes the user's name, date of the posting tweet, count of retweets, count of replies, link of publication, mentioned users, and hashtags. The tweets were collected from March 6 to April 8, 2021, during the second round of the Ecuadorian election campaign.

3.2 Tweets Preprocessing

From Fig. 1, the second step corresponds to the tweets preprocessing. Since there is no defined writing pattern to be employed on Twitter, it is necessary to clean the content of the tweets in order to standardize them. Furthermore, all emoticons were eliminated once it is analyzed the lexicon of the Spanish language. Figure 2 shows the five sub-steps followed to perform the cleaning of the tweets.

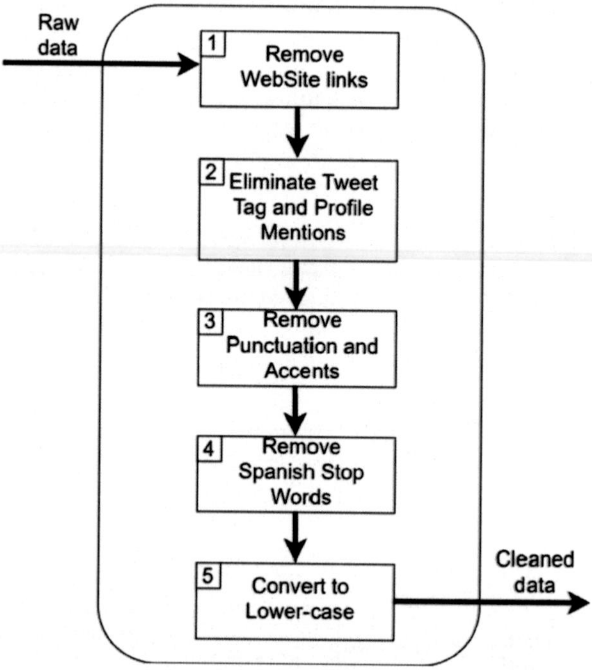

Fig. 2. Internal sub-steps of tweets preprocessing

From Fig. 2, the first sub-step removes website links while the second sub-step eliminates tweet tags and profile mentions. Later, the third and fourth sub-steps remove punctuation, accent marks, and all Spanish stop words, respectively. It is worth noting that stop words play a very small role in the sentiment analysis process of the tweets, and, for that reason, they must be removed. Finally, the fifth sub-step converts the text of the tweets into lowercase words completing the tweets preprocessing step.

3.3 Tweets Stemming

Once the texts from the tweets have been preprocessed, their words pass through a stemming process that reduces inflection in words to their root forms, e.g., the words "retrieval", "retrieved", and "retrieves" are reduced to the root "retrieve". In this paper, it was implemented the stemming algorithm for the Spanish language given by [1].

3.4 Tweets Sentiment Analysis

This step corresponds to the sentiment analysis of the tweets after the stemming. Sentiment analysis allows us to identify and extract subjective information from text or sentences in order to classify them as positive, negative, or neutral.

The sentiment analysis is based on a text classifier supervised learning algorithm, which is trained with a training database (G_i, α_i), where G_i represents the feature vector and $\alpha_i = [-1, +1]$ determines the class index corresponding to the ith tweet. The training of the text classifier is performed by computing the discriminant function $f(\cdot)$ and applying the thresholds t_1 and t_2, as follows:

$$\begin{cases} f(G_i) < t_1 & \text{then } c_i \text{ is negative} \\ t_1 \leq f(G_i) \leq t_2 & \text{then } c_i \text{ is neutral} \\ f(G_i) > t_2 & \text{then } c_i \text{ is positive} \end{cases} \tag{1}$$

In this paper, the Naive Bayes Supervised Learning (NBSL) algorithm is employed as text classifier, since it is the most common one. In order to train the NBSL algorithm, the R package *syuzhet* [14] is employed as the training dataset. This package provides a useful dictionary of terms named NRC Emotion Lexicon [15], which has approximately 14000 words in 105 different languages, including Spanish.

NBSL Algorithm. During the training phase, both the prior probabilities of each class $P(c_i)$ and the likelihood of different features g_i for each class $P(g_i|c_i)$ are learned. Later, by considering that all features g_i are independent, the NBSL algorithm determines the probability $P(c_i|g_i)$, which represents that a given feature set G_i belongs to a particular class label. Thus, c_i corresponds to the greatest probability value obtained by:

$$\underset{c_i}{\operatorname{argmin}} \left(P(c_i \mid G_i)\right) \tag{2}$$

Thus, the NBSL algorithm is able to classify each word as a positive, negative, or neutral sentiment.

3.5 Data Visualization

The last step of Fig. 1 corresponds to the data visualization, which is the representation of the preprocessing data or information into a graph, chart, or other visual formats.

4 Results

This section aims to analyze the results achieved after applying the methodology proposed in Sect. 3 to the tweets posted by the candidates Arauz and Lasso.

Figure 3 shows the Twitter sources, that is, from where the tweets were originated. It is evident that the principal source for both candidates was the Web App followed by the mobile application for iPhone. Furthermore, it is noted that the candidate Arauz has used in a lower percentage Media Studio and mobile application for Android, which demonstrates more resource sources in the social network Twitter.

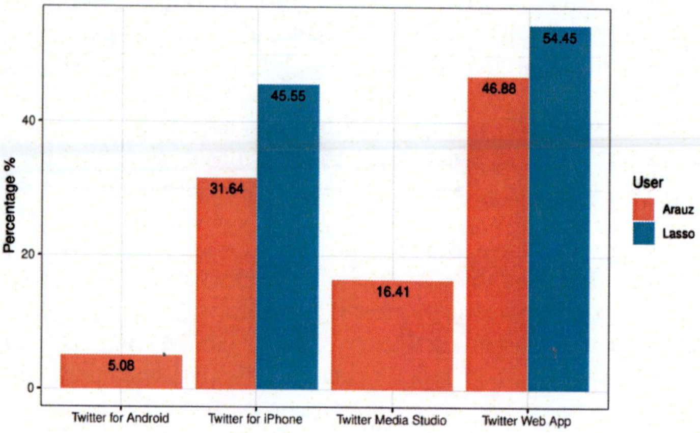

Fig. 3. Twitter sources used by each candidate

Figure 4 analyzes the percentage of organic, retweets, and replies tweets of both candidates. It is observed that 76.95% of the tweets produced by the candidate Arauz were organic, which means that the candidate has produced most of his tweets whereas the organic tweets produced by the candidate Lasso were just 7.99% of the whole. Indeed, the majority of tweets, 91.55%, from candidate Lasso were replies to tweets where he was mentioned, which shows a better interaction between the candidate Lasso and Twitter's users, this is also corroborated by its low percentage of retweets in comparison with the candidate Arauz, since nobody likes a Twitter account that exclusively retweets, for example, without any individual content.

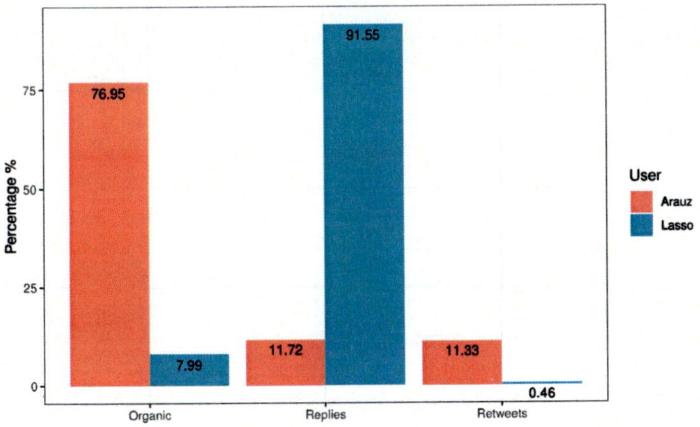

Fig. 4. Kind of the tweets published by each candidate

In Fig. 5, the frequency of the tweets of both candidates during the election campaign is shown. From this figure, it is clearly inferred that both candidates have published their tweets at a different rate since the candidate Lasso has had about 8 times as many tweets as the candidate Arauz, which shows the importance that the candidate Lasso gave to the Twitter social network in order to interact with its users.

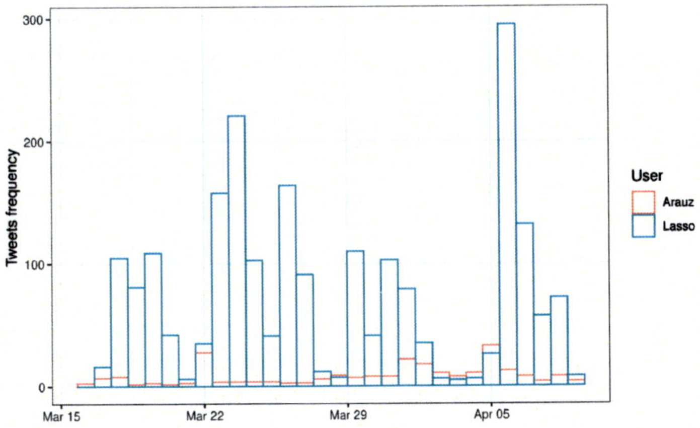

Fig. 5. Tweets frequency for each candidate

Once the sources, kind, and frequency of both candidates' tweets were shown, hereinafter in this paper, the context of each tweet of each candidate will be analyzed. Thus, Fig. 6 shows a frequency comparison between the most common words used in the tweets of each candidate. Hence, in this figure, words that are near to the red line are employed with approximately equal frequencies in their tweets by both candidates whereas words away from the red line are employed much more by one candidate compared to the other. It can be visualized that both candidates have employed common words like: *jóvenes* (young), *país* (country), *compromiso* (commitment), which indeed represents typical words used in presidential speech. On the other hand, it is noted that candidate Arauz has used a greater variety of words in the tweets, which is an expected result since 76.95% of his tweets were organics. The candidate Lasso have employed in a greater frequency the words *vamos* (let's go) and *gracias* (thanks), which can be associated with the content of a reply tweet since 91.55% of the tweets published by the candidate Lasso were reply tweets.

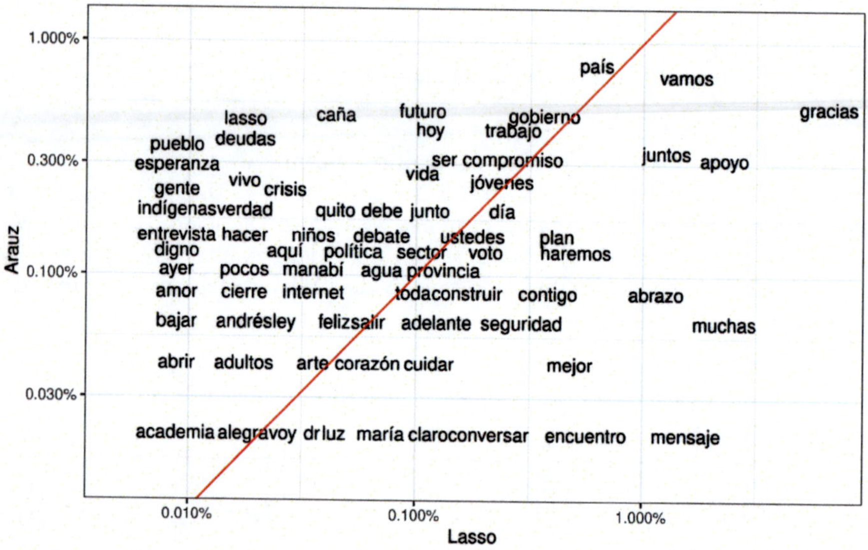

Fig. 6. Frequency comparison of the words used by each candidate in their tweets

In order to compare the usage of the words of both candidates, Fig. 7 shows the Logarithmic Odds Ratio (LOR), which determines the words that are more or less likely to come from each candidate. Thereby, it is counted the number of times each candidate has used each word in their tweets, and words used more than 5 times are retained. Later, it is computed the LOR for each word by using the following expression:

$$\mathrm{LOR} = \ln\left(\frac{\left[\frac{n+1}{total+1}\right]_{\mathrm{Arauz}}}{\left[\frac{n+1}{total+1}\right]_{\mathrm{Lasso}}}\right) \tag{3}$$

In Eq. 3, n indicates the number of times that a word is employed by each candidate and *total* represents the total words for each candidate. From this figure, the candidate Arauz has tweeted with more frequency the word *andres presidente* (president Andres). On the other hand, the candidate Lasso has employed with more frequency the word *cambio* (change) and *encontremonosparalograrlo* (let's meet to achieve it), the slogan of his presidential campaign.

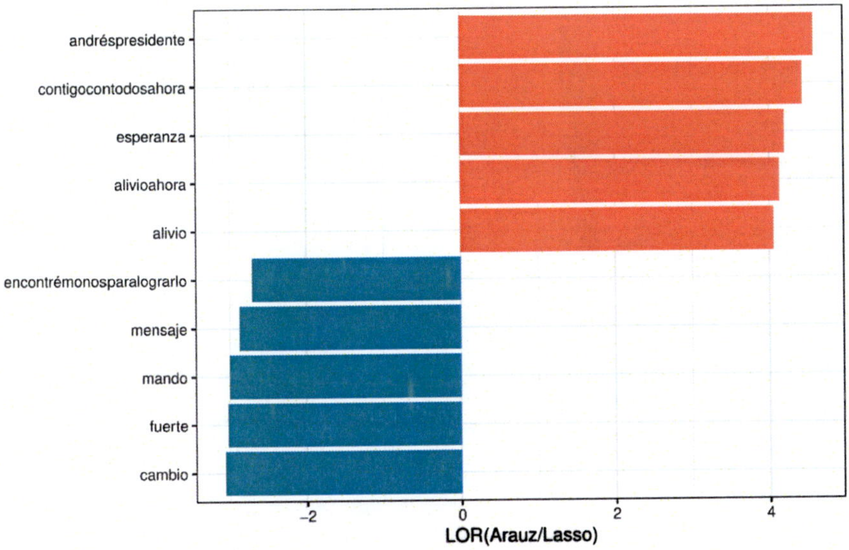

Fig. 7. Logarithmic Odds Ratio (LOR) comparison of the words used by both candidates

Figure 8a and 8b show the top 5 hashtags used by the candidate Lasso and Arauz, respectively. For the candidate Lasso, the hashtag dominant was *encontremonosparalograrlo* (let's meet to achieve it), whereas the candidate Arauz used both *contigocontodosahora* (with you with everyone now) and *andres presidente* (president Andres) as the predominant hashtags.

Finally, Figs. 9 and 10 show the comparison of the distribution of sentiments and emotions of both candidates by using the NRC Emotion lexicon. Thus, in Fig. 9, the tweets' texts of each candidate are associated with two sentiments, which are negative and positive, and, in Fig. 10, they are associated with eight emotions, which are anger, fear, anticipation, trust, surprise, sadness, joy, and disgust.

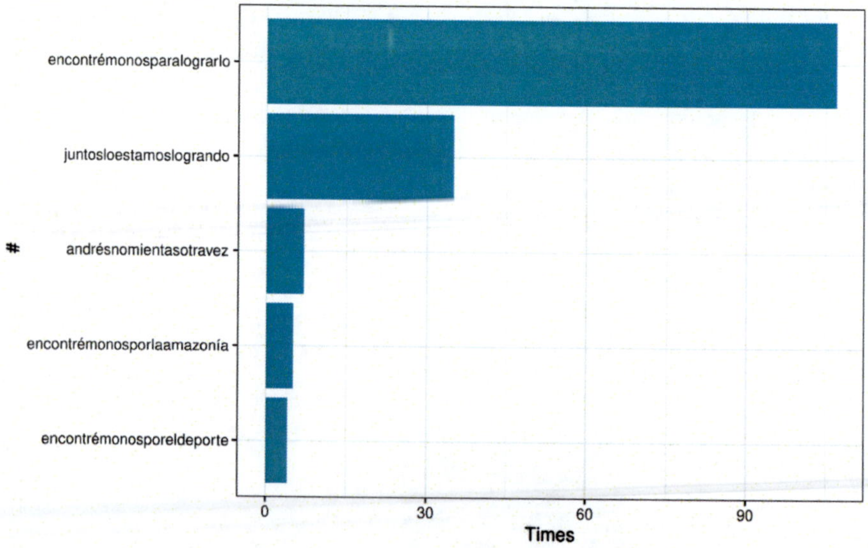

(a) Top 5 hashtags used by the candidate Lasso

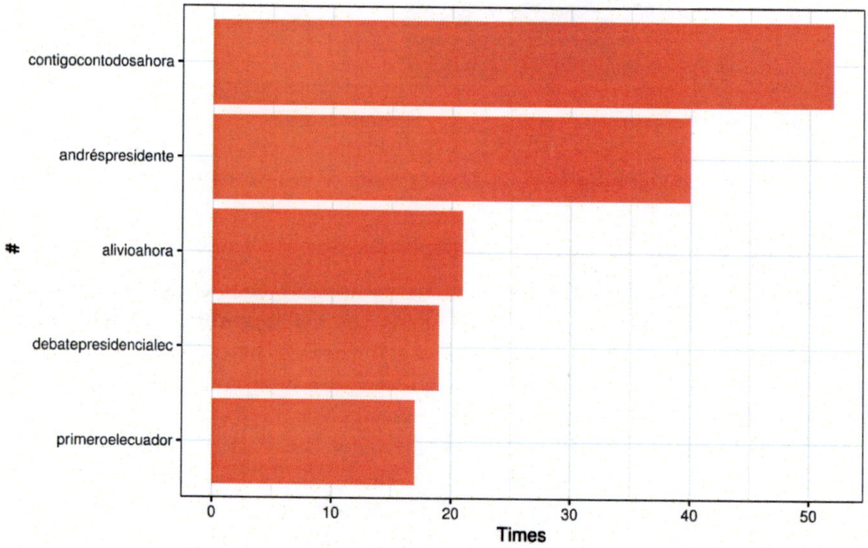

(b) Top 5 hashtags used by the candidate Arauz

Fig. 8. Top 5 hashtags

From Fig. 9, it is noted that the tweets of both candidates have reflected positive sentiments. On the other hand, from Fig. 10, it can be observed that Arauz's tweets have associated trust, fear, and sadness as emotions, whereas Lasso's tweets have been influenced by trust, joy, and anticipation as their principal emotions.

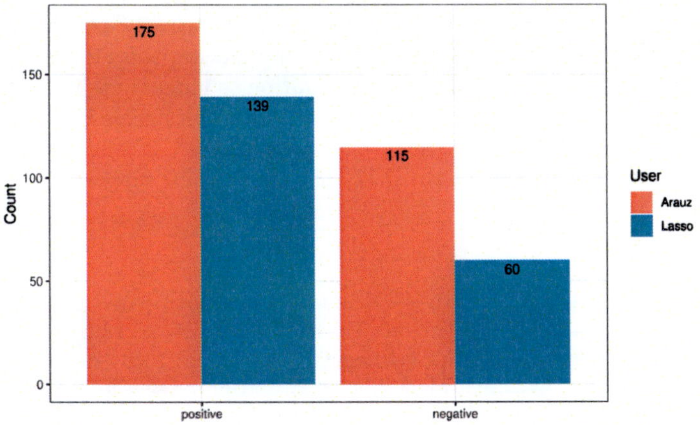

Fig. 9. Distribution of sentiments categories for both candidates

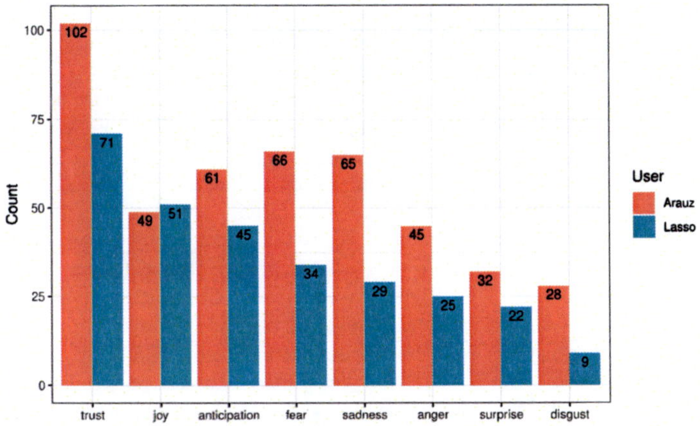

Fig. 10. Distribution of emotion categories for both candidates

5 Conclusion

This paper presented a process of text cleaning and sentiment analysis to study the context of messages collected from the online social network Twitter corresponding to the second round of 2021 Ecuadorian presidential elections, seeking a relationship and comparison between the tweets posted by both candidates, Arauz and Lasso.

For this purpose, several stages of sentiment analysis were performed, such as extraction, preprocessing, stemming, and polarity sentiment analysis of the tweets. Besides, exploration of tweets information was performed, which allowed determining the words that are tweeted more and less frequently, as well as the words that are most likely to be tweeted from each of the candidates' Twitter accounts.

Moreover, it was also possible to analyze and extract the sentiments and emotions present in the text written by the candidates, and all the results were presented in graphs of frequency and distribution to make it easier for the reader to visualize and understand the analysis and the results achieved.

Acknowledgments. Thanks to the ISTER by the financial support.

References

1. Yaqub, U., Chun, S.A., Atluri, V., Vaidya, J.: Sentiment based analysis of tweets during the US presidential elections. In: Proceedings of the 18th Annual International Conference on Digital Government Research, pp. 1–10 (2017)
2. Choy, M., Cheong, M., Laik, M.N., Shung, K.P.: US presidential election 2012 prediction using census corrected Twitter model. arXiv Prepr arXiv12110938 (2012)
3. Bansal, B., Srivastava, S.: On predicting elections with hybrid topic based sentiment analysis of tweets. Procedia Comput. Sci. **135**, 346–353 (2018)
4. Eder, N., Jenny, M., Müller, W.C.: Manifesto functions: how party candidates view and use their party's central policy document. Elect. Stud. **45**, 75–87 (2017)
5. Wagner, M., Johann, D., Kritzinger, S.: Voting at 16: turnout and the quality of vote choice. Elect. Stud. **31**(2), 372–383 (2012)
6. Hendricks, J.A., Denton, R.: Political campaigns and communicating with the electorate in the Twenty-First Century. Commun. How Barack Obama used New Media Technol. Win White House, 1-18 (2010)
7. Gilmore, J.: Ditching the pack: digital media in the 2010 Brazilian congressional campaigns. New Media Soc. **14**(4), 617–633 (2012)
8. Cerón-Guzmán, J.A., León-Guzmán, E.: A sentiment analysis system of Spanish tweets and its application in Colombia 2014 presidential election. In: 2016 IEEE International Conferences on Big Data and Cloud Computing (BDCloud), Social Computing and Networking (Socialcom), Sustainable Computing and Communications (Sustaincom)(BDCloud-Socialcom-Sustaincom). IEEE, pp. 250-257 (2016)
9. Baker Al Barghuthi, N., E. Said, H.: Sentiment analysis on predicting presidential election: twitter used case. In: Brito-Loeza, C., Espinosa-Romero, A., Martin-Gonzalez, A., Safi, A. (eds.) ISICS 2020. CCIS, vol. 1187, pp. 105–117. Springer, Cham (2020). https://doi.org/10.1007/978-3-030-43364-2_10
10. Ansari, M.Z., Aziz, M.B., Siddiqui, M.O., Mehra, H., Singh, K.P.: Analysis of political sentiment orientations on twitter. Procedia Comput. Sci. **167**, 1821–1828 (2020)
11. Eluniverso.com (2021) Elecciones en Ecuador: Redes sociales del candidato, partido político y hasta las páginas de apoyo serán consideradas para el gasto electoral — Política — Noticias — El Universo. In: Eluniverso.com. https://www.eluniverso.com/noticias/2021/01/08/nota/9344974/candidatos-gasto-electoral-redes-sociales-ecuador-facebook/. Accessed 25 Aug 2021
12. Comissão Nacional de Eleições (2021). https://resultados.cne.gob.ec/. Accessed 25 Aug 2021
13. Primicias (2021) Segunda vuelta: la campaña sucia reinó en las redes sociales. In: Primicias. https://www.primicias.ec/noticias/politica/segunda-vuelta-campana-sucia-redes-sociales. Accessed 25 Aug 2021
14. Jockers, M.L.: Syuzhet: extract sentiment and plot arcs from text (2015). Accessed 21 Oct 2015
15. Mohammad, S.M.: Word affect intensities. arXiv Prepr arXiv170408798 (2017)

Spectroscopy Technologies for Biological Samples Throughout History

Talita Cristina Colomeu[1](✉), Marina de Gea Neves[1], and Euclides Lourenço Chuma[2]

[1] Fiberwork Optical Communication, Campinas, São Paulo, Brazil
{talita.colomeu,marina.neves}@fiberwork.net
[2] iTech-Instituto de Inovações Fotônicas, Campinas, São Paulo, Brazil
euclides.chuma@ieee.org

Abstract. The evolution of biotechnological processes has been facing a huge demand for developing low-cost, non-invasive, and highly accurate platforms for biological samples. Thus, biospectroscopy has emerged to meet this demand. Important applications arise from biomedical spectroscopy in the areas of identification of proteins, microorganisms, and others. Biological samples such as saliva, serum, blood, and urine are composed of proteins and metabolic that can be used to identify different diseases. Compounds present in biological fluids can be an excellent source of identification of viruses, bacterial infection, cancer, metabolic diseases, among others. In the last few years, the biospectroscopy has been introduced in biological process, acting in the band from 1800 to 900 cm^{-1}, also known as "biofingerprint" (biological fingerprint), has a high density of information about important biomolecules, and is used to identify different nucleic acids and other biological constituents. In this review, we describe the main spectroscopy techniques and how they work. It is also presented works about the potential application of spectroscopic techniques for different biological samples, focusing on ultraviolet-visible, infrared (NIR and MIR), fluorescence, and Raman spectroscopy.

Keywords: Biospectrocopy · Infrared · UV-Vis · Fluorescence · Raman · Biological fluids

1 Introduction

The study of spectroscopy began around the 1600s when Isaac Newton exerted influence on this science. Newton was responsible for discovering the colors of the rainbow when he focused light through glass split. Years later, the science of spectroscopy became the study of the interaction between electromagnetic radiation and matter as a function of wavelength [1]. Spectroscopy techniques operate at different ranges of wavelengths and depending on spectral range, different spectroscopic transitions can be excited. In other words, under certain conditions, the materials absorb the radiation and emit and scatter energy [2].

Y. Iano et al. (Eds.): BTSym 2021, SIST 207, pp. 269–275, 2023.
https://doi.org/10.1007/978-3-031-04435-9_25

Currently and in the recent past, diverse types of spectrophotometers have been developed for particular wavelength ranges, using different detectors, light sources, filters, and wavelength-section. These spectrophotometers are generally coupled to optical fibers covering different wavelength ranges, which include Near Infrared (NIR), MID infrared (MIR), ultraviolet-visible (UV-Vis), Raman, and fluorescence spectroscopy. In recent years, the miniaturization of spectrometers became reality with USB-powered devices and micro-opto electro-mechanical systems (MEOMS), leading the spectrometers to reduce sizes and weight [2].

In the last few years, spectroscopy has been introduced in biological processes, called biospectroscopy. The 1800 to 900 cm^{-1}, also known as "biofingerprint" (biological fingerprint), has a high density of information about important biomolecules and is used to identify different nucleic acids and other biological constituents [3]. In this context, spectroscopy can be a useful tool for analyzing biological samples.

Searching the Pubmed scientific database through the years to verify the evolution of Infrared, UV-Vis, Raman, and Fluorescence Spectroscopy showed an increase of applications for biological samples in the 2000 s (Fig. 1 and Fig. 2).

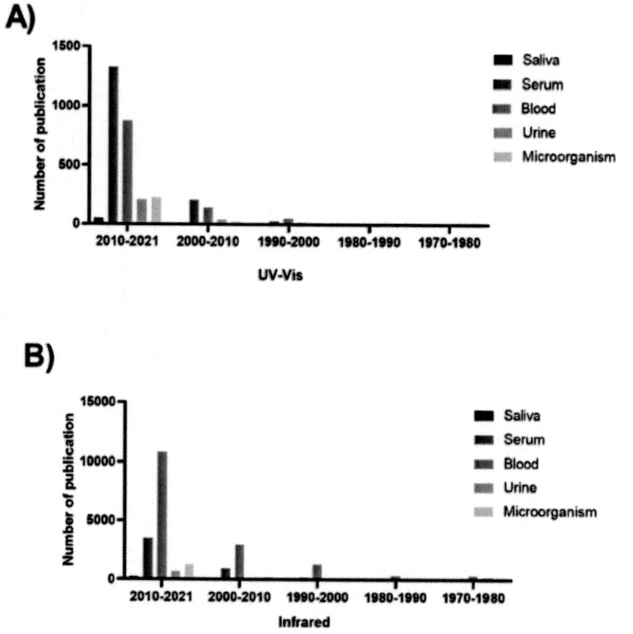

Fig. 1. Demonstrative of the techniques A) UV-Vis and B) Infrared published in scientific articles from 1970 to 2021.

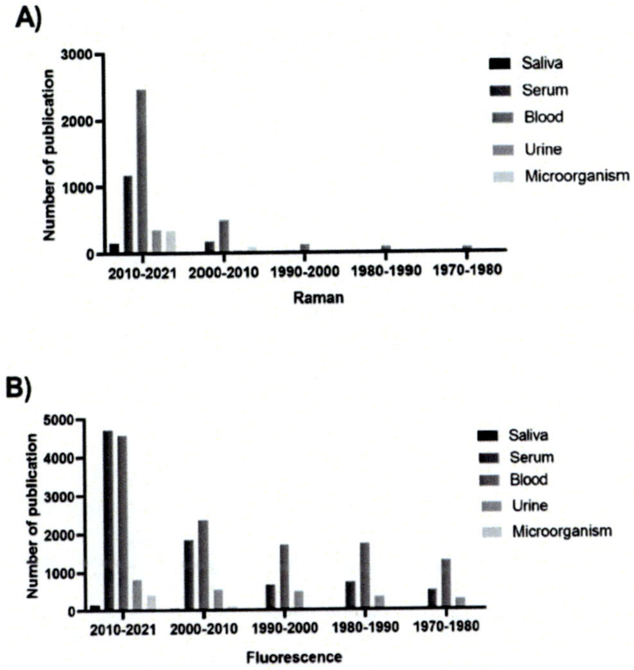

Fig. 2. Demonstrative of the techniques A) Raman and B) Fluorescence published in scientific articles from 1970 to 2021.

In addition, Pubmed scientific database for the years 2020 to 2021, with themes: saliva, serum, blood, urine, and microorganism for Infrared, UV-Vis, Raman, and Fluorescence Spectroscopy showed that Infrared and Fluorescence Spectroscopy was widely used for blood analysis. It is also possible to realize that other spectroscopy techniques are being investigated for different biological samples (Fig. 3).

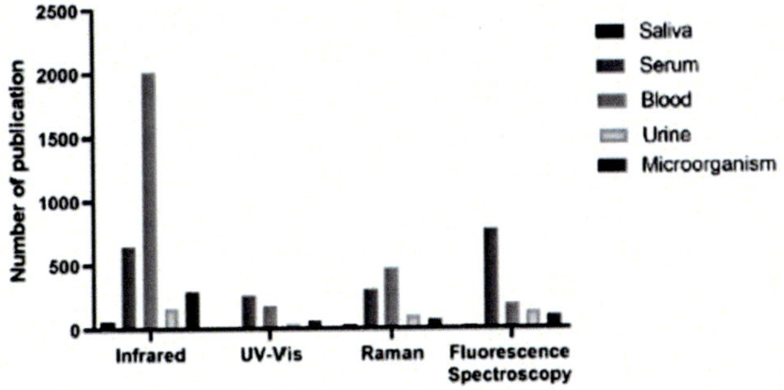

Fig. 3. Spectroscopy techniques employed for the biologicals of saliva, serum, blood, urine, and microorganism in the years 2020 and 2021.

Given the works that have been published over the years, we realize that spectroscopy can be an agile, easy, and sensitive tool for biological samples.

2 Ultraviolet-Visible (UV-Vis) Spectroscopy

UV-Vis spectroscopy corresponds to the absorption of ultraviolet-visible radiation (200–780 nm) and usually results from the excitation of bonding electrons. This excitation act in electron transitions between orbitals, involving higher energy, bigger than in infrared spectroscopy. As a result, the typical band contains many closely spaced lines corresponding to vibrational and rotational states.

With UV-Vis spectroscopy, it is possible to detect individual electron transfers between orbitals or atomic, ionic, and molecular bands. This technique is very used in analytical chemistry for quantitative determination, such as transition metal ions, highly conjugated organic compounds, and biological macromolecules.

Conventional UV-Vis spectrum is collected by measuring the transmitted light intensity of a narrow bandpass, which scans the wavelength in the time domain. Ultracompact spectrometers have been developed based on MEMS (micro-electromechanical systems) and image sensors suitable for applications with portable devices, such as tablets and smartphones [3].

UV-Vis was applied with aim of analyzing the phosphate levels in unstimulated saliva from patients with type 2 Diabetes Mellitus to support already existing results regarding the increased risk of dental caries in this population. The results of this study showed that phosphate levels from patients with type 2 Diabetes Mellitus are significantly lower compared to individuals without Diabetes Mellitus [4]. Heidari, A. and Gobato, R. compared malignant and benign human cancer cells after synchrotron radiation and analyzed by UV-Vis. The results showed that malignant human cancer cells have gradually transformed to benign human cancer cells under synchrotron radiation with the passage of time [5].

3 Fluorescence Spectroscopy

Fluorescence spectroscopy investigates fluorescence from molecules. The term fluorescence refers to a type of luminescence, which can be defined by the light emission from the molecule. Generally, the energy during the fluorescence is normally lower than its absorption energy for a determinate molecule, thus, the emitted light has longer wavelengths than those used for excitation [6].

The conventional spectrometers or fluorimeters contain Uv-Vis light, used as photon source, like a xenon lamp, lasers, and LEDs, a sample holding cell, monochromator/filters to select a specific wavelength (both the excitation beam and the fluorescence result), and detector. Simple instruments often use a bandpass filter to choose the excitation wavelength [7].

Fluorescence detection is a high-sensitive technique that is capable to determine very low concentrations. Because the fluorescence measured signal has zero background, the technique is considered more sensitive when compared to light absorption techniques [7]. Many biological compounds, such as proteins, enzymes, vitamins, primary and secondary metabolites from microbial growth have excitation by visible or near-UV, rendering fluorescence spectroscopy a tool for bioprocess monitoring.

Serum samples with or without the dengue or chikungunya viruses were investigated using molecular fluorescence spectroscopy. The technique proved to be a useful, specific, and sensitive tool able to identify spectral variations between the samples [8].

Maurice, J et.al. demonstrated that non-invasive monitoring of gut function is feasible using transcutaneous fluorescence spectroscopy of orally administered contrast agents. Through this technique, the absorption of intestinal cells was verified [9].

4 Infrared Spectroscopy

Infrared spectroscopy (IR) works using the measurement of the interaction of infrared radiation with matter and can be by absorption, emission, or reflection. In IR spectroscopy, the intensity and position of the vibrational band are well defined and the atoms participating in the chemical bond vibrate and change the electric dipole moment of the molecule, working as a system of oscillators [10].

IR can be split into three regions: NIR – near-infrared (780 to 2500 nm), MIR – mid-infrared (2500 to 50.000 nm), and FIR - far-infrared (50.000 to 100.000 nm) [3]. As mentioned before, the biofingerprint region is located between 1800 to 900 cm^{-1} (5.556 to11.111 nm), and because of that, this short revision will be focused on the MIR technique.

The spectrum MIR contains information about fundamental transitions ($\Delta_V = \pm 1$) and, in general, shows narrow and well-defined bands being used for characterization and identification of organics and analytic quantification.

The MIR technique can nowadays be easily attached to sampling accessories such as attenuated total reflectance (ATR). The ATR-FTIR has been used to identify Chronic Kidney disease (CKD) in the saliva. Rodrigues et al. (2019) demonstrated that several components showed different expressions between CKD patients and control subjects [11]. In 2018, Santos et.al. used the technique ATR-FTIR in blood samples for the identification of dengue, chikungunya, and zika. FTIR spectroscopy coupled to chemometric tools such as GA-LDA (genetic algorithm - linear discriminant analysis), PCA-LDA (principal component analysis – linear discriminant analysis), SPA-LDA (successive projections algorithm - linear discriminant analysis), and multivariate classification algorithms could be a powerful and sensitive strategy to verify biochemical variations caused by the presence of different viruses in the blood [12].

5 Raman

Raman Spectroscopy is a fast, non-destructive, and “green” analysis (no solvent is needed) that provides detailed information about chemical and molecular structure. This technique is based on the inelastic scattering of monochromatic light when interacting with a molecule. Raman can provide information about rotational, vibrational, and other transitions in molecules for which the polarizability changes on excitation [13].

Generally, the laser focuses on the sample and is collected through a lens. It is necessary to filter the elastic scattered radiation (Rayleigh scattering) from the rest of the scattering before the light is dispersed in a grating/spectrograph. After this, the photons can be directed to the detector according to wavenumbers [14]. Usually, multichannel detectors are used in Raman spectroscopy, such as CCD (charge couple device), and more recently, CMOS (complementary metal-oxide-semiconductor).

The Raman technique is based on inelastic scattering, which occurs when the energy of an incident photon is different from the energy of the scattering photon, resulting in a wavelength shift. The inelastic scattering could be split into two classes. Whether the energy of the incident photon is higher than the energy of the scattering photon, the phenom is called stokes. If the energy of the incident photon is lower than the energy of the scattering photon, the phenom is called anti-stokes [3].

Raman spectroscopy can use laser wavelengths from ultraviolet to near-infrared. The choice of laser has a direct impact on sensitivity, spatial resolution, and fluorescence suppression into measurements. Because of the laser ranges, the Raman technique can be used for different samples. Raman spectroscopy is being used in biological samples, because several techniques have recently been developed, such as surface-enhanced (SER), Fourier transform (FT-Raman), and two-dimensional (2D) Raman, which favor methodologies for biological samples [13].

The Raman technique was a viable alternative for analyzing analytes from fractions with high and low molecular weight in human serum for diagnostic purposes [15]. In another work, Zhu, W. et.al. used Raman Spectroscopy for evaluated creatinine (CDK) concentration in human urine and showed capable of effectively detecting CDK using a Raman spectrometer in optimized Au-sol SERS substrate [16].

6 Conclusion

In conclusion of this short review, the use of UV-Vis, fluorescence, infrared, and Raman spectroscopies, applied to biological samples, has been shown a very powerful tool in science life. An increasing number of papers using spectroscopy and biofluids is observed in the last 10 years. Is important to highlight some spectroscopies, such as infrared and Raman, are non-destructive, fast, and do not need solvents or molecular reagents, resulting in a huge advantage compared to other techniques as chromatography and molecular tests.

Because spectroscopy can result in a huge amount of data, it is needed the use of capable tools for the treatment of data, such as artificial intelligence (AI), statistical, and chemometric. The accelerated growth of spectroscopy applications, in the last decades, could be directly related to the advances in instrumentation, computer processing, and AI. A plenty of algorithms (ANN – artificial neural network, SVM – support vector machine, PCA – principal component analysis, MCR- multivariate curve resolution, LDA, GA, SPA, and others) are already been used to develop models and treat spectroscopy databases.

The application in the biological area is enlarging, especially their utilization in monitoring, medical diagnostics, and biophotonic devices for the management of human diseases. Opportune conditions of the current market improve the popularization of spectroscopy into biological samples and are expected the development of new researches in this field.

References

1. Martins, R.A., Silva, C.C.: As pesquisas de Newton sobre a luz: Uma visão histórica. Revista Brasileira de Ensino de Física **37**(4), 1–32 (2015)
2. Tkachenko, N.V.: Optical Spectroscopy Methods and Instrumentations, 1st edn. Elsevier, Oxford (2006)
3. Skoog, D.A., Holler, F.J., Crouch, S.R.: Principles of Instrumental Analysis, 7th edn. Cengage, Boston (2007)
4. Meidyawati, R., Adiyasa, J., Djauharie, N.: Phosphate concentration in unstimulated saliva of patients with type 2 Diabetes Mellitus. J. Int. Dental Med. Res. **12**(3), 1182–1188 (2019)
5. Heidari, A., Gobato, R.: Human cancer cells and tissues with the passage of time under synchrotron radiation. Parana J. Sci. Educ. **4**(6), 18–33 (2018)
6. Horiba (2021), http://www.horiba.com. Accessed 01 Sep 2021
7. Lakowicz, J.R.: Principles of Fluorescence Spectroscopy, 2nd edn. Springer, New York (1999)
8. Santos, M.C.D., Monteiro, J.D., Araújo, J.M.G., Lima, K.M.G.: Molecular fluorescence spectroscopy with multi-way analysis techniques detects spectral variations distinguishing uninfected serum versus dengue or chikungunya viral infected samples. Nature **10**(13758), 1–13 (2020)
9. Maurice, J., Lett, A.M., Skinner, C., Lim, A., Richardson, M., Thomas, A.P.: Transcutaneous fuorescence spectroscopy as a tool for non-invasive monitoring of gut function: first clinical experiences. Nature **10**(16169), 1–13 (2020)
10. Stuart, H.B.: Infrared Spectroscopy: Fundamentals and Applications: Analytical Techniques in the Sciences.1nd edn. John Wiley & Sons Ltd, England (2004)
11. Rodrigues, R.P.C.B., Aguiar, E.M.G., Cardoso-Sousa, L., Caixeta, D.C., Guedes, C.C.F.V., Siqueira, W.L.: Differential molecular signature of human saliva using ATR-FTIR spectroscopy for chronic kidney disease diagnosis. Braz. Dent. J. **30**(5), 437–445 (2019)
12. Santos, M.C.D., Nascimentos, Y.M., Monteiro, J.D., Alves, B.E.B., Melo, M.F., Paive, A.A.P.: ATR-FTIR spectroscopy with chemometric algorithms of multivariate classification in the discrimination between healthy vs. dengue vs. chikungunya vs. zika clinical samples. Analytical Methods, **10**(1280), 1–23 (2018)
13. Ferraro, J.R., Nakamoto, K.: Introductory Raman Spectroscopy: Academic Press, 2nd edn. Elsevier Science, California, USA (1994)
14. Stavelley, L.A.K.: The Chracterization of chemical purity organic compounds. 1nd edn. Pure and Applied Chemistry, Great Britain (1971)
15. Parachalil, D.R., Bruno, C., Bonnier, F., Blasco, H., Chourpa, I., McIntyre, J.: Raman Spectroscopic screening of high and low molecular weight fractions of human serum Analyst. Royal Soc. Chem. **144**(7), 4295–4311 (2019)
16. Zhu, W., Wen, B.Y., Jie, L.J., Tian, X.D., Yang, Z.L., Radjenovic, P.M.: Rapid and low-cost quantitative detection of creatinine in human urine with a portable Raman spectrometer. Biosensors Biolectron. **154**(112067), 1–8 (2020)

Technology for Electric Bus in the Brazilian Scenario: Focus on the Adoption of National Components

Lilian Regis Laraia(✉), Yuzo Iano, Ricardo Takahira, Luiz Vicente Figueira de Mello Filho, Gabriel Gomes de Oliveira, and Gabriel Caumo Vaz

School of Electrical and Computer Engineering, University of Campinas, Campinas, Brazil
lilian.laraia@gmail.com, yuzo@unicamp.br, luizvicente.mello@icloud.com, oliveiragomesgabriel@ieee.org

Abstract. According to the Brazilian National Association of Public Transportation (ANTP), in 2018, about 28% of the urban mobility in the Brazilian cities was made by public transportation, especially buses. For that reason, the supply market of this segment is constantly innovated. The objective of this work is to present a Technological Roadmap (TRM) for electric buses in Brazil, focusing on the adoption of national components in a timeframe of 15 years, from 2018 to 2033. The approach applied in this study is based on wide documentary research of the available data in scientific databases, interaction with specialists in this field, morphological analyses, decision matrices, and Delphi research to develop scenarios that allowed confirming and qualifying the estimate of the technologies adoption. The results pointed out that most of the components are already nationalized, bringing many benefits, as such availability, price, and other positive aspects, which make it possible to improve and qualify the system as a whole. Besides, this work approaches basic concepts about electric buses, ITS, and IIS to strengthen the perspective of the sector concerning economic development within a sustainable and intelligent ecosystem.

Keywords: Electric bus · Brazilian market · Intelligent Transportation System (ITS) · Intelligent Information System (IIS) · Brazilian components

1 Introduction

Electric vehicles have gained space in the international market, helping diminish the dependency on fossil fuels and reduce greenhouse gas emissions. Considering the whole transport sector, in 2009, those emissions reached a worrying level, representing 25% of all global emissions caused by energy consumption [1,2].

Y. Iano et al. (Eds.): BTSym 2021, SIST 207, pp. 276–285, 2023.
https://doi.org/10.1007/978-3-031-04435-9_26

With technological advancement and government incentives, such as Agenda 2030, which was established by the United Nations (UN) to propose goals concerning the development of smart sustainable systems and cities to a better world, it is natural that, in the current century, actions are taken to accomplish the established objectives. In this sense, there is a constant transformation in urban mobility, mainly in the adoption of more efficient, intelligent, and sustainable vehicles [3–5].

This article presents the results of a Technological Roadmap (TRM) for heavy electric vehicles in Brazil, focusing on adopting national components for buses in a timeframe of 15 years, from 2018 to 2033. The approach applied in this study is based on documentary research of the available data, interaction with specialists, morphological analyses, decision matrices, and Delphi research to assess scenarios. This study was drawn on PROMOB-e, a project on electric mobility conducted in partnership with the Brazilian Ministry of Economics and the German Ministry of Cooperation and Economic Development through the Deutsche Gesellschaft für Internationale Zusammenarbeit (GIZ) GmbH.

2 Background

2.1 Electric Bus

In the current century, electric buses are growing on an unimaginable scale, and all the concerned parties, i.e., automobile manufactures, companies related to this segment, governments, and the population, can take advantage of it. With the introduction of electric buses, studies point out benefits in the environmental, financial, and ergonomic aspects, making this theme very relevant [6].

According to the Chinese News Agency, in 2019, 99% of the electric buses with battery in the world was installed in China, with more than 421,000 buses on its roads, which represents 17% of the total Chinese bus fleet [7].

The electric buses can store electricity or be powered by an external source. Most of the buses that store electricity have an onboard battery, from where the electric motors achieve the necessary energy. There are other kinds of storage, such as the gyroscope, which stores the energy of the steering wheel. When the electricity is not stored onboard, it is provided through contact with aerial wires or conductors on the ground [8,9].

2.2 Intelligent Transportation System

The concept of Intelligent Transportation System (ITS) was established in the early 2000 s s and applied worldwide, in all kinds of transportation systems, either terrestrial, aquatic, or aerial. This concept has been receiving crescent attention in the academic and industrial fields [10].

ITS integrates information and communication data technologies and applies them in the transportation field, aiming to develop an integrated system of people, roads, and vehicles and establish a transportation management system that is totally functional, precise, and efficient [11,12].

It is worth mentioning that Intelligent Transportation Systems work with information and control technologies, which are the core of their functions. In this way, it is necessary to approach the Intelligent Information Systems, which are closely related to the main theme of this work.

2.3 Intelligent Information System

The Intelligent Information System (IIS) is characterized as the next generation of Information Systems since it joins two technologies: Technology Integration and Database [13]. The IIS acquires and establishes knowledge, exhibiting and intelligent behavior and making it possible to communicate better with users and other systems, which allows a more efficient and faster solution for potential problems [14].

For any IIS to meet its purpose, information must be available whenever it is necessary. That means that the computing systems used to store data and process information and the security controls that protect them must work properly [14].

2.4 Roadmap

The Roadmap is a technique of flexible planning that supports strategic planning, combining goals in the short and long terms with specific technologies. It is a plan applied in new products or processes and can include technology forecasting to identify the appropriate emerging technologies. Besides, it is a known technique to help manage diffuse innovation front-end [15–18].

The expectation is that the roadmap techniques may help companies withstand turbulent situations and plan more holistically to include non-financial goals and focus on more sustainable development. It is worth highlighting that the roadmaps may be combined with other corporate forecast methods to facilitate systemic changes [19,20].

As shown in Fig. 1, the first layer represents the strategic guidance for the sales of buses in the timeframe of 15 years in the context of applications for propulsion systems in 2018, 2023, 2028, and 2033.

3 Methodology

The objective of this section is to illustrate in Fig. 2 the flowchart of the methodology applied in this work, aiming to describe briefly and clearly the model adopted by the authors to achieve the previously proposed results.

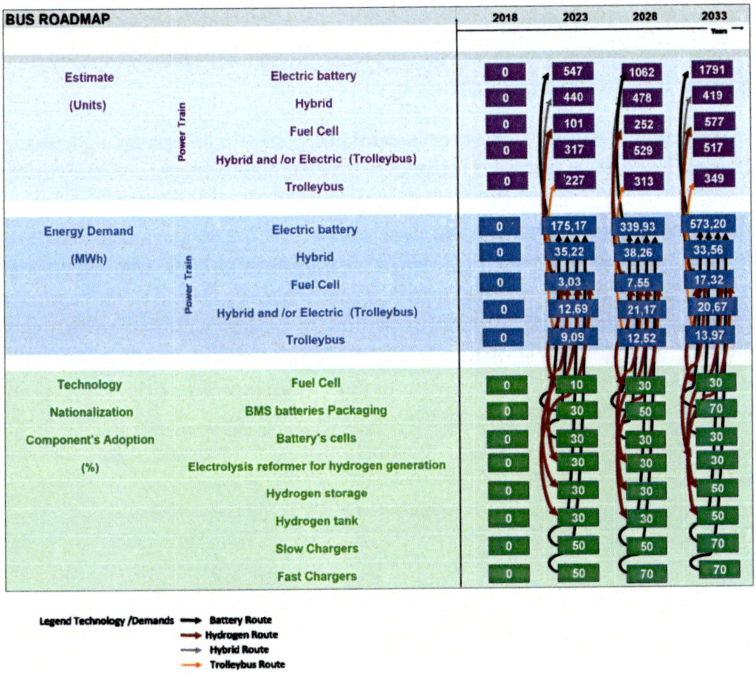

Fig. 1. Technology roadmap.

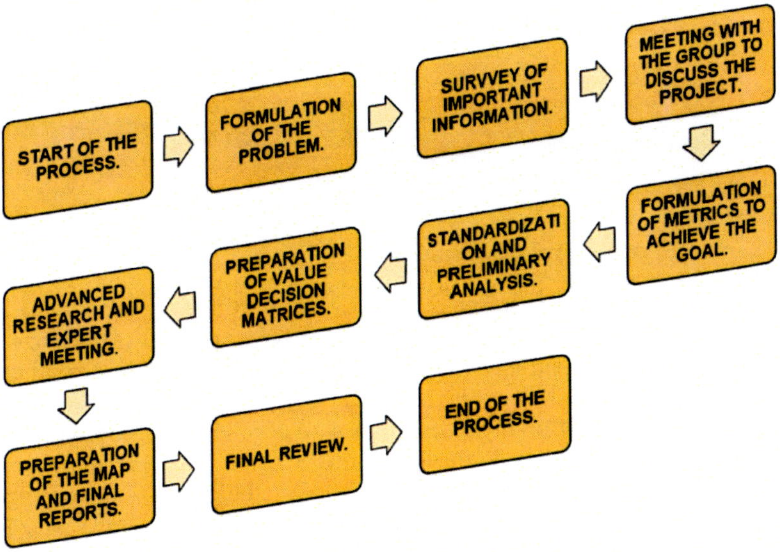

Fig. 2. Flowchart of the work.

4 Results Analysis

4.1 Delphi Research Results

The development began with the recognition of the components through interviews with the bus assembler industries. With the content of the interviews, it was possible to identify all the components of these vehicles with a preliminary classification of the imported components. This step, combined with field data search, generated inputs for the Delphi research. The research was conducted in two rounds of questionnaires. This process allowed each respondent to reflect before answering an upcoming round of questionnaires, avoiding a common consensus. It was used for the following purposes: to adjust electrified bus sales scenarios; estimate the evolution of the adoption of national components in vehicles sold in 2023, 2028, and 2033; evaluate some policy initiatives and incentives; obtain all the elements of a decision matrix that would allow verifying the consistency of the item results.

The return rates of external surveys, which have external respondents to the entity that promotes the survey, are between 10% and 15%, as shown in Table 1. The rate obtained is within the general expectation, although it is believed that it could be higher if the research was not applied near the end of the year.

Table 1. Results obtained on the participation of respondents for Delphi Research

	Delphi step 1	Delphi step 2
Questionnaires sent	375	342
Questionnaires answered	55	40
Return time	12 days	10 days
Return rate	14,7%	11,7%

4.2 Merit Figures and Assessments

The following issues were considered: reduction in particulate matter emissions - MP and NOx (versus Diesel Euro 5); scalability of infrastructure with the fleet; cost of supply infrastructure; acquisition cost; autonomy; shelf-life; total cost of ownership (TCO); availability; reliability.

Concerning all the aforementioned issues, the following questionnaire was answered by the respondents: From the perspective of a bus dealership company, how would you rate the relative importance of the above questions to judge the technological options when equipping your fleet with new units. Use a linear scale from 1 (minimum importance) to 7 (maximum importance).

Figure 3 presents the results achieved from the research.

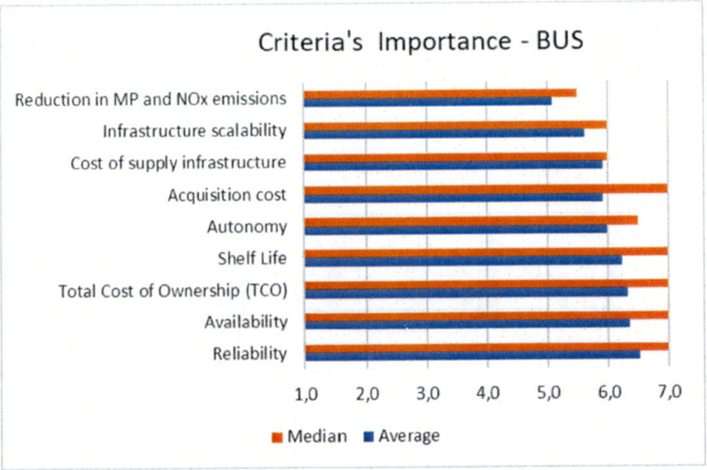

Fig. 3. Assessment of merit figures.

4.3 Evolution of Electric Bus Sales Forecasts and Participation of Each Technological Option

Figure 4 shows the prospects for the evolution of estimated bus sales for the 15 years starting from 2018 in thousands of vehicles, and Fig. 5 considers the probable percentage participation of each technological option. In the opinion of the experts participating in the work, there will be a growth for the technology of electric batteries, whereas hybrids, hybrids and/or electric, and trolleybus will gradually reduce their participation. For the fuel cell the forecast is for growth, but with less participation when compared to other technologies.

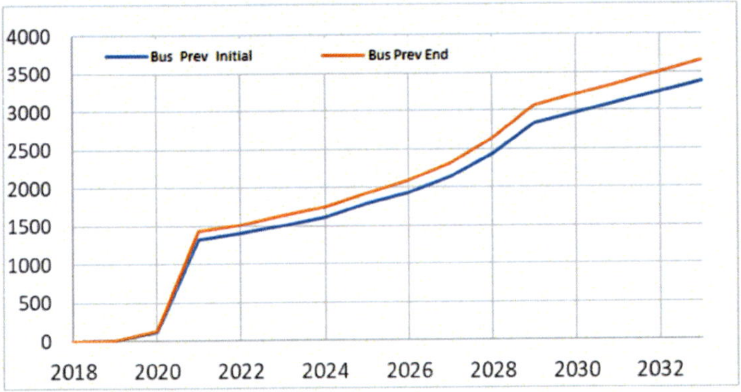

Fig. 4. Evolution of electric bus sales forecasts.

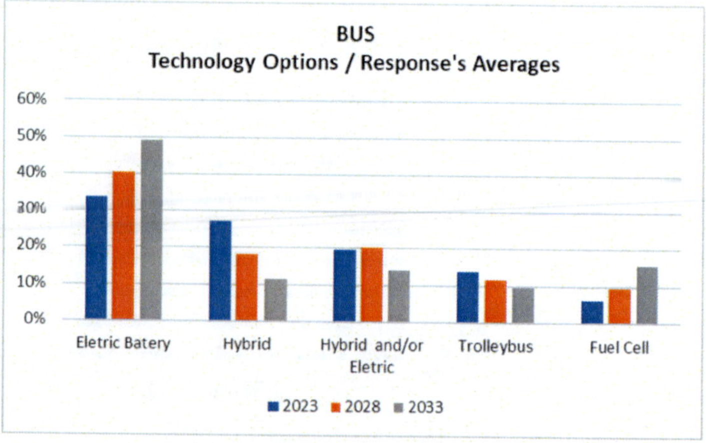

Fig. 5. Evolution of the participation of technological options for buses.

4.4 Results for the Adoption of National Components

The deployment of the propulsion system allowed us to identify that the vast majority of the technologies of its components are already nationalized. During the Delphi research process, it was possible to confirm these facts and qualify the estimates of adoption of the technologies for the next 15 years (Fig. 6).

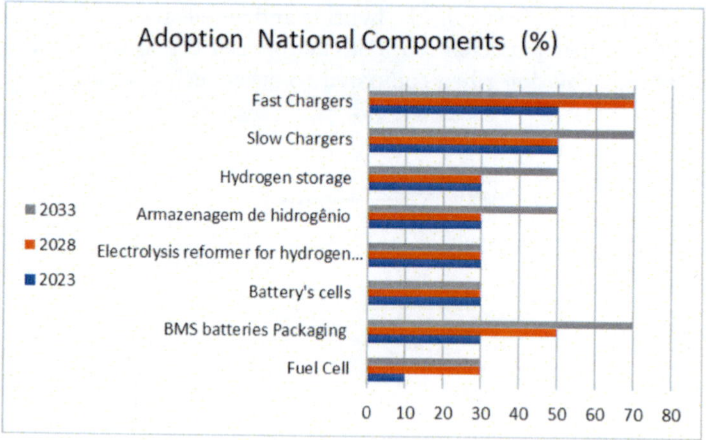

Fig. 6. Results for the adoption of national components.

5 Discussion

As previously mentioned, currently, in the Brazilian scenario, there is already a sustainable and efficient strategy for the maintenance of automobile parts to serve the national market, especially the buses, which are the main means of transport in the daily mobility of people in the Brazilian cities.

However, when the national scenario is compared with other countries, especially the developed ones, it is visible that there is scope for improvement. The international market goes beyond the parts maintenance and also seeks intelligent and efficient ways to meet the goals established by governments and, in global scope, the United Nations. The objective is to transform the current urban spaces into sustainable ecosystems for the next generations.

For that reason, the Technological Roadmap is extremely relevant for the scenario approached in this study. It allows developing strategic long-term planning, combining technological solutions with a unique and specific character, aiming to supply the national market of automobile parts, especially buses parts to serve the services of public transportation. Moreover, considering the current scenario, where the national and international markets constantly seek innovation, a behavior boosted by global competition, incentives should be given by the governments, aiming to achieve an increasingly intelligent and sustainable system.

6 Conclusion

It is important to highlight that this article approaches the Brazilian national market of public transportation parts with a growth perspective for the next years. In general, the results achieved were plausible from the point of view of logic and comparison, which were formalized in this work. It is worth mentioning that data from national and international literature were considered in the development of the study.

An interesting point is that the results of the decision matrices initially represent the attractiveness of technological options nowadays. Consequently, there should be a level of compatibility between them and the relative participation planned for 2023. This fact proved distorted in the case of the buses. It is also possible that the respondents might have "embedded" a change forecast in the activities of attraction in the short term, between the date of the research and 2023.

In this work, it was observed a study that demonstrates an efficient supply of public transportation parts, but the national market still cannot serve efficiently and strategically the buses that have been recently introduced into the national territory, which are the electric vehicles. This is a concern since there is a perspective of a considerable increase in the adoption of the new technology.

Finally, this article points out the perspective of effective use of the methodology to serve the national market, which needs a revision since it does not have, currently, know-how and capability of meeting the demands of an innovative branch, i.e., electric buses, which have been developing and being applied worldwide.

References

1. Mobilize.org.br (2021) Ônibus híbridos: como funcionam as estações de recarga rápida. https://www.mobilize.org.br/noticias/10260/onibus-eletricos-como-funcionam-as-estacoes-de-recarga-rapida.html. Accessed 21 July 2021
2. Repositorio.ipea.gov.br (2021), http://repositorio.ipea.gov.br/bitstream/11058/5328/1/td_2123.pdf. Accessed 21 July 2021
3. Yan, M., et al.: Enhancing the transmission grid resilience in ice storms by optimal coordination of power system schedule with pre-positioning and routing of mobile DC deicing devices. IEEE Trans. Power Syst. **34**(4), 2663–2674 (2019)
4. Galus, M.D., et al.: Integrating power systems, transport systems and vehicle technology for electric mobility impact assessment and efficient control. IEEE Trans. Smart Grid **3**(2), 934–949 (2012)
5. Sofia, D., Gioiella, F., Lotrecchiano, N., Giuliano, A.: Cost-benefit analysis to support decarbonization scenario for 2030: a case study in Italy. Energy Policy **137**, 111137 (2020)
6. de Oliveira, G.G.: Estudo do transporte público da cidade de Campinas, utilizando conceito de smart city e equipamentos específicos, para a coleta precisa de dados e melhoria deste segmento no conceito ergonômico (2021)
7. Bloomberg.com, Bloomberg - Are you a robot? (2021), https://www.bloomberg.com/news/articles/2018-04-23/electric-buses-are-hurting-the-oil-industr. Accessed 20 July 2021
8. Bloomberg.com (2021) Bloomberg - Are you a robot? https://www.bloomberg.com/news/articles/2019-05-15/in-shift-to-electric-bus-it-s-china-ahead-of-u-s-421-000-to-300. Accessed 20 July 2021
9. American-automobiles.com (2021) Account Suspended. http://www.american-automobiles.com/. Accessed 20 July 2021
10. Moisio, L., Hillmansen, S.: Intelligent Transport System-RRUKA Special Issue. IET Intell. Transp. Syst. **10**(1), 1 (2016)
11. Hou, Z., Zhou, Y., Du, R.: Special issue on intelligent transportation systems, big data and intelligent technology. Transp. Plan Technol. **39**(8), 747–750 (2016). https://doi.org/10.1080/03081060.2016.1231893
12. De Brucker, K., Macharis, C., Verbeke, A.: Two-stage multi-criteria analysis and the future of intelligent transport systems-based safety innovation projects. IET Intell. Transp. Syst. **9**(9), 842–850 (2015). https://doi.org/10.1049/iet-its.2014.0247
13. Jarašūniene, A.: Research into intelligent transport systems (ITS) technologies and efficiency. Transport **22**(2), 61–67 (2007). https://doi.org/10.1080/16484142.2007.9638100
14. Nkambou, R., Mizoguchi, R., Bourdeau, J.: Advances in intelligent tutoring systems. Springer Science & Business Media (2010)
15. Alexander, P.: Creating a technology road map. In: Entrepreneur (2021). https://www.entrepreneur.com/article/83000. Accessed 20 Aug 2021
16. Rohrbeck, R., Kallehave, P.: The role of corporate foresight in promoting sustainability. UN Glob Compact Int Yearb (2012)

17. Petrick, I.J., Echols, A.E.: Technology roadmapping in review: a tool for making sustainable new product development decisions. Technol. Forecast Soc. Change **71**(1), 81–100 (2004). https://doi.org/10.1016/S0040-1625(03)00064-7
18. Oliveira, M.G., Rozenfeld, H.: Integrating technology roadmapping and portfolio management at the front-end of new product development. Technol. Forecast Soc. Change **77**(8), 1339–1354 (2010). https://doi.org/10.1016/j.techfore.2010.07.015
19. Phaal, R., Farrukh, C.J.P., Probert, D.R.: Technology roadmapping-a planning framework for evolution and revolution. Technol. Forecast Soc. Change **71**(1), 5–26 (2004). https://doi.org/10.1016/S0040-1625(03)00072-6
20. Alexander, P.: Creating a technology road map. In: Entrepreneur (2021). https://www.entrepreneur.com/article/83000. Accessed 20 Aug 2021

ICT Infrastructure Usability Model in the Sustainability of Computational Equipment

José Fortunato Zuloaga-Cachay[1](✉), Gilberto Carrión-Barco[2], Denny John Fuentes[2], Jessica Karin Solano-Cavero[1], and Gisella Luisa Elena Maquen-Niño[2]

[1] Universidad Católica Santo Toribio de Mogrovejo, Chiclayo, Perú
{jzuloaga,jsolano}@usat.edu.pe
[2] Universidad Nacional Pedro Ruiz Gallo, Lambayeque, Perú
{gcarrion,dfuentesad,gmaquenn}@unprg.edu.pe

Abstract. The interaction of technology users in an educational environment requires the application of virtual environments that favor student learning. This process can be carried out through the use of technological resources, which provide, in a timely manner, physical and logical support to the pedagogical processes that are executed in the teacher-student interaction. Technological resources are not exempt from failures due to the very nature of their technological infrastructure, making it necessary to apply monitoring methods with reliable standards. This led to the proposal of an ICT infrastructure usability model supported by the strategic management dimension, the risk management dimension, and the incident management dimension, in order to improve the sustainability of computer equipment, because it takes into account the functionality and useful life of hardware and software with a systemic approach and strategic structuring techniques, establishment of risk and monitoring judgments of the service provided in the Pedagogical Innovation Classrooms (PIC), and because, also, it is based on a need to adequately renew and maintain computer equipment to advantage the teaching-learning process in the environment of regular basic education.

Keywords: Usability of ICT infrastructure · Sustainability of computer equipment · Pedagogical innovation classrooms

1 Introduction

According to the technical standard ISO 9241 [1], usability is the ratio that establishes the extent to which products, systems, services, environments, and facilities can be used by people of various abilities, capable of achieving in a given context, the objectives set for a certain purpose. On the other hand, usability is framed in quality processes, without whose attributes it would not be possible to determine whether a technological product is aligned with reality [2].

Y. Iano et al. (Eds.): BTSym 2021, SIST 207, pp. 286–293, 2023.
https://doi.org/10.1007/978-3-031-04435-9_27

The Information and Communication Technologies (ICT) infrastructure is made up of physical equipment such as computers and logical equipment such as computer programs that are necessary to manage, process, convert and store information through network cabling and wired connection [3].

The usability of ICT resources leads to establishing the functionality and useful life of the physical, logical, and connectivity parts, through monitoring and systemic control that guarantees a high impact interaction with the end-user [4].

With the world increasingly digitized, it has been necessary to explore the relationship between different types of use of ICT at school and home, student attitudes towards ICT, and academic performance. For this, mathematical models have been applied to evaluate the usability of ICT in the school and family environment, finding positive results between the use of ICT and academic performance from a perspective of home entertainment, also showing a greater availability of the usability of computers per student relative to higher levels of academic achievement. These results highlight the need to consider and include ICT as an input in the educational production function. In addition, public policies must consider the interaction between ICT and the educational performance of low-achieving students [5].

From another perspective, the Ministry of Economy and Finance of Peru indicates that there are no policies to manage compliance with information technology service levels aimed at seeking user satisfaction and that there are limited formal procedures and use of good practices for the service of Information Technologies [6].

Regarding ICT in the School, it is established that the principal is responsible for the proper use and pedagogical purposes of the technological infrastructure that the Educational Institution has, which must be found in adequate and properly protected environments. The principal of the EI designates a teacher in charge of technological resources who has the technical training to carry out maintenance routines to preserve computer equipment, supervise the installation, and ensure the maintenance of the computer and communication equipment in the Pedagogical Innovation Classrooms (PIC) [7], using strategies for the due control of assets and the monitoring of the functionality of the ICT equipment [8].

Given that the problematic situation shows a lack of a form of ICT infrastructure usability, the following research problem arises. In what way could modeling of the ICT infrastructure usability process improve the sustainability of computing equipment in pedagogical innovation classrooms?

The importance of this research work consists in improving the sustainability of computational equipment in the PIC Classrooms of Educational Institutions, through the modeling of the ICT infrastructure usability processes to determine the relationship between the functionality management processes and useful life of hardware and software, as well as risk prevention processes in infrastructure and computing equipment.

The main objective is the modeling of the ICT infrastructure usability process that takes into account the functionality and useful life of hardware and software to improve the sustainability of computational equipment in the PIC Classrooms of the Educational Institutions of the Province of Lambayeque.

The proposed hypothesis is if the ICT infrastructure usability process is modeled taking into account the functionality and useful life of hardware and software, then, it contributes to improving the sustainability of computational equipment in the PIC Classrooms of the Educational Institutions of the Province of Lambayeque, Peru.

The alternative hypothesis (H_A) is, if the ICT infrastructure usability process is modeled taking into account the functionality and useful life of hardware and software, then it contributes to improving the sustainability of computational equipment in the PIC classrooms of the Educational Institutions of the Province of Lambayeque, on the other hand, the null hypothesis (H_O) states that the usability process doesn´t contribute to the improvement of the sustainability of the computational equipment. The result of the contrast can be seen in Fig. 3.

The IT infrastructure usability modeling involves the description of models for hardware and software systems by simulating the behavior of the system based on functional logic for a set of input data variations, taking into account the failure approach to evaluate the model [9].

The theoretical bases that support the research are based on reference frameworks accepted in the IT field, such as the Spanish Association for Standardization and Certification [10], which are good practices in Information Technology activities that have allowed the development of various technical standards and standardized models that position the standards in terms of their application following their regulatory framework such as ITIL, CMMI, ISO/IEC 27001, among others.

2 Methodology

The research is descriptive with a bibliographic-documentary field design because it takes into account the functionality and useful life of hardware and software with a systemic approach and techniques of the Balanced Scorecard. The research is proactive in nature since it is based on a need or gap to renew and properly maintain the computational equipment of the Pedagogical Innovation Classrooms (PIC) of the Educational Institutions of the Province of Lambayeque.

At the theoretical level, research methods such as synthetic analytical, inductive deductive, historical-logical, holistic-systemic, systematization, and empirical methods are used for the characterization of the theoretical and historical antecedents of the usability process of ICT infrastructure and the construction of the theoretical and practical contributions.

At the empirical level, for the characterization of the current state of the functionality of the computational equipment of the Pedagogical Innovation Classrooms of the Educational Institutions of the Province of Lambayeque, techniques, such as surveys, interviews, observation guides, documentary analysis, and the corroboration of the feasibility and scientific-methodological value of the research results (expert criteria), were applied.

The population is made up of 50 PIC classroom teachers from the Educational Institutions of the Province of Lambayeque and an expert teacher in Information and Communication Technologies. The sample size is 37 PIC Classroom Teachers (PICT) calculated by sampling for proportions [11].

We worked with the survey to collect information on the current state of computational usability and its relationship with the sustainability of computational equipment in the Pedagogical Innovation Classrooms. The survey was applied to the teachers in charge of the PIC Classroom. The instrument used consists of three dimensions of strategic management, risk prevention, and incidents, with eighteen items that were applied with virtual forms.

The theoretical review of usability models led to a modeling proposal under five approaches to the usability of ICT infrastructure oriented to the sustainability of computational equipment.

3 Results and Discussion

As a result of the survey of PICT teachers (Fig. 1), the factors that affect the process of ICT infrastructure usability in the PIC environment were determined, being the lack of maintenance of computer equipment responsible for 68% of the equipment failures. Likewise, the malfunction is due to the fact that 22% of the equipment is obsolete or was damaged by electrical problems.

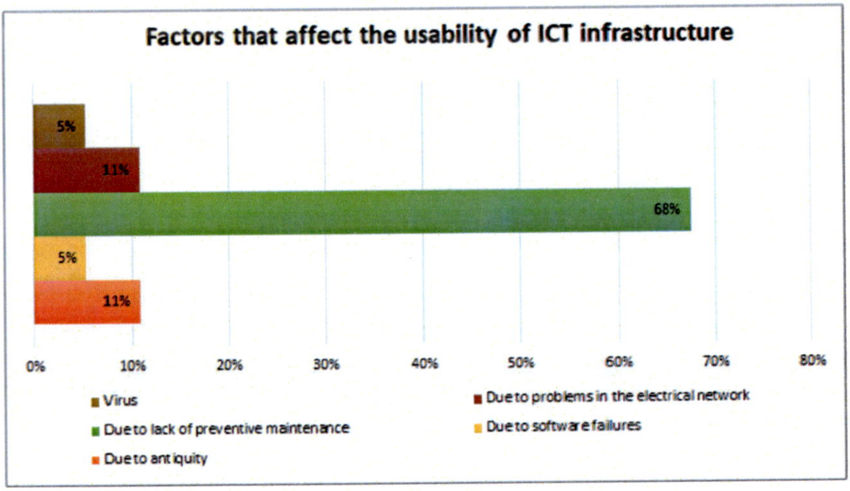

Fig. 1. Factors that harm the usability of ICT infrastructure

Based on these results, the ICT infrastructure usability model (Fig. 2) is proposed to improve the sustainability of computational equipment. The epistemological construction of the ICT infrastructure usability process contextualized to the PIC classrooms of the education sector was established under the systemic, failure, change, risk reduction, and monitoring approaches, considering three dimensions called: strategic management dimension, risk management dimension, and incident management dimension, which are materialized in the interrelationships that are achieved in the sustainability of the computing infrastructure.

The analysis of the sustainability of computational equipment through the systemic approach in organizations that become more complex involved objectives, resources, and the environment [12] with an application of the principles of the General Theory of Systems [13] and the changes of the Soft Systems Methodology [14], aimed at management control systems, reorganization of institutions and analysis and design of information systems.

In the same way, the failure approach is related to incident management since its main objective is the resolution of incidents to restore computer service in the shortest possible time [15], even when there are factors that are difficult to predict such as breakage of parts or machine elements [16].

The technological evolution of computer systems has a change approach that is clearly described by Moore's Law [17]. Likewise, the change is related to the useful life of computer equipment, which is shortened with the accelerated increase in the supply of new equipment, mainly due to the process of artificial obsolescence that entails the classification of an artifact as obsolete even when it is not [18].

All the activities concerning risk management have the objective of developing a strategy to effectively deal with risks. This entails the consideration of three fundamental aspects: avoiding risk whenever possible, monitoring risk, and, finally, managing risk and establishing contingency plans [19].

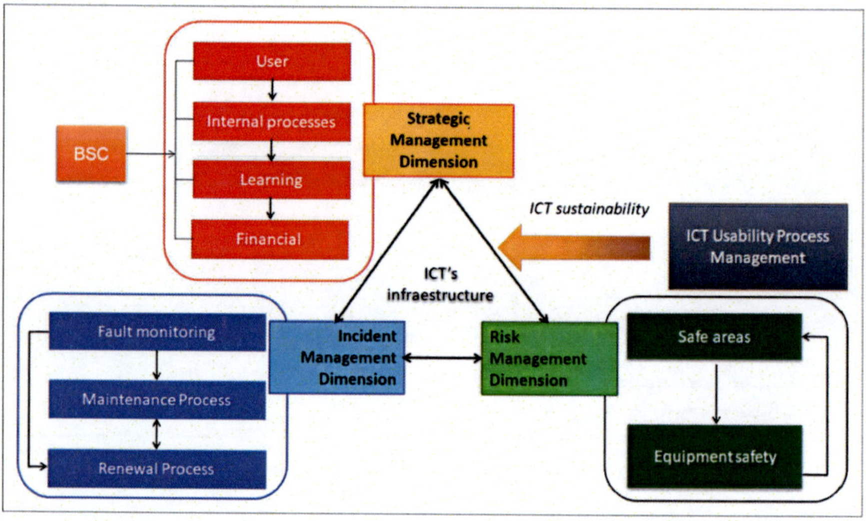

Fig. 2. ICT infrastructure usability model (prepared by the authors)

The monitoring approach is aimed at reviewing and evaluating the operation of the architecture of computer equipment, both of the physical components and the evaluation of software applications through the procedure of computer management, security measures, the use of its resources, and the essential consumer goods for the proper management of the computer room [20] and whose result of the evaluation will be carried out by issuing a report and/or professional opinion indicating the situation in which they are developed and use these computational resources [21].

In Fig. 2, the relationship between ICT sustainability in PIC classrooms and the available technological infrastructure is observed, which improves with the management of the ICT usability process. This approach is supported by three dimensions that contribute to the improvement of sustainability for the benefit of technology users.

The Risk Management Dimension allows determining, analyzing, assessing, and classifying the risk that occurs in the dynamics of the usability of the ICT infrastructure, to implement mechanisms that allow it to be controlled.

The Incident Management Dimension allows the prompt restoration of the service due to the malfunction of the hardware and software of the ICT infrastructure. Likewise, it allows the continuation of the service with an adequate preventive maintenance schedule and without the occurrence of a previous failure.

The application of the ICT infrastructure usability model was positive to understand by PICT teachers, who favorably rated the model applied in the context of PIC classrooms.

The statistical test for two usability samples assuming equal variances (Table 1) was done by contrasting the hypothesis with the Student's t-test and with a significance level of $\alpha = 0.01$.

The tool used was found to establish statistically that usability 2 is significantly higher than usability 1 since the second variable contemplates the use of ICT infrastructure with standardized methods that favor the sustainability of computational equipment.

Table 1. T-test for two usability samples assuming equal variances

	Usability 1	Usability 2
Average	32.5945946	64.375
Variance	33.46997	56.5535714
Observations	37	8
Pooled variance	37.2277656	
Hypothetical difference of means	0	
Degrees of freedom	43	
T statistic	-13.3587415	
P (T < = t) one tail	3.3276E-17	
A critical value of t (one-tailed)	2.41625013	
P (T < = t) two tails	6.6551E-17	
A critical value of t (two-tailed)	2.69510208	

Decision: it can be seen in Table 1 that the critical value for a tail is equal to 2.42 and, taking into account that the Student t-test acquires the t value equal to 13.36, which falls in the rejection region of the null hypothesis (Fig. 3); then, at a 99% confidence level, the null hypothesis (Ho) is rejected and the alternative hypothesis (HA) is accepted.

Consequently, the proposed ICT infrastructure usability model (Fig. 2) is comprehensive and dimensionally encompasses all aspects related to the usability process in

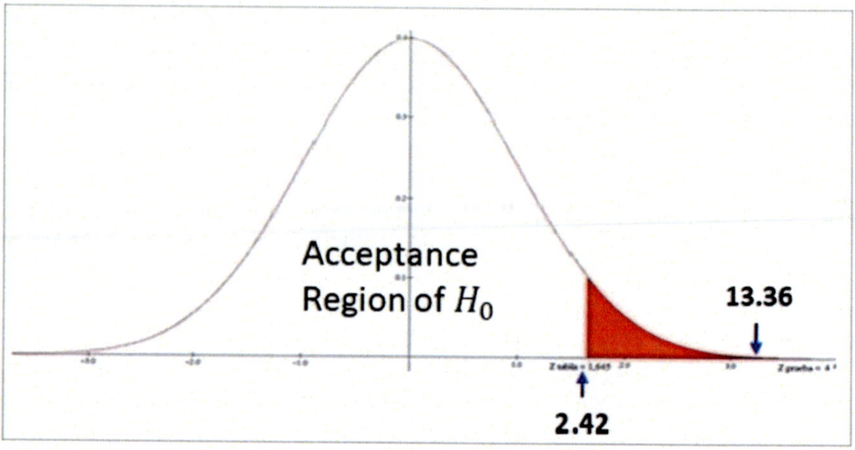

Fig. 3. Alternative hypothesis acceptance region

pedagogical innovation classrooms, taking into account the functionality and useful life of hardware and software under the failure approach [9] and supported by theoretical bases whose reference frameworks are accepted in the IT field [10].

4 Conclusions

The ICT infrastructure usability process was characterized, considering the conceptual and procedural approaches framed in the management of the functionality and useful life of hardware and software, however, it can be observed that there are limitations in the proposals adapted to the process of teaching-learning with technological resources, carried out in the PIC and/CRT classrooms, the same ones that require, to guarantee the sustainability of their computer equipment, the implementation of standardized maintenance and renewal procedures.

The ICT infrastructure usability model was developed based on the systemic, failure, risk reduction, and monitoring approaches and structured based on the dimensions of strategic management, to improve the performance of PIC classrooms, risk management, to minimize damage to the ICT infrastructure and incident management, to standardize maintenance processes and monitoring of failures.

The relevance of the model was corroborated, through the assessment carried out by the PIC teachers, implying to recognize that the proposal has the potential to contribute to the sustainability of the computational equipment of the PIC classrooms of educational institutions.

References

1. International Organization for Standardization (1998) ISO 9241–11. Switzerland
2. Paz Espinoza, F.A.: Método para la evaluación de usabilidad de sitios web transaccionales basado en el proceso de inspección heurística. Pontificia Universidad Católica del Perú (2017), http://hdl.handle.net/20.500.12404/9903

3. Aprende en línea (2021) Las TIC como apoyo a la educación. http://aprendeenlinea.udea.edu.co/lms/investigacion/mod/page/view.php?id=3118. Accessed 30 Mar 2021
4. Silva, F., Analide, C.: Computational sustainability and the PHESS platform: using affective computing as social indicators. Futur. Gener. Comput. Syst. **92**(1), 329–341 (2019)
5. Gómez-Fernández, N., Mediavilla, M.: Exploring the relationship between Information and Communication Technologies (ICT) and academic performance: a multilevel analysis for Spain. Socio-Econ. Plan. Sci. **77**(1) (2021)
6. Ministerio de Economía y Finanzas (2015) RM 010_2015EF Plan Estratégico de Tecnologías de la Información 2015–2016 del Ministerio de Economía y Finanzas, Perú
7. Gerencia Regional de Educación Lambayeque (2011) Funciones del docente responsable del AIP- CRT, Chiclayo Perú
8. Ministerio de Educación (2016) RM-627–2016 Normas y orientaciones para el desarrollo del año escolar 2017 en Instituciones Educativas y programas educativos de la Educación Básica, Perú
9. Sinha, S., Kumar Goyal, N., Mall, R.: Early prediction of reliability and availability of combined hardware-software systems based on functional failures. J. Syst. Architect. **92**(1), 23–38 (2019)
10. Asociación Española de Normalización (2009) ISO/IEC 20000 Guía completa de aplicación para la gestión de los servicios de tecnologías de la información. Madrid – España
11. Rodriguez, A., García, M.: Estadística II. Sistema Universidad Abierta y Educación a Distancia, México (2013)
12. Rodriguez, A.M.: Centros de cómputos y administración de proyectos. Instituto Politécnico Nacional, México (1991)
13. Checkland, P.: Pensamiento de sistemas, práctica de sistemas. 1st edn. Editorial Limusa S.A, México (1993)
14. Wilson, B.: Systems: Concepts, Methodologies and Applications. John Wiley and Sons Ltd., Chichester (1984)
15. Ríos, S.: ITIL v3 Manual íntegro. Biable Management, Excellence and innovation, Sevilla, España (2014)
16. Navarro, L., Pastor, A.C., Mugaburu, J.M.: Gestión integral de mantenimiento. 2nd edn. Marcombo, S.A., Barcelona (2009)
17. Barros, A.: 50 años de la Ley de Moore (2015). https://www.alejandrobarros.com/50-anos-de-la-ley-de-moore/. Accessed 09 May 2021
18. Vega, O.A.: Efectos colaterales de la obsolescencia tecnológica. Revista Facultad de Ingeniería, UPTC **21**(32), 55–62 (2012)
19. Sánchez, J.S., Chalmeta, R., Coltell, O., Monfort, P., Campos, C.: Ingeniería de proyectos informáticos: Actividades y procedimientos, 1st edn. Universidad Jaume I, Servicio de Comunicacion y Publicacione, Castelló, España (2003)
20. Muñoz, C.: Auditoria en Sistemas Computacionales. 1st edn. Pearson Educación, México (2002)
21. Aguirre, J.: Auditoria en Informática. Sistema Universidad Abierta Educación a Distancia, México (2005)

Challenges and Perspectives of the Furniture Market in Brazil

Luiz Vicente Figueira de Mello Filho[1](✉) and Italo Galvani[2]

[1] Recliners Industrial Ltda, Piracicaba, Brazil
luizvicente.mello@icloud.com
[2] Italo Galvani Soluções Industriais, São Paulo, Brazil
italo@italogalvani.com.br

Abstract. Traditionally the furniture market in Brazil is owned by family businesses and of national origin, with few exceptions. Based on this premise and the high demand promoted, especially in the Covid-19 pandemic, the market has adapted to the growing demand for residential furniture. Therefore, a conceptual restructuring needs to be carried out due to the great variability of products in the same factory, driven by the target audience, which goes beyond the traditional concepts of factory standardization and enables a new way of thinking according to the demand for products offered by catalog and/or bespoke. In addition, as they are family businesses, there is an inherent risk of closing activities due to the lack of family motivation of the following generations linked to the lack of knowledge to update manufacturing processes. This article concludes that the variability of products offered in each factory is a high option. New concepts must be adopted from handcrafted to manual transition. And, with positional and functional factory configurations, ensuring high efficiency and quality concerning the degree of difficulty, associated with the characteristics of furniture dealerships. To get success, the companies must be directed towards the sustainable production chain to the companies involved.

Keywords: Furniture · Manufacturing · Efficiency

1 Introduction

1.1 The History of Furniture in Brazil

The first piece of furniture began with the arrival of Dom João VI, who established the capital of Portugal in Brazil. Among other novelties, the arrival of furniture, initially all imported from Europe, gradually promoted a new concept of family comfort [1].

In the 1920s, the Patent factory was devoted to the production of beds with metal frames, popularizing this furniture in the country, which until then had the use of hammocks.

During and between the first two world wars, it was the moment that the greatest development of furniture took place, as imports were suspended, creating an internal

Y. Iano et al. (Eds.): BTSym 2021, SIST 207, pp. 294–300, 2023.
https://doi.org/10.1007/978-3-031-04435-9_28

need for this demand. Most of the artisans were of European origin and already had skills in making furniture [1].

In the 1960s, both the architecture and industrial design faculties began to professionalize furniture with a foundation [1].

Also in this decade, furniture manufacturers sought out the so-called designers, who signed the furniture collections, adding value.

Over time, the first generation of furniture manufacturers was replaced by the heirs, which is a new challenge given the context of remaining competitive with a high range of furniture offered in the market.

Some furniture manufacturing hubs have also emerged, including the cities of Maringá, in the state of Paraná, in Ubá, in the state of Minas Gerais, and in Porto Alegre, in the Rio Grande do Sul [2].

This fierce competition for an expressive number of manufacturers in the market has tended the market to open up the range of furniture possibilities even more offered in the catalog, in order to have a greater quantity of orders.

On the one hand, it was advantageous, but from the perspective of production engineering, there was a retreat from manufacturers to workshops, from production in line, in cell, or functional, to positional production, that is, through artisans in custom-made furniture.

1.2 The Challenge of the Furniture Market

The first challenge is to maintain the manufacture of furniture in view of the variability offered by each furniture configuration, with the number of variations exceeding 10,000 combinations among competitors being usual to the same products family when it comes to the crossing of dimensions, coatings, and finishes.

Make production batches more flexible, which, due to the high number of combinations, tend to be small-scale and often "single" batches, that is, to manufacture a single product with those characteristics.

The equipment used in manufacturing, over the years, has shown itself to be more and more capable of meeting such flexibility with smaller setups and more robust software, with the objective of better use of materials and available labor.

Based on this variability offered, keeping prices competitive in the market and maintaining earnings in line with a profit in the face of unfavorable cash flow.

The furniture market requires a settlement of invoices in days and the market requires a receipt in splits for months.

In order for organizations to organize themselves systemically, one of the ways established by senior management is in the management system known as the Balanced Scorecard.

This balanced system is organized into four distinct perspectives, which are: finance, customers, internal processes, and learning and growth [3], which were internally adapted as financial results, markets and customers, internal processes, and competencies.

2 The Balanced Scorecard

The Balanced Scorecard provides a necessary framework for translating the strategy into operational terms, which enables each employee to understand the importance of their activity in the general context of the company [3].

Each of the perspectives has performance indicators that interact with each other and can leverage the actions defined in the strategic planning, that is, the results of the indicators of the base formed by the competence perspective, influence internal processes, which affect markets and customers, and consequently, the financial results.

This methodology also meets the five production performance objectives, which are: quality in the manufacture of the product, the speed that is executed in the representative number of variations, the reliability of product delivery, the flexibility of each step of production, and, consequently, in the advantages regarding the elaboration of costs, which can be managed according to the first four performance indicators [4].

Perspectives are presented in the following subsections.

2.1 Learning and Growth Perspective

From this perspective, learning and growth are directly related to the prerequisites in training for hiring, with the continuous promotion of identifying new talents in different areas, in training in specific conditions such as fabric cutting software, and consistent use with less waste. Encouragement of changes in the manufacturing and/or assembly area for employees, and recycling of current or newly promoted ones to exercise the specific function of the position, which will provide the means to achieve the goals measured by the indicators.

For this, depending on the employee's education level and willingness to grow in the company, a career plan with two operational levels per area is sought, with a sector leader accompanied by employees. The choice of the leader is essential for the success in the progress of each sector. The example in the work of each leader, how to respect the schedule, commitment in the execution of tasks, encouragement in the search for continuous improvement. As a result, the spontaneous output turnover should be low when the factory is located in a region absent from the furniture industry.

With the identification of engaged employees, it is possible to mitigate work accidents, deviations in processes, and absenteeism, which are indicators of this perspective.

2.2 Internal Process Perspective

In order for the processes to be carried out normally in manufacturing, there are four management macro-processes, which support and maintain the rigor of quality in the production of furniture.

The management macro processes follow the sequence presented in Fig. 1.

Fig. 1. The management macro processes in the furniture market.

The first macro-process aims to guarantee the order for furniture after the purchase is made and is the responsibility of the commercial area. The area of costs and production planning and control (PPC), which supports this macro-process, has the essential function of offering the products that best suit the market's needs.

The development macro process is responsible for the development of new products aiming at possible standardizations in the manufacture of furniture, recalling the challenge of product variability. This macro process has the largest number of support in the operation, which is the driver of sales due, mainly, to the definition of the degree of difficulty in manufacturing and which will impact the delivery time. It is also responsible for preparing the technical datasheets for the products.

Third, the manufacture/invoice macro process, which plays a fundamental role in ensuring that designed products are effectively fulfilled within the pre-established deadlines.

The manufacturing macro process, to be successful, must use the Lean Manufacturing concept in the elaboration of processes, that is, it seeks to eliminate the seven categories of waste [5], which are: excess production; movement; wait; transport; unnecessary stocks; unnecessary processing, which is done exactly what is defined in the corrections, which is constantly avoided due to macro processes and defined instructions.

Finally, the delivering macro process, which means not only in the act of delivering but must be verified by the production control, tested, especially when using reclining devices. The quality of the sewing and the tailoring used to dress the structure is essential to deliver a product that generates repurchases and boosts the macro sale process.

The indicators of internal processes are related to the definition of each stage and the degree of difficulty in manufacturing, which will contribute to the balance and equation of different programmed products, with different production times according to each area. As there are different products to be made, a more elaborate product in the initial assembly is not necessarily in the tapestry and vice versa. Therefore, product development, when well designed, minimizes production imbalances.

2.3 Market and Customer Perspective

The market and customer perspective is the checker if the competence and internal processes perspectives are executed under the monitoring of indicators, i.e., if there is a base work in hiring, training, and recycling professionals, there is consequently the execution of work related to internal processes.

The main indicators in this perspective are the effective sale of the furniture produced, at the level of complaints regarding the products delivered, which will affect possible operational errors. The treatment of deviations in these indicators will not necessarily have an isolated action, it is often the initiative of a reflection by the manager, via indicators that are related, on how the basic works were performed to cause some hypothetical deviation, generating an action plan.

For this reason, each stage and layer of management has a vision based on the design of the macro processes, so that the indicators guide each professional in charge of managing a team and making a decision with tangible parameters and with greater assertiveness in guiding operational professionals.

2.4 Financial Results Perspective

The financial results follow the same principle from the perspective of markets and customers. That is, as there is a definition in the processes, the manager has control of how the labor must be made available so that they can serve the production according to a certain period.

When there is a large variability of products, a minimum stock of raw material is essential, especially in the stock of fabrics in catalogs, as they directly impact production time. There are also commodity items, such as wood, foam, fiber, staple, glue, which are used independently of the product to be produced.

The financial perspective is the consolidation of all layers of the Balanced Scorecard and is the closing of all stages and management of the complexity of the furniture market. At this stage, it is verified if the product development was well executed, if the technical file has a definition of the raw materials used and if the commercial area is practicing competitive prices and with profitability for the company's sustainability.

The indicator that represents this planning is the average operating cost due to the variability of products and the complexity of different production methods and with different execution times. For this indicator to be assertive, the challenge is to seek less variation in time and complexity for each product developed.

To support this indicator, variable costs must also be monitored, such as consumption of water, electricity, and disposal of solid waste, resulting from the production of furniture.

All day-to-day control with the financial results makes it possible to have rigor, control, and assertiveness in planning the budget for the following year.

3 Results and Discussions

The four perspectives presented in chapter 2 for the progress of the processes can be translated by standard strategic planning in the mobile market, as shown in Fig. 2.

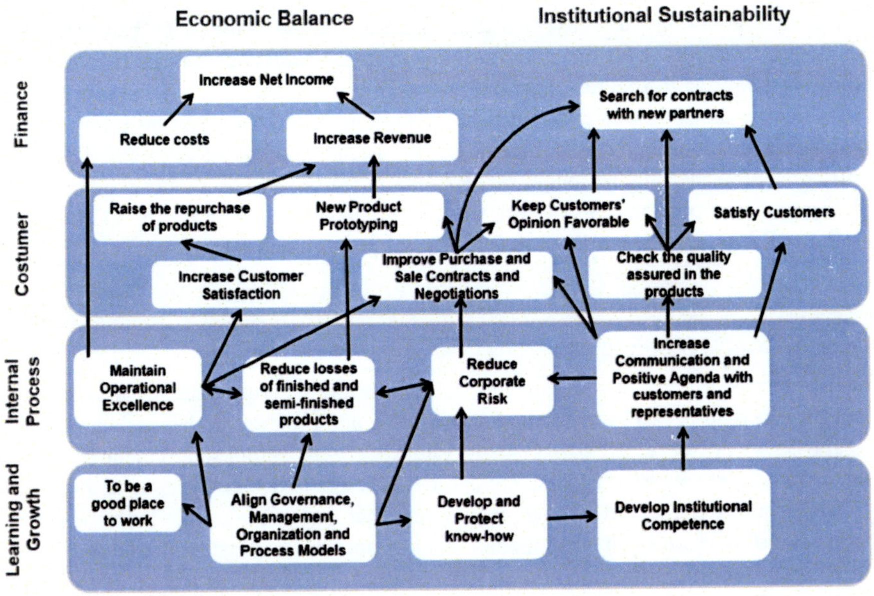

Fig. 2. The proposed strategic map given the complexity of the furniture market.

Figure 3 shows the direction with the four perspectives and each block the step to be completed, with the basic steps in the perspective of competence, going through the definition of internal processes, working with the market and customer, and ending with sustainability financial.

For the market and customer, as well as the last stage, the financial one, a matrix is proposed that has as its principle the rationalization of decision-making in face of the complexity of order variation.

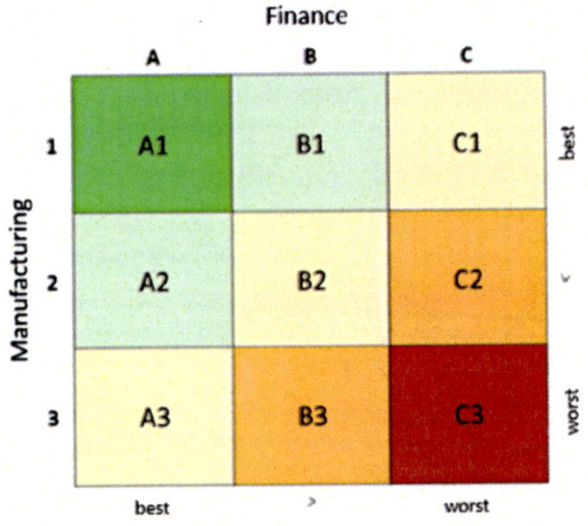

Fig. 3. Matrix of the degree of difficulty in relation to store customers.

Figure 3 shows how assertiveness in the face of what is available in the orders can collaborate to supply with the minimum stock, in the understanding of receipt after billing is issued and the degree of difficulty in each product assumed for manufacturing.

Indirectly, the action of what must be in stock is taken, the payment terms of the raw material concerning receipts. In addition, the equalization of the degree of difficulty contributes to formatting, no longer being taken by feeling and becoming rational.

There is a proportionality between the quadrants, as follows: A1 = 5 × C3; A2 = B1; A3 = B2 = C1; B3 = C2; A1 = 2 × A2; A1 = 3 × A3 e A1 = 4 × B3.

4 Conclusions

For the success of traditional companies in the furniture market, systemic thinking and management by processes are essential to maintain and progress in the market.

The Balanced Scorecard management system encompasses all stages in the furniture market and directs the top management at the key points in dedicating time and energy.

The Balanced Scorecard management system encompasses all stages in the furniture market and directs the top management at the key points in dedicating time and energy.

The production matrix in relation to financial health provides guidance in decision-making, collaborating with the company's sustainability.

Investments in technology through sophisticated numerical control equipment become the solution to offer the flexibility of products in single batches. The first perspective of the Balanced Scorecard contributes to training professionals in this new market need.

References

1. Santos, M.C.L.: Modern Furniture in Brazil. 1st edition, SENAC (2017)
2. Brainer, M.S.C.P.: Setor Moveleiro: aspectos gerais e tendências no Brasil e na área de atuação do BNB. Caderno Setorial ETENE (2018)
3. Kaplan, S.R., Norton, D.P.: A Estratégia em Ação – Balanced Scorecard, 32nd edn. Elsevier (1997)
4. Slack, N., Chambers, S., Johnston, R.: Administração da Produção. Atlas, 3rd edition. São Paulo (2009)
5. Ohno, T.: O Sistema Toyota de Produção, reimpression, Bookman (1997)

Study of Depression, Academic Stress and Demographic Variables in the Academic Performance of Students of a Public University Using Multivariate Methods, Lima - Peru

Félix Bartolo Gotarate(✉) and Olga Lidia Solano Dávila

Facultad de Ciencias Matemáticas, Universidad Nacional de Mayor de San Marcos, Lima, Peru
{fbartolog,osolanod}@unmsm.edu.pe

Abstract. In this paper, we studied some characteristics, such as demographic variables, depression, and college stress, which are associated or not with university achievement, in students of a faculty of basic science at a non-private university. A transactional, descriptive, and explanatory study was used on a sample, of probabilistic type, with 280 participants. The data were gathered during the second half of 2017, applying the Beck depression test, Barraza academic stress test, and the variables age, stages of life, sex, professional jobs, civil status, and socioeconomic status. For the statistical analysis, the multiple logistic regression model was used to determine if the variables study depression, component for academic stress, and demographic influence academic performance. The results show that the variable that influences university achievement is depression and the characteristics that do not associate university achievement are age, stages of life, gender, professional school, civil status, socioeconomic level, and that dimension for college stress.

Keywords: University achievement · Depression · University stress · University students · Logistic regression

1 Introduction

Depression in university students can be considered as a deterioration of their mental health, having negative consequences on their academic performance.

The student performs various academic activities, however, sometimes they may feel tired or have a lack of concentration, lack of energy, and have no interest in performing the tasks assigned to them in the courses they are enrolled in [1, 2]. It is estimated that depression affects more than three hundred million people in the world. Depression is a public health problem, especially when it is moderate to severe, it can disrupt activities at work and college, sometimes even leading to suicide. On the other hand, depression is one of the main mental disorders worldwide [3].

Depression is a mental disorder characterized by a combination of sadness, disinterest in tasks, and decreased energy, it also mentions that other symptoms are distrust

Y. Iano et al. (Eds.): BTSym 2021, SIST 207, pp. 301–307, 2023.
https://doi.org/10.1007/978-3-031-04435-9_29

and lack of self-confidence, unjustified sad mood, desire to commit suicide, decreased concentration, as well as insomnia and anxiety to eat [2].

Stress is defined as the physiological reaction of the body in which various defense mechanisms come into play to cope with a situation that is perceived as threatening or in greater demand [4]; when it is originated in the context of a university process, this is called university stress. For a university institution, it is significant to know the degree of stress in its pupils, as well as to know how university stress has been related to depression in its pupils, how stress has been related to mental diseases, incurable illness [5], heart diseases [6], failures in the immune system [7], complementary school failure [8], and poor academic performance [9]. The purpose of this work is to show a model to explain academic achievement based on sociodemographic variables, depression, and academic stress in students enrolled in the second academic semester of 2017, at the faculty of basic science.

This work has the full consent of the National University of San Marcos, on behalf of the university representatives Edinson Montoro Alegre, Dean of a faculty of basic science at a non-private university. Additionally, the authors Félix Bartolo and Olga Solano were in charge of managing the information collected in the survey.

2 State-Of-The-Art

There are several studies about depression, and we present some below. Students with severe depression have academic problems [10]. On the other hand, female students have greater depressive symptoms than men, and the risk of failing subjects is higher in those who have symptoms of depression [11, 12].

In [13], it was evidenced that 38.7% of students present some degree of depression, and among male pupils, 56% present some degree of depression.

In other research related to stress, it has been found that the main academic stressors were an overload of tasks and work, evaluations, and the type of work requested by teachers [14, 15].

On the other hand, the association between the stress at university variable and the academic achievement variable is positive, because the pupils presented conservative stress and obtained a basic academic performance [16].

Likewise, stress does not significantly affect academic performance, however, it interferes with some of the academic activities carried out by students [17, 18].

Anxiety has also a significant role in doing an exam. It is necessary to meditate on the distress suffered by students until they have taken the exam [19].

In academic performance, a high degree of stress alters the triple response system: (a) At a cognitive level, students believe they do not know anything before the exam, that their minds are going to go blank, that they will ask questions that they have not studied, that they will lose the scholarship if they fail the exam, etc. (b) on a motor level, students need to go to the bathroom without urinary elimination, crunch their fingers, bite their nails, snort, smoke excessively, etc., and (c) physiologically, they suffer from tachycardia, sweating, general tremor, meteorism, diarrhea, etc. The disturbance of these three levels of answers has an inverse relationship with university achievement. In some cases, it decreases the students' marks in the exams, and in other cases, the students do not show up for the exam or leave the classroom minutes before [18].

3 Methodology

A transactional, descriptive, and explanatory perspective was used, since this study aims to explain university performance from sociodemographic variables, depression, and academic stress, in students of a faculty basic sciences at a non-private university [20].

The authorization was obtained from the faculty of basic sciences to carry out this study. The pupils signed an informed document. The list of students enrolled in the undergraduate was provided by the head of registration of the faculty of basic sciences.

3.1 Selected Sample

The population in this research was all undergraduate students enrolled in the faculty of basic sciences of a non-private university in the academic period 2017-II. Stratified sampling was used [21], where each stratum was a professional career of the faculty. In total, four professional careers were considered. Students were randomly selected for each career path. Considering a degree of significance of 0.05, an estimation error of 0.0415, the percentage of students with depression in each career professional, and the information of the students enrolled in the faculty of basic sciences [21], the sample was 280 participants.

3.2 Variables and Instruments Used

The variables considered in the research were sex, professional careers, marital status, socioeconomic status, depression, and levels of depression in students.

In the questionnaire used, the following students' data were considered: test of depression [22, 23] and test of academic stress [24, 25]. The instruments were applied during the months of October and November of 2017. The value of Cronbach's alpha coefficient for the stress test was 0.85, and, for the depression inventory, it was 0.85, which is considered good reliability.

4 Results

For the statistical analysis of the research [26], academic performance was used as the response variable and the independent variables were depression, stress at university, and demographic and social variables.

Table 1 shows us that the important variable in the regression equation is depression, with a significance degree of 0.05. Likewise, having minimal depression increases the association to pass courses, compared to having a level of severe depression, with an OR = 2.041 (IC95%: 1.543:2.701).

Also, Table 2 shows that the variables that are not important in the regression equation are age, status of life, gender, professional school, civil status, socioeconomic level, stressors, reactions to stress, and coping tactics for college stress, at a significance degree of 0.05 ($p < 0.05$).

Table 1. Logistic regression equation estimated with important variables influencing academic performance – Semester 2017-II

Variable	ß	Wald	Df	p-value	Odds Rati(OR)	IC 95% exp(ß)
Level of depression[a]		26.473	3	0.000		
Level of depression (1)	0.713	24.942	1	0.000	2.041	1.543 2.701
Level of depression (2)	0.318	0.939	1	0.332	1.375	0.722 2.618
Level of depression (3)	-0.405	0.592	1	0.442	0.667	0.237 1.873

[a] Level of depression (1): minimal, level of depression (2): mild, and level of depression(3); moderate

Table 2. Logistic regression equation estimated with not-important variables influencing academic performance – Semester 2017-II

Variable	Wald	Df	p-value
Age	0.871	1	0.351
Stages of life[a]	0.032	2	0.984
Stages of life (1)	0.009	1	0.926
Stages of life(2)	0.011	1	0.915
Gender	2.841	1	0.092
Professional school[a]	3.598	3	0.308
Professional school (1)	0.115	1	0.735
Professional school (2)	2.468	1	0.116
Professional school (3)	2.274	1	0.132
Civil status	0.003	1	0.958
Socioeconomic degree[c]	1.357	2	0.507
Socioeconomic degree(1)	0.004	1	0.949
Socioeconomic degre(2)	0.124	1	0,725
Stressors	0.694	1	0.405
Reactions to stress	0.261	1	0.609
Coping strategies	0.005	1	0.944

[a] Stages of life(1): Adolescence, stages of life(2): Youth
[b]Professional school (1): mathe, professional career (2): statistics, and professional career (3): operations research
[c]Socioeconomic degree(1): medium, socioeconomic level(2): low

5 Discussion

The results show that depression influences university achievement. This is similar to various studies which indicate that students with depression have academic problems [10–12]. Likewise, in [27] they determined that there is an important association between depression and regular or low university achievement. Students having depression have disinterest in academic activities, less ability to concentrate, and loss of self-esteem and confidence in themselves [2]. In our study, the dimensions of academic stress are not significant so they do not influence academic performance, but it is present in a moderate way.

The most overwhelming university demands revolve around academic work, and in a short time to be able to fulfill them, such as the evaluation of the professors, reloading of works and tasks, the work model they require, as well as the limited time so that they can do it and not understanding the topics covered in the classroom. Which are considered alarming for learners. This is consistent with some research, in [16] they reported that students in the sixth semester in education at the Autonomous University of Yucatán show that academic stress does not significantly harm academic performance, however, it intervenes in some of the academic activities carried out by students. And in [28] they indicated that there is no association between stress and violence with academic achievement. Likewise, the intensity of student stress is not associated with college achievement [29], thus in [30], they concluded that there is no significant association between the level of stress and university performance.

6 Conclusions

In this article, we present a statistical model that explains university achievement according to the features' demographic variables, depression, and academic stress. According to this, we conclude that the variable that influences university achievement is depression and the features that are not related to achievement are age, stages of life, gender, professional school, civil status, socioeconomic level, stressors, reactions to stress, and coping strategies for college stress ($p > 0.05$).

It is recommended that the professors of the different courses do not overload the students with works, assignments, exhibitions, and practical qualifications simultaneously, because these activities are thus considered to result in depression. It is suggested that teachers coordinate their exams so that they are not taken simultaneously on the same day.

References

1. García-Herrera, J., Nogueras, E.V.: Guías de autoayuda para la depresión y los trastornos de ansiedad. Sevilla: Servicio Andaluz de Salud (2013). https://consaludmental.org/publicaciones/Guiasautoayudadepresionansiedad.pdf. Accessed 11 Oct 2020
2. OMS (2020) La depresión según la Organización Mundial de la Salud. https://www.who.int/es/news-room/fact-sheets/detail/depression. Accessed 11 Oct 2020

3. Pardo, G.A., Sandoval, A.D., Umbarila, D.Z.: Adolescencia y depression. Revista Colombiana de Psicología **13**, 17–32 (2004). https://www.redalyc.org/pdf/804/80401303.pdf. Accessed 11 Oct 2020
4. Selye, H.: The Stress of Life. McGraw-Hill, New York (1956)
5. Rich, A.R., Scovel, M.: Causes of depression in college students: a crosslagged panel correlation analysis. Psychol. Rep. **60**(1), 27–30 (1987)
6. Lowe, G., Urquhart, J., Greenman, J., Lowe, G.: Academic stress and secretory immunoglobulin. Psychol. Rep. **87**, 721–722 (2000)
7. Verdhara, K., Nott, K.: The assessment of the emotional and immunological consequences of examination stress. J. Behav. Med. **19**, 467–478 (1996)
8. McDonald, A.S.: The prevalence and effects of test anxiety in school children. Educ. Psychol. **21**, 89–101 (2001)
9. Zeidner, M.: Test Anxiety: The State of the Art. Plenum Press, New York (1998)
10. Gutiérrez, J.A., Montoya, L.P., Toro, B.E., Briñón, M.A., Rosas, E., Salazar, L.E.: Depresión en estudiantes universitarios y su asociación con el estrés académico. CES Medicina. **24**, 7–17 (2010)
11. Fouilloux, C., Barragán, V., Ortiz, S., Jaimes, A., Urrutia, M.E., Guevara-Guzmán, R.: Síntomas depresivos y rendimiento escolar en estudiantes de Medicina Salud Mental. Salud Mental **36**(1), 59–65 (2013). https://www.redalyc.org/pdf/582/58225671008.pdf. Accessed 11 Oct 2020
12. Riveros, Q.M., Hernández, V.H., Rivera, B.J.: Niveles de depresión y ansiedad en estudiantes universitarios de Lima Metropolitana. Revista de Investigación en Psicología, **10**(1), 91–102 (2007). https://doi.org/10.15381/rinvp.v10i1.3909. Accessed 11 Oct 2020
13. Rosas, M., Yampufé, M., López, M., Carlos, G., Sotil, A.: Niveles de depresión en estudiantes de Tecnología Médica. Anales de la Facultad de Medicina **72**(3), 183–6 (2011). https://revistasinvestigacion.unmsm.edu.pe/index.php/anales/article/view/1065. Accessed 11 Oct 2020
14. Zárate, N.E.: Estrés académico en estudiantes universitarios: medidas preventivas. Revista de la Alta tecnología y la Sociedad **9**(4), 92–8 (2017). http://www.academiajournals.com/revista-alta-tec-y-sociedad. Accessed 11 Oct 2020
15. Toribio-Ferrera, C., Franco-Bárcenasa, S.: Estrés Académico: El Enemigo Silencioso del Estudiante. Salud y administración **3**(7), 11–18 (2016). http://www.unsis.edu.mx/revista/doc/vol3num7/A2_Estres_acad.pdf. Accessed 11 Oct 2020
16. Valencia, J.A., Moreno, Y.V.: Relación entre el estrés académico y el rendimiento académico de estudiantes de sexto semestre de psicología de la Corporación Universitaria Minuto de Dios.Facultad de Ciencias Humanas y Sociales. Programa de psicología (2015). https://repository.uniminuto.edu/bitstream/handle/10656/4185/TP_ValenciaJennifer_2015.pdf?sequence=1&isAllowed=y. Accessed 11 Oct 2020
17. Domínguez, R.A., Guerrero, G.J., Domínguez, J.G.: Influencia del estrés en el rendimiento académico de un grupo de estudiantes universitarios. Educación y ciencia, **4**(43), 31–40 (2014). http://www.educacionyciencia.org/index.php/educacionyciencia/article/viewFile/313/pdf_12. Accessed 11 Oct 2020
18. Pérez, M.A., Martín, A., Borda, M., Del Río, C.: Estrés y rendimiento académico en estudiantes universitarios. Cuadernos de medicina psicosomática y psiquiatría de enlace, 67–68, 26–33 (2003). http://www.editorialmedica.com/editorialmedica_publicacion_numero_detalle.php?nid=159&t=cuadernos&y=2003&m=Julio-Agosto-Sept/Oct-Nov-Diciembre%202003&num=13. Accessed 11 Oct 2020
19. Álvarez, J., Aguilar, J-M., Lorenzo, J.J.: La Ansiedad ante los exámenes en estudiantes universitarios: Relaciones con variables personales y académicas. Electron. J. Res. Educ. Psychol. **10**(1), 333–354 (2012). https://www.redalyc.org/pdf/2931/293123551017.pdf. Accessed 11 Oct 2020

20. Hernández, T., Mendoza, C.P.: Metodología de la investigación, 6a edn. McGraw Hill, México (2018)
21. Scheaffer, R., Mendenhall, W., Ott, L.: Elementos de muestreo, 6ª Paraninfo Cengage Learning, Madrid (2007)
22. Beck, A.T., Ward, C.H., Mendelson, M., Mock, J., Erbaugh, J.: An inventory for measuring depression. Arch. Gen. Psychiatry **4**, 561–571 (1961)
23. Beck, R., Taylor, C., Robbins, M.: Missing home: sociotropy and autonomy and their relationship to psychological distress and homesickness in college freshmen. Anxiety Stress Coping **16**, 155–162 (2003)
24. Barraza, A.: Un modelo conceptual para el estudio del estrés académico. Revista Electrónica de Psicología Iztacala **9**(3), 110–129 (2006). http://www.psicologiacientifica.com. Accessed 10 Oct 2020
25. Barraza, A.: Propiedades psicométricas del Inventario SISCO del Estrés Académico. Revista Psicología Científica, **9**(10) (2007). http://www.psicologiacientifica.com/sisco-propiedades-psicometricas/. Accessed 10 Oct 2020
26. Hosmer, D.W., Lemeshow, S., Sturdivant, R.X.: Applied Logistic Regression. Wiley, New Jersey (2013)
27. Franco, C., Gutierrez, S., Perea, E.: Asociación entre depresión y bajo rendimiento académico en estudiantes universitarios. Psicogente **14**(25), 67–75 (2011). https://www.redalyc.org/pdf/4975/497552358007.pdf. Accessed 10 Oct 2020
28. Barahona, L., Amemiya-Hoshi, I., Sánchez-Tejada, E., Oliveros-Donohue, M., Pinto-Salinas, M., Cuadros-Tairo, R.: Asociación entre violencia, estrés y rendimiento académico en alumnos de medicina del primer y sexto años de una universidad pública, 2017. Anales de la Facultad de Medicina **79**(4), 307–311 (2018). https://doi.org/10.15381/anales.v79i4.15635. Accessed 10 Oct 2020
29. Evaristo-Chiyong, T., Chein-Villacampa, S.: Estrés y desempeño académico en estudiantes de Odontología. Odontol. Sanmarquina **18**(1), 23–27 (2015). https://revistasinvestigacion.unmsm.edu.pe/index.php/odont/article/view/11336/10211. Accessed 10 Oct 2020
30. Caldera, J.F., Pulido, B.E.: Niveles de estrés y rendimiento académico en estudiantes de la carrera de Psicología del Centro Universitario de Los Altos. Revista de Educación y Desarrollo 77–82 (2007). http://www.cucs.udg.mx/revistas/edu_desarrollo/anteriores/7/007_Caldera.pdf. Accessed 10 Oct 2020

Transformative School Digital Sustainability Model to Improve Teacher Professionalism and Identity in Primary Education

Lucia Isabel Chamán Cabrera[1(✉)], Silvia Juana Saavedra Díaz[2], Luis Alberto Chunga Pajares[2], and Consuelo Ivonne Del Castillo Castro[1]

[1] Pedro Ruiz Gallo National University, Lambayeque, Peru
{lchaman,cdelcastilloc}@unprg.edu.pe
[2] César Vallejo University, Chiclayo, Peru
sisavedrad@ucvvirtual.edu.pe, lchungap@ucv.edu.pe

Abstract. In the transformation of education, it is essential that the teacher conceives their pedagogical practice as a way to strengthen their professional identity and exercise their teaching work in accordance with the trends and challenges of education, showing good preparation, developing research, using technology, and reflecting on their teaching practice. In order to improve the development of Teacher Professional Identity (TPI), the transformative school digital sustainability model was applied. The design was pre-experimental of an explanatory type, and a questionnaire was applied to 31 teachers from a school in Peru. Taking into account the dimensions of teaching performance established in the Framework for Good Teaching Performance (MBDD, from the Spanish Marco de Buen Desempeño Docente), the extent of the development of teaching professionalism and identity was chosen. By applying the model, it was possible to reverse the problem identified in the pre-test, increasing the high level to 67.74%.

Keywords: Digital model · Teaching identity · Professionalism · Transforming school

1 Introduction

The teacher has a leading role in the educational process; therefore, he must have good scientific and technological preparation. Currently, their pedagogical mediation must be carried out in a different way from the face-to-face modality, making it possible to engage with important ones, to develop learning skills virtually with their students, assuming new challenges, empowering themselves in the management of technological tools and the digital competence through the development of online courses, self-training, workshops, study circles, etc.

The development of professionalism must be linked to the evaluation of teachers, taking into account the educational context in which the teacher develops

Y. Iano et al. (Eds.): BTSym 2021, SIST 207, pp. 308–313, 2023.
https://doi.org/10.1007/978-3-031-04435-9_30

their pedagogical work [1]. In Iran, TPI is carried out by designing local programs that are tailored to the needs of teachers where contexts can facilitate the construction of the identity of teachers, not only those who are in service but also those who are in training [2].

We need an education in digital communication media for the 21st century; that is why the incorporation of technology contributes to achieving continuous improvement, being reflected in instructing in teaching and learning. Along the same lines, we agree with David Buckingham's statement, mentioning that if digital influences life, then digital must also influence teachers' professional practices. It is important to transform schools with leading professionals in technological knowledge who carry out innovations, and this research constitutes a permanent practice in their pedagogical work.

2 State-of-the-Art

In current times, digital technologies are immersed in various fields of life, highlighting their relevance in educational innovation through their integration, resulting in a range of different ways in which students and teachers access, engage, and create knowledge [3]. Likewise, their teaching methods must be redefined with leadership models and interaction channels through digitization towards the improvement of the sustainable development of their teaching [4]. Considering innovative education, the teacher must develop critical, creative, digital, and socio-emotional skills so that they feel committed to playing an active role in social changes for a sustainable world [5].

Also, it is vitally important to promote a transforming school, using digital tools and technology not only for pedagogical practice but also for teacher training [6] and thus promote the strengthening of TPI, generating virtual spaces for reflection and use of digital technology for the sustainable development of their teaching work, with research and innovation as a constant in their teaching practice. In an investigation carried out in Malaysia, professional qualifications, experience, and development were found to be part of the teaching identity, having a direct impact on the curriculum [7]. Likewise, in Denmark, dimensions of the ethical perception of professionalism were analyzed, highlighting its importance and applicability in the formulation of policies, professional training, and education [8].

Furthermore, in Australia, the professional proposes particular standards related to the identity, the practice, as well as the vocation and training of educators [9]. Also, the importance of involving educational actors in the planning and dissemination of activity plans was demonstrated, highlighting the importance of collaborative work [10]. Finally, professional qualities are considered in adapting the content of educational work to the personal characteristics and educational needs of students, highlighting the importance of technology in professional teacher training [11].

3 Proposal

The transformative school digital sustainability model is sustainable, and its purpose is to improve the TPI of Basic Education by making use of technology; it is based on the four pillars of education of this century published by Delors, such as: learning to do and learning to be, learn to know and learn to live together, taking into account the humanistic approach to education. This model presents a profile of the teacher of the digital transforming school who must be a leader, who promotes the strengthening of their interpersonal skills, their continuous improvement at a personal and institutional level, demonstrating solid knowledge, skills, and attitudes, in addition to being a critical, reflective, and practical researcher, who conceives research as a constant practice in the development of knowledge. In the pedagogical aspect, the teacher must be a manager and generator of emerging pedagogical skills, an expert in content and learning environments, using technology in the student's environment and their relationship with the family. The teacher must also carry out pedagogical innovation projects that allow the achievement of the learning in the students and collaborate to the progress of the educational quality.

This model focuses on Siemens' connectivism, making it necessary to apply this new orientation in the social sphere represented by the generation of value through human intelligence networks to generate knowledge, setting itself up in a new context where technology fulfills a relevant role [12]. From this perspective, professional learning communities play a very important role in this model as they seek to increase the efficiency and opportunity offered by the electronic space, offer and generate new educational spaces, channels of exchange, and participation; In addition, learning networks that are online learning environments will be taken into account, allowing them to be competitive, sharing and collaborating information. In this regard, the learning networks are intended to enrich the learning experience in the aspects of distance and face-to-face modality.

Ten sessions were applied with various strategies where the teacher was strengthened in their professional skills through collaborative work, participating in workshops and virtual learning circles. The following technological tools were applied: Regarding the collaborative work of the teacher, they were trained in the use of Office 365, Google drive; for communication: Zoom, Microsoft Teams, Meet; also to share files: Dropbox, Mega, WeTransfer; also for the development of the sessions: Easyclass, Remind, Socrative, Projected, Thing Link, Kahoot; in the same way for the elaboration of contents: Ted-Ed, CK-12, Academic Google, EduClipper; for creating activities: Storybird, Ardora, Hopscotch; also for creating mind maps: Mindomo, Mindmeister and for creating infographics: Visual.ly, Infogram. Concerning the tools to strengthen investigative competencies, training was provided in the use of EduBirdie, to detect plagiarism, as well as bibliographic managers: Zotero, Mendeley. It should be noted that all the digital tools in which the teachers were trained are free. The model can be seen in Fig. 1.

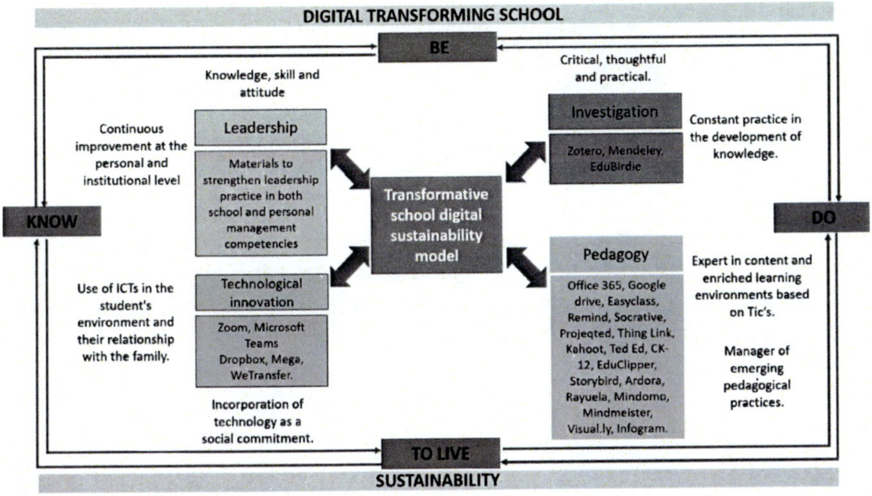

Fig. 1. Transformative school digital sustainability model to improve teacher professionalism and identity in primary education.

4 Results and Discussion

In Table 1, we can see that 45.16% of teachers present a regular level in the pre-test, and after applying the sustainable digital model, 67.74% of teachers obtained a high level; which indicates that after the application of the sustainable digital model of a transforming school, the TPI increased.

A pre-test consisting of a questionnaire on the development dimension of professionalism and teacher identity was applied that refers to their participation in improvement studies, training, research, in which it was observed that 22.58% have a high-performance level and after applying the sustainable digital model of a transforming school, the performance increased to 67.74%, demonstrating the validity of the model.

Table 1. Development of teacher professionalism and identity

Development of teacher professionalism and identity	Pre test		Post test	
	fi	%	fi	%
Low	10	32.26	2	6.45
Regular	4	45.16	8	25.81
High	7	22.58	21	67.74
Total	31	100	31	

Taking into account the results, we affirm that, for professional improvement, it is necessary to carry out workshops, courses on disciplinary content of the different areas or subjects to be taught, research experiences, and internships as an opportunity for professional improvement [13].

For the TPI, we must apply a model that strengthens their teaching skills in the digital world, which coincides with the Holistic model that promotes teacher development, with the constant motivation of digital pedagogical content and practices, forming a reflective-practical educator, incorporating technology into their pedagogical practice and their professional training for their continuous improvement, achieving a quality education [14]. The teacher must not only acquire technological skills but must adapt their role to their professional development.

5 Validation

A seventeen-item questionnaire was developed based on four dimensions of the MBDD - MNEDU 2012: Teaching for student learning, preparation for student learning, participation in the school's management articulated to the community, development of professionalism, and teaching identity. In addition, each indicator considers three items: Promotes their professional training, undertakes postgraduate studies, participates in training and specializations. After preparing the items, it is validated through the judgment criteria of 3 experts, giving conformity to the questionnaire; the reliability was obtained through Cronbach's Alpha, equivalent to 0.976.

The Transformative school digital sustainability model was applied, previously validated by expert judgment.

6 Conclusions

The pre-test shows that the level of TPI is between low and regular; this could be verified through the application of the questionnaire on the dimensions of teacher performance, validated by the judgment of 3 experts; in addition to this, it has a high degree of reliability.

In the post-test, when the Transformative school digital sustainability model was applied, it generated an increase in the TPI by 67.74%. (high level), checking its validity and viability.

The proposed model allowed to strengthen the competencies of teachers in leadership, research, pedagogy, and technological innovation through online specialization courses, supporting workshops through their continuous improvement, both personal and institutional. In addition, it seeks that the teacher is critical, reflective, practical, and experienced in learning environments and content supported by technological tools.

References

1. Manso, J.: Una mirada a la evaluación del profesorado en perspectiva internacional. Cuadernos de pedagogía **50**(504), 98–102 (2019)
2. Haghighi, F., Chalak, A., Heidari Tabrizi, H.: Assessing pre-service teachers' professional identity construction in a three-phase teacher education program in Iran. Asian-Pacific J. Second Foreign Lang. Educ. **5**(1), 19(2020). https://doi.org/10.1186/s40862-020-00100-3
3. Howard, S.K., Schrum, L., Voogt, J., Sligte, H.: Designing research to inform sustainability and scalability of digital technology innovations. Educ. Technol. Res. Dev. **69**(4), 2309–2329 (2021). https://doi.org/10.1007/s11423-020-09913-y
4. Sá, M.J., Serpa, S.: The COVID-19 pandemic as an opportunity to foster the sustainable development of teaching in higher education. Sustainability **12**(20), 8525 (2020)
5. Triviño-Cabrera, L., Chaves-Guerrero, E.I., Alejo-Lozano, L.: The figure of the teacher-prosumer for the development of an innovative, sustainable, and committed education in times of COVID-19. Sustainability **13**(3), 1128–1146 (2021). https://doi.org/10.3390/su13031128
6. Makarova, E.: Blending pedagogy and digital technology to transform educational environment. Int. J. Cogn. Res. Sci. Eng. Educ. **6**(2), 57–65 (2018). https://doi.org/10.5937/ijcrsee1802057M
7. Masnan, A.H., et al.: The concept of professional identity: Kindergarten teachers' professionalism requirement in Malaysian preschool curriculum. Int. J. Eval. Res. Educ. (IJERE) **10**(1), 126–134 (2021). https://doi.org/10.11591/ijere.v10i1.20849
8. Ribers, B., Balslev, G., Jensen, C.: Education, collaboration and pedagogical phronesis: essential dimensions in professional learning and development. Professional Development in Education (2021). https://doi.org/10.1080/19415257.2021.1902835
9. Tyler, M.A., Dymock, D.: Constructing a professional identity in VET: teacher perspectives. Res. Post-Compulsory Educ. **26**(1), 1–18 (2021). https://doi.org/10.1080/13596748.2021.1873404
10. Arinaitwe, D.: Practices and strategies for enhancing learning through collaboration between vocational teacher training institutions and workplaces. Empirical Res. Vocational Educ. Train. **13**(1), 1–22 (2021). https://doi.org/10.1186/s40461-021-00117-z
11. Alizhanova, K., Mikailov, D.: Technology of formation of legal competence of future bachelor of pedagogical education. Revista Publicando **5**(18 (Octubre-Noviembre-Diciembre)), 1–11 (2018)
12. Bates, A.W.T.: 2.6 Conectivismo. In: Enseñar en la Era Digital, pressbooks.com (2021)
13. Evans, P.K., Craig, C.J.: Enhancing preservice teacher preparation through formal and informal learning experiences. In: Stokes, D.W. (ed.) Preparing Teachers to Teach the STEM Disciplines in America's Urban Schools, Advances in Research on Teaching, vol. 35, Emerald Publishing Limited, pp. 65–83 (2021). https://doi.org/10.1108/S1479-368720210000035005
14. Mon, F.M.E., Quintero, L.J.C., Segura, J.A.: Un modelo holístico de competencia docente para el mundo digital. Revista interuniversitaria de formación del profesorado: RIFOP **32**(91), 105–116 (2018)

Emerging Digital Transformation Model for Teaching Strategies to the Achievement of Meaningful Learning in Education 4.0

Consuelo Ivonne Del Castillo Castro[1](✉), Luis Alberto Chunga Pajares[2], Lucia Isabel Chamán Cabrera[1], and Silvia Juana Saavedra Díaz[2]

[1] Pedro Ruiz Gallo National University, Lambayeque, Perú
{cdelcastilloc,lchaman}@unprg.edu.pe
[2] César Vallejo University, Chiclayo, Perú
lchungap@ucv.edu.pe, sisavedrad@ucvvirtual.edu.pe

Abstract. Several decades ago, the new technologies burst into the classrooms of education. Since then, studies have shown the benefits of the use of ICT in learning processes; however, a large percentage of teachers do not implement the use of digital tools in their subjects, making classes to be developed with orthodox methods such as rote learning or repetition; without achieving meaningful learning. For that reason, some students do not develop adequate critical thinking, a good analysis of information with reflective moments that lead them to build new ideas. In this investigation, a pre-experimental design was used and the type of investigation was explanatory. 39 students of the third cycle of a Peruvian university participated. In the pre-test, 48.7% obtained a regular level in the achievement of their learning, and in the post-test, the results improved significantly, reaching 81.1% in the good level. This reveals that after the application of the emerging digital transformation model for teaching strategies in Education 4.0, students had greater skills and abilities to achieve significant learning, demonstrating the validity of the proposed model.

Keywords: Digital transformation · Teaching strategies · Education 4.0

1 Introduction

In the last decades, technological development has been decisive in the way our society has changed, especially in the activities related to human beings, from housework, work, education, commerce, etc. Information and Communication Technologies (ICTs) are considered, in the XXI century, as sources of information and knowledge management, but above all as resources for Teaching-Learning, becoming trend the emergence of each emerging technology and at the same time, contributing and greatly improving education, providing teachers and students new ways and tools to achieve their goal with a higher level of quality in teaching that exceeds the imagined standards.

In 2020, due to the pandemic caused by COVID-19, education moved from face-to-face to virtual classes, despite the resistance of some teachers to the use of technology.

Y. Iano et al. (Eds.): BTSym 2021, SIST 207, pp. 314–321, 2023.
https://doi.org/10.1007/978-3-031-04435-9_31

However, the pandemic has forced us to make the leap that many authors announced that "had to be made at some point". Despite not being fully prepared and trained for that purpose, new models, approaches, and strategies have helped this process emerge [1].

2 State-of-the-Art

We live in a society marked by technological advances and these improve greatly the quality of life of human beings. We can say that success lies in choosing the right technology, and, in the future, the professional who best selects the tools and technologies for their professional practice will have a competitive advantage.

Universities are no strangers to these changes and technological advances, but making the leap to transformation and digital literacy is not an easy task. It is said that digital transformation is achieved when we bring creativity and innovation through the use of digital technologies, generating significant changes in an environment of interest [2], and, on the other hand, digital literacy is the set of skills necessary for the generation of knowledge from the analysis and interpretation of information [3].

We must evaluate whether there is a digital balance in the achievement of learning, and there should be a variation related to the design of learning models and thus ensure the achievement of competencies [4]. However, there are studies in which we can see obstacles for this digital transformation to occur properly, especially due to the lack of participation of teachers to provide pedagogical support [2]. In this sense, to ensure and promote digital transformation in universities, we must develop competencies in both students and teachers: both users of technology must acquire these skills and knowledge and use them effectively in the learning process [5].

Emerging technologies are considered scientific agents that are constantly evolving, and, in the educational field, they are one of the trends with the greatest contribution [6]. When combined with best practices and innovative pedagogical procedures it is known as Education 4.0, which proposes four main components for its achievement in higher education: competencies, learning models, information and communication technologies, and technological infrastructure [7].

It has been demonstrated that the use of emerging technologies allows the development of skills and improves both learning and academic performance in university students. This has been evidenced during the pandemic due to COVID-19, having a high technological impact on universities [1].

David Ausubel defines meaningful learning as that type of learning in which the student relates new information with the knowledge he already has in the cognitive structure; modifying and reconstructing both information, that is, new knowledge is conditioned to the structure of existing knowledge (constructivist approach). In this sense, authentic inquiry experiences are related to the achievement of significant learning, in their performance and self-perception [8], and if we combine these with emerging technologies, we can have effective and efficient results.

3 Methodology

The design used in this work was pre-experimental with pre-test and post-test, since the degree of control is minimal and we only worked with a single study group. The type of research was explanatory, since we intended to establish the causes of the facts; in the same way, it is applicative, since it allowed us to confront the theory with the problem under study.

The population consisted of 39 students ranging from 18 to 27 years old; having as inclusion criteria: that they are students of the career of business administration of a Peruvian university, who are studying the third cycle, and are enrolled in a particular subject and have actively participated in the emerging digital transformation model. On the other hand, the 39 students made up our sample group, since this research work focuses on a single study group.

To measure meaningful learning in education 4.0, a 15-item questionnaire was constructed based on Marzano's [9] 5 dimensions: Perception and skills; acquiring and processing knowledge, deepening and refining knowledge, meaningful use of knowledge and habits of mind, combined with the components of Education 4.0 [7]. Each dimension was divided into 4 items, the fourth item being the technological factor. The elaborated instrument was validated through the judgment of experts, who gave their agreement. Likewise, the statistical process of Crombach's Alpha was performed to see its reliability, having as a result 0.888, thus determining that the instrument has a good degree of reliability.

4 Proposal

The model of emerging digital transformation for teaching strategies proposed aims to increase the levels of depth of knowledge achieving significant learning in students of a particular subject, where the teacher makes use of summaries, mind maps, case studies, and discussions in their classes supported by technological tools and infrastructure, and thus improve the learning of the course at a university in Peru, enhancing the achievement of significant learning with the support of emerging technologies.

In addition, the technological basis of the proposal is based on the four core components of Education 4.0, which allowed teachers and students to take advantage of emerging technologies and infrastructure to improve procedures in higher education:

- Competencies: It considers the development of soft and hard professional competencies, designing educational models that allow facing challenges through the development of knowledge and efficient management of the "Know-how".
- Learning methods: It incorporates new learning methods that consider the use of technologies, principles, strategies, and pedagogical procedures in higher education, which allow the generation of knowledge and the transfer of information and resources.
- Information and communication technologies: Implementation of emerging ICTs, indispensable tools to contribute to new teaching models, working collaboratively and enhancing the learning process, especially when combined with teaching-learning method. Two aspects are considered: technology-based and tools and platforms.
- Infrastructure: Improve learning processes through the use of new infrastructure, adaptation of innovative infrastructure: virtual and physical, in response to current challenges and needs.

Education 4.0 promotes self-learning based on constructivist and humanistic principles, encourages metacognition and self-reflection as well as the understanding of the learning process itself [7]. The model can be seen in Fig. 1.

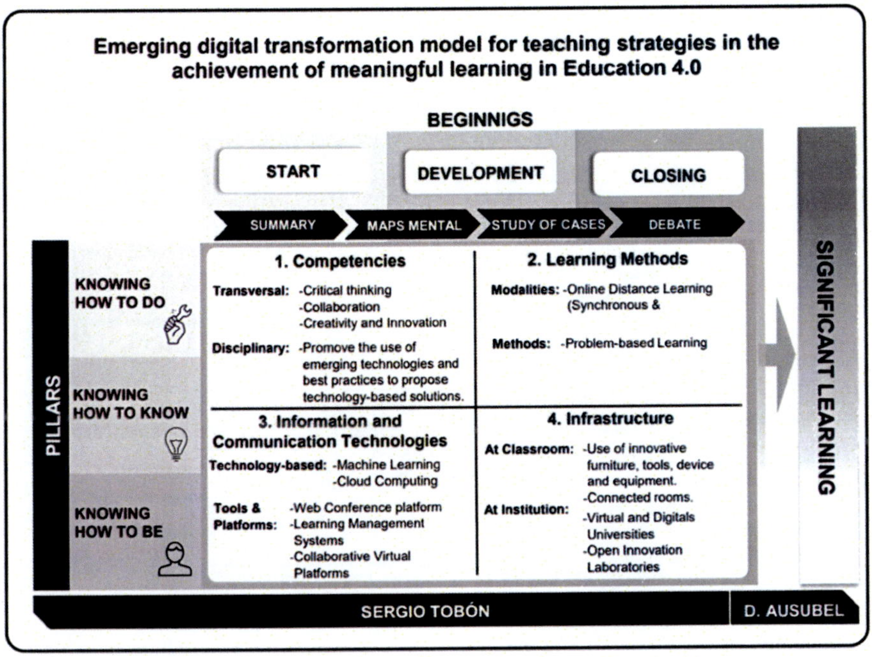

Fig. 1. Emerging digital transformation model for teaching strategies in achieving meaningful learning in Education 4.0.

These pillars are based on Sergio Tobón's three knowledge and the combination of these allows guaranteeing optimal results with significant learning [10], they go in a staggered way where students must first know how to be; managing to generate their own motivation facing the studies, assume challenges and interest in a well-done job. Subsequently, students must know how to know; they must understand the problem in order to provide solutions and finally develop the know-how, executing a series of specific procedures to reach a solution. In each of these three areas of knowledge, appropriate technologies and tools will be used to help to enhance each of the activities.

This model is based on David Ausubel's theory of meaningful learning, in which students integrate new knowledge into their cognitive structure; integrating previous knowledge with new knowledge [11]. This learning is approached as a constructive and non-receptive process in which students investigate, experiment, interact, reflect, and communicate their acquired learning to their teachers [12]. Ausubel proposes that for learning to be meaningful, logical and psychological significance is required. This model is also based on the theory of John Dewey and the experimental school. Dewey related theory to practice, stating that learning is the product of lived or experienced situations. This experimental school makes the student be exposed to situations that cause expectation and leads them to reflect, seek information to provide solutions to the posed problems.

Table 1. Technological tools and platforms for proposal

Strategy	*Tools*		
Summary	Parafrasist	Resoomer	LinguaKit
	Online tools for summarizing documents and long texts, provide a simple and easy interface. They handle synonyms and allow you to paraphrase and create new texts.		
Maps Mental	MindMeister	LucidChart	Bubble.us
	Tools for creating mind maps online. They allow collaborative work in real-time, addition of notes, version history, embedding of labels and icons, etc. Some use brainstorming, they are quite intuitive.		
Study of Cases	Wikispaces	Documento360	Google Sites
	Collaborative tools of public scope, free or protected, available to the entire educational community. They include hosting services to create workspaces in the form of wikis.		
Debate	Meet & Zoom		Google drive
	Applications for video conferencing and online meetings. They allow collaboration between teams from anywhere and offers cloud-based services and tools that allow you to save, share and collaborate on documents in the cloud.		

Source: Own elaboration

On the other hand, in this model, four sessions were applied, and the following strategies were used: summaries, mind maps, case studies, and debates. For their application, a variety of tools were made available to the students for their selection. In the summaries, students had to read the text, select main ideas and analyze. Also, when making mind maps, they had to organize the information, make the graphic resource using keywords. As for the case study, students had to select the information, interpret the case to give a logical solution. Finally, in the debate, it was necessary to search for information, prepare logical arguments and define their ideas to debate.

The tools that were available to the students for each of the strategies proposed are shown in Table 1.

Each of these sessions was based on the principles of beginning, development, and closing, using virtual tools for videoconferencing, likewise, moodle was used as virtual classroom support.

5 Results and Discussion

A questionnaire was applied as a pre-test and post-test to measure meaningful learning in Education 4.0; as we can see in Table 2, 48.7% of the students have a regular level in the pre-test in terms of their learning, compared to 82.1% of students who achieved a good level in the post-test. This leads us to think that after the application of the emerging digital transformation model for teaching strategies in Education 4.0, students had better skills to achieve knowledge in a given subject supported by the strategies and technological tools made available to them.

Table 2. Meaningful learning in education 4.0.

Levels of meaningful learning	PRE-TEST		POST-TEST	
	fi	%	fi	%
Low	3	7.7	0	0
Regular	19	48.7	7	17.9
High	17	43.6	32	82.1
TOTAL	39	100.0	39	100.0

Likewise, in Fig. 2, we can see how the low and regular levels decrease from the pre-test to the post-test, the opposite case with the good level that increases from 43.6% to 82.1% respectively, demonstrating that the levels of significant learning are improved by applying the proposed model, demonstrating the validity of the proposal.

Our results are similar to those obtained in a Colombian university. They made a model based on strategies and technological tools to optimize educational quality. The authors conclude that the model contributed to the development of creative and reflective thinking, strengthening collaboration, interaction, innovation, and entrepreneurship of students; making the achievements of meaningful learning in Education 4.0 occur effectively [13]. Similarly, the research [7] work was done based on three cases where the Education 4.0 model was applied to be used as designs in educational innovation projects; having as results that in the applied cases positive effects were shown, such as the generation of meaningful learning, as well as the creation of innovative solutions and the use of digital resources.

For these emerging digital transformation models in education 4.0 to become a reality, not only technological innovations and changes in methodologies are necessary, but also the profound transformation of the actors in the education system, which is closely related to leadership and culture change.

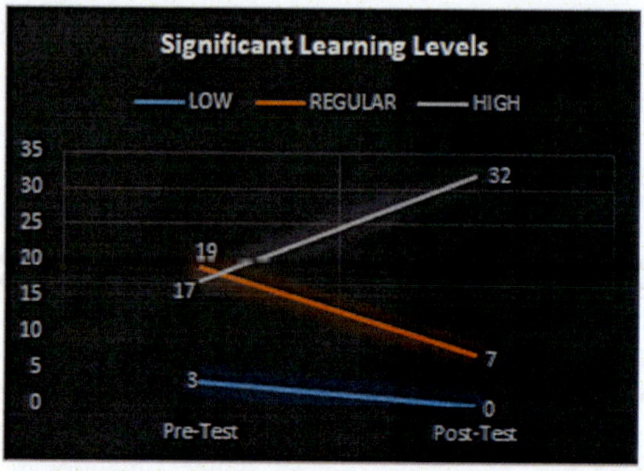

Fig. 2. Improvement of significant learning levels from pre-test to post-test

6 Conclusions

The level of meaningful learning in education 4.0 was determined by employing a questionnaire that was validated through expert judgment and has a good degree of reliability. Initially, in the pre-test, regular levels of meaningful learning were obtained. Subsequently, the emergent digital transformation model for teaching strategies was applied, generating good results in the post-test, since they changed to a high level. This model gave the students digital strategies to improve their learning by combining the various tools with the three learning skills proposed by Sergio Tobón: Knowing how to be, knowing how to know, and knowing how to do.

To achieve these good learning levels, the characteristics of the students must be taken into account; that is to say, they must seek to generate autonomy and intrinsic motivation; likewise, we must involve all educational agents. This will allow deep learning, where the student will achieve a certain level of personal significance, obtaining an adequate level of understanding.

It is important to banish those resistances and negative attitudes towards pedagogical innovations on the part of university teachers; on the contrary, we must train teachers with a constructivist profile, that is, teachers who are focused on student learning and provide them with all the technological resources to achieve their academic objectives.

With the results found, we can affirm that our model achieves a digital balance in the achievement of learning; thus achieving the achievement of competencies.

References

1. Tawafak, R.M., Malik, S.I., Alfarsi, G.: Impact of technologies during the COVID-19 pandemic for improving behavioral intention to use e-learning. Int. J. Inf. Commun. Technol. Educ. **17**(3), 137–150 (2021). https://doi.org/10.4018/IJICTE.20210701.oa9
2. Monteiro, A.R., Leite, C.: Digital literacies in higher education: skills, uses, opportunities and obstacles to digital transformation. Revista de Educación a Distancia **21**(65), 1–20 (2021). https://doi.org/10.6018/RED.438721

3. Reyes, C.E.G., Avello, R.: Alfabetización digital en la educación. Revisión sistemática de la producción científica en SCOPUS. Revista de Educación a Distancia **21**(66), 2–21 (2021)
4. Bergdahl, N., Nouri, J., Fors, U.: Disengagement, engagement and digital skills in technology-enhanced learning. Educ. Inf. Technol. **25**(2), 957–983 (2020). https://doi.org/10.1007/s10639-019-09998-w
5. Farias-Gaytan, S., Aguaded, I., Ramirez-Montoya, M.-S.: Transformation and digital literacy: systematic literature mapping. Educ. Inf. Technol. 1–21 (2021). https://doi.org/10.1007/s10639-021-10624-x
6. Cantero, C.L., Oviedo, G.B., Barboza, W.F., Feria, M.V.: Emerging technologies in the teaching-learning process: towards the critical thinking development. Revista Electronica Interuniversitaria de Formación del Profesorado **23**(3), 83–98 (2020). https://doi.org/10.6018/REIFOP.435611
7. Miranda, J., et al.: The core components of education 4.0 in higher education: three case studies in engineering education. Comput. Electr. Eng. **93**, 107278 (2021). https://doi.org/10.1016/j.compeleceng.2021.107278
8. Wu, X.B., Sandoval, C., Knight, S., Jaime, X., Macik, M., Schielack, J.F.: Web-based authentic inquiry experiences in large introductory classes consistently associated with significant learning gains for all students. Int. J. STEM Educ. **8**(1), 1–18 (2021). https://doi.org/10.1186/s40594-021-00290-3
9. Cejas, M.F., Rueda, M.J., Cayo, L.E., Villa, L.C.: Formación por competencias: Reto de la educación superior. Revista de Ciencias Sociales **25**(1), 94–101 (2019). https://doi.org/10.31876/rcs.v25i1.27298
10. Carranza, M.R., Caldera, J.F.: Percepción de los Estudiantes sobre el Aprendizaje Significativo y Estrategias de Enseñanza en el Blended Learning. REICE Revista Iberoamericana sobre Calidad Eficacia y Cambio en Educación **16**(1), 73–88 (2018). https://doi.org/10.15366/reice2018.16.1.005
11. Flores, V.: Aprendizaje significativo con estrategia de enseñanza activa para un curso de proyecto software. Una experiencia en el norte de Chile. Ingeniare Revista chilena de ingeniería **29**(1), 120–128 (2021). https://doi.org/10.4067/s0718-33052021000100120
12. Vega, L.Y., Botero, L.E.: Formación profesional inmersa en la transformación digital con el fin de mejorar la calidad en la educación. CULTURA EDUCACIÓN Y SOCIEDAD **12**(1), 37–46 (2020). https://doi.org/10.17981/cultedusoc.12.1.2021.03
13. Karim, S., Prima, E.C., Utari, S., Saepuzaman, D., Nugaha, M.G.: Reconstructing the physics teaching didactic based on Marzano's learning dimension on training the scientific literacies. J. Phys: Conf. Ser. **812**, 012102 (2017). https://doi.org/10.1088/1742-6596/812/1/012102

Study of Malnutrition in Children Under Five in Peru Using the Exploratory Spatial Data Analysis

Bryan Martin González Cusi and Olga Lidia Solano Dávila(✉)

Facultad de Ciencias Matemáticas, Universidad Nacional de Mayor de San Marcos, Lima, Peru
{bryan.gonzalez,osolanod}@unmsm.edu.pe

Abstract. This research presents the application of the Exploratory Spatial Data Analysis to assess the type of spatial pattern that follows the cases of patients with malnutrition in children under 5 years who were attended in the Ministry of Health of Peru (MINSA, after its Spanish abbreviation) in 2017. The results demonstrated that malnutrition cases in children under 5 years were non-homogeneously distributed throughout Peru in 2017. The boxplot identified that a larger number of malnutrition cases in girls under five years was presented in the departments of Piura, Loreto, and Ucayali; and in the case of the boys was in Piura, Cajamarca, and Ucayali. The rates map evidenced that a higher rate of malnutrition was presented in the departments of Piura, Cajamarca, Loreto, Ucayali, Madre de Dios, and Ayacucho. The LISA indexes found the presence of significant clusters in departments with a high demand for this disease as in Loreto, as well as significant clusters in departments with less demand as in Lima, Moquegua, and Puno. The Moran's Index indicated that the spatial dependence is positive and almost nil.

Keywords: Exploratory Spatial Data Analysis · Malnutrition · LISA map · Moran's Index

1 Introduction

In 2017, chronic malnutrition affected 6.4% of children under 5 years who were attended in health establishments of DIRIS (Spanish acronym for Integrated Health Networks Directorates) Lima Norte-Peru, which, in accordance with the ENDES (Spanish acronym for Demographic and Family Health Survey) 2017, the said value was below the national average which was 12.9%. In the past years, child malnutrition has been reduced; however, it still affected 13.1% of children under 5 years in 2016; it reaches 26.5% in rural areas and 7.9% in urban ones [1]. Districts with a higher proportion of chronic malnutrition in children under five years are Ancon, Puente Piedra, and Rimac with 8.2%, 7.2%, and 6.8%, respectively [2].

The objective of this paper was to identify the spatial pattern and the departments with a higher incidence of chronic malnutrition in children under 5 years in Peru.

Y. Iano et al. (Eds.): BTSym 2021, SIST 207, pp. 322–333, 2023.
https://doi.org/10.1007/978-3-031-04435-9_32

The present document is organized as follows: introduction, theoretical framework related to the study; subsequently, the research methodology, then, the results of the Exploratory Spatial Data Analysis (ESDA), and finally, the discussion and conclusions.

2 State-of-the-Art

2.1 Exploratory Spatial Data Analysis

The Exploratory Spatial Data Analysis (ESDA) is a group of techniques to describe and visualize the spatial scatter, discover patterns of spatial association, identify the spatial atypical observations, and suggest other ways of spatial heterogeneity [3, 4]. According to the spatial trend, the ESDA graphics methods are the thematic map, dynamic map, conditional plots, Moran's scatter plot, and Multivariate Moran's dispersion plot. While, according to the spatial atypical trend, the graphic methods are the boxplot, percentile maps, cartogram, LISA's graphics, and multivariate LISA's graphics [5–7].

2.2 Exploratory Analysis of Global Spatial Dependence

The spatial dependence or autocorrelation involves the existence of a functional relationship between what occurs in a determined point of the space and what happens in close or neighbor places. That is to say; a variable will be spatially auto-correlated when the values observed in a point or region depend on the values observed in neighbor regions producing a certain geographical continuity in the distribution of this variable, for example, on a map. In the ESDA grid, the most important techniques of the spatial association are the LISA's maps and the Moran's scatter plot as well as the multivariate versions [5–8].

Moran's Univariate. The Moran's univariate is a scatterplot where each observation is associated with its value within the same variable and the corresponding average values of the neighbor observations where the Moran's index statistic is the slope of a regression line. This scatterplot divides the type of spatial association into four categories: two for positive spatial autocorrelation (high and low values of a univariate surrounded by high and low values, respectively) and two for negative spatial autocorrelation (high values surrounded by low values, and vice versa) [5, 6].

2.3 Rate Maps and Empirical Bayesian Smoothing

Rate maps are commonly used to analyze the spatial scatter of the risk that a certain event may occur when the data are organized according to the count of areas. However, an important problem associated with the use of rates is their high instability to express the risk of rare events in regions with small populations. Alternatively, empirical Bayesian smoothing rates are presented to use information from the whole region or neighbor to estimate the risk that may occur in each area [9, 10]. A simple method is implemented to calculate the Bayesian empirical estimations, which do not assume any specific distribution. Being z_i, a_i, and r_i the number of events, the area, and the rate of the i area, respectively:

$$\hat{\theta}_i = C_i r_i + (1 - C_i)\hat{m} \tag{1}$$

$$\hat{m} = \frac{\sum_{i=1}^{n} z_i}{\sum_{i=1}^{n} a_i} \tag{2}$$

$$S^2 = \sum_{i=1}^{n} \frac{z_i (r_i - \hat{m})^2}{z} \tag{3}$$

$$C_i = \frac{S^2 - \hat{m}/\overline{z}}{S^2 - \hat{m}/\overline{z} + \hat{m}/z_i} \tag{4}$$

Therefore, the empirical Bayesian rate is a weighted average between the raw rate of the town and the global rate of the region. If the location has a significant population, its rate will demonstrate little variability and it will be practically maintained without changes [9, 10].

2.4 Spatial Non-Homogeneous Poisson Process

The Spatial non-homogenous Poisson process is obtained replacing the constant intensity, λ, of a Poisson process with a variable intensity function λ (x).

A non-homogeneous Poisson process is defined by the following properties:

1. The number of events in a region A, N(A) followed by a Poisson distribution with a $\int A$ λ(x)dx mean for some non-negative function λ (x).
2. Given N(A) = n, the n events form a random sample of the distribution in A with a distribution function of proportional probabilities to λ (x).
3. For two disjointed regions A and B, the random variables N(A) and N(B) are independent [11, 12].

2.5 Complete Spatial Randomness Tests

Generally, the null model is the appropriate one in the homogeneous Poisson process for a dot pattern. The basic task when analyzing a dot pattern is to find evidence against the complete spatial randomness (CSR). If it is suspected that the intensity (number of cases per area) is non-homogeneous, the Chi-Square statistic (X^2) is a non-parametric statistic test to verify this suspicion. This test is based on quadrant counts where if it was a homogeneous Poisson punctual process, it would be expected that the regions with equal areas have around the same number of dots.

The method involves dividing the observation window W into subregions named quadrants, B_1, …, B_m of area a_1, …, a_m (equal areas) respectively, saying that $B_1 \cup \ldots \cup B_m = W$. We count the number of dots that fall in each square, $n_j = (X \cap B_j)$ for $j = 1, \ldots, m$ [11, 12]. In a statistical test for homogeneity that consists of counting the quadrants, and at the same time that the dot process is Poisson. Where, if the null hypothesis is not rejected, it involves that the n_{js} is the implementation of independent Poisson random variables, with an expected value of $\mu_j = \rho a_j$ [8]. The following hypotheses are considered: H_0: The intensity is homogeneous (CSR), and H_1: The intensity is non-homogeneous (it is a non-homogeneous Poisson process). Considering the homogeneity test, the uniformity Chi-Square test is applied. Given that the number of dots $= \sum n_j$ and the total area of the window $a = \sum a_j$, then, the estimated intensity is $\hat{\rho} = n/a$ and the expected count in the quadrant B_j is $e_j = \hat{\rho} a_j = n a_j / a$ [11, 12].

The test statistic is:

$$X^2 = \sum_j \frac{(n_j - e_j)^2}{e_j} = \sum_j \frac{(n_j - \hat{\rho} a_j)^2}{\hat{\rho} a_j} \sim X^2_{(m-1)} \quad (5)$$

where:

n_j: Number of dots in quadrant j.

a_j: The area of quadrant j.

n: Sum of dots of all the quadrants.

a: Sum of the areas of all the quadrants.

This is a very useful test to identify the type of pattern of the studied population. On the other hand, only if the intensity is non-homogeneous, it is recommended to use the LISA's graphic, because, if the intensity is homogeneous, the LISA's graphic will not identify any significant cluster.

2.6 Exploratory Analysis of the Local Spatial Dependence

Both, Moran's index statistic and Moran's scatter graphic are procedures that allow us to study the dependence phenomenon or the spatial autocorrelation from a global point of view. Therefore, they cannot find the existence of data groups or sets which are away from the general trend of a variable. The Moran's scatterplot is able to detect certain local aspects of a distribution pointing out those dots which are away from the mean (greater than 2) but the existence of special concentrations or the absence of concentrations of high/low values of such variable cannot be determined.

For that, LISA's graphics of local spatial dependence have been designed, which offer an indication of the concentration degree of similar values (high or low) regarding each geographical unit [5, 6, 8].

LISA Map. The LISA (Local Indicator of Spatial Association) map represents the locations with significant values in certain statistical indicators of local spatial association, which are included in the GeoDa program.

It is a statistic that, different from the Moran's index mentioned above, is not globally calculated for all the observations of the map but acquires a different value for each one of them.

Within the geographical area, this statistic calculates the concentration level of the low/high values of a univariate for each of the observations of the sample. To calculate the statistical significance level of rejection of the null hypothesis of the absence of similarity or difference in values of a geographical location, it is feasible to prepare an inference for each value of this statistic. In such a way, the existence of hot spots or spatial atypical are detected whose intensity, whether is higher or lower, depends a lot on the significance related to the mentioned statistics [5–8].

2.7 Global Spatial Association Measure

Spatial autocorrelation is a clustering or dissemination of values of a certain variable within a map. That is to say, the spatial autocorrelation evidences the level of objects such as activities in a geographical unit that are similar to them but in neighbor geographical units [12]. The global spatial autocorrelation analysis makes a joint examination of all the units that involve the sample to determine if the spatial units are randomly distributed or if, on contrary, they are distributed as per a determined pattern [13]. The most used tool to evaluate the autocorrelation is the Moran's index, being this a transformation of a non-spatial correlation measure to a spatial environment and is generally performed in spatial units where there is information in the form of ratios or intervals [5, 6]. One of the most used formulas is:

$$I = \left(\frac{n}{\sum_{i=1}^{n}(y_i - \overline{y})^2}\right)\left(\frac{\sum_{i=1}^{n}\sum_{j=1}^{n} w_{ij}(y_i - \overline{y})(y_j - \overline{y})}{\sum_{i=1}^{n}\sum_{j=1}^{n} w_{ij}}\right) \tag{6}$$

where:

n: The number of areas.

y_i: The value that takes the variable that is being considered in area i.

$\overline{y}$: The average value that takes the variable in the region of the study.

w_{ij}: The weights of the W matrix.

y_j: The value that takes the variable which is being considered in area j.

The value that takes the Moran's index may vary between ±1, where if values are zero, they indicate the presence of a spatial cluster with territorial units of very different values, and if the values are greater than zero, they indicate the presence of a spatial cluster with territorial units of very similar values which may be low or high. If the value obtained with Moran's index is 0 or very close to 0, this indicates that the analyzed variables are spatially independent; therefore, any autocorrelation or spatial dependence would not exist between these variables.

3 Methodology

The type of study is descriptive since the spatial pattern of malnutrition in all the children of Peru under five years attended in Outpatient Consultation of MINSA in 2017 will be determined and analyzed.

The type of design is non-experimental, because the variables are not manipulated, and cross-sectional, because the disease is diagnosed in children under 5 years at a particular time, in 2017.

The population of the study was composed of children under five years with malnutrition and attended in Health Centers and Hospitals of the Ministry of Health of Peru

(MINSA). The data were provided by MINSA. For the ESDA processing and analysis, the GeoDa software version 1.14 was used. GeoDa is a free and open-source software that works as an introduction to spatial data science. It is designed to facilitate new knowledge from the data analysis through the exploration and modeling of spatial patterns. It was developed by Dr. Luc Anselin and his team. The software provides an easy-to-use graphic interface for exploratory spatial data analysis (ESDA) methods, such as spatial autocorrelation statistics for aggregate data (many thousands of records) and basic spatial regression analysis for dots and polygon data (dozens of thousands of records). To work with big data in GeoDa, units of area should be aggregated first [14]. From its initial launch in February 2003, the number of GeoDa users has increased exponentially to more than 360.000 (September 2020). This includes laboratory users in universities like Harvard, MIT, and Cornell. The user community and the press received the program with enthusiasm calling it an "analytic tool of great importance", a "very classy software piece", and an "exciting development" [14, 15].

For the creation of the map of Peru, as well as the limitation of its departments, in the Geoda software, the latitude and longitude variables of each department were previously added in our database (shape). For the case map, we followed the natural breaks criterion whose objective is the greater homogeneity in each interval to minimize the variance in each group and the greater difference among specified intervals by coloring with the same color the departments that belong to the same interval for easy identification. For the boxplot, we started with the box diagrams whose construction begins with the calculation of quartiles and the mean of one variable, therefore, the geographical units – whose data in one variable corresponds to the mean, the atypical values, both low and high, and the interquartile range – can be represented with different colors in the boxplot. In the rate map, the raw rate obtained from the division of the total cases of malnutrition of each department between the total of children under 5 years of each department was previously calculated. Then it was grouped in intervals and the departments belonging to the same interval were colored to identify them easily. For the Empirical Bayesian Smoothing case, the rate was obtained through the formulas mentioned in section 2.3 of this paper, and then, it was also grouped in intervals. Finally, for the Moran's index and the LISA's Graphic, we had to create a Weights Manager through the weights contiguity (neighbors) method, and the LISA's graphic and the Moran's index were obtained using the raw rate previously calculated and the Moran's formula shown in section 2.7.

4 Results

The ESDA results will be shown in this section explaining the study variable through a colored map and following a creating intervals criterion. For this research, we prepared a Case Map using the natural breaks criterion and the boxplot. Both represent the univariate spatial scatter.

In Figure 1 and Table 1, a larger number of cases of malnutrition in girls and boys is presented in the departments of Piura (3121) and Ucayali (3006), respectively, in comparison to the other departments. This makes us think that maybe the spatial scatter of malnutrition in girls and boys under 5 years is not homogeneously distributed in Peru.

In Figure 2 and Table 1, it is observed that the malnutrition in girls in the departments of Piura, Loreto, and Ucayali with 3121, 2307, and 2808 cases respectively are highlighted in dark red since in these places the number of cases of this disease is very high and they are considered as outliers departments. For the malnutrition cases in boys, the departments of Piura, Cajamarca, and Ucayali with 2579, 1849, and 3006 cases respectively are considered as outliers departments due to a large number of malnutrition cases.

After using the Raw Rate, it is observed that the departments with higher rates of malnutrition in children of 5 years are: Piura, Cajamarca, Loreto, Ucayali, Madre de Dios, and Ayacucho with 0.032, 0.029, 0.037, 0.104, 0.021, and 0.026 rates respectively which are shadowed in dark orange.

Nevertheless, since the population of the department was very large, the EBS effect was almost nil, and values very similar to the Raw Rate were obtained. (See Figure 3, Table 1).

To prove that the intensity of the malnutrition has a homogeneous scatter, it is proposed the alternative hypothesis where the malnutrition intensity has a non-homogeneous scatter (it is a non-homogeneous Poisson process), the test statistical value was 69805.614, the critical value to a significance level of 5% resulted in 36.415, the nil hypothesis is rejected and it is concluded that the intensity of the malnutrition in children under 5 years in Peru arises from a non-homogeneous Poisson process; that is to say, it is non-homogeneously spatially scattered.

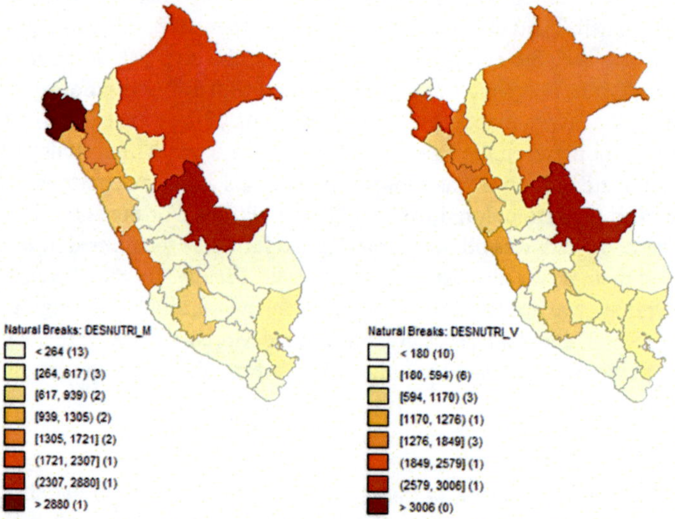

Fig. 1. Case Map of malnutrition in girls (left) and boys (right) under five years in Peru, MINSA – 2017

After having demonstrated that the spatial scatter of children under five years with malnutrition in Peru is non-homogeneously spatially scattered, we will locate the departments with high rates of malnutrition disease using LISA maps.

In Fig. 4, left side, we can observe that Moran's index obtained a value very close to zero (0.069) in the malnutrition case with which it can be concluded that the spatial

Table 1. Table of cases and rates of malnutrition in children under 5 years of age in Peru MINSA – 2017

Departments	Cases and rates of malnutrition		
	Cases in girls	Cases in boys	Rates (%)
Amazonas	303	293	0.015
Ancash	617	810	0.015
Apurimac	220	301	0.015
Arequipa	173	159	0.003
Ayacucho	645	791	0.026
Cajamarca	1721	1849	0.029
Callao	120	93	0.003
Cusco	164	180	0.003
Huancavelica	124	133	0.008
Huánuco	261	326	0.009
Ica	188	176	0.005
Junín	176	169	0.003
La Libertad	1056	1276	0.014
Lambayeque	939	594	0.014
Lima	1305	1170	0.004
Loreto	2307	1410	0.037
Madre de Dios	157	146	0.021
Moquegua	11	7	0.001
Pasco	96	133	0.01
Piura	3121	2579	0.032
Puno	264	205	0.005
San Martin	396	364	0.009
Tacna	17	20	0.002
Tumbes	158	107	0.012
Ucayali	2880	3006	0.104

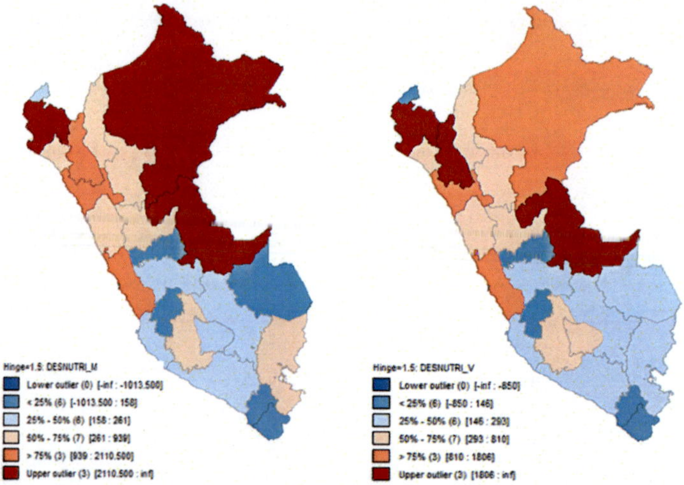

Fig. 2. Box map of the distribution of malnutrition cases in girls (left) and boys (right) under 5 years in Peru – MINSA, 2017

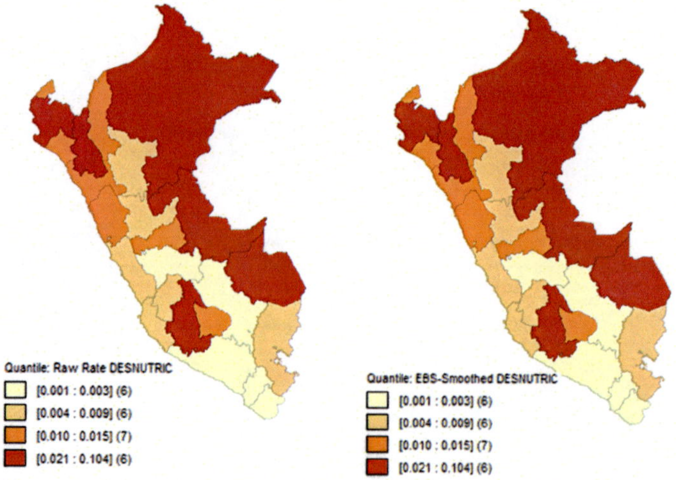

Fig. 3. Rate Map of malnutrition (left) and Empirical Bayesian Smoothing (right) in children under 5 years in Peru – MINSA, 2017

dependence is positive and almost nil for malnutrition in children under five years in Peru.

Four different types of clusters were detected. They are (HL), which indicates that the department presents high (H) values and is surrounded by other departments with low (L) values. Also, it is observed the presence of HH, LL, and LH, which were previously mentioned, stands out the HH label that is highlighted in red and comprised of the department of Loreto. This area represents an important concern since there are areas with a very high number of malnutrition cases (see Fig. 4, right side).

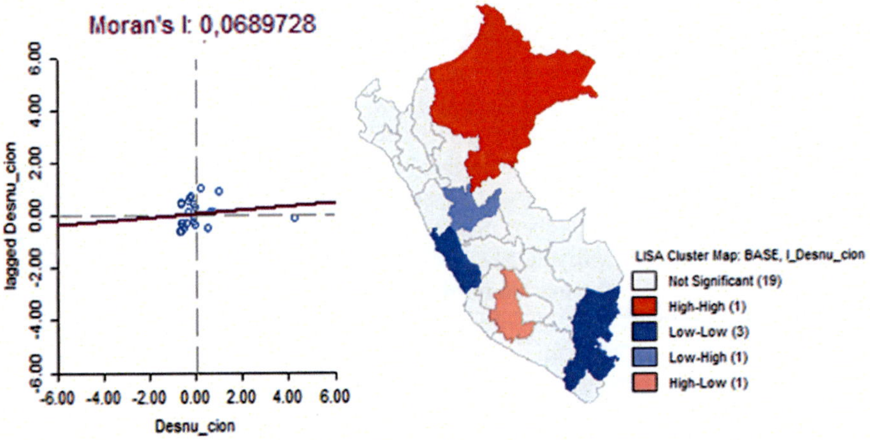

Fig. 4. Moran's Scatterplot (left) and detection of spatial clusters (right) for malnutrition cases in children under 5 years in Peru MINSA – 2017

5 Discussion

The boxplot identified that the departments with a very large number of patients, girls and boys, with malnutrition attended in Outpatient Centers in MINSA in 2017 correspond to Piura and Ucayali. Likewise, the boxplot identified that the departments with larger numbers of cases in girls with malnutrition were Piura, Cajamarca, and Ucayali; and, in the case of boys, they were Piura, Loreto, and Ucayali. On the other hand, the rates map identified that the highest malnutrition rate is in the departments of Piura, Cajamarca, Loreto, Ucayali, Madre de Dios, and Ayacucho. Furthermore, the LISA map identified the department of Loreto as the most significant cluster. In 2016, according to an ENDES study, it was identified that the departments with higher prevalence in this age group were: Huancavelica, Cajamarca, Ucayali, Pasco, and Loreto. Piura and Ayacucho were the departments that obtained prevalence of 15.3% and 18.9%, respectively [16], which also need attention from the authorities. In 2012, according to the ENDES, children with chronic malnutrition in departments with higher prevalence were in Huancavelica, Cajamarca, Loreto, Apurímac, Huánuco, and Ayacucho; Ucayali and Piura also stood out with 24.7% and 21.7% of prevalence, respectively [17]. On the other hand, there is good news from Unicef, which says that the chronic malnutrition in children under five years was reduced from 30% in 2000 to 12% in 2019 [18] throughout Peru. However, there are gaps to shorten in most of the departments. Thus, in the first semester of 2019, the prevalence in Huancavelica was 30.1% and in Lima 9.1%, which is a very significant difference [19], representing a public health problem. On the other side, chronic malnutrition negatively affects the individual during their life, limits the development of the society, and hinders the elimination of poverty [20].

6 Conclusions

Chronic malnutrition is still a nationwide public health problem. The ESDA identified the departments with the most cases of malnutrition, and the departments that stand out are Piura, for girls, and the Ucayali, for boys.

With the box map, the outlier departments were identified due to their large numbers of cases of malnutrition, with most of the girls being in the departments of Piura, Loreto, and Ucayali, and most of the boys in the departments of Piura, Cajamarca, and Ucayali.

The departments with the highest rates of malnutrition were also identified, such as Piura, Cajamarca, Loreto, Ucayali, Madre de Dios, and Ayacucho. On the other hand, it was demonstrated that the morbidity of malnutrition in children under 5 years in 2017 is non-homogenous spatially scattered throughout Peru. Likewise, the LISA map discovered the existence of significant clusters of departments with high malnutrition rates where the cluster composed of the department of Loreto stands out. Moran's index indicated that the spatial dependence is positive and almost nil.

References

1. MINSA. Plan Nacional para la reducción y control de la anemia Materno Infantil y la Desnutrición Crónica Infantil en el Perú: 2017-2021 (2017). http://bvs.minsa.gob.pe/local/MINSA/4189.pdf. Accessed 27 Aug 2021
2. MINSA. Análisis de Situación de Salud 2018 (2018). http://www.dirislimanorte.gob.pe/wp-content/uploads/2019/01/asis-2018.pdf. Accessed 27 Aug 2021
3. Anselin, L.: Local indicators of spatial association-LISA. Geogr. Anal. **27**(2), 93–115 (1995). Accessed 27 Aug 2021
4. Haining, R., Wise, S, Signoretta, P.: Providing scientific visualization for spatial data analysis: criteria and an assessment of SAGE. J. Geogr. Syst. **2**(1), 121–140 (2000). https://link.springer.com/article/10.1007%2FPL00011450. Last accessed 28 Aug 2021
5. Chasco, C.: Análisis exploratorio de datos espaciales al servicio del geomarketing. In: III Seminario sobre Nuevas Tecnologías en la Investigación, el Marketing y la Comunicación, Instituto Lawrence R.Klein, Universidad Autónoma de Madrid (2003)
6. Chasco, C.: Métodos gráficos del análisis exploratorio de datos espaciales. Anales de economía aplicada, Departamento de Economía Aplicada, Universidad Autónoma de Madrid, Madrid (2003). https://www.asepelt.org/ficheros/File/Anales/2003%20-%20Almeria/asepeltPDF/93.PDF. Accessed 27 Aug 2021
7. Acevedo, I., Velásquez, E.: Algunos conceptos de la econometría espacial y el análisis exploratorio de datos espaciales. Ecos de Economía **12**(27), 9–34 (2008). https://pdfs.semanticscholar.org/342a/b77d86bb065f2391a84ca9e416919f8636b2.pdf?_ga=2.84691353.1480765035.1567562032-1771610990.1566949864. Accessed 27 Aug 2021

8. Vilalta, C.J.: Cómo enseñar autocorrelación espacial. Economía Sociedad y territorio **1**(18), 323–333 (2005). https://www.redalyc.org/pdf/111/11101804.pdf. Accessed 27 Aug 2021
9. Marshall, R.: Mapping disease and mortality rates using Empirical Bayes Estimators. J. R. Stat. Soc. Ser. C Appl. Stat. **40**(2), 283–294 (1991). https://doi.org/10.2307/2347593. Accessed 27 Aug 2021
10. Fuentes, I.: Inferencia estadística en procesos puntuales espaciales. Aplicación al análisis de incendios forestales en Galicia (2008). http://eio.usc.es/pub/mte/descargas/ProyectosFinMaster/Proyecto_296.pdf. Accessed 27 Aug 2021
11. Baddeley, A.: Analysing spatial point patterns in R (2008). https://training.fws.gov/courses/references/tutorials/geospatial/CSP7304/documents/PointPatterTutorial.pdf. Accessed 27 Aug 2021
12. Quispe, B.: Modelos Estadísticos en Procesos Puntuales Espaciales Poisson para evaluar la distribución Espacial de los hechos delictivos en Lima, Perú (2016). http://repositorio.lamolina.edu.pe/bitstream/handle/UNALM/2806/E10-Q8-T.pdf?sequence=1&isAllowed=y. Accessed 27 Aug 2021
13. Goodchild, M.: Spatial analytical perspective on geographical information systems. Int. J. Geogr. Inf. Syst. **1**(4), 327–334 (1987). http://www.geog.ucsb.edu/~good/papers/95.pdf. Accessed 27 Aug 2021
14. Geoda. Introducción a la ciencia de datos espaciales (2021). https://geodacenter.github.io/
15. Rodriguez, R.: Software de análisis geoestadístico GEODA (2017). https://risharkygis.wordpress.com/2017/03/23/software-de-analisis-geoestadistico-geoda/. Accessed 21 Aug 2021
16. MINSA. Plan nacional para la reducción y control de la anemia materno infantil y la desnutrición crónica infantil en el Perú: 2017-2021 (2017). https://anemia.ins.gob.pe/rm-ndeg-249-2017-minsa. Accessed 27 Aug 2021
17. MINSA. Plan nacional para la reducción de la desnutrición crónica infantil y prevención de la anemia en el país 2014 - 2016 (2014). http://bvs.minsa.gob.pe/local/MINSA/3514.pdf. Accessed 27 Aug 2021
18. Agencia Peruana de Notícias. Perú redujo en 21% desnutrición crónica en dos décadas (2019). https://andina.pe/agencia/noticia-peru-redujo-21-desnutricion-cronica-infantil-dos-decadas-769961.aspx. Accessed 21 Aug 2021
19. Instituto Nacional de Salud. Informe Gerencial Estado Nutricional de Niños y Gestantes que acceden a Establecimientos de Salud. 2019 primer semestre (2019). https://web.ins.gob.pe/sites/default/files/Archivos/cenan/van/informes/2019/Informe%20Gerencial%20SIEN%20HIS%20I%20Semestre%202019-c.pdf. Accessed 27 Aug 2021
20. Sanchez, A.: Evolución de la desnutrición crónica en menores de cinco años en el Perú **29**(3) (2012). https://rpmesp.ins.gob.pe/rpmesp/article/view/377/2496. Accessed 27 Aug 2021

eHealth: A Focus on Software Architectures for Brazilian Public Health

Caroline Luiz Gonçalves, Luciana Pereira, and Alessandra Cristina Santos Akkari(✉)

Graduate Program in Engineering and Innovation Management, Federal University of ABC, Santo Andre, Brazil
caroline.goncalves@aluno.ufabc.edu.br, {luciana.pereira, alessandra.akkari}@ufabc.edu.br

Abstract. There is constant pressure to increase the quality of healthcare services and reduce costs, which drives the use of digital technologies and an optimized management process for a large volume of data. The existence of different archetypes, which have several legacy systems and which, in many cases, are difficult to be used, makes the environment even more challenging for healthcare organizations to create innovative solutions. This article aimed to map the software architectures related to Brazil's Unified Health System (in Portuguese, Sistema Único de Saúde - SUS), using a systematic literature review as a procedure. To this end, an exploratory and descriptive study with a qualitative approach was developed based on the collection and analysis of secondary data. Among the results, seven architectures stood out in the context of Brazilian public health, as well as efforts to develop a flexible and comprehensive software architecture were observed, with less bureaucracy, correction of flaws in real-time, and prioritizing a responsive and resolutive system. This study contributed both to professionals and researchers by providing a document that brings together the existing architectures, highlighting the main points of each one and identifying the existing gaps.

Keywords: Software architecture · eHealth · Public health · Brazil · SUS

1 Introduction

An unprecedented increase in the use of digital data and the consequent growth of cyberspace are transforming the way human beings live. Therefore, it becomes necessary to rethink systems linked to the different areas of the economy to optimize indicators and allow alignment with new attributes and techniques, such as cloud and big data [1, 2]. From this perspective, the healthcare segment and health care systems stand out, which, in the face of global facts that challenge public health, such as population aging [3]; increase in the number of individuals with chronic diseases [4]; and of public health emergencies, such as the COVID-19 pandemic [5], assume global relevance, requiring a holistic approach to the application of digital technologies with new social practices in the context of eHealth.

Y. Iano et al. (Eds.): BTSym 2021, SIST 207, pp. 334–341, 2023.
https://doi.org/10.1007/978-3-031-04435-9_33

Regarding the development of health information systems, the World Health Organization suggests that countries work in an integrated manner in international networks to ensure the standardization and quality of data, in addition to providing dissemination and easy access to information. To carry out these activities, it is necessary to use all the new technologies and innovations that can provide better communication between the participants, in addition to providing digital medical records and telemedicine, increasingly contributing to the provision of better and cheaper treatments for patients [6]. In addition, Grol [7] states that the number of new programs and technologies that are currently available reflects an intense scientific development and results in a large amount of knowledge that quickly becomes obsolete, demanding an efficient integration of organizations and health systems for internalizing so many changes and innovations.

In fact, in recent years, there has been intense dissemination of information technologies in the medical field, so that the global health technology market, between 2015 and 2020, grew 141%, reaching more than 14 billion dollars of investments in 2019. Given this scenario, Brazil is the seventh-largest market for health technologies in the world and first in Latin America [8], motivating the choice of this country as a study model for this paper.

The Brazilian Ministry of Health has made a constant effort to improve health systems to reduce costs and increase the quality of services provided. To this end, in 1991 the SUS Department of Informatics (DATASUS) was created. Through DATASUS it is possible to use digital tools to manage resources and information. Until 2020, DATASUS was responsible for the creation of more than 200 auxiliary systems for national health. However, many of them are not integrated, making interoperability between them difficult and hampering many activities that could be performed through new data science technologies [9].

To assimilate this set of systems and innovations, the mapping of the main software architectures associated with health information systems becomes strategic. Several works in the literature point out the importance of qualifying the value and impact of systems by evaluating the software architecture so that its absence leads to sub-optimal design decision-making. In addition, an inaccurate architectural design generates significant flaws, pointing out that architecture plays an important role in software development, especially in the healthcare field [10–13].

Software architecture is related to a process of conceiving, expressing, defining, documenting, communicating, and certifying the correct implementation and maintenance during the life of a software [14]. The benefits of the architecture consist of easier programming, increased code quality, reduced work development time [15].

Therefore, this article aimed to map the main software architectures employed in the context of Brazilian public health, with a focus on the Unified Health System, known as SUS in Portuguese, being the largest unified healthcare system in the world. This study contributes both to scientists who want to find relevant studies and gaps that exist and to specialists seeking to understand software architectures already available in the literature.

2 Methodology

A systematic literature review was conducted, and the method was compressed into phases, following the Systematic Research Flow protocol [16], including search strategy, database consultation, and standardization and selection of documents.

Regarding search strategy, "Software Architecture", "Health", and "SUS" keywords were defined to understand the software architectures built for the Brazilian health system.

Subsequently, the CAPES Periodical Portal was selected as a database. Based on a search procedure, the keywords were combined, and the relationship between them was identified. This process was carried out in July 2021 and there was no time or language restriction for publications.

In the fourth phase on standardization and selection of documents, the files were analyzed under three aspects: (i) title, (ii) keywords, and (iii) abstract. The total number of selected documents was 305, but 283 articles were eliminated for not presenting a research theme related to the topic investigated, resulting in the remaining 22 scientific documents that were analyzed according to their content and level of contextualization. Finally, there was analysis, synthesis, and consolidation of the results.

3 Results and Discussion

3.1 Architecture for Standardizing and Building Tools – Software Development Methodology (MDS)

Some measures were taken by the Informatics Department of the Brazilian Unified Health System (DATASUS) to make the computational architecture of its information systems more adequate, including [17]:

- *Electronic Government Interoperability Standards (ePing)* – to increase interoperability between the various systems of the Federal Government.
- *Minimum Data Set* – mandatory adoption by all health establishments, it consists of a communication bus for the collection and storage of care data.
- *e-SUS Primary Care* – to restructure and integrate Primary Care information at the national level, facilitating the daily tasks of professionals.
- *Digital Health Strategy in Brazil (digiSUS)* – to improve all areas of national health by providing accurate and secure information [17].
- *Simplified Data Collection* – software used only to store the data entered by health agents.
- Citizen's Electronic Medical Record – to computerize the flow of care, where professionals directly use an electronic medical record.
- *Health Information System for Primary Care (SISAB)* – processing and dissemination of data related to primary care, helping in the analysis of information.

Prepared in line with the DATASUS Project Management Methodology, MDS is focused on the software engineering process and aims to provide a tool for the software

Table 1. Milestones of each phase of MDS.

Project proposal	Document for carrying out the project's feasibility analysis
Initiation	Document with the scope of the iteration or demand
Elaboration	Description of the result to be built or modified
Construction	System operating in an approval environment
Transition	Established system operating in the production environment

development and maintenance teams of the SUS IT department. According to Table 1, each phase of the development or maintenance project has a milestone [17].

The use of MDS allows standardizing the construction of software, aligning work processes, and carrying out adequate documentation. As a result, there is an improvement in governance, reflecting an increase in effectiveness, productivity, and retention of knowledge in the organization [17].

In addition, MDS was built on the principles of the Rational Unified Process (RUP), which refers to a collection of best practices in software engineering and uses a sequence of incremental steps, so that each new version of the system is improved accordingly. incremental compared to the previous version. Regarding management, the RUP software lifecycle consists of four sequential phases. At the end of each phase, an examination is carried out to verify the completion of the objectives, where, upon reaching them, the project moves to the next phase [17].

3.2 Architecture for SUS Health Care Terminals

It was identified a software architecture to be applied to healthcare terminals. Due to the varied conditions of healthcare terminals throughout the country, some requirements were imposed on the architecture such as ease of use in different places, ease of operation, security, and privacy of the information collected. Table 2 describes the requirements of this architecture [18].

3.3 Architecture for the Electronic Citizen Record – OpenPEC

Brazil has more than 40 thousand Basic Health Units spread across its national territory and less than half of them use electronic medical records because of the high computerization costs, lack of infrastructure, and lack of financial resources. A possible cause for this is that the medical records have closed source code, that is, they are restricted to fixed functions in which their viewing and editing are not possible, making it difficult to adjust to the peculiarities of users [9].

With the proposed architecture, OpenPEC seeks to establish a software architecture for the Electronic Citizen's Record and follows a worldwide trend of free software development, where any developer has the possibility to execute, copy, distribute, study, change, and improve the service offered. OpenPEC stores all exported files on its own basis, which permits users to always access them [9].

Table 2. Requirements of the architecture applied in the Health Care Terminals.

Basic utilities	Terminals must allow the collection of information from users online and offline and must have volatile memory (RAM) with the capacity to support several simultaneous operations. In addition, it needs to contain non-volatile memory to store service data records, codes, and programs
Usability	The interface must be simple, and the terminals must have good ergonomics and ensure comfort
Electromechanical	The healthcare terminal must have a compact and robust physical structure to be used in different environments, also to have high Mean Time Between Failures (MTBF) values
Reliability	To prevent hackers from misusing patient information, the system needs to be reliable, continuously operating, reboot-free, and resistant to viruses and other forms of attacks or invasion. Furthermore, it must have simple and low-cost maintenance

OpenPEC has two advantages in its use, including as it is a microservices architecture, the software is divided into several services, each with a specific activity, providing scalability and allowing a high growth of the program, in addition to faster development and greater simplicity in creation, testing, and deployment. Also, as it is free software, which allows any programmer, when identifying an error, to correct the code almost instantly, resulting in a reduction in financial resources and time invested [9].

3.4 Architecture to Support the Activities of the SUS Community Health Agents' Program

The service performed by the Community Health Agents (CHAs) is of great relevance for the progress of the Community Agents Program. However, during the mapping of the data collection process carried out by the CHAs in the community, some problems were identified, such as the redundant performance of activities, because the data collected from the patients were written on paper and then manually entered into the computer, which could generate inconsistencies; incompatibility of data with standards for the electronic representation of patient record information; difficulties in monitoring the communities due to the mismatch; and disappearance of clinical data. Therefore, there is a lack of use of technologies that help in carrying out daily tasks combined with a mobile system that helps data collection and monitoring of families [19].

To solve these problems, an architecture that uses the concepts of Ubiquitous and Mobile Computing was proposed, helping CHAs carry out their daily activities in different environments in the context of the Family Health Program. The main gain in using the architecture in these patterns is the optimization of the information collection process and broad monitoring of families through the application of context awareness techniques. Thus, it was possible to improve the process and obtain results of efficiency and effectiveness in the daily routine, in addition to gaining interoperability of information. After performing tests, it was found that the results obtained were satisfactory [19].

3.5 Architecture for Publication and Distribution of Information Systems for the SUS – iSalud

The work of building the iSalud architecture aimed to specify, implement, and validate an architectural model that allows the developed tools to be shared among several health centers, reducing the redundancy of creation. One of the pillars on which the project was based was the development of an internet-based software platform, since, in addition to allowing the greater reach of users, it has as a competitive advantage its operation on various devices such as computers, cell phones, and tablets. Therefore, an easy distribution mechanism for all telehealth centers is obtained. The repository also transmits constant updates to users, informing them about new applications and Ministry of Health campaigns [20].

Two tests were carried out to verify the viability of the architecture regarding web and mobile applications so that two locations around the world were selected as remote servers. As a result, it was found that all mobile applications were successfully deployed on virtual devices and, for web applications, it was found that 24.8% of the requests were not answered in good time, in the worst evaluation scenario. The more instances of applications are running, the more mechanisms are required for the applications to function normally. Therefore, in a period of greater use, these mechanisms may be insufficient, affecting the availability of applications [20].

3.6 Healthcare Reference Software Ecosystem Architecture – Aiki Ecosystem Architecture

Few articles in the literature demonstrate studies carried out on Software Ecosystem architectures with a focus on the health field. It is a little-explored topic that can positively impact the quality and efficiency of services provided to the community [21].

Siqueira's work [21] proposes rules for the creation and adoption of the health information systems in health organizations; establishes parameters for the interaction between the health information systems and the people that make up the environment; and proposes a Software Ecosystem architecture consistent with the Brazilian reality, requiring an extensive study of the environment to be implemented [21].

This architecture was developed in four Brazilian hospitals, in Sergipe. Among the environmental factors found, the following can be cited: the high complexity of the environments, the reduced number of beds and employees, and the services offered. As a result, it was observed the dependence on the development of applications by external organizations and the difficulty in the administration and evolution of the applications, where Siqueira's work contributed a lot to the solution of these problems [21].

3.7 Architecture for the Health Domain of the *Hospital Das Clínicas* of the Faculty of Medicine of the University of São Paulo

The SIGEAN architecture was built using the Archi 2 tool, which uses the Archimate language to carry out the architectural description and allows the semi-formal modeling of software architectures. Interviews were carried out with some of the stakeholders to better understand the problems to be solved [8].

After the analysis, it was observed that three systems are needed to address user requirements, including the electronic medical record system, the integrated intraoperative support system, and the postoperative support system. In addition, the requirements were expressed with the stakeholders through feature questions which the system should present to end-users [8].

All decision-making on the architectural design that met the Architecturally Significant Requirements (RAS) were defined. The architecture description presented a set of architectural views using the Archi tool. As the last step developed, all documentation was registered, representing, and describing the views and tracing the relationships between views [8].

4 Conclusion

In line with the increasing complexity of technologies and health systems, there were efforts in the literature to propose a flexible and comprehensive software architecture that allows the inclusion of solutions for the integration of technologies already developed. Furthermore, the trend of open-source code was identified so that when detecting possible errors or failures in integration, software developers can solve the problems almost in real-time, reducing bureaucracy in the process and producing speed in the progress of obtaining a unified health system increasingly responsive and resolute.

References

1. Fukuda, K.: Science, technology and innovation ecosystem transformation toward society 5.0. Int. J. Prod. Econ. **220** (2020). https://doi.org/10.1016/j.ijpe.2019.07.033
2. Zhang, Y., Qiu, M., Tsai, C.W., Hassan, M.M., Alamri, A.: Health-CPS: healthcare cyber-physical system assisted by cloud and big data. IEEE Syst. J. **11**, 88–95 (2017). https://doi.org/10.1109/JSYST.2015.2460747
3. Reher, D.S., et al.: The COVID-19 pandemic in an aging world (2020). https://doi.org/10.31235/OSF.IO/BFVXT
4. ECONOMIC POLICIES IN THE AGE OF GLOBALISATION EPOG STUDENTS AND ALUMNI ASSOCIATION POLICYBRIEF 1. AGING AND RISING HEALTHCARE COST (2019)
5. Sarti, T.D., Lazarini, W.S., Fontenelle, L.F., Almeida, A.P.S.C.: What is the role of primary health care in the COVID-19 pandemic? Epidemiologia e Serviços de Sa-úde **29**, e2020166 (2020). https://doi.org/10.5123/S1679-49742020000200024
6. World Health Organization: Health 2020: a European policy framework and strategy for the 21st century (2013). https://apps.who.int/iris/handle/10665/326386. Accessed 6 Nov 2021
7. Grol, R., Wensing, M.: Implementation of change in healthcare. Improving Patient Care 1–20 (2020). https://doi.org/10.1002/9781119488620.CH1
8. Garcés, L., Oliveira, B., Arenas, C.: Arquiteturas de software para o domínio da saúde, pp. 1–47 (2020)
9. Girotto, V.V., Reis, R.S., da Conceição, A.F.: ARQUITETURA DE REFERÊNCIA PARA O PRONTUÁRIO ELETRÔNICO DO CIDADÃO (OPENPEC). Temas em Saúde 20 (2020). https://doi.org/10.29327/213319.20.3-13

10. Plaza, A.M., Díaz, J., Pérez, J.: Software architectures for health care cyber-physical systems: a systematic literature review. J. Softw. Evol. Process **30**, e1930 (2018). https://doi.org/10.1002/SMR.1930
11. Carroll, N.: Key success factors for smart and connected health software solutions. Computer **49**, 22–28 (2016). https://doi.org/10.1109/MC.2016.340
12. Dybå, T., Dingsøyr, T.: Empirical studies of agile software development: a systematic review. Inf. Softw. Technol. **50**, 833–859 (2008). https://doi.org/10.1016/J.INFSOF.2008.01.006
13. Bowers, J., May, J., Melander, E., Baarman, M., Ayoob, A.: Tailoring XP for large system mission critical software development. In: Wells, D., Williams, L. (eds.) XP/Agile Universe 2002. LNCS, vol. 2418, pp. 100–111. Springer, Heidelberg (2002). https://doi.org/10.1007/3-540-45672-4_10
14. Júnior, A.A.C., Misra, S., Soares, M.S.: ArchCaMO - a maturity model for software architecture description based on ISO/IEC/IEEE 42010:2011. In: Misra, S., et al. (eds.) ICCSA 2019. LNCS, vol. 11623, pp. 31–42. Springer, Cham (2019). https://doi.org/10.1007/978-3-030-24308-1_3
15. Kannengiesser, U., Müller, H.: Multi-level, viewpoint-oriented engineering of cyber-physical production systems: an approach based on industry 4.0, system architecture and semantic web standards. In: Proceedings - 44th Euromicro Conference on Software Engineering and Advanced Applications, SEAA 2018, pp. 331–334 (2018). https://doi.org/10.1109/SEAA.2018.00061
16. Pagliosa, M., Tortorella, G., Ferreira, J.C.E.: Industry 4.0 and lean manufacturing: a systematic literature review and future research directions. J. Manuf. Technol. Manag. **32**, 543–569 (2021). https://doi.org/10.1108/JMTM-12-2018-0446/FULL/PDF
17. Ministério da Saúde (n.d.) Metodologias – MDS – DATASUS. https://datasus.saude.gov.br/metodologias-mds/. Accessed 6 Nov 2021
18. Hexsel, R.: A Arquitetura dos Terminais de Atendimento à Saúde (2003)
19. Maran, V., Paz, L., Machado, A., Boufleur, R., Augustin, I.: Uma Arquitetura Móvel para Auxílio as Atividades do Programa de Agentes Comunitários de Saúde do SUS (2013)
20. Morais, S.D.: iSalud: uma arquitetura para publicação e distribuição de sistemas de informação para o SUS (2015)
21. Siqueira, O.M.P.: Aiki ecosystem architecture: uma proposta de arquitetura de ecossistema de software de referência para a área de saúde (2017)

Graphene Geology and the Fourth Industrial Revolution

Augusto Gonçalves Nobre(✉), Lara Poliny Nogueira da Silva, and Fábio Ramos Dias de Andrade

Institute of Geosciences, University of São Paulo, São Paulo, Brazil
{augusto.goncalves,lara.poliny.silva,dias}@usp.br

Abstract. In the first three industrial revolutions, the role played by Geology was restricted to the supply of energy and mineral resources. The popularization of the Fourth Industrial Revolution (Industry 4.0) has been hampered by the technological limitation of traditional raw and processed materials. In this sense, nanotechnology appears as a potential alternative to provide materials with properties that allow wide implementation of Industry 4.0 initiatives. The hitherto unprecedented technological properties of graphene made it emerge as an exponent raw material for high-tech devices. Among the materials currently studied, it is the one that has thus far presented the most alternatives for being obtained directly from geological precursors. Since the last decade, 'geological thinking' has been included in the contemporary conception of the production of advanced materials to help in the issue of popularizing the enabling technologies of Industry 4.0, through the abundant and cheap availability of nanomaterials. Two forms of geological thinking in this regard are (i) the production of graphene and its derived materials from untreated mineral precursors, and (ii) the extraction of nano minerals directly from geological deposits.

Keywords: Graphene · Industry 4.0 · Mineral nanotechnology

1 Introduction

The Fourth Industrial Revolution, also called Industry 4.0, started at the beginning of the 21st century and has as its main approach the implementation of intelligent production processes. Among the enabling technologies of Industry 4.0, there are the Internet of Things (IoT), Cloud Computing, Big Data, Advanced Robotics, Additive Manufacturing, Digital Manufacturing, Digital Security, and Systems Integration [1].

The success of that industrial revolution will only be fully achieved with the help of new high-performance materials and nanotechnology. These enabling technologies will require the exchange of massive digital data at high speed with more efficiency in energy generation, storage, and transport and better mechanical, electrical, and electronics performance, properties which are not attainable by traditional metallic, ceramic, polymeric, and composite materials.

As in all industrial revolutions, Geology plays a fundamental role in Industry 4.0, but specifically, in this phase, it goes beyond providing mineral, water, and energy

Y. Iano et al. (Eds.): BTSym 2021, SIST 207, pp. 342–348, 2023.
https://doi.org/10.1007/978-3-031-04435-9_34

resources. This time, Earth Sciences will be able to play an even greater role, helping chemists, physicists, and engineers, in trans and interdisciplinary projects, planning new technologies.

As the costs to produce several high-performance and nanotechnological materials are high, due to the use of synthetic raw materials with high purity and structural rigor, it is difficult to quickly popularize devices that will allow the global emergence of Industry 4.0 [2]. Thus, to favor the development and economic production of materials with new properties, it is important to integrate communication from mining and ore processing to the product manufacturing industry, as the popularization of Industry 4.0's enabling technologies requires integration among the different production sectors in order to reduce waste and improve cost-benefit rates [3].

In this scenario, this work presents a review and bibliographic compilation of Graphene Geology as an example of how geological thinking should be participative in the Fourth Industrial Revolution. Graphene is related to graphite and is a strategic mineral resource available in nature as nanoplatelets. Natural graphene can be used by the nanotechnology industry for the production of high-performance materials in cases where analytical purity and the absence of crystalline defects in the raw material are not required.

2 Industrial Revolutions and the Consumption of Mineral Resources

The First Industrial Revolution began in the 18th century, with a sequence of inventions, such as the steam engine and the automated loom, which gave rise to a new production system in England. The energy sources changed from mainly vegetable substances (e.g. firewood) or animals (e.g. whale oil) to mineral coal [4]. England, the cradle of the First Industrial Revolution, had available coal and iron to build, and the workforce, from the growing rural exodus, to operate steam engine systems that emerged in urban areas.

The Second Industrial Revolution started in the mid-19th century and was marked by a significant increase in industrial production and productivity and was made possible by new materials and energy resources [4]. The new cycle of technological innovations had as examples the more efficient and cheaper production of steel by the Bessemer-Mushet and Siemens-Martin processes [5] and the ammonia synthesis by the Haber-Bosch process [6]. Regarding new energy sources, this revolution introduced petroleum (oil for combustion engines) and electric energy, obtained through hydroelectric plants. In particular, petroleum products had their outbreak in this industrial phase, with Ford's Model T line production [7], which used gasoline as fuel. Another extremely important use of this resource is the production of polymeric materials.

With the end of World War II, the Third Industrial Revolution was initiated, which has as its pillar the rapid technological development, favored by rapid society digitalization. This phase was marked by the introduction of robotics, biotechnology, and electronics, with important advances in computing and telecommunications [8]. This revolution was marked by profound changes in all sectors of the economy, as all production became more mechanical and technological, with the replacement of human forces by robots, machines, and computers. The devices of the Third Industrial Revolution demand much more energy and diversity of mineral resources for their production. For example, approximately 40 minerals are used to build a computer, including quartz, silver, aluminum, and gold. This means that mining and metallurgical processes were even more fundamental for that revolution [9].

In the context of expanding globalization and greater historical demand for natural resources, both in diversity and volume, there was the emergence of an environmental agenda, including sustainable development guidelines. It is noteworthy that in the first three revolutions, the natural resources that made technological development possible were available in nature, unlike what occurs in Industry 4.0 [10].

The scenario of human development before the Fourth Industrial Revolution is one of over-digitalization of society and progressive automation of production processes, with high demand for digital data storage and transfer capacity [11].

Currently, Industry 4.0 is in progress – the term was used for the first time in 2012 in Germany, a pioneer country in the implementation of such a revolution [12]. This revolution differs from the previous ones because it is based on several simultaneous enabling technology changes that must take place in a highly integrated manner in the production processes. The enabling technologies of this industrial revolution are the Internet of Things (IoT), Cloud Computing, Big Data, Advanced Robotics, Additive Manufacturing, Digital Manufacturing, Digital Security, and Systems Integration.

IoT can be defined as a network of physical pieces of equipment, which are able to connect to the internet [1]. IoT is being developed to improve life quality, connecting the various domestic devices directly through the internet. On the other hand, in an industrial environment, it contributes to more efficient manufacturing by promoting the integration of information technology with mechanical and electronic components in a cyber-physical environment.

Cloud Computing is crucial in the Fourth Industrial Revolution. There are many types of cloud computing but all of them do the same tasks. It provides numerous resources in the computer system, like online data storage, and it can be noticed that Cloud Computing is also important in the management of Big Data.

Big Data is a term used to define the concentration of all data in a digital environment, as a result of many processes taking place in devices, electronic equipment, business management systems, and social networks [1].

Advanced Robotics systems are making the industrial process more efficient since the robots have increasing mobility and adaptability. Advanced Robotics will save time and perform many tasks simultaneously, making them more economic.

Additive Manufacturing is expressed in 3D printing, so this type of manufacturing uses a computer to allow manipulating products with complex geometries. As a result, it can simplify the fabrication process, improve performance, and spend less on raw materials.

Digital Manufacturing is manufacturing using a computer system. This mechanism is applied to a factory's production procedures and supply chain, but it can also be used as a simulation tool for the manufacturing process. Digital manufacturing implies reducing errors, increasing efficiency, and reducing costs.

Digital security is essential for the new Industry because, as the Fourth Industrial Revolution advances, the digital space becomes more interconnected, and important decisions, fabric operations, and business management tend to be made on the internet. Therefore, Digital Security will be essential for providing security and privacy in the digital environment.

System Integration in Industry 4.0 can be interpreted as the integration between different systems (in physical and virtual environments, connected or not to the internet) that should work together. The aim is to enhance product quality, save time and increase work efficiency.

The enabling technologies of the Fourth Industrial Revolution demand greater computational capabilities, with more chemical and mechanically resistant materials of easy repair and which reject programmed obsolescence – considering the good environmental practices of the circular economy [2]. For the first time in the history of industrial revolutions, there are no natural materials available that would allow the integrated popularization of these enabling technologies.

Although there are several prototypes, both academic and industrially approved, and all enabling technologies are computationally ready, the widespread use of these technologies is unfeasible so far, as the materials currently available do not provide the necessary properties or there is a cost to provide these properties that would allow massive production of devices bearing the enabling technologies of Industry 4.0. It is currently believed that the solution to enable Industry 4.0 technologies is nanotechnology.

Given the interdisciplinary context of Industry 4.0 and the objective of making nanomaterials available on large scale, it is necessary to integrate the study of raw materials with stakeholders involved in the production process of devices carrying the enabling technologies. It is in this context that we will work with the example of Graphene Geology.

3 Graphene in the Geological Context

Graphene aroused the interest of the scientific and technological community, as it is the first material of atomic thickness obtained by mankind and with unprecedented properties (chemical, mechanical, electrical, optical, etc.) in material sciences. These properties have made graphene research develop at a relentless pace in the early 21st century [13] and have made it an exponent of potential materials for Industry 4.0.

Based on graphene structure, other two-dimensional materials with potentially interesting properties matching those seen in graphene can be envisaged, from minerals bearing van der Waals forces, fostering the search for new nanomaterials from mineral precursors [14]. This demonstrates that graphene, with its planar structure composed of covalent bonds and van der Waals interactions that form the graphite structure, is a milestone in the search for new materials in the Fourth Industrial Revolution.

The first mention of Graphene Geology in the literature dates back to 2011 [15]. The authors worked on the morphology of polycrystalline structures of graphene, but that did not directly address the natural occurrence of graphene or its production from mineral precursors.

The use of natural graphite in graphene production started in 2012; it was reported that graphene obtained from natural graphite will inevitably inherit the compositional impurities of the latter [16]. In 2020 researchers presented the structural influence that different types of graphite have on the preparation of graphene oxide, reaching analogous conclusions about the inheritance of crystal defects in the obtained nanomaterial [17]. The production of graphene nanosheets from untreated natural graphite for application in Li-ion batteries was tested promisingly in 2018 [18], demonstrating that despite the intrinsic compositional heterogeneities and the presence of crystalline defects, the properties of nanomaterials obtained from natural precursors can be of interest to the high-tech industry.

For the first time in 2015, the influence of the geological and thermodynamic context of natural graphite formation was studied in respect to a modified Hummers method for obtaining graphite oxide that would be potentially used to obtain graphene oxides [19].

Alongside the research groups that seek to produce nanomaterials from mineral precursors most simply and cheaply possible, there are also researchers engaged in the search for nano minerals that are naturally available in a geological context [20]. This group presented the possibility and technological potential of extracting two-dimensional nanomaterials from nanoplatelets of talc and graphite in geological heterostructures.

Thus, it is noteworthy that, over the last decade, an increasing interest in integrating 'geological thinking' into the chain of nanomaterials production, in particular, to graphene, took place. Graphene is a prominent example of a material with super properties potentially capable of helping in Industry 4.0's enabling technologies dissemination.

4 Discussion: The Inclusion of Geological-Graphene Thinking in the Fourth Industrial Revolution

Graphene is the most prominent example of a nanomaterial that can be obtained from natural resources and that combines properties such as electrical superconductivity at room temperature, high hardness, and flexibility. These features place graphene as one of the main potential raw materials of devices based on Industry 4.0 enabling technologies.

The impurities in composition and structural defects of natural precursors imply changes in the properties of the obtained nanomaterials when compared with those obtained from synthetic raw materials. In this sense, mineral nanotechnology needs to focus on the property and application of materials, given the abundant availability of raw materials at low cost, which differs from the analytical purity and compositional homogeneity of synthetic raw materials used in academic nanotechnology.

'Geological thinking' has already been included in the production of nanomaterials in the last decade. In this sense, every step from the mineral deposit to the final manufactured product must be considered.

5 Conclusion

In all industrial revolutions, Geology played a fundamental role, but in the first three, its function was fundamentally limited to the supply of mineral and energy resources.

With the arrival of the Fourth Industrial Revolution and the inherent difficulty in popularizing the devices of its enabling technologies, due to the technical limitation of materials available so far, it is necessary to rethink the production chain. Nanotechnology is a likely key to the production of materials with properties that will enable the wide dissemination of Industry 4.0.

Nanotechnology has been introducing devices with properties hitherto unprecedented in the history of science and technology, but it depends mostly on synthetic materials and obtaining processes with extremely controlled conditions. Those issues make the process more expensive and limit production to relatively small volumes of nanomaterials.

Geology is the science that studies the composition, structure, and processes that form natural resources. Bringing geology closer to the materials sciences and the manufacture of high-tech devices has been shown to be relevant in the integration of production processes for Industry 4.0, since the production of traditional nanomaterials has not been able to democratically popularize devices around the world.

To contribute to the massive and low-cost production of nanomaterials, 'geological thinking' has been included in the production chain of nanotechnology for nearly a decade. Graphene is the most prominent example of a material whose production has been integrating the participation of mineral precursors in its production, given its enormous scientific, technological, and commercial appeal.

In this sense, there are two paths that scientific advances have been showing:

- Production of graphene and its derived materials, such as graphene oxide and reduced graphene oxide directly from untreated natural raw materials;
- Extraction of nano minerals directly from geological deposits.

In the first case, it was observed that the obtained nanomaterials, despite having worse properties than their analogs produced by synthetic raw material, are efficient in producing materials abundantly and can be potentially employed in various high-tech uses demanded by Industry 4.0.

On the other hand, planar nano minerals available directly in rocks often stabilize in heterostructures and require future technological characterization to verify the possibility of directly using these natural materials in devices.

It is concluded that the integration of 'geological thinking' in the production chain of high-tech devices, for example, those carrying nanomaterials, should be strongly encouraged, as despite being recent, it has demonstrated the possibility of making materials with promising properties available abundantly and cheaply, which will favor the popularization of Industry 4.0 initiatives around the world.

References

1. Da Costa, M.B., Dos Santos, L.M.A.L., Schaefer, J.L., Baierle, I.C., Nara, E.O.B.: Industry 4.0 technologies basic network identification. Scientometrics **121**(2), 977–994 (2019). https://doi.org/10.1007/s11192-019-03216-7
2. Nobre, A.G., Martínez, J.A.E., Florêncio, O.: Mineral nanotechnology in circular economy. In: Iano, Y., Saotome, O., Kemper, G., Mendes de Seixas, A.C., Gomes de Oliveira, G. (eds.) BTSym 2020. SIST, vol. 233, pp. 220–226. Springer, Cham (2021). https://doi.org/10.1007/978-3-030-75680-2_26

3. Santos, B., Alberto, A., Lima, T., Charrua-Santos, F.: Industry 4.0: challenges and opportunities. RevistaProdução e Desenvolvimento **4**(1), 111–124 (2018)
4. Barca, S.: Energy, property, and the industrial revolution narrative. Ecol. Econ. **70**(7), 1309–1315 (2011). https://doi.org/10.1016/j.ecolecon.2010.03.012
5. Barraclough, K.: "Puddled steel": a forgotten chapter in the history of steelmaking. J. Iron Steel Inst. **209**(10), 785–789 (1971)
6. Erisman, J., Sutton, M., Galloway, J., Klimont, Z., Winiwarter, W.: How a century of ammonia synthesis changed the world. Nat. Geosci. **1**(10), 636–639 (2008). https://doi.org/10.1038/ngeo325
7. Wilson, J., Mckinlay, A.: Rethinking the assembly line: organisation, performance and productivity in Ford Motor Company, 1908–27. Bus. Hist. **52**(5), 760–778 (2010). https://doi.org/10.1080/00076791.2010.499425
8. Greenwood, J.: The Third Industrial Revolution: Technology, Productivity, and Income Inequality. American Enterprise Institute, Washington, D.C. (1997)
9. Bleiwas, D., Kelly, T.: Obsolete computers, "gold mine," or high-tech trash? Resource Recovery from Recycling. US Geological Survey, Denver (2001)
10. Maynard, A.: Navigating the fourth industrial revolution. Nat. Nanotechnol. **10**(12), 1005–1006 (2015). https://doi.org/10.1038/nnano.2015.286
11. Colombo, A., Karnouskos, S., Kaynak, O., Shi, Y., Yin, S.: Industrial cyberphysical systems: a backbone of the fourth industrial revolution. IEEE Ind. Electron. Mag. **11**(1), 6–16 (2017). https://doi.org/10.1109/MIE.2017.2648857
12. Bartodziej, C.: Technologies and functions of the concept Industry 40. In: Bartodziej, C.J. (ed.) The Concept Industry 40. BestMasters. Springer, Wiesbaden (2017). https://doi.org/10.1007/978-3-658-16502-4_4
13. Geim, A.: Graphene: status and prospects. Science **324**(5934), 1530–1534 (2009). https://doi.org/10.1126/science.1158877
14. Frisenda, R., Niu, Y., Gant, P., Muños, M., Castellanos-Gomez, A.: Naturally occurring van der Waals materials. NPJ 2D Mater. Appl. **4**(1), 1–13 (2020). https://doi.org/10.1038/s41699-020-00172-2
15. Yakobson, B., Ding, F.: Observational geology of graphene, at the nanoscale. ACS Nano **5**(3), 1569–1574 (2011). https://doi.org/10.1021/nn200832y
16. Ambrosi, A., Chua, C., Khezri, B., Sofer, Z., Webster, R., Pumera, M.: Chemically reduced graphene contains inherent metallic impurities present in parent natural and synthetic graphite. Proc. Natl. Acad. Sci. **109**(32), 12899–12904 (2012). https://doi.org/10.1073/pnas.1205388109
17. Hou, D., Li, K., Ma, R., Liu, Q.: Influence of order degree of coaly graphite on its structure change during preparation of graphene oxide. J. Materiomics **6**(3), 628–641 (2020). https://doi.org/10.1016/j.jmat.2020.04.009
18. Simón, M., Benítez, A., Caballero, A., Morales, J., Vargas, O.: Untreated natural graphite as a graphene source for high-performance Li-ion batteries. Batteries **4**(1), 13 (2018). https://doi.org/10.3390/batteries4010013
19. Wong, C., Sofer, Z., Pumera, M.: Geographical and geological origin of natural graphite heavily influence the electrical and electrochemical properties of chemically modified graphenes. Chem. Eur. J. **21**(23), 8435–8440 (2015). https://doi.org/10.1002/chem.201500116
20. Nobre, A., Martínez, J., Terence, M., Florêncio, O.: The action of shear zones in the natural availability of graphite nanoplatelets: the example of the metadolomites of the Itaiacoca Group and the mica schist of the Dom Silverio Group. Braz. J. Anim. Environ. Res. **3**(4), 3108–3118 (2020). https://doi.org/10.34188/bjaerv3n4-031

The BFS Method in a Cloud Environment for Analyzing Distributed Energy Resource Management Systems

Adolfo Blengini Neto[1(✉)], Yuzo Iano[1], Gabriel Gomes de Oliveira[1], Gabriel Caumo Vaz[1], Fabiana Silva Podeleski[1], Higor de Paula Kolecha[2], and Marcius F. H. de Carvalho[2]

[1] School of Electrical and Computer Engineering, State University of Campinas, Campinas, Brazil
blengini@mail.com, yuzo@unicamp.br, oliveiragomesgabriel@ieee.org

[2] Pontifical Catholic University of Campinas, Campinas, Brazil
marcius@puc-campinas.edu.br

Abstract. This article presents a cloud environment for analyzing Distribution Energy Resource Management System (DERMS) through the Backward Forward Sweep (BFS) method. The software architecture is modeled in five layers: (i) DERMS clients, (ii) standardization from the network data, (iii) integration with API (Application Programming Interface) available in a cloud environment, (iv) normalization and caching, and (v) processing of Power Flow algorithm. The BFS cloud environment provides a Load Flow API for supporting Energy Resource Management of Distribution Systems. The use of different methods through APIs makes it possible to choose the most adherent method to the network topology, load models, constraints flexibility, and generation limits. The proposal provides benefits such as computational time, accessibility, availability, flexibility, ease, and cost-effectiveness. The BFS in the cloud is tested on Taiwan Power Company 83-bus systems showing scalability, performance, and convergence in two tests scenarios considering 50 load flow transactions per second and 100 load flow transactions per second.

Keywords: Power flow · Backward forward sweep · Distribution energy resources · Distribution management systems

1 Introduction

Distribution energy systems are subject to a sharp growth in demand that overloads the distribution networks, requires more available capacity from feeders, increases losses, and reduces system efficiency [1]. Besides the increasing demand in modern power grids, generation and control are continuously moving towards the edge of the grid, closer to the end customers [2]. Furthermore, the deregulation and privatization of power systems encourage the inclusion of Distributed

Y. Iano et al. (Eds.): BTSym 2021, SIST 207, pp. 349–362, 2023.
https://doi.org/10.1007/978-3-031-04435-9_35

Generation (DG) as the main element of Distribution Energy Resources (DER) [3]. Through the integration of an increasing DER within the management software solutions, new opportunities have arisen for real-time management of the resources in the distribution network, enabling new computational solutions through Distribution Energy Resource Management System (DERMS).

A DERMS is a software-based solution that allows a distribution system operator to have real-time visibility and control of the distribution network and its distributed assets to manage operation more effectively [2]. This proposal provides a Load Flow (LF) API for DERMS. The LF model as a service in a cloud environment offers benefits such as computational time, convergence, accessibility, availability, flexibility, ease, and cost-effectiveness [4].

There are numerous LF techniques available in the literature for the analysis of power distribution systems. Sarfi et al. [5] presents a survey of the state of the art in distribution system where optimization approaches are based on searching for a configuration with minimum losses and classified accordingly to the solution methods as based upon a blend of heuristic and optimization methods and making use of heuristics only. In the earlier stage of research, Newton-Raphson (NR) and fast-decoupled [6] method of power flow calculation were the most popular LF techniques according to the trend in transmission systems [7]. These techniques exhibit good convergence characteristics for well-conditioned transmission systems [8,9], but do not offer the same performance for distribution systems, as discussed in the literature [10].

The high R/X ratio distribution system characteristic deteriorates the diagonal dominance of the Jacobian matrix, and NR methods are not suitable for this purpose [11]. To overcome the computational burden of the NR approaches, recent literature indicates Backward Forward Sweep (BFS) approach [12], which considers that the radial characteristic of the distribution system is capable of providing good LF solutions with some application constraints [11,13,14]. This proposal allows utility partners and researchers to evaluate existing and future DERMS and other utility management system applications in a realistic laboratory environment. In brief, the main contributions of this work are:

- An alternative to the dedicated LF software system for traditional approaches of non-linear equations based on the NR nodal formulation;
- The cloud architecture allows for scalability and high performance, ensuring real-time responses for DERMS;
- The BFS model is available in the API model as a microservice, considering the elements of distribution networks with their capacities and voltage profile.

To achieve the above aims, the paper is organized in the following way: Sect. 2 gives a description of the Solution Layers, Sect. 3 presents numerical results, and Sect. 4 outlines the conclusions.

2 Description of Solution Layers

The software architecture is modeled in layers and groups. Layered architectures focus on grouping relative functionality within an application with different layers stacked vertically from start to end. The features applied in each layer are related to a common rule or responsibility, as shown in Fig. 1. Groups are considered the boundaries of the solution, i.e., clients request environment or DERMS systems and the proposed cloud environment model. The proposed architecture is divided into five layers, namely client or DERMS systems, input and standardization of network data, integration with microservice API environment, normalization and caching, and processing of the Current Flow algorithm.

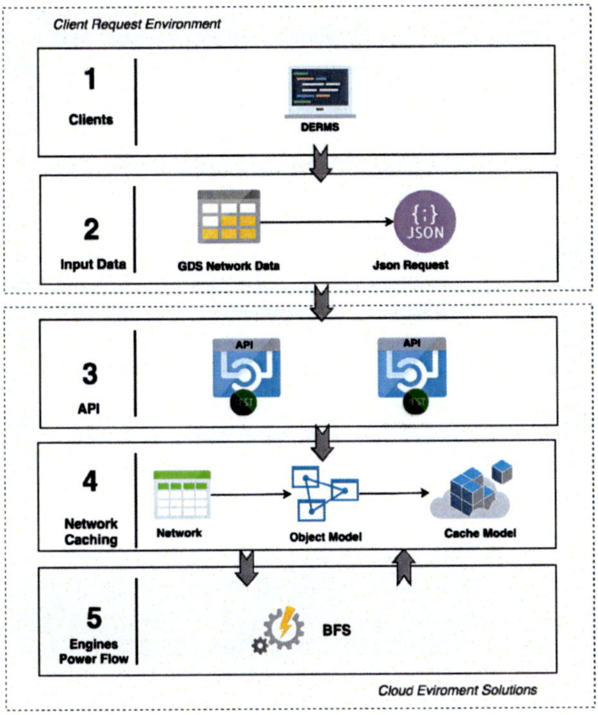

Fig. 1. Software architecture model.

2.1 DERMS

The DERMS solution consists of DMS (Distribution Management System) and ADMS (Advanced Distribution Management System) systems, DER resources, and models for simulation of network operations (loss analysis, network loading, restoration, and planning maneuvers) [2].

DERMS is a relatively new concept for power distribution. The objectives and functionalities consist mainly of organizing, managing, optimizing, and effectively controlling the DER resources and the distribution network in order to maximize the economic benefits of the network and propose more reliability of network operation and quality of service [15].

In the proposed solution, DERMS and Input Data are considered the client of the cloud solution in layer one, using Internet standard REST API communication protocols. Data from the distribution network of DERMs systems (input data) are normalized to the JSON (JavaScript Object Notation) standard.

2.2 Input Data

This layer consists of establishing a standard or layout for the solution's distribution network input data, in order to facilitate the integration with DMS and ADMS systems. To exemplify the input model, a 5-bus distribution network composed of feeder, branch, source bus, destination bus, active power (P), reactive power (Q), voltage limit (V), resistance (R), section reactance (X) and line state (NC - Closed or NO - Opened), and type (G-Generation L-Load or S-Switch) is shown in Fig. 2 and Table 1.

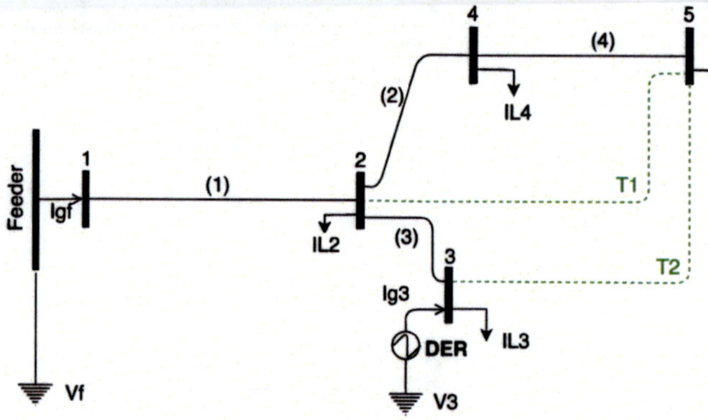

Fig. 2. 5-Bus distribution system with DER [8].

The JSON format is chosen to input the data into an API layer. This is a simple data format used for data exchange. The structure of the JSON content follows the syntax structure for JavaScript [16]. The example of Fig. 3 shows the JSON format of line 1 (branch 0) of Table 1.

Generally, a JSON string contains either an array of values or an object, which is a set of name/value pairs. An object is surrounded by a pair of curly brackets and contains a comma-separated list of name-value pairs. A name-value pair consists of a field name (in double quotes), followed by a colon (:), followed by the field value (in double quotes).

Table 1. 5-Bus distribution system with DER - data

Feeder	*Branch*	*Source*	*Destination*	*P*	*Q*	*R*	*X*	*State*	*Type*
A	0	0	1	0.00	0.00	0.01	0.01	NC	G
A	1	1	2	2.50	2.70	0.07	0.10	NC	L
A	2	2	4	3.30	0.90	0.11	0.11	NC	L
A	3	4	5	1.30	0.90	0.11	0.11	NC	L
A	4	2	3	3.35	1.55	0.08	0.11	NC	L
A	5	2	5	0.00	0.00	0.04	0.04	NO	S
A	6	3	5	0.00	0.00	0.08	0.11	NO	S

```
{
    "label" : "A",
    "branch" : 0,
    "source" : 0,
    "destination" : 1,
    "p" : 0.0,
    "q" : 0.0,
    "v" : 0.0,
    "r" : 0.01,
    "x" : 0.01,
    "state": "NC",
    "type": "G"
}
```

Fig. 3. JSON request example branch 0.

The advantage of a JSON message is that it can contain both structured and semi-structured data. By supporting the exchange of information in the form of JSON objects, the boundary layer 2 (input data) of DERMS systems and layer 3 (API solution) becomes simple and standardized and allows for different DERM systems to connect to the API published in the cloud.

2.3 API

In integrated DMS operation with DER and microgrids, the goal is not only to allow mutual sharing of data, functionality, and task performance but to achieve better coordination in how different functions operate in relatively independent systems. The integration of DERM System with Cloud System Solutions is implemented with API model with JSON format, as shown in Fig. 4.

The architecture of the proposed solution exports two Web microservices API, BFSController and FeederInfoController. The BFSController is responsible for performing, with POST requests, the LF calculation for the BFS model, using the JSON structure in the format of Fig. 3 as an input parameter. The FeederInfoController microservice is responsible for GET requests for analyzing the feeder's characteristics, such as DER number, buses identification, and whether there is mesh formation in the feeder. The documentation for both microservices is available in Swagger API (inputs and outputs).

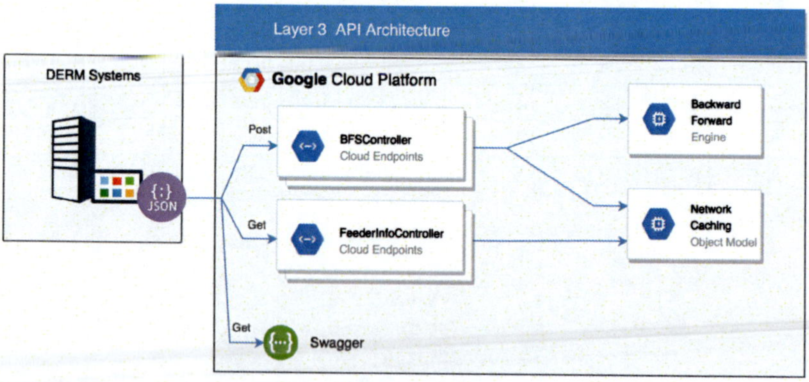

Fig. 4. API architecture layer.

The API documentation is available in the Swagger tool, through HTTP requests to bfs-controller and feeder-info-controller API, as shown in Fig. 5.

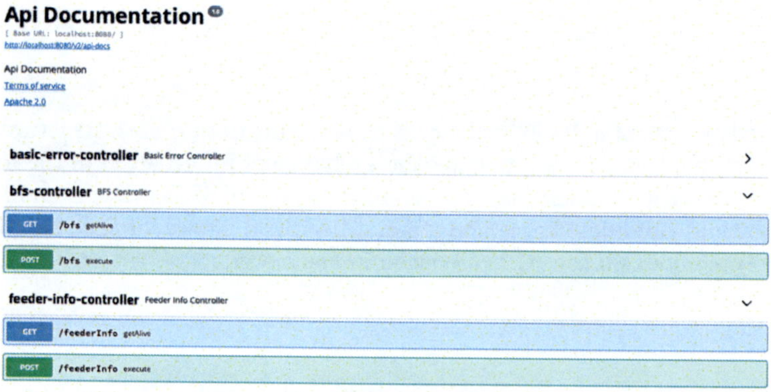

Fig. 5. Swagger API documentation.

2.4 Network Caching

The distribution network modeling process is one of the main steps in optimization this power flow process. It is possible to have standardized access to the input data, abstraction of the real network elements within the object-oriented model, flexibility to access element properties, and reuse of objects in other implementations of power flow algorithms. This approach uses a concept that models the energy network elements in the form of objects in memory [14]. The modeling defines a cache data structure for each distribution system element, allowing better exploitation of its resources by current flow calculation engines. The distribution network modeling process is one of the main steps in optimization.

The use of object-oriented programming in the load flow process presents an interesting idea: the representation of each element of the distribution network as an object or class. This representation seeks to approximate the system being created to what is observed in the real world, with characteristics and actions, as observed in reality [13].

The logical model of objects of the radial distribution network is shown in Fig. 6. The model represents the real elements of the network: feeders, lines, buses, and their relationships. The Node-Edge or Bus-Branch relationship represents the heart of the data structure. Through this, it is possible to organize the cache network data in the form of graphs. To perform this processing, the "Father-Child" technique is used, where the section or branch is composed of the source node and destination node, and each section has a referenced parent section, as shown in Fig. 7.

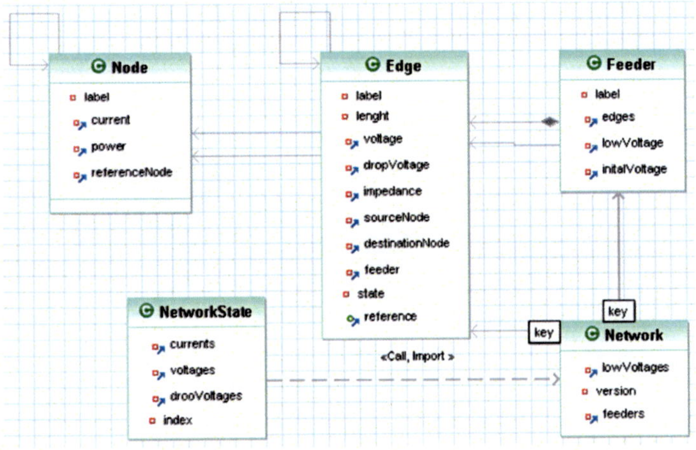

Fig. 6. Object model of distribution network.

In this manner, in-memory connectivity of the network in the form of objects is created with the input data for the distribution network, allowing high performance in object queries in real-time.

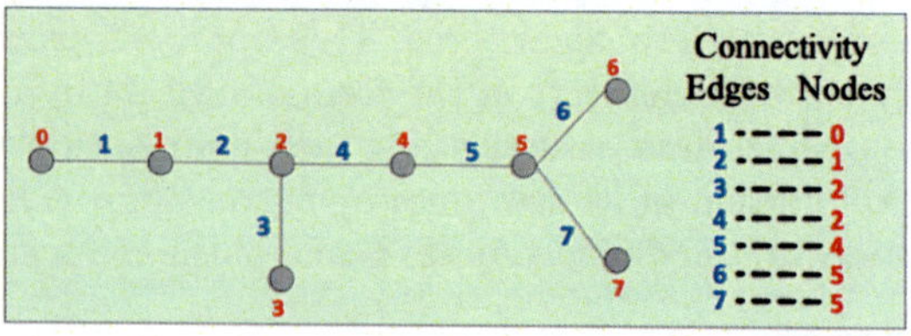

Fig. 7. Graph structure for cache model [13].

2.5 Load Flow Engines

The solution considers a current flow implementation with the Backward-Forward Sweep model as the heart of the proposed solution, supporting DERMS planning and management. The BFS algorithm has the advantage of lower computational effort compared to Newton-Raphson time and Gauss-Seidel methods [17] and better convergence in radial distribution networks. The BFS model is represented in five steps, as shown in Fig. 8:

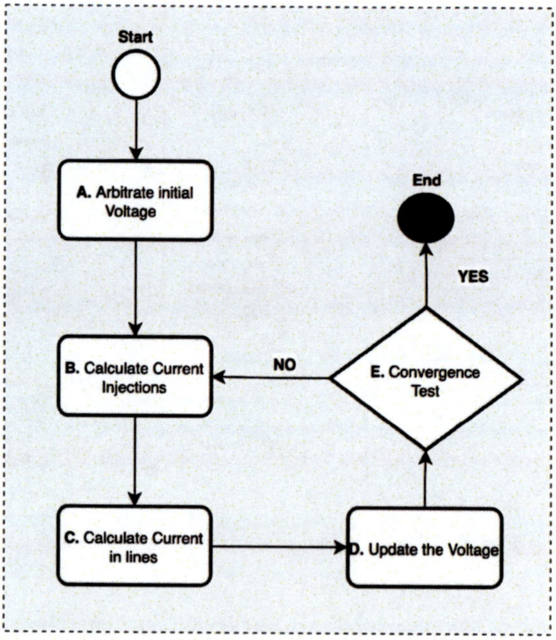

Fig. 8. Backward-forward sweep model [17]

A - Arbitrate Initial Voltage. The initial voltage vector (real) to all buses is arbitrate equal 1.0 p.u. as Eq. (1). The iteration counter is set to $p = 0$.

$$V_k,\ where\ k = 1, ..., NB \tag{1}$$

B - Calculate Current Injections. With a new voltage, the current injected in the load buses is recalculated according to Eq. (2).

$$I_k = \left(\frac{S_k}{V_k}\right)^* - {Y_h}^{sh} * V_k\ where\ k = 1, ..., NB \tag{2}$$

C - Calculate Current in Lines. Calculate the currents in the loops connecting the k and m buses from the terminal lines to the root bus feeder with Eq. (3).

$$I_{km} = I_m + \sum_{j \in Fm} I_{mj} \tag{3}$$

D - Update Bus Voltages. Update bus voltages of the distribution network from feeder to all terminal buses according to Eq. (4).

$$V_m = V_k - Z_{km} * I_{km} \tag{4}$$

E - Convergence Comparison. Finally, the convergence comparison is performed. The algorithm returns to Step B if there is a ΔV component of the bus vector V greater than ϵ, as Eq. (6), or, if ΔV is lower or equal to ϵ, the model is finalized, as Eq. (5).

$$[\Delta V_k, k = 1, ..., NB] \leq \epsilon \tag{5}$$

$$[\Delta V_k, k = 1, ..., NB] > \epsilon \tag{6}$$

All nomenclatures presented in the equations are described in Table 2.

The initial voltage vector (real) for all buses is arbitrarily equal to 1.0 p.u. The iteration counter is set to $p = 0$. As shown in Step B, the proposed model is iterative. For each iteration performed in Step C, a new bus voltage vector is determined. With a new voltage, the injected current in the load buses is recalculated.

The load can be represented by a constant impedance model, constant current model, or constant power model. If all the system loads are represented by a constant current load model, the process is not iterative, because the current is independent of the bus voltage. In this case, the algorithm converges after one iteration ($p = 1$).

Table 2. Nomenclature

kV	- Kilo-volt
MVA	- Mega-volt Ampere
NB	- Number of buses
I_r	- Active Current
I_i	- Imaginary Current
S_k	- Apparent Power k bus
Y_h^{sh}	- Admittance of the shunt element connected to bus k
I_{km}	- Current line k-m
I_m	- Injection Current m bus
V_k	- Initial voltage vector
V_m	- Voltage m bus
Z_{km}	- Impedance k-m

3 Numerical Results and Discussions

The proposed API Architecture Layer is coded using Java version 16 and Spring Framework 2.5.3. The tests were executed in Google Cloud Platform Intel Core i5 2.4 GHz processor, with 4 GB RAM. The developed approach is investigated on the balanced three-phase network. This section highlights the performance of the approach in terms of accuracy and convergence rate.

3.1 83-Bus Taiwan Power Company Distribution Network

To demonstrate the robustness proposed in this work, computer simulations were carried out using the algorithm developed for a constant power model in the Taiwan Power Company (TPC) distribution network [18], using the balanced three-phase network of 11.4 kV for voltage as input parameter, 11 feeders(A-B-C-D-E-F-G-H-I-J-K), 83 tap-changers with normal-closed state, 13 tap-changers with normal-open state, 28.35 MW of total active load, and 20.70 MVAr of total reactive load, as shown in Fig. 9.

Taiwan Power Company (TPC) Distribution Network Test with convergence factor $\epsilon = 0.0001$. The initial voltage at buses and feeder are set at $V_k = 1.0$ p.u. and $V_m = 1.02$ p.u. The result of the algorithm is represented in Fig. 10 with the following data: voltage profile (module p.u.), buses, and feeders. Convergence in power flow calculation with the BFS method was achieved in all network feeders, but with different iterations, as shown in Table 3.

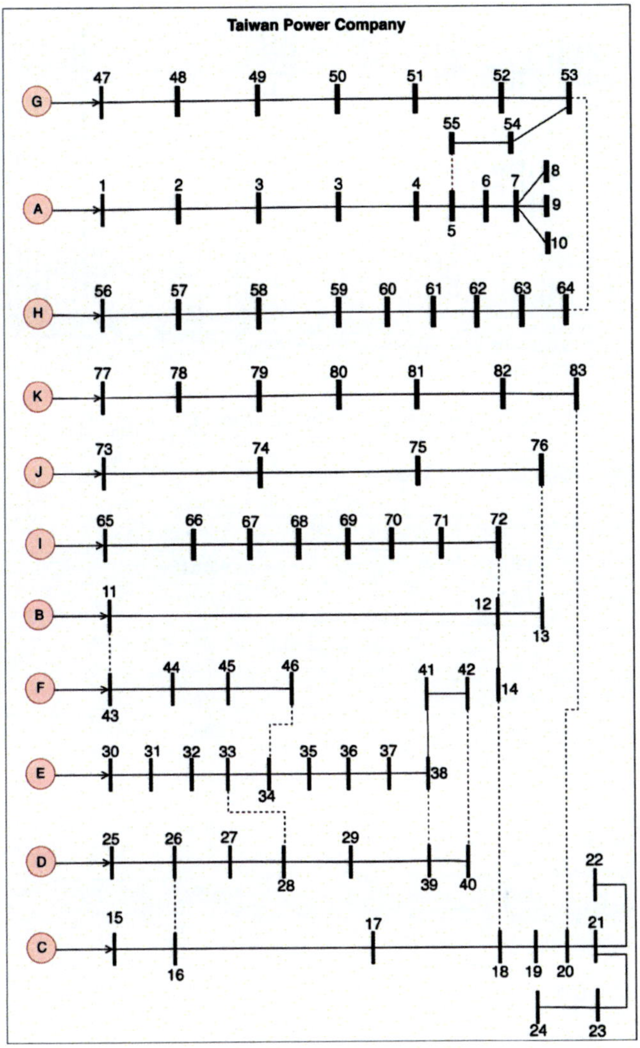

Fig. 9. 83-Bus Radial Network Taiwan Power Company [18]

To prove the high availability, performance, and multiple accesses in the cloud API infrastructure, two stress tests were simulated in the solution: the first test simulates a DERM system sending 50 simultaneous requests to the energy flow calculation API, and the second test simulates 100 DERMs requests per second. In both scenarios, the API solution remained stable, with scalability in the Google Cloud Platform of microservices up to 100 available images (configured limit) and with CPU processing times as shown in Table 3.

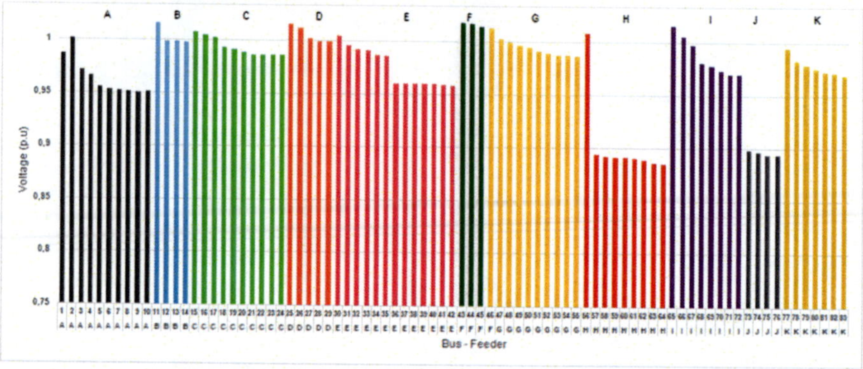

Fig. 10. Voltage (p.u) - Taiwan Power Company [18]

Table 3. Convergence time and transactions per second (TPS) the TPC 83-bus RDS

Feeder	Iterations	50 DERMS TPS CPU time (s)	100 DERMS TPS CPU time (s)
A	5	0.028	0.033
B	2	0.004	0.011
C	4	0.029	0.037
D	2	0.007	0.015
E	4	0.028	0.035
F	2	0.004	0.009
G	4	0.028	0.032
H	7	0.031	0.048
I	4	0.028	0.038
J	7	0.031	0.044
K	4	0.028	0.032

4 Conclusion

This solution can be integrated with the DERM system to assess the performance of new DMS applications centrally and add new APIs to the microservices catalog. The use of the cloud solution enables the reductions of the operational costs, the complexity, and time-to-market of the deployment compared to the on-premise model and especially the availability and scalability of APIs and scenarios of high volume requests.

This platform includes the following benefits: increased efficiency, reliability, performance, resilience with real-time DERM operations, centralized and switching studies, and operations for distributed energy systems. In addition, it is possible to incorporate new APIs for optimizing operations in DERMs aiming at studying network switching, minimizing losses, self-healing, and other power flow algorithms implemented in the solution cloud environment. Sources, network examples, and DERM simulators are available at https://github.com/blengini/ocf-api.

For future work, another API will be developed to consider distribution network optimizations and minimize the injected current and transmission losses with a linear or quadratic function.

References

1. Serban, A.C., Lytras, M.D.: Artificial intelligence for smart renewable energy sector in Europe-smart energy infrastructures for next generation smart cities. IEEE Access **8**, 77364–77377 (2020)
2. Nowak, S., Tehrani, N., Metcalfe, M.S., Eberle, W., Wang, L.: Cloud-based DERMS test platform using real-time power system simulation. In: 2018 IEEE Power Energy Society General Meeting (PESGM). IEEE, pp. 1–5 (2018)
3. Sultana, U., Khairuddin, A.B., Aman, M.M., Mokhtar, A.S., Zareen, N.: A review of optimum DG placement based on minimization of power losses and voltage stability enhancement of distribution system. Renew. Sustain. Energy Rev. **63**, 363–378 (2016)
4. Nayak, S.K., Panda, S.K., Das, S., Pande, S.K.: A renewable energy-based task consolidation algorithm for cloud computing. In: Singh, A.K., Tripathy, M. (eds.) Control Applications in Modern Power System. LNEE, vol. 710, pp. 453–463. Springer, Singapore (2021). https://doi.org/10.1007/978-981-15-8815-0_40
5. Sarfi, R.J., Salama, M.M.A., Chikhani, A.Y.: A survey of the state of the art in distribution system reconfiguration for system loss reduction. Electr Power Syst Res **31**(1), 61–70 (1994)
6. Stott, B., Alsac, O.: Fast decoupled load flow. IEEE Trans. Power Appar. Syst. **3**, 859–869 (1974)
7. Tinney, W.F., Hart, C.E.: Power flow solution by Newton's method. IEEE Trans. Power Appar. Syst. **11**, 1449–1460 (1967)
8. Ghatak, U., Mukherjee, V.: An improved load flow technique based on load current injection for modern distribution system. Int. J. Electr. Power Energy Syst. **84**, 168–181 (2017)
9. Yang, N.-C., Chen, H.-C.: Decomposed Newton algorithm-based three-phase power-flow for unbalanced radial distribution networks with distributed energy resources and electric vehicle demands. Int. J. Electr. Power Energy Syst. **96**, 473–483 (2018)
10. Yang, J.A.M., Ganesh, S.: Power flow analysis for radial distribution system using backward/forward sweep method. Int. J. Electr. Comput. Electron. Commun. Eng. **8**(10), 1540–1544 (2014)
11. Rupa, J.A.M., Ganesh, S.: Power flow analysis for radial distribution system using backward forward sweep method. Int. J. Electr. Comput. Energ. Electron. Commun. Eng. **8**(10), 1621–1625 (2014)
12. Chen, T.-H., Yang, N.-C.: Loop frame of reference based three-phase power flow for unbalanced radial distribution systems. Electr. Power Syst. Res. **80**(7), 799–806 (2010)
13. Aravindhababu, P., Ganapathy, S., Nayar, K.R.: A novel technique for the analysis of radial distribution systems. Int. J. Electr. Power Energy Syst. **23**(3), 167–171 (2001)

14. Neto, A.B., Carvalho, M.F.H., Mota, L., Pezzuto, C.: Object-oriented load flow modeling and implementation for distribution power systems. IEEE Lat. Am. Trans. **15**, 439–444 (2017). https://doi.org/10.1109/TLA.2017.7867593
15. Jianhui, W., Chen, C., Xiaonan, L.: Guidelines for implementing advanced distribution management systems-requirements for DMS integration with DERMS and microgrids. United States (2015). https://doi.org/10.2172/1212266
16. Dusan, P.: JSON integration in relational database systems. Int. J. Comput. Appl. **108**(5), 0975–8887 (2017)
17. Pereira, C.A.N.: Alocação Ótima de reguladores de tensão em redes de distribuição de energia elétrica (2009)
18. Esmaeilian, H.R., Fadaeinedjad, R.: Distribution system efficiency improvement using network reconfiguration and capacitor allocation. Int. J. Electr. Power Energy Syst. **64**, 457–468 (2015)

Characterization of the Behavior of LoRa Networks in a Fluvial Environment in the Rio Negro

Danilo Frazão and Edgard Silva(✉)

School of Technology, Amazonas State University, Manaus, Brazil
{dsf.eng17,elsilva}@uea.edu.br

Abstract. Few articles study the transmission and analysis of data captured in rivers that cross a city using LoRa radiofrequency. LoRa is a network that was specially developed for the construction and development of Internet of Things services. Among its main features, there are long-range, low power, and wireless data transmission. This work presents a characterization of Lo-Ra networks in a fluvial environment in the Rio Negro (bridge) and a semi-urban environment. Where the transceiver will cross the bridge over the Rio Negro, crossing the other bank and continuing its route in a semi-urban region. The Sect. 1 is an introduction, then a brief description of the LoRa network is made. In the Sect. 3, the physical layer is addressed. In the following section, a description of the methodology of this article is made and finally, the results obtained and a conclusion were verified.

Keywords: LoRa · Smart cities · Rio Negro · IoT

1 Introduction

The Rio Negro is the longest blackwater river in the world and the second-largest in terms of water volume. It bathes three countries: Venezuela, Colombia, and Brazil. The objective of this work is to characterize the transmissions of the LoRa network in a fluvial environment on the bridge that crosses the Rio Negro from one bank to the other, passing through a semi-urban region and returning to the fixed point of the receiver. For this purpose, a testbed with sensors, transceiver, and receiver is implemented to verify the capacity of the network. The stretch of river where the test platform will be built is in front of the city of Manaus, a city located in the middle of the Amazon rainforest and the capital of the state of Amazonas in Brazil.

Smart Cities are those that typically use wireless sensor networks (WSN), and Internet of Things technologies, to extract data and provide services that contribute to a better quality of life for citizens. In addition, the public administration uses these data to reduce costs and increase efficiency [1]. These sensors can also be used within river regions that bathe a city [2]. So far, there is still no information about the performance of the LoRa technology in the Amazon environment, especially in the Rio Negro river region. Due to the peculiar characteristics of Manaus, bathed by the Rio Negro, the monitoring of climatic and river conditions in this river is extremely interesting.

Y. Iano et al. (Eds.): BTSym 2021, SIST 207, pp. 363–369, 2023.
https://doi.org/10.1007/978-3-031-04435-9_36

2 What is LoRa?

It is Semtech's modulation technology for low-power long-distance networks (LPWANs). The name LoRa comes from the abbreviation of the words "long" and "range". LoRa-based devices require little power and their battery can last up to 10 years. In addition, this technology allows you to demodulate signals better than using FSK and surpasses the range of transmissions via Wi-Fi, Bluetooth, and cell phone. In urban areas, the range can be a few kilometers and, in rural areas, 10–15 km [3].

LoRa is the physical layer, and devices based on this technology are capable of converting data into radio frequency (RF) signals that are sent and received over the air, using spread spectrum chirp communications and frequency bands licensed for use by ANATEL. Within the 900 MHz band with a width of 26 MHz, it goes from 902 to 928 MHz. The advantage of using chirp technology is that it is very resistant to interference in adverse radio frequency environments [3].

LoRaWAN refers to the asynchronous network software protocol that uses the physical layer described in the previous paragraph. It is managed by a non-profit association called the LoRa Alliance [4]. With this technology, it is possible to create flexible networks on a large scale that are ideal for saving costs and preserving battery life. These networks are perfect where there are a large number of devices that capture low amounts of data with small power requirements and require long-range communication [3].

3 The LoRa Physical Layer

3.1 Overview

The Semtech corporation owns the LoRa modulation technology and the physical layer is not fully open. So this section only describes the open parts. The physical layer uses a radio modulation technique called Chirp Spread Spectrum (CSS). The Industrial Scientific and Medical (ISM) frequency bands are sets of unlicensed bands reserved internationally for industrial, scientific, and medical development. The LoRa technology has sensitivity on the order of −130 dBm, operates in the ISM bands of 433–868 or 915 MHz (used in this work), but it depends on the region where it is deployed [5].

3.2 Physical Layer Parameters

Three parameters are available for customizing the LoRa modulation: Bandwidth (BW), Spread Factor (SF), and Code Rate (CR). These parameters influence the effective bit rate of the modulation, its resistance to interference noise, and its ease of decoding [5].

The scattering factor (SF) will define the number of chirps that will compose a LoRa symbol, with 2SF chirps for a symbol. And it also directly sets the number of bits per symbol. Values for this factor are between 7 and 12. The higher the value for this parameter, the higher the signal-to-noise ratio (SNR), improving sensitivity and range. However, this causes the baud rate to halve, increasing the packet transmission time and consuming more power [5]. Figure 1 displays three frequency spreading factors as a function of time.

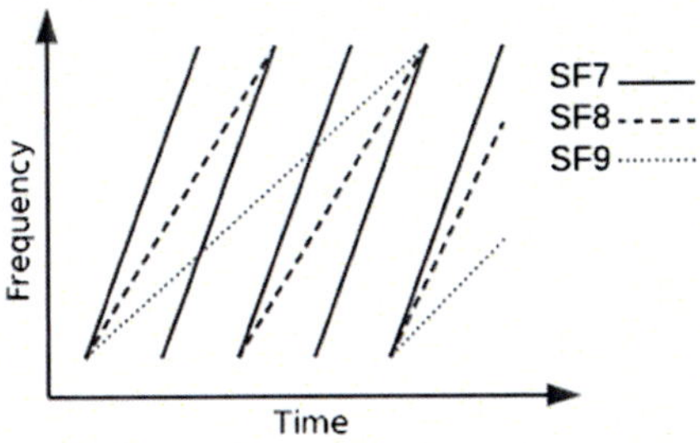

Fig. 1. Variation of symbol duration as a function of mirroring factor (Source: Modified from [6]).

Figure 2 gives an example of a LoRa transmission in the variation of frequency (on the ordinate axis) over time (on the abscissa axis), where bandwidth (BW) is represented by the vertical arrow. When the maximum frequency of the band is reached, it returns to normal and starts increasing again from the minimum frequency. The position of this discontinuity in frequency is what encodes the transmitted information. A chirp is a signal that varies continuously, which can increase (upchirp) or decrease (downchirp) its frequency. So 2SF chirps form a LoRa symbol that covers the entire frequency band and usually starts with several ascending chirps. Since there are 2SF chirps in a symbol, a symbol can effectively encode SF bits of information. Generally, depending on the region, the bandwidth parameters that can be chosen are 125, 250, and 500 kHz values. The higher the bandwidth, the higher the data rate with less airtime, but with lower sensitivity due to additional noise integration [5].

LoRa includes a progressive error correction code which is code rate (CR). Responsible for performing a direct correction in each data transmission, it is represented in Eq. (1) with n ∈ {1, 2, 3, 4}. The code rate can assume the values of 4/5, 4/6, 4/7, and 4/8. A 4/5 code rate is less tolerant of interference than a signal transmitted with a 4/8 code rate. However, it decreases the useful bit rate, according to Eq. (3), which will be discussed later, as it increases the transmission time [5].

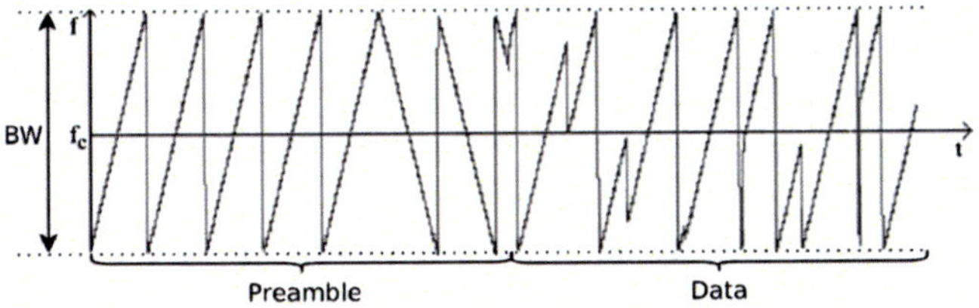

Fig. 2. Frequency variation over time of a sample signal emitted by a LoRa transceiver. The center frequency of the channel is fc and the bandwidth is BW [5].

$$CR = \frac{4}{(4+n)} \tag{1}$$

Equation (2) relates the duration of a symbol (TS) in the air, in seconds, to the bandwidth and the scattering factor [5].

$$T_S = \frac{2^{SF}}{(BW)} \tag{2}$$

Equations (3) and (4) allow calculating the useful bit rate (Rb) given in bits/second, where the SF directly represents the number of bits and the TS, as already mentioned, is in seconds [5].

$$R_b = \frac{SF}{T_S} \times CR \tag{3}$$

$$R_b = SF \times \frac{BW}{2^{SF}} \times CR \tag{4}$$

4 Methodology

4.1 Transceiver

It is agreed that this device will transmit every 1 s. The data to be sent by the transceiver are Counter, GPS Time, Latitude, Longitude, and Altitude. The counter is an integer-type variable that starts at zero and adds one unit to each transmission. Latitude and longitude are being returned in degrees and decimals of degrees. Altitude is being returned in meters. In total, we have a 19-byte size data structure to be transmitted. The two devices (transceiver and receiver) will be configured with a code rate (CR) of 4/5, bandwidth (BW) of 125 MHz, spread factor (SF) of 7 to 12, and radiofrequency of 915 MHz. The main materials used are ESP32 LoRa v2, Steelbras AP3900 Antenna, GPS Neo-6M v2.

4.2 Receiver

The data to be received is the same data that will be sent by the transceiver, already mentioned in Sect. 4.1. In addition to these data, other information will be shown on the ESP32 display: RSSI and Packet Size (Byte). The Arduino IDE (on a notebook) will display the time this information was received on the serial monitor. So, in total, there are nine variables listed below that will be saved in the Arduino IDE installed on the notebook: Schedule (Arduino IDE), RSSI, Packet Size, and data received by the transceiver. Successfully received packages will be stored in a notebook in the Arduino IDE for analysis, obtained results, and conclusions. It is possible to know the maximum transmission and reception range through the GPS location, and through the "Counter" variable, a survey must be made of how many packets were lost and how many were successfully received. The receiver must have the same settings defined on the transceiver for a transmit and receive synchronization.

The original antenna that came with the microcontrollers (ESP32 LoRa v2) has a maximum range of only 350 m, so that, it was replaced by Steelbras AP3900 antennas that operate at the same frequency of 900 MHz and have an estimated range of 6.5 km. The main materials used are ESP32 LoRa v2 and Antenna Steelbras AP3900.

4.3 Testbed

The trajectory that the transceiver will travel is represented on the map in Fig. 3 and will be from point A to point B. The receiver can be fixed (stationary) at point A. The transceiver will preferably be in constant motion and is transported by a vehicle from point A to point B and returning from point B to point A. It will perform this trajectory six times in each scattering factor (SF) from 7 to 12. The distance between the two points, which form a straight line, is around 7.9 km. The package delivery rates for each scattering factor are presented in Fig. 4.

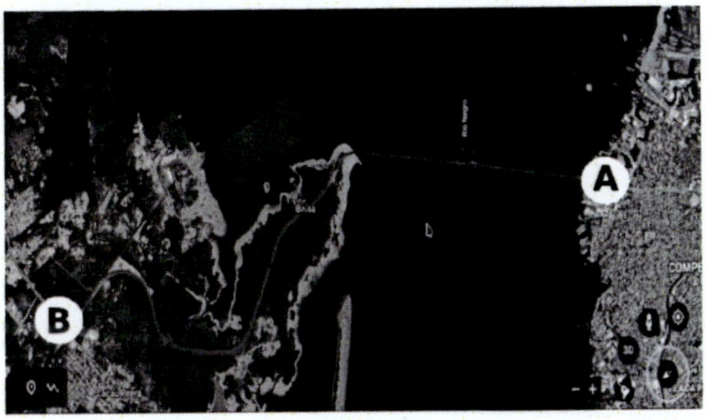

Fig. 3. Map of the trajectory that the transceiver will travel.

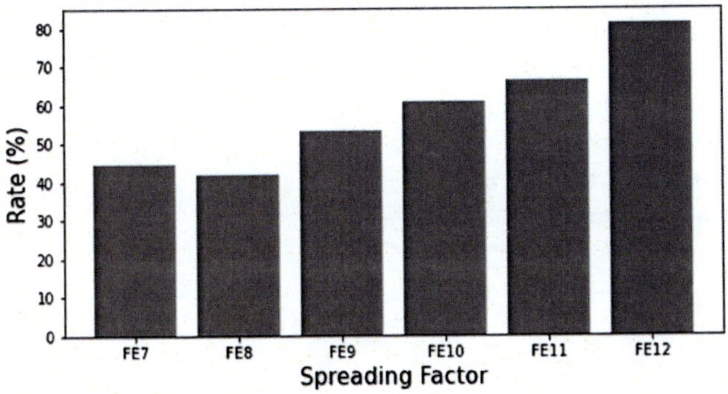

Fig. 4. Package delivery rate (percent). These are the packages that were successfully delivered.

5 Results

The Rio Negro bridge, being a metallic structure, influenced the reduction of the antenna's sensitivity. This can be seen in the graph in Fig. 5, where the RSSI is a function of the distance in meters. As the transceiver moves away from the receiver, the received signal strength is reduced. Around 4,000 m away, an improvement in the RSSI signal is seen. This is because at this point there is a target area (without obstacles) which is at the end of the bridge on the way to this route and at the beginning of the bridge on the way back. The remaining five scattering factors had almost the same result as this graph. The signal managed to reach the maximum range of the route, which is 7.9 km. But LoRa receivers offer a sensitivity on the order of −130 dBm [4] and the RSSI on the chart has come to around −115 dBm. Thus, greater transmission/reception distances may be obtained.

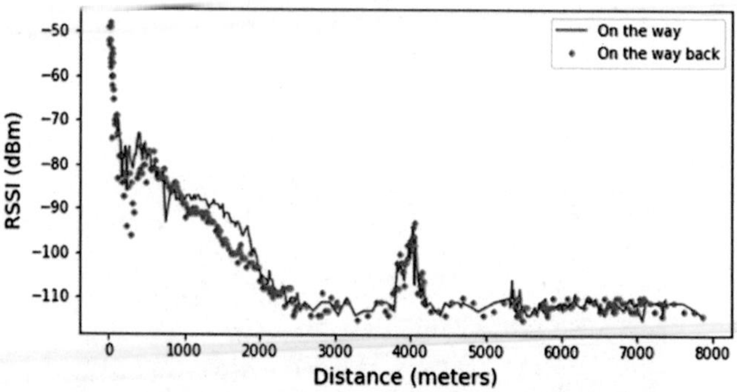

Fig. 5. Spread factor 7 on the way to and from the course

The byte-per-second rate is similar to the bits-per-second. Then, from the collected data, the average throughput in bytes/s for the six scattering factors from seven to twelve was extracted. Equation (3) shows that the useful bit rate (bits/s) is reduced as the spreading factor increases. In the graph in Fig. 6, this event happens from the value seven to eight and from the value ten to twelve. But in values from nine to ten this rate remains constant. And in the scattering factor from 8 to 9, there is an increase that shows to have an inverse logic of Eq. (3) in this interval. It is necessary to consider that in this experiment the devices were relatively moving away in the outward journey, and approaching the return journey.

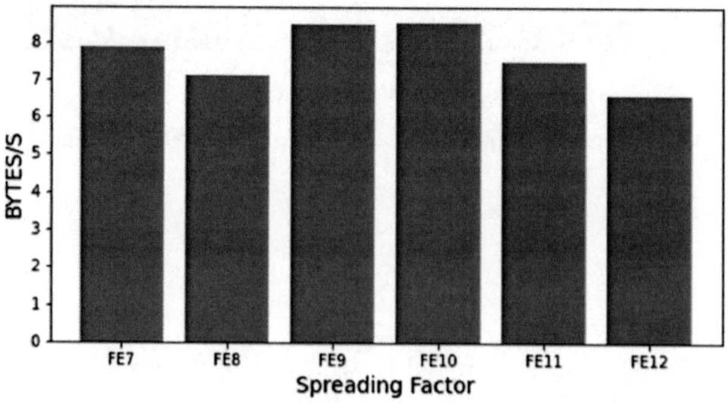

Fig. 6. Average packet throughput in bytes per second

6 Conclusion

This article provides an experimental methodology on the LoRa physical layer different from the works of [1, 2]. Because the transceiver moves relatively far away from the receiver on the way out, it gets closer on the way back, and data is captured while it is in motion. Another difference is the fact that the testbed is carried out in a fluvial and semi-urban environment, with an equatorial climate characteristic of the Amazon region. Equation (3) worked very well in the article by [4] that performed the experiments at fixed points, without movement. But the average throughput result of this article, given in bytes/s, shows that this equation did not have a very approximate application due to the fact that increasing scattering factors did not have a constant reduction in their rate of bytes per second for all scattering factors. The graph of the rate of packets that were successfully delivered had a result very close to theory resulting in greater range and sensitivity with increasing scattering factors. The signal could have a range greater than 7.9 km.

Acknowledgments. This study was financed in part by Fundação de Amparo à Pesquisa do Estado do Amazonas – FAPEAM. The foundation provides fellowship to the first author.

References

1. Garcia-Font, V., Garrigues, C., Rifa-Pous, H.: A comparative study of anomaly detection techniques for smart city wireless sensor networks. Sensors **16**(6), 868 (2016)
2. Guibene, W., Nowack, J., Chalikias, N., Fitzgibbon, K., Kelly, M., Prendergast, D.: Evaluation of LPWAN technologies for smart cities: river monitoring use-case. In 2017 IEEE Wireless Communications and Networking Conference Workshops (WCNCW), pp. 1–5. IEEE (2017)
3. Portal do Desenvolvedor LoRa (n.d.). https://lora-developers.semtech.com/. Accessed 20 Sept 2021
4. LoRa Alliance official website (n.d.). https://lora-alliance.org/. Accessed 20 Sept 2021
5. Augustin, A., Yi, J., Clausen, T., Townsley, W.M.: A study of LoRa: long range and low power networks for the internet of things. Sensors **16**(9), 1466 (2016)
6. Bedaque, D.: Testes de Desempenho de Enlaces Ponto a Ponto da Tecnologia Lo-Ra. Instituto Federal de Santa Catarina (Campus São José) - Engenharia de Tele-comunicações (2016)

Utility Pole Detector for Inspection of Power Distribution Networks Using Faster R-CNN

Júlio César Pereira(✉), Rangel Arthur, and Luis Augusto Libório

School of Technology, University of Campinas, Limeira, Brazil
j159576@dac.unicamp.br, rangel@ft.unicamp.br,
l167994@g.unicamp.br

Abstract. Electricity users make power companies better monitor their services to keep distribution networks up and running. To monitor working electrical distribution systems an important tool is the visual inspection. Infrared thermography and image processing techniques are used to improve the reliability of this visual inspection to avoid several electrical failures. The target devices for visual inspections are located on poles. Therefore, the focus is on the correct detection of poles to automate the inspection from these devices. The research proposed by this work aims to build a utility pole detector as the first step of an automated diagnostic system for network inspection. The database composed of images collected from Google Street View was used to train detectors using Faster R-CNN network, YOLO, and Mask R-CNN. The Faster R-CNN detector presented 93,59% of detections when applied to a video and 14,62% of wrong detections. The research is still ongoing, and the results can still be improved.

Keywords: Utility pole · Deep learning detector · Faster R-CNN · Power distribution networks

1 Introduction

Electric power distribution networks face many challenges due to the growing need for energy worldwide. These networks are often hazardous environments, so automating these systems can reduce costs, improve efficiency, and lower operational risk [1].

Visual inspection, one of the oldest monitoring techniques, is one of the most used in predictive maintenance routines because of its low operating cost and ease of use. However, several manifestations may indicate defects outside the field of vision. It may lead to power interruption and danger to lives [2].

2 Researches Related

Several applications were developed for pole detection using different techniques.

One of the works used Bag of Visual Words with different feature extractors together with an SVM (Support Vector Machine) classifier to identify the poles for inspection

Y. Iano et al. (Eds.): BTSym 2021, SIST 207, pp. 370–377, 2023.
https://doi.org/10.1007/978-3-031-04435-9_37

of the electrical network [3]. The algorithm using the Fast-SIFT technique achieved a precision of 96.25% and an inference time of 23 s to classify 240 images.

The objective of [4] was to map the location of the poles using aerial images using the method of Deep Learning ATSS (Adaptive Training Sample Selection). The author used 1057 images of 5619x5946 pixels. Images were divided into smaller images of 256x256 pixels, thus generating 99473 images. The technique showed 91.3% accuracy with an IoU (Intersection over Union) threshold greater than 0.5.

Also, for mapping, [5] uses Google Street View images to find pole crossheads and estimate their location using RetinaNet-101. It was used 500 images containing 937 poles. With the IoU threshold greater than 0.3, it obtained 0.78 precision.

Another research proposes the extraction of poles and electrical networks for 3D measurements using Hough transforms and 2D Gaussian functions to simulate underground networks [6]. According to the author, the accuracy of the extraction of the poles was 68.97% although the recognition rate was 100%.

The detection of poles using UAV (Unmanned Aerial Vehicles) using Radon transform is proposed in [7]. The research achieved as results the identification of 70% of the poles, identification of 95% of elements like poles, and only 5% of lost detections.

The use of unsupervised learning networks to detect anomalies in electrical distribution networks, more specifically in poles, is made in [8]. The authors propose a model that consists of a generic CNN network to learn the parameters of telephone poles. The second component is an inference model that visually describes the model's behavior. The third component is responsible for proving the data points and the result of the interpolation near those points. This is done using a GAN network for probabilistic data analysis. The objective is to create an architecture that improves the human being's ability to understand the decision-making of models.

The detection of electrical distribution networks in real-time using UAVs is also done in [9]. For this, the YOLACT algorithm is used. The study uses 600 images in 512 x 288 pixels format. The model features 36.4 mAP (mean Average Precision) at a rate of 37.2 frames per second using a Resnet-50-FPN backbone to identify crossheads, isolators, transformers, and primary wiring.

A model for detecting normal and faulty poles using only a computer with CPU, based on YOLO-Lite and classification model using SPP network (Spatial Pyramid Pooling) was presented in [10]. The model achieved 75.8% of accuracy, 57.33% of recall at 9 frames per second using images in 224 x 224 pixels.

The main objective of [11] is the detection of bird nests on poles of the electrical distribution network. The pole detection was used to filter the wrong detections of the nests. The model uses two SSD (Single Shot Detector) detectors, one for each class. The pole detector performance was 84.21% using 4105 images for training and 795 for testing.

The use of a self-organizing map (SOM) and an SVM is proposed by [12]. A CNN network was used to extract the characteristics, the SOM network does the clustering and the SVM is responsible for the classification. The process was used to recognize disasters in electrical distribution networks.

This article seeks to present a system capable of performing the inspection of the electricity distribution network automatically. The system uses a vehicle with cameras (optical and thermographic) traveling through the city streets to visualize the network and its elements. The images captured by the cameras are sent to a software responsible for identifying the network elements. The network diagnosis is given according to the thermographic image evaluation. The elements detection is done by convolutional neural network techniques such as YOLOv4 [13, 14], YOLOv5, Faster-RCNN [15], and Mask-CNN [16] for performance comparison.

3 Faster R-CNN

Faster R-CNN is composed of two modules (Fig. 1). The first uses RPN (Region Proposal Network) instead of Sliding Windows. The second module is a Fast R-CNN detector. RPNs generates from the input images, rectangles with probable objects, and a score as output. RPNs are designed to predict candidate regions with a wide range of scales and aspect ratios [15]. The function of RPN is to show the Fast R-CNN detector where to look. The Fast R-CNN detector extracts the characteristics of the candidate regions using RoiPool and performs the classification and regression of the bounding boxes.

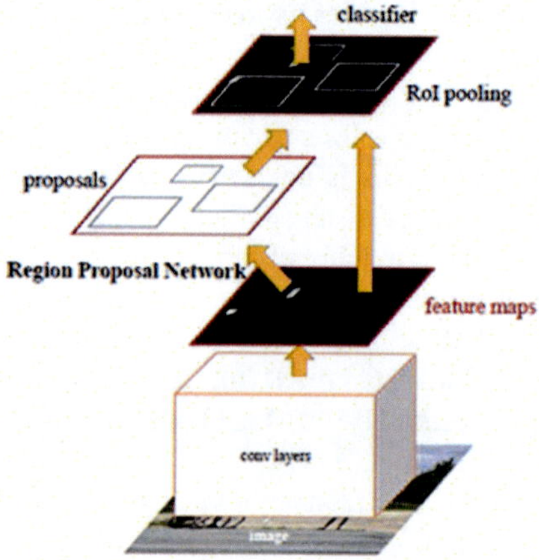

Fig. 1. Faster R-CNN network representation [15].

The Mask R-CNN network is also composed of two stages: the first is the RPN and the second has, in parallel with the prediction of the class, the output of masks for each ROI (Region of Intrest) [15].

4 Proceedings

4.1 Image Acquisition

The entire structure was developed using the free environment of Google Collaboration (Google Colab) which allows the execution of commands in Python programming language and has GPUs and TPUs for use.

The first step of the pole detection project was to build an image database containing poles. As there is no database available from which images of poles used in electrical distribution networks could be obtained, it was decided to use the Google Cloud Platform service, more specifically the Street View Static API. Street View Static allows images to be obtained from the Google Street View platform through addresses or geographic location. Along with the API, the Python package "google-streetview" (https://rrwen.github.io/google_streetview/) was used to automate the collection of images through code. The API together with the "google-streetview" package allows collecting images from an address, or geographic location, with different points of view, and with a maximum size up to 640 × 640.

At first, a Street View Python module was used to acquire street images in the City of Limeira, Brazil. Around 1000 640 × 640 pixel images were acquired at four different angles every 5 m of street. Figure 2 shows images collected from a stretch of street.

Fig. 2. Images collected on Google Street API

4.2 Labeling

To train the detectors it is necessary to label the images. The label gives the model a location and the name of the object to be located. The annotation process generated 2575 poles in 960 images. Among these images, 70% were used for training, 20% for validation, and 10% for testing. The label model used was the bounding box and the annotations were performed using the VGG Annotator. A sample of labeled images is shown in Fig. 3.

Fig. 3. Sample of labeled images.

4.3 Training

After acquiring the images and labeling the poles, the dataset was used as input for the Faster R-CNN network. No pre-processing was performed on the images before entering the network. The network was trained for 25 thousand epochs and presented the results shown in Fig. 4.

Fig. 4. Detector applied to test dataset

5 Experiments and Results

Every city may have different patterns of electricity distribution network elements, primarily due to the age and size of the network, or different companies responsible for the networks and different equipment suppliers. The same happens when we extend to different cities. A set of images containing 94 poles from a neighboring city was selected for the experiment. The main objective was to verify the detector's performance when applied in a different city. We notice difficulties in the inspection due to urban noise (trees, houses, and other obstacles), but also due to the difference in the composition of the lighting on the poles. In the region chosen for inspection, the poles have two rods that support two lamps, while the trained model only had one rod. Even with these differences, the models were able to identify more than 70% of the poles. Testing in a different city showed us the need for a diversified dataset to make the model appliable in different regions.

In Table 1 we can observe that the Faster R-CNN correctly identifies a larger number of poles but makes more identification errors (identifies some residential and commercial entrance poles and pole shadows). The Yolo technique identified fewer poles, but it makes fewer identification errors. The Mask-RCNN model, despite the number of 78 identifications, identified poles that were more distant and that were not considered in the selection of this test, failing to identify other selected ones. The Mask network still presented a high number of errors in relation to the other models.

Table 1. Preliminary detector results in a different city

	Detected	Not detected	Wrong detections
Faster R-CNN	73	21	13
YOLOv5	70	24	9
Mask R-CNN	78	27	36
YOLOv4	66	28	6

The model must be able to identify the elements with the camera in motion since the system is for street inspection. A video was recorded using Google Street View circling the streets of the city where the model was trained. The video was recorded at a rate of 30 frames per second with a resolution of 1280 × 720 pixels. There are 156 poles selected as possible to be identified, although there are other poles in the video that are further away. The video was analyzed by the detectors, and we obtained the results shown in Table 2. The Faster R-CNN detector obtained the best detection results (93.59%), the detector based on YOLOv5 identified 84.62%, YOLOv4 identified 81.41% and Mask R-CNN detected 80,13% of the poles.

Most of the wrong identifications were from pole shadows and standard residential and commercial entry poles that are also elements of the electrical distribution network. Some of the unidentified poles were due to urban noise, such as trees, buildings, houses, and vision obstructions. In addition, the image capture angle influences the identification of the pole, so a more detailed study of the dataset, the image capture position, and the use of a set of cameras will be carried out.

Detecting a network element that has problems should be done as quickly as possible. The element, however, can take a longer time to be identified, due to the number of elements in a pole and the technique used for detection. It is necessary to detect the pole

Table 2. Detector performance applied to a video

	Detected	Not detected	Wrong detections
Faster R-CNN	146 (93,59%)	10 (6,41%)	25 (14,62%)
YOLOv5	132 (84,62%)	24 (15,38%)	16 (11,19%)
YOLOv4	127 (81,41%)	29 (18,59%)	13 (9,29%)
Mask R-CNN	125 (80,13%)	31 (19,87%)	56 (30,94%)

the moment the inspection vehicle passes through it. So, detection techniques that have shorter video inference times and efficiently perform detection should be applied. As standard for the experiment, a personal notebook without the use of a graphics processing unit (GPU) was used. Among the presented techniques, the one that presented the worst performance in real-time video analysis was the Mask R-CNN model. The inference time was about 15 s for each frame. The Faster R-CNN technique performed slightly better than the Mask network, with inference varying between 8 and 10 s. This long period makes it impossible to use these models to detect poles using the equipment available for the experiment. The technique that presented the highest processing speed was YOLO, which manages to act at different frame rates while maintaining a detection pattern. The inference time was about 0.450 s. The use of machines with GPU can enable the use of Mask R-CNN and Faster R-CNN techniques, improving the inference time.

6 Conclusions

The process of choosing an element detection technique must consider the network's ability to identify the elements, but also the inspection process that will be carried out. In this study, the network that showed the best result in identifying the poles was the Faster R-CNN network, but it has a high inference time that can make it impossible to locate the poles in real-time. The same applies to the Mask R-CNN network which has a structure like Faster R-CNN.

In terms of inference speed, YOLOv5 and YOLOv4 networks have a great advantage over the others, but they had a lower capacity to identify the poles.

Adjustments will be made so that the efficiency in identifying the poles increases, as this is the first step in the process of identifying and inspecting the elements of the network using thermographic cameras.

A more detailed study of the parameters of all networks, cameras, and hardware will be made in order to choose the one that will best perform under the inspection conditions found by the energy company.

References

1. Menendez, O., Cheein, F.A.A., Perez, M., Kouro, S.: Robotics in power systems: enabling a more reliable and safe grid. IEEE Ind. Electron. Mag. **11**(2), 22–34 (2017)

2. Gorchinski, A.L., Amarante, H.M.S., Kovaleski, J.L., Matos, E.A.Á., Francisco, A.C.: Termografia em redes de distribuição de energia elétrica. In: III Congresso Brasileiro de Engenharia de Produção – Conbrepro, pp. 1–12 (2013)
3. Cabello, F.C., Iano, Y., Arthur, R., Dueñas, A., León, J., Caetano, D.G.: Automatic detection of utility poles using the bag of visual words method for different feature extractors. In: CAIP 2017. LNCS, vol. 10425, pp. 116–126. Springer. Cham (2017). Doi: https://doi.org/10.1007/978-3-319-64698-5_11
4. Gomes, M., et al.: Mapping utility poles in aerial orthoimages using atss deep learning method. Sensors **20**(21), 6070 (2020)
5. Zhang, W., Witharana, C., Li, W., Zhang, C., Li, X., Parent, J.: Using deep learning to identify utility poles with crossarms and estimate their locations from google street view images. Sensors **18**(8), 2484 (2018)
6. Gao, Z., et al.: Utility pole extraction processing from point cloud data from 3D measurement and its applications. In: 2020 11th International Conference on Awareness Science and Technology (iCAST), pp. 1–6. IEEE (2020)
7. Sharma, H., Adithya, V., Dutta, T., Balamuralidhar, P.: Image analysis-based automatic utility pole detection for remote surveillance. In: 2015 International Conference on Digital Image Computing: Techniques and Applications (DICTA), pp. 1–7. IEEE (2015)
8. Lee, D., Nam, J., Choi, H.J.: Deep electric pole anomaly detection and unsupervised description generation. In: Proceedings - 2020 IEEE International Conference on Big Data and Smart Computing (BigComp), pp. 535–537. IEEE (2020)
9. Vemula, S., Frye, M.: Real-time powerline detection system for an unmanned aircraft system. IEEE Trans. Syst. Man Cybern. (SMC), 4493–4497 (2020)
10. Zhang, S., Chen, B., Wang, R., Wang, J., Zhong, L., Gao, B.: Unmanned Aerial Vehicle (UAV) vision-based detection of power line poles by CPU-based deep learning method. In: 9th IEEE International Conference on Cyber Technology in Automation, Control and Intelligent Systems (CYBER), pp. 1630–1634. IEEE (2019)
11. Ju, M., Yoo, C.D.: Detection of bird's nest in real time based on relation with electric pole using deep neural network. In: 34th International Technical Conference on Circuits/Systems, Computers and Communications (ITC-CSCC), pp. 1–4. IEEE (2019)
12. Bochkovskiy, A., Wang, C.Y., Liao, H.Y.M.: Yolov4: Optimal speed and accuracy of object detection. arXiv preprint arxiv:2004.10934 (2020)
13. Wang, C.Y., Bochkovskiy, A., Liao, H.Y.M.: Scaled-yolov4: scaling cross stage partial network. In: Proceedings of the IEEE/CVF Conference on Computer Vision and Pattern Recognition, pp. 13029–13038. IEEE (2021)
14. Wang, R., Chen, B., Han, J., Li, C.: The intelligent recognition of the disaster damage of distribution network detected by UAV. In: Proceedings of 2019 IEEE 4th Advanced Information Technology, Electronic and Automation Control Conference (IAEAC), pp. 1708–1711. IEEE (2019)
15. Ren S, He K, Girshick R, Sun J.: Faster R-CNN: towards real-time object detection with region proposal networks. IEEE Trans. Pattern Anal. Mach. Intell. **39**(6), 1137–1149 (2017)
16. He, K., Gkioxari, G., Dollár, P., Girshick, R.: Mask R-CNN. IEEE Trans. Pattern Anal. Mach. Intell. **42**(2), 386–397 (2020)

Lean-Agile Education: A Bibliometric Analysis

Julia Yumi Ito, Franciane Freitas Silveira,
and Alessandra Cristina Santos Akkari(✉)

Graduate Program in Engineering and Innovation Management, Federal University of ABC, Santo Andre, Brazil
julia.ito@aluno.ufabc.edu.br, {franciane.silveira, alessandra.akkari}@ufabc.edu.br

Abstract. There are several studies on the application of Lean Thinking, as well as the Agile Mindset, in the literature, including the sectors of healthcare, technology, agriculture, industry, finance, among others. However, few works focus on the joint exploration of both themes, especially in the education field. This article aimed to develop a bibliometric analysis on Lean-Agile Education, mapping the global scientific production on this topic and focusing on the process of education management in the digital age. Therefore, an exploratory and descriptive study was developed with a quantitative approach based on the collection and treatment of secondary data. Between 2000 and 2021, only 34 articles were identified on the Web of Science database, organized into three categories according to management, pedagogical, and curricular content perspectives, mostly found in English and published in the US and Europe, with three publications from the Aalto University (Finland) affiliation. It was observed that this discipline does not follow the main bibliometric laws (Lotka's Law, Zipt's Law, and Bradford Law) with only 6 authors with more than one publication, high dispersion of journals responsible for publications, and frequency of words without constant proportion within the field of study, with emphasis on "software", "lean", "agile", and "education". Therefore, this bibliometric analysis pointed out trends for the area, indicating that Lean-Agile Education is a contemporary topic and still scarce in number and diversity of publications with an opportunity to go beyond new studies.

Keywords: Lean · Agile · Education · Management · Lean-agile education

1 Introduction

The digital age is a characterization of a global environment connected by digital technologies, involving devices, processes, people, networks, and real-time data sharing, fostering a social, environmental, economic, and cultural transformation [1]. This technological advance also promotes the adaptation of organizations and professionals due to continuous and agile changes in the market and society's behavior. Thus, a traditional management model, vertical, bureaucratic, mechanistic, and not flexible, does not follow the market dynamics, nor does it have the speed to respond to new demands of different natures [2].

Y. Iano et al. (Eds.): BTSym 2021, SIST 207, pp. 378–385, 2023.
https://doi.org/10.1007/978-3-031-04435-9_38

According to this multifaceted scenario, new philosophies emerge to solve complex problems in the organizational environment, highlighting the agile approach. Agility is the ability to face dynamic and turbulent environments, promoting multidisciplinary teamwork with flexibility and adaptation. The Agile mindset is related to values and principles that can be applied in the management of small projects or strategically through an organizational systemic view, being directly related to the prioritization and allocation of resources in value streams [3]. In the education field, the literature points to Agile Education connected to four values, including (i) people are more important than processes and tools (human-centric); (ii) contact and interaction between people, from the beginning of the construction of an action/solution, is more relevant than a well-structured plan to be presented only at the end; (iii) delivering small improvements and experiences daily, (iv) and adjusting routes is more interesting than taking a long time to deliver a big change that may generate value to people [4].

Agile was influenced by Lean concepts and practices, which is a philosophy focused on systemic optimization, considering the effectiveness and efficiency of resources. Lean Thinking when applied to the educational sector is called Lean Education, which, in addition to the continuous improvement, focuses on generating value for the individual and reducing waste of human potential, exploring continuous learning [5]. Lean Education is also based on five principles: (i) creating value from the user perspective, including both teachers and students; (ii) engaging and interacting with everyone involved in the value stream, including government agencies and the community for problem-solving and innovation; (iii) continuous flow based on skills such as a sense of responsibility; (iv) flow pulled by the needs of students and teachers through interaction and collaboration within a human-centric approach; (iv) and search for continuous improvement regarding quality of products and services offered to the community [5–7].

Therefore, the concepts of Lean and Agile can contribute to the educational area and have their effects enhanced from a joint application. This article aimed to carry out a bibliometric study on Lean-Agile Education, being of interest to map global scientific productions regarding a contemporary theme to assess the main authors, keywords, trends in the area, among other parameters.

2 Methodology

For the development of the bibliometric analysis, a systematic and quantitative search to evaluate the global scientific production on Lean-Agile Education was carried out, based on the management of information available in bibliographic online databases.

This study included the analysis of scientific articles extracted from the Web of Science Core Collection (WoSCC), the largest worldwide-used citation database [8].

The searches were performed using an advanced search model, without restrictions on language, academic disciplines, areas of study, types of documents, methods, or names of journal/periodic initially to identify the total number of related publications available, resulting in little scientific documentation in earlier data tests. The keywords Lean, Agil* (agile variants such as agile, agility, among others), Educa* (considering variants such as education, educational, among others), and school were used, standardized with boolean AND and OR algorithms as follows: TS = (Lean AND Agil* AND (Educa* OR

SCHOOL); Timespan: 2000–2021; Indexes: SCI-EXPANDED, SSCI, A&HCI, CPCI-S, CPCI-SSH, ESCI. Where: Topic (TS) = default search field (title, abstract, authors' keywords, and Keywords Plus); AND = Boolean operator "E"; OR = Boolean operator "OR".

The period defined as the search filter was between 2000 and 2021 (the month of August as the 2021 cutoff, when this analysis was carried out). This start period, from the year 2000, is justified by the historical milestone of the publication by Dahlgaard and Østergaard [5, 9], referring to the first publication on Lean Education, with an application of Lean Thinking in Higher Education.

Search results were exported to the software VosViewer [10], for network and overlay visualization and analysis. The VOSviewer parameters used for analysis were: occurrence analysis as type of analysis, the unit of analysis was words in titles, abstracts, and keywords (ignoring structured abstract labels and copyright statements), minimum number of 1 occurrence of a term, selecting the 60% most relevant (default choice), full counting as counting method, circles labels visualization and the association strength method of normalization of strength of the links between items.

The software Biblioshiny (graphical interface of the R Bibliometrix package) [11] was used to evaluation about authors, sources, and citations, further to the classical laws of bibliometrics. Regarding Bradford Law, papers were analyzed to identify the main periodics; (ii) Lotka's Law, to investigate the productivity of related authors, and (iii) Zipt's Law, to identify the words most used by authors [12, 13].

3 Results and Discussion

From the research procedure presented in Sect. 3, only 57 results were identified on Web of Science (Fig. 1). There was a curve of academic interest in studies related to Lean-Agile Education, with a peak of publications between 2016 and 2018, as well as a

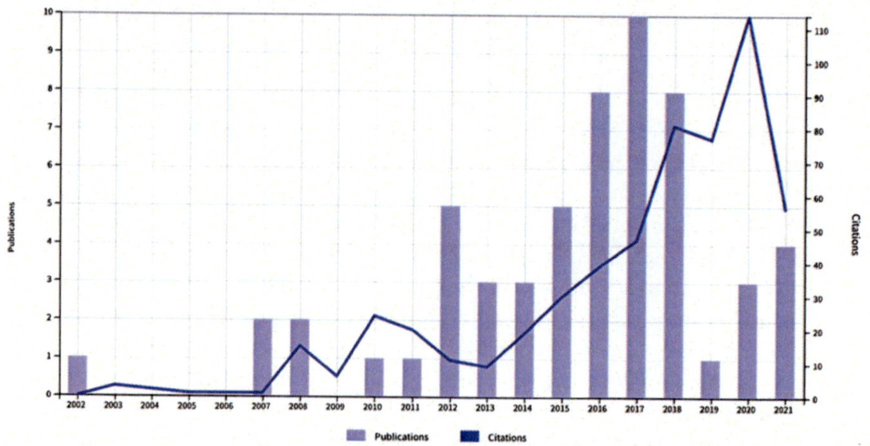

Fig. 1. Publications and citations on Lean-Agile in education, between 2000 and 2021, according to Web of Science.

progressive increase in citations from 2013, suggesting that it is a contemporary theme and there are still many contributions to be made in the area.

Although all studies demonstrated the joint application of Lean Thinking and Agile Mindset, it was observed that the focus diverged among 4 perspectives categorized as: (i) 6 results grouped by the managerial perspective, contemplating application of Lean Thinking and Agile Mindset from a systemic view of education management; (ii) 12 results grouped by pedagogical perspective, adopting resources from both concepts to facilitate the teaching-learning process (Lean and/or Agile learning); (iii) 16 results grouped by the perspective of curriculum content, referring to the presence of attributes of both methodologies in the curriculum of schools, universities, institutes, and companies as a field of study; (iv) 23 results grouped by a general perspective and focusing on other areas, so these studies only touched the educational aspect. All documents in category (iv) were excluded from this bibliometric analysis due to not being related to the direct application of Lean-Agile in education, resulting in 34 articles.

Table 1. Words most frequently used in abstracts, titles, and keywords

	Abstract	Title	Keywords	Keywords Plus	Total
Software	58	8	4	1	71
Agile	48	9	8	1	66
Engineering	42	9	4	0	55
Lean	37	10	6	0	53
Education	35	11	4	1	51
Development	39	4	3	2	48
Learning	42	5	0	0	47
Students	40	0	0	2	42
Project	39	0	0	0	39
Research	37	0	0	0	37
Industry	0	5	0	0	5
Educational	0	4	1	0	5
Practices	0	4	0	0	4
Innovation	0	0	4	0	4
Kanban	0	0	3	0	3
Scrum	0	0	3	0	3
Serious games	0	0	3	0	3

Following Zipt's Law, the most common words ("basic lexicon") in this field of study were ordered in Table 1, according to Abstract, Title, Keywords, and Keywords Plus, and a constant pattern was not identified in the frequency of words [12]. The words lean (53), agile (66), and education (51) were often expected because they were part of the initial search query and are short words, easier to repeat. It was observed that the word software (71) was the most used, once Agile is original and usually in the context of software development, contemplating different practices and tools already used by Lean, such as kanban. These words reflect the three perspectives of results (managerial, pedagogical, and curriculum), considering software, engineering, and development more related to curriculum perspective; learning and serious game more pedagogical, and research and innovation more related to managerial scope.

Figures 2 and 3, from VosViewer software, show word co-occurrence networks that point to the relationship of research themes. In Fig. 2, it is observed two keywords groups with closer relationship; one refers to the Lean-Agile methodology connecting keywords like agile, lean, kanban, scrum; the second refers to Agile methods in the education context.

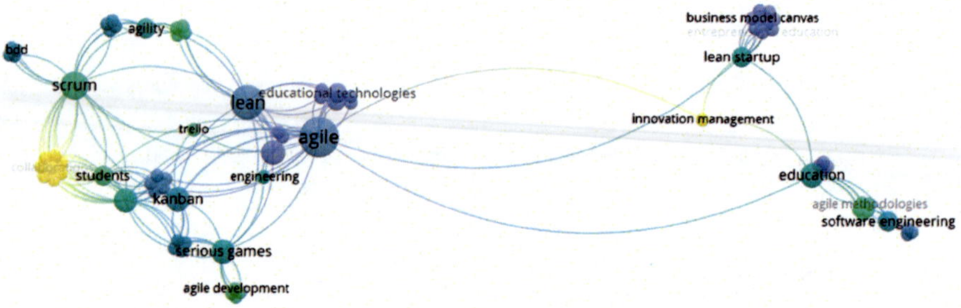

Fig. 2. Word co-occurrence network for Lean-Agile Education.

The size of nodes can reflect the frequency of keywords and the higher frequency of keywords, the larger size of the node. In addition, the thickness of the connecting line between two nodes demonstrates the relevance of the connection, the thicker the line, the more often these two words connect. Therefore, besides lean and agile, scrum stands out with thicker lines, referring to the agile framework most applied by the market and, also simpler to experiment, reflecting in studies about Agile Education that use Scrum as a teaching resource [3, 4]. Words in purple are more present in older publications, before 2012, while words in yellow are more recent (after 2012) and are locally close to clusters. Innovation management, for example, is a new expression in this field of study, with fine lines.

Figure 3 shows a zoom in the main words used in the most recent studies (yellow keywords) that represent the collaboration between industry and education in Lean-Agile methodologies. The connection between scrum and students brings collaborations, experience, and curriculum for the network. These relationships are directly related to what is expected for Lean-Agile Education.

To assess the distribution of researchers, the productivity law created by Alfred Lotka was analyzed. Only six authors had more than one publication in this field of study (from 2000 to 2021). Thus, an almost homogeneous group of authors and little connection with this theme was identified, with 92.4% of the authors responsible for the publication of only one paper, while the other 6 authors wrote two documents in the same period. As a result, this field of study does not comply with Lotka's Law, in which approximately 9 authors were expected to produce two publications, 4 authors to three publications, and about 60% of authors produced only one publication [13].

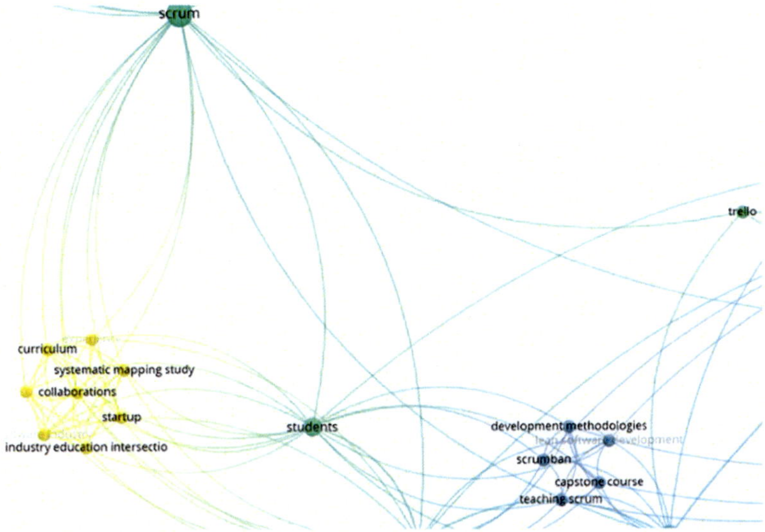

Fig. 3. Zoom of the word co-occurrence network for Lean-Agile Education standing out as the main words used in the most recent studies.

The ten main authors' affiliations (Table 2) represent 61.9% of the affiliations according to all analyzed articles. Aalto University(Finland) stands out with 3 papers and others institutions in Table 2 with 2 publications over 21 years; other 26 affiliations appeared just once. Considering all affiliations mentioned, US(12), Finland(4), Spain(3), Greece(3), India(3), Germany(2), Austria, Brazil, Colombia, Czech Republic, England, Estonia, Japan, New Zealand, Norway, China, Portugal, Russia, Serbia, Slovenia, Thailand and Turkey (1 mention each) are countries in this field of study.

US and Finland stand out in the education field, with innovative management models [14], while India is the only developing country with more than one affiliation related, suggesting that educational intervention is a huge opportunity for research in emerging countries [15, 16]. It is noticeable that UK universities, references in Lean Education [17], still do not present a similar and innovative concept connected with Lean-Agile Education, according to the analysis carried out in this article.

Regarding publishers, most publications (85.3%) occurred in proceedings papers and only 5 are journal articles. Therefore, as identified by the Biblioshiny software, the three main publishers in this field of study are '12th International Technology, Education and Development Conference (INTED)', '2016 IEEE Frontiers in Education Conference (FIE)' and 'Journal of Systems and Software', with two publications each.

This field of study does not meet the Bradford Law [13], which identifies a group of publishers focused on the research field, responsible for the concentration of publications.

Table 2. Top 10 authors' affiliation according to publications on Lean-Agile Education

Affiliation	Country	Item	%
Aalto University	Finland	3	8.8%
American Society for Engineering Education	US	2	5.9%
International Institute of Information Technology Hyderabad	India	2	5.9%
Purdue University	US	2	5.9%
Purdue University System	US	2	5.9%
Purdue University West Lafayette Campus	US	2	5.9%
University of Eastern Finland	Finland	2	5.9%
University of Minnesota System	US	2	5.9%
University of Minnesota Twin Cities	US	2	5.9%
University of Thessaly	Greece	2	5.9%

With these three main publishers with a maximum of two publications, it is visible that publications on this topic are dispersed distributed by publishers.

4 Conclusion

It was found that the joint application of Lean and Agile is still reduced in the education field and it is contemporary, predominantly in the last 10 years, with only 34 results globally in the Web of Science referring to specific applications in higher education. It was identified the predominance of studies related to Lean and Agile in technology industries and manufacturing companies, connected with the origin of both concepts and dissemination of practices and tools. Therefore, this analysis points to the opportunity for the development of new studies and dissemination of Lean-Agile Education, being a starting point for academic and practical applications in this context of the evolution of the digital age, mainly by developing countries.

More recent articles have pointed out the connection between Agile methodologies, such as Scrum, and education topics, such as curriculum development. It was also found that this field of study does not comply with the classical laws of bibliometrics. One of the limitations of this study is since only one database was used, being of interest for future studies to search in other scientific databases.

References

1. Valamede, L., Akkari, A.: Lean 4.0: A new holistic approach for the integration of lean manufacturing tools and digital technologies. Int. J. Math. Eng. Manage. Sci. **5**, 851–868 (2020)
2. Sturdy, A., Wright, C., Wylie, N.: Managers as consultants: the hybridity and tensions of neo-bureaucratic management. Organization **23**(2), 184–205 (2016)

3. Stachowiak, A., Oleśków-Szłapka, J.: Agility capability maturity framework. Procedia Manufacturing **17**, 603–610 (2018)
4. López-Alcarria, A., Olivares-Vicente, A., Poza-Vilches, F.: A systematic review of the use of agile methodologies in education to foster sustainability competencies. Sustainability **11**(10), 2915 (2019)
5. Hines, P., Holweg, M., Rich, N.: Learning to evolve: a review of contemporary lean thinking. Int. J. Oper. Prod. Manag. **24**(10), 994–1011 (2004)
6. Gupta, S., Sharma, M.: Lean services: a systematic review. Int. J. Product. Perform. Manag. **65**(8), 1025–1056 (2016)
7. Stone, K.: Four decades of lean: a systematic literature review. Int. J. Lean Six Sigma **3**(2), 112–132 (2012)
8. Hui, J., et al.: A bibliometric analysis of international publication trends in premature ejaculation research (2008–2018). Int. J. Impot. Res. **33**, 86–95 (2021)
9. Dahlgaard, J., Østergaard, P.: TQM and lean thinking in higher education. The Best on Quality: Targets. Improvements, Syst. **11**(1), 203–226 (2000)
10. Van Eck, N., Waltman, L.: Software survey: VOSviewer, a computer program for bibliometric mapping. Scientometrics **84**(2), 523–538 (2009)
11. Aria M, Cuccurullo C (2017) Bibliometrix: An R-tool for comprehensive science mapping analysis. Journal of Informetrics 11(4):959–975, Elsevier, USA
12. Araújo, C.: Bibliometria: evolução histórica e questões atuais. Em questão **12**(1), 11–32 (2006)
13. Ferreira, A.: Bibliometria na avaliação de periódicos científicos. DataGramaZero-Revista de Ciência da Informação **11**(3), 1–9 (2010)
14. Sahlberg, P.: Finnish Lessons 3. 0: What can the World Learn from Educational Change in Finland? 3rd ed. Teachers College Press, Columbia University, New York (2021)
15. International Statistical Institute: ISI webpage (2020). https://www.isi-web.org/capacity-building/developing-countries. Accessed 29 Oct 2021
16. Ganimian, A., Murnane, R.: Improving education in developing countries: lessons from rigorous impact evaluations. Rev. Educ. Res. **86**(3), 719–755 (2016)
17. Sfakianaki, E., Kakouris, A.: Lean thinking for education: development and validation of an instrument. Int. J. Qual. Reliability Manage. **36**(6), 917–950 (2019)

Glioma Brain Tumor Segmentation in Four MRI Modalities Using a Convolutional Neural Network and Based on a Transfer Learning Method

Nazanin Tataei Sarshar[1], Ramin Ranjbarzadeh[2(✉)], Saeid Jafarzadeh Ghoushchi[3], Gabriel Gomes de Oliveira[4], Shokofeh Anari[5], Mahboub Parhizkar[6], and Malika Bendechache[7]

[1] Department of Engineering, Islamic Azad university Tehran North Branch, Tehran, Iran
ab.tataee@gmail.com

[2] Department of Telecommunications Engineering, Faculty of Engineering, University of Guilan, Rasht, Iran
ramin.ranjbarzadehkondrood2@mail.dcu.ie

[3] Faculty of Industrial Engineering, Urmia University of Technology, Urmia, Iran
s.jafarzadeh@uut.ac.ir

[4] School of Electrical and Computer Engineering, State University of Campinas, Campinas, Brazil
oliveiragomesgabriel@ieee.org

[5] Department of Accounting, Economic and Financial Sciences, Islamic Azad University, South Tehran Branch, Tehran, Iran
shokofehanarii@gmail.com

[6] Department of Basic Science, Faculty of Basic Science, Central Tehran Branch, Islamic Azad University, Tehran, Iran
mah.parhizkar.sci@iauctb.ac.ir

[7] School of Computing, Faculty of Engineering and Computing, Dublin City University, Dublin, Ireland
malika.bendechache@dcu.ie

Abstract. Accurate segmentation of brain tumors from magnetic resonance imaging (MRI) images is still a challenging task for many applications in the medical domain. To gain a better segmentation outcome, it is vital to employ more information from brain tissue, which is available in different modalities. So, in this study, to segment brain tumors in MRI images, a new four different feature extraction Convolutional Neural Network (CNN) architecture is utilized. Also, we used four image modalities, including T1, T2, T1-c, and FLAIR. Weights and biases from a ResNet-50 network were imported to our structure to increase the segmentation accuracy. We designed two main building blocks to extract additional features from each modality. Our model uses both local and global features by defining two different sizes of patches. The proposed framework was evaluated on the BRATS 2018 dataset and showed that our architecture has a competitive dice score (0.9223, 0.8993, and 0.9211 for Enhanced, Whole, and Core tumor areas, respectively).

Y. Iano et al. (Eds.): BTSym 2021, SIST 207, pp. 386–402, 2023.
https://doi.org/10.1007/978-3-031-04435-9_39

Keywords: Brain tumor · Inception structure · Image segmentation · MRI · Transfer learning

1 Introduction

Cells are the basic and smallest biological block of the human body that could be grown and divided through a process, which is identified as cell division. In other words, new cells are normally created when the body requires them wherever old cells are dead or damaged. Every genetic change that is interfered with this regular process could be the reason for cancer. Two major factors that cause cancers are internal factors, including inherited mutation, hormones, and immune conditions as well as environmental acquired factors. Specifically, cancer is a disease in which cells are grown and divided uncontrollably in the body. These abnormal cells would form tumors, which are lumps of tissue. There are variant types of cancers, which are named based on organs or tissues where they begin [1–3].

A brain tumor (BT) is an abnormal mass in the brain that is graded with respect to the speed of growth and how likely it would grow back when treatment is terminated. The brain tumor is divided into benign (non-cancerous) with low grade (1–2) or malignant (cancerous) by high grade (3–4) based on the nature of the constructed cells. The tumor would originate in the brain tissue or spread to another part of the brain (metastasis) [4,5]. It is essential to mention that two main types of cancerous tissue are primary tumor, which is initiated within the brain, and secondary tumors, which occur when cancer cells spread uncontrollably from another organ to the brain and are also known as a metastatic brain tumors. There are different symptoms of brain tumors, including headaches, seizures, problems with vision, vomiting, and mental changes that modify depending on the part of the involved brain besides difficulty for walking or speaking. Unconsciousness happens by progress in diseases. Patients frequently experience strict functional impairment and cognitive dysfunction while the medical outcomes are improved. Therefore, patients with this type of cancer are made greater dependent and hopeless than others.

Brain tumors in children are the reason for one-quarter of overall cancer deaths according to research by the National Brain Tumor Foundation (NBTF) in the United States. Moreover, the number of brain tumor mortality is increased significantly to around 300% in the last three decades. Primary brain tumors are originated in neuroepithelial tissue that would be affected in any part of the brain. This type of tumor usually occurs in the posterior cranial cavity for children while it is initiated in the anterior two-thirds of the cerebral hemisphere for adults [6]. However, the most common primary brain tumors are Glioma (50.4%), meningioma (20.8%), pituitary adenoma (15%), and nerve sheath tumors. Gliomas are the most prevalent deadly malignant tumors along with various grades. In other words, the rate of survival in patients would be determined by the grades of Gliomas. Particularly, High-Grade Glioma (HGG) and Low-Grade Glioma (LGG) are two main categories of Gliomas, and HGG is more aggressive and low infiltrative in comparison to LGG. The condition of the tumor has a critical role in diagnosis and treatment [7,8].

Medical imaging is considered a central way to diagnose tumors. In fact, extracting meaningful and accurate information from medical images while minimizing possible error, is the primary objective of image processing in this field. Computed tomography (CT) and magnetic resonance imaging (MRI) are usually used as a diagnosis of brain tumors along with medical examination. In addition, confirmation of the result is performed by a biopsy which is taken a small part of tissue for testing in the laboratory. This procedure is the only way to determine surely the presence of a tumor in an area of the body. Nevertheless, magnetic resonance images (MRI) and other forms of medical imaging modalities have a vital role in the early detection of brain tumors and beginning the treatment. Great image resolution and speed of acquisition in MRI are the cause of the dominance of this modality usage [9].

In the medical imaging field, computer-aided diagnosis (CAD) is considered as an assistant for radiologists and doctors to interpret the medical images. It is a computer-based system that combines elements of artificial intelligence and computer vision to developed methods for the detection and pattern recognition of images [10,11]. The output of CAD analysis with sufficient performance in terms of sensitivity and specificity could be taken into account as a "second opinion" in making diagnostic decisions. The other point to mention is that MRI result in automated medical diagnostic systems is usually better than computed tomography (CT) due to providing more contrast between diverse soft tissues of the body and high-resolution spatial images [7]. Brain tumor detection and segmentation is not only the most challenging task in medical image processing but also is a time-consuming issue because of variety in form of shape, size, location, and image intensities. Thus, an automatic computer-aided system for brain tumors is identified as a desirable method to improve the speed of detection and segmentation. Accurate segmentation of MRI within CAD applications has a key role to provide more information on the soft tissue to detect brain tumors. Generally, the tumor core is visible in T2 and T1-c modalities, the necrotic parts can be identified employing contrast-enhanced T1 modality, the whole tumor is mainly observable in T2 and FLAIR (Fluid-attenuated inversion recovery) images, and T1 is usually utilized for identifying healthy tissues [7,12]. Four different modalities used in this study are Flair, T1, T1C, and T2, which are indicated in Fig. 1 from left to right.

Recently, deep learning models have demonstrated a high ability in computer-based applications. These architectures have obtained outstanding outcomes in the field of detection, segmentation, recognition, and classification tasks. Basically, the current automated segmentation techniques can be labeled into two major classes:

1. Hand-crafted approaches, which need to define a powerful strategy to extract some pre-defined features by an expert person. These kinds of segmentation techniques, by employing some explored features of the input image (or images), can be trained as a discriminative structure. The Hand-crafted

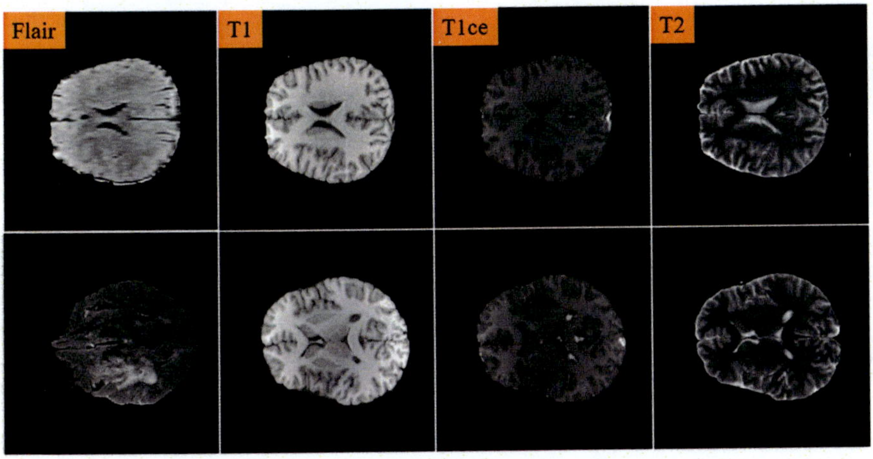

Fig. 1. Eight samples of four MRI modalities from two patients.

pipelines normally work with many classifiers including clustering techniques [13], Random Forest [5], Nested Neural Network [14], and Sparse Representation [15].

2. Deep learning (DL) algorithms, which are capable of exploring key textural information and classifying them automatically. This ability routes in the fact that these neuron-based models mimic the behavior of the brain.

Recently, deep learning (DL) methods have become the attractive field of artificial intelligence that outperforms hand-crafted computer vision strategies in many applications. Zeineldin et al. [12] suggested a modular decoupling pipeline based on two connected core parts that employs an encoding and decoding relationship. For extracting spatial information and to get the full-resolution probability map, an encoder and a decoder structure were used, respectively. A combination of a pre-trained convolution-base of VGG16 (transfer learning) and a fully connected classifier based on the U-net proposed by [16] for tumor segmentation. Nema et al. [17] suggested a residual cyclic unpaired encoder-decoder network (RescueNet) employing mirroring and residual principles. Their model employs an unpaired adversarial training approach for segmenting the different parts of the brain tumor in an MRI scan. Zhou et al. [18] developed an efficient 3D residual neural network (ERV-Net), which has less GPU memory consumption and computational complexity. They used a combination of the Cross-entropy loss and Dice loss to overcome the issues of data imbalance and model convergence. Also, to develop the efficiency of ERV-Net and diminish GPU memory, a 3D ShuffleNetV2 block is employed as an encoder.

Owing to the characterization of the CNN structures to extract more sufficient features of each tissue, in this study, we develop a CNN-based pipeline to amend the accuracy of semantic tumor segmentation in MRI images. Figure 1 demonstrates four MRI modalities that have been employed in this study. The main contributions of this paper are: (1) Suggest a multi-route feature extraction model; (2) Employ a transfer learning method to bring learned weights and biased to the model; (3) Address the issue of multi modalities input images for brain tumors; (4) Better segmentation outcomes are achieved.

This study is structured as follows. In Sect. 2.1, a pre-processing strategy of four MRI modalities is explained. In Sect. 2.2, the augmentation of data is produced. In Sect. 2.3, the proposed architecture of the CNN model is described. In Sect. 3, results and discussion are described. The experiment and concluding remarks are explained in Sect. 4.

2 Material and Methods

In this part, the procedure of our pipeline is demonstrated in detail.

2.1 Pre-processing

Accurate segmentation of brain tumors from magnetic resonance imaging (MRI) images is still a challenging task for many applications in the medical domain. To gain a better segmentation outcome it is vital to employ more information from brain tissue available in different modalities. So, in this study, to segment brain tumors in MRI images, a new four different feature extraction Convolutional Neural Network (CNN) architecture is utilized. Also, we used four image modalities including T1, T2, T1-c, and FLAIR. Weights and biases from a ResNet-50 network were imported to our structure to increase the segmentation accuracy. We designed two main building blocks to extract additional features from each modality. The proposed framework was evaluated on the BRATS 2018 dataset and showed that our architecture has a competitive dice score (0.9223, 0.8993, and 0.9211 for Enhanced, Whole, and Core tumor areas, respectively).

Due to the difference of modalities, devices, and patients, anisotropic characteristics of the MRI volumes (especially for the FLAIR modality), normalization is a core building block for the training of a deep learning model. Also, it is perceptible that the volume resolution is not consistent across subjects. To address the mentioned issues, a pre-processing normalization strategy is utilized to accelerate training and prevent over-fitting [27]. To perform the normalization, it is necessary to obtain an image with zero mean and unit variance. Because of the variance in intensity profile between modalities, the normalization needs to be implemented independently for each modality, as follows.

$$Z = \frac{x - \mu}{\sigma} \tag{1}$$

where σ is the standard deviation, x is the current pixel intensity, and μ is the mean of the intensity of each slice. The results of the normalization are presented in Fig. 2.

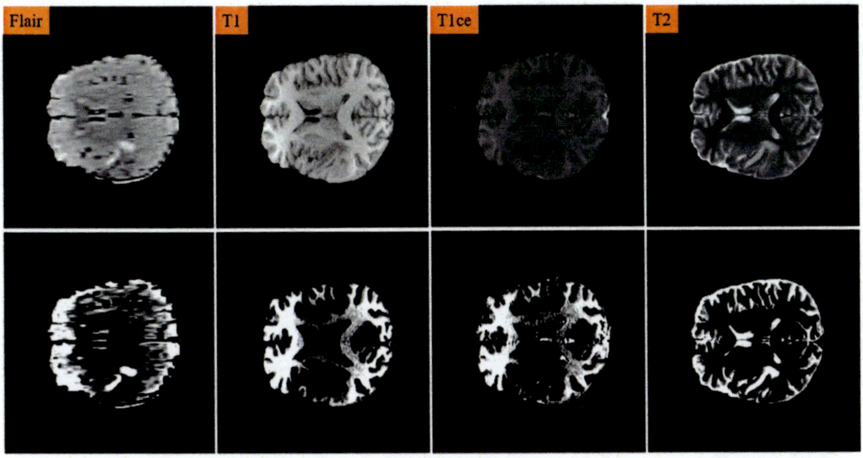

Fig. 2. The results of applying the normalization approach to the BRATS 2018 dataset. First row and second row indicate the original and normalized images, respectively.

2.2 Data Augmentation

We employed the random scaling deformations, brightness transformation, mirror transformation, random elastic deformations, vertical flipping, gamma correction, random rotations, Gaussian noise, and random intensity variation strategies of data augmentation in this study [19]. Owing to the natural symmetrical shape of the brain, vertical flipping, which mirrors all pieces along the horizontal axis is applied. Since there is a high-intensity variation between MRI scans, we employ a random intensity variation method. As demonstrated in [20], the standard deviation of the FLAIR slices is much bigger than the mean (by almost a factor of 10), we perform the intensity variation strategy on the normalized dataset.

2.3 Proposed Structure

Although noteworthy progress has been achieved in the field of brain tumor segmentation and numerous DL pipelines have been implemented, there is still room for improvement [21–24]. Firstly, the position, shape, orientation, and dimension of gliomas differ meaningfully among patients, making them difficult to segment. Secondly, as the tumor regions obtained in MRI images are minor parts inside the brain, there is a class imbalance between the background region and the tumor regions [7,25]. In this study, to address the mentioned problems, a new pipeline is suggested, which used a pre-trained ResNet-50 architecture on ImageNet at the beginning of the model [26,27]. The proposed model is indicated in Fig. 3.

Generally, in a deep convolutional neural network, numerous layers are stacked to train the hyperparameters, and the network learns some low/mid/high-level features at the end of the model [7]. The architecture of the ResNet-50 comprises 5 steps, including convolution layers (Conv layers) and identity blocks, and comprises over 23 million trainable parameters. The ResNet-50 structure includes 50 layers deep and capable of classifying images into 1000 object categories, such as pencil, keyboard, and many animals. Also, to improve the efficiency of the segmentation process, our model greatly benefited from four images modality T1, T1-c, T2, and FLAIR.

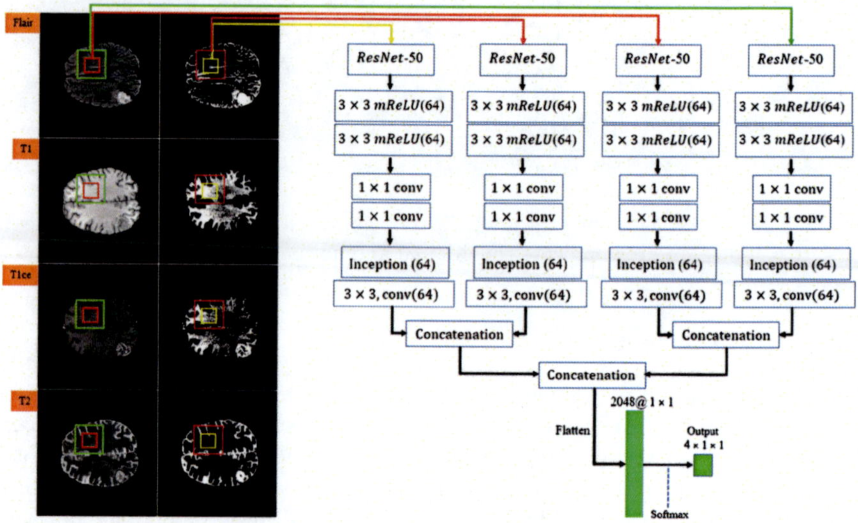

Fig. 3. Our implemented model using four MRI modalities and their corresponding Z-score normalized images.

Our pipeline contains four MRI modalities and their corresponding Z-score normalized images. As various modalities undeniably contain complementary and detailed information, our investigational outcomes imply that our multi-modality strategy is effective in improving segmentation performance. In order to extract more unique details from the input images, we implemented our structure based on a cascade feature extraction model (four paths). Each route is responsible for exploring features inside a specific patch with a specific center of each pixel. It means for each pixel inside the image we can consider a local and a global patch. The area around the pixel is considered as a local vicinity and global neighborhood. In this work, the size of global and local patches is 128 × 128 and 25 × 25, respectively. Inside each modality, we defined two patches for extracting both global and local key information. In order to overcome the drawbacks of large patches that are difficult for neural networks to learn local information, we use both small and semi-large patches. Our model extracts all key features from 4 routes and then these features are concatenated.

In this study, for extracting high-level visual features from each modality (low-level and features were explored by ResNet-50 model) and performance improvements, a 3×3 new.mReLU block is implemented. The output of each ResNet-50 model in four routes is fed to the 3 × 3 new.mReLU block. To obtain a better dice score, two 3 × 3 new.mReLU blocks are used sequentially.

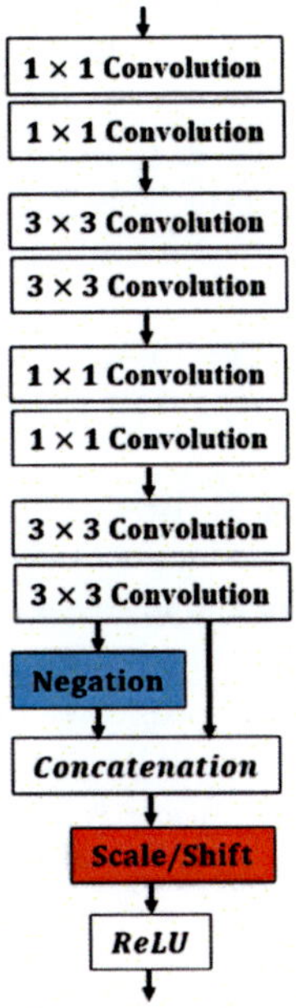

Fig. 4. Our proposed 3 × 3 new.mReLU.

Our proposed 3 × 3 new.mReLU pipeline is clarified in detail in Fig. 4. The capacity intensification of the system to learn multiple patterns using intermediate layers in a convolutional neural network has become the source of motivation for this architecture [28,29]. It has been observed that to recognize a broad varying-scale tumor, employing stacking up new.mReLU architecture is more powerful than a simple linear chain of convolution layer. Taking advantage of this property, 8 Conv layers with small receptive fields have been employed, and they are followed by a negation layer. We used small 3 × 3 filters to overcome the overfitting issue and allow the utilization of deeper architectures. The small kernel in convolution layers is chosen for detecting key features in a small area, which leads to differentiating between areas of the tumor more accurately [29]. We applied two convolutional layers before the negation layer to enhance the feature extraction precision by using the experimental technique. Another important layer is a 1 × 1 Conv layer (bottleneck layer) that considerably reduces the computation procedure. Although a 1 × 1 kernel is not able to learn any spatial patterns inside the input image, patterns across the depth (cross channel) of the image can be learned [30,31]. Consequently, not only does a 1×1 Conv layer (bottleneck layer) provide a dimension reduction technique, but it also provides the additional benefit of enabling the network to learn more [32,33].

The negation layer is responsible to multiply -1 into the output of the former layer for increasing the possibility of obtaining the distinct features. The output of the last two convolutional layers and the negation layer are fed into a concatenation layer. By applying a concatenation layer, all feature maps tend to be paired with their reverse sign. This strategy can effectively improve the generalization ability of the structure and avoid overfitting. Moreover, we employed the Scale/Shift layer to the output of the concatenation layer to apply some trainable weights and biases for better labeling one pixel to one of the predefined classes [31,33]. Since the tumor segmentation efficiency relies heavily on considering all differences in all four modalities, the proposed 3×3 new.mReLU block plays a key role in our pipeline. In order to cope with the issues of tumor volume variations, the proposed new.mReLU block has indicated the improved accuracy of the final system output.

To achieve a better dice score in brain tumor segmentation and diminish the overall number of false positives and false negatives, it is being essential to survey all regions with a variety of dimensions and shapes. To meet this, a new inception structure to obtain the multi-scale nature of tumor segmentation tasks maximally is proposed [33,34]. Our proposed inception block in high detailed descriptions is represented in Fig. 5, which is intended to clarify overall concepts. The core idea behind our new inception block has been inspired by some of the new works in detection and recognition fields [30,31]. The proposed new inception structure is responsible for maintaining the sparsity of the pipeline and enhancing the computational efficiency by decreasing the number of feature filters in each layer. In this structure, firstly, three parallel bottleneck layers (1×1 convolutional layers) with fewer neurons have been applied for nonlinear dimensionality reduction of the input feature maps produced in the former layer. Then, in each path, a convolutional layer with a different receptive field is applied to cover all possible states of tumor shape. Different receptive fields enable the model to learn different spatial patterns at various scales. Next, the output of the first two feature extraction layers is concatenated. This procedure is followed by a bottleneck layer. Then the output of the previous stage is concatenated with the third feature extraction layer. Finally, a bottleneck layer is applied again.

This procedure is implemented for all 4 feature extracting routes from 2 local and 2 global patches. For the next step, all extracted feature maps need to be concatenated and then flattened. As in the last Conv layer in CNN pipelines, the receptive fields do not explore the whole spatial dimension of the input image, the obtained key features by the former convolutional layer are related to a part of the input image. Accordingly, at least one Fully-Connected layer (FC) needs to be employed in such a scenario [35,36]. The FC layers allow the system to learn the non-linear features in an input image. The number of neurons employed in the suggested architecture for the FC layer is 2048. Softmax regression (also called Multinomial Logistic) layer is the latest layer for classification tasks in a

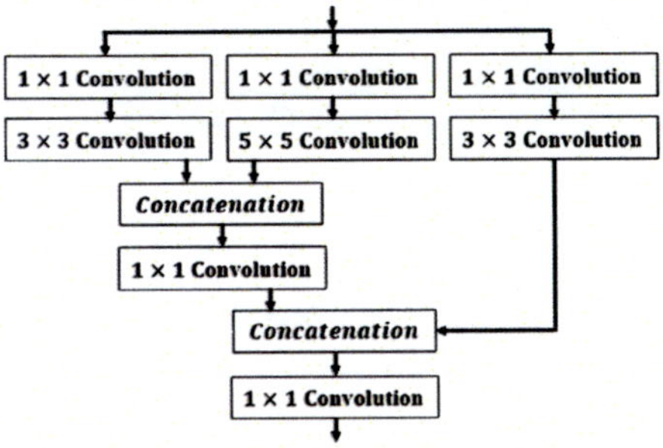

Fig. 5. Our proposed inception architecture.

CNN-based system that is responsible for recognizing one class from the others. Finally, by employing the Softmax layer, the classification of each selected patch for its center is accomplished. It means that the center of each patch is labeled into one predefined class.

As in most cases, only 2% of whole brain tissue covers brain tumors, the significant problem of dealing with the BRATS datasets can be considered as the data imbalance. Additionally, convergence issues can be observed in loss functions employed to overcome the imbalanced issue [17,37].

To our best knowledge, one of the popular and most commonly utilized loss functions for segmentation and classification is Cross-entropy. However, it cannot reach a high level of accuracy when dealing with an imbalanced dataset. Moreover, the Dice loss function aims to optimize the similarity between ground truth and segmentation outcomes of neural networks. Consequently, to compensate for the lack of Cross-entropy we can use Dice loss to deal with the imbalanced data. However, it is prone to convergence problems in shallow models.

To overcome the issues of data imbalance and model convergence, a combination of the Cross-entropy loss and Dice loss has been used in this study. To diminish the impact rooted in imbalanced data for the segmentation task, a multi-class Dice loss function was employed. Moreover, the Softmax function is employed for normalizing the output of the Fully Connected (FC) layer before applying this loss function. The proposed Dice loss function is defined as [18]:

$$Loss_{Dice} = -\frac{2}{|C|}\sum_{j\in C}\frac{\sum_i P_i^j GT_i^j}{\sum_i P_i^j + \sum_i GT_i^j} \tag{2}$$

where i indicates the number of voxels in the training patch, j is one of all the classes C in the BRATS 2018 data, P implies the probability of Softmax acquired by model, and GT demonstrates the ground truth.

While data is exceedingly imbalanced, convergence problems can easily occur in Dice loss. By formulating the Dice loss function as Eq. (3), its differentiable formula can be obtained using Eq. (4), as follow:

$$Loss_{Dice} = \frac{2 \times P \times GT}{P + GT} \quad (3)$$

$$Loss_{Dice} = \frac{2 \times GT^2}{(P + GT)^2} \quad (4)$$

Equation (4) implies that, by applying the small number of P and GT, a larger gradient will be occurred and lead to an unstable condition during the training procedure. To overcome this issue, a Cross-entropy loss function is introduced. Finally, the sum of Dice and Cross-entropy loss functions are utilized as the total fusion loss as [18]:

$$Fusion_{Loss} = -\frac{2}{|C|} \sum_{j \in C} \frac{\sum_i P_i^j GT_i^j}{\sum_i P_i^j + \sum_i GT_i^j} - \frac{1}{num} \sum_{i \in num} \sum_{j \in C} GT_i^j \log(P_i^j) \quad (5)$$

where num indicates the number of voxels in a patch.

3 Experiments

3.1 Evaluation Indexes

The scores employed to assess the performance of our strategy and the given references are Dice similarity, Specificity, and Sensitivity. Dice score (Eq. 6) is a criterion to evaluate how closely our architecture matched the training dataset's hand-annotated ground truth segmentation that is achieved by overlapping predicted labels with actual labels [7,24].

$$DICE = 100\% * \left(\frac{2 \times TP}{2TP + FP + FN}\right) \quad (6)$$

where TP, FP, TN, and FN represent true positive, false positive, true negative, and false negative, respectively. Sensitivity (also known as a true positive) is the representative of tumor pixels that have been correctly distinguished and is given by:

$$Sensitivity = \frac{TP}{TP + FN} \quad (7)$$

Specificity is the representative of non-tumor pixels that have been distinguished precisely and is given as:

$$Specificity = \frac{TN}{TN + FN} \quad (8)$$

3.2 Results and Discussion

In this study, we employed four different MRI modalities from the BRATS 2018 dataset for training, validation, and testing the proposed structure [6]. This public dataset includes Multimodal MRI images and patients' clinical data and is skull-stripped and resampled to $1 \times 1 \times 1$ mm isotropic resolution. Also, the T2-weighted, T1-weighted, FLAIR, and T1 with gadolinium-enhancing contrast (T1-c) with the size of $240 \times 240 \times 155$ have been utilized for both testing and training phase that clinically captured in an anticipated manner in several institutions. Labeling of the training samples are conducted by medical experts and are comprised of 3 dissimilar brain tumor including 1) the tumor core, 2) the enhancing tumor, and 3) the edema. The procedure of image obtaining is accomplished utilizing different scanners, magnetic field strengths, and various protocols.

The number of 210 HGG cases and 75 cases with LGG of the dataset has been employed to train the proposed model. Additionally, 20% of the training data is reserved for evaluating the accuracy of the pipeline in the testing phase. Furthermore, we employed 66 cases of the dataset for the validation phase.

The experimental results were obtained using Python3 with the Keras library over CUDA 9.0 and CuDNN 5.1 on Intel Core I7- 3.4 GHz, 32 GB RAM, 15 MB Cache, and GPU 1080Ti NVIDIA computer under a 64-bit operating system. The validation of our suggested strategy was verified by assessing the outcomes of different state-of-the-art models on the BRATS 2018 dataset.

To have a clear understanding for visual comparison purposes and to evaluate the semantic tumor segmentation performance, five state-of-the-art structures were implemented (Multi-task Learning [38], RescueNet [17], ERV-Net [18], HOG [39], and Fusion of features [40]).

Table 1 indicates quantitative evaluations that emphasize there is a small difference between the proposed strategy and Multi-task Learning [38] and RescueNet [17] in all indexes, but our pipeline attains more efficiency. From Table 1, it can be observed that our approach, Multi-task Learning [38], and RescueNet [17] methods are superior to other methods in Dice. Our pipeline exhibits better segmentation performance than Multi-task Learning [38] and RescueNet [17] with enhancing the sensitivity score by 2% for Enh, 3% and 0.2% for Whole, and 1% for Core. The mean Specificity of the Multi-task Learning [38] and RescueNet [17] techniques are very similar to our technique; however, these models have a significantly different Sensitivity. Also, it is observed that the Fusion of features [40], HOG [39], and ERV-Net [18] structures gained low sensitivity scores than the other three techniques with a significant margin.

Figure 6 visually shows outcomes obtained by the proposed technique on the BRATS 2018 dataset. Owing to the employ of transform learning (ResNet-50) and applying two new building blocks namely inception and new.mReLU, the proposed structure was enabled to extract more unique contextual information of the tumor that leads to a better segmentation outcome. The mean of dice indexes of our approach for the Core, Whole, and Enhanced regions are 0.9211, 0.8993, and 0.9223, respectively.

Table 1. BRATS 2018 tumor labeling outcomes.

Model	Dice Score (mean)			Sensitivity (mean)			Specificity (mean)		
	Enh.	Whole	Core	Enh.	Whole	Core	Enh.	Whole	Core
Multi-task Learning [38]	0.9213	0.8939	0.9122	0.9369	0.9362	0.9221	0.9815	0.9820	0.9822
RescueNet [17]	0.9220	0.8962	0.9194	0.9383	0.9627	0.9201	0.9808	0.9782	0.9715
ERV-NET [18]	0.8181	0.8912	0.8662	0.8520	0.9139	0.8558	0.8847	0.9088	0.9134
HOG [30]	0.8838	0.8066	0.8075	0.9025	0.8829	0.8531	0.9124	0.9073	0.9035
Fusion of features [40]	0.8573	0.8485	0.8818	0.8613	0.9287	0.8825	0.9233	0.9185	0.9359
Proposed method	**0.9223**	**0.8993**	**0.9211**	**0.9513**	**0.9641**	**0.9318**	**0.9834**	**0.9826**	**0.9859**

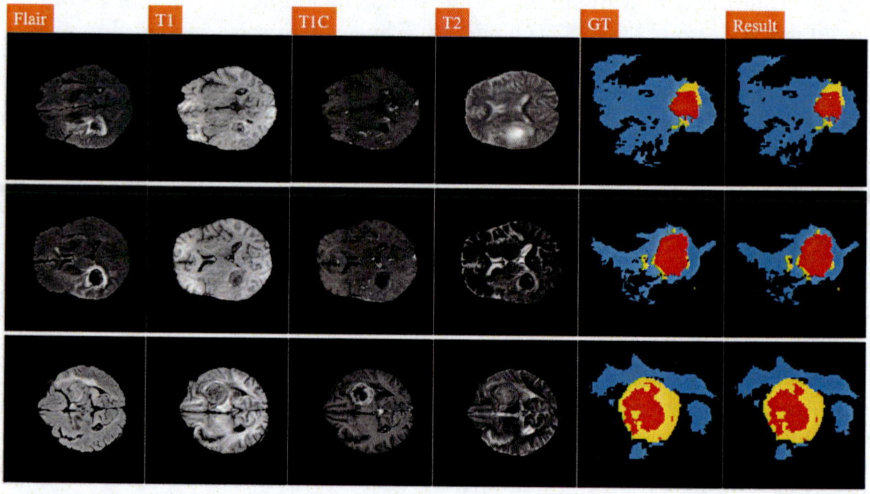

Fig. 6. Visual comparison of segmentation using suggested pipeline and Ground Truth (the blue, red, and yellow are edema, core, and enhanced areas, respectively).

Figure 7 indicates the difference between the proposed technique and three state-of-the-art models in the BRATS 2018 dataset. As can be seen, the HOG method [39] can roughly recognize the borders of the tumor area. This model under-segment the Edema regions and over-segment the Enhanced tumor area. The segmentation performance of Multi-task Learning [38] and RescueNet [17] models based on the Dice indices with regard to tumor Enh and Core regions are superior to ERV-Net [18], HOG [39], and Fusion of features [40]. As indicated in Fig. 7, employing the ERV-Net [18] pipeline reaches less segmentation accuracy compared to Multi-task Learning [38] and RescueNet [17] models. The RescueNet and Multi-task Learning models under-segment the Core and Edema areas, respectively. The RescueNet approach indicates more core matching with the GT compared to the HOG but still has insufficient accuracy, especially in Whole areas. The ERV-Net, HOG, and Fusion of features methods were unable to recognize the Whole and Enhanced regions properly. The proposed model and RescueNet are excellent in segmentation (the size and shape of the final segmented volume) in all regions, but ours demonstrate more accurately matching with the ground truth.

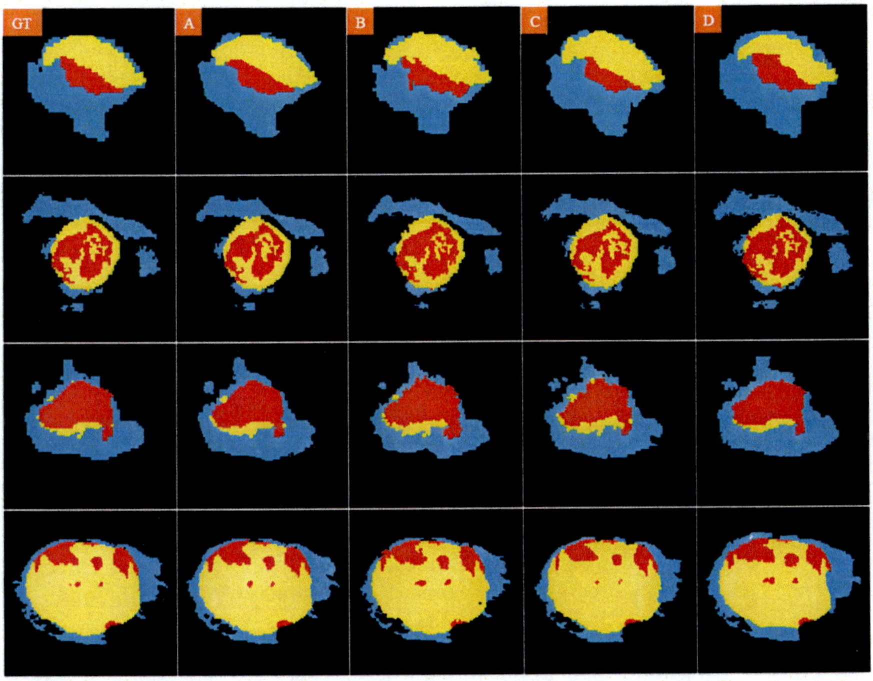

Fig. 7. Visual comparison between our model and the other strategies. The red, yellow, and blue colors represent core, enhanced, and edema, respectively. GT is the ground truth. Segmentation using the (a) Ours, (b) Multi-task Learning [38], (c) RescueNet [17], and (d) HOG [39].

4 Conclusion

In this study, we developed a new cascade feature extraction based on a CNN architecture that benefits from the characterization of the four MRI modalities. To achieve a significantly better outcome, a learned ResNet-50 (the transformed learning) has been employed to bring the learned weights and biases into the structure rather than random selection. Besides, we observed that normalization and bias correction is able to increase the accuracy of the segmentation vividly. Also, we suggested the 3×3 new.mReLU and inception blocks that enable the proposed model to get broad varying-scale tumor detection. We also developed a new loss function, that is the sum of Dice and Cross-entropy loss functions and utilized as the total fusion loss.

Acknowledgments. The author Malika Bendechache is supported, in part, by Science Foundation Ireland (SFI) under the grants No. 13/RC/2094_P2 (Lero) and 13/RC/2106_P2 (ADAPT).

References

1. Razmjooy, N., Razmjooy, S.: Skin melanoma segmentation using neural networks optimized by quantum invasive weed optimization algorithm. In: Razmjooy, N., Ashourian, M., Foroozandeh, Z. (eds.) Metaheuristics and Optimization in Computer and Electrical Engineering. LNEE, vol. 696, pp. 233–250. Springer, Cham (2021). https://doi.org/10.1007/978-3-030-56689-0_12
2. Do, N.-T., Jung, S.-T., Yang, H.-J., Kim, S.-H.: Multi-level seg-unet model with global and patch-based X-ray images for knee bone tumor detection. Diagnostics **11**(4), 691 (2021)
3. Ranjbarzadeh, R., Saadi, S.B.: Automated liver and tumor segmentation based on concave and convex points using fuzzy c-means and mean shift clustering. Measurement **150**, 107086 (2020)
4. Hu, A., Razmjooy, N.: Brain tumor diagnosis based on metaheuristics and deep learning. Int. J. Imaging Syst. Technol. **31**(2), 657–669 (2021)
5. Lefkovits, L., Lefkovits, S., Szilágyi, L.: Brain tumor segmentation with optimized random forest. In: Crimi, A., Menze, B., Maier, O., Reyes, M., Winzeck, S., Handels, H. (eds.) BrainLes 2016. LNCS, vol. 10154, pp. 88–99. Springer, Cham (2016). https://doi.org/10.1007/978-3-319-55524-9_9
6. Bakas, S., et al.: Identifying the best machine learning algorithms for brain tumor segmentation, progression assessment, and overall survival prediction in the BRATS challenge. arXiv Prepr arXiv:1811.02629 (2018)
7. Ranjbarzadeh, R., Kasgari, A.B., Ghoushchi, S.J., Anari, S., Naseri, M., Bendechache, M.: Brain tumor segmentation based on deep learning and an attention mechanism using MRI multi-modalities brain images. Sci. Rep. **11**(1), 1–17 (2021)
8. Yaqub, M., Feng, J., Zia, M.S., Arshid, K., Jia, K., Rehman, Z.U., Mehmood, A.: State-of-the-art CNN optimizer for brain tumor segmentation in magnetic resonance images. Brain Sci. **10**(7), 427 (2020)
9. Zhang, W., et al.: ME-Net: multi-encoder net framework for brain tumor segmentation. Int. J. Imaging Syst. Technol. **31**, 1834–1848 (2021)
10. Razmjooy, N., Estrela, V.V.: Applications of image processing and soft computing systems in agriculture. IGI Global (2019)
11. Abdel-Maksoud, E., Elmogy, M., Al-Awadi, R.: Brain tumor segmentation based on a hybrid clustering technique. Egypt Inform. J. **16**(1), 71–81 (2015)
12. Zhang, D., Shou, Y., Xu, J.: A mapreduce-based approach for shortest path problem in road networks. J. Ambient Intell. Humaniz. Comput., 1–9 (2018)
13. Zeineldin, R.A., Karar, M.E., Coburger, J., Wirtz, C.R., Burgert, O.: DeepSeg: deep neural network framework for automatic brain tumor segmentation using magnetic resonance FLAIR images. Int. J. Comput. Assist. Radiol. Surg. **15**(6), 909–920 (2020). https://doi.org/10.1007/s11548-020-02186-z
14. Zhuge, Y., Krauze, A.V., Ning, H., Cheng, J.Y., Arora, B.C., Camphausen, K., Miller, R.W.: Brain tumor segmentation using holistically nested neural networks in MRI images. Med. Phys. **44**(10), 5234–5243 (2017)
15. Li, Y., Jia, F., Qin, J.: Brain tumor segmentation from multimodal magnetic resonance images via sparse representation. Artif. Intell. Med. **73**, 1–13 (2016)

16. Naser, M.A., Deen, M.J.: Brain tumor segmentation and grading of lower-grade glioma using deep learning in MRI images. Comput. Biol. Med. **121**, 103758 (2020)
17. Nema, S., Dudhane, A., Murala, S., Naidu, S.: RescueNet: an unpaired GAN for brain tumor segmentation. Biomed. Signal Process. Control **55**, 101641 (2020)
18. Zhou, X., Li, X., Hu, K., Zhang, Y., Chen, Z., Gao, X.: ERV-Net: an efficient 3D residual neural network for brain tumor segmentation. Expert Syst. Appl. **170**, 114566 (2021)
19. Wang, F., Zhong, S., Peng, J., Jiang, J., Liu, Y.: Data augmentation for EEG-based emotion recognition with deep convolutional neural networks. In: Schoeffmann, K., Chalidabhongse, T.H., Ngo, C.W., Aramvith, S., O'Connor, N.E., Ho, Y.-S., Gabbouj, M., Elgammal, A. (eds.) MMM 2018. LNCS, vol. 10705, pp. 82–93. Springer, Cham (2018). https://doi.org/10.1007/978-3-319-73600-6_8
20. Benson, E., Pound, M.P., French, A.P., Jackson, A.S., Pridmore, T.P.: Deep hourglass for brain tumor segmentation. In: Crimi, A., Bakas, S., Kuijf, H., Keyvan, F., Reyes, M., van Walsum, T. (eds.) BrainLes 2018. LNCS, vol. 11384, pp. 419–428. Springer, Cham (2019). https://doi.org/10.1007/978-3-030-11726-9_37
21. Valizadeh, A., Jafarzadeh Ghoushchi, S., Ranjbarzadeh, R., Pourasad, Y.: Presentation of a segmentation method for a diabetic retinopathy patient's fundus region detection using a convolutional neural network. Comput. Intell. Neurosci. (2021)
22. Ranjbarzadeh, R., et al.: Lung infection segmentation for COVID-19 Pneumonia based on a cascade convolutional network from CT images. Biomed. Res. Int. (2021)
23. Razmjooy, N., et al.: Computer-aided diagnosis of skin cancer: a review. Curr. Med. imaging **16**(7), 781–793 (2020)
24. Aghamohammadi, A., Ranjbarzadeh, R., Naiemi, F., Mogharrebi, M., Dorosti, S., Bendechache, M.: TPCNN: two-path convolutional neural network for tumor and liver segmentation in CT images using a novel encoding approach. Expert Syst. Appl., 115406 (2021)
25. Iqbal, S., Ghani Khan, M.U., Saba, T., Mehmood, Z., Javaid, N., Rehman, A., Abbasi, R.: Deep learning model integrating features and novel classifiers fusion for brain tumor segmentation. Microsc. Res. Tech. **82**(8), 1302–1315 (2019)
26. Tian, Z., Shu, M., Lyu, P., Li, R., Zhou, C., Shen, X., Jia, J.: Learning shape-aware embedding for scene text detection. In: Proceedings of the IEEE/CVF Conference on Computer Vision and Pattern Recognition, pp 4234–4243 (2019)
27. He, T., Huang, W., Qiao, Y., Yao, J.: Text-attentional convolutional neural network for scene text detection. IEEE Trans. Image Process. **25**(6), 2529–2541 (2016)
28. Naiemi, F., Ghods, V., Khalesi, H.: A novel pipeline framework for multi oriented scene text image detection and recognition. Expert Syst Appl. **170**, 114549 (2021)
29. Hong, S., Roh, B., Kim, K.-H., Cheon, Y., Park, M.: PVANet: Lightweight deep neural networks for real-time object detection. arXiv Prepr arXiv:1611.08588 (2016)
30. Mansour, R.F., Aljehane, N.O.: An optimal segmentation with deep learning based inception network model for intracranial hemorrhage diagnosis. Neural Comput. Appl. **33**(20), 13831–13843 (2021). https://doi.org/10.1007/s00521-021-06020-8
31. Alom, M.Z., Hasan, M., Yakopcic, C., Taha, T.M., Asari, V.K.: Inception recurrent convolutional neural network for object recognition. Mach. Vis. Appl. **32**(1), 1–14 (2021). https://doi.org/10.1007/s00138-020-01157-3
32. Shang, W., Sohn, K., Almeida, D., Lee, H.: Understanding and improving convolutional neural networks via concatenated rectified linear units. In: International Conference on Machine Learning, PMLR, pp 2217–2225 (2016)

33. Szegedy, C., Vanhoucke, V., Ioffe, S., Shlens, J., Wojna, Z.: Rethinking the inception architecture for computer vision. In: Proceedings of the IEEE Conference on Computer Vision and Pattern Recognition, pp. 2818–2826 (2016)
34. Szegedy, C., et al.: Going deeper with convolutions. In: Proceedings of the IEEE Conference on Computer Vision and Pattern Recognition, pp 1–9 (2015)
35. Coşkun, M., Uçar, A., Yildirim, Ö., Demir, Y.: Face recognition based on convolutional neural network. In: 2017 International Conference on Modern Electrical and Energy Systems (MEES), pp. 376–379. IEEE (2017)
36. Mahmood, A., et al.: Deep learning for coral classification. In: Handbook of Neural Computation. Elsevier, pp 383–401 (2017)
37. Wu, S., Li, H., Quang, D., Guan, Y.: Three-plane-assembled deep learning segmentation of gliomas. Radiol. Artif. Intell. **2**(2), e190011 (2020)
38. Weninger, L., Liu, Q., Merhof, D.: Multi-task learning for brain tumor segmentation. In: Crimi, A., Bakas, S. (eds.) BrainLes 2019. LNCS, vol. 11992, pp. 327–337. Springer, Cham (2020). https://doi.org/10.1007/978-3-030-46640-4_31
39. Arunkumar, N., Mohammed, M.A., Mostafa, S.A., Ibrahim, D.A., Rodrigues, J.J.P.C., de Albuquerque, V.H.C.: Fully automatic model-based segmentation and classification approach for MRI brain tumor using artificial neural networks. Concurr. Comput. Pract. Exp. **32**(1), e4962 (2020)
40. Sharif, M., Amin, J., Raza, M., Yasmin, M., Satapathy, S.C.: An integrated design of particle swarm optimization (PSO) with fusion of features for detection of brain tumor. Pattern Recognit. Lett. **129**, 150–157 (2020)

Aztec Curve: Proposal for a New Space-Filling Curve

Diego Ayala(✉), Daniel Durini, and Jose Rangel-Magdaleno

Instituto Nacional de Astrofísica, Óptica y Electrónica, Puebla, Mexico
diegoayala002@gmail.com, {ddurini,jrangel}@inaoep.mx

Abstract. Different space-filling curves (SFCs) are briefly reviewed in this paper, and a new one is proposed. A century has passed between the inception of this kind of curve, and, since then, they have been found useful in computer science, particularly in data storage and indexing due to their clustering properties, being Hilbert curve the most well-known member of the family of fractals. The proposed Aztec curve, with similar characteristics to Hilbert's curve, is introduced in this paper, accompanied by a grammatical description for its construction. It yields the possibility of creating bi-dimensional clusters, not available for Hilbert or Peano curves. Additional to this, a case of application on the scope of Compressed Sensing is implemented, in which the use of Hilbert curve is contrasted with the Aztec curve, having a similar performance, and positioning the Aztec curve as viable and a new alternative for future exploitation on applications that make use of SFC's.

Keywords: Space-filling curve · Fractal · Memory-efficient clustering

1 Introduction

It is considered that the first space-filling curve (SFC) was presented in 1890 by G. Peano, who described a continuous curve that goes through every coordinate point of the unit square [1]. The following year D. Hilbert proposed his own curve [2]. Both curves pass through the center points of all inner squares that are enclosed by the outer square unit and exhibit self-similar patterns at smaller scales, being each scale referred to as an 'order' and represented as $n \in \mathbb{R}^+$. As there is no upper limit for n, the patterns can be replicated an infinite number of times without exceeding the boundary of the square unit, as illustrated in Fig. 1.

Initially, space-filling curves were considered no more than mathematical curiosities, being even called "mathematical monsters" by Goldschager in 1981 [3], but, around the time, a variety of curves had been proposed. Great attention was drawn to these mathematical artifacts due to their intricate and elegant recursive structure. They belong to the family of fractal curves discussed by B. Mandelbrot [4], who coined and popularized the term "fractal".

Y. Iano et al. (Eds.): BTSym 2021, SIST 207, pp. 403–411, 2023.
https://doi.org/10.1007/978-3-031-04435-9_40

Fig. 1. The first, second, and third-order of a) Peano curve and b) Hilbert curve.

Later, in 1988, the so-called Generic Space-Filling Heuristic (GSFH) was formulated by Bartholdi and Platzman [5], which consisted of using SFCs to translate a combinatorial problem from the Euclidean space to the unit interval, which yields an easier solution. The authors applied this heuristic to solve the Travelling Salesman Problem (TSP), which is well-known in the realm of Computer Sciences, exhibiting solutions close to the optimal. Since then, SFCs have been proven to have important locality properties, which are useful to arrange and access neighboring elements inside data structures [3]. The following examples illustrate some direct applications of the SFCs relevant to date: improving the quality partitioning in parallel computing [6], defining the "S2Cells" path (which is a key library used by GoogleMaps) based on the Hilbert curve [7] to identify the regions ranging from millimeters to kilometers on Earth's surface (Fig. 2a), designing more efficient planar resonant antennas using the physical path of the Hilbert curve [8], or developing a set of Lumped Element Kinetic Inductance Detectors (LEKID) used inside the NIKA2 instrument of the 30 m IRAM telescope located at Pico Valeta in Spain (Fig. 2b) [9].

Not all SFC's tend to cross have a correspondence to adjacent sub-squares, as it is the case, for example, of the so-called Lebesgue's curve [3], nor are they restrained to square-like figures, as it is shown in the case of the Gosper curve that fills hexagonal shape figures instead [3]. Moreover, different SFC paths, as it is shown in Fig. 1, do not necessarily divide the square unit following the same rate of sub-divisions. For example, the Peano curve is based on iterations that cover 9^n squares, whilst the Hilbert covers 4^n squares.

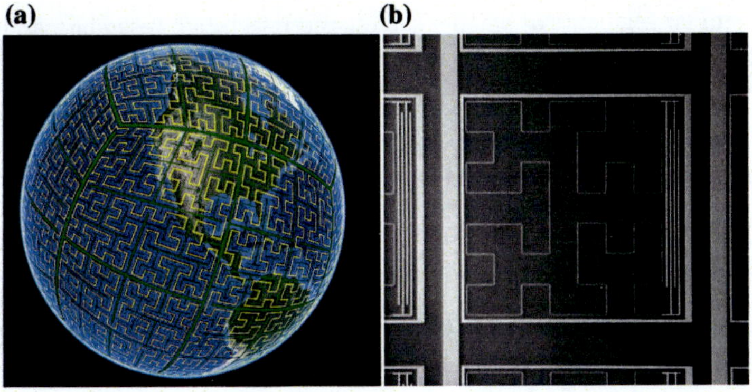

Fig. 2. Applications of the Hilbert curve on a) S2Cells projected onto the sphere [7] and b) a front-illuminated microstrip pixel for the 260 GHz on NIKA2 [9].

2 Aztec Curve Proposal

In this work, we propose another curve that we named the "Aztec curve", as it was partially inspired by the ornaments discovered on the Quetzalpapálotl Palace located in the archaeological site of Teotihuacán in Mexico (see Fig. 3).

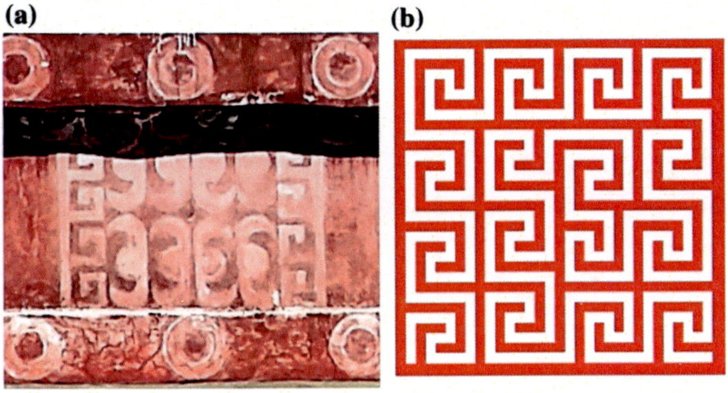

Fig. 3. a) Detail on Quetzalpapálotl Palace. b) Second-order Aztec curve.

This curve is a unit interval $[0,1]$ that has a one-to-one mapping onto the square unit $[0,1]^2$, and is defined as follows: the square unit is subdivided into sixteen sub-squares, each of them with sides quarter of the length of the father square. A curve fills each sub-square via rotations or reflections of itself. The sixteen curves are connected through the operators of rotation, and reflection in a way that preserves its continuity through all the sub-squares. The curve is chosen along with the operators in order to fill the defined square unit, grows into self-similar segments at smaller scales, and if iterated, resembles a meander-shape figure.

2.1 Grammar-Based Description

To describe the iterations of the proposed Aztec curve, the grammar description defined by M. Bader [3] will be used. The unit square is divided into 16 sub-squares, which will be populated by line segments that follow the path of the first-order curve A_1, which is the original pattern (see Fig. 4a). A set of non-terminal symbols $\{A,B,C,D\}$ represents the following: A the n-order path, B a reflection and 90° clockwise rotation of A, C is a 180° rotation of B, and D is a a 180° rotation of A. The terminal symbols $\{\uparrow, \downarrow, \leftarrow, \rightarrow\}$ describe the translations between the sub-squares. A grammatical description of how to construct an n-order Aztec curve (for $n = k + 1$), is given as:

$$A_{k+1} \Leftarrow B_k \uparrow B_k \uparrow B_k \uparrow A_k \rightarrow A_k \rightarrow A_k \rightarrow A_k \downarrow C_k \downarrow D_k \leftarrow D_k \uparrow B_k \leftarrow C_k \downarrow C_k \downarrow C_k \rightarrow A_k \rightarrow A_k \quad (1)$$

$$B_k = R_{O,-90}(refl(A_k)) \quad (2)$$

$$C_k = R_{O,180}(B_k) \quad (3)$$

$$D_k = R_{O,180}(A_k) \quad (4)$$

Equation (1) for $k = 1$, describes the second-order curve A_2, which is equivalent to a sequence of first-order non-terminal symbols (defined in (2), (3), (4)) placed on a path by the terminal symbols (arrows); on Fig. 4 the above-mentioned proceeding is visually represented.

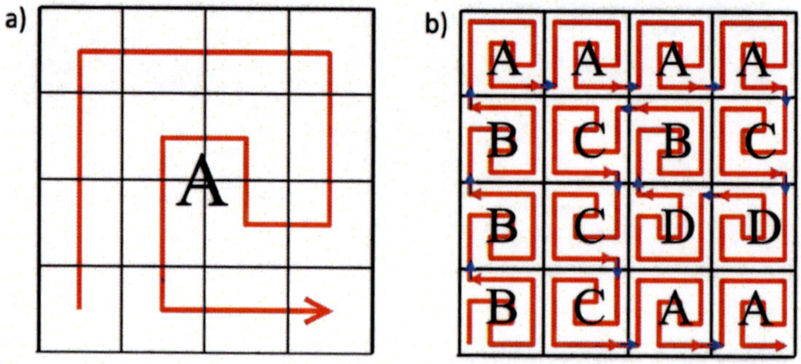

Fig. 4. a) First-order Aztec curve, b) Second-order Aztec curve is constructed by rotations and flips of its predecessor.

The next iteration would be a third-order Aztec curve (depicted on Fig. 5), which is thus constituted by a sequence of second-order replacements of $\{A2,B2,C2,D2\}$ spatially arranged by the same terminal symbols shown in (1). This procedure can be repeated infinite times, producing self-similar patterns at smaller scales, a property known from fractals.

Fig. 5. Third-order Aztec curve.

2.2 Clustering Properties

Although the number of sub-squares filled by the Aztec curve equals 4^{2n}, it can contain various rectangular sub-sets smaller in size, each one housing a continuous path that crosses all the adjacent sub-squares, this will be referred to as clusters.

If each sub-square is considered to be a memory cell, then it becomes apparent that the Aztec curve can be used to store clusters of 4^{2n} size, moreover its structure allows the existence of internal rectangular clusters that, under other paths, would rather be non-connected. These subsets with different proportions are shown in Fig. 6.

Therefore the Aztec curve has a correspondence in the 2D-space that harbors sets of clusters consisting of 4^{2n} sub-squares, 4^{2n-1} sub-squares (which in conjunction with the 4^{2n} conform a set that have a one-to-one correspondence to the set available to Hilbert curve), and $9 \times 16^{n-1}$ sub-squares. Hence, sets of clusters that are not continuous for the Hilbert or Peano curves are contained within the Aztec curve as a continuous path, which makes it suitable to use as an alternative to index and access data as well as other applications.

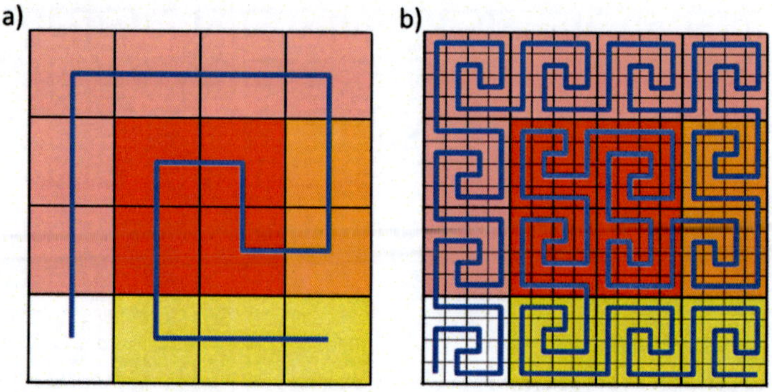

Fig. 6. Rectangular subsets covering a continuous path of 4^{2n-1} (dark orange), $9 \times 16^{n-1}$ (yellow), $2 \times 3 \times 16^{n-1}$ (light orange), and $3 \times 4 \times 16^{n-1}$ (pink) sub-squares, are overlayed on a) first-order Aztec curve and b) second-order Aztec curve.

3 Case of Application

A case of application of the Aztec curve is proposed on the scope of Compressed Sensing (CS), which is an advanced signal processing technique proposed in 2006 by Donoho [10], and whose principle is that a signal sparse in nature can be efficiently reconstructed by a number of measurements much smaller than the length of the digitally sampled signal [11]. Likewise, a digital signal can be compressed and reconstructed from a set of sparse non-zero coefficients which are computed through a Matching Pursuit (MP) algorithm, proposed by S. Mallat et al. [12], decomposing the signal to a linear combination of coefficients with a series of wavelets from a fixed dictionary, selecting waveforms that has the best fit compared to the original signal, and calculating the weight of this waveforms as a few sparse coefficients. If the MP algorithm is considered as a black-box it can process a 2D image through the use of an SFC converting it to a 1D signal, the original image can be reconstructed through a linear combination of the output coefficients with the known dictionary.

A test is conducted over the publicly available MNIST database [13], which consists of 28-by-28 pixel images of white handwritten numbers on a black background. The image on Fig. 7 is padded with black pixels until it is 32-by-32 in size, then it is arranged as a 1D vector column-by-column (thorough a vertical raster scan), and fed to the stock MATLAB's Orthogonal Matching Pursuit (OMP) algorithm, which was proven as a slight improvement over MP [14], afterward it calculates the sparse coefficients based on the Haar wavelet dictionary, and finally the image is reconstructed as the linear combination of the sparse coefficients with the corresponding wavelets.

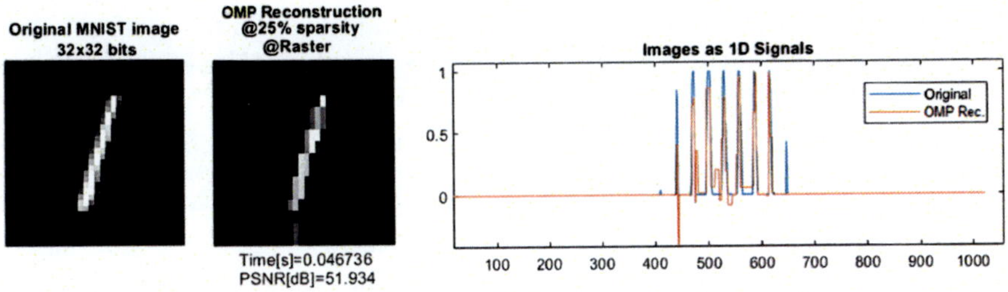

Fig. 7. An image from the MNIST database processed through the OMP algorithm.

The sparsity value is often an input to MP algorithms and limits the number of non-zero coefficients, which are only a fraction of the total length of the signal, consequently, an image is not reconstructed as a copy of the original, and it can be considered as a lossy compression. It can be observed in Fig. 7 that some vertical artifacts are introduced. The padding of black pixels permits these images to be 1024 pixels in size, which enables them to be rearranged as a one-dimensional vector by either Hilbert or Aztec curves, which in turn may cause an impact on the performance of the OMP algorithm. To asset this, a batch of 100 images is subject to a similar test, but the rearrangement of the one-dimensional vector is made through 4 different methods: Hilbert curve, Aztec curve, Zig-Zag scan, and Raster scan. To provide a statistical insight the processing time and the Peak signal-to-noise ratio of the batches are represented as box plots in Fig. 8.

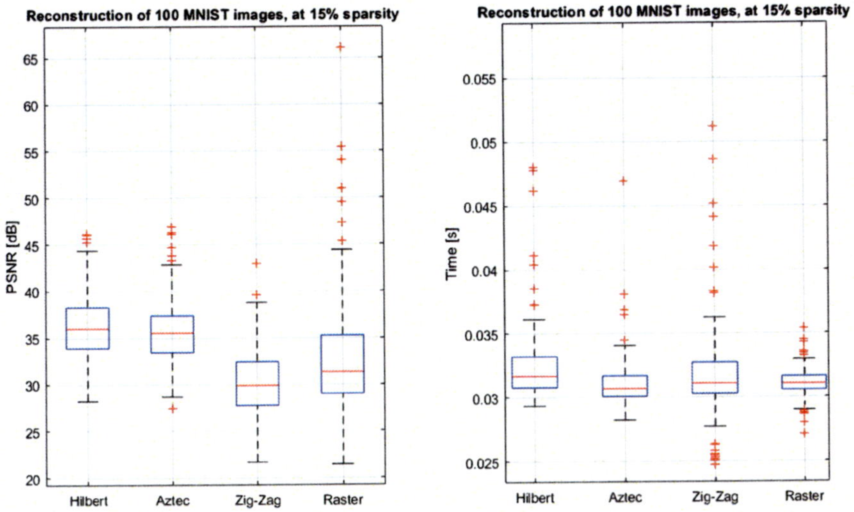

Fig. 8. Box-plots from batches of 100 MNIST database images reconstructed by OMP algorithm under different scanning methods; each image from the batch was fed through the following methods: Hilbert curve, Aztec curve, Zig-Zag scan, and Raster scan.

From the box-plots of Fig. 8, it can be seen that both SFC's yield a better PSNR than Zig-Zag, and raster scans, being the median of the batch rearranged by Hilbert curve meagerly greater than that of the Aztec curve; on the other hand, there seems to be a slight reduction in processing time when the Aztec curve is used, exhibiting a lower median, and overall more consistent times when compared to Hilbert curve. In consequence, the Aztec curve is not only an alternative to the Hilbert curve but closely matches the advantages on performance, in this particular application of Compressed Sensing.

4 Conclusions

A new SFC called Aztec curve was presented that is designed to fill the defined square units and remain self-similar at reduced scales of observation, as do also the well-known Hilbert and Peano curves. The Aztec curve may offer a new set of possibilities that can outperform similar space-filling curves if applied to different applications, such as, for example, those used in Compressed Sensing. The clustering properties of the Aztec curve allow for the connection of sets of adjacent sub-squares in a continuous uninterrupted manner, which might be used in efficient data storage or accessing applications, especially in two-dimensional arrays.

Acknowledgments. D. A. acknowledges INAOE and its staff, for their academic excellence even on the face of the current world pandemic, and under the support of CONACYT's scholarship (CVU:1016578). Also special thanks to Alexandra Elbakyan and Aaron Swartz, for making research accessible to any person in this world, regardless of place of birth, economic level, or affiliation.

References

1. Peano, G.: Sur une courbe, qui remplit toute une aire plane. Math. Ann. **36**(1), 157–160 (1890). https://doi.org/10.1007/BF01199438
2. Hilbert, D.: Ueber die stetige Abbildung einer Line auf ein Flächenstück. Math. Ann. **38**(3), 459–460 (1891). https://doi.org/10.1007/BF01199431
3. Bader, M.: Space-Filling Curves. Springer, Berlin Heidelberg (2013). https://doi.org/10.1007/978-3-642-31046-1
4. Mandelbrot, B.B., Freeman, W.H.: The Fractal Geometry of Nature. Einaudi paperbacks, Henry Holt and Company (1983)
5. Bartholdi, J.J., Platzman, L.K.: Heuristics based on spacefilling curves for combinatorial problems in Euclidean space. Manage. Sci. **34**(3), 291–305 (1988). https://doi.org/10.1287/mnsc.34.3.291
6. Borrell, R., et al.: Parallel mesh partitioning based on space filling curves. Comput. Fluids **173**, 264–272 (2018). https://doi.org/10.1016/j.compfluid.2018.01.040
7. Google LLC (n.d.) S2 Cells. https://s2geometry.io/devguide/s2cell_hierarchy. Accessed 1 Feb 2016
8. Vinoy, K.J., Jose, K.A., Varadan, V.K., Varadan, V.V.: Hilbert curve fractal antenna: a small resonant antenna for VHF/UHF applications. Microw. Opt. Technol. Lett. **29**(4), 215-219 (2001). https://doi.org/10.1002/mop.1136

9. Adam, R., et al.: The NIKA2 large-field-of-view millimetre continuum camera for the 30 m IRAM telescope. A&A **609** (2018). https://doi.org/10.1051/0004-6361/201731503
10. Donoho, D.L.: Compressed sensing. IEEE Trans. Inf. Theory **52**(4), 1289–1306 (2006). https://doi.org/10.1109/TIT.2006.871582
11. Stanković, L., Sejdić, E., Stanković, S., Daković, M., Orović, I.: A tutorial on sparse signal reconstruction and its applications in signal processing. Circ. Syst. Sign. Process. 1–58 (2018). https://doi.org/10.1007/s00034-018-0909-2
12. Mallat, S.G., Zhang, Z.: Matching pursuits with time-frequency dictionaries. IEEE Trans Sign. Process **41**(12), 3397–3415 (1993). https://doi.org/10.1109/78.258082
13. LeCun, Y., Cortes, C., Burges, C.J.C.: MNIST handwritten database. http://yann.lecun.com/exdb/mnist/ (1998). Accessed 28 Aug 2021
14. Pati, Y.C., Rezaiifar, R., Krishnaprasad, P.S.: Orthogonal matching pursuit: recursive function approximation with applications to wavelet decomposition. In: Proceedings of 27th Asilomar Conference on Signals, Systems and Computers, vol. 1, pp. 40–44 (1993)

Evaluation of Adjustment Models for Taking Body Temperature with a Thermal Imaging Camera in the Context of COVID-19

Gabriel Cruz, Raul Mamani, Henry Torres, Victor Salcedo, Cristian Benites, Reencarnacion Quispe, Franklin Cardeñoso, and Jorge L. Arizaca-Cusicuna(✉)

Universidad Nacional de San Antonio Abad del Cusco, Cusco, Peru
{141025,120297,120686,130111,130351,111213,franklin.cardenoso, jorge.arizaca}@unsaac.edu.pe

Abstract. In the context of the COVID-19 health emergency, it is necessary to have devices that help identify symptoms that indicate whether a person has COVID-19. As one of the main symptoms is fever, which can be identified by measuring body temperature, different non-contact measurement methods are being widely used as an alternative to traditional contact thermometers. However, readings with thermographic cameras present limitations in terms of high dependence on the environment. For this reason, this article aims to validate the temperature measurements of a thermographic camera by comparing different models using as a reference standard the readings of a calibrated medical infrared thermometer. For this purpose, 463 measurements were analyzed using an infrared thermometer and a thermographic camera simultaneously. As significant differences were observed between the measurements made using statistical analyses with ($p < 0.05$), models were developed establishing weighting and compensation criteria to obtain similar readings between the values measured from the thermal images and the infrared thermometer. For uncertainty estimation, linear and non-linear regression models such as artificial neural networks were tested, selecting the best model that allows reducing the variation among the readings.

Keywords: Thermal imaging · Body temperature · Thermographic camera · Artificial Neural Networks · COVID-19

1 Introduction

COVID-19 has caused an unprecedented health crisis worldwide, forcing people to take prevention and detection measures to mitigate the spread of the virus. According to studies developed by WHO [1], among the most important symptoms presented by a person with COVID-19 is high body temperature, fever

Y. Iano et al. (Eds.): BTSym 2021, SIST 207, pp. 412–419, 2023.
https://doi.org/10.1007/978-3-031-04435-9_41

(usually from 37.5 °C to more) being very important to measure this indicator for the location of infected people, being an alternative to the Covid-19 tests.

The ideal temperature measurement should be noninvasive and quick, as well as provide a safe physical distance between the operator and the person in care, but above all, it must be accurate enough to identify elevated temperatures. Traditional clinical thermometers present high reliability, but their reading procedures are time-consuming and cumbersome [2]. Then there are digital clinical thermometers that have a display for simple reading, but still require contact and relatively long waiting times. Finally, there are infrared thermometers and thermographic cameras, whose main advantage is that they allow almost instantaneous and non-contact temperature measurements, but their disadvantage is that they measure skin temperature, which calls into question their reliability; however, studies on the accuracy of temperature measurement using infrared sensors have shown that they offer reliable measurements, while the reliability of readings with thermographic cameras is still pending [3,4].

Thermal imaging cameras detect and measure the infrared radiation emitted by an object, which provides a non-contact method to determine the surface temperature of a person's skin in real-time, but this measurement is affected by external factors, such as the data processing in the camera application, the transmission coefficient of the medium between the sensor and the person, the average radiant temperature of the measurement environment, and, of course, the distance between the thermal sensor and the person [4,5].

The article presents an analysis of temperature measurements from a thermographic camera and a calibrated medical infrared thermometer to validate the readings of the thermographic camera. The proposed system applies a criterion of weighting and compensation to ensure that the reading of the thermographic camera values is similar to those obtained by an infrared thermometer, improving the detection of people with high temperatures.

2 Materials and Methods

For temperature measurements, a thermographic camera was installed at the entrance of the campus of the National University of San Antonio Abad del Cusco at a height of 2.20 m above ground level. These measurements are sent via a wireless link to central monitoring where all the data is stored. For the implementation of the models, in total, measurements from 463 subjects were used to build the models. In this context, it should be taken into account that the measures were taken in an outdoor environment without prior acclimatization of the subjects.

The temperature taking procedure begins at the entrance door, where the temperature is taken on the person's forehead with the infrared thermometer at a distance of no more than 2 cm, immediately after which the subjects record their temperature with the thermographic camera in the database, the camera processes the image at a distance between 1.0 m and 2.0 m. The technical specification of the devices are as follow: the infrared thermometer used is a JPD-FR202

Jumper thermometer, with 0.95 of emissivity, 1 s for latency, measurement distance from 1 cm up to 5 cm, temperature range from 34.9 °C to 42.2 °C, accuracy ± 0.2 °C and resolution of 0.1 °C; on the other hand, the thermographic camera with a temperature range from −15 °C to 550 °C, with the accuracy of ± 0.3 °C and 1920 × 1080 and 256 × 192 of image and thermal resolution respectively was used.

With the data obtained, a paired samples comparison is performed, using a p value equal to 0.05 for the statistical significance of the measurements. From the analysis, it is obtained a P value equal to 0.000010374, so the null hypothesis is rejected, indicating that there are differences between the readings of the infrared thermometer and the thermographic camera. Similarly, a regression analysis [6] is performed using a double square model, obtaining a correlation coefficient equal to 0.316815, indicating a relatively weak relationship between the two measurements. Figure 1 shows the correlation and regression equation between body temperature measurement with the infrared thermometer and the thermographic camera [7].

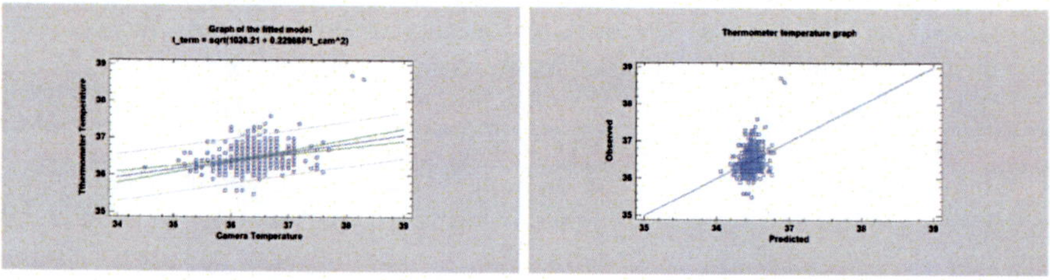

(a) Regression (b) Prediction

Fig. 1. Double square regression.

To validate the obtained readings, we used three models, the first model using a double square regression, the following model with compensation of the aforementioned regression, and the last one with a neural network with two hidden layers. The first model follows the regression with a double square model, which gives us the following equation:

$$T_{term} = \sqrt{1026.21 + 0.229868 * T_{cam}^2} \tag{1}$$

where T_{term} is the temperature measured with the infrared thermometer and T_{cam} is the temperature measured with the thermographic camera.

The second model uses the regression equation plus compensation factors due to an uncertainty factor presented by the measurements with the thermographic camera, this uncertainty factor is due to the reading distance, the environmental conditions of reading, and the homeostatic mechanisms of the body thermoregulation of people [8]. We can estimate this uncertainty factor by first analyzing the importance of the distance between the camera and the subject; studies have shown a small difference of 0.2 °C between measurements taken at 0.2 m and 2.5 m [8,9]. Then we have that, according to ISO GUM [10], the standard combined uncertainty due to several factors (measurement, instrument, environmental influence, and operator) with a fixed threshold, is estimated to range from 0.20 °C (for a subject at rest in an indoor environment and after a suitable stabilization time) to 0.31 °C (for a subject immediately after being in an outdoor environment and without thermal stabilization). In that sense, we have that the compensated temperature can be calculated as [9]:

$$T_{comp} = T_{obj} + a * D + b \tag{2}$$

where, T_{comp} is the compensated temperature, T_{obj} is the temperature obtained by the thermographic camera with regression, D is the separation distance between the sensor and the subject in meters, a is the estimated coefficient related to distance in °C, such as low thermal image resolution and out of focus at long distances, and b is the estimated coefficient in °C for other uncontrolled factors such as hardware calibration errors and environmental factors. For our analysis, we will consider the distance coefficient $a = 1$, the coefficient for uncontrolled factors $b = 0.31$, and the separation distance $D = 1$.

Finally, a third model based on an Artificial Neural Network (ANN) was trained in a supervised manner to find a relationship between the input and output data (camera measurement and infrared thermometer measurement) so that it can be used to adjust the temperature measured by the thermographic camera and thus obtain values that are as close as possible to those obtained with the infrared thermometer [11]. The type of ANN used in this research corresponds to the Multilayer Perceptron (MLP), which is a type of network composed of multiple fully connected layers, each one with a given number of neurons. Additionally, these layers are divided into three types: The input layer, the hidden layers, and the output layer [12].

Using the temperatures read as a training and validation dataset, an MLP network with two hidden layers was trained and optimized using a predefined loss function and an optimization algorithm. Additionally, it is required to define certain hyperparameters in advance, such as the learning rate, the batch size, and the number of epochs to perform the training of the proposed network. It is worth mentioning that the correct choice of these hyperparameters will largely influence the learning of the trained network.

For this work, the constructed artificial neural network consists of an input layer and an output layer with one neuron each and two hidden layers of 32 and 16 neurons respectively. On the other hand, the nonlinear activation used in the hidden layers was the rectified linear unit (ReLU) due to its low computational cost and better performance in training based on the gradient descent optimization algorithm. This activation function is defined through the following equation:

$$y_x = \max(0, x) \tag{3}$$

where x is the input and y_x is its corresponding output.

A total of 463 data were used for the training data, which were divided into the following proportions: 75% for training and 25% for validation. Additionally, the following training criteria were used: the Stochastic Gradient Descent (SGD) was used as the optimizer, and the Mean Squared Error (MSE) was used as the loss function.

To choose the best model, different parameters were varied within the architecture. The following table summarizes the combinations made (Table 1):

Table 1. Combinations of neural network parameters.

Parameter	Min	Max
Learning rate	10^{-4}	10^{-2}
Hidden layers	1	5
Neurons	4	64

3 Results and Discussion

For the evaluation of the data, we use the statistical tool of Equivalence Tests and the Hypothesis Test. The former will allow us to test whether the two measurements are equivalent, in this case, the infrared thermometer body temperature measurement readings with the thermography camera measurements obtained after the three methods mentioned above. In equivalence testing, the null hypothesis is that the population mean differs from a target value or the mean of another population. The second; hypothesis testing; will allow us to evaluate the probability associated with the null hypothesis (H_0) that there is no effect or difference between the measurements obtained with the infrared thermometer and the thermographic camera. The p-value obtained reflects the probability of rejecting H_0 being true, the null hypothesis is a statement of "textitno difference" between the two readings. The rejection of the null hypothesis and the low correlation obtained are explained by the uncertainty factor presented by the thermographic camera measurements, for which we use the three methods already mentioned.

The results of the first method were obtained with the model using only regression, as shown in Fig. 2. This model is using a double squared regression, and we see that the results show a slight variation of the mean as presented in Fig. 2, however, Fig. 1 highlights that as the real values with the thermographic camera increase, the predicted values are maintained in the mean, this is, this model only presents an adequate behavior for measurements below 36.7 °C, but for measures above this value this model does not present a correct approximation, this detail was also observed by [8] and by [9] for this kind of models. The behavior of this model is due to this model does not take into account the different external factors that influence the processing of thermal images.

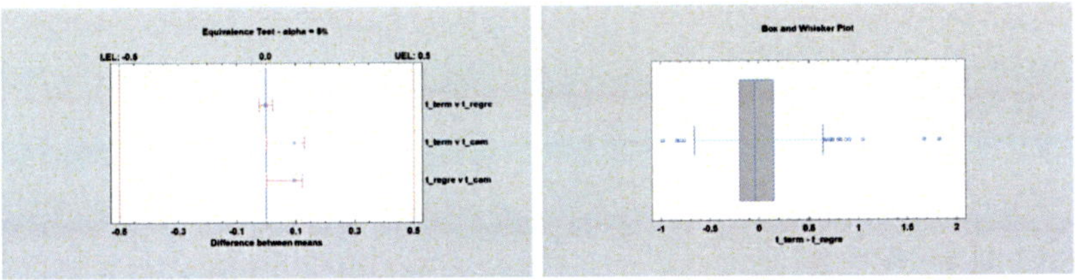

(a) Equivalence Tests (b) Boxplot

Fig. 2. Regression model.

The results of the second method, using the regression plus compensation model, are shown in Fig. 3. This model uses a compensation factor to improve the values from the previous model and performs a double analysis measurement procedure. First, a temperature measurement is performed using the first method (regression equation), then, the resulting prediction is evaluated. If the subject's temperature is below a threshold set at 36.7 °C the measure is labeled as valid, otherwise, a second analysis is performed using the compensation equation. The results are shown in Fig. 3, where we observe that it presents a greater variation of the mean compared to model 1, improving the response to temperatures above 36.7 °C, however the resulting p-value rejects the null hypothesis.

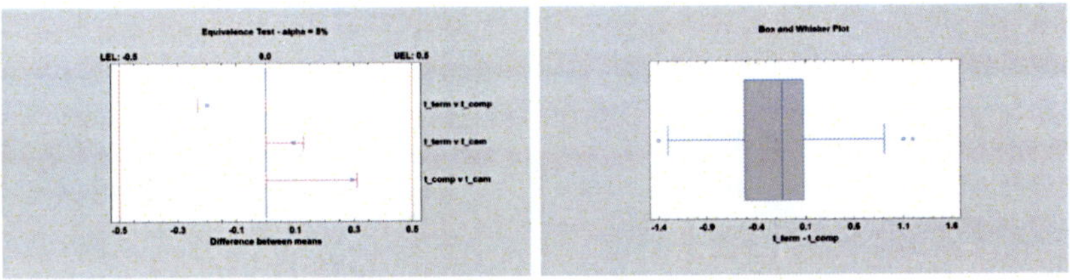

(a) Equivalence Tests (b) Boxplot

Fig. 3. Regression plus compensation model.

Finally, the results of the third method, using the neural network, are shown in Fig. 4. The model built with the neural network presents the results shown in Fig. 4, where we see that the variation of the mean is improved compared to the second model, likewise, the p-value obtained indicates that this time the null hypothesis is not rejected, which means that there is a significant relationship between the data measured with the infrared thermometer compared to the data adjusted using the neural network.

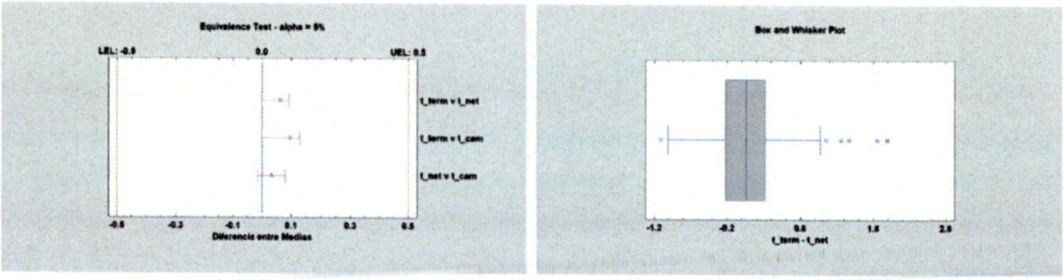

(a) Equivalence Tests (b) Boxplot

Fig. 4. Neural network model.

4 Conclusions

According to the statistical analysis performed between the readings of an infrared thermometer and a thermographic camera, a hypothesis test was run for the three models. For this reason, to perform non-contact body temperature measurements in the context of the current COVID-19 health emergency, differences were found between the values obtained. Under this context, the present work considered three different models that were implemented and evaluated to be able to decrease these variations in temperature readings. These three models correspond to the following: a double squared regression model obtained a p-value of 0.0029364, a double squared regression model with compensation with a p-value of 0, and a prediction model using an MLP network with a p-value of 0.136971.

Despite the good fit obtained with the first model, it was found that, for high temperatures, the model presents a stationary prediction tendency. Although this model was improved by using a compensation factor, the results obtained were not as expected in statistical terms for the analysis of the hypothesis test. Finally, a third model, using an MLP network with two hidden layers, was trained to perform the adjustment and found to perform better than the models presented above. Thus, it is possible to affirm that by using a simple ANN model it is possible to reduce the variation between the measurements of both devices obtaining a more reliable fit between the different values, which was verified by the respective hypothesis test.

Acknowledgements. This article was made thanks to the financing of Universidad Nacional de San Antonio Abad del Cusco for research projects by contract N° COVID-06-2020-UNSAAC and resolution R-536-2020-UNSAAC.

References

1. WHO, Novel Coronavirus (2019-nCoV) The WHO report. https://apps.who.int/iris/bitstream/handle/10665/330762/nCoVsitrep23Jan2020-eng.pdf. Accessed 20 May 2021
2. Childs, C.: Body temperature and clinical thermometry. Handb. Clin. Neurol. **157**, 467–482 (2018). https://doi.org/10.1016/B978-0-444-64074-1.00029-X

3. Zaproudina, N., Varmavuo, V., Airaksinen, O., Närhi, M.: Reproducibility of infrared thermography measurements in healthy individuals. Physiol. Meas. (2008). https://doi.org/10.1088/0967-3334/29/4/007
4. Lahiri, B.B., Bagavathiappan, S., Jayakumar, T., Philip, J.: Medical applications of infrared thermography: a review. Infrared Phys. Technol. (2012). https://doi.org/10.1016/j.infrared.2012.03.007
5. Fernández-Cuevas, I., Bouzas-Marin, J., Arnais-Lastras.: Classification of factors influencing the use of infrared thermography in humans: a review, infrared physics and technology (2015). https://doi.org/10.1016/j.infrared.2015.02.007
6. Somboonkaew, A., Prempree, P., Vuttivong, S., Wetcharungsri, J., Porntheeraphat, S., Chanhorm, S.: Mobile-platform for automatic fever screening system based on infrared forehead temperature. OECC - 2017 (2017). https://doi.org/10.1109/OECC.2017.8114910
7. Padilla-Raygoza, N., Ruiz-Paloalto, M.L., Díaz-Guerrero, R., Olvera-Villanueva, G., Maldonado, A., Raygoza-Mendoza Mdel, P.: Correlación de mediciones de temperatura corporal con 3 termómetros: Ótico, cutáneo y digital. Enferm. Clin. (2014). https://doi.org/10.1016/j.enfcli.2014.01.005
8. Dell'Isola, G.B., Cosentini, E., Canale, L., Ficco, G., Dell'Isola, M.: Noncontact body temperature measurement: uncertainty evaluation and screening decision rule to prevent the spread of COVID-19. Sensors **21**, 346 (2021). https://doi.org/10.3390/s21020346
9. Chin, J.W., Wong, K.L., Chan, T.T., Suhartono, K., So, R.H.: An infrared thermography model enabling remote body temperature screening up to 10 meters. In: 2021 IEEE/CVF Conference on Computer Vision and Pattern Recognition Workshops (CVPRW) (2021). https://doi.org/10.1109/cvprw53098.2021.00429
10. International Organization for Standardization Geneva (2008) Evaluation of measurement data - Guide to the expression of uncertainty in measurement. isbn = 9267101889. issn = 00099147. http://www.bipm.org/en/publications/guides/gum.html
11. Lin, J-W., Lu, M-H., Lin, Y-H.: A thermal camera based continuous body temperature measurement system. In: 2019 IEEE/CVF International Conference on Computer Vision Workshop (ICCVW) (2019). https://doi.org/10.1109/iccvw.2019.00208
12. Haykin, S.S.: Neural Networks and Learning Machines. Prentice Hall (2009). ISBN: 9780131471399

Characterization of the Electromagnetic Propagation in Inca Walls of the Archaeological Site of Rumiwasi Using ARMA Processes in the Band from 800 MHz to 1 GHz

Henry L. Davila-Andrade(✉) and Jorge L. Arizaca-Cusicuna

Universidad Nacional de San Antonio Abad del Cusco, Cusco, Peru
{114155,jorge.arizaca}@unsaac.edu.pe

Abstract. This article describes the modeling of electromagnetic propagation in the frequency domain for the most representative types of walls existing in the archaeological zone of Rumiwasi in the band from 800 MHz to 1 GHz using autoregressive moving-average (ARMA) process, which is considered important since it can simulate a variety of systems in different fields of engineering. This was done from the S_{21} parameter, which roughly represents the transfer function of channel communication that was obtained using a channel sounder in the frequency domain. The calculation of the parameters of the ARMA processes is presented with the use of the *statsmodels* library in Python because it provides simple commands to obtain statistical models for each wall. This is based on the Akaike information criterion to acquire the optimal order for each model. On the completion of this process, the transfer functions are achieved. Its precision is calculated with the graphic analysis of the model's comparison and measured values, which also appear in the graphic analysis of residuals. As a result, there is a good approximation to the models previously obtained.

Keywords: ARMA process · Inca wall · Electromagnetic propagation · Statsmodels · Python

1 Introduction

Autoregressive Moving-Average (ARMA) models are widely used for the computational simulation of various systems because they make easier the replication of the measurements as close as possible to the measured values. One application of these models is the behavioral representation of a communication channel in the frequency domain, having a minimum number of parameters [1].

The objective of this work is to model the propagation of electromagnetic waves in the band from 800 MHz to 1 GHz in two of the most representative walls of the archaeological zone of Rumiwasi located in the province and department

Y. Iano et al. (Eds.): BTSym 2021, SIST 207, pp. 420–427, 2023.
https://doi.org/10.1007/978-3-031-04435-9_42

of Cusco, district of San Sebastian, in such a way the behavior of these walls can be simulated for future works related to the modeling of a wireless channel, simulation of propagation conditions of a wireless communication system, and related topics.

In previous works, the RMS delay time and the RSSI were characterized in this archaeological site providing tools, so that in the future a wireless network of sensors can be optimally designed using LoRa technology as part of the project "Análisis y caracterización experimental de un modelo de canal inalámbrico y su uso en áreas de conservación arqueológica" [2], which, as a complementary part, aims to generate statistical models of channels in such a way that they can be applied for the simulation of other types of technologies.

Currently, there are many ways to obtain the parameters of an ARMA process, but the simplest ones, which is gaining popularity, are the tools provided in the *statsmodels* library that belongs to Python [3], since it provides us with simple methods to calculate the parameters of the statistical models, we make use of this in the present work, using the Akaike information criterion to obtain the optimal order of the model. The parameters for each wall are shown, in addition to their comparison graphic with the prediction of the models.

2 Theory

2.1 Autoregressive Moving-Average (ARMA) Process of a Communication Channel

The frequency response of a wireless channel can be interpreted as the output of an Autoregressive Moving-Average (ARMA) process. Assuming this process of order (M, K), the frequency response will have the following form:

$$\begin{aligned} H(f_n, x) + a_1^* H(f_{n-1}, x) + \cdots + a_M^* H(f_{n-M}, x) \\ = v(f_n) + b_1^* v(f_{n-1}) + \cdots + b_k^* v(f_{n-K}) \end{aligned} \tag{1}$$

where $a_1, \cdots, a_m$ and $b_1, \cdots, b_K$ are called ARMA parameters, x is the position in which the experimental measurements are carried out, $v(f_n)$ is white noise source of zero mean and variance σ_v^2. An extended version of the ARMA processes of order (M, N) used for the identification of its parameters in terms of the z-transform is the following:

$$H(z) = \frac{X(z)}{W(z)} = \frac{G(1 + \beta_1 z^{-1} \cdots + \beta_M z^{-M})}{1 + \alpha_1 z^{-1} \cdots + \alpha_N z^{-N}} \tag{2}$$

where $X(z)$ and $W(z)$ are the z-transforms of $x(k)$ and $w(k)$; the ARMA process is the $x(k)$; $w(k)$ is the white noise source of zero mean and variance σ. This approximation can only be made considering that the frequency response of the channel is a stationary time series.

2.2 Model Order Selection

There are different criteria for choosing the model order, but in this work, we will use the Akaike Information Criterion (AIC) because it is widely used as a tool in the modeling of wireless communication channels [1].

Akaike Information Criteria (AIC). For the Autoregressive Moving-Average (ARMA) process of order (p, q), the normalized Akaike criterion [4] has the following form:

$$AIC(p,q) = ln(\hat{\sigma}_v^2) + \frac{2(p+q)}{N} \tag{3}$$

where $\hat{\sigma}_v^2$ is the maximum likelihood estimate of σ_v^2. The value for which $AIC(p,q)$ is minimum is the optimal order of the model.

3 Methodology and Technique

The considered communication channel has, in a large part of its trajectory, Inca walls, where a channel probe was used in the frequency domain proposed in [5–7] consisting of two model HG908P-NF patch antennas with 8 dBi gain and the Fieldfox Handheld Analyzer N9914A established as a vector network analyzer. The S_{21} parameters were obtained in a frequency range from 800 MHz to 1 GHz. The block diagram is shown in Fig. 1.

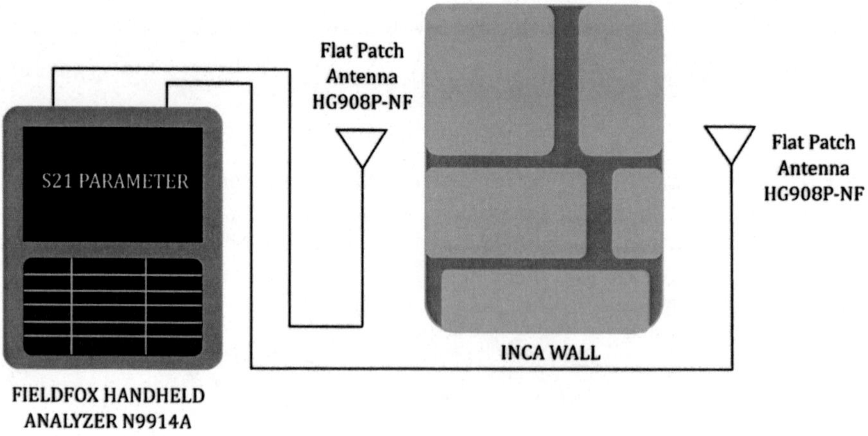

Fig. 1. Block diagram of the measurement system

This was done for the most representative Inca walls. Two different types of them were found in the archaeological site of Rumiwasi. Both are 1 m thick and

(a) Test Wall 1 (b) Test Wall 2

Fig. 2. S_{21} parameter measurement

different in geometry and material. They are shown in Fig. 2, and the process is about taking 4 profiles of the S_{21} parameter for each wall at the same height as the antennas (99 cm) with a distance of 60 cm from the wall to each antenna. By averaging these profiles, achieving the S_{21} parameter for each wall is possible. Then, it is necessary to import this data and calculate the parameters using the Python distribution Anaconda version 2.0.3. The Jupyter Notebook environment was used for having the necessary commands executed from the *statsmodels* library version 0.12.2.

Using each S_{21} average parameter in dBm, we proceeded to convert them to milliwatts, so later the measured signal was normalized. We used this approach because the transfer function in milliwatts will be expressed as the product of terms instead of the additions and subtractions that usually happen in dBm. It was also normalized because the values in milliwatts were tiny. Once this was done, the order of the model was calculated using the Akaike information criterion (AIC) with the command *arma_order_select_ic*, adding later *res.aic_min_order* to then specify an ARMA model. For example, *model=ARMA (data,order=(p,q))* with the command *model.fit()*, which uses the exact maximum likelihood through the filter of Kalman, to calculate the model parameters of each Inca wall. In addition, a graphic analysis of the residuals is shown for each model using the command *residuals = pd.DataFrame (model_fit.resid)*, which shows us the residuals and their density.

4 Results

4.1 Test Wall 1

The calculated parameters for this wall can be summarized as a transfer function that follows:

$$H_{60cm} = \frac{0.0797 * (1 - 0.4799 * z + 0.7109 * z^2)}{1 - 1.2626 * z + 0.8448 * z^2 - 0.7324 * z^3 + 0.1915 * z^4} \quad (4)$$

A comparison of measured data versus model prediction (4) is shown in Fig. 3. A good approximation of the ARMA process to the measured data is evident. The graphs of the residual errors and their distribution are shown in Fig. 4. We can see that according to Fig. 4(b) and values obtained using residuals *residuals = pd.DataFrame(model_fit.resid)* the errors have approximately a Gaussian behavior, which indicates that the data analysis was carried out correctly, with mean $\mu = 0.000242$. This reflects a small variation in the model prediction, the measured data, and a standard deviation of $\sigma^2 = 0.042467$.

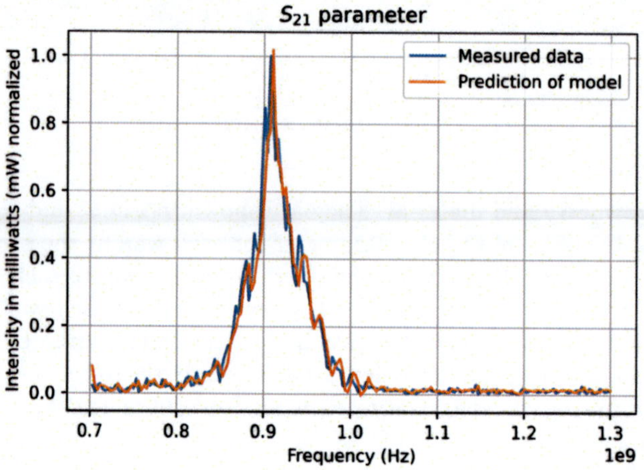

Fig. 3. Comparison of the prediction model with the data obtained

4.2 Test Wall 2

The calculated parameters can be summarized in a transfer function, as in the following equation:

$$H_{60cm} = \frac{0.2318 * (1 - 0.9161 * z + 0.4280 * z^2)}{1 - 1.1497 * z + 0.3219 * z^2 + 0.3233 * z^3 - 0.4158 * z^4} \quad (5)$$

A comparison of measured data versus prediction of model (5) is shown in Fig. 5. A good approximation of the ARMA process to the measured is evident. The graphs of the residual errors and their distribution are shown in Fig. 6. We can see that, according to Fig. 6(b) and values obtained using residuals *residuals = pd.DataFrame(model_fit.resid)*, the errors have approximately a Gaussian behavior, which indicates that the data analysis was carried out correctly with

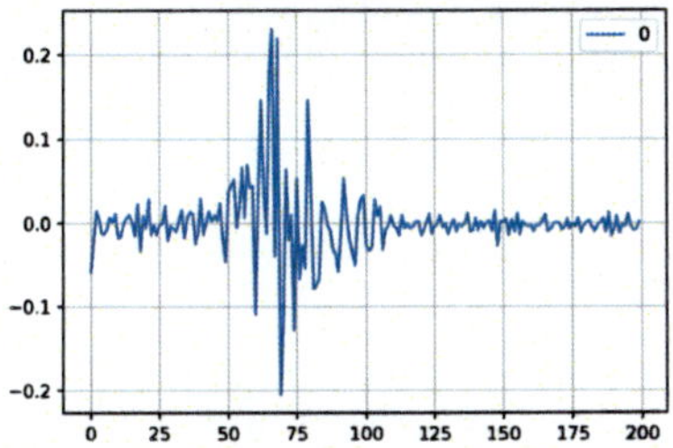

(a) Residues of model

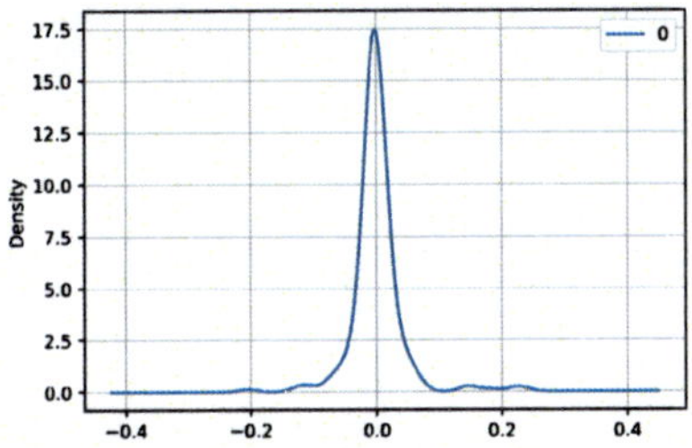

(b) Density of residues of model

Fig. 4. Description of the residual errors of the model

mean $\mu = 0.003280$. This reflects a small variation between the model prediction, the measured data, and a standard deviation of $\sigma^2 = 0.135413$. It should be noted that this wall presents high attenuation compared to the previous wall that was shown with an analysis of the RSSI in prior work [2].

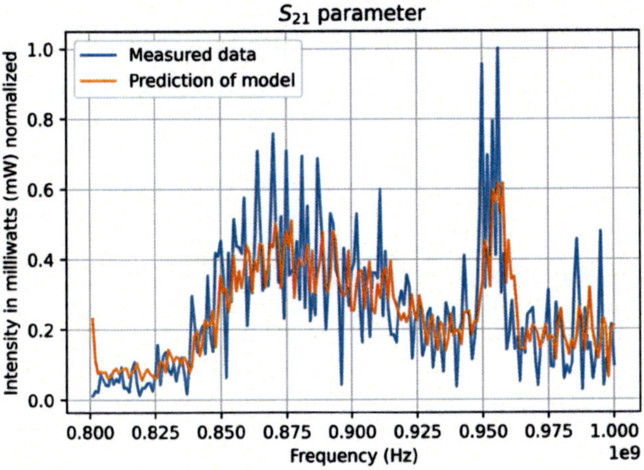

Fig. 5. Comparison of the prediction model with the data obtained

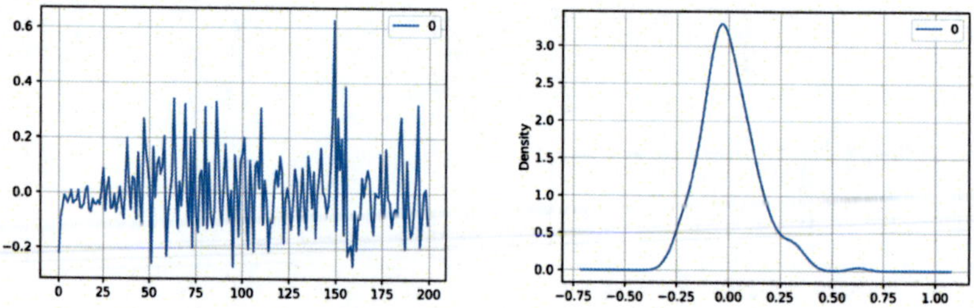

(a) Residues of model (b) Density of residues of model

Fig. 6. Description of the residual errors of the model

5 Conclusions

The most representative walls of the Rumiwasi archaeological zone were characterized, obtaining, in each of them, a transfer function in terms of the z-transform. The data predicted by the obtained models have a good approximation to the measured data but with a slight bias that is shown in the graph of the residual errors of each model.

The S_{21} parameters can be approximated as a stationary time series because the same average behavior will be obtained, regardless of the time used for tests performance, the channel probe in the frequency domain allows us to obtain a greater amount of channel parameters, either the RMS delay time or the ARMA model of the communication channel compared to a channel sounder in the time domain.

Acknowledgments. This article was made thanks to the financing of CONCYTEC, FONDECYT, and WORLD BANK for research projects and also to the Ministry of Culture for the permits granted to carry out tests at the archaeological site of Rumiwasi. Also thank Alessandra M. Sota Gonzales for the traslation of this work.

References

1. Ghassemzadeh, S., Jana, R., Rice, C., Turin, W., Tarokh, V.: Measurement and modeling of an ultra-wide bandwidth indoor channel. IEEE Trans. Commun. **52**(10), 1786–1796 (2004). https://doi.org/10.1109/TCOMM.2003.820755
2. Davila-Andrade, H., Arizaca-Cusicuna, J.: Measurements and outdoor propagation channel characterization for Rumiwasi archaeological site at 920 MHz. In: Proceedings of the 6th Brazilian Technology Symposium (BTSym 2020), pp. 949–956. Springer International Publishing (2021). https://doi.org/10.1007/978-3-030-75680-2_103
3. Seabold, S., Perktold, J.: Statsmodels: Econometric and statistical modeling with python. In: 9th Python in Science Conference (2010). https://conference.scipy.org/proceedings/scipy2010/pdfs/seabold.pdf

4. Akaike, H.: A new look at the statistical model identification. IEEE Trans. Autom. Control **19**(6), 716–723 (1974). https://doi.org/10.1109/TAC.1974.1100705
5. Howars, S., Pahlavan, K.: Measurement and analysis of the indoor radio channel in the frequency domain. IEEE Trans. Instrum. Measur. **39**(5), 751–755 (1990). https://doi.org/10.1109/19.58620
6. Howars, S., Pahlavan, K.: Autoregressive modeling of wide-band indoor radio propagation. IEEE Trans. Commun. **40**(9), 1540–1552 (1992). https://doi.org/10.1109/26.163575
7. Chen, L., Ostertag, T., Loschonsky, M., Reindl, L.M.: Measurement of mobile radio propagation channel in ruins. In: 2010 IEEE International Conference on Wireless Communications, Networking and Information Security, pp. 252–256, Beijing (2010). https://doi.org/10.1109/WCINS.2010.5541724

Check for updates

Change Management in a Brazilian Family-Owned Business in the Window and Door Frames: A Case Study

Gabriela Vieira Passos[1](✉), Maria Thereza de Moraes Gomes Rosa[1], Marlucy de Godoy Ricci[1], Daniela Helena Pelegrine Guimarães[2], and Alessandra Cristina Santos Akkari[3]

[1] Science and Technology Center, Universidade Presbiteriana Mackenzie, Campinas, Brazil
gabrielavieirapassos@gmail.com, gomes.mtms@gmail.com, marlucy.ricci@mackenzie.br

[2] Lorena Engineering School (EEL-USP), Lorena, Brazil
dhguima@usp.br

[3] Federal University of ABC (UFABC), São Paulo, Brazil
alessandra.akkari@ufabc.edu.br

Abstract. The aim of this work was to unveil how a family-owned business impacts change management within a Brazilian family business in the door and window frame industry, located in the São Paulo State. In order to meet that goal, a unique case study was carried out along with a quantitative analysis. A structured questionnaire with close-ended questions was chosen for collecting data, which was composed of 35 questions devoted to factors hindering business changes and 4 questions on sociodemographic issues. The Likert scale of 5 points was chosen. Data were collected by using Google Forms and processed at BioEstat 5.4 software. Qualitative variables were presented by means of absolute and relative frequency distributions, whereas quantitative variables were displayed by using measures of central tendency and dispersion. This study analyzed the factors inhibiting internal organizational change and identified the employee's awareness of communication and information issues, organization strategy, and internal/external environments. In conclusion, there are some change hindering factors according to the data from the examined sample, such as information not reaching the employees and business units on time, difficulty in adjusting strategy to the external environment, and poor capabilities in handling the pressures arising from it.

Keywords: Change management · Family business · Change hindering factors

1 Introduction

Family businesses play a key role as a driving force for growth and national development, according to the Brazilian Institute of Geography and Statistics (IBGE) and the Brazilian Micro and Small Enterprises' Support Service (SEBRAE). Literature data support that

Y. Iano et al. (Eds.): BTSym 2021, SIST 207, pp. 428–435, 2023.
https://doi.org/10.1007/978-3-031-04435-9_43

those companies are liable for 65% of the Brazilian Gross Domestic Product (GDP). Those companies correspond to 90% of the businesses in the country and employ 75% of its economically active population [1].

According to a Global Research on Family Business from PwC [2], family businesses may have a competitive advantage over non-family ones if they pursue an ambitious goal and explore their values as core assets. It was also pointed out that 64% of the studied companies achieved a higher performance in sales and operations, while 34% of them reached a double-digit rise within 12 months before that study. Data evidenced the expectancy of growth in those companies was high, reaching 79% in Brazil and 84% in the world, which represents a positive scenario in disruptive times.

Family companies face the duality of preserving their own identities, while they adapt to actual or future new demands [3]. A recent study indeed suggests that these types of organizations face the uneasiness between tradition and the quest for innovation [4]. Organizational change management enables structuring the transition of organizations and people from the present state to the desired state, providing means to carry out a strategy. It is a framework capable of preparing employees to accept and comprehend changes within their environment [5].

Mapping organization change may contribute to the execution of management systems, which in turn provides a competitive advantage to organizations by means of improving operations and knowledge about the company. The promotion and implementation of continuous process improvement in those companies will depend on the high administration to handle such a change [6]. When the change is integrated into the organizational culture, there is the possibility to reach new profitable business models, which might overcome expectations [7].

Since family management may be a hindering factor for internal organizational changes, the main contribution from this work was to substantiate a study able to allow family businesses to identify weaknesses within their family models. This work provided the concerned parties an opportunity for enhancing their knowledge of their business and some possible solutions for eventual problems identified. The theoretical relevance of this research is justified by the search for understanding the concepts behind the family business, change management, and organizational change, as well as the articulation among them and mutual influence. From a methodological perspective, this work sheds some light on factors such as those promoting or hindering change in this type of organization. Moreover, it was conceived to boost the scientific study on family management associated to change management, since the research interest on those subjects is still rather low.

2 Methodology

The present study was characterized as an exploratory and descriptive search with a quantitative approach based on collecting primary data. The exploratory approach was carried out by bibliographic search. The descriptive research method was based on observation with the aim of unveiling if there are factors hindering change within a family business. A structured questionnaire with close-ended questions was chosen to examine which components should be observed and analyzed in depth by the statistical method.

Data were collected from a unique case study within a Brazilian door and window frame industry. The proposal herein was to understand the employee's awareness about internal changes in this organization.

The instrument for data collection was a structured questionnaire with close-ended questions developed and validated by Neiva and Domingos [8], which was based on the work published by Burke and Litwin (1992) and research data concerning fifteen experts from private and public organizations.

Based on a sample composed of 1,246 individuals for instrument validation, this work aimed to analyze their awareness about company changes. Besides, factors promoting or hindering change in the organization were split in this research, which was performed from bibliographic search and information from eight experts in organizational studies.

The instrument in this study was composed of 35 questions to identify factors hindering change and 4 sociodemographic questions (gender, age, time, and position in that company) devoted to analyzing the employees' awareness of internal organizational changes. The Likert scale with 5 points was considered herein, with a variation from totally disagree (0) up to totally agree (4).

This instrument was then applied to a family business in a self-administered mode, once the questionnaire was sent online along with an invitation letter and a free and informed consent (TCLE). Data were processed electronically: the questionnaire was built in Google Forms and sent to the company by e-mail. The population or universe in this research refers to 100 employees from that family business chosen as a case study. Non-probability sampling was chosen since the questionnaire was sent to respondents by sampling based on accessibility because the goal was to reach the highest number of respondents possible.

The research analysis in this work is a family business, which is already in the market for 28 years. This organization has an area of 16,000 m^2 and about 100 employees. It is located within one of the most industrialized regions of the São Paulo State, around 450 km away from its capital, São Paulo, in Brazil. Its main economic activity is the fabrication of door and window frames, with a monthly production of approximately 35,000 pieces.

Qualitative variables were presented by means of absolute and relative frequency distribution, whereas quantitative variables were processed by using central tendency and dispersion. The normality test was performed according to the Shapiro-Wilk test, which determines whether there is a normal distribution. The distribution of qualitative variables was analyzed by the chi-square test, which examines the relationship between actual and observed data. Nonetheless, when the variance was restricted to $npq < 5$, the G test was applied for proportions expected to be equal, as recommended by Ayres et al. [9]. An alpha error was previously fixed at 5% for rejecting the null hypothesis. Data were collected by applying a questionnaire, which was compiled within an electronic spreadsheet (Excel software) and later processed at BioEstat, version 5.4—free Brazilian software with 210 statistical applications.

3 Results and Discussion

Data collected from 26 respondents, all coworkers at a Brazilian family-owned business in the industry of door and window frames, were analyzed. Their age ranges from 25 to

44 years old (38.5%) and 57.7% are from the feminine gender. Those who have already worked in the company for over 8 months represent 53.8%, whereas 26.9% of them work at the company from 8 months up to 1 year. This is a limiting factor for this research, once the COVID-19 pandemic was an important variable since it was not possible to completely understand the normal routine of the company. The most frequent position among the respondents was from assistant functions (30.8%).

Questions from this research were classified in factors with a common subject and were complementary, such as Information and Communication (questions 5–15), External Environment (questions 16, 17, 21, 24, 25, 32–39), Internal Environment (questions 22, 23, 26–31), Organizational Strategy (questions 18–20) and Organizational Capacity for Change (all the questions).

In terms of analyzing the factors: Information and Communication (A), External Environment (B), Internal Environment (C), Organizational Strategy (D), and Organizational Capacity for Change, a distribution was performed and, when analyzed by the Shapiro-Wilk normality test, resulted in a p-value = 0.0091, 0.0089, 0.0094, 0.0087 and 0.0097, respectively, which does not belong to a Gaussian distribution, as shown in Fig. 1 (A, B, C, and D) and Fig. 2. That curve shows the central values characterized by the median and the region within the first and the third quartile.

The average of the first factor (A), of 49.3 $\pm$ 18.8 points, means poor capacity in terms of information and communication. The smallest punctuation was observed for the Receptionist function, which reached only 22.7 points.

In (B), an average of 70.6 $\pm$ 4.2 points indicates that the company is well adjusted to the external environment. The "Owner" category performed very well in this subject, scoring 78.2 points, which is justifiable since they are in charge of the strategic level concerning the analysis of the external environment. Employees from the operational level, such as installers, interns, and assistants might not be aware of how the company performs such an analysis.

In (C), an average of 59.5 $\pm$ 4.6 points evidences the company has a satisfactory capacity to adjust to the internal environment. The highest scores were found for the categories of manager (65.6 points) and cutter (68.8 points).

In (D), an average of 64.1 $\pm$ 7.5 means the company has a satisfactory organizational strategy. The category "Manager" scored the highest: 75.0 points. Finally, the Organizational Capacity for Change (Fig. 2) assessment resulted in an average of 60.9 $\pm$ 4.4, which means the company has a satisfactory organizational capacity for change. The category "Owner" scored the highest: 68.8 points.

In order to examine the questions hindering internal changes in the family business, an evaluation was carried out for the items, from 0 to 100, according to their performance.

For the Information and Communication factor, the most influential questions were 7 and 13, with the worst (42.3 points) and best (57.7 points) performance, respectively. The former stated that "the process of communication reaches all the organizational units equally" and the latter "the formal communication resources match well to the needs of employees). Therefore, question #7 would be a change hindering factor concerning internal change, which means that communication may not reach all the organization units equally. And question #13 would be a facilitator since formal communication reached the highest score within this category.

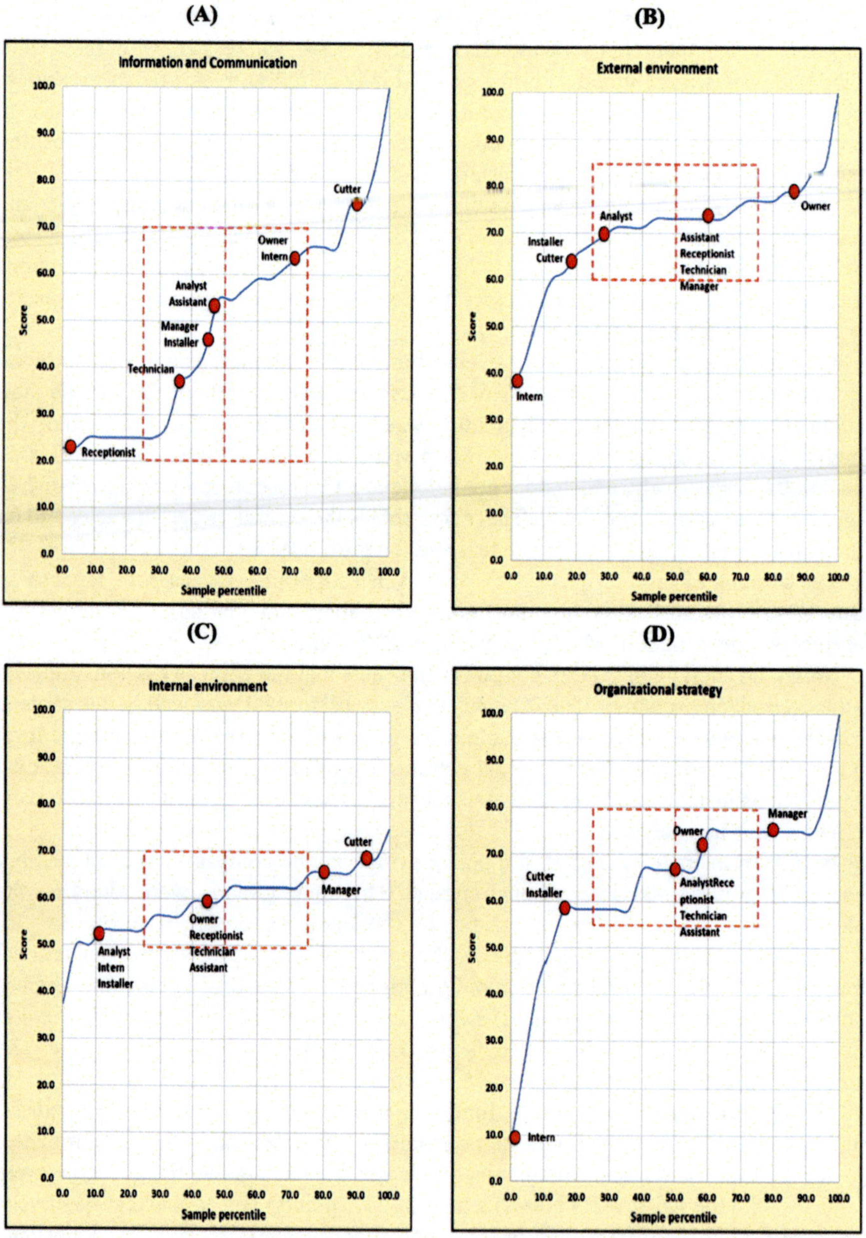

Fig. 1. Evaluation of information and communication factor (A), external environment (B), internal environment (C), and organization strategy (D) for change management inside a family business, based on data from n = 26 employees from an industry of door and window frames in Campinas, Brazil, 2021. Source: Plotted by the author by using BioEstat software, 2021.

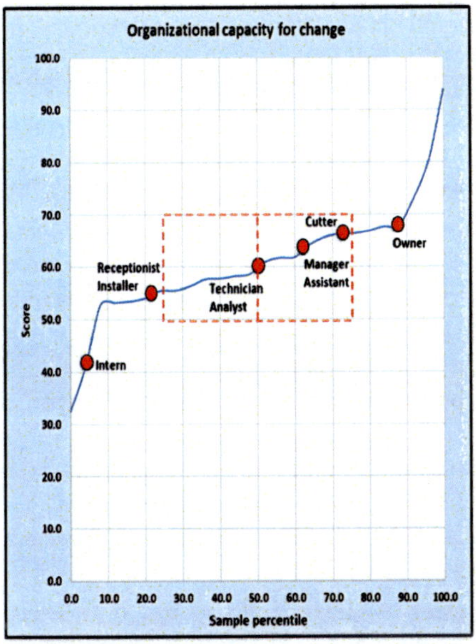

Fig. 2. Evaluation for the organizational capacity for change in a family business, based on data from n = 26 employees from an industry of door and window frames in Campinas/SP, Brazil, 2021. Source: plotted by the author by using BioEstat software, 2021.

In terms of the External Environment, the most influential questions were: facilitator (question #39–78.8 points) and hinderer (question #16–58.7 points). The former declares "the organization works within an extremely competitive environment", while the latter says "the organization has mechanisms to evaluate, periodically, the adequacy of its strategy to the demands from the external environment". It is understood that the employees are aware of the competitive external scenario in which the company thrives to survive, even though they do not believe the company has enough mechanisms to evaluate its strategy at the expense of external demands.

In terms of Internal Environment, the most influential questions were #26 and #27. The former has the following statement: "some groups in the organization were willing to resist to change" and the latter "some groups in the organization were willing to fight for change". Question #26 provided the lowest score (28.8 points) between the items since the scale is 0 for "totally disagree" and 1 for "disagree". Therefore, such a low score was thought of being a facilitator, once it was understood they were not willing to resist to changes. Question #27 (73.1 points) was also considered a facilitator since it was found the respondents were willing to make changes.

Finally, concerning organization strategy, the most influential questions were #18 and #19, with the worst (55.8 points) and best performance (69.2 points), respectively. The former had the statement "the organizational strategy was decided after an ample internal debate", while the latter stated "the organization strategy is based on the organization's vision of the future". Therefore, it was then assumed that the organization's future statement was the most influential in terms of organizational strategy and the internal debate would be subject to improvements.

It is understood that the company, according to the evaluation from the respondents, has some weaknesses. Regarding information and communication, there is essentially a hinderer associated with not matching communication goals within the entire organization hierarchy. The respondents do not believe the information was spread fast enough to all organizational units. Nonetheless, most of the respondents agreed upon that the company is in search of new ways to solve the issue. Such results indicate the need to further research in order to find means to improve communication and information dissemination.

The information and communication factor discussed herein is at odds with the results published by Neiva and Domingos [8]. That work considered it was a facilitator to change since the studied companies had a fast communication flow, objective, and accessibility to all employees. Conformity was observed in this work in terms of difficulty in handling changes from the external environment, even though the identification that the organizational structure is based on the external environment has already been found only for non-familiar organizations.

The choice of employees was made in such a way as to reach the highest number of respondents inside the organization. There was no discrimination from the researchers towards respondents in terms of their education level or function. It is possible that the employees from the operational level are not aware of the organizational strategy or how the company acts towards the external environment.

Nevertheless, the company is responsible for discovering why most of the respondents believed the organization should adopt a different way to structure its processes, which mechanisms could be adopted to enhance the lag in terms of responding to the external environment, and recognize the employees who supported internal changes.

4 Conclusions

From the description of the employees' profile at the company and their awareness of the challenges faced by the managers, it was possible to identify the results suggesting that the factors hindering change inside the chosen organization would be the dissemination of information not fast enough to all employees and organizational units, the difficulty in adjusting its strategy towards the external environment and its capacity to handle the pressure arising from such an environment. Despite the regular average scores found herein, the employees had expectancy for a higher sponsorship for making changes and a stronger alliance to make them come true.

In this respect, it was concluded that the aims of this research were reached, as long as it was understood that the family management system may have weaknesses, such as in terms of information and communication among the employees and in providing resources and instruments to encourage changes.

This study had its sample size as the main limitation since the reduced sample size allowed only the identification of results for that population. Besides, the length of time the employee has worked for the company impacted their vision about the company at pre-pandemic times. Considering future research, it is suggested the application of the questionnaire to companies with a larger size and number of employees, especially if that participation is primarily motivated by the high administration. Moreover, the study can also be applied to companies from different sectors, and finally comparing the data.

References

1. Cabral, J.: (2020) Empresas familiares são maioria no país. http://www.ceisebr.com/conteudo/empresas-familiares-sao-maioria-no-pais.html. Accessed 20 Sept 2020
2. PWC: (2020) Pesquisa global sobre empresas familiares. https://www.pwc.com.br/pt/estudos/setores-atividades/pcs/2018/pesquisa-empresas-familiares-18.pdf. Accessed 27 Sept 2020
3. Suddaby, R., Jaskiewicz, P.: Managing traditions: a critical capability for family business success. Fam. Bus. Rev. **33**(3), 234–243 (2020)
4. Erdogan, I., Rondi, E., Irmak, A.F.: Managing the tradition and innovation paradox in family firms: a family imprinting perspective. Entrep. Theory Pract. **44**(1), 20–54 (2019)
5. Ryerson University (2021) http://mtpinnacle.com/pdfs/ChangeManagementGuide_FINAL.pdf. Accessed 17 Sept 2021
6. Francisco, R.P., Machado, R.L.: A mudança eficaz: usando metodologias para auxiliar a liderança na implementação de novas visões ou estratégias. In: Encontro Nacional de Engenharia de Produção. Resumo... Porto Alegre, ABEPRO, Porto Alegre (2005)
7. Sanchez, W.: Gestão da Mudança. Senac, São Paulo (2020)
8. Neiva, E.R., Domingos, S.G.: Validação de instrumento para avaliação da capacidade organizacional para a mudança. In: Encontro da ANPAD, ANPAD, Rio de Janeiro (2011)
9. Ayres, M., et al.: Bioestat 5.3 aplicações estatísticas nas áreas das ciências biológicas e médicas. 5th ed. IDSM, Belém (2007)

Physical, Chemical and Rheological Properties of Blueberry Jam

Daniela Helena Pelegrine Guimarães[1](✉), Gisele Letícia Alves[2], Amanda Faria Querido[3], Pedro Felipe Arce Castillo[1], and Maria Thereza de Moraes Gomes Rosa[4]

[1] Lorena Engineering School (EEL-USP), Lorena, Brazil
{dhguima,parce}@usp.br
[2] Adolfo Lutz Institute, Central Laboratory of Public Health, São Paulo, Brazil
giseleleticia@uol.com.br
[3] Agronomy Department, Taubate University, Taubaté, Brazil
[4] Mackenzie Presbyterian University, Campinas, Brazil

Abstract. Blueberry stands out for its functional properties, which provide antioxidant potential to the fruit. Manufacturing jams are an alternative for the processing, utilization, and consumption of fruits. However, factors as the kind of pectin, sugar index, and the addition of soluble fibers may interfere with bioactive compounds' stability during the thermal process and storage. The proposal of this study includes the analysis of the effect of storage time on the physical, chemical, and texture properties of blueberry jams (traditional and light), immediately after processing and after 30, 90, and 120 days of storage. Results were analyzed using ANOVA and regression analysis. Soluble solids content showed no significant interaction with time for light products. Longer cooking times and a higher sugar concentration resulted in the instability of anthocyanins during storage. Time was of influence on most texture properties of the conventional products. For light jams, time exerted influence on hardness and adhesiveness. Physical, chemical, and rheological properties, on the other hand, were predicted by mathematical simulation by using molecular descriptors + smile codes with several architectures of artificial neural networks. The optimal configuration 10-30-10-3 was the architecture that obtained the lowest deviations for the dependent variables (hardness, water activity, and adhesiveness).

Keywords: Blueberry jam · Storage · Texture · Artificial neural networks

1 Introduction

1.1 Blueberry Background

Blueberry is a nutritious and very popular fruit containing high levels of anthocyanins, which contribute to the high antioxidant capacity and other biological activities of the fruit, including a disease-free and healthy life [1, 2, 3].

Y. Iano et al. (Eds.): BTSym 2021, SIST 207, pp. 436–442, 2023.
https://doi.org/10.1007/978-3-031-04435-9_44

However, blueberry is a perishable product where its industrialization emerges as an attempt to, at least, reduce waste [4]. In this context, manufacturing jams are an alternative to add value to the fruits; in addition, a product is achieved with pleasant sensorial properties and long shelf life [5, 6, 7].

Food, industrialized or not, remains in constant biological activity manifested by changes (chemical, physical, microbiological or enzymatic) that can lead not only to loss of nutrients but also to the alteration of other attributes, such as pH, acidity, texture, and viscosity [8].

Jam texture is an essential quality attribute that determines its acceptability. This property influences its appearance, taste, sensorial impressions in the mouth and hands [9]. That is, the consistency of a jam must be such that when extracted from its flask, it is capable of maintaining the semi-solid state, with a smooth texture and without resistance to cutting [10].

Despite the availability of some studies involving the stability of phenolic compounds in blueberry jams [11, 12, 13], there are few studies on the effect of storage time on the rheological (or texture) properties of this type of product.

2 Material and Methods

2.1 Materials

For the processing into jam, blueberries of the *Rabbiteye* species (*Climax* and *Powder Blue* varieties) were pulped by pulp machine. Conventional products were processed according to Santos et al. [14]. For the *Powder Blue* variety, a synthetic HM pectin was used, and for the *Climax* variety, HM pectin was extracted from passion fruit peel because, according to Guimarães et al. [13], these products presented a greater acceptance. *Light* formulations were prepared according to the scientific literature [15, 16].

2.2 Analytical Methods

In order to verify the effect of the storage on the characteristics of the jams, they were evaluated according to Lutz [17], regarding pH (Method 4.7.1), acidity (Method 4.7.2), soluble solids (Method 315/IV), non-reducing and reducing sugars (Method 309/IV), anthocyanins [18], water activity (Alpax, 650 AW Water activity meter), and rheological properties, at the first day of storage (immediately after processing) and after 30, 90, and 120 days. The rheological parameters were measured in a texturometer (Texture Analyser, TA- TX2) with a 1.0 cm^2 probe (TA3/100) and distance, pre-test, test, and post-test speeds of 30 mm, 2, 1, and 1 mm/s, respectively. The results, obtained from the force × time curve, were calculated by the Texture Expert Software Version 1.22. The analyzed parameters were hardness and adhesion. These analyses were performed in triplicate and the results were the means of the three replicates.

2.3 Statistical Analysis

The results of the physical, chemical, and texture properties were analyzed using the following statistical methodologies:

- Analysis of Variance (ANOVA), to evaluate the level of significance of the main effects and interactions, according to Ferreira [19];
- Regression analysis to evaluate changes during storage in relation to time. The results that did not have a significant effect according to ANOVA, were not plotted.

3 Results and Discussions

In Table 1, the effect of time on the physical and chemical properties, based on the ANOVA summary, is shown for the four products, where: FA: jam with synthetic high methoxyl pectin (HM) and *Powder Blue* fruit variety; FB: light jam with *Powder Blue* fruit variety and synthetic low methoxyl amidated (LM) pectin; FC: jam with high methoxyl pectin (HM) extracted from passion fruit and *Climax* fruit variety; FD: light jam with *Climax* fruit variety and synthetic low methoxyl amidated (LM) pectin.

From the experimental results of the variables analyzed in Table 1, the regression models were established at the 5% level of significance (Table 2). The adequacy of these models can be verified by the determination coefficients (R^2). In the present work, the low value of R^2 for time and soluble solids content in the jam made with fruit of the *Climax* variety and HM pectin extracted from passion fruit peel (FC), indicates that the adjusted model was not adequate for the experimental data.

According to the results presented in Table 1, pH and acidity showed no significant interaction with time for all jams (since $F_{tab} = 3.24$). With respect to the soluble solids content, there is a significant interaction with the time only for the conventional products (FA and FC), indicating a tendency to crystallization at the end of storage [20]; this tendency was more evident when synthetic pectin HM was used, since the hydration level of the pectin extracted from the passion fruit is high. Regarding the water activity, the storage time influenced only the product prepared with the *Climax* variety and pectin extracted from passion fruits (FC); in this case, sucrose hydrolysis occurred, as evidenced by the increase in the reducing sugar content.

The reduction of the anthocyanin content as storage time progressed was not significant in the conventional product prepared with synthetic HM pectin (FA). In this case, according to the results presented by Guimarães et al. [13], the longer cooking process (when compared to FB and FD), in addition to the higher concentration of sucrose (when compared to FC), promoted higher losses of this pigment and the initial anthocyanin content of the product immediately after processing was therefore lower. This way, the greater proportions of anthocyanins in FB, FC, and FD intensified possible pigment interactions with ascorbic acid, oxygen, or light during the storage. Howard et al. [12] analyzed the jam processing and storage effects on blueberry polyphenolics and concluded that sugar-free versions retained higher levels of anthocyanins than sugar jams late during storage. However, in the present work, the light products were prepared with 50% of the sugar amount of conventional jams. This amount of sugar was enough to produce polymers of degradation products that diminish their stability. These results corroborate the experimental results from Melgarejo et al. [21], which show that anthocyanins in pomegranate jam are quite unstable during storage at 25 °C. Although the processing into jam decreased the anthocyanin content, the jam can still be considered a source of this phenolic compound due to the high content of anthocyanin even after processing, according to Sellappan et al. [22]. The authors obtained results for the anthocyanin content in berries in the range of 12.70 to 197.34 mg/100 g and these values classified blueberries and blackberries as rich sources of this pigment.

Table 3 shows the ANOVA results for the rheological parameters and based on these the regression models were set at a level of 5% of significance (Table 4).

Table 1. Summary of ANOVA of the physical and chemical characteristics of blueberry jam during storage.

	pH	Acidy	NRS	RS	Anthocyanins	SS	WA
FA	0.03	2.39	1.35	0.09	1.17	16.47*	2.29
FB	0.08	0.07	1.76	0.33	23.44*	1.43	0.10
FC	0.22	2.29	0.21	5.91*	53.65*	3.52*	11.64*
FD	0.58	1.91	1.76	1.04	41.51*	0.17	0.97

NRS: Non Reducing Sugars; RS: reducing Sugars; SS: Soluble Solids; WA: Water Activity.

Table 2. Regression equations with coded variables and determination coefficients for the physical and chemical analyzes of blueberry jam.

Variable response	Time (days)	Product	Estimating model	R^2
RS	120	FC	$y = 64.79 - 0.083x$	0.75
Anthocyanins	120	FB	$y = 53.09 - 0.117x$	0.92
Anthocyanins	120	FC	$y = 39.25 - 0.136x$	0.96
Anthocyanins	120	FD	$y = 73.52 - 0.251x$	0.95
SS	120	FA	$y = 47.85 + 0.046x$	0.89
SS	120	FC	$y = 52.11 - 0.009x$	0.08
WA	120	FC	$y = 0.73 - 0.001x$	0.85

From the results presented in Table 3, a significant effect of time on adhesiveness can be observed for all products. With respect to hardness, this parameter was not significantly influenced at time 0 (jam just prepared) with the *Climax* fruit variety and pectin extracted from passion fruit peel (FC), with consequent maintenance of gel stability. In relation to the other formulations, Table 4 shows that the hardness reduced throughout the storage (as evidenced by the negative angular coefficient), which can be explained

Table 3. Summary of ANOVA of the rheological properties of blueberry jam during storage.

		F_{cal}	
Product	F_{tab}	Hardness	Adhesiveness
FA	3.240	13.083*	95.474*
FB	3.240	24.147*	79.775*
FC	3.240	0.199	7.234*
FD	3.240	17.982*	5.272*

Table 4. Regression equations with coded variables and determination coefficients for the rheological properties of blueberry jam.

Variable response	Time (days)	Product	Estimating model	R^2
Hardness	120	FA	$y = 1187.50 - 2.767x$	0.87
Hardness	120	FB	$y = 650.31 - 1.822x$	0.92
Hardness	120	FD	$y = 613.38 - 1.086x$	0.90
Adhesiveness	120	FA	$y = 86.29 + 0.583x$	0.78
Adhesiveness	120	FB	$y = 34.42 - 0.146x$	0.96
Adhesiveness	120	FC	$y = 68.41 - 0.260x$	0.78
Adhesiveness	120	FD	$y = 33.05 - 0.104x$	0.56

by the discontinuity of the structure of the jam, related to a possible syneresis. The same effect was observed by Dias et al. [23], who observed that the firmness of the jam made from the banana (*Musa spp.*) peel had its value reduced throughout the storage, at different temperatures.

Regarding conventional jams, a great difference was observed in the rheological properties between the product prepared from the *Powder Blue* (FA) and *Climax* (FC) varieties. It should be noted that, in the latter case, the jam was made with the pectin extracted from the passion fruit peel, which was highly hydrated (with a moisture content of around 90%) during processing. This way, a partial breakdown of the molecules may have occurred, with a consequent decrease in molecular weight, as well as greater demethoxylation of the esterified groups [24].

The storage resulted in a significant increase in the adhesiveness of the traditional jam prepared from the *Powder Blue* (FA) fruit, possibly due to the increase of the soluble solids content since, according to Rahman and Al Farsi [25], the adhesiveness increases when the moisture content of the gel decreases, which would explain the occurrence in the present work, as an increase in the soluble solids content was observed (with consequent reduction of the moisture content). The opposite effect was observed in the traditional jam made with the *Climax* variety, where a reduction of the solids content was observed during storage with a consequent reduction of the adhesiveness. This observation corroborates by Oliveira et al. [8], who reported a tendency for soluble solids content to increase during the storing of fruit jams.

4 Conclusion

Regarding the physical, chemical properties of the jam, the storage time had a significant influence on the quality of the product prepared with the *Climax* variety and pectin extracted from passion fruit, with a decrease in water activity, anthocyanins, reducing sugars and soluble solids content.

Regarding the rheological parameters of the gels, the conventional product prepared with synthetic HM pectin was the most influenced throughout storage, where one can observe that time interfered with the hardness and adhesiveness of the final product.

An artificial neural network model was used to predict the physical, chemical, and rheological properties of four types of blueberry jams using time (days), % water, % citric acid, % glucose, % sucrose, % anthocyanin, % HM pectin, % LM pectin, % xanthan fim, pH, acidity (%), and molecular descriptors as independent variables. The ANN model can be used to model the physical, chemical, and rheological properties of blueberry jams by using molecular descriptors. For different values of the independent variables, the physical, chemical and rheological properties of blueberry jam can be predicted with good accuracy with the optimum architecture found of four layers, represented as 10-30-10-3 (where 10 represents the number of independent variables, 30 and 10 represent the node numbers of the hidden layers, and 3 represents the number of dependent variables), returning overall average absolute deviations below 0.06, 0.07 and 0.13% for the three dependent variables: hardness (g), water activity and adhesiveness (mJ), respectively.

Acknowledgment. Fapesp (2007/59114-4).

References

1. Castagnini, J.M., Zapata, L.M., Quinteros, C.F., Noceti, A.: Multiple response optimization of blueberry juice depectinization. Cienc **47**, 1–8 (2017)
2. Zou, Y., Hou, X.: Sonication enhances quality and antioxidant activity of blueberry juice. Food Sci. Tech. **37**, 599–603 (2017)
3. Avram, A.M., Morin, P., Brownmiller, C., Howard, L.R., Sengupta, A., Wickramasinghe, R.: Concentrations of polyphenols from blueberry pomace extract using nanofiltration. Food Bioprod. Process **106**, 91–101 (2017)
4. Goldmeyer, B., Penna, N.G., Melo, A., Rosa, C.S.R.: Características físico-químicas e propriedades funcionais tecnológicas do bagaço de mirtilo fermentado e suas farinhas. Rev Bras Frut **36**, 980–987 (2014)
5. Lespinard, A.R., Bambicha, R., Mascheroni, R.H.: Quality parameters assessment in kiwi jam during during pasteurization. Modelling optimization of the thermal process. Food Bioprod. Process **90**, 799–808 (2012)
6. Schafaschek, T.P., Hickel, E.R., Pereira, H.L., Oliveira, C.A.L., Toledo, V.A.A.: Performance of Africanized honeybee colonies settled by queens selected for different traits. Acta Sci Animal Sci **38**, 91–100 (2016)

7. Curi, P.N., et al.: Characterization of different native american physalis species and evaluation of their processing potential as jam in combination with brie-type cheese. Food Sci. Technol. **38**, 112–119 (2017)
8. Oliveira, E.N.A., Santos, D.C., Rocha, A.P.T., Gomes, J.P.: Development, characterization and stability of traditional umbu-caja jam. Rev. Bras. Frut. **36**, 640–651 (2014)
9. Abu-Gahannan, N., Crowley, H.: The effect of low temperature blanching on the texture of whole processed potatoes. J. Food Eng. **74**, 335–344 (2006)
10. Menezes, C.C., Borges, S.V., Cirillo, M.A., Ferrua, F.Q., Oliveira, L.F., Mesquita, K.S.: Physical and physico-chemical characterization of different formulations of guava preserve (Psidium guajava, L.) from Pedro Sato cultivar. Food Sci. Tech. **29**, 618–625 (2009)
11. Brownmiller, C., Howard, L.R., Prior, R.L.: Processing and storage effects on monomeric anthocyanins, percent polymeric color, and antioxidant capacity of processed blueberry products. J. Food Sci. **73**, H72–H79 (2008)
12. Howard, L.R., Castrodale, C., Brownmillers, C., Mauromoustakos, A.: jam processing and storage effects on blueberry polyphenolics and antioxidant capacity. J. Agri. Food Chem. **58**, 4022–4029 (2010)
13. Guimarães, D.H.P., Alves, G.L., Querido, A.F.: Geleia de mirtilo (Blueberry): análises dos parâmetros sensoriais e do efeito do armazenamento nas propriedades físicas e químicas. RBTA. **5**, 19–25 (2014)
14. Santos, P.R.G., et al.: Geleia de cagaita (Eugenia dysenterica DC.): desenvolvimento, caracterização microbiológica, sensorial, química e estudo da estabilidade. Adolfo Lutz Institute. **71**, 281–290 (2012)
15. Lago, V.E.S., Ramin, P., Umsza, M.A.G., Santos, G.V., Silva, R.: Chemical and sensory characteristics of pulp and peel "cajá-manga" (Spoldias cytherea Sonn.) jelly. Food Sci. Tech. **31**, 398–405 (2011)
16. Granada, G.G., Zambiazi, R.C., Mendonça, C.R.B., Silva, E.: Caracterização física, química, microbiológica e sensorial de geleias light de abacaxi. Food Sci. Tech. **25**, 629–635 (2005)
17. Lutz, A. Normas Analíticas do Instituto Adolfo Lutz. Débora Rebocho, São Paulo (1985)
18. Lima, V.L.A.G., Melo, E.A., Maciel, M.I.S., Lima, D.E.S.: Avaliação do teor de antocianinas em polpas de acerola congelada proveniente de frutos de 12 diferentes aceroleiras (Malpighia emarginata). Food Sci. Tech. **23**, 101–103 (2003)
19. Ferreira, D.F.: Análise Multivariada. UFLA, Lavras (1996)
20. Queiroz, M.B., Braúna, I.R.L., Fadini, A.L., Kieckbusch, T.G.: Solubilidade de equilíbrio de sacarose em misturas multicomponentes de açúcares. Braz. J. Food Tech. **11**, 70–77 (2008)
21. Melgarejo, P., Martínez, R., Hernández, F.C.A., Martínez, J.J., Legua, P.: Anthocyanin content and color development of pomegranate jam. Food Biopro. Process **89**, 477–481 (2011)
22. Sellappan, S., Akoh, C.C., Krwer, G.: Phenolic compounds and antioxidant capacity of Georgia-grown blueberries and blackberries. J. Agr. Food Chem. **50**, 2432–2438 (2002)
23. Dias, C.S., Borges, S.V., Queiroz, F., Pereira, P.A.P.: Effects of temperature on the physical, physicochemical, chemical and microbiological alterations during of the storage of jelly made from Cv. Prata banana (*Musa* spp.) peel. Adolfo Lutz Institute. **70**, 28–34 (2011)
24. Bobbio, F.O., Bobbio, P.A. Introdução à Química dos Alimentos. Varella, São Paulo (2003)
25. Rahman, M.S., Al Farsi, A.S.: Instrumental texture profile analysis of the date flesh as a function of moisture content. J. Food Eng. **66**, 505–511 (2005)

Flipped Classroom to Improve the Learning of Mathematics in Students of the Public University of Lambayeque

Diana Mercedes Castro Cardenas[1(✉)], Segundo Francisco Segura Altamirano[1], and Merly Liliana Yataco Bernaola[2]

[1] Pedro Ruiz Gallo National University, Lambayeque, Peru
{dcastroc,sseguraal}@unprg.edu.pe

[2] San Luis Gonzaga National University, Ica, Peru
merly.yataco@unica.edu.pe

Abstract. Usually, in the Mathematics courses in Engineering careers at universities, there are many disapproved students, and even those who pass lack adequate mastery of the developed topics. We consider that the causes of this problem are the lack of motivation that, along with traditional methods, do not favor student learning. For this reason, we propose the Flipped Classroom, with constructive alignment to improve students' performance in these courses that will be measured using the SOLO Taxonomy, which allows evaluating different levels of learning. We use a quantitative approach with a pre-experimental design, applying a pre-test, followed by the traditional method, and, then, a post-test. In the second part of the course, we used the Flipped Classroom and compared the test results. We noticed a significant difference between the final grades of exam 1 (traditional method) and exam 2 (Flipped Classroom method), considering that the average value of the Flipped Classroom grade (14.7 ± 0.279) is much higher than the average value of the score when we used the traditional method (12.5 ± 0.279). We then affirm that when we use Flipped Classroom, the students achieve a better performance than the traditional method.

Keywords: Flipped classroom · Student performance · Traditional method · Constructive alignment · SOLO Taxonomy

1 Introduction

In more than 16 years of experience as a teacher of public university education (most of them developing Mathematics courses at different schools of Engineering), it was possible to see that the common denominator of the students is their little interest in these courses and the consequent lack of commitment that, in the end, affects negatively their final grades and also their performance in subsequent courses to which they are not sufficiently prepared.

Y. Iano et al. (Eds.): BTSym 2021, SIST 207, pp. 443–450, 2023.
https://doi.org/10.1007/978-3-031-04435-9_45

The relationship between engagement in mathematics courses (Specifically Discrete Mathematics) and academic performance was studied in [1] and [2], who found that these variables are strongly correlated. In addition, they indicate that students who perform adequately use technology to learn and have found higher levels of engagement and academic performance. This is partly a consequence of a wrong perception of the student whose activities possibly seek to emphasize skills that have little to do with educational objectives [3].

In our reality, a possible cause of the little interest of the students is due to the fact that, in a traditional development of the courses, the center is still the teacher, which in general does not manage to commit to the learning of the students, due to the inappropriate use of new tools that include new hardware and software technologies [4,5]. Despite the fact that the majority, if not all of our students belong to the so-called Millennial generation, who demonstrate great mastery in the management of technological gadgets and their applications but little tolerance for reading [6], they can facilitate methodological strategies based on technological means. In addition, [7] and [8] present strategies to achieve level 5 learning (Taxonomy Only), evaluating the competencies of students in the first cycles of Sciences and Engineering, where it is shown that the majority have a good perception of the implementation of this new model, which also includes new forms of evaluation that allow adequate and even personalized monitoring of the learning path followed by the students.

Besides that, the use of student-centered methodologies, which have shown a high level of approval on the part of students [9], allows teachers to design different activities and evaluations before, during, and after the classes, which have proven to be effective in improving academic performance [10].

One of the promising active learning strategies is “Flipped Classroom” or “Inverted Class”, and its variants, which consist basically in inverting the activities developed at class and home, i.e., what was done during class is now done at home, and what was done at home is now done in class [11]. But, in reality, it means much more than that.

To better understand the term, we describe how the inverted class “looks” on a day-to-day basis. First, the video seen before is discussed in a few minutes. It is necessary to train students to watch the videos effectively, which is achieved by eliminating all distractions while watching the video. We also indicate that they can “pause” or “back off” their teacher. In addition, in moments of pause, they write down the key points of the lesson, thus summarizing his learning. This method enables students to come to class with the appropriate questions. We can use these questions to evaluate the effectiveness of our videos since if each student has a similar question, it is clear that we do not teach that topic well, and that particular video must be corrected.

2 Methodology

In the present investigation, we use a quantitative approach, where the observation units are the results of the students' exams, the Mathematical Analysis II course, and the exam questions. Variables were measured, such as autonomous learning and the degree of knowledge acquired; according to the SOLO taxonomy, we used the answers of the surveys, questionnaires, and the answers of the exams in the different stages. The research design is pre-experimental (Fig. 1), since it is a single group, with pre-test and post-test. It consisted of applying a pre-test where we used the SOLO taxonomy to measure the level of knowledge reached in the subject, carrying out traditional classes, and applying a post-test where we used the SOLO taxonomy to measure the level of knowledge achieved in the subject after the classes. In the following week of the course, we evaluated with a pre-test before the topic, applied Inverted Class, and evaluated the level of knowledge using the SOLO taxonomy. We compared the pre-tests and the post-tests to corroborate the hypothesis's validity that if the Inverted Class method is applied, students achieve a higher level of learning.

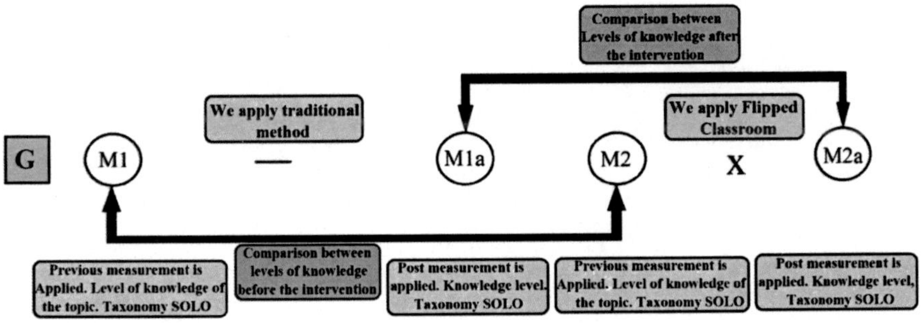

Fig. 1. Pre-experimental research design

In Fig. 2, we show more details, with the SOLO taxonomy. We classify both the learning outcomes of the students and the materials and evaluations in order to control the degree of student learning, starting with Multistructural learning (Level 1), which allows the generalization only in terms of a few and limited independent aspects, continuing with Relational learning (Level II), where it generalizes within the received context or experience using related aspects, up to the Extended Summary (Level III), using the deduction and induction to make generalizations to unexperienced situations [12].

In the School of Electronics of the Pedro Ruiz Gallo National University, there are 147 students in the first three cycles, and we have 40 students enrolled in the Mathematical Analysis II course.

As mentioned above, we applied a pre-test to students of the Mathematical Analysis II course of the Electronics Engineering Program. We focused on week 1, with the traditional method, where we evaluated whether the student "knew the concept of the antiderivative and the definition of the indefinite integral". In week 2, we applied the Flipped Classroom, with the performance "Know the integration technique of change of trigonometric variable and integration by parts".

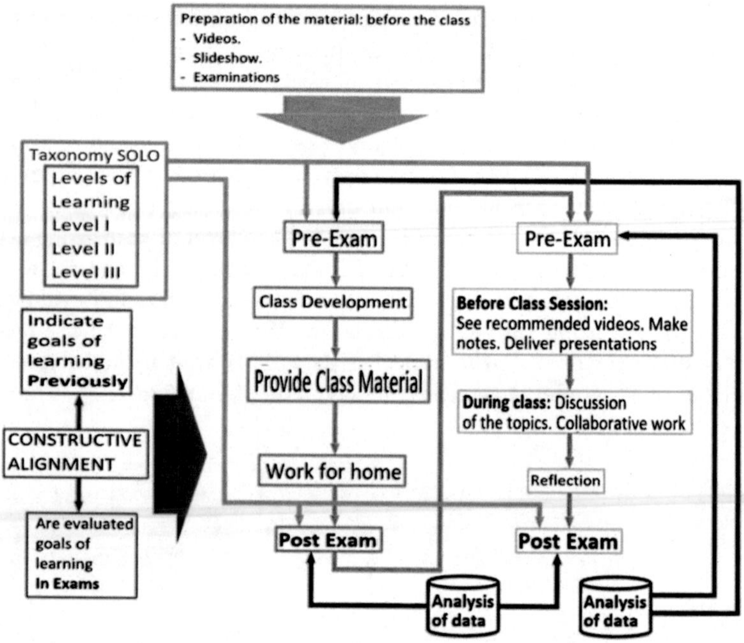

Fig. 2. Pre-experimental research details

3 Results

In this research, we generate summary statistics from the analysis of the results of the applied tests, which we compare to measure the statistical difference so that we can infer the influence of the method used and offer recommendations to serve for later analysis. Since the results of the exams are independent of each other, in addition to being the same size, we have chosen to use the ANOVA Test.

We use the null hypothesis (Ho): There is no significant difference in the students' performance in the tests carried out. We also have an alternative hypothesis (H1): We will achieve a significant difference in students' performance in the exams taken. We assume that the test results are consistent for each assessment. It is a non-directional (2-tailed) hypothesis test. As a first assumption, we seek to perform the test at a 5% significance level. We use the ANOVA test to check if the samples belong to the same distribution function. We check whether the results of the examinations of each applied method have the same distribution.

We have tested homogeneity and normality and verified that there are no extreme observations that could affect the results obtained; we performed ANCOVA where we verified that the difference is statistically significant between the final marks of exam 1 (applying the Traditional model) and exam 2 (applying Flipped Classroom), since $F(1, 77) = 30.105$, $p \leq 0.0001$.

An analysis of which means differ has also been carried out, with Bonferroni adjustment, to identify the ranges of difference of the means, in such a way that it is determined that the average of the final grade of exam 2, where we used Flipped Classroom, (14.7 ± 0.279) is higher if we compare with the average of the final grade of exam 1, where we used the traditional method, (12.5 ± 0.279). We evaluated the Flipped Classroom and the learning achievements of the students at the different levels proposed in this research, classified the results of exams 1 and 2 in the different levels, indicated the strategy used (whether it is the traditional method or Flipped Classroom), and considered the methodology used and the levels of learning as independent variables as dependent variables, respectively. We performed ANCOVA and found a statistically significant interaction between the methodology used and the levels considered in this research, with $F(2, 233) = 8.056$, $p = 0.000414$.

We evaluated the effect of a variable at each learning level and vice versa. We analyzed the effect of the methodology used in each level of learning and executed three tests. Now the significance level will be $0.05/3 = 0.0166667$. We indicated that the strategy used was statistically significant in levels 1 and 2 with $p = 0.00000231$ and $p = 0.0000000565$. However, it was not statistically significant at level 3 ($p = 0.767$).

4 Discussion

From the results of the averages and the median of the results of the exams, it can be observed that although in both pre-exams there is a final failing grade, in the exams, there is already a passing grade that in its average value exceeds 12.0 mark. In addition to the results in each of the levels in the exams, it can be seen that the contribution of the grade from level 3 to the final grade in both pre-exams is small, but in the exams at the end of each strategy, their contribution is similar and in some cases higher than that of the other levels.

In Fig. 3, we graphically observe the significant statistical difference, where we use an analysis of differences of means, with Bonferroni adjustment, where we identify the ranges of difference of the means, with the result that the mean of exam 2, after the application of the Flipped Classroom Method with a value of 14.7 ± 0.279, which is higher than the average grade of exam 1 (where we used the traditional method). We affirm that the Flipped Classroom allows students to achieve greater learning than the traditional method.

It has been essential to understand how the application of the Flipped Classroom Method affects the learning achievements of the students in the three levels considered. For this, the examinations proposed consisted of three sections of differentiated questions that allowed the evaluation of the learning achievement in each of the mentioned levels. In this way, it has been possible to collect the evaluations taking into account the levels of learning and the methodology used, which has made it possible to evaluate the effect of the independent variable on the independent variable considered in the levels of learning.

We found a statistically significant interaction between the methodology used and the levels considered in this research, with $F(2, 233) = 8.056$, $p = 0.000414$.

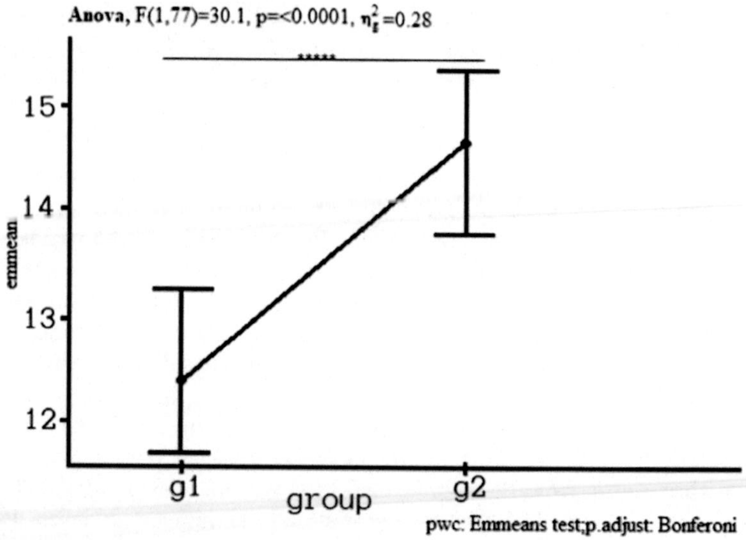

Fig. 3. Difference ranges of average values

In other words, the effect of the methodology used in the final grade also depends on the levels of learning.

It can be observed in the results (See Fig. 4), that the effect of the strategy used was statistically significant at levels 1 and 2 with $p = 0.00000231$ and $p = 0.0000000565$, respectively. However, it was not statistically significant at level 3 ($p = 0.767$). These results were verified (See Fig. 4) by observing the ranges of difference of the means. There is a statistically significant difference in levels 1 and 2, with a trend that level 2 is lower in both exams. However, there is no statistically significant difference at level 3 since the ranks overlap.

We consolidate that in the final marks of the exams applied at the end of each methodology considered in the studies, the average Flipped Classroom Method is higher than the average of the Traditional Method. This difference is statistically significant, although in similar studies, such as [13], where the authors find that the knowledge achieved by students using the Flipped Classroom method is similar to those achieved using the methods based on readings.

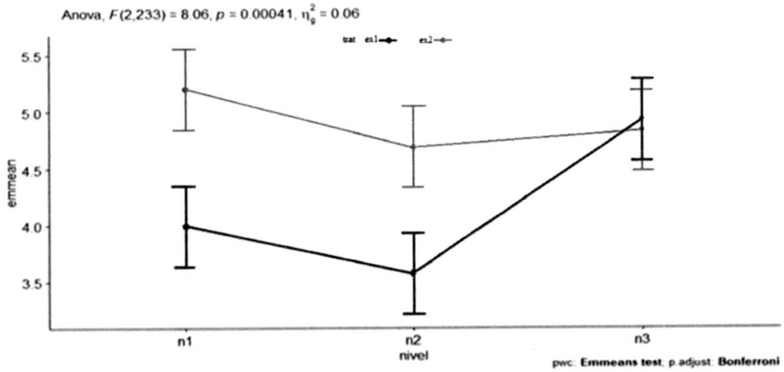

Fig. 4. Difference ranges of the means at the three levels

5 Conclusions

We found a significant difference between the final grades of exam 1 (where we used the traditional method) and exam 2 (where we used the Flipped Classroom method); we noticed the difference considering that the average value of the Flipped Classroom grade (14.7 ± 0.279) is much higher than the average value of the note where we use the traditional method (12.5 ± 0.279). We then affirm that when we use Flipped Classroom, the students achieve a better performance than the traditional method since the final grades resulting in level 3 are higher than those in the traditional method.

The applied methodology affects student achievement at the levels considered in this research since the findings determine that the effect of the strategy used was statistically significant at levels 1 and 2 with $p = 0.00000231$ and $p = 0.0000000565$, respectively. However, it is not statistically significant at level 3 ($p = 0.767$).

The students achieved more significant learning with the Flipped Classroom method since the averages of the pre-exams are similar. However, the difference between the marks of their corresponding post-exam in the case of the Flipped Classroom is more significant, which means that the students achieved further learning with this method.

It is advisable to apply the Flipped Classroom method in all university Calculus classes, since it allows the students to develop their exercises in class with the teacher's supervision, which helps them detect their errors, improving learning and disclosing the benefits they present, both the Flipped Classroom, Constructive Alignment, and Taxonomy Only in the achievements of the different levels of student learning. Finally, it is also advisable to foster workshops that allow the development of digital skills of teachers in software and hardware tools, which allow them to develop multimedia content for their different learning sessions.

References

1. González, T., García, A.: Estudio de los factores de estudiantes y aulas que intervienen en el "engagement" y rendimiento académico en Matemáticas Discretas. Revista Complutense de Educación **31**(2), 195–206 (2020). https://doi.org/10.5209/rced.62011
2. Fernández, M., Gaytán, P.: Experiencia del aula invertida para promover estudiantes prosumidores del nivel superior. RIED. Revista Iberoamericana de Educación a Distancia **22**(2), 245–263 (2019). https://doi.org/10.5944/ried.22.2.23065
3. Margolis, E.: The Hidden Curriculum in Higher Education. 1er ed. Routledge, Taylor & Francis Group, New York (2001)
4. Luzardo, M., Sandia, B., Aguilar, A.: Conocimiento y frecuencia del uso de las tecnologías de información y comunicación en la práctica educativa. Variables sociodemográficas de los docentes en la Universidad de Los Andes. Revista Cubana de Educación Superior **39**(1), 45–62 (2020)
5. Ruiz, A.: ACUTIC: cuestionario para el estudio de la actitud, el conocimiento y el uso de TIC en profesores de educación superior. Investigar con y para la sociedad, **3**, 1733–1744 (2015). ISBN 978-84-686-6906-9
6. Roehl, A., Reddy, S., Shannon, G.: The flipped classroom: an opportunity to engage millennial students through active learning strategies. J. Family Consum. Sci. **105**(2), 44–49 (2013)
7. Sologuren, E., Nuñez, C., Gonzáles, M.: La implementación de metodologías activas de enseñanza-aprendizaje en educación superior para el desarrollo de las competencias genéricas de innovación y comunicación en los primeros años de Ingeniería. Cuaderno de Pedagogía Universitaria **16**(32), 19–34 (2019)
8. Uribe, R., Sologuren, E., Matamoros, A.: Instalación e implementación de las competencias comunicativas en la Facultad de Ciencias Físicas y Matemáticas de la Universidad de Chile. In: XXIX Congreso Chileno de Educación en Ingeniería, Santiago de Chile (2016)
9. Gilboy, M., Heinerichs, S., Pazzaglia, G.: Enhancing student engagement using the flipped classroom. J. Nutrition Educ. Behav. **47**(1), 109–114 (2015). https://doi.org/10.1016/j.jneb.2014.08.008
10. Rodríguez, L., Díaz, P.: Estrategias de las universidades españolas para mejorar el rendimiento en matemáticas del alumnado de nuevo ingreso. Aula Abierta **43**(2), 69–76 (2015). https://doi.org/10.1016/j.aula.2015.01.002
11. Bergmann, J., Sams, A.: Flip Your Classroom: Reaching Every Student in Every Class Every Day, 1er edn. International Society for Technology in Education, Washington DC (2012)
12. Biggs, J., Collis, K.: Evaluating the Quality of Learning: The SOLO Taxonomy, 1er edn. Academic Press Inc London LTD, United Kingdom (1982)
13. Scott, C., Green, L., Etheridge, D.: A comparison between flipped and lecture-based instruction in the calculus classroom. J. Appl. Res. Higher Educ. **8**(2), 252–264 (2016). https://doi.org/10.1108/JARHE-04-2015-0024

Simulation of Hail Damage in the Amylaceous Corn Production Acobamba, Huancavelica

William Huamán-Tovar, Roberto Chuquilín-Goicochea, Arcadio Sánchez-Onofre, and Ruggerths De La Cruz-Marcos(✉)

Universidad Nacional de Huancavelica, Acobamba, Peru
{arcadio.sanchez,ruggerths.delacruz}@unh.edu.pe

Abstract. During the 2018–2019 agricultural season, in the town of Choclococha, province of Acobamba—Huancavelica, located at 3440 m above sea level, hail damage was simulated 90 days after planting to evaluate the effects on the yield of amylaceous corn under rainfed conditions. A randomized block design with 7 treatments was used: T1 (no defoliation, Control), T2 (defoliation in 25% of the laminae of each leaf), T3 (defoliation in 50% of the laminae of each leaf); T4 (defoliation in 75% of the laminae of each leaf), T5 (defoliation in 100% of the laminae of each leaf), T6 (total tearing of the laminae of the leaf) and T7 (total tearing of the laminae and rib break-age); all in triplicate. A significant effect was found in the components of corn yield ($p < 0.05$), except for plant height, at 120 days after planting. The decrease in dry grain yield ranged from 13.53 to 61% with respect to the witness treatment. A significant negative linear relationship was found between the percentage of defoliation and the number (90.66%) and weight (90.25%) of grains per ear, and yield (90.25%). The results would allow estimates to make coherent decisions to compensate for agricultural insurance and the implementation of policies related to food security.

Keywords: Hail damage · Amylaceous corn · Leaf area index

1 Introduction

Eighty percent of the world's food production is produced by smallholder farmers, who account for 75% of farms holdings and comprise 60% of the agricultural labor force worldwide [1]. Most of these farmers are food insecure due to their vulnerability to factors, including climatic trends and resources, domestic shocks, and seasonality [2]. The province of Acobamba belongs to the Huancavelica region in Peru (12°50′27″S 74°34′14″W). Its agricultural production is based on corn, beans, and peas and presents a high risk, as a high Andean area, due to a variety of adverse climatic factors such as drought, frost, wind, hail, and soil salinity [3].

Frosts are of great importance in the highlands of the Andes, especially in the southern part of Peru and Bolivia, with significant diurnal temperature fluctuations and night frosts up to 200 days per year [4]. In recent years, the occurrence of sudden low temperatures (called frosts) is a major natural disaster recognized by the Civil Defense Institute in

Y. Iano et al. (Eds.): BTSym 2021, SIST 207, pp. 451–457, 2023.
https://doi.org/10.1007/978-3-031-04435-9_46

Peru since 2006 [5]. One of the frost events is hail and has an effect on maize due to the loss of foliage area [6], better known as defoliation. There is an effect of defoliation on corn grain yield. In fact, defoliations have been used to simulate hail damage [7]. An optimal understanding of the seasonal factors, such as solid precipitation (hail), that influence food insecurity would allow the systematic incorporation into development programs of measures to help meet the needs of farming families [8]. For this reason, this study has been implemented to simulate the effects of solid precipitation on the yield of amylaceous maize in an Andean area such as Acobamba, Huancavelica.

2 Materials and Methodology

2.1 Materials

Ccarhuay ecotype amylaceous maize seed, commonly used in the province of Acobamba, belongs to the harvest of the previous season (2017–2018). Decomposed cow manure was used as fertilizer. The field was located in the town center of Choclococcocha, Acobamba, where in the last three agricultural seasons amylaceous corn was planted. The experiment was conducted in the fourth consecutive year of maize planting.

2.2 Field Preparation

The field was prepared with agricultural machinery, plowing was done with a disk plow, and then manual weeding was done manually using a pickaxe. The field was demarcated with plaster to separate the blocks and experimental units, and wooden stakes were placed at the corners of the blocks (day 1).

2.3 Sowing, Fertilization, and Cultural Practices

Sowing was carried out in furrows manually, using picks, spaced 0.70 m apart. The seeds were deposited at the bottom of the furrow, three seeds were deposited 0.40 m apart (day 2). Fertilization was carried out in two moments. The first moment was at sowing, first with decomposed manure, it was applied to the bottom of the furrow between the spaces of the seeds at a rate of 50 g. Then, 50% of a mixture of chemical fertilizer was applied at a dose of 100 – 100 – 80 NPK (urea, di-ammonium phosphate, and potassium chloride) in the form of a continuous stream at the bottom of the furrow and covered with a light layer of soil, prior to seed placement. The second moment was in the hilling, in which 50% of the remaining mixture was applied. Weeding was done manually, removing the soil slightly and removing the weeds that were present in the field. The weeds identified were: yuyo or turnip (*Brassica silvestris*), barley (*Avena sativa*), and kikuyo (*Pennisetum sp*). A single manual hilling was carried out when the plants were approximately 40 to 50 cm high. This was done to provide the plants with support and greater anchorage (day 42). Harvesting was carried out when the corn had finished its physiological maturity, the cobs were expected to mature and lose moisture naturally, an indicator of this was that the cobs showed a decay parallel to the stalk (day 120).

2.4 Hail Damage Simulation

The simulation of hail damage (solid precipitation) was carried out manually on 68 plants sing scissors to simulate leaf damage, cutting the leaves of the plants and tearing them, as previously established in the experimental design (day 90). Table 1 shows the distribution of the treatments applied.

Table 1. Experimental Treatments.

Treatment	Description
T1	Witness
T2	Defoliation in 25% of the laminae of each leaf
T3	Defoliation in 50% of the laminae of each leaf
T4	Defoliation in 75% of the laminae of each leaf
T5	Defoliation in 100% of the laminae of each leaf
T6	Total tearing of the laminae of each leaf of the total number of leaves of the plant, without breakage of the central rib
T7	Total tearing of the laminae of each leaf of the total number of leaves of the plant, with breakage of the central rib

2.5 Data Collection

Data were collected using the following techniques [9]:

Plant Height at Maturity (cm). At maturity, five plants were randomly selected from each plot. Their height was measured from the soil surface to the tip of the panicle/flag leaf with the help of a metric rod and the average height was calculated. Their height was measured from the soil surface to the tip of the panicle/flag leaf with the help of a metric rod and the average height was calculated.

Corn Cob Length (cm). Five plants were randomly selected; their cobs were measured, recorded, and averaged.

Corn Cob Diameter (cm). Five plants were randomly selected; their cobs were measured, recorded, and averaged.

Number of Grains per Major Cob Perimeter. Five plants were selected at random; the grains in their cobs were counted, recorded, and averaged.

Number of Grains per Cob. Five plants were selected at random; the grains in their cobs were counted, recorded, and averaged.

Grain Weight per Cob (g). Five plants were selected at random; the grains of their cobs were weighed, recorded, and averaged.

Corncob Weight (g). Five plants were randomly selected; their corncob weights were weighed, recorded, and averaged.

Grain Yield (kg.ha-1). After sun drying, the cobs were threshed manually and the yield per plot was recorded and then converted to kg.ha-1 using Eq. 1:

$$Yield = \frac{Yield\ per\ plot(kg) \times 10000}{plotsize(m^2) \times 1000} \tag{1}$$

2.6 Statistical Analysis

A randomized complete block design with three replications was used. For the analysis of variance, a Tukey test was applied. The software used was MS Excel.

3 Results and Discussion

Table 2 shows the analysis of variance for the yield components of amylaceous corn. There was no effect of blocks. It can be observed that corn plant height did not show significant differences among the treatments generated for the hail damage simulation ($p > 0.05$). This is because the simulations were performed 90 days after planting. The height of the plant is an important component that helps determine the growth achieved during the growing season [9]. Plant height can be influenced by the level of fertilization and the availability of nutrients in the soil, the interaction with environmental factors such as water (rainfall) and climate; in addition to cultural practices such as weeding and hilling that were performed at appropriate times; plant crowding and greater intraspecific competition for resources is another factor that increases competition between plants for nutrient absorption and sunlight interception [10].

Table 2. Analysis of variance for yield components.

Yield component	p
Plant height at maturity	2,9961
Cob length	0,000*
Cob diameter	0,000*
Number of grains per greater perimeter of cob	0,000*
Number of grains per corncob	0,000*
Grain weight per corncob	0,000*
Weight per corncob	0,005*
Grain yield	0,000*

**Significance at 0,05.*

The dry cob diameter of amylaceous corn showed significant differences by the hail damage simulation ($p < 0.05$). Treatment T1 (control or without defoliation) presented a greater diameter (11.10 cm) and was statistically different from treatment T5 (6.98 cm), treatment T6 (8.03 cm), and treatment T7 (6.98 cm) (Table 3). It should be noted that the leaf area index of the plants was also affected by the level of simulated hail damage, which in turn affected the light interception capacity; this explains the variability of the maize cob sizes obtained in the treatments [11].

Plant defoliation and leaf tearing as a simulation of hailstorm effects (solid precipitation) had negative effects on ear size, grain filling, and consequently on grain yield. The lowest yield was recorded in the T7 treatment with 2239.98 kg/ha, 61% less than the control (5743.41 kg/ha).

Table 3. Tukey's test for yield components.

Component	T1	T2	T3	T4	T5	T6	T7
Cob length (cm)	$11{,}18^{a}$	$10{,}33^{ab}$	$9{,}89^{abc}$	$9{,}58^{bcd}$	$8{,}96^{cde}$	$8{,}03^{f}$	$6{,}98^{f}$
Cob diameter (cm)	$7{,}19^{a}$	$6{,}94^{ab}$	$6{,}91^{abc}$	$6{,}79^{abcd}$	$6{,}33^{bcde}$	$6{,}07^{de}$	$4{,}97^{f}$
Number of grains per greater perimeter of cob	$14{,}03^{a}$	$12{,}33^{ab}$	$11{,}88^{bc}$	$11{,}71^{bcd}$	$11{,}66^{bcde}$	$11{,}49^{bcdef}$	$5{,}56^{g}$
Number of grains per corncob	$211{,}33^{a}$	$184{,}13^{b}$	$177{,}33^{bc}$	$174{,}18^{bcd}$	$148{,}02^{e}$	$116{,}67^{f}$	89^{g}
Grain weight per corncob (g)	$104{,}62^{a}$	$90{,}46^{b}$	$87{,}13^{bc}$	$85{,}52^{bcd}$	$71{,}41^{e}$	$52{,}43^{f}$	$43{,}20^{f}$
Weight per corncob (g)	$7{,}66^{a}$	$6{,}83^{ab}$	$6{,}80^{ab}$	$6{,}54^{ab}$	$6{,}51^{ab}$	$5{,}52^{b}$	$5{,}05^{b}$
Grain yield (kg/ha)	$5743{,}41^{a}$	$4966{,}09^{b}$	$4783{,}23^{b}$	$4694{,}85^{b}$	$3920{,}29^{c}$	$2878{,}37^{d}$	$2239{,}98^{e}$

Means in the same row that do not share a common superscript letter are significantly different

Something very interesting that could be found with the data collected is that a linear regression could be made between the percentage of simulated leaf damage (0, 25, 50, 50, 75, and 100%) and the yield components of amylaceous corn that were significant, i.e., for plant height, it was not performed (Table 4). The model for cob length was the one that was able to explain 97.10% of the variation of this yield component, being the highest of all. The number (90.66%) and weight (90.25%) of grains per cob and yield (90.25%) presented linear models with a considerable R2. In this regard, Hernández and Carreño state that the growth rate is highly dependent on the radiation that the foliage can intercept and the efficiency of conversion of this radiation into the dry matter, which in turn depends on the available resources and the genotype, and that the plant canopy that receives the highest proportion of incident radiant energy will have a higher photosynthetic efficiency than the plants that do not receive the same amount of radiation. Therefore, it will have an increased photosynthetic efficiency [11].

Based on the theory of photosynthetic efficiency by plant utilization of solar radiation, Díaz-López et al. state that total dry matter production is the result of the efficiency of crop foliage in intercepting and utilizing the available solar radiation during the growth cycle [12].

Table 4. Correlation of percent simulated leaf damage (0, 25, 50, 50, 75, and 100%) with yield components of amylaceous corn.

Component	Liner model	R^2 (%)
Cob length	$Y = -0{,}0208X + 11{,}03$	97,10
Cob diameter	$Y = -0{,}0074X + 7{,}20$	87,38
Number of grains per greater perimeter of cob	$Y = -0{,}0214X + 13{,}39$	73,18
Number of grains per corncob	$Y = -0{,}5463X + 206{,}31$	90,66
Grain weight per corncob	$Y = -0{,}2854X + 102{,}1$	90,25
Weight per corncob	$Y = -0{,}0103X + 7{,}38$	76,86
Grain yield	$Y = -15{,}67X + 5605{,}1$	90,25

However, this efficiency can be influenced by the amount of solar radiation, the ability of leaves to photosynthesize, leaf area index, plant architecture, respiration, among others, which are summarized in internal growth factors related to the genotype and external factors related to the environment and management practices. Yield results of amylaceous corn were directly influenced by the percentage of simulated defoliation since corn plants had lower photosynthetic efficiency due to insufficient leaf area. The results will also help to estimate crop damage and losses to make more coherent decisions in a real situation, to compensate damages due to hailstorms through possible agricultural insurance, among other applications.

4 Conclusions

The effect of foliar damage by hailstorm simulation was manifested in cob size, the number of grains, and yield. The percentage of defoliation of amylaceous corn plants produced negative effects on the components of dry grain yield, with considerable determination coefficients, the variation in the decrease was in the range of 13.5% to 61.0%, with respect to the witness treatment.

References

1. Donatti, C.I., Harvey, C.A., Martinez-Rodriguez, M.R., Vignola, R., Rodríguez, C.M.: Vulnerability of smallholder farmers to climate change in Central America and Mexico: current knowledge and research gaps. Clim. Dev. **11**, 264–286 (2019)
2. Blackmore, I., Rivera, C., Waters, W.F., Iannotti, L., Lesorogol, C.: The impact of seasonality and climate variability on livelihood security in the Ecuadorian Andes. Clim. Risk Manag. **32**, 100279 (2021)
3. García-Parra, M., Zurita-Silva, A., Stechauner-Rohringer, R., Roa-Acosta, D., Jacobsen, S.E.: Quinoa (Chenopodium quinoa Willd.) and its relationship with agroclimatic characteristics: A Colombian perspective. Chil. J. Agric. Res. **80**, 290–302 (2020)
4. Imfeld, N., et al.: A combined view on precipitation and temperature climatology and trends in the southern Andes of Peru. Int. J. Climatol. **41**, 679–698 (2021)
5. Young, K.R., León, B.: Natural hazards in Peru: causation and vulnerability. Dev Earth Surf Process **13**, 165–180 (2009)
6. Trappeniers, G., Ledent, J.F., Fayt, O., Nijs, A.: Effects of simulated hail damage on the yield of forage maize. J. Agron. Crop Sci. **168**, 13–19 (1992)
7. Carter, P.R.: Clipping corn following early-season frost. J. Prod. Agric. **3**, 394–398 (1990)
8. Below, T.B., Schmid, J.C., Sieber, S.: Farmers' knowledge and perception of climatic risks and options for climate change adaptation: a case study from two Tanzanian villages. Reg. Environ. Change **15**(7), 1169–1180 (2014). https://doi.org/10.1007/s10113-014-0620-1
9. Abuzar, M.R., et al.: Effect of plant population densities on yield of maize. J. Anim. Plant Sci. **21**, 692–695 (2011)
10. Sibonginkosi, N., Mzwandile, M., Tamado, T.: Effect of plant density on growth and yield of maize [Zea mays (L.)] Hybrids at Luyengo, Middleveld of Eswatini. Asian Plant Res. J. **3**, 1–9 (2019)
11. Hernández Córdova, N., Soto Carreño, F.: Determinación de índices de eficiencia en los cultivos de maíz y sorgo establecidos en diferentes fechas de siembra y su influencia sobre el rendimiento. Cultiv. Trop. **34**, 24–29 (2013)
12. Díaz-López, E., Morales-Rosales, E.J., Franco-Mora, O., Domínguez-López, A.: Atenuación de luz, radiación interceptada y rendimiento de maíz en función del fósforo. Terra Latinoam **29**, 65–72 (2011)

Energy Startup Business Model: An Analysis of Scientific Publications and Trends for the Future

Larissa de Souza Lemos Farali(✉), Franciane Freitas Silveira, and Alexandre Acácio de Andrade

Graduate Program in Engineering and Innovation Management, Federal University of ABC, Santo Andre, Brazil
{larissa.lemos,franciane.silveira,aacacio}@ufabc.edu.br

Abstract. Whether shaped by fluctuating energy and fuel prices, energy transition, environmental and security issues, aspects of technological innovations, or goals to improve access to energy, nowadays, attention regularly turns to ways to improve energy pathways. However, whatever is understood about change in the energy system, especially concerning new business models provided by startups, is still in its infancy. Therefore, this bibliometric search includes articles from the last 5 years, retrieved from Thompson Reuters Web of Science TM, and allows obtaining a complete view of citation, co-citation, and keyword analysis. Other emerging search streams were evaluated using Histcite and VOSviewer software. The results are divided into Bibliometric Survey and Citation Analysis, Co-citation Analysis, and Research Flow in the field of energy business model for startups. Regarding the first, an increase in related research was identified, beyond 2018 and 2020. Regarding the analysis of co-citation, a grouping about the frequency of cited authors identified five clues, in addition, the three most articles were identified. Finally, a specific analysis of the startups' business model pointed out three major contribution strands of startups in (1) the dynamism of the traditional market, (2) in socio-environmental strategies, and (3) in the strategy facilitators: business model and technologies. This suggests that the literature on energy business models, specifically startups, seeks to achieve the energy transition through responsible innovations and technologies. The article aims to contribute to those interested in the topic and present suggestions for future studies.

Keywords: Innovation · Startups · Energy · Business model · Energy transition

1 Introduction

The perspective of the energy transition as a key issue for climate change emission targets in environmental, economic, and social terms places the energy sector as central to the fulfillment of the Paris Agreement, a treaty in which 195 nations unite to reduce global warming that happened at the 21st Conference of the Parties (COP21) of the United Nations Framework Convention on Climate Change [1]. In fact, the IEA [2]

Y. Iano et al. (Eds.): BTSym 2021, SIST 207, pp. 458–468, 2023.
https://doi.org/10.1007/978-3-031-04435-9_47

estimates that three-quarters of greenhouse gas emissions expected efforts to reduce CO2 emissions after the Paris Agreement is signed will come, directly or indirectly, from the energy sector. In this article, we will understand the energy transition as the current process resulting from global efforts from the beginning of the 21st century to achieve a low-carbon economy [2, 3, 4]. This is because the energy transition is not something punctual but consists of broader terms that cover the entire history [5].

Because of imminent changes in the energy industry, utility companies face a major challenge to find different strategies to create, deliver and capture value from renewable energy technologies. In fact, according to Richter's study (2013), the innovation capability of the traditional business model organizations will be crucial to mastering the energy transition [6]. Additionally, the author who investigated based on the German market that such companies run the risk of losing market share if they do not innovate their business models, especially regarding business models for the small generation of renewable electricity.

We can understand a business model (BM) as "the logic of how an organization creates, delivers, and captures value" [7, 8]. This concept can be concise in three key questions [9, 10, 11]: (1) Strategies and connections that the organization adopts to create value to the customer; (2) The way to deliver this value, which involves the entire supply chain and relationship with stakeholders, and (3) How the organization monetizes these connections.

An important source of innovation in BM is startups [12]. There is not just one and commonly accepted definition of "startup" [13, 14]. For some authors [13], startups are defined as enterprises made-up to insert innovation on the market. Ries [15] explain startups as a venture intended to launch a product or service in market circumstances of countless ambiguity. In the same way, Blank [16] pointed out that the main characteristic and goal of a startup is to discover an easily exponential business model.

Understanding how businesses are reinventing themselves when developing new technologies and BMs in the framework of the energy transition is essential, not only from a social and environmental outlook but also from an economic perspective. In fact, it is noted a substantial growth of literature on the energy transition and BM (separately). But, in addition to all this, a comprehensive view that research progresses on BM in the context of the energy transition is not identified in the literature, especially related to the business model of energy startups. In this way, this study aims to obtain an overview of authors, journals, countries, and critical areas of the BM in the energy sector, with an emphasis on startups through bibliometric and network analysis techniques, systematizing the articles of a domain in specific.

Finally, this article aims to contribute to the academic and professional community by systematizing and rationalizing the literature on energy transition for the contemporary body, mainly interested in innovation, business, and a low carbon economy.

2 Materials and Methods

The present work is based on bibliometrics. Bibliometric analysis is a statistical tool for mapping the status of a particular area of academic interest [17, 18, 19]. This method meets the expectations of the research, as it allows insights from a vast body of literature [20].

The purpose of using this methodology is to gain insights into the energy sector's business model from publications of peer-reviewed journal articles indexed from the year 2015, date marked by the 21st Conference of the Parties (COP21) of the Framework Convention of Nations United on Climate Change (UNFCCC) in Paris, where the first universal agreement on climate change and global warming was assumed [1]. The data for this research were extracted from the central collection Web of Science TM (WoS) database due to its scope and because it is a recommendation for studies involving the areas of business and management domains [21].

Figure 1 synthesizes the main steps of the data collection and the research methodology. The flow referring to the data collection presents the filters that were performed to refine the data, in the end only 191 articles that help answer the objective of this study were selected. The second process presented, the research methodology, was applied to analyze the interconnection of the selected articles to provide insights into the BM scenario in the energy sector. The defined string was: "energy" OR "energy transition" AND "startup" OR "start up" OR "small business" OR "new ventures" AND "business model" OR "canvas" OR "innovation". Being AND = Boolean operator "AND" and OR = Boolean operator "OR".

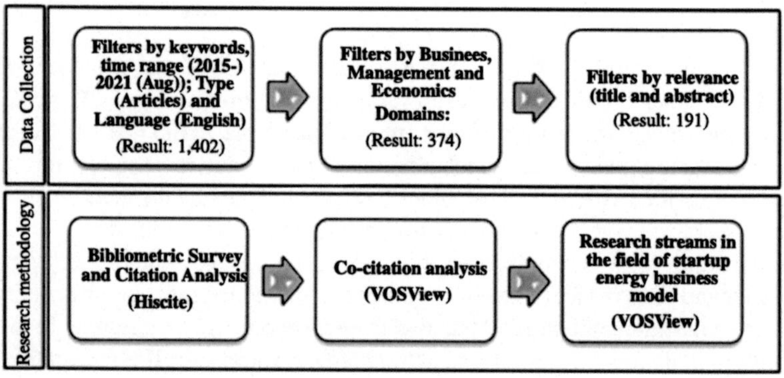

Fig. 1. Research methodology flows

For bibliometric analysis, two software were used, Histcite and VOSviewer. Regarding Histcite, it is a system that generates maps of bibliographic collections resulting from the WoS database, allowing the analysis of the relevance of selected studies [22]. Then, VOSviewer was used to correlate such studies and provide author clues and keywords. The VOSviewer, a grouping algorithm software and item similarity location in the network [23], an analysis of co-citation between authors and keywords was established and demonstrated in the sections below.

3 Results and Discussion

3.1 Overview

Using the methodology, 191 articles published from 482 authored journals were obtained by 90 authors associated with 313 institutions in 42 countries with citations of 11,180 references.

3.2 Bibliometric Survey and Citation Analysis

In this section, citations of published articles are analyzed by year, the most cited articles, authors, and countries. This, to identify the measure of influence [19, 23] given the correlation between authors and their disciplines. This analysis was carried out in the Histcite system to measure the productivity and influence of large WoS data sets [19].

With regards to the number of publications per year, Fig. 2 summarizes the articles chronologically by year from 2015 to 2021 (August). The trend shows a growing interest in the topic with a falling point in 2018 and 2020, probably due to the lack of public incentives in renewable energy [24, 25] from 2017 to 2018 and the advent of the crisis caused by Covid-19 respectively.

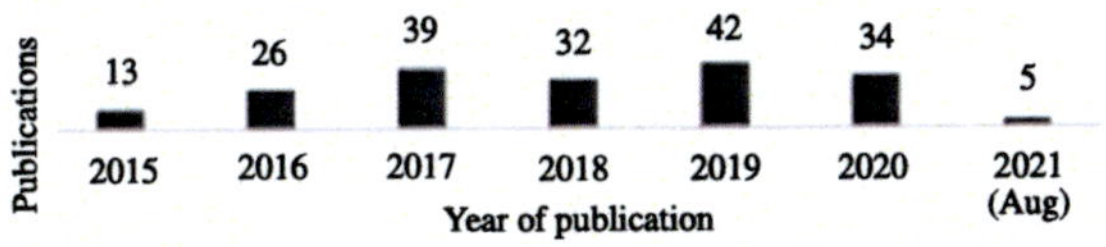

Fig. 2. Volume publications on energy business model between 2015 and 2021 (August)

The second point analyzed was the most relevant articles based on the volume of citations in the WoS database. Of the 193 referenced articles, the average citation rate is 16% with one citation per article. 8% of the articles were never cited and 46% were cited from 1 to 10 times. To identify the most relevant, the minimum limit of 60 citations per article was established. As a result, 10 articles were identified as shown in Table 1. In short, the most relevant articles are between 2015 and 2017 pertaining to the themes of sustainability, sharing economy, and innovations, indicating a trend in these themes in general. The most relevant with 195 citations is "A literature study for DEA applied to energy and the environment", published in the journal Energy Economics by Sueyoshi et al. (2017) which analyzes publications from 1980 to 2010 on energy and the environment through statistical methodology [26]. This article points out that technological innovation is a great driver for solving economic and commercial issues.

The most contributing and most influential authors were the third analyzed point. Related to the identification of the most eminent authors, an analysis of publication and citation frequency was performed. Table 2 shows the classification of the 5 authors who contributed the most to these two factors, but it is evident that there is no correlation between the volume of publications (6) and the frequency of citations, as the rankings have different authors. Socacool BK was the highest contributing author in terms of publication, but the top authors achieved a higher citation score (both 219). Benjamin

Table 1. Most cited articles in energy business model.

Rank	*Title and Journals*	*Authors and Publication year*	*No citation*
1	A literature study for DEA applied to energy and environment (Energy Economics)	Sueyoshi *et al. (2017)*	195
2	Commercial orientation in grassroots social innovation: Insights from the sharing economy (Ecological Economics)	Martin *et al. (2017)*	115
3	Does FDI influence renewable energy consumption? An analysis of sectoral FDI impact on renewable and non-renewable industrial energy consumption (Energy Economics)	Doytch and Narayan (2015)	104
4	Governing sustainability transitions through business model innovation: Towards a systems understanding (Research Policy)	Bolton and Hannon (2016)	103
5	Electric Vehicles with a Battery Switching Station: Adoption and Environmental Impact (Management Science)	Avci and Girotra (2015)	98
6	Capturing value from big data—a taxonomy of data-driven business models used by start-up firms International Journal of Operations & Production Management)	Hartmann *et al. (2016)*	89
7	Business, model innovation in electricity supply markets: The role of complex value in the United Kingdom (Energy Policy)	Hall and Roelich (2016)	75
8	Public financing of innovation: new questions (Oxford Review of Economic Policy)	Mazzucato and Semieniuk (2017)	64
9	Does corporate R&D investment affect firm environmental performance? Evidence from G-6 countries (Energy Economics)	Alam *et al.* (2019)	64
10	Business models for distributed energy resources: A review and empirical analysis (Energy Policy)	Burger and Luke (2017)	60

Sovacool serves as an academic at the University of Sussex Business School in the UK researching and advising on issues of renewable energy and energy efficiency, large scale energy infrastructure policies [27]. Professor Toshi Sueyoshi is also dedicated to risk and policy analysis as well as management science and data envelopment analysis [28].

Table 2. Analysis of publication frequency and citation frequency

Rank	*Author/Publication frequency*	*Author/Citation frequency*
1	Sovacool BK (6)	Sueyoshi T (219)
2	Rizzi F (3)	Yuan Y (219)
3	Frey M (3)	Goto M (195)
4	Brink T (3)	Budd L; Martin CJ; Upham P (115)
5	Gasbarro F (3)	Doytch N and Narayan S (104)

Eventually, a country by publication was carried to include the classification of countries in terms of published research on Business Model in the energy sector, and the synthesis is presented in Table 3. Only 3 out of 43 (7%) countries published more than 20 articles. Leading the contribution in the volume of articles and in the number of citations the UK is significantly the largest contributor. It is noted that the most relevant countries in the study of the business model in the energy sector are European, apart from the USA. One factor that highlights Europe from the energy sector is the government's interest in developing economic and environmental liberalization policies for the development of an efficient internal market [29].

Table 3. Most Contributing Country

Rank	*Country*	*Documents published*	*Citations*
1	UK	38	1,005
2	USA	31	744
3	Germany	26	466
4	Italy	17	165
5	Denmark	16	219

3.3 Co-citation Analysis

To better understand the 191 publications in the BM area of the energy industry, a systematic co-citation analysis using VOSviewer was performed. From the perspective of authorship, of these 191 selected articles, a total of 11,118 authors were identified. In order to present the highlights of this analysis, only authors with at least 20 citations were selected; as a result, 25 authors appear, as shown in Fig. 3 below, which shows the network of co-cited authors.

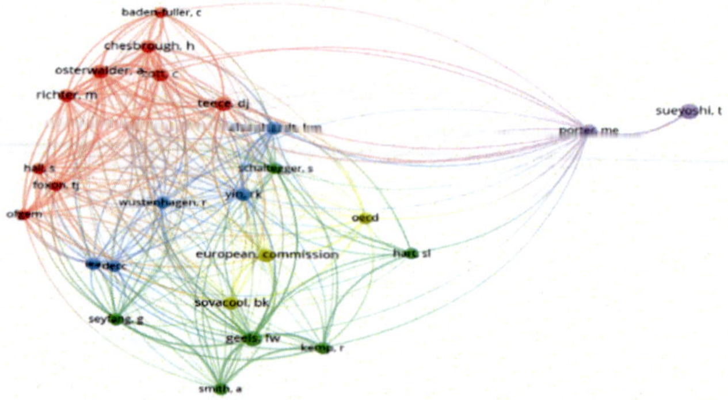

Fig. 3. Co-citation analysis for frequently cited authors

Five clusters represented in different colors were employed. A second deepening analysis of this result was established. The cluster in the "red" at the top emphasizes co-authoring largely related to managing strategies for innovative business models. Likewise, the "blue" cluster represents the research method design through the case study, emphasizing the dynamism, complexity, and practice orientation of the BM study. Correspondingly, the cluster in "yellow" represents the co-authorships related to approaches on human and social aspects. In turn, the "green" cluster represents the interconnection of social and technical aspects of organizations. Finally, the "purple" cluster addresses the economic and environmental issues of energy sector policies. In general, most of the co-cited authors, apart from the last one, are grouped, pointing to a strong relationship of domains under BM in the energy sector.

The most cited reference (22), "Business Models, Business Strategy and Innovation," by Teece (2020), a Long-range planning journal, which conceptualizes business models and explores their specifications with business strategy, innovation management, and economic theory [8]. Next, the second most co-cited article (22) was published at the Academy of management review, 1986, entitled "Building theories from case study research" by Eisenhardt K, who theorizes about the case study methodology, indicating that a relevant part of studies on the topic of BM in the energy sector is carried out through exploratory case studies that are empirically based on data collection and analysis [30]. Finally, the third most cited article is "Innovation of the business model for sustainable energy: German utilities and renewable energy" by Richter (2013), which is in line with the two previous ones and analyzes a case study of large-scale successful German utilities that fail in small-scale renewable generation concluding that as enablers of business model innovation, they are crucial to meeting an extreme energy transition needed for global decarbonization [6], this last one was published at Energy Policy journal that also published 30% of the most cited articles in BM in the energy sector as shown in Table 1. Through understanding relationships, insights into future lines of research can emerge. Additionally, an analysis of the co-knowledge of words by two authors was performed. Such systematization seeks to identify the most recurrent words within two 191 selected articles as well as how they are schematized. Aiming to simplify the analysis, the selected forms are subject to the words with a minimum occurrence of 6 appearances; As a result, 55 keywords were identified as presented in the network analysis Fig. 4.

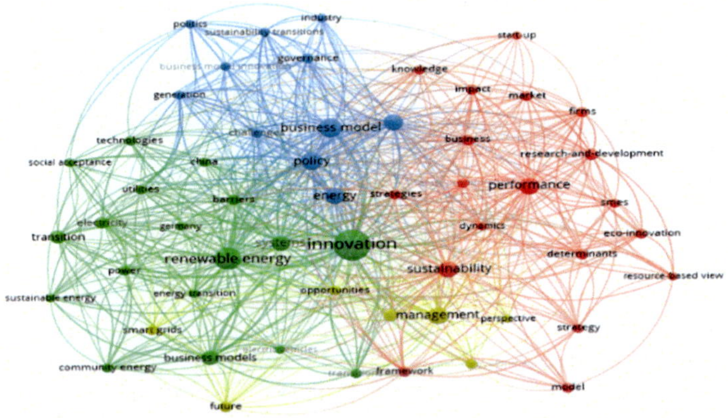

Fig. 4. Co-occurrence of author's keywords

Figure 4 shows "Innovation", "Business Model", and "Renewable Energy" as pillar themes used in the literature since they are the most recurrent words and therefore the three largest circles. That indicated what Richter pointed out as the challenges the public-private sectors to innovate in the way they create, deliver and capture value to provide people with cleaner energy and contribute to a low-carbon economy [6]. Then, the keyword "Innovation" (main node) is the one that has more relationship with the other words, indicating that your applications are the BM driver. Here are also presented clusters that group the keywords based on a central theme. The "red" cluster addresses issues related to the configuration of the energy market, with the main "Performance" and "Sustainability" being the most notorious words. The "blue" cluster is related to the strategies with "Business Model", "Politics" and "Technologies" as the most relevant words. "Yellow", in turn, permeates the trends in the sector, with "Management", "Energy Efficiency" and "Future" being the most frequent words. Concluding the analysis by clusters, the "green" cluster covers socioenvironmental themes such as "Innovation", "Renewable Energies" and "Systems". All these clusters have the potential to guide future BM research in the energy sector.

3.4 Research Streams in the Field of Startup Energy Business Model

To assess the research flows in which the energy sector BM and startups are correlated, a cut was made in the VOSviewer by selecting the existing links with the word "startup". In other words, from the data presented in Fig. 4, only the words that have a link with the word "startup" were selected, as shown in Fig. 5. Taking as a premise that the larger the circle size, the more the word is mentioned, we can observe that the study of energy startups is still emerging. From the selected period (2015–2021 Aug) most articles that directly present the term "startup" in their titles, keywords, and/or abstract are from the year 2019.

On the other hand, non-academic literature such as the metadata of the technology company Crunchbase, which is one of the most complete commercial information platforms on private and public companies, indicates that since 2015 there has been growing market interest in energy startups [31]. Despite the largest number of links being related to the energy market, the "red" cluster, pointing to the contribution of startups in the dynamism of the traditional market, that is, to companies (or concessionaires) already consolidated in the market; the most relevant clusters are related to socio-environmental strategies, "green" and the strategy enablers in "blue": the business model and technologies. While not a keyword, all these interlinked terms point to the importance of innovation ecosystems as a driving force for social, environmental, and economic development.

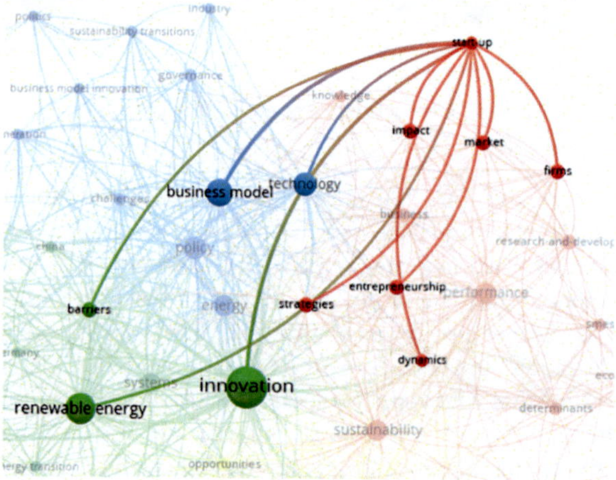

Fig. 5. Co-occurrence of author's keywords related to "startup" keyword

4 Implications and Future Research Directions

Aiming to obtain an overview of authors, journals, countries, and critical areas of the BM in the energy sector, with a focus on startups research, this work performed bibliometric research divided into two phases: data collection and research methodology.

The results are divided into Bibliometric Survey and Citation Analysis, Co-citation analysis, and Research streams in the field of startup energy business model. Regarding the first one, with the help of the Hiscite software, a growth in research related to the topic of BM in the energy sector in recent years was identified, except for 2018 and 2020. The article also scored the 12 most cited files related to the topic as well as the top

5 most frequent authors in terms of publication number and citation frequency. Sovacool BK and Sueyoshi T respectively. Regarding the analysis of co-citation, a grouping concerning the frequency of cited authors identified five clues, in addition, the three most cited articles were identified. Still on the co-citation aspect, an analysis of the most frequent keywords was conducted, identifying a strong relationship between the words "innovation", "business model" and "renewable energies" with the research theme, this suggests the Electric sector faces opportunities and challenges in guaranteeing innovative business models mainly focused on renewable energies and, consequently, a low carbon economy. Finally, an analysis specifically of the startups business model pointed out three main streams contribution of startups in the dynamism of the traditional market, socio-environmental strategies, and the strategy facilitators. This suggests that the literature on energy business models, specifically startups, seeks to achieve the energy transition through sustainable innovations and technologies.

We believe that this article has contributed to the academic, organizational, and government community that is interested in the topic with the points pointed out throughout the article, and for future research that deepens quantitatively (with other databases or complementary keywords) and qualitatively with the deepening through case studies, interviews, and systematic literature reviews.

References

1. United Nations Climate Change (2015) The Paris Agreement. https://unfccc.int/sites/default/files/english_paris_agreement.pdf. Acessed 6 Aug 2021
2. International Energy Agency (2021) Net Zero by 2050. https://www.iea.org/reports/net-zero-by-2050. Acessed 15 Sept 2021
3. Huisingh, D., Zhang, Z., Moore, J.C., Qiao, Q., Li, Q.: Recent advances in carbon emissions reduction: policies, technologies, monitoring, assessment and modeling. J. Clean. Prod. **103**, 1–12 (2015)
4. Olmos, L., Ruester, S., Liong, S.J.: On the selection of financing instruments to push the development of new technologies: application to clean energy technologies. Energy Policy **43**, 252–266 (2012)
5. Smil, V.: Energy transitions: history, requirements, prospects, ABC-CLIO (2010)
6. Richter, M.: Business model innovation for sustainable energy: German utilities and renewable energy. Energy Policy **62**, 1226–1237 (2013)
7. Osterwalder, A., Pigneur, Y.: Business model generation: a handbook for visionaries, game changers, and challengers, Wiley, Vol.1 (2010)
8. Teece, D.J.: Business models, business strategy and innovation. Long Range Plan. **43**(2–3), 172–194 (2010)
9. Chesbrough, H.: Business model innovation: opportunities and barriers. Long Range Plan. **43**(2–3), 354–363 (2010)
10. Osterwalder, A.: The business model ontology a proposition in a design science approach. Université de Lausanne, Switzerland (2004)
11. Johnson, M.W., Christensen, C.M., Kagermann, H.: Reinventing your business model. Harv. Bus. Rev. **86**(12), 57–68 (2008)
12. Hunt, R.A.: Entrepreneurial tweaking: an empirical study of technology diffusion through secondary inventions and design modifications by start-ups. European Journal of Innovation Management (2013)

13. Ries, E., Dillard, S., Eisenmann, T.R.: Hypothsis driven entrepreneurship: the lean startup, Harvard Business School Cases (2011)
14. Paternoster, N., Giardino, C., Unterkalmsteiner, M., Gorschek, T., Abrahamsson, P.: Software development in startup companies: a systematic mapping study. Inf. Softw. Technol. **56**(10), 1200–1218 (2014)
15. Ries, E.: The lean startup: how today's entrepreneurs use continuous innovation to create radically successful businesses, Currency, pp. 25–32 (2011)
16. Blank, S., Dorf, B.: The startup owner's manual: the step-by-step guide for building a great company. Wiley, pp. 32–39 (2020)
17. Garfield, E., Sher, I., Torpie, R.: The use of citation data in writing the history of science. Institute for Scientific Information, Philadelphia PA. (1964). https://doi.org/10.21236/ad0466578
18. Small, H.: Co-citation in the scientific literature: a new measure of the relationship between two documents. Journal of the American Society for Information Science **24**(4), 265–269 (1973). https://doi.org/10.1002/asi.4630240406
19. Dzikowski, P.: A bibliometric analysis of born global firms. J. Bus. Res. **85**, 281–294 (2018). https://doi.org/10.1016/j.jbusres.2017.12.054
20. Strozzi, F., Colicchia, C., Creazza, A., Noe, C.: Literature review on the 'Smart Factory' concept using bibliometric tools. Int. J. Prod. Res. **55**(22), 6572–6591 (2017). https://doi.org/10.1080/00207543.2017.1326643
21. Hausberg, J.P., Liere-Netheler, K., Packmohr, S., Pakura, S., Vogelsang, K.: Research streams on digital transformation from a holistic business perspective: a systematic literature review and citation network analysis. J. Bus. Econ. **89**(8–9), 931–963 (2019). https://doi.org/10.1007/s11573-019-00956-z
22. Garfield, E.: From the science of science to scientometrics visualizing the history of science with histcite software. J. Informet. **3**(3), 173–179 (2009)
23. Van Eck, N.J., Waltman, L.: Software survey: VOSviewer, a computer program for bibliometric mapping. Scientometrics **84**(2), 523–538 (2010). https://doi.org/10.1007/s11192-009-0146-3
24. Dudin, M., Zasko, V., Dontsova, O., Osokina, I.: The energy politics of the european union and the possibility to implement it in post-soviet states. Int. J. Energy Econ. Policy **10**(2), 409 (2020)
25. International Energy Agency (2020) Energy Storage. https://www.iea.org/reports/energy-storage. Acessed 15 Sept 2021
26. Sueyoshi, T., Yuan, Y., Goto, M.: A literature study for DEA applied to energy and environment. Energy Econ. **62**, 104–124 (2017)
27. University of Sussex (2021) Professor Benjamin Sovacool. https://profiles.sussex.ac.uk/p373957-benjamin-sovacool. Acessed 23 Sept 2021
28. New Mexico Tech (2021) Professor Toshi Sueyoshi Phd. https://www.nmt.edu/academics/management/faculty/tsueyoshi.php. Acessed 23 Sept 2021
29. Organisation for Economic Co-operation and Development (2013) European Union energy regulation https://read.oecd-ilibrary.org/governance/international-regulatory-co-operation-case-studies-vol-2/european-union-energy-regulation_9789264200500-4-en#page6. Acessed 5 Sept 2021
30. Eisenhardt, K.: Building theories from case study research. Acad. Manag. Rev. **14**(4), 532–550 (1989)
31. Crunchbase (2021) Energy Startups. https://www.crunchbase.com/hub/energy-startups Accessed 24 Sept 2021

Effect of Immersion Time in NaCl and Blanching Time on the Texture and Sensory Acceptability of Preserves Based on Ginger (*Zingiber Officinale*) Surplus in Syrup

Betty Yolanda Roque-Velásquez, Verónica Náthaly Jácobo-Domínguez, and Cesia Boñón(✉)

Universidad Privada del Norte, Trujillo, Peru
cesia.bonon@upn.edu.pe

Abstract. Peru exports ginger on a large scale. However, what does not meet export quality standards is considered as surplus and can be used in the development of new products, giving it an added value Therefore, the objective of this work was to evaluate the effect of immersion time in NaCl solution and blanching time on the texture and sensory acceptability of preserves based on the ginger surplus in syrup. We worked with a 3 x 3 x 3 experimental design of 9 samples with 3 replicates, in immersion times in NaCl solution (1.3% m/v) of 0, 15, and 30 min, without and with blanching times of 15, 30, and 40 min, with ginger samples in transverse cuttings of approximately 2 mm. The texture of the samples of ginger in syrup was measured in a Stable Micro Systems Texture Analyzer (model TA-HD plus). The pre-treatment of immersion in NaCl solution helped in the change of texture, observing that a longer immersion time allows obtaining a more compact product, with better results at a time of 30 min. In blanching, the best results were obtained with 15 min of treatment. When evaluating the effect of the combination of immersion time in NaCl and blanching, it was determined that it is better to work with medium and long times, since the product presents better texture, which agrees with the sensory analysis where the sample with the best acceptability had 15 min of immersion in NaCl and 30 min of blanching.

Keywords: Ginger · Ginger surplus · Preserved ginger · Texture · Sensory acceptability

1 Introduction

Ginger (*Zingiber officinale*) is a plant of the Zingiberaceae family that grows as a horizontal rhizome [1]. Peru is one of the countries where the cultivation and export of ginger are growing considerably, positioning itself as the second supplier of ginger to two of the most important markets in the world, Europe and the United States [2], being Junín the region with the main production area [3].

Ginger has shown various health effects, as reported in several studies. For example, it has shown positive effects in counteracting high plasma cholesterol levels [4] and in

Y. Iano et al. (Eds.): BTSym 2021, SIST 207, pp. 469–478, 2023.
https://doi.org/10.1007/978-3-031-04435-9_48

improving insulin sensitivity and glucose homeostasis. It has also shown antioxidant, anti-inflammatory, anticancer, and antimicrobial activity [5], besides its effect in reducing pain and inflammation in patients with osteoarthritis, muscle discomfort, and arthritis [6].

Regarding its chemical composition, ginger contains essential oils (2.5–3.0%) and non-volatile spicy substances. The main components of the fraction of spicy substances are gingerols, with 25%. High content of gingerols and an intense spicy flavor are considered as indicators of freshness and quality [7], an attribute that can be evaluated through sensory analysis.

The sensory analysis allows the characterization and analysis of acceptance or rejection of a food product by the taster and/or consumer, according to the sensations experienced from the moment it is observed and after it is consumed [8].

In a study carried out by [9] on ginger-producing farms in the Cuyani-Pichanaki micro-basin (Junin, Peru), it was determined that ginger was considered for disposal when there were warts, deformed rhizomes or white tips, burned, or more than 50% peeled. In this regard, [10] mentions that all those gingers that do not present a good appearance, shiny surface, and are not as smooth as possible are discarded. Special care is taken to discard those that have a fibrous appearance and many branches. The excess of ramifications is not a safety problem, it is perceived negatively by customers, therefore, it is a criterion to control. The author [9] refers in the same study mentioned above, that the highest percentage of discard of ginger obtained was 30% and that it is sold in the national market at lower prices.

In this sense, since the existing literature does not report the development of alternative products based on ginger surplus, the objective of this study was to evaluate the effect of immersion time in NaCl solution and blanching time on the texture and sensory acceptability of preserves based on the ginger surplus in syrup, considering NaCl for its preservative effect and its contribution in masking undesirable flavors [11], in this case, spiciness. We worked with 9 samples of 3 replicates each, with immersion times in NaCl solution of 0, 15, and 30 min, and blanching of 0, 15, 30, and 40 min.

2 Materials and Methods

2.1 Experimental Design

An experimental design was used to determine the number of samples needed for the analysis. Tables 1 and 2 show the variables and the experimental design applied.

Table 1. Experimental design variables.

Variable	Factor
Dependent	• Compressive strength (Newton)
Independent	• Immersion time in NaCl solution (min)
	• Blanching time at 90 °C (min)

The formula used was:

$$N = a \times b \times r \quad (1)$$

Where a and b are the levels of the independent variables, and r is the number of repetitions (3), obtaining: $N = 3 \times 3 \times 3 = 27$ experimental tests.

Table 2. Two-factor design applied.

Immersion time in NaCl solution (min)	Blanching time at 90 °C (min)		
	15	30	40
0	Y_{111}	Y_{211}	Y_{311}
	Y_{112}	Y_{212}	Y_{312}
	Y_{113}	Y_{213}	Y_{313}
15	Y_{121}	Y_{221}	Y_{321}
	Y_{122}	Y_{222}	Y_{322}
	Y_{123}	Y_{223}	Y_{323}
30	Y_{131}	Y_{231}	Y_{331}
	Y_{132}	Y_{232}	Y_{332}
	Y_{133}	Y_{233}	Y_{333}

2.2 Preparation of Preserves Based on Ginger Surplus

The method proposed by Guevara was followed for the preparation of preserves based on the ginger surplus in syrup [12]:

The ginger was weighed and selected, discarding those samples that presented textural defects (fibrous) and advanced maturity. The raw material was washed with water to eliminate any foreign particles and the peel was removed manually by using stainless steel knives. Subsequently, the rhizomes were cut transversally (longitudinally) to obtain samples of approximately 2 mm thick. Samples were immersed in NaCl solution (13 g of NaCl in 1000 ml of solution), at periods of 0, 15, and 30 min, followed by blanching at 90 °C for 15, 30, and 40 min. Then, the product was bottled in 125 ml jars, and the syrup solution with 6% sugar and 0.4% CMC in 1 L of water at a temperature above 85 °C was added. Afterward, exhausting and sealing were carried out. The sealed jars were sterilized in an autoclave at 121 °C for 15 min. Finally, the finished product was measured for pH (OAKTON, United States), NaCl percentage (THERMO ORION, United States), and °Brix (ATAGO, United States).

2.3 Texture Profile Analysis

The Texture Profile proposed by Bourne [13] was used to determine the force required to compress 2 mm of a ginger slice and to evaluate the effect of NaCl immersion time and blanching time. A Stable Micro Systems texturometer (model TA-HD plus) (Stable Micro Systems, UK), in an environment at 25 °C, a 5 kg load cell and Exponent software (ANAME S.L., Spain) were used to export the data obtained in the texturometer, by using the following measurement parameters: Test mode: Compression, Pre-Test Speed: 5.00 mm/s, Test Speed: 0.50 mm/s, Post-Test, Speed: 10.00 mm/s, Target Mode: Distance, Distance: 2.00 mm, Trigger Type: Auto (Force), Trigger Force: 10.00 g, Break Mode: Off, Stop Plot At: Start Position, Tare Mode: Auto, Advanced Options: On.

2.4 Statistical Analysis

Analysis of variance (ANOVA) and Multiple Range Test (Tukey's HSD) were performed to determine significant differences between treatments at significance levels of 5%. The Statgraphics Centurion XV statistical package (Statistical Graphics Corporation, Rockville, USA) was used for these analyses. All experimental tests were performed in triplicate.

2.5 Statistical Sensory Analysis

The products were evaluated with a nine-point hedonic scale. Samples were coded with three numbers and served on plates. The sensory attributes evaluated were taste, color, appearance, and texture, by an untrained team of 30 panelists. The sample used of 30 panelists was based on the references of Meilgaard et al. [14], who, in their study, used the Friedman's Statistical Analysis, which is a nonparametric test equivalent to ANOVA test, for repeated measures in the nonparametric version. The method consists of arranging the data by rows or blocks, replacing them by their respective order. When sorting, the existence of identical data must be considered.

3 Results and Discussion

3.1 Physicochemical Analyses

Table 3 shows the results of physicochemical analyses of ginger surplus preserves.

Table 3. Physicochemical parameters (pH, % NaCl, and °Brix) of ginger-based preserves. Nomenclature: 0–15 indicates 0 min of immersion in NaCl solution and 15 min of blanching at 90 °C.

	0–15	0–30	0–40	15–15	15–30	15–40	30–15	30–30	30–40
pH	5.38	5.38	5.39	5.39	5.38	5.49	5.38	5.40	5.38
% NaCl	0.84	0.84	0.85	0.87	0.87	0.90	0.93	0.93	0.95
°Brix	18.60	19.20	19.60	18.7	19.00	19.8	18.4	18.8	19.00

3.2 Compressive Strength of Preserves Based on Ginger Surplus

Figure 1 shows the compressive strength required to deform ginger slices treated at different times of immersion in NaCl solution and blanching. The highest compressive strength was obtained in sample 30–40 (30 min of immersion in NaCl solution and 40 min of blanching at 90 °C). On the contrary, the sample with the lowest result was the 30–15 sample, i.e. 30 min of immersion in NaCl solution and 15 min of blanching at 90 °C.

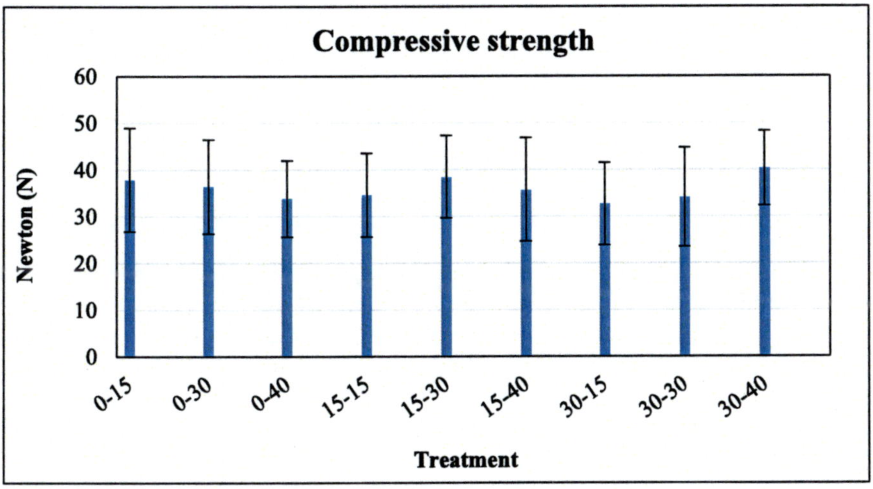

Fig. 1. Compression force required to deform ginger slices in syrup (sample thickness: 2 mm). Nomenclature: 0–15 indicates 0 min of immersion in NaCl solution and 15 min of blanching at 90 °C.

The texture is among the parameters to be taken into account during the blanching process, which is an important quality attribute and can be considered as an indicator of deterioration during processing and handling of foods, which undergo changes during the application of heat treatments, due to gelatinization of starches and solubilization of pectic substances, causing the loss of firmness in the vegetable tissue [15]. During processing, when moist heat is applied, plant texture is altered, mainly due to the effect on the cell wall. This organized structure breaks down and causes changes in permeability and increases the flexibility of the tissues [16].

Figure 1 shows a certain loss in firmness of the samples without immersion in NaCl as blanching time elapses, so that, at 15 min of blanching, a better texture was obtained than at longer times. This result agrees with that reported by Latorre et al. [17], who mention that the effects of both temperature and blanching times cause a decrease in firmness, because the application of heat treatments (especially by immersion in water) causes various changes in cellular structures, mainly the disorganization of the cell membrane and changes in the polymers of the cell wall, causing a softening of the tissues.

The use of additives such as glycerol and a mixture of sucrose and NaCl to preserve the texture of dehydrated vegetables has been previously evaluated [18]. Salts such as sodium or potassium chlorides and sulfates, without modifying the pH of blanching water, have also been used to improve the texture of vegetables such as Brussels sprouts, peas, and green beans, or calcium chlorides or citrates to improve the texture of cauliflower and potatoes [19]. Although the evaluation of texture in ginger-based preserves has not been reported, the previously mentioned studies support the results obtained (Fig. 1), since, in the sample without immersion in NaCl solution, there is a decrease in compressive strength as the blanching time increases. On the contrary, when ginger pieces are immersed in NaCl solution, compressive strength improves as the immersion time increases. This behavior would indicate that the presence of cations in the water contributes to the increase of plant stiffness, due to their interaction with the peptide substances, creating a mesh structure originated by the bridges between plant molecules and ions [20].

3.3 Chewiness Test Applied to Preserves Based on Ginger Surplus

Figure 2 shows the graph obtained from the chewiness tests. The gain and distribution of NaCl in the food depends on the salting method applied, as well as on the thickness of the samples, the food/solution ratio, the composition of the mixture of solutes as dehydrating agents, time, and temperature, among other factors [21]. Likewise, the immersion time depends on the NaCl concentration, degree of agitation, size, and thickness of the food, although, if it is not very thick, most of the NaCl will probably penetrate during the first 3 or 4 min [22]. In this study, 3 immersion times in NaCl solution were used: 0, 15, and 30 min. Although the cutting thickness used was 2 mm, the part of the ginger called rhizome has an irregular structure between fibrous and soft (fleshy) parts, so that blanching times do not have the same effect on its structure [23]. For this reason, the samples were subjected to longer immersion times than those reported in previous studies. As shown in Figs. 1 and 2, samples that were not immersed in NaCl solution (immersion time = 0 min) lost firmness with blanching, while samples that were subjected to 15 and 30 min of NaCl immersion show an increase in firmness.

Blanching consists of exposing the treated product to a humid, caloric medium, generally between 70 and 100 °C, maintaining that temperature for a certain time, and then cooling the product rapidly to a temperature close to room temperature [24]. According to the results shown in Figs. 1 and 2, the samples without NaCl immersion showed a firmer texture with the shortest blanching time (0–15), in agreement with previous similar studies [25].

It has been previously reported that blanching in combination with the addition of sodium chloride and calcium acetate improved firmness in fermented vegetables [26], which is consistent with the results shown in Figs. 1 and 2, where the combination of NaCl addition and blanching resulted in a firmer texture as the blanching time increased.

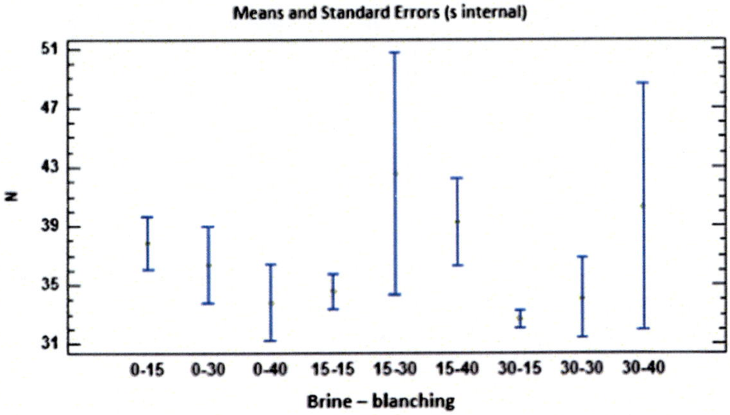

Fig. 2. Results of the chewing test of preserves based on ginger surplus samples. Nomenclature: 0–15 indicates 0 min of immersion in NaCl and 15 min of blanching at 90 °C.

3.4 Sensory Acceptability of Preserves Based on Ginger Surplus

Figure 3 shows the net diagram with the results obtained from the hedonic scale of preserves based on the ginger surplus in syrup, in order to determine preference. The results show a lower acceptance for sample 0–15, which was not immersed in NaCl solution and was treated with 15 min of blanching.

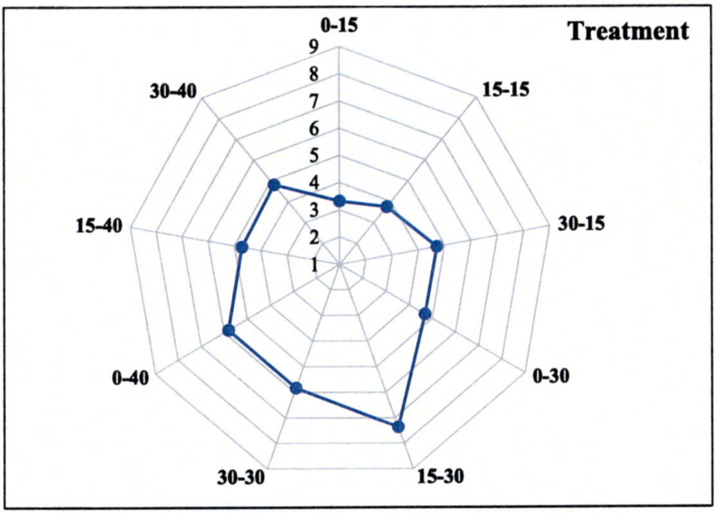

Fig. 3. Net diagram of preferences based on the hedonic scale. Nomenclature: 0–15 indicates 0 min of immersion in NaCl solution and 15 min of blanching at 90 °C.

According to Doellin [27], NaCl gives its own flavor to a food product and has important effects in enhancing and modifying the flavor of other ingredients. For example, it reduces the sensation of bitterness. This effect is possibly related to the decrease in the activity of water, increasing the concentration of other compounds in the medium and, therefore, their perception.

NaCl also contributes to masking flavors [28], so, in this study, it was used to mask the spiciness of ginger, so that it does not cause rejection at the time of consumption, which is consistent with the results shown in Fig. 3, where the samples with lower acceptability were those without immersion in NaCl solution, i.e., these samples were characterized by a greater spicy sensation.

On the other hand, the sample with the highest acceptability was the one immersed in NaCl solution for 15 min and with a blanching time of 30 min (15–30).

Sensorially, compounds present in ginger, such as flavonoids and carotenoids, are responsible for giving bitterness and spiciness [29], in addition to its known antioxidant capacity [5]. Thus, the application of the pre-treatment of immersion in NaCl solution, besides contributing to improving the texture profile (Figs. 1 and 2), helped reduce the spicy sensation, especially in the sample with 15 min of immersion in NaCl (15–30), as shown in Fig. 3.

4 Conclusions

The results obtained show that there are no significant differences between the different NaCl immersion and blanching times on the texture profile of preserves based on ginger (*Zingiber officinale*) surplus in syrup. However, the sensory evaluation shows higher acceptability in the samples pre-treated by immersion in NaCl solution. Regarding the combined effect of NaCl immersion time and blanching time, greater sensory acceptability was obtained with the sample immersed for 15 min in NaCl solution and with 30 min of blanching at 90 °C (sample 15–30), as it was evaluated as having the best flavor and the most compact texture by the panelists.

Acknowledgment. This study was supported by the Universidad Privada del Norte, Project UPN-20191005 "Desarrollo de nuevos productos con potencial de exportación a partir de descarte industrial de jengibre (kión)".

References

1. Acuña, O., Torres, A.: Aprovechamiento de las propiedades funcionales del jengibre en la elaboración de condimento en polvo, infusión filtrante y aromatizante para quema directa. Revista Politécnica. **29**(1), 60–69 (2010). https://bibdigital.epn.edu.ec/handle/15000/4343
2. ADEX Asociación de Exportadores (2020). http://www.adexperu.org.pe/. Accessed 29 Sept 2021
3. SENASA Servicio Nacional de Sanidad Agraria (2020). https://www.senasa.gob.pe/senasacontigo/tacna-senasa-certifica-ocho-toneladas-de-kionpara-su-exportacion-holanda/. Accessed 28 Sept 2021
4. Tzeng, T., Liu, I.: 6-gingerol inhibits rosiglitazone-induced adipogenesis in 3T3-L1 adipocytes. Phytother. Res. **28**(2), 187–192 (2014). https://doi.org/10.1002/ptr.4976

5. Vinothkumar, R., Sudha, M., Nalini, N.: Chemopreventive effect of zingerone against colon carcinogenesis induced by 1,2-dimethylhydrazine in rats. Eur. J. Cancer Prev. **23**(5), 361–371 (2014). https://doi.org/10.1097/CEJ.0b013e32836473ac
6. Arablou, T., Aryaeian, N., Valizadeh, M., Sharifi, F., Hosseini, A., Djalali, M.: The effect of ginger consumption on glycemic status, lipid profile and some inflammatory markers in patients with type 2 diabetes mellitus. Int. J. Food Sci. Nutr. **65**(4), 515–520 (2014). https://doi.org/10.3109/09637486.2014.880671
7. Ebrahimzadeh, V., Malek, A., Javadivala, Z., Mahluji, S., Zununi, S., Ostadrahimi, A.: A systematic review of the anti-obesity and weight lowering effect of ginger (*Zingiber officinale* Roscoe) and its mechanisms of action. Phytother. Res. **32**(4), 577–585 (2018). https://doi.org/10.1002/ptr.5986
8. Stone, H., Bleibaum, R., Thomas, H.: Descriptive analysis. In: Sensory Evaluation Practices, 4th ed., pp 233–289. Elsevier, Oxford (2012). https://doi.org/10.1016/B978-0-12-382086-0.00006-6
9. Refulio, P.B.A.: Procesamiento de jengibre fresco orgánico para exportación. Bach Tesis, Universidad Nacional Agraria La Molina, Lima, Perú (2018)
10. Maraví, L.J.Y.: Caracterización de fincas productoras de kion, piña y plátano en la microcuenca Cuyani-Pichanaki (Junin, Perú). Bach Tesis, Universidad Nacional Agraria La Molina, Lima, Perú (2018)
11. Matthews, L., Strong, P.: Taste masking of diclofenac sodium using microencapsulation. J. Microencapsul. **19**(1), 45–52 (2002). https://doi.org/10.1080/02652040110055612
12. Universidad Nacional Agraria La Molina (2020). https://docplayer.es/67947537-Elaboracion-de-fruta-en-almibar.html. Accessed 28 Sept 2021
13. Bourne, M.C.: Texture profile analysis. Food Technol. **32**(62–66), 72 (1978)
14. Meilgaard, M., Civille, G., Carr, T.: Sensory evaluation techniques. Fifth edition. CRC Press Taylor & Francis Group, Boca Raton, Florida (2015). https://doi.org/10.1201/b19493
15. Mendoza, R., Herrera, A.: Cinética de inactivación de la enzima peroxidasa, color y textura en papa criolla (*Solanum tuberosum* Grupo *phureja*) sometida a tres condiciones de escaldado. Información Tecnológica. **23**(4), 73–82 (2012). https://doi.org/10.4067/S0718-07642012000400009
16. Szczesniak, A.: Texture is a sensory property. Food Qual. Prefer. **13**(4), 215–225 (2002). https://doi.org/10.1016/S0950-3293(01)00039-8
17. Latorre, V., Pantoja, D., Mejía-España, M., Osorio, A., Hurtado, A.: Evaluación de tratamientos térmicos para inactivación de enzimas en jugo de fique (*Furcraea gigantea* Vent.). Biotecnología en el Sector Agropecuario y Agroindustrial. **11**(1), 113–122 (2013). http://www.scielo.org.co/pdf/bsaa/v11n1/v11n1a14.pdf
18. Hombre, R., Castro, E.: Parámetros mecánicos y textura de los Alimentos. Universidad de Chile, Santiago, Chile (2007). Disponible en: http://www.repositorio.uchile.cl/handle/2250/121381
19. Sanjuan, N., Benedito, J., Clemente, G., Mulet, A.: The influence of blanching pretreatments on the quality of dehydrated broccoli stems. Food Sci. Technol. Int. **6**(3), 227–234 (2000). https://doi.org/10.1177/108201320000600305
20. Fennema, O.: Química de los alimentos. 2ª edición. Editorial Acribia S.A., Zaragoza (2000)
21. Martínez, O., Salmerón, J., Guillén, M., Casas, C.: Sensorial and physicochemical characteristics of salmon (*Salmo salar*) treated by different smoking processes during storage. Food Sci. Technol. Int. **13**(6), 477–484 (2007). https://doi.org/10.1177/1082013207087816
22. Rodríguez, D., Barrero, M., Kodaira, M.: Evaluación física y química de filetes de bagre (*Pseudoplatystoma* sp.) salados en salmuera empacados al vacío y almacenados en refrigeración. Archivos Latinoamericanos de Nutrición (ALAN). **59**(2), 206–213 (2009)
23. Maistre, J.: Las plantas de especias. Blume, Barcelona (1969)

24. Fellows, P.: Food Processing Technology: Principles and Practice, 3rd edn. CRC Press, Oxford (2009)
25. Drake, S., Swanson, B.: Energy utilization during blanching (water vs steam) of sweet corn and subsequent frozen quality. J. Food Sci. **51**(4), 1081–1082 (1986). https://doi.org/10.1111/j.1365-2621.1986.tb11242.x
26. Andersson, A., Gekas, V., Lind, I., Oliveira, F., Öste, R., Aguilfra, J.: Effect of preheating on potato texture. Crit. Rev. Food Sci. Nutr. **34**(3), 229–251 (1994). https://doi.org/10.1080/10408399409527662
27. Doellin, L.: Análisis de los alimentos. pág.5–7,22–28 (2005)
28. Gutiérrez, C., Salazar, A., González, D.: Variaciones físico-químicas de la sardina (*Sardinella aurita*) durante el proceso de salado-madurado en seco. Memoria de la Fundación La Salle de Ciencias Naturales **163**, 109–119 (2005). http://flasa.msinfo.info/portal/bases/biblo/texto/memoria/men_2005_65_163_109-118.pdf
29. Ares, G., Barreiro, C., Deliza, R., Gámbaro, A.: Alternatives to reduce the bitterness, astringency and characteristic flavour of antioxidant extracts. Food Res. Int. **42**(7), 871–878 (2009). https://doi.org/10.1016/j.foodres.2009.03.006

Detection of Deforested Areas in the Peruvian Amazon Using Satellite Images

Ricardo Carbajal[1], Alejandro Alvarez[1], Joshi Lopez[1], Miguel Saturno[1], and Victor Murray[1,2,3](✉)

[1] Department of Electrical Engineering, Universidad de Ingenieria y Tecnologia, Lima, Peru
{ricardo.carbajal,alejandro.alvarez,joshi.lopez, miguel.saturno}@utec.edu.pe, vmurray@ieee.org

[2] Department of Electrical and Computer Engineering, University of New Mexico, New Mexico, USA

[3] School of Engineering and Applied Sciences, Harvard University, Massachusetts, USA

Abstract. Deforestation brings many problems to humanity, so studies started to appear to help reduce the damages done by the root of the problem and restore the health of our forests. In this paper, we present a method that can help identify deforested areas with the intention that deforestation may be monitored automatically using satellite imagery in the future. We analyze satellite images from the Peruvian Amazon in three separate timelines to view change over time. First, we preprocessed the images to remove haze and cloud using the dark channel prior method. Then, we segment the deforested area in the clean images by converting the RGB images to HSV format and later filtering the color corresponding to forested areas. Then we produce three binary images from the HSV conversion. Finally, these images are merged to obtain a single image that shows the change of deforestation over time. The method is compared with methods based on NDVI and NDTS detecting more deforested regions.

Keywords: Deforestation · Satellite imaging · Haze and cloud removal

1 Introduction

The Amazon has great importance for various reasons. It provides essential resources to local communities by storing a significant amount of carbon dioxide, and the biodiversity it contains helps us maintain a habitable ecosystem [1,2]. However, for the past decades, the health of the Amazon rainforest has been deteriorating for several different reasons. One of the main reasons is deforestation; this is very concerning for several reasons. For instance, fewer forested areas mean a decrease in the ability of the forest to hold more CO_2, which can lead to a higher yield of emitted greenhouse gases per year [3].

Y. Iano et al. (Eds.): BTSym 2021, SIST 207, pp. 479–486, 2023.
https://doi.org/10.1007/978-3-031-04435-9_49

To help keep the Amazon alive, it is of utmost importance to develop monitoring methods and systems that can help us identify the areas of interest affected by illegal deforestation practices. One particular example is a work done in [4], in which the authors analyze forest health using normalized difference vegetation index (NDVI) calculations followed by the time-domain (NDTS) information processing. As follow-up work, we present a different approach that increases the detection of the deforested areas. First, we preprocess the images to remove noise from clouds and reduce haze using the dark channel prior method to improve color contrast. Then we use color segmentation methods to filter out non-vegetated areas; this will give us a binary image showing where deforestation occurred. We did this for images taken in three separate timelines to analyze the change of deforested areas over time.

In the following sections, we describe the materials and methods used here. Next, we present the results and discussions and, finally, the conclusions.

2 Methodology

This section breaks down the steps used for the data acquisition from the Sentinel Hub at the EO Browser [5]. We describe the image preprocessing used to filter the haze using the dark channel. Then, we remove the clouds using averaging filters and a threshold. Finally, we apply an image segmentation method by converting the image to HSV (hue, saturation, and value) format and filtering the dark green areas to distinguish deforested areas and produce a binary image for each selected year. Then, we merge the three binary outputs to produce an RGB (red, green, and blue) image in which we can analyze the change of deforestation over the selected timeline.

2.1 Data Acquisition

The images used for this study were acquired from the Sentinel-Hub's EO Browser[1], a private repository collecting data since the year 2000. The satellite Sentinel-2 was selected for its high spatial resolution ($10\,m^2$/pixel, $20\,m^2$/pixel, and $60\,m^2$/pixel for different bands) [5]. However, the Level-1C of the Sentinel does not provide atmospheric correction, and this is why it is necessary to apply a cloud noise removal process and a haze removal process using the dark channel prior method (to be described in Subsect. 2.2).

The region for this study is located at 8.71°S, 74.32°W, in Ucayali, Peru (see Fig. 1). As a reference, it is located within the conservation area of Imiría[2]. Each image has a 2500×1183 pixels resolution and a projected resolution of 10 m/pixel. We use data from the year 2017 to 2020. For our purposes, we only analyze the bands B04, B03, and B02, which correspond to the RGB channels.

[1] Available at apps.sentinel-hub.com/eo-browser.
[2] More information about Imiría available at old.sernanp.gob.pe/sernanp/noticia.jsp?ID=191.

(a) (b) (c)

Fig. 1. Satellite images from the region of Ucayali. (a) 2019. (b) 2018. (c) 2017.

2.2 Cloud Noise Removal

For the preprocessing part, we start with an atmospheric correction to eliminate the noise from the clouds. In this case, all the rivers and clouds were viewed as noise in the images. There are images with very dense clouds, so a threshold was used to detect them. The removal is achieved because the clouds have relevant characteristics to be detected very quickly. In Fig. 2a, one can see the image used as input, which has two types of noise identified, dense clouds and haze. In Fig. 2b, it is seen how a threshold technique identifies these clouds. Moreover, finally, in Fig. 2c, one can see the image already masked.

(a) (b) (c)

Fig. 2. Cloud masking process. (a) Input image. (b) Binarization using a threshold. (c) Masked image.

As stated before, Level-1C does not provide atmospheric correction, and the haziness of the image can prevent the full binarization of the deforested area. Haze is considered any kind of fog, mist, or atmospheric phenomenon that can reduce the contrast and degrade the colors in an image [6,7]. We can remove the haze using the dark channel prior method based on the work presented in [6] to solve that problem. This method assumes that hazy images are composed of the scene radiance, the transmission map, and atmospheric light. Mathematically a hazy image can be denoted as:

$$I(x) = J(x)t(x) + A\left(1 - t(x)\right), \tag{1}$$

where $I(x)$ is the hazy image, $J(x)$ is the scene radiance or the haze-free image, A is the global atmospheric light, and $t(x)$ is the transmission map.

To recover the haze-free image, we can rewrite (1) as:

$$J(x) = \frac{I(x) - A}{max(\tilde{t}(x), t_0)} + A, \quad (2)$$

where t_0 is a correcting parameter to ensure the division by 0 does not occur. Before recovering the haze-free image, it is required to enhance the transmission. Then the refined transmission map is denoted as $\tilde{t}(x)$, as seen in (2).

Several methods are used to refine the transmission map. For our goals, we use the guided filter derived from [8]. The haze removal process is summarized as follows:

1. Acquire the dark channel of the input image.
2. Use the resulting dark channel to estimate the atmospheric light and the transmission map.
3. Use a refinement method to filter the transmission map.
4. Use (2) to obtain the haze-free image.

For example, Fig. 3 shows the result of the combination of the cloud and haze removal process on an image of a deforested area in Ucayali in 2019.

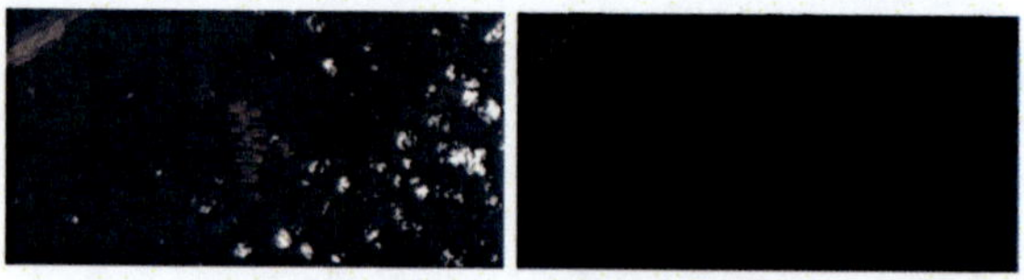

(a) (b)

Fig. 3. Cloud noise removal process. (a) Input cloudy and hazy image. (b) Cloud filtered image.

2.3 Deforestation Analysis

For the analysis, we use color segmentation in the HSV color space to take advantage of the color domain transformation focusing on hue. We are interested in the hue specifically for our purposes, and we know that the hue ranges from 0° to 360°. To obtain the desired output, we filter the dark green areas to be left with non-vegetated areas. In our case, the hue range selected is from 120° to 220°; this means we are filtering green, cyan, and blueish colors. We chose this range because the forested areas in our images have a dark green with blueish tones. Figure 4 shows the output of filtering the color of the vegetated areas in 2019.

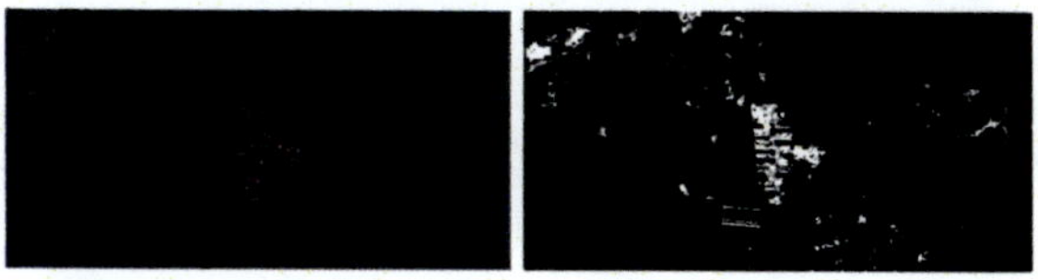

(a) (b)

Fig. 4. Image segmentation of the deforested area. (a) Image atmospherically corrected. (b) Binary output of (a).

3 Results

After performing the segmentation process for each selected image, we can merge the binary images into a single RGB image in which the images of the years 2019, 2018, and 2017 correspond to the red, green, and blue channels, respectively. Figure 5 shows the resulting image, in which we can observe the change in deforested areas from one year to another. The bright red spots indicate that the green and blue channels have intensity values of 0. This means that for the years 2018 and 2017, those areas were previously filled with trees zones. So a bright red indicates that in the year 2019, there was an increase in deforestation. White spots indicate that there was a deforested zone that started in 2017 and remained until 2019. Yellow spots are areas that still had vegetation in 2017, were cut in 2018, and remained deforested in 2019. Also, in Fig. 5, we observe a summary of the meaning of each color in the resulting image.

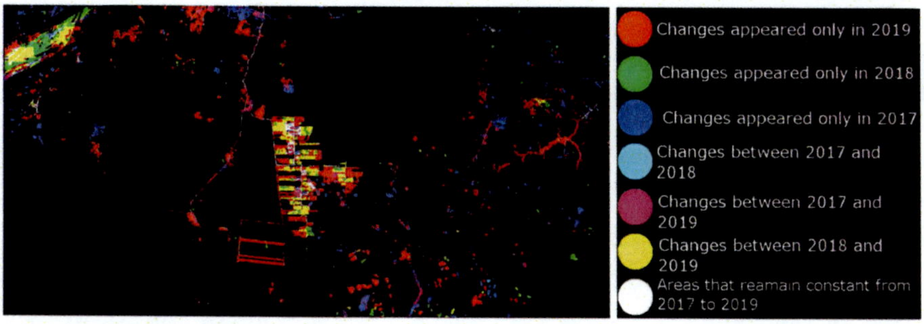

(a) (b)

Fig. 5. Output image with the interpretation of each color. (a) Image showing the change in deforestation overtime. (b) Interpretation of each color.

As mentioned above, each image is 2500 × 1183 pixels with a resolution of 10 m/pixel, which would be equivalent to a total of 29575 ha in total land. As shown in Table 1, the percentages of deforestation were obtained each year concerning the total hectares of land. The total number of hectares that were deforested per year is also presented. The obtained results show a notable increase between the deforested hectares in 2017 and 2018–2019, respectively.

Table 1. Percentage and area of deforested vegetation.

Year	Deforested (%)	Total in hectares
2017	1.39	413.38
2018	1.92	569.78
2019	4.49	1330.20

Now, we compare the results of our methods with the methods presented in [4]. For this comparison, we will use the same general area they used in their study, located at the coordinates 8.37°S, 75.14°W. Figures 6 and 7 present the input images and the output image of this area. We can see in Fig. 7 the change over time in the region selected using the color defined in Fig. 5. In Table 2, we compare the results found in [4] and our results.

The images that we gathered from the Sentinel-2 have the exact resolution as before: 2500×1183 pixels. However, in this case, the spatial resolution is 15 m/pixel, which gives us an area of 66543 ha. The images presented in [4] have a resolution of 2500×1197 and a spatial resolution of 14 m/pixel, which yields 58653 ha. At first glance, the results do not seem to differ too much from one another. So it would seem that to analyze deforested areas, both methods can help bring some insight into the scale of the deforested areas and the areas where illegal logging occurs more frequently.

However, it is also necessary to consider some areas of improvement for our methods. For instance, to correctly identify deforested zones, we need to improve the color contrast of the hazy images. We used the single image haze removal using the dark channel prior method based on the work done in [6]. The subsequent study could use other methods that could potentially improve further the color contrast. In addition, we can still experiment with other color segmentation techniques. The reason being some deforested areas could still be detected as green areas by our methods. According to the Food and Agriculture Organization for the United Nations (FAO), a large portion of forests can be used for cattle-raising [9]. These cattle-raising areas are sometimes covered with grass as a food supply [9,10]. This is why our methods might not be able to segment these zones as deforested accurately in some cases.

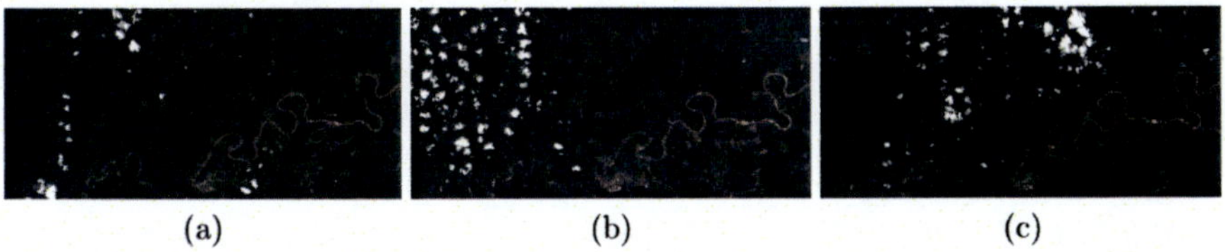

Fig. 6. Satellite images from the region of Ucayali. (a) 2020. (b) 2019. (c) 2018.

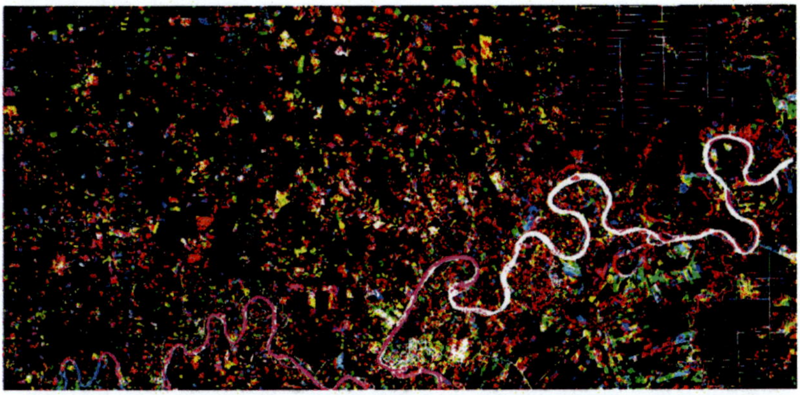

Fig. 7. Output image showing change over time (see color legend in Fig. 5).

Table 2. Result comparison of non-vegetable zones calculated by hectares.

Year	Total area in hectares	
	Work in [4]	Proposed method
2018	4208.59	4310.20
2019	5058.27	5918.60
2020	8246.23	9395.70

4 Conclusions

Deforestation is a severe problem that we find in the Amazonian forest. As the results show, the deforested area gradually increases over time. Therefore we have to develop solutions as soon as possible to solve the problem. The method presented detects deforestation using the RGB channel from Sentinel Hub; we can report that this method allows us to analyze deforestation only using the primary RGB channels to have a wider variety of satellites that we can use. Also, it is an easy method to implement. As future work, a web application that reports the areas that were deforested would be implemented.

References

1. Harris, N., Gibbs, D.: Forests absorb twice as much carbon as they emit each year. Technical report, World Resources Institute (2021). www.wri.org/insights/forests-absorb-twice-much-carbon-they-emit-each-year
2. Carrington D (2018) What is biodiversity and why does it matter to us? Technical report, the Guardian. https://www.theguardian.com/news/2018/mar/12/what-is-biodiversity-and-why-does-it-matter-to-us
3. Ellwanger, J.H., et al.: Beyond diversity loss and climate change: Impacts of amazon deforestation on infectious diseases and public health. Anais da Academia Brasileira de Ciências 92 (2020)
4. Velayarce, D., Alvarez, M., Guevara, D., Murray, V.: Analysis of deforestation in Ucayali-Peru using satellite imagery from sentinel-2. In: Iano, Y., Saotome, O., Kemper, G., Mendes de Seixas, A.C., Gomes de Oliveira, G. (eds.) BTSym 2020. SIST, vol. 233, pp. 308–316. Springer, Cham (2021). https://doi.org/10.1007/978-3-030-75680-2_35
5. L1C S (2021) About sentinel-2 l1c data. https://docs.sentinel-hub.com/api/latest/data/sentinel-2-l1c/
6. He, K., Sun, J., Tang, X.: Single image haze removal using dark channel prior. IEEE Trans. Pattern Anal. Mach. Intell. **33**(12), 2341–2353 (2011). https://doi.org/10.1109/TPAMI.2010.168
7. Markchom, T., Lipikorn, R.: Thin cloud removal using local minimization and logarithm image transformation in hsi color space. In: 2018 4th International Conference on Frontiers of Signal Processing (ICFSP), pp 100–104 (2018). https://doi.org/10.1109/ICFSP.2018.8552064
8. He, K., Sun, J., Tang, X.: Guided image filtering. In: Daniilidis, K., Maragos, P., Paragios, N. (eds.) ECCV 2010. LNCS, vol. 6311, pp. 1–14. Springer, Heidelberg (2010). https://doi.org/10.1007/978-3-642-15549-9_1
9. Faoorg (2007) Ganadería y deforestación. http://www.fao.org/3/a0262s/a0262s.pdf
10. Hance, J.: Cattle driving big forest loss in peru's 'under-appreciated' amazon. Mongabay Series: Global Forest Reporting Network (2016). https://news.mongabay.com/2016/07/cattle-driving-big-forest-loss-in-perus-under-appreciated-amazon/

Micropropagation of *Myrcianthes Ferreyrae* (McVaugh) "Arrayán": an Endemic Species from Lomas De Atiquipa

Ana Caldas Araujo, Fiorella Medina Medina(✉), Patricia Viveros Flores, and Herbert Lazo Rodriguez

Universidad Nacional de San Agustín de Arequipa, Arequipa, Peru
{fmediname,hlazor}@unsa.edu.pe

Abstract. *Myrcianthes ferreyrae* is a species endemic to Peru, paleoendemic of Lomas de Atiquipa, and classified as critically endangered. This plant is exposed to strong anthropic pressure, climate change, and limited conservation policies. Therefore, it has been proposed to develop a methodology using biotechnological techniques to facilitate its propagation *in vitro*. Explants taken from the district of Atiquipa were selected from acclimatized plants from a greenhouse. Explants were disinfected with 15% (v/v) NaOCl (sodium hypochlorite) for 15 min and other group with 0.1% (w/v) $HgCl_2$ (mercury dichloride) for 10 min. For the establishment stage, two culture mediums were used, MS (Murashige and Skoog) and WPM (Woody Plant Medium), with 30 g/L of sucrose and 4g/L of phytagel. Nodal segments from the greenhouse plants were used for the multiplication stage. As WPM was the best culture medium on the establishment stage, four treatments were carried out on this medium supplemented with BAP (6-Benzylaminopurine) in increasing concentrations (1, 2, 4 and, 8 mg/L). The results show a high disinfection efficiency (100%) and no significant difference between both disinfectants. Hence, the least aggressive and toxic disinfectant was chosen. WPM culture medium was determined as the best establishment medium, with a survival rate of 30%. For the multiplication stage, the optimal treatment was 4 mg/L BAP, with an average of 3.67 shoots/node and an average length of 8 mm.

Keywords: *Myrcianthes* · Nodes culture · Conservation · Micropropagation

1 Introduction

Genus Myrcianthes have 14 species, and 6 of them are endemic. *Myrcianthes ferreyrae* (McVaugh) McVaugh, known as "arrayán" is one of the most important species and constitutes a registered paleoendemism only for the Arequipa region [1]. This species has a narrow altitudinal range of distribution (400–900 masl). Small forests are formed on the top of slopes with hill vegetation [2]. In this ecosystem, *M. ferreyrae* naturally captures water from the mist, retains moisture in aquifers, produces fruits, and can be used in landscaping works [3, 4].

Y. Iano et al. (Eds.): BTSym 2021, SIST 207, pp. 487–493, 2023.
https://doi.org/10.1007/978-3-031-04435-9_50

"Arrayán" is threatened due to the fragmentation of its habitat because of the introduction of cattle that stops its natural regeneration and the indiscriminate felling of trees. These factors are affecting their abundance and distribution. For those reasons, this species was classified as in critical hazard by Peruvian legislation (DS N° 43–2006-AG) and categorized as critically endangered by the International Union for Conservation of Nature (IUCN) [5].

Background of propagation of *Myrcianthes* and related genus, either by seeds or cuttings, shows little success because it has recalcitrant seeds and presents high oxidation during rooting [6]. Likewise, *in vitro* micropropagation is rarely used because it has complications in all stages of culture. For example, *Myrcianthes coquimbensis* [6], *Myrcianthes pungens* [7]; *Campomanesia rufa* [8], and *Psidium guajava* [9–11].

Therefore, it is necessary to use more sophisticated methodologies for multiplication, such as plant tissue culture. This powerful biotechnological tool is used for the preservation of species because it allows a rapid massive propagation to recover threatened populations [11, 12]. Nevertheless, no information is currently available on a protocol for *in vitro* micropropagation of *M. ferreyrae*. For this reason, the objective of this work was to micropropagate "arrayán" from nodal segments to contribute to their conservation.

2 Material and Methods

The research was developed in the Plant Physiology and Biotechnology Laboratory, Faculty of Biological Sciences, Universidad Nacional de San Agustin de Arequipa. This study consisted of three stages, disinfection, establishment, and multiplication.

2.1 Plant Material

Plants with approximately one year of life were collected from the Atiquipa district (15′47′LS, 74′16′LW), with a license from SERFOR-Arequipa (N° D000230–2020-MINAGRI-SERFOR-DGGSPFFS). The collected plants were transferred to the greenhouse of the Faculty of Biological Sciences of UNSA to acclimatize for six months.

2.2 Explant Disinfection

Side branches of approximately 15 cm with 5 to 6 nodes each were cut from the acclimatized plants. The segments were washed for 30 min under a stream of water. Two disinfection treatments were tested; T1: nodes were soaked in a solution of sterile distilled water (SDW) + 0.3 ml/100 ml of Tween 20 for 10 min. After that, the nodes were transferred to 70% ethanol (v/v) for 1 min, then to a solution of 15% (v/v) Sodium Hypochlorite (NaOCl) (from commercial bleach with 4% active ingredient) for 15 min. Finally, rinse three times with sterile distilled water for 5 min each time. T2: the nodes were soaked in 20% (v/v) Isodine for 10 min, then transferred to 0.1% (v/v) $HgCl_2$ for 10 min and finally rinsed 5 times with SDW. The nodes were seeded in MS medium and the disinfection rate was evaluated after one week. The experimental unit was a culture vessel containing approximately 15 ml of culture medium with one explant; 15 replications were made per treatment. Data was transformed with the formula $\sqrt{}(x + 1)$.

2.3 Establishment of Nodes

The establishment of explants was tested in MS basal medium [13] and WPM. Both supplemented with sucrose 3% (w/v), citric acid 350 mg/L, and ascorbic acid 300 mg/L. The pH was adjusted to 5.8 with HCl (0.1N) and with KOH (0.1N). The gelling agent was 0.4% (w/v) Phytagel. The media was sterilized in an autoclave at 15 lb pressure and 121 °C for 20 min. One explant with one leaf per culture container was placed under aseptic conditions [14]. 20 repetitions were carried out and incubated at 23 ± 2 °C, with a photoperiod of 16 h light and 8 h dark with the illumination of 800 lx (40W fluorescent lamps) for 30 days [6, 14]. Finally, the survival rate of the explants was evaluated. The data was transformed with the formula $\sqrt{(x + 1)}$.

2.4 Multiplication

The explants were cultured in basal WPM medium supplemented with Benzyl amino purine (BAP) at 1, 2, 4, 8 mg/L to study their effect on the multiplication of shoots. A change of medium was made after 3 weeks and then after 6 weeks. The number of shoots per explant and their length (mm) were evaluated [10]. Each treatment had 15 replications; the data were subjected to ANOVA analysis of variance and Tukey's specificity test ($p < 0.05$) [15].

3 Results and Discussion

3.1 Explant Disinfection

For in vitro culture of plant tissues, one of the problems that affect the establishment stage is contamination by microorganisms, mainly with woody species. This study found that surface disinfection of nodal segments of *M. ferreyrae* was successfully achieved with sodium hypochlorite or mercury dichloride when the explants were acclimatized in a greenhouse for 6 months (100% of success). Though when samples were brought from the field, disinfection was less efficient, which lead to low survival rates due to high contamination (data not shown). This has been borne out in *Psidium guajava* and *Psidium friedrichsthalianum* [16]. Despite the efficiency of HgC12 compared to NaOCl in the disinfection of explants in tissue culture of many *Myrtaceae* woody plants, its removal from the tissue by rinsing is difficult and could affect the explant establishment [10]. This work reports the successful disinfection of viable nodal segments using NaOCl 15% (v/v). Since HgC12 is a powerful disinfectant but corrosive and an environmental pollutant, NaOCl was the best option as it is less toxic and just as effective as HgC12. In addition, it is recommended to keep the donor plants in a greenhouse as detailed in the methodology.

3.2 Establishment of Nodes

WPM medium presented a better survival response (30%) compared to MS medium (0%) presenting significant differences ($p < 0.05$) (Fig. 1). The low survival rate of the nodal segments of *M. ferreyrae* was due to strong oxidation produced by the release of phenols that oxidized the culture medium. Probably it would be associated with the high concentration of NH4NO3 and NO3K salts in the medium, causing greater browning, as observed in MS medium with a high concentration of salts [17].

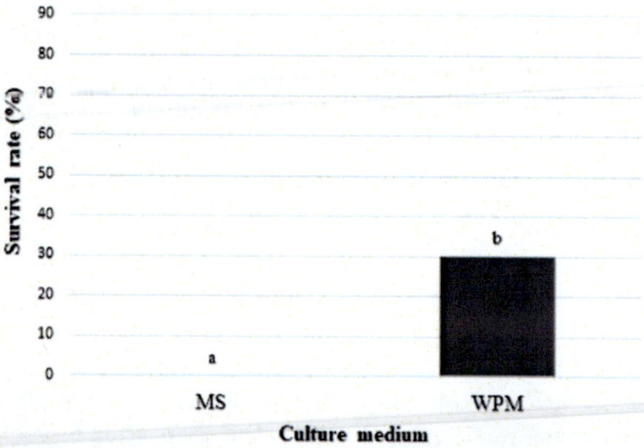

Fig. 1. Survival rate of explants in WPM culture medium: (Woody Plant Medium) and MS: (Murashige Skoog basal medium). Values are expressed as percentage, n = 15

Myrtaceae family are rich in phenolic compounds, mainly in woody plants such as *M. coquimbensis* [6] and varieties of *P. guajava* [17]. These compounds hinder the initial establishment of the *in vitro* culture. Application of antioxidant substances, such as citric and ascorbic acid, amino acids such as arginine, glutamine, asparagine, L-cysteine, Phloroglucinol, or PVPP reduces explants oxidation [18, 19]. However, the application of antioxidants such as citric acid and ascorbic acid in the medium was not enough. That's why another effective and straightforward technique was added: inverting the explant in the medium [20]. It was found that inversion of nodal (Fig. 2) segments under aseptic conditions in the WPM medium for 7 days and its subsequent subculture in normal position, decreased the release of phenolic compounds to the medium. With this procedure, wound healing was obtained with a survival rate of 100%.

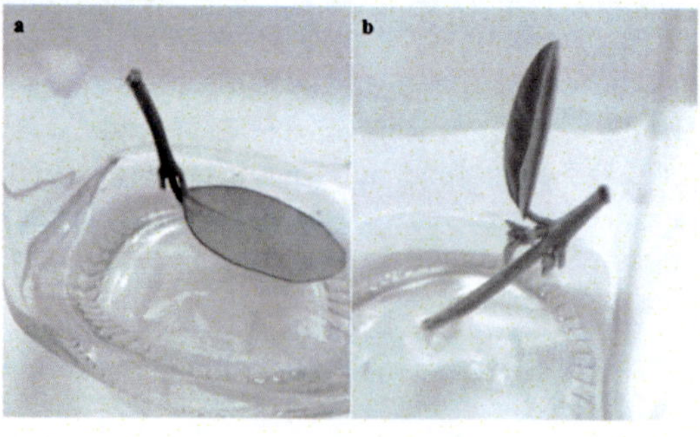

Fig. 2. Explants *of M. ferreyrae* in the establishment stage. After 1 day (a) and after 1 week (b) of culture.

3.3 Multiplication

The results show the highest survival rate (100%) in the treatment with BAP 4 mg/L and the lowest (68%) with BAP 8 mg/L (Fig. 3). However, there are no significant differences ($p > 0.05$). Treatment with BAP 4 mg/L showed (Fig. 4) significant differences ($p < 0.05$) in the highest number of shoots (3.67) and the lowest with BAP 1 mg (L (2.17). Measurement of shoots growth showed significant differences ($p < 0.05$). The response was inversely proportional to the BAP concentration (Fig. 5). Treatment with a lower concentration (1 mg/L) shows a greater growth, with 12.5 mm in length, and lower growth at a higher concentration 8mg/L, with 8.2 mm in size. The best treatment was BAP 4 mg/L, with more remarkable survival and an average of 3.67 shoots per node and with an average length of 8 mm.

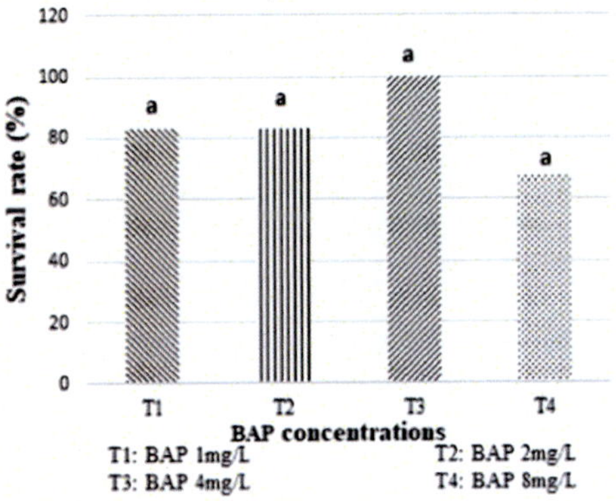

Fig. 3. Survival rate of explants in different BAP concentrations node (values expressed as the average, n = 15) BAP: 6-Benzylaminopurine.

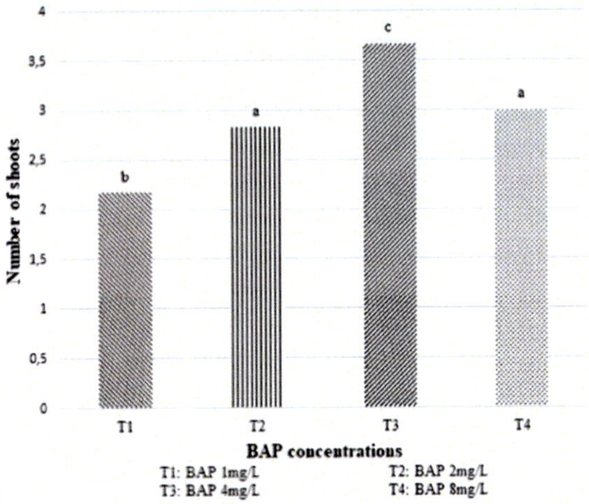

Fig. 4. Effect of different BAP concentrations on the number of shoots per node (values expressed as the average, n = 15) BAP: 6-Benzylaminopurine

How the results show, the number of shoots increased with a higher concentration of BAP in the culture medium, but the length of the shoots decreased inversely to the concentration of cytokine. This is contrary to what was found in *Psidium guajava* [9, 16, 17], *M. pungens* [7], and *Campomanesia rufa* [8], where it was reported that they sprouted in low concentrations of BAP (0.2 to 2 mg/L). However, in other varieties of *Psidium guajava* [10, 17] and *M. coquimbensis* [6, 17], the best responses were found with higher concentrations of BAP (2 mg/L at 8 mg/L), which are like the results found in *M. ferreyrae* (4 mg/L). The responses can probably be related to the concentration of mineral salts in the culture medium which are high in MS and low in WPM, the nature of the explant and genotype of the species [18, 19].

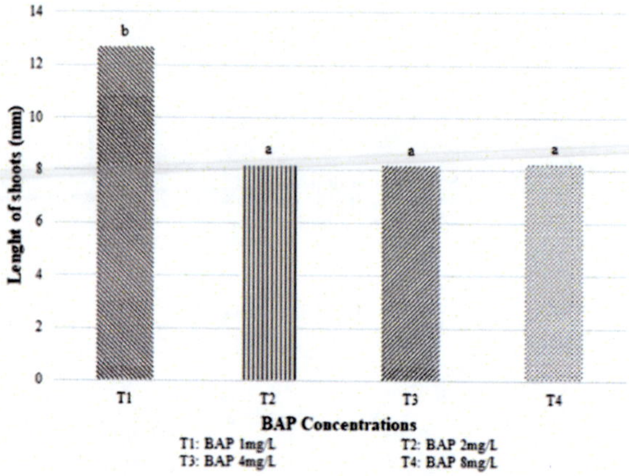

Fig. 5. Effect of different BAP concentrations (6-bencilaminopurina) on length of shoots. values expressed as the average, n = 15

Cytokines are widely used in propagation due to their participation in different processes of growth and development. However, in some species, the application of high concentrations of cytokines causes a decrease in growth or oxidation of the medium [7, 9]. It has been reported that using BAP in the micropropagation of plants of the *Myrtaceae* family promotes shoot induction, presenting a better response than other cytokines such as 2iP and Kinetin [6], and BA and TDZ in *Campomanesia rufa* [8]. Another reaction *in vitro* culture is associated with the elongation of stems (Fig. 3), greater stem elongation was evidenced with 1 mg/L BAP (12 mm), which is concordant as reported in *P. guajaba* [21].

4 Conclusion

It was possible to establish an efficient protocol for the micropropagation of *Myrcianthes ferreyrae*, which will serve to continue research on its rooting and acclimatization in *ex vitro* conditions, contributing to its conservation. Likewise, we seek to promote the use of this species in landscaping work.

Acknowledgments. We appreciate the financial support was provided by Universidad Nacional de San Agustín de Arequipa through contract N° TT-0138–2016. We sincerely thank Ph.D. Luis Martínez Manchego for his guidance in the statistical analysis of this investigation.

References

1. Arias, C., Torres, J.: Dinámica de la vegetación de Lomas del Sur del Perú: Estacionalidad y Productividad primaria caso: Lomas de Atiquipa (Arequipa). Zo áridas **6**, 55–76 (1990)
2. Kawasaki, L., Holst, B.K.: Myrtaceae endémicas del Perú. Rev peru biol **13**, 463s–468s (2006)
3. Brack, A.: Ecología de un país complejo, en gran geografía del perú naturaleza y hombre. In: Dourojeanni, M. (ed.) Flora-Ecología tomo II. Lima, pp. 175–319 (1986)
4. Fundación Manuel Bustamante. Cambio climático en el Perú. Regiones del Sur. 125 (2010)
5. Gonzáles, G., Llerena, G., Villasante, J.: Algunos aspectos sobre el estado de conservación de Myrcianthes ferreyrae (M.C. Vaught, 1958) especie arbórea endémica y amenazada del Perú. In: Memoria Segundo encuentro de Investigadores Ambientales. Ministerio del Ambiente, Dirección General de Investigación e Información Ambiental, Perú, pp. 1–6 (2013)
6. Saldías, G., Velozo, J.: Estudio de la propagación de Myrcianthes coquimbensis (Barnéoud) Landrum et Grifo por semillas y esquejes. Gayana Bot **71**, 17–23 (2014)
7. de Souza, L., Fior, C.S., de Souza, P.V.D., Schwarz, S.F.: Disinfestation of seeds and in vitro multiplication of guabijuzeiro from apical juveniles segments (Myrcianthes pungens O.Berg) D. Legrand. Rev Bras Frutic **33**, 691–697 (2011)
8. Sant'Ana, C.R., de Paiva O.R., dos Reis, M.V.: In vitro propagation of Campomanesia rufa: An endangered fruit species. Cienc e Agrotecnologia **42**, 372–380 (2018)
9. Amin, M.N., Jaiswal, V.S.: Micropropagation as an aid to rapid cloning of a guava cultivar. Sci. Hortic. (Amsterdam) **36**, 89–95 (1988)
10. Papadatou, P., Pontikis, C.A., Ephtimiadou, E., Lydaki, M.: Rapid multiplication of guava seedlings by in vitro shoot tip culture. Sci. Hortic. **45**, 99–103 (1990)
11. Rajasekharan, P., Sahijram, L.: In vitro conservation of plant germplasm. In: Bir Bahadur, et al. (eds.) Plant Biology and Biotechnology: Volume II: Plant Genomics and Biotechnology, pp.417–443. II (2015)
12. Medina, F.M., Rodríguez, H.L.: Establishment of an in vitro micropropagation protocol for cantua volcánica (Polemoniaceae): a potential ornamental plant of Peru. Idesia **36**, 203–208 (2018)
13. Murashige, T., Skoog, F.: A revised medium for rapid growth and bio assays with tobacco tissue cultures. Physiol. Plant **15**, 474–497 (1962)
14. Le Roux, J.J., Staden, J.V.: Micropropagation of eucalyptus species. HortSceince **26**, 199–200 (1991)
15. Reyes, P.: Diseño de Experimentos Aplicados, 2da ed. México (1982)
16. Ocampo, F., Núñez, V.M.: Propagación in vitro de Psidium guajaba mediante organogénesis directa a partir de segmentos nodales. Corpoica Cienc y Tecnol Agropecu **8**, 22 (2007)
17. Singh, K.K., Singh, S.P.: A review: micropropagation of guava (Psidium Spp.). Hortic. Int. J. **2**, 462–467 (2018)
18. Ibáñez, C., Vargas, M., Jofré, E.: Establecimiento de protocolos de propagación de cuatro especies nativas de alto valor ecológico de la región árida y semiárida de Chile. Simiente **83**, 134–136 (2013)
19. Pierik, R.: In vitro culture of Higher Plants. MatinusNijhoff, The Neatherlands (1991)
20. Varshney, A., Anis, M.: Trees: Propagation and Conservation. Springer, India (2014)
21. Loh, C.S., Rao, A.N.: Clonal propagation of guava (Psidium guajava L.) from seedlings and grafted plants and adventitious shoot formation in vitro. Sci. Hortic (Amsterdam) **39**, 31–39 (1989)

Perceived Effectiveness of Gender-Based Virtual Education in Times of COVID-19 Pandemic in University Higher Education

Carlos Valdivia[1], Roger Alarcón[1], Jessie Bravo[1](✉), Janet Aquino[1], Nilton Germán[1], and Fiorella Li[2]

[1] School of Computer Engineering and Informatics, Pedro Ruiz Gallo National University, Lambayeque, Peru
{cvaldivias,ralarcong,jbravo,jaquino,ngerman}@unprg.edu.pe
[2] School of Statistics, Pedro Ruiz Gallo National University, Lambayeque, Peru
fli@unprg.edu.pe

Abstract. The COVID-19 pandemic caused many processes to be digitized, one of them being education. With regard to university education, it was carried out virtually, in such a way that students were affected by the confinement and the coupling to a new educational reality. The objective was to analyze the perception of the effectiveness of the students based on their gender in the teaching - virtual learning process, under a quantitative, non-experimental, cross-sectional, and descriptive approach. A survey structured in five dimensions: psychological, technological, pedagogical, academic activities, and communication, was applied to a sample of 1,348 students from a public university. The results show that there is a significant difference between the perception of effectiveness by gender, taking into account the career groups. In technology, women perceive high effectiveness compared to men, in pedagogical terms, men have a better perception than women, academic activities and communication are perceived in a similar way in both genders, finally, psychologically, women were more affected than men.

Keywords: Gender · COVID-19 · Higher education · Effectiveness

1 Introduction

In many countries, the adaptation of education towards a virtual modality has generated in different actors, mainly students and teachers, being part of the change to successfully carry out this teaching-learning process.

As mentioned by Vialart [1] in Cuba, despite the experience they have in the virtual modality, they have presented difficulties related to the connectivity of students and teachers, which directly influences the didactics to be used to conduct the process when considering the varied and unequal means of access. In Colombia, Ojeda et. al. [2] found that students consider digital tools and learning platforms useful, but consider the experience developed from face-to-face processes such as communication much more significant. In the study carried out by Gazca [3], a diagnosis of the consequences

Y. Iano et al. (Eds.): BTSym 2021, SIST 207, pp. 494–501, 2023.
https://doi.org/10.1007/978-3-031-04435-9_51

of COVID-19 in higher education was carried out, measuring the perception in the technological, training, social, economic, and health fields that have been presented in the academic community, concluding that the latent threat has to see that if the pandemic is not resolved in the short term, it can generate uncertainty for proper academic planning. According to Moser et. al. [4], the abrupt shift from face-to-face contexts to remote learning is fundamentally different from planned online learning, and understanding the nature of this shift may influence the future of online and remote teaching.

In Peru, the national emergency due to COVID-19 stopped the school year in March 2020 and many of the universities had to adapt quickly to this new modality of virtual education, but will the perceived effectiveness be the same with respect to the virtual modality based on gender?

Given this, the purpose of this study is to analyze the perception of the effectiveness of students based on their gender in the teaching process - virtual learning.

2 State of the Art

The COVID-19 pandemic has caused many changes in all sectors of life, especially in the field of education, where different strategies were used to seek the effectiveness of the teaching-learning process. As Klein et al. [5] point out, although the success of distance learning requires good planning and the development of technological and human resources, there was no time for much preparation in the current situation, the courses had to be transferred to online formats, which resulted in a variety of synchronous, asynchronous or mixed activities.

According to Moreno et al. [6], in a study in Mexico, students positively value the "quality of teaching role and practice", where the instrument assesses the perception of the effort and dedication of teachers during the pandemic, their willingness to adapt and incorporate new tools to guarantee student learning; Likewise, a high evaluation received the "quality and accessibility of ICT", where men and women reported that they had adequate electronic devices and few problems of connection or access to the Internet. In the investigation carried out by Espinoza et al. [7], it was found that the effectiveness of the virtual learning environment is mediated by internal factors such as inadequacies in the organization of content and activities and external factors such as the limited technological preparation of the tutor, as well as the non-availability by some students of the technological equipment and internet connectivity. Likewise, for Asgari et al. [8], regarding online education for engineering, problems that had a negative influence were identified such as logistics/technical, learning/teaching challenges, and lack of sufficient practical training, where more than half of the students had difficulty maintaining their attention afterward of attending multiple sessions online.

The pandemic and confinement also affected the psychological aspect of the students, as pointed out by González et al. [9], who found out that stress (31.92%) and anxiety (40.3%) affected their mental health and influenced their academic performance, however, for men these episodes have been a little more bearable at home and women showed more anxiety, sadness, or depression than men. Finally, another factor that influenced the physical and psychological health of university students was the strenuous load of activities and evaluations.

In the study made by Tang et al. [10], there is a significant difference between men and women with respect to five contributing factors: technological availability, student control, self-efficacy of online communication, self-directed learning, and motivation to learn, with particularity in communicative self-efficacy and self-directed learning. It was also found that the motivation of the students is greater due to their enthusiasm with the use of technology and communication tools for learning, however, online learning is more attractive for women than for men. Regarding the perception of online learning, Warfvinge et al. [11] identified that men were more negative for the online experience, while women seemed more able to benefit from the shift to online learning. In the same way, Butnaru et al. [12] applied regression coefficients, which yielded results where students who have a computer or laptop and access to the internet to access their classes online, perceive their courses as more effective than those who do not have that technology. In general, women perceive online education as more effective, compared to men who perceive it as less effective.

3 Methodology

The population consisted of 11,600 students enrolled in the 2020 academic year of a Peruvian public university. A probabilistic sample of 1,348 students was used, with a confidence level of 95% and a margin of error of 2.51%. The research was carried out with a non-experimental, cross-sectional, descriptive, and correlational quantitative approach. A questionnaire was applied, validated with a Cronbach's alpha of 0.83 in a pilot test of 100 respondents, with good internal consistency. The instrument was designed using Google Forms, it had 34 questions on a Likert scale, distributed in five dimensions: technological (5), pedagogical (10), psychological (5), academic activities (10) and communication (4), and three levels according to the average of their scores: high, medium, and low. The 28 professional careers were formed into five groups: engineering, economics, medical sciences, social sciences, and pure sciences. The survey was delivered to students by institutional email and published in the virtual classroom. For descriptive and inferential analysis, SPSS version 26 software was used, using descriptive statistical measures and the Chi-Square statistical test.

4 Results and Discussion

In the **descriptive analysis**, for the students' perception of the effectiveness of the virtual modality in their teaching-learning process, results were obtained by dimensions and gender, as well as an analysis by career groups.

Regarding the technological dimension, Table 1 describes the levels of effectiveness used for the analysis, where it is observed that only 8.7% of the students identified it as effective at a high level, 90.5% at the medium level. and 0.8% at the low level, this shows that it is necessary to reinforce the mastery of technological tools so that they can be used to the maximum and that the Internet connection is a problem that affects the development of virtual classes.

Table 1. Definition of levels of effectiveness for the technological dimension.

Level	Rank	Description	%
Low	0–10	Students have little command of technological tools and do not have an adequate Internet connection	0.8%
Medium	11–20	Students have a regular command of technology tools and have an occasional adequate Internet connection	90.5%
High	21–25	Students are proficient in technological tools and have an adequate Internet connection	8.7%

Analyzing the effectiveness based on the gender of the students and the career groups, it was observed that there is a significant difference in the results obtained. In the group of economic sciences, medical sciences, social sciences, and pure sciences, men perceived a high level of effectiveness; on the other hand, in the engineering group, women perceived it with a high level of effectiveness.

For the pedagogical dimension, Table 2 describes the levels of effectiveness, where it is observed that 17.14% of students identified it as effective at a high level for their teaching-learning process, 82.49% moderately effective, and 0.37% not so effective. This shows that teachers have done their best to adapt to virtual education quickly.

Table 2. Definition of levels of effectiveness for the pedagogical dimension.

Level	Rank	Description	%
Low	0–15	The students perceived in the teachers a low mastery of their courses, a not so effective methodology, outdated content, inadequate use of technologies, and received little motivation	0.37%
Medium	16–40	The students occasionally perceived in the teachers an adequate mastery of the courses and their methodology; up-to-date content, they used technologies well and regularly received motivation	82.49%
High	41–50	Students frequently perceived in teachers a good command of their courses, correct methodology, updated content, adequate use of technologies, and they always received motivation	17.14%

At the level of professional career groups, it was found that there are differences in the perception of the effectiveness of the pedagogical dimension by gender, where men who study in medical sciences considered it to be highly effective, those who study in engineering, economics, social sciences, and pure sciences with average effectiveness. On the other hand, for women who study some engineering, they perceived it as highly effective, in medical sciences with medium effectiveness, and in economics with low effectiveness.

In the dimension of academic activities, Table 3 describes the levels of effectiveness considered, it can be seen that 16.1% of students were located in the high level, while

83.9% were located in the medium level, also finding that none had a perception at the low level, which shows the need to better organize and plan virtual courses, according to the activities and scheduled times.

Table 3. Definition of levels of effectiveness for the academic activities dimension.

Level	Rank	Description	%
Low	0–15	The students consider that the number of activities was excessive, the time allocated was not adequate, not all the activities were graded and they rarely received feedback	0.00%
Medium	16–40	The students consider that the number of activities was moderately adequate with adequate time, not all the activities were graded and some of them received feedback	83.90%
High	41–50	The students consider that the number of activities and time were adequate, they were scored and had good feedback	16.10%

In addition, considering the groups of professional careers, it was detected that there are differences in the perception of effectiveness in the dimension of academic activities by gender, where women corresponding to the engineering group had a higher perception in the medium level in opposition to men who perceived it at the high level. On the other hand, women who study medical sciences or pure sciences have a greater perception at the medium and high levels, and men from the social sciences and economic sciences groups perceived greater effectiveness at the same levels.

Regarding the communication dimension, Table 4 describes the levels of effectiveness considered, where it is observed that 44.14% of the students identified it as effective at a high level, 52% at the medium level, and 3.86% at the level low. This shows that the means and communication tools used were not totally effective.

Table 4. Definition of levels of effectiveness for the communication dimension.

Level	Rank	Description	%
Low	0–10	The students perceive that the means and communication tools used were not enough and that the teacher rarely answered their queries	3.86%
Medium	11–15	The students perceive that the communication means and tools were occasionally sufficient and the teacher occasionally absolved inquiries	52.00%
High	16–20	The students perceive that the communication means and tools used were effective and that the teacher frequently cleared their queries	44.14%

Analyzing the gender of the students based on the career groups, the men who study economics and pure sciences had a high perception of effectiveness, while in engineering careers their effectiveness was medium. Regarding women in engineering, medical sciences, and social sciences, their perception of effectiveness was high.

For the psychological dimension, Table 5 describes the levels of effectiveness considered, where 16.39% of students perceived it as influential at a high level for their teaching-learning process, 79.01% moderately influential, and 4.60% not so influential. This shows the need to implement a student psychological support program to reduce the impact caused by the pandemic.

Table 5. Definition of levels of effectiveness for the psychological dimension.

Level	Rank	Description	%
Low	0–10	Students have rarely experienced an episode of stress, anxiety, irritability, or depression	4.60%
Medium	11–20	Students have occasionally experienced an episode of stress, anxiety, irritability, or depression	79.01%
High	21–25	Students have frequently experienced an episode of stress, anxiety, irritability, or depression	16.39%

At the level of professional career groups, it was found that there are differences in the perception of the effectiveness of the psychological dimension by gender. About men, in social sciences and pure sciences, they perceived the effectiveness with a high level and in economics with a medium level; in medical sciences, the perception was higher at all levels. In the case of women, in economics, they perceived it with a high level and in engineering, the perception was higher at all levels.

The inferential analysis determines the relationship between gender and each of the five dimensions of the analysis, for which the Chi-square statistical test was used. The results are observed in Table 6. It can be concluded that there is statistical evidence that gender is related only to the psychological dimension (Sig. 0.027) with which it has been verified that the new form of virtual education has generated episodes of stress, anxiety, irritability, or depression in students, both men and women, thus affecting their academic performance.

Table 6. Results of the Chi-Square statistical test.

Dimension	Chi square	gl	Sig
Psychological	7,219	2	0,027
Technological	0,391	1	0,532
Pedagogical	1,824	1	0,177
Communication	0,461	2	0,794
Academic activities	0,2	1	0,655

In the discussion of results, González et al. [9] showed the influence of the psychological aspect on the academic performance of students, indicating that women were more affected than men, our research also concludes with the same results. Tang et al. [10] found that there is a significant difference in the use of technology and communication tools. In our results, there is no such difference in the use of technologies, and greater effectiveness of communication tools was obtained. According to Warfvinge et al. [11] and Butnaru et al. [12], men were more negative in the online experience, in our results, the men have a good perception of this virtual modality because the teachers showed mastery of their courses, acceptable methodologies, and used the technology correctly, likewise, they consider academic activities with an average perception in both genders.

For Moreno Candil et al. [6] the students positively value the role and the teaching practice. Similarly, in our results, the students perceived the effort that the teachers are developing in this new virtual modality in adapting to the use of technologies. Espinoza Freire et al. [7] found inadequacies in the organization of the contents and activities, likewise, in our results, it was perceived that the amount of programmed academic activities was excessive even though the time allocated to the activities was adequate.

5 Conclusions

From the study carried out, it is concluded that the perceived effectiveness in students on virtual education based on gender, at a global level, is very similar in the dimensions analyzed, however, in the analysis at the level of career groups it allows differentiating different levels of effectiveness.

Women are more affected than men psychologically, but their perception of technology is high in engineering careers, unlike men. In the pedagogical dimension, men who study medical sciences perceive it more effectively and women in engineering. At the level of academic activities, women who study in medical sciences and pure sciences perceive it more effectively, as do men in economics and social sciences.

It is statistically evidenced that there is a significant difference in the relationship between gender and the psychological dimension, but its influence is not evidenced in the rest of the dimensions.

References

1. Vidal MNV. Estrategias didácticas para la virtualización del proceso enseñanza aprendizaje en tiempos de COVID-19. Rev Cuba Educ Médica Super **34**(3) (2020)
2. (2021) In: Revistaespacios.com. https://revistaespacios.com/a20v41n42/a20v41n42p07.pdf. Accessed 22 Nov 2021
3. Gazca Herrera, L.A.: Implicações do coronavírus covid-19 nos processos de ensino no ensino superior. RIDE Rev Iberoam para la Investig y el Desarro Educ **11**(21) (2020)
4. Moser, K.M., Wei, T., Brenner, D.: Remote teaching during COVID-19: implications from a national survey of language educators. System **97**, 102431 (2021)
5. Klein, P., et al.: Studying physics during the COVID-19 pandemic: student assessments of learning achievement, perceived effectiveness of online recitations, and online laboratories. Phys. Rev. Phys. Educ. Res. **17**(1), 10117 (2021)

6. Moreno Candil, D., Bravo Verdugo, E.R., Burgos Dávila, C.J., Grijalva Verdugo, A.A.: Educación superior y pandemia: estudio piloto del cuestionario de experiencia educativa durante pandemia. Rev. Digit Investig en Docencia Univ. **15**(1) (2021)
7. Espinoza Freire, E.E., Granda-Ayabaca, D.M., Villacres Arias, G.E.: Educación a distancia en tiempos de COVID-19 en la carrera de Enseñanza Básica de la Universidad Técnica de Machala. Transformación **17**(2), 224–245 (2021)
8. Asgari, S., Trajkovic, J., Rahmani, M., Zhang, W., Lo, R.C., Sciortino, A.: An observational study of engineering online education during the COVID-19 pandemic. PLoS ONE **16**(4), e0250041 (2021)
9. González, N., Tejeda, A., Espinosa, C., Ontiveros, Z.: Impacto psicológico en estudiantes universitarios mexicanos por confinamiento durante la pandemia por Covid-19. Preprint **644**(1), 17 (2020)
10. Tang, Y.M., et al.: Comparative analysis of Student's live online learning readiness during the coronavirus (COVID-19) pandemic in the higher education sector. Comput. Educ. **168**, 104211 (2021)
11. Warfvinge, P., Löfgreen, J., Andersson, K., Roxå, T., Åkerman, C.: The rapid transition from campus to online teaching–how are students' perception of learning experiences affected? Eur. J. Eng. Educ. 1–19 (2021)
12. Butnaru, G.I., Nită, V., Anichiti, A., Brînză, G.: The effectiveness of online education during covid 19 pandemic—a comparative analysis between the perceptions of academic students and high school students from Romania. Sustainability **13**(9), 5311 (2021)

A Proposal for a Process Framework to Guide the Digital Transformation in the R&D Center of a Brazilian Chemical Industry

Talison Alvarenga dos Santos(✉), Anderson Orzari Ribeiro, and Franciane Freitas Silveira

Graduate Program in Engineering and Innovation Management, Federal University of ABC, Santo Andre, Brazil
{talison.santos,anderson.ribeiro, franciane.silveira}@ufabc.edu.br

Abstract. Despite the known advantages of digital transformation, many chemical industries have failed to advance digital maturity. Recent studies have shown that the main difficulties lie in the implementation process of digital technologies. In fact, digital projects are started within companies, but most of them do not go beyond the pilot phase. So, how should the digital transformation process be conducted to mitigate implementation difficulties? This article presents a proposal of a process framework to be used as a guideline during the digital transformation implementation, specifically for the Research and Development (R&D) center of a Brazilian large chemical company. Therefore, prescriptive research was developed with a qualitative approach based on the Design Science Research method resulting in the artifact (process framework) design. The proposed framework is composed of three phases, Preparation, Elaboration, and Application, aiming to support the identification, prioritization, and execution of digital projects inside the sector, improving the likelihood of their implementation effectiveness.

Keywords: Digital transformation · Maturity model · Technology roadmap · Design Science Research · Chemical industry · R&D center

1 Introduction

The Digital Transformation (DT) subject has been quite discussed throughout the industries due to its potential gains such as cost reduction, process control enhancement, data-driven solutions, efficiency improvement, etc. [1]. In the chemical industry, especially in Research and Development (R&D) area, the digital transformation brings many opportunities to improve the research capacity. For example, Artificial Intelligence (AI) could be used to predict chemical properties before the bench experiments, the Internet of Things (IoT) could improve the pilot process control and safety, and so on [2].

However, despite the advantages, digital transformation lags in the chemical industry. In fact, pilot projects are started in the companies, but most of them (approx. 75%)

Y. Iano et al. (Eds.): BTSym 2021, SIST 207, pp. 502–511, 2023.
https://doi.org/10.1007/978-3-031-04435-9_52

fail during the rollout process [3]. In other words, the unsuccess regards to the implementation process, not to the technologies themselves. According to some studies [3–6], it is possible to enumerate some failure causes, for example, conduction of the projects in a siloed way reducing the technologies robustness throughout the sector; lack of a concrete value proposition to justify the investment due to an unclear digital strategy; neglection of intermediary steps to advance in digital maturity; lack of knowledge about digital transformation by the employees and the digital project leaders; poor communication about the theme and others [3–6].

Based on the above, the guiding question of the research is: How should the process of conducting the digital transformation in the R&D area be to maximize the effectiveness of technological solutions implementation?

The main objective of this paper is to propose a process framework to guide the execution and control of digital transformation at a chemical company's R&D center. In the future, the framework will be applied and assessed at a Brazilian leader chemical company, specialized in surfactants, in order to validate its applicability.

The Design Science Research (DSR) method was chosen to operationalize the research once the objective to be achieved is the development of an artifact. The DSR can be defined as a set of analytic and perspective techniques to execute an investigation where knowledge is used to design, create, and validate models applying rigorous scientific techniques [7].

2 Digital Transformation in Chemical Industry R&D Center

Digital Transformation (DT) refers to the disruptive improvement process that introduces changes in information management, computing, communication, and connectivity technologies that impact organizational operations, structure, and business strategy. It is important to state that the DT is more than just applying technology in the current process, but it also encourages the organization to rethink the way they work and deliver value to internal or external customers [8].

At the heart of digitalization's discussions are data. In the modern world, data is considered one of the most valuable assets of companies [9]. New digital technologies are improving the whole data path from the collection (e.g., using online sensors) until the availability (e.g., integrated systems giving ease access to data). At the R&D centers, the data have a protagonist role, once the scientists rely on data to raise knowledge that can be used to develop innovative products, for example. So, we can infer that the DT has great potential in Chemical Industry R&D centers.

Some examples of potential digital technologies applications at R&D centers are listed in Table 1. These technologies may increase the sector's efficiency and effectiveness. For example, the High-Throughput Experimentation technique enables the laboratory to run a large number of experiments at once, which speeds up the research and data collection. Properties predication by AI can reduce the number of required experiments for research, reducing the laboratory costs [2].

Table 1. Potential technologies application examples at chemical R&D

Technology	Potential application examples
New robots generation	- High-throughput Experiments (HTE) - Clean in place - Sample transportation
Artificial Intelligence (AI)	- Physicochemical properties prediction - Product Design - Process control and optimization
Internet of Things (IoT)	- Process automation - Equipment communication for maintenance - Device control
Virtual reality	- Security training - Accident investigation - Equipment setup and maintenance

2.1 Challenges to Digital Transformation in Chemical Industries

According to the research conducted by the consultant Michael Williams and presented at ARC Industry Forum in 2019, there is still less than 25% of chemical industries moving beyond the pilot phase of digitalization projects. Indeed, the pilot projects find technically viable solutions, but they end up not evolving in the business case that would allow the demand for more investments, reducing their roll out throughout the organization [3].

In his presentation, Williams identified some of the top challenges, which include: Lack of a concrete value proposition to justify investment; Islands of information prevalent in existing infrastructure; Lack of seamless integration between Internet of Things (IoT) and Operational Technologies (OT) solutions; Cybersecurity issues. [3].

Also, Parviainen *et al.* state the importance of a well-defined overall digitalization strategy to the transformation success: *"The transformation did not succeed because organizations did not change mindsets and processes or build a culture that could foster the change. Lack of an overall digitalization strategy and competing priorities were the most typical obstacles for digitalization, together with security concerns and insufficient technical skills* [4]."

Therefore, we can infer that a structured plan is essential to enhance the success likelihood of DT. That said, this paper aims to propose a process framework to guide the R&D center of a Brazilian chemical company in conducting the DT inside the area. The framework should assist in the identification, prioritization, and execution of digital projects in order to increase the sector's digital maturity.

3 Method—Design Science Research

For the development of the proposed framework, prescriptive research with a qualitative approach was carried out based on the company's internal information and external references as scientific articles and correlated companies' whitepapers.

It was chosen Design Science Research (DSR) as the applied method once the main objective is to build, test, and validate an artifact (process framework). Some guidelines should be met when using DSR [10]. Table 2 exposes the applicable requirements and how they are fulfilled at present work:

Table 2. Compliance with DSR requirements of the current work

DSR requirements	Research elements
Problem relevance	DT can improve the performance of the chemical R&D sector, affecting the competitiveness of the industry [5]
Problem class	The relatively low success of digital transformation at Chemical Industry R&D Centers [3–6]
Artifact design	The artifact is the proposed process framework (step-by-step guide)
Artifact validation	Observational: Predefined questions survey with R&D leadership and experts regarding framework applicability
Research contribution	A proposal to help the Chemical R&D sectors more effectively conduct the Digital Transformation process
Operating method	The process suggested by Vaishnavi and Kuechler (2007) [11]

In this article, only the design of the framework will be presented. Its application and validation will be carried out in the future at the R&D center of a Brazilian company, which has 8 laboratories, around 150 employees, and operates in various segments such as Oil and Gas, Home and Personal Care, Paints, and Coatings.

3.1 Operating Method

The chosen operating DSR method is that proposed by Vaishnavi and Kuechler (2007) [11] (Fig. 1). The first step, Awareness of the Problem, consists of deeply understanding the problem to be solved. The researcher should expand their knowledge about the problem looking for related information as to its nature, involved actors, limitations, past projects, and so on. In this case, a committee was formed to discuss the potential failure root causes of digital projects within the sector. At least one experienced employee from each laboratory participated in the meetings, which were carried out with open-ended questions.

Second, starts the Suggestion step. At this moment, the researcher uses their creativity to suggest (abduction) a solution based on the state of theory. For the present research, a process framework is suggested to be used as a guideline to conduct the digital transformation process, from the identification to the execution of digital projects.

After that, comes the third step: Development. This is when the researcher describes what and how the artifact will be designed.

Then, the artifact goes through the Evaluate step. In this case, the framework will be validated by a group of R&D leaders and experts through a survey of predefined

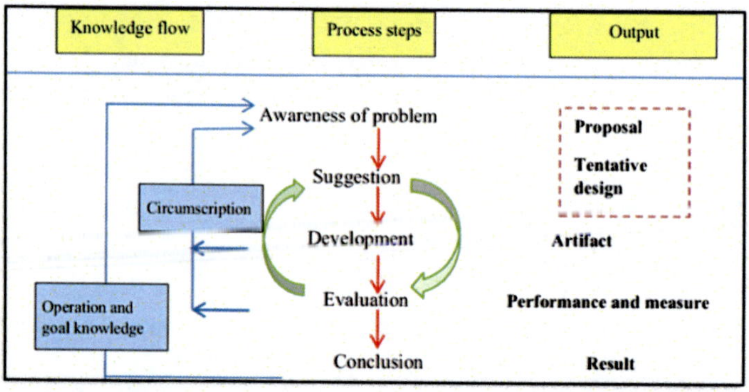

Fig. 1. DSR operating method adapted from Vaishnavi and Kuechler (2007) [11]

questions. The main objectives are to verify whether the framework has the potential to serve as a guideline during the DT process and to raise feedbacks to improve it.

Finally, the last step is the Conclusion, when the researcher will summarize the research's findings and generalize their importance, discuss ambiguous data, and recommend further research. In the next section, the artifact development process will be explained in more detail.

4 Artifact Development

This section will explain the artifact (process framework) development and how it will be applied at the R&D center during the Digital Transformation process.

The framework is divided into three phases: Preparation, Elaboration, and Application (Table 3). In the first phase, the main objective is to understand the present and future needs of the company's R&D sector. The collected information will be useful to identify digitalization opportunities. The second phase consists of the Maturity Map and Technology Roadmap elaboration. The tools are designed using the information collected during the first phase. Finally, in the third phase, the tools are applied to guide the DT process.

Table 3. Proposed framework composed of three phases and seven steps.

Digital transformation framework guide		
Preparation	Elaboration	Application
1. Preliminary 2. Information collection 2.1. Processes 2.2. Pain points 2.3. Strategy	3. MM issue 3.1. Levels 3.2. Dimensions 4. TRM issue 4.1. Tech prospection	5. Auto-assessment (MM) 6. Project prioritization (TRM) 7. Project execution

4.1 Preparation

Preliminary: Team Building and Communication Plan Development. There is a preliminary step before going through information gathering with the R&D teams. The objective is to define some parameters that will be considered during the DT process. The main parameters to be defined are the project team(s), the sponsor(s), involved areas, key employees to be interviewed, and the communication plan.

Information Collection: Process Mapping, Pain Points, and Strategy. At this point, the project team (or commission) will start the information collection aiming to figure out how the sector works and what the employees' needs and pains points are. Thus, an initial discussion team will be defined containing at least one representative from each area. During the conversations, other people may be requested to join the meetings, dynamics, or even private interviews.

The first task is to map the sector's processes using the SIPOC tool [12] (SIPOC stands for supplier, input, process, output, and customer) seeking to understand how different processes or laboratories are interconnected. After that, the commission will be able to ask for the operational pain points in each specific process, so the problems can be compiled, analyzed, and categorized.

The information collection of strategy is done similarly, that is, it will be based on meetings, group dynamics, and interviews. The difference is that top leadership participants will be prioritized because they may have a broader vision of the sector and are able to talk about how the area is mobilized to deliver value.

The strategy will be evaluated in two domains: external and internal [13]. The external domain is related to the posture of the sector in the environment outside the company, that is, how the competition for products and services occurs, relationship with suppliers and customers, positioning in value chains, etc. It is worth pointing out that depending on the structure of the R&D center, some laboratories may not have external activities.

The internal domain, on the other hand, refers to the way in which the area is mobilized as a team to generate value. For certain experimental work, for example, the sector may choose to do it in-house or outsource it, and this choice may directly affect the roadmap to be developed. Both domains will be captured by the project team to elaborate the tools.

4.2 Elaboration

Once the Preparation phase is completed, the project team has already obtained important information that will be used to build a customized plan for the sector under evaluation. It is noteworthy that everything that is raised must be constantly revalidated (iterative cycle) with the project's key people in order to increase the accuracy of the information and ensure that the important items are being considered. Two tools have been considered in the framework: Maturity Map (MM) and Technology Roadmap (TRM).

Many researchers have used the MM related to Industry 4.0 assessment for different segments [14–18]. The tool is commonly used as an instrument to conceptualize and measure the maturity of an organization or a process regarding some specific target state such as digital transformation [14]. The TRM is a useful tool to visually plan the implementation of systems, allowing the project team to identify the path to be taken with the intermediary projects signalized [19].

Thus, the two tools applied together have the potential to coordinate the identification, prioritization, and execution of digital projects. Next, it will be explained in more detail how the tools will be issued.

Issue of Maturity Map (MM). For the present work, it is not applied a specific methodology for MM construction, but the adaptation of different models to suit the project's objectives [14–18]. It contains two perpendicular axes: maturity Levels and Dimensions.

The maturity levels definition was made by inspiration of the Deloitte model (2019) [16]. To characterize the levels, two factors are taken into account: the intensity of digital technologies application and their impact on the way in which the sector operates. The first factor is more intuitive and is often considered the only one to assess digital maturity. However, the digital transformation is also about remodeling the way of acting with the technology support, so the second factor is necessary [16]. Therefore, the proposed levels are:

- Level 1. Passive—None or little application of basic digital technologies aimed only at meeting certain internal or market demands;
- Level 2. Explorer—Application of traditional technologies to automate some tasks, without changes in the way of acting;
- Level 3. Starter—Application of some digital technologies to increase performance capabilities, but still without major changes in the way of acting;
- Level 4. Becoming digital—Application of digital technologies increasing synchronization between different areas, causing changes in the way of internal and external performance;
- Level 5. Being digital—Operation and performance model completely optimized using digital tools through a new way of working, differently from the previous ones.

The definition of dimensions can be made from a prefixed set, as proposed by Schumacher et al. (2016) [14], or from a tailor-made set, which takes into account the main strategic pillars of the company for example [18]. The present study will consider as dimensions the sector's main macro processes, mapped on the Preparation phase.

Due to the specificity of the object of study of this work, some dimensions considered strategic for the R&D sector will be suggested. However, it will be up to the team leading the project to evaluate this set and may use it in its entirety, make modifications or even create a new set. The suggested dimensions are Data and Analytics; Sample Management; Knowledge Management; Laboratory Management; Integration; Safety.

Issue of Technology Road Map (TRM). The TRM aims to answer the three main questions: 1) Where are we going? 2) Where are we now? and 3) How will we get there? [19]. However, in the framework presented in this paper, questions 1 and 2 were considered during the construction of the Maturity Map. Therefore, the TRM will have the role of identifying the technologies needed to fulfill each MM's scenarios, that is, it will answer the question: how do we reach each scenario?

The used structure is the same as the MM, which on the first column contains dimensions and the levels are on the first row (Table 4). The technologies or systems are arranged in boxes and the arrows indicate a dependency relationship, that is, the systems downstream the arrow depend on the implementation of the upstream systems to be implemented. Table 4 illustrates the TRM, the blue boxes are representing the required systems for each level per dimension.

For the technological prospecting process, the IT team will be directly involved, due to their experience in the subject. Both internal and external sources will be used to survey potential technological solutions to meet the maturity map scenarios [19].

Internally, systems developed in other sectors may be used as a reference for R&D. In addition, technology experts from sectors such as engineering, IT, industrial automation, and others will be consulted for the subject discussion. Externally, the team will search for solutions in different channels. Such as direct search on the internet, participation in seminars, fairs, and workshops of technology providers, conducting benchmarks with other companies, contacting institutions and associations that encourage industry and technology development, etc.

Table 4. TRM structure representation. Blue boxes represent digital systems and arrows represent dependencies. Elaborated by the author.

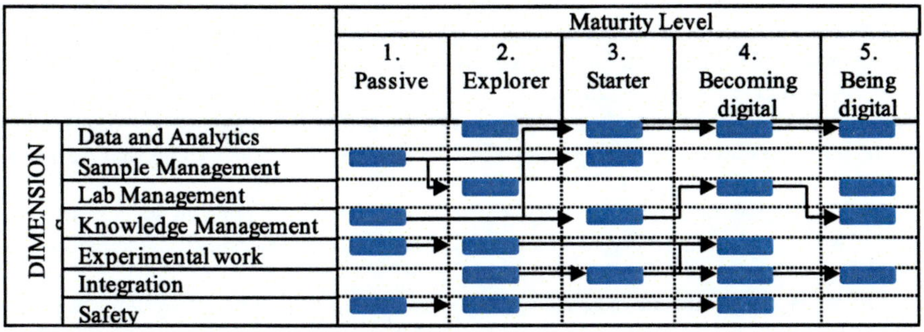

4.3 Application

The final phase consists of the application of the tools by R&D areas. Firstly, each area must run an auto-assessment for maturity level (step 5). Then, managers will be able to define which dimension(s) will be prioritized to be advanced at the moment. So, the system implementation projects can be identified using the TRM. In this way, it is possible to select the projects that are applicable in more than one area, increasing their relevance and destinated resources (step 6).

In addition, it will be clearer the sequence of projects that must be adopted for the Digital Transformation process to move forward. This can contribute to the coordination of initiatives across the sector, enhancing the feasibility and effectiveness of digital projects. Finally, the selected projects can be executed by the responsible teams.

5 Final Considerations

Chemical industries have failed in most digital initiatives, slowing the advancement of digital maturity in areas such as the R&D. Recent studies show that problems range from proposing projects to their implementation. Therefore, a structured process framework was proposed to serve as a guideline for the R&D center to conduct the area's digital transformation process. The applied method was Design Science Research since the main objective was to prescribe an artifact (process framework).

The designed process framework comprises three phases: Preparation, when relevant information is gathered, such as people involved, sector macro processes, employees' pain points, and defined strategies; Elaboration, when the MM and TRM tools are issued; and Application, finally when the tools are applied within the R&D areas. It has the potential to support the area in identifying, prioritizing, and executing digital projects, in order to improve the implementation success rate.

This article covered up to the framework design stage. The application and evaluation steps will be carried out in the future in the chemical R&D center of a large Brazilian company.

References

1. Abdallah, Y., Shehab, E., Al-Ashaab, A.: Digital transformation challenges in the manufacturing industry. In: Shafik, M., Case, K. (eds) Advances in Manufacturing Technology XXXIV, pp 9–14. IOS Press, Online (2021)
2. Kalinowski, M. et al.: Towards lean R&D: an agile research and development approach for digital transformation. In: 46th Euromicro Conference on Software Engineering and Advanced Applications (SEAA), pp. 132–136. IEEE Xplore, Portorož (2020)
3. Hill, D.: Digital transformation lags in chemical industry (2019). https://www.chemicalprocessing.com/articles/2019/digital-transformation-lags-in-chemical-industry/. Accessed 20 Jul 2020
4. Parviainen, P., Tihinen, M., Kääriäinen, J., Teppola, S.: Tackling the digitalization challenge: how to benefit from digitalization in practice. Int. J. Inf. Syst. Proj. Manag. **5**, 63–77 (2017)
5. Peleg, L.: Digital transformation & industry 4.0 in the chemical industry (2021). https://www.precog.co/blog/digital-transformation-industry-4-0-in-the-chemical-industry. Accessed 10 Jul 2021
6. Mckinsey: Unlocking success in digital transformations (2018). https://www.mckinsey.com/business-functions/organization/our-insights/unlocking-success-in-digital-transformations. Accessed 10 Jul 2021
7. Dresch, A., Lacerda, D., Miguel, P.: A distinctive analysis of case study, action research and design science research. RBGN **17**(56), 1116–1133 (2015)
8. Rosalia, R., Wahba, K., Milevska-Kostova, N.: How digital transformation can help achieve value-based healthcare: Balkans as a case in point. The Lancet Regional Health - Europe **4**, 100100 (2021)
9. The Economist: The world's most valuable resource is no longer oil, but data (2017). https://www.economist.com/leaders/2017/05/06/the-worlds-most-valuable-resource-is-no-longer-oil-but-data. Accessed 12 Jul 2021
10. Mettler, T.: Maturity assessment models: a design science research approach. Int. J. Soc. Syst. Sci. **1**(2), 81–98 (2011)

11. Vaishnavi, K., Kuechler, W.: Design science research methods and patterns: Innovating information and communication technology, 2nd edn. CRC Press, Boca Raton (2007)
12. Brown, C.: Why and how to employ the SIPOC model. J. Bus. Contin. Emer. Plan. **12**, 198–210 (2019)
13. Henderson, J., Venkatraman, H.: Strategic alignment: leveraging information technology for transforming organizations. IBM Syst. J. **32**, 4–16 (1993)
14. Schumacher, A., Erol, S., Sihn, W.: A maturity model for assessing industry 4.0 readiness and maturity of manufacturing enterprises. Procedia CIRP **52**, 161–166 (2016)
15. Gracel, J., Łebkowski, P.: The concept of industry 4.0 related manufacturing technology maturity model. Decis. Making Manufac. Serv. **12**, 17–31 (2019)
16. Delloite: Digital maturity model to help ease the transition: achieving the next frontier of chemicals excellence (2019). https://www2.deloitte.com/content/dam/Deloitte/us/Documents/energy-resources/us-er-achieving-the-next-frontier-of-chemicals-excellence.pdf. Accessed 20 Oct 2020
17. Bibby, L., Dehe, B.: Defining and assessing industry 4.0 maturity levels - case of the defence sector. Prod. Plan. Control **29**, 1–14 (2018)
18. Stefan, L., et al.: Concept for an evolutionary maturity based industrie 4.0 migration model. Procedia CIRP **72**, 404–409 (2018)
19. Rezgui, Y., Zarli, A.: Paving the way to the vision of digital construction: a strategic roadmap. J. Constr. Eng. Manag. **132**, 767–776 (2006)

Identification of Soil Characteristics for the Extraction of Ch'ampa as Earth Block Masonry

Yasmani Vitulas Quille(✉)

Departamento de Ingeniería Civil, Universidad Nacional del Altiplano de Puno, Puno, Peru
yvitulas@unap.edu.pe

Abstract. At present, many houses in the Peruvian highlands do not have the necessary infrastructure to face climatic phenomena, even though there are pre-Hispanic vernacular houses as ancestral legacies that allow them to resist this inclemency, among them are the *Putucos* of Puno (Taraco and Samán), Therefore, the objective of this research is to study and characterize the identification criteria for the extraction zones and the ideal soil that can be used for extraction of *Ch'ampa* as earth block masonry, the methodology used is a preliminary study of the characteristics of extraction to blocks or natural grass in a traditional way, tests to determine the limits of consistency, fine content, soil classification, and granulometric curve. According to the findings obtained, the Liquid limit has an average value of 42.33%, the Plastic limit has a value of 32.33%, and the Plasticity index has a result of 10.00%, having a considerable variation between the LP with reference to the adobe, has a natural humidity of 3.80% (sample dried outdoors for 20 days), specific weight of solids of 2.587 gr/cm^3, a wet density of 1.385 gr/cm^3, and a dry density of 1.350 gr/cm^3, being the predominant soil type ML (Classification USCS) and its contrast with other materials through the NRCS Soil triangle. These results give us a first idea of the difference with other traditional construction materials with earth, looking for new environmentally friendly technologies that can recover ancestral technologies, which for some time now have been in the process of extinction.

Keywords: Ancestral · Physical characteristics · Blocks of natural grass · Ch'ampa · Cultural identification · Putuco · Vernacular

1 Introduction

We must start from the concept of eco-technologies, as a methodology that favors a harmonious relationship with the environment without neglecting the social and economic values of the interested parties [1], and, among the most used materials worldwide, we have *adobe*, *tapia*, *CEB*, and *quincha*, which are traditional and are also widely studied throughout the world.

Y. Iano et al. (Eds.): BTSym 2021, SIST 207, pp. 512–520, 2023.
https://doi.org/10.1007/978-3-031-04435-9_53

However, there are no current studies of the *Ch'ampa*, being a material predominantly from the *Colla* culture of the Peruvian and Bolivian highlands, material that is extracted directly from the ground [2–4], but there are many investigations on the traditional adobe as ancestral knowledge of Colombia [5], as the heritage of Najran [6], recovering it as a lesson of traditional architecture of Spain [7] and even Europe [8] and Africa [9]. There are many standards and regulations that have included the earth blocks as a building material [10], even filling part of the gap in the thermal comfort standards for Libya [11], even considering researches of comparison between thermal and resistant aspects with earth construction elements [12,13], in addition to its mechanical behavior [14], and we could mention many more, but none addresses the *Ch'ampa* as a traditional material, which is worth studying since its use cannot remain unnoticed in the existing information base. This is the main reason why we have chosen to present this first study on the identification of soil characteristics for the extraction of *Ch'ampa* as earth block masonry.

2 *Ch'ampa* Extraction

In order to have a clear idea of this type of raw earth block masonry, named by George Squier (1864) as natural turf blocks [3], this study presents the characteristics that make this traditional material called *Ch'ampa*, which is used in the very particular construction of rural dwellings in the *Andes* between Peru and Bolivia.

2.1 Extraction Zones

The *Ch'ampa* or *T'epe* (native name) is used to build traditional pre-Hispanic dwellings called *Putukus* (Bolivia) [15] or *Putucos* (Peru) [4,16], easily reaching 50 years of duration and even exceeding 80 years without collapse, being resistant to erosion.

For its extraction, the selected area should be identified well in advance and avoid using it as an agricultural area, reserving the land for at least five years (see Fig. 1), in which wild plants called *Ch'iji* and *Quemello* sprout and thanks to their roots, the block of earth obtains its resistance properties.

Fig. 1. Extraction and stacking area for the drying process, the difference between a current extraction area and an old extraction area (3 years old), which is in the process of restoration and growth of wild vegetation called *Ch'iji*.

2.2 Method of Extraction

It is done with the use of traditional tools in the form of semi-curved shovels. The masonry of earth blocks is usually 40 × 50 × 12.5 cm and two shovels must be used. To extract units of 50 × 70 × 15 cm, three shovels are necessary (see Fig. 2).

Once the blocks are extracted, they must be shaped to give them a regular prismatic form that can be stacked. For the smaller blocks, an extraction yield of 600 units/day to 650 units/day is achieved, and for the larger ones, it is only possible to extract between 350 units/day to 400 units/day, requiring at least three to four days to complete the work necessary for the construction of a *Putuco* type house.

Fig. 2. Extraction of *Ch'ampa* with the aid of two blades (left) and three blades (right), which depend on the required constructive characteristics.

2.3 *Ch'iji* and *Quemello* Plant Matter Content

The roots of the wild grasses called in Quechua *Ch'iji* and *Quemello* (*Muhlenbergia fastigiata, Muhlenbergia ligularis*), spread inside the earth block both vertically and horizontally forming a natural mesh and providing a natural property of adherence, making the *Ch'ampa* resistant to the effects of erosion [17].

This species can be found in Colombia, Peru, Bolivia, Argentina, and Chile, between 3300 m (above sea level) and 4300 m (above sea level) [3].

The grass formed by the stem and leaves together can reach 25 cm in height, it has been found in some places, the grass crawls on the outer surface of the soil, this phenomenon is because when they manage to grow more than 25 cm, their stems cannot withstand their weight because they are thin and crawl along the ground, forming numerous additional roots.

The diameter of the roots has been measured. In the case of *Ch'iji*, the horizontal roots reach an average of 5.59 mm and the vertical roots 1.28mm. For *Quemello* the horizontal roots reach an average of 4.40 mm and the vertical roots 0.97mm.

Regarding the root content based on the weight of the sample (Wm), the weight of roots (Wr) has a mean of 10% (9.847%), considering that there is a linear relationship between the weight of the sample and the weight of the roots contained (see Fig. 3).

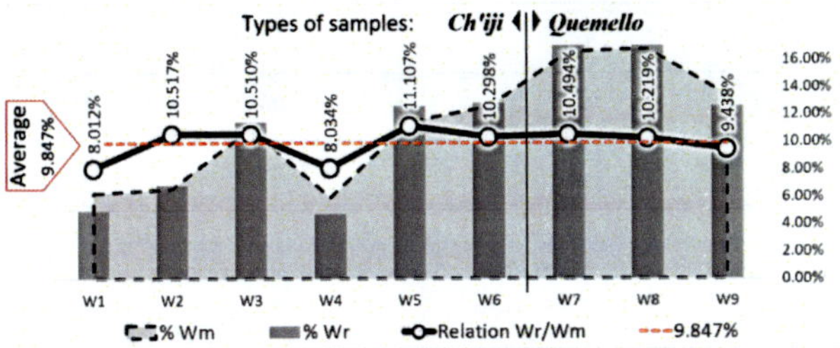

Fig. 3. Ratio between sample weight (%Wm) and root weight (%Wr), in addition to the Wr/Wm ratio in % of the total weight of the samples.

2.4 Density and Consistency Limits

In additional tests, according to ASTM D2216-10, a moisture content ranging from 3.5% to 4.1% has been obtained, considering that the earthen masonry blocks are traditionally dried in the open air and with the help of the sun for a period of 15 to 20 days. The specific weight of the solids, according to ASTM D854-10 with the help of the pycnometer, has an average of 2586.67 kg/m^3, and the volumetric unit weight, according to ASTM D7263-9, has an average of 1350.0 kg/m^3.

To determine the consistency limits, ASTM D4318-10 was used to obtain the liquid limit (LL) and plastic limit (PL), obtaining an average value of 42.33% for LL and 32.33% for LP. (see Table 1). In addition, its silt content varies between 61% to 63%, clay between 19% to 22% and sand between 5% to 9%, which can be classified as an ML soil by USCS (Low plasticity silt) and ASTM (Silt, rock dust, slightly plastic sandy, or clayey silts) (see Fig. 4).

Table 1. Soil index properties.

Property	Value	Standard deviation	Coefficient of variation
Liquid limit	42.33%	2.3094	5.46%
Plastic limit	32.33%	2.3011	7.12%
Plasticity index	10.00%	–	–
Humidity	3.80%	0.0030	7.89%
Specific gravity of solids	2.587 gr/cm^3	0.0306	1.18%
Wet density	1.385 gr/cm^3	0.0250	1.81%
Dry density	1.350 gr/cm^3	0.0100	0.74%

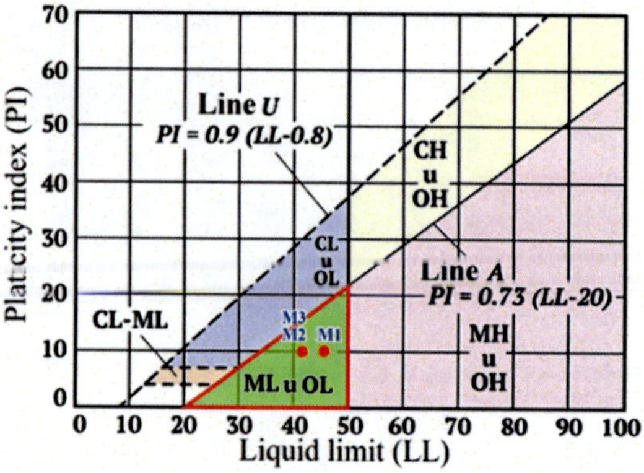

Fig. 4. CASAGRANDE Plasticity Chart - ASTM.

2.5 Granulometric Analysis

It is developed based on the ASTM D422-63 standard, using different sieve sizes to analyze the particle size, and, secondly, the densimeter was used.

The tests were carried out with 25 different diameters or sieves, in which the 3" to #4 sieve let through 100%, the #10 sieve let through between 98.70% to 99.90%, and finally, the #200 sieve let through between 90.60% to 96.20%, which indicates that a more thorough analysis of the sample is necessary. In the case of smaller diameters, the densimeter test was developed as a complement, since these diameters are not normally analyzed, finishing the control on the #200 mesh, where the 0.038 mm sieve passes between 73.04% to 78.06%, the 0.011 mm sieve passes between 44.82% to 55.24%, and the smallest sieve of 0.001 mm only passes between 17.32% to 20.91% (see Fig. 5).

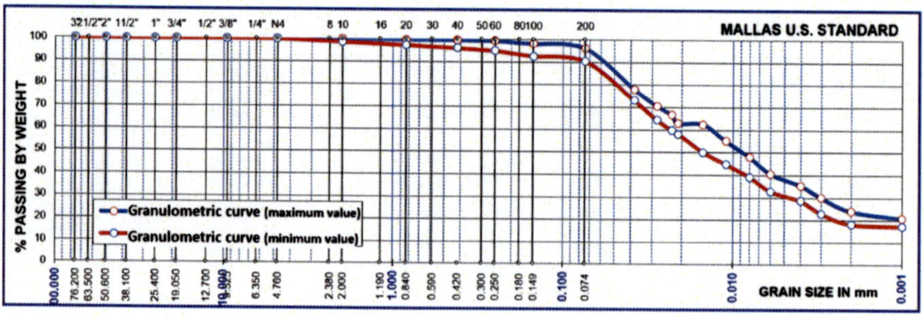

Fig. 5. Maximum and minimum sieving sieve analysis curve, the two curves represent the maximum and minimum values of all tests, generating an upper and lower limit for the variation of the results.

3 Discussion of Results

In the case of the content of organic components in the form of roots, it was determined that the *Ch'ampa* has a proportion to 10% of weight, it should be considered that this quantity incorporated is natural and depends on the period of years that the soil is allowed to regenerate the wild plants, which is approximately no less than five years.

Regarding the limits of consistency (LL and PL), which are compared with other factors and according to the Mann-Whitney U correlation, it was determined that the promised value of LL and PL of *Ch'ampa* are similar to other raw earth materials, except for the PL between *Ch'ampa* and *adobe*, since they are different, giving this earth block masonry a greater capacity for deformability and adaptation to small deformations due to settlements, which are very common in the construction of *Putucos.*

According to Table 2, it can be seen that *Ch'ampa* has a higher silt content and very little sand, which makes this material more workable, in addition to its LL and PL being similar to *tapial* because it is a naturally formed material.

Table 2. Soil index properties.

Material	Ref.	Silt	Clay	Sand	Gravel	LL	PL	PI
Ch'ampa	This study	61–63%	19–22%	5–9%	–	42.3%	32.3%	10.0%
	[18]	59–63%	11–20%	5–15%	–	44.3%	32.9%	11.4%
Adobe	[19]	20%	15%	65%	–	27.8%	19.8%	8.0%
	[20]	7.1%	46.5%	46.4%	–	28.0%	17.0%	11.0%
	[21]	10.0%	41.8%	48.2%	–	32.0%	22.0%	10.0%
	[22]	15–25%	10–20%	55–70%	–	30.1%	19.8%	10.3%
	[23]	10%	72%	18%	–	44.0%	25.0%	19.0%
Tapial	[24]	50.85%		47.40%	1.75%	47.0%	27.0%	20.0%
	[25]	65%		40%	25%	31.0%	20.0%	11.0%
	[26]	34%		47%	19%	26.0%	18.0%	8.0%
	[27]	15–25%	10–20%	55–70%		31.0%	20.0%	11.0%
CEB	[28]	54.95%		44.60%	0.45%	27.6%	15.5%	12.1%
	[29]	–	–	–	–	52.4%	45.7%	6.7%
	[30]	22–53%	22–53%	55–75%	–	–	–	–

Presenting these results in the NRCS Soil Triangle (see Fig. 6), we can see the comparison of the soil considerations that make up the *Ch'ampa*, in contrast to both *adobe* and *tapial*, which gives us a clear idea of the difference between these two materials and clearly shows the advantages of the traditional material called *Ch'ampa*, which, as indicated at the beginning of this study, has few antecedents and studies with great depth, and it is necessary to deepen these studies, as was done in both *adobe* and *tapial.* The data are obtained using the Bouyoucos method.

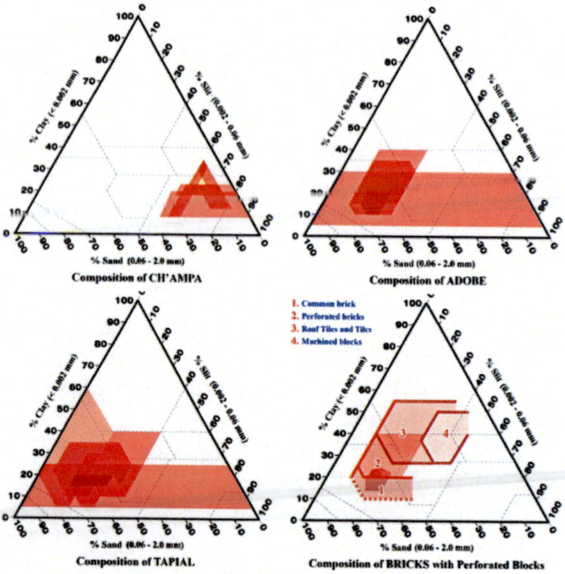

Fig. 6. Granulometric characteristics according to the NRCS Soil Triangle. For the case of *Ch'ampa* and other similar materials (considering the contributions of [31–33]).

4 Conclusions

Ch'ampa is a natural earth block masonry that is used in the construction of handmade houses, preferred over other materials such as *adobe* and *tapial*. According to the present study, this construction is validated with physical studies that classify the soil as type ML (Classification USCS), an average liquid limit of 42.33%, plastic limit with a value of 32.33% and a plasticity index of 10%, the block dried in the open air for 20 days has a natural humidity of 3.80%, its specific weight is 2.587 gr/cm^3, a humid density of 1.385 gr/cm^3 and a dry density of 1.350 gr/cm^3.

According to its internal composition, a higher percentage of silt can be seen in the *Ch'ampa*, which varies between 61% to 63%, above the *adobe* and *tapial* (average of 20%) and gives this traditional material, ideal characteristics to its thermal behavior and work-ability. Regarding the clay, the *Ch'ampa* varies between 19% to 22%, which is very similar to the *adobe* and *tapial*, finally the sand has a variation between 5% to 9%, which gives *Ch'ampa* greater stability and compensates for its fragility with the natural incorporation of roots such as *Ch'iji* in a proportion of 10% of the total weight, an aspect it does not share with any other type of earth construction material, despite the incorporation of natural artificial components in both *adobe* and *tapial*.

The Soil Triangle according to the texture of the NRCS of the U.S.D.A., gives us a graphic vision of the composition of the soil based on the percentages of silt, clay, and sand content, thus filling an existing void in the study of *Ch'ampa*.

These characteristics make *Ch'ampa* earth block masonry suitable for the construction of traditional houses called *Putuco*, so the area of extraction should be properly identified, in addition to being traditionally used in the high Andean Altiplano of Puno city in Peru.

References

1. Ortiz-Moreno, J.A., Masera-Cerutti, O., Fuentes-Gutierrez, A.F.: La ecotecnología en México. Imagia, p. 126, Mexico (2014)
2. Gamio-Felipa, A.L., González-García, Mra.N., Verdú-Vázquez, M.A.: Comparision of the physical properties of earth construction units: adobe, tapial and champa. The importance of its use in the peruvian highlands. In: 4th CITE, pp. 284–285. E.T.S. of Edification (UPM), Spain (2019)
3. Vitulas-Quille, Y.T., Reynoso-Machaca, K.B.: Ancestral Construction and Mystical Conception of the Putucos of the Peruvian Highlands, 5th edn., pp. 65–73. Antioquia Research Institute, Chapter (2020)
4. Marussi-Castellan, F.: Arquitectura vernacular: Los Putucos de Puno, 1st edn. University Research Center Ricardo Palma, Peru (2021)
5. Rivera-Salcedo, H., Valderrama-Andrade, O.M., Daza-Barrera, A.A., Plazas-Jaimes, G.S.: Adobe como saber ancestral usado en construcciones autóctonas de Pore y Nunchía, Casanare (Colombia). Rev. Arquit. **23**(1), 74–85 (2021)
6. Al-Sakkaf, Y.K., Abdullah, G.: Soil properties for earthen building construction in Najran City. Saudi Arabia. C Mater Contin **67**(1), 127–140 (2021)
7. Villacampa-Crespo, L., García-Soriano, L., López-Manzanares, F.V., Mileto, C.: Constructive techniques of the past for a sustainable future. The case of traditional earthen architecture in Aragon (Spain). Int. J. Latest Trends Eng. Technol. **11**(3), 30–36 (2018)
8. Duarte-Carlos, G., Alcindor, M., Correia, M.: Arquitectura tradicional de tierra en Europa: un patrimonio de entramado y encestado, adobe, tapia y pared de mano. In: Anales del Instituto de Arte Americano e Investigaciones Estéticas "Mario J. Buschiazzo", vol. 48, no.2, pp. 239–256 (2018)
9. Sturm, Th., Ramos, L.F., Lourenço, P.B.: Characterization of dry-stack interlocking compressed earth blocks. Rev. Mater. Struct. **48**(1), 3059–3074 (2015)
10. Cid, J., Mazarrón, F.R., Cañas, L.: The earth building normative documents in the world. Inf. Rev. Constr. **63**(523), 159–169 (2011)
11. Gabril, N.: Thermal comfort and building design strategies for low energy houses in Libya: Lessons from the vernacular architecture, Ph.D. thesis University of Westminster (2014)
12. Cuitiño-Rosales, M.G., Rotondaro, R., Esteves, A.: Análisis comparativo de aspectos térmicos y resistencias mecánicas de los materiales y los elementos de la construcción con tierra. Rev. Arquit. **22**(1), 138–151 (2020)
13. Wieser, M., Onnis, S., Meli, G.: Desempeño térmico de cerramientos de tierra alivianada. Posibilidades de aplicación en el territorio peruano. Rev. Arquit **22**(1), 164–174 (2020)

14. Castillo-Valencia, W., Areiza-Palma, G., Coral-Moncayo, H.: Comportamiento físico mecánico de la tapia por pandeo y conexión de esquina, Caso Teatro Imperial de Pasto, Inge CUC **14**(2), 81–96 (2018)
15. Escalante-Moscoso, J.F.: Arquitectura prehispánica en los Andes Bolivianos, 2nd edn. Producciones CIMA, Bolivia (1994)
16. Vitulas-Quille, Y.T., Reynoso-Machaca, K.B.: Etnoingeniería de los Putucos de Puno: el predominio de la sabiduría Colla. 1st edn., pp. 59–66. Universidad Nacional del Altiplano (2013)
17. Peterson, P.M., Giraldo-Cañas, D.: The genus Muhlenbergia (Poaceae: Chloridoideae: Cynodonteae: Muhlenbergiinae) in Colombia. Rev. Caldasia **31**(2), 269–302 (2009)
18. Gamio-Felipa, A.L.: Comparison of the construction system of Putucos with other construction systems in earth. Rev. An. Edif. **3**(2), 1–8 (2017)
19. Rojas-Tirado, F.D.: Diseño de un bloque de adobe compactado con adición de viruta de madera para mejorar la resistencia a la compresión, Thesis UCV (2020)
20. Laime-Ancalle, G.: Diseño de elaboración del adobe incorporando la fibra de totora para reforzar las propiedades físico y mecánico, Thesis UCV (2020)
21. Mantilla-Calderón, J.C.: Variación de las propiedades físico mecánicas del adobe al incorporar viruta y caucho, Thesis UNC (2018)
22. Carhuanambo-Villanueva, J.T.: Propiedades mecáinicas y físicas del adobe compactado con adición de viruta y aserrín, Thesis UPN (2016)
23. Rivera-Torres, J.C., Muñoz-Díaz, E.E.: Structural characterization of materials used in construction systems with soil material, The adobe. Rev. Int. Des. Nat. Accid. Infraestr. Civil **5**(2), 135–148 (2005)
24. Nardez-Giraldo, M.F.: Análisis del comportamiento físico mecánico de muros en tapia pisada con inclusión de biomasa residual, Thesis UPBSB (2018)
25. Tantaleán-Altamirano, C.J.: Desarrollo de un diseño estructural por esfuerzos de trabajo (DET) en viviendas de tapial de 1 piso, Thesis PUCP (2018)
26. Chuqui-Paucar, W., Challco-Ruelas, R.A.: Evaluación de las propiedades mecánicas de muros tipo tapial para viviendas económicas con presencia de hiladas de mortero de cemento – arena, Thesis UNSAA (2018)
27. Tejada-Schmidt, U., Mendoza-Garcia, A., Torrealva-Dávila, D.: Uso del tapial en la construcción. SENCICO (2016)
28. Medina-Saucedo, O.: Diseño de bloque de tierra comprimida estabilizado con cemento, como material sostenible, para el uso en la construcción de viviendas rurales en la provincia de San Martín, Thesis UNSM (2021)
29. Cañola, H.D., Builes-Jaramillo, A., Medina, C.A., González-Castañeda, G.E.: Bloques de tierra comprimida (BTC) con aditivos bituminosos. Rev. TecnoLógicas **21**(43), 135–145 (2018)
30. Donkor, P., Obonyo, E.: Earthen construction materials: assessing the feasibility of improving strength and deformability of compressed earth blocks using polypropylene fibers. Rev. Mater. Des. **83**(1), 813–819 (2015)
31. Blanco-Montero, C., Morales-Pereyra, P.: Mejora de las condiciones de habitabilidad básica en el marco del Programa Oxlajuj Tz'ikin. Fase III Cooperación Galega. File ASF/ASIAPRODI, Guatemala (2010)
32. Gaete-Cruz, M.: Estrategias morfológicas de diseño sismo resistente para la construcción contemporánea en tierra portante, Thesis PUCCH (2010)
33. Álvarez-Rozo, D.C., Sánchez-Molina, J., Corpas-Iglesias, F.A., Gelves, J.F.: Características de las materias primas usadas por las empresas del sector cerámico del área metropolitana de Cúcuta (Colombia). Boletín la Soc. Española de Cerámica y Vidrio **57**(6), 247–256 (2018)

Social Impacts of Artificial Intelligence and Mitigation Recommendations: An Exploratory Study

Vanessa Marques Paes(✉), Franciane Freitas Silveira, and Alessandra Cristina Santos Akkari

Federal University of ABC (UFABC), São Bernardo do Campo, Brazil
{vanessa.paes,franciane.silveira,alessandra.akkari}@ufabc.edu.br

Abstract. With all the changes and adaptations of the industrial sector to migrate to a more technological and connected environment, society also needs to adapt to a new format of work and lifestyle. Therefore, the challenge of evaluating the social impact related to Artificial Intelligence (AI) becomes evident, since the speed of this transformation leads society to rethink the man-machine interaction, as well as the impacts related to the new forms of work that emerge every day. From this perspective, this study aimed to develop a systematic review of the literature and evaluate qualitatively the main social impacts linked to the use of artificial intelligence, as well as the main mitigation actions, using as database the Web Of Science and Scopus. Nine social impacts and twelve mitigation actions were mapped. Among the impacts, bias and discrimination growth due to the use of AI (26%) stand out, followed by the possibility of autonomous systems hurting humans (16%), and then the impact related to the violation of data, privacy, and freedom (15%). It was evident that the recommendation considered to be the most comprehensive refers to increasing the breadth and depth of public debate regarding the risks of AI application. In second place is the recommendation to create laws, principles, and regulations for the use of AI. The results of this study point to the protection of life and human ethics, which are topics that need to be widely investigated for the dissemination of knowledge, and to enable the sustainable development of AI.

Keywords: Industry 4.0 · Digital technologies · Artificial intelligence · Social impacts

1 Introduction

Artificial Intelligence (AI) refers to the ability of a system to interpret internal and external data correctly, using pattern recognition, developing a learning format, and using that learning to achieve goals in a flexible, adaptable, and autonomous way [1]. With all the changes and adaptations of the industrial sector to migrate to a more technological and connected environment, society also needs to adapt to a new format of work and lifestyle. The use of AI is able to provide opportunities for fairer decisions [2], favor

Y. Iano et al. (Eds.): BTSym 2021, SIST 207, pp. 521–528, 2023.
https://doi.org/10.1007/978-3-031-04435-9_54

the world economy [3], and enable human development in an unprecedented way [4]. Therefore, the challenge of evaluating the social impact related to Artificial Intelligence becomes evident, since the speed of this transformation leads society to rethink the man-machine interaction, as well as the impacts related to new forms of work that arise every day [5].

Therefore, technological advances must be accompanied by discussions about the resulting social impacts. The understanding of the social impacts of AI in diverse environments is a necessary precursor to the sustainable development of technologies, enabling technological benefits and preserving humans from possible social impacts and implications [6].

Thus, the objective of this paper was to develop a systematic review of the literature seeking to identify the main social impacts linked to the use of Artificial Intelligence and to point out the main actions that aim to mitigate or eliminate such impacts.

2 Methodology

The objective of the systematic literature review is to evaluate, in a qualitative way, a set of selected academic publications. The data analysis was performed in three main stages: i) planning, in which the need for the research was identified; ii) development of the protocol of the planned review; iii) conduction of the systematic review, performed by searching the publications, selecting the articles according to the established criteria, extracting the data, validating the quality of the publications, and synthesizing the data [7].

Based on the results of the 690 publications selected from the Web of Science and Scopus databases, the publications were selected using the criteria of keywords that contained the expressions.

- Scopus:

```
TITLE-ABS-KEY = ("ARTIFICIAL INTELLIGENCE" OR ai) AND
(ethics * OR morals *) AND (soci *)
```

Where:

- TITLE-ABS-KEY = standard search fields (title, abstract and keywords)
- AND = Boolean operator "AND"
- OR = Boolean operator "OR".

```
TS = (("Artificial Intelligence" OR ai) AND (ETHICS * OR
morals *) AND soci *)
```

Where:

- TS = standard search field (title, abstract, and keywords)
- AND = Boolean operator "AND"
- OR = Boolean operator "OR".

The RSL was developed based on the recommendations of the international PRISMA protocol, which consists of a 27-item checklist that describes and suggests the processes of systematic review [8].

After obtaining 690 publications, the eligibility criteria were applied, divided into three filters. The first filter was the analysis of titles. The second filter was the analysis of the abstracts and conclusions. In this process, the abstracts of the publications were analyzed in order to evaluate, in a more in-depth way, the objective and the results of the article. The last filter used was the complete analysis of the publication. In this step, 175 articles were read in their entirety to qualitatively identify the main social impacts of the use of AI.

3 Results and Discussion

The evaluated publications were categorized according to their affinity with the themes addressed in this study. For this, in a first categorization, the publications were classified according to the area of AI use (Fig. 1), where twelve categories were defined: government and politics, society, leisure, information technology, industry, agriculture, services, transportation and logistics, commerce, military, health, and education. Among the most researched areas regarding the use of AI, the areas of technology (28%), society (20%), and industry (10%) stand out. The first three categories together represent 58% of the themes addressed in this research, showing how technology is linked to human life, from work to the city where they live. According to Fig. 1, a shapeless profile for the application of AI is observed, pointing out the relevance of the topic for the areas of technology and society, highlighting the importance of AI in the social context, since society has the most diverse and complex interfaces with human daily life.

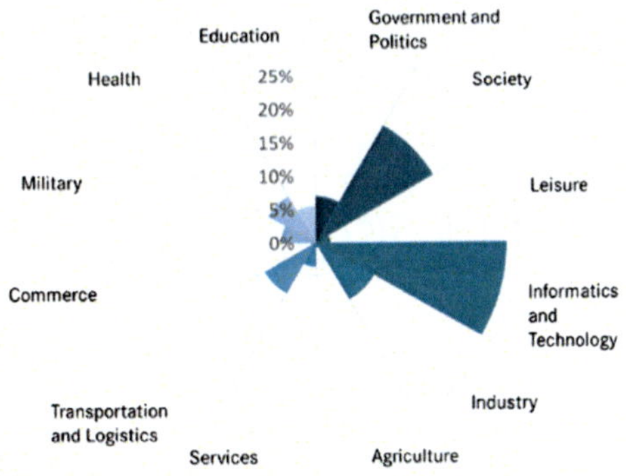

Fig. 1. Main areas that use AI.

3.1 Main AI-Related Impacts

In the sequence, the results were classified according to the most recurrent social impact categories in the selected database, as presented in Table 1. As main impacts identified, bias and discrimination due to the use of AI stand out (n = 26%), followed by the impact of the possibility of autonomous systems hurting humans (n = 16%) and the impact related to the violation of data, privacy, and freedom (n = 15%). The top three impacts when added together represent 57% of all those mentioned in this study.

Table 1. Main Social Impacts on the selected base

Social Impact	Amount mentioned in the 175 publications
Bias and discrimination	72
Risk of Injury	45
Data Breach/Privacy & Liberty	42
Usurpation of jobs by automation	35
Lack of transparency	32
Reduced Autonomy/Responsibility	25
Injustice	21
Over-dependence on technology	7
Environmental Impacts	2

The decision process used by AI systems has the potential to present biased choices, either because it acts from criteria that will generate forms of bias or because it is based on the history of choices [9].

Poorly designed intelligent systems can cause moral, psychological, and physical harm. For example, the use of predictive policing tools may cause more people to be arrested or physically harmed by the police [10].

The risks associated with the use of AI are still unpredictable and unprecedented [11], and there are already several examples that show AI has made discriminatory decisions against minorities [12], reinforced social stereotypes in Internet search engines and enabled data breaches [13], and eliminated jobs in various types of companies [14].

In situations in which the development and use of AI are not explained to the user, or in which the decision processes do not provide the criteria or steps that constitute the decision, the use of AI becomes inexplicable and the ability of the human being to understand and trust the technology becomes challenging [15].

AI is providing more and more solutions for complex activities, and by taking advantage of this process, people are becoming able to perform a greater number of activities more quickly and accurately. However, the result of this innovation is enabling choices that were once exclusively human responsibility to be made by AI systems [16]. Furthermore, the production process of these devices requires raw materials such as nickel, cobalt, and lithium in such high quantities that the Earth may soon no longer be able to sustain them in sufficient quantities [17].

Thus, as in a traditional problem-solving process, understanding the risks and correctly describing the difficulties of a given situation is the initial step in developing an appropriate mitigation or elimination solution [18]. Thus, knowing the impacts of AI is essential to outline an adequate plan of action that allows the sustainable development of this new technology. This route is feasible if the appropriate measures are taken [19].

3.2 Recommendations for Eliminating or Mitigating Impacts

In the last step, the main recommendations for eliminating or mitigating the social impacts of AI were analyzed. The objective was to identify the actions that have greater relevance in the evaluated publications, as well as to evaluate which actions may have greater scope and effectiveness in relation to mitigating the impacts.

In the third phase of the study, the recommendations for the mitigation of social impacts were classified into twelve categories, according to the most cited in the evaluated publications, as shown in Table 2.

Table 2. Main mitigation recommendations

Recomendações de Mitigação dos Impactos Sociais em IA	Frequency in the selected base
Breadth and depth of public debate	85
Laws, principles, regulations, or policies to ensure ethical deployment in AI	71
Standardization of the AI language, translating ethical principles into prototypes	42
Multidisciplinary cooperation between AI researchers and programmers and ethicists	37
Preservation of human decision-making processes for certain types of decisions	22
Joint programs and/or training	18
Increased accessibility of correct information to the public in the form of fact sheets and statements of ethical values on reliable web pages	10
Alternative countries for major AI, ethics, and governance research conferences	4
Translation of the main studies and reports	4
Universal basic income	3
Creation of legally deregulated, or special, zones for research and development in robotics	1
Applying fees to the use of AI	1

It was evident that the recommendation considered as most comprehensive refers to "increasing the breadth and depth of public debate regarding the risks that AI can provide." In second place is the recommendation to "create laws, principles, and regulation of the use of AI," which also received mention in all impact categories. In third place is the recommendation to "standardize AI language" with percentage results above 9%. These top three recommendations add up to 66% of the total recommendations identified in the base. One of the recommendations with the lowest occurrence is the application of "fees or taxes for companies that use AI systems to replace humans", with a percentage of 0.3%. One of the desirable contexts for the use of AI is that certain decisions remain under human responsibility, especially when they are related to the evaluation of people's capabilities, for example in university admissions or job selection processes [20]. Encouraging research and training focused on AI will help create a supportive and developing network of researchers, fostering cooperation and understanding among the general public as research advances [21]. Developing methodologies so that society has access to the right information regarding the development of systems that use AI is a way to bring the public closer to the creation process and ensure that stakeholders are informed about the latest updates [9].

Encouraging global participation in AI conferences is a way to optimize the performance of experts and to involve countries that are still in the early stages of the technology [15]. One of the mitigation proposals is the development of a universal basic income, aiming to foster a way of solidarity and dignified survival for the less fortunate population and to ensure the process of transition from the present to the future [16].

Creating legally unregulated zones for AI research and development is a controlled way to overcome legal and even moral impasses so that thorough testing can take place and that, through the evidence and conclusions obtained, developers and researchers can find potential for improvement and even prevent undesirable impacts [22]. The idea of taxes on the use of robots emerged a few years ago as a way to try to exercise control concerning this technology in replacing people with machines [14].

4 Conclusion

The need to evaluate the social impacts of Artificial Intelligence is a topic on the rise; however, the evolution of this discussion has not reached the same speed as the technological development in AI. The public debate about the social impacts of AI is in its early stages and needs to be developed in a balanced way, taking into account opportunities and risks, to enable initiatives that foster innovation but also exercise control over the potential risks. This study aimed to develop a systematic review of the literature between the years 2000 and 2020 and to evaluate, in a qualitative way, the main social impacts linked to the use of artificial intelligence, as well as the main mitigation recommendations, nine main social impacts were mapped, in which bias and discrimination resulting from the use of AI stand out (n = 26%), followed by the impact of the possibility of autonomous systems harming human beings (n = 16%) and, in sequence, the impact related to the violation of data, privacy, and freedom (n = 15%). The three main impacts when added together represent 57% of all those mentioned in this study and point to the protection of life and human ethics, which have been critical issues since the beginning of humanity, and which need to be widely researched for the dissemination of knowledge and to enable the sustainable development of AI.

References

1. Shadowen, A.N.: Ethics and Bias in Machine Learning: A Technical Study of What Makes Us "Good". City University of New York (2017). https://academicworks.cuny.edu/cgi/viewcontent.cgi?article=1042&context=jj_etds. Accessed 1 Oct 2020
2. Kaufman, D.: Inteligência Artificial e os desafios éticos. PAULUS Rev. Comun. da FAPCOM **5**, 9 (2021). https://doi.org/10.31657/rcp.v5i9.453
3. Corinne, C.: Governing artificial intelligence: ethical, legal and technical opportunities and challenges. Philos. Trans. R. Soc. A Math. Phys. Eng. Sci. **376**, 20180080 (2018). https://doi.org/10.1098/rsta.2018.0080
4. Cath, C., Wachter, S., Mittelstadt, B., Taddeo, M., Floridi, L.: Artificial intelligence and the 'Good Society': the US, EU, and UK approach. Sci. Eng. Ethics **24**(2), 505–528 (2017). https://doi.org/10.1007/s11948-017-9901-7
5. Morrar, R., Arman, H., Mousa, S.: The Fourth Industrial Revolution (Industry 4.0): A Social Innovation Perspective. Technol. Innov. Manag. Rev. **7**(11), 12–20 (2017). https://doi.org/https://doi.org/10.22215/timreview/1117
6. Hagerty, A., Rubinov, I.: Global AI Ethics: A Review of the Social Impacts and Ethical Implications of Artificial Intelligence (2019)
7. Wahono, R.: A systematic literature review of software defect prediction: research trends, datasets, methods and frameworks. J. Softw. Eng. **1** (2015)
8. Moher, D., Liberati, A., Tetzlaff, J., Altman, D.G.: The PRISMA Group (2009) preferred reporting items for systematic reviews and meta-analyses: the PRISMA Statement. PLoS Med **6**(7), e1000097. Doi:https://doi.org/10.1371/journal.pmed.1000097
9. Vesnic-Alujevic, L., Nascimento, S., Pólvora, A.: Societal and ethical impacts of artificial intelligence: critical notes on European policy frameworks. Telecomm Policy **44**(6), 101961 (2020). https://doi.org/10.1016/j.telpol.2020.101961
10. Morley, J., Floridi, L., Kinsey, L., Elhalal, A.: From what to how: an initial review of publicly available ai ethics tools, methods and research to translate principles into practices. Sci. Eng. Ethics **26**(4), 2141–2168 (2019). https://doi.org/10.1007/s11948-019-00165-5
11. Kratsas, G., Truby, J.: Regulating sovereign wealth funds to avoid investment protectionism. J. Financ. Regul. **1**(1), 95–134 (2015). https://doi.org/10.1093/jfr/fju002
12. Buolamwini, J., Gebru, T.: Gender shades: intersectional accuracy disparities in commercial gender classification. In: Proceedings of the 1st Conference on Fairness, Accountability and Transparency, in Proceedings of Machine Learning Research 81, pp. 77–91 (2018). https://proceedings.mlr.press/v81/buolamwini18a.html. Accessed 1 Oct 2020
13. Kay, M., Matuszek, C., Munson, S.A.: Unequal representation and gender stereotypes in image search results for occupations. In: Proceedings of the 33rd Annual ACM Conference on Human Factors in Computing Systems, pp. 3819–3828. ACM, New York (2015). https://doi.org/10.1145/2702123.2702520
14. Pham, Q., Madhavan, R., Righetti, L., Smart, W., Chatila, R.: The impact of robotics and automation on working conditions and employment [ethical, legal, and societal issues]. IEEE Robot. Autom. Mag. **25**(2), 126–128 (2018). https://doi.org/10.1109/MRA.2018.2822058
15. Munoko, I., Brown-Liburd, H.L., Vasarhelyi, M.: The ethical implications of using artificial intelligence in auditing. J. Bus. Ethics **167**(2), 209–234 (2020). https://doi.org/10.1007/s10551-019-04407-1
16. Floridi, L., et al.: AI4People—an ethical framework for a good AI society: opportunities, risks, principles, and recommendations. Mind. Mach. **28**(4), 689–707 (2018). https://doi.org/10.1007/s11023-018-9482-5
17. Kaplan, A., Haenlein, M.: Rulers of the world, unite! the challenges and opportunities of artificial intelligence. Bus. Horiz. **63**(1), 37–50 (2020). https://doi.org/10.1016/j.bushor.2019.09.003

18. Gick, M.L.: Problem-Solving Strategies. Educ. Psychol. **21**(1–2), 99–120 (1986). https://doi.org/10.1080/00461520.1986.9653026
19. Truby, J.: Governing artificial intelligence to benefit the UN sustainable development goals. Sustain. Dev. **28**(4), 946–959 (2020). https://doi.org/10.1002/sd.2048
20. Krupiy, T.: A vulnerability analysis: theorising the impact of artificial intelligence decision-making processes on individuals, society and human diversity from a social justice perspective. Comput. Law Secur. Rev. **38**, 105429 (2020). https://doi.org/10.1016/j.clsr.2020.105429
21. ÓhÉigeartaigh, S.S., Whittlestone, J., Liu, Y., Zeng, Y., Liu, Z.: Overcoming barriers to cross-cultural cooperation in AI ethics and governance. Philos. Technol. **33**(4), 571–593 (2020). https://doi.org/10.1007/s13347-020-00402-x
22. Pagallo, U.: When morals ain't enough: robots, ethics, and the rules of the law. Mind. Mach. **27**(4), 625–638 (2017). https://doi.org/10.1007/s11023-017-9418-5

Total Phenolic Content and Antioxidant Activity of a Macha Macha Beverage (*Vaccinium Floribundum* Kunth)

Domingo Utus-Ramires[1], Pedro Arteaga-Llacza[2], Ruggerths De La Cruz-Marcos[1], Angélica Miranda-Jara[2], Helí Miranda-Chávez[3], Miguel Ángel Quispe-Solano[4], and Roberto Chuquilín-Goicochea[1](✉)

[1] Universidad Nacional de Huancavelica, Acobamba, Peru
ruggerths.delacruz@unh.edu.pe, robercarl33@gmail.com
[2] Escuela de Ingeniería Agroindustrial, Universidad Nacional Federico Villarreal, San Miguel, Peru
amiranda@unfv.edu.pe
[3] Universidad Católica de Trujillo Benedicto XVI, Trujillo, Peru
amiranda@uct.edu.pe
[4] Facultad de Ingeniería en Industrias Alimentarias, Universidad Nacional del Centro del Perú, Huancayo, Peru

Abstract. The effect of pH, temperature, and pasteurization time on the total anthocyanin and phenol contents, and antioxidant capacity in a macha macha beverage (*Vaccinium floribundum* Kunth) was evaluated. pH (3 and 4), temperature (80 and 85 °C), and time (20 and 40 min) were studied using a complete 2^3 factorial design. All three variables have shown a significant effect ($p < 0.05$) on the content of anthocyanins and total phenols, while only time and temperature influenced the antioxidant capacity. It was found that: the anthocyanin content increases with higher pH, while it decreases with time and temperature; the total phenol content decreases with temperature, while it increases with pH and time; and the antioxidant capacity increases with temperature while it decreases with pasteurization time. A pH of 4.0, a temperature of 85 °C, and a time of 20 min allow greater retention of anthocyanins, phenols (9.7642 mg of gallic acid equivalent/mL), and antioxidant capacity (2.5968 μmol Trolox/mL) in a macha macha beverage. This beverage may have nutraceutical properties and should be optimized with further research.

Keywords: Polyphenols · Cyanidin · Differential pH · Folin · Anthocyanin

1 Introduction

Macha macha is a fruit obtained from a deciduous perennial shrub belonging to the *Ericaceae* family, which is endemic to the South-American Andes and grows between 1600 and 4000 m above sea level [1]. Although its nutraceutical properties have been

Y. Iano et al. (Eds.): BTSym 2021, SIST 207, pp. 529–534, 2023.
https://doi.org/10.1007/978-3-031-04435-9_55

studied due to its high antioxidant content, this wild species is still under-exploited but has shown potential for agricultural production and agro-industrial use. The genus *Vaccinium* includes industrial crops such as *V. corymbosum*, *V. ashei*, and *V. angustifolium*, considered worldwide as "superfruits" due to their high phenol, anthocyanin, and antioxidant contents [2–4].

In its ripe and raw stage, the macha macha fruit shows high antioxidant activity (1203 ± 94 mg Trolox/100 g) [5], and total polyphenol content of 66.83 mg of gallic acid per 100 g of dry matter [6], according to studies on Ecuadorian fruits. Studies on its antioxidant potential concluded that macha macha is a good candidate for a dietary supplement [7].

Phenolic compounds in fruits and their beneficial effects on health have been trending in recent years. Berries and their derived products show several positive effects due to their high phenol content and chemopreventive activity against degenerative diseases, including oxidative stress, inflammation, and carcinogenesis [8]. Phenols are strong antioxidants against reactive oxygen species [9], especially found in berries of the genus *Vaccinium* [3].

Alarcón-Barrera et al. [10] found that *Vaccinium floribundum* Kunth attenuates oxidative damage in adult human dermal fibroblasts, showing that this fruit may represent a relevant source of bioactive compounds with promising benefits for human health.

The promising agro-industrial potential of the macha macha fruit presents, however, challenges regarding the maximum retention of its bioactive compounds during its processing. For this reason, the present study aimed at evaluating processed products for their possible health benefits. Total phenol thermal stability and antioxidant capacity were studied during the processing of a macha macha fruit beverage, by controlling pH, time, and pasteurization temperature.

2 Materials and Methods

2.1 Materials

Macha macha fruits (*Vaccinium floribundum* Kunth) were collected in the Acoria district, Huancavelica, Peru.

2.2 Determination of Total Phenol Content

Total phenol content was determined applying the Folin-Ciocalteau method [11]; results were given in gallic acid equivalent (GAE)/mL of beverage. First, the beverage was diluted 1/100 with absolute ethanol and water (50:50) and acidified with 0.01% chlorohydric acid. 0.5 mL of the solution was added to 2.5 mL of 1:10 diluted Folin-Ciocalteau reagent. The solution was vortex-mixed and incubated at room temperature for 5 min. 2 mL of a 7.5% sodium carbonate aqueous solution was added to the solution. The solution was vortex-mixed, then kept at 45 °C for 125 min. The solution absorbance was read at 765 nm using the Shimadzu UV-visible, model UV 160 U spectrophotometer (Kyoto, Japan), and 1 cm disposable cells. All reads were repeated four times.

2.3 Determination of Antioxidant Activity

The DPPH free radical method [4] was applied. IC50 values (μg/mL) of the beverage was obtained from 1:2, 1:5 and 1:1 (v/v) dilutions in a stock solution at 0.54 mg/mL.

2.4 Statistical Analysis

A complete 2^3 factorial design with three repeats was used, for a total of 24 experiments. The independent variables were pH (3 and 4), temperature (80 and 85 °C), and pasteurization time (20 and 40 min). The dependent variables were total phenol content and antioxidant activity. Analysis was done using MS Excel and Minitab v.17 (trial version).

3 Results and Discussion

The factorial design variance analysis (Table 1) confirms the effect of the three selected variables: pH, temperature, and time, as well as their interactions. Conventional thermal food processing continues to be the most adopted technology in the food industry around the world because it significantly inactivates enzymes and microorganisms, which guarantees food safety and prolongs the food products' shelf life. However, it also results in the instability of bioactive compounds, such as phenols, and affects their antioxidant activity [12].

Table 1. Variance analysis of total phenols and antioxidant activity in the macha macha beverage.

Effects	Total phenols	Antioxidant activity
	p	
pH	0.031*	0.076
Temperature	0.002*	0.000*
Time	0.005*	0.031*
pH × temperature	0.683	0.002*
pH × time	0.000*	0.000*
Temperature × time	0.385	0.004*
pH × Temperature × time	0.945	0.183

* Significance level at 0,05

The total phenol content was significantly ($p < 0.05$) influenced by pH, temperature, and pasteurization time, as well as the pH × time interaction (Table 1). The heating temperature and duration have a great influence on total phenol stability. Santhirasegaram et al. [13] reported that heat treatment significantly lowers total phenol content in juices (37.8% reduction).

Variation of total phenol content associated with the variables in this study is shown in Fig. 1. Total phenol content in the beverage increases with higher pH and longer time

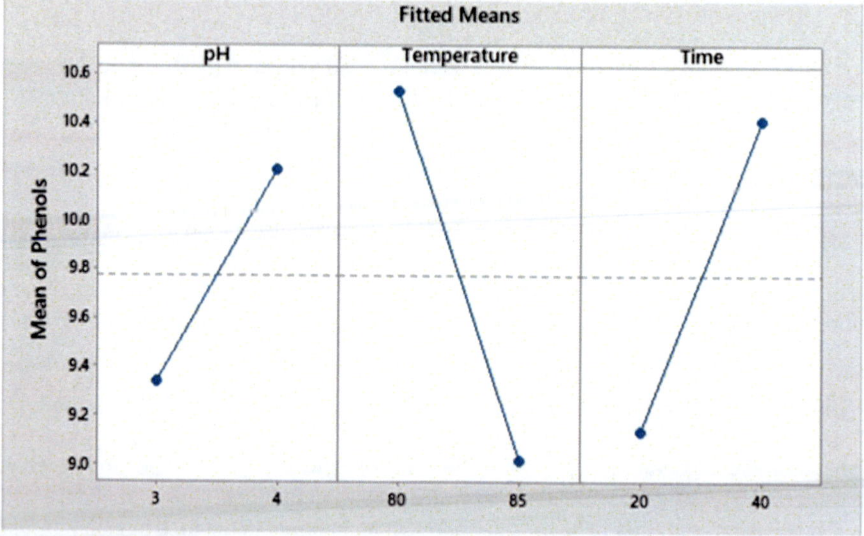

Fig. 1. Main effects plot for total phenol content in macha macha beverage.

but decreases with higher pasteurization temperature. During food thermal processing and storage, the temperature has a great influence on phenols. Strict temperature control is important to maintain phenol levels and stability, as it has been shown that they can undergo epimerization at high temperatures [14].

Liu et al. [15] reported that the total phenolic content decreased by 20.21% when heated at 70 °C for 30 min. Similarly, the total phenol content decreased significantly when the temperature was 90 °C during the processing period. Phenol instability is due to two mechanisms: epimerization and autoxidation. Epimerization is one of the most important mechanisms that lead to the instability of phenols [14], it occurs easily with a variety of factors such as temperature, pH, and metal ions in the food system. Autoxidation occurs in the presence of oxygen, forming peroxides and hydroperoxides [14]. In this study, the oxygen is removed during the beverage pasteurization, and the packing occurs under vacuum conditions, so it can be deduced that there is no autoxidation. pH is the principal factor affecting phenol stability in fruits and vegetables. In general, lower pH values of a solution are associated with greater phenol stability. For example, the total phenol content of the extract from the millet seed pod fraction remained constant at low to almost neutral pH ($\leq$6.5) but decreased as alkalinity increased (pH of 10) [16]. According to Fig. 1, a pH of 4, a temperature of 80 °C, and a time of 40 min are recommended to retain the highest total phenol content.

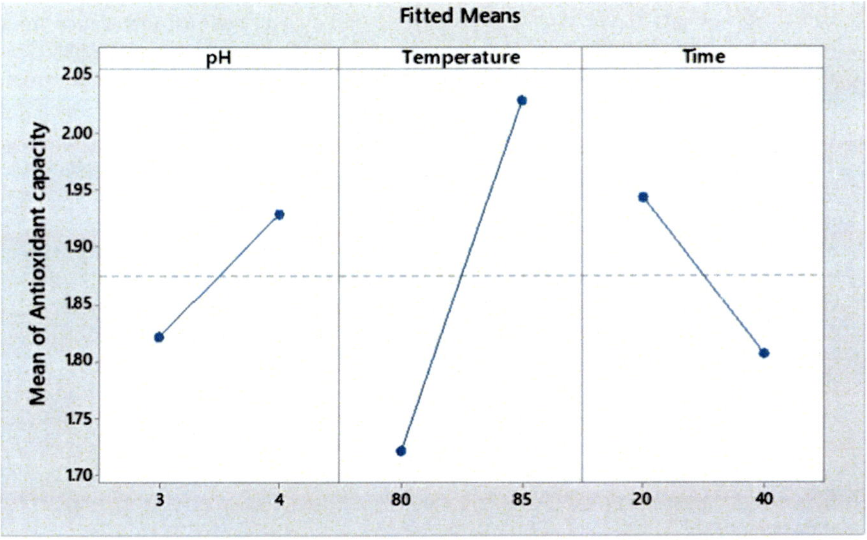

Fig. 2. Main effects plot for antioxidant activity in macha macha beverage.

The antioxidant capacity was significantly influenced ($p < 0.05$) by temperature, time, and the interactions pH × temperature, pH × time, and temperature × time (Table 1). This result is compatible with Claudio et al. [17], who observed that time, temperature, and their interaction would influence the antioxidant activity of *V. floribundum*.

The beverage antioxidant capacity increases with higher pH and temperature but decreases with longer pasteurization time (Fig. 2). According to Fig. 2, a pH of 4, a temperature of 85 °C, and a time of 20 min are recommended to retain the highest antioxidant capacity. Increased antioxidant activity may be related to cell membrane disruption and enzyme inactivation, among others [18].

4 Conclusions

pH, temperature, and pasteurization time of a macha macha beverage (*Vaccinium florbundum* Kunth) have a significant effect on its total phenol content, higher pH, and longer time but lower temperature associate with higher total phenol content. Temperature and pasteurization time of a macha macha drink (*Vaccinium florbundum* Kunth) have a significant effect on its antioxidant capacity, higher temperature but shorter time associate with higher antioxidant activity. Finally, our results indicate that a pH of 4.0, a temperature of 85 °C and a time of 20 min increase total phenol and antioxidant capacity retentions in a macha macha beverage.

References

1. Cobo, M.M., Gutiérrez, B., Torres, A.F., de Torres, M.L.: Preliminary analysis of the genetic diversity and population structure of mortiño (Vaccinium floribundum Kunth). Biochem. Syst. Ecol. **64**, 14–21 (2016)

2. Grace, M.H., et al.: Hypoglycemic activity of a novel anthocyanin-rich formulation from lowbush blueberry, Vaccinium angustifolium Aiton. Phytomedicine **16**, 406–415 (2009)
3. Martín-Gómez, J., Varo, M.Á., Mérida, J., Serratosa, M.P.: Influence of drying processes on anthocyanin profiles, total phenolic compounds, and antioxidant activities of blueberry (Vaccinium corymbosum). LWT **120**, 108931 (2020)
4. Wang, Y., Ye, Y., Wang, L., Yin, W., Liang, J.: Antioxidant activity and subcritical water extraction of anthocyanin from raspberry process optimization by response surface methodology. Food Biosci **44**, 101394 (2021)
5. Vasco, C., Riihinen, K., Ruales, J., Kamal-Eldin, A.: Phenolic compounds in Rosaceae fruits from Ecuador. J. Agric. Food Chem. **57**, 1204–1212 (2009)
6. Castillo Páez, M.A., De Janon Brito, M.J.: Caracterización y microencapsulación de compuestos bioactivos del mortiño (Vaccinium floribundum Kunth), en la sierra norte del Ecuador. Universidad De Las Américas, Quito (2018)
7. Prencipe, F.P., Bruni, R., Guerrini, A., Rossi, D., Benvenuti, S., Pellati, F.: Metabolite profiling of polyphenols in Vaccinium berries and determination of their chemopreventive properties. J. Pharm. Biomed. Anal. **89**, 257–267 (2014)
8. Jiang, T. et al.: The anti-inflammatory potential of protein-bound anthocyanin compounds from purple sweet potato in LPS-induced RAW264.7 macrophages. Food. Res. Int. **137**, 109647 (2020)
9. Agati, G., Azzarello, E., Pollastri, S., Tattini, M.: Flavonoids as antioxidants in plants: location and functional significance. Plant Sci. **196**, 67–76 (2012)
10. Alarcón-Barrera, K.S., et al.: Wild Andean blackberry (Rubus glaucus Benth) and Andean blueberry (Vaccinium floribundum Kunth) from the highlands of Ecuador: nutritional composition and protective effect on human dermal fibroblasts against cytotoxic oxidative damage. J. Berry Res. **8**, 223–236 (2018)
11. Li, Q.: Chen X (2021) Impact of food additive titanium dioxide on the polyphenol content and antioxidant activity of the apple juice. LWT **154**, 112574 (2022)
12. Brasili, E., Chaves, D.F.S., Xavier, A.A.O., Mercadante, Z., Hassimotto, N.M.A., Lajolo, F.M.: Effect of pasteurization on flavonoids and carotenoids in Citrus sinensis (L.) Osbeck cv. 'Cara Cara'and 'Bahia'juices. J. Agric. Food Chem. **65**, 1371–1377 (2017)
13. Santhirasegaram, V., Razali, Z., Somasundram, C.: Effects of thermal treatment and sonication on quality attributes of Chokanan mango (Mangifera indica L.) juice. Ultrason. Sonochem. **20**, 1276–1282 (2013)
14. Deng, J., Yang, H., Capanoglu, E., Cao, H., Xiao, J.: Technological aspects and stability of polyphenols. In: Polyphenols: Properties, Recovery, and Applications, pp 295–323. Elsevier, the Netherlands (2018)
15. Lin, D., et al.: An overview of plant phenolic compounds and their importance in human nutrition and management of type 2 diabetes. Molecules **21**, 1374 (2016)
16. Chethan, S., Malleshi, N.G.: Finger millet polyphenols: optimization of extraction and the effect of pH on their stability. Food Chem. **105**, 862–870 (2007)
17. Claudio, R., Nájera, J., Santacruz, S.: Evaluación de la actividad antioxidante de Vaccinium floribundum sometido a tratamiento térmico. Vitae **19**, S352–S354 (2012)
18. Dewanto, V., Wu, X., Liu, R.H.: Processed sweet corn has higher antioxidant activity. J. Agric. Food Chem. **50**, 4959–4964 (2002)

Evaluation of the Yield, Sensory Characteristics, and Digestibility of Expanded Quinoa

Noemi Quispe-Boza[1], Pedro Arteaga-Llacza[2], Ruggerths De La Cruz-Marcos[1], Angélica Miranda-Jara[2], Helí Miranda-Chávez[3], Miguel Ángel Quispe-Solano[4], and Roberto Chuquilín-Goicochea[1](✉)

[1] Universidad Nacional de Huancavelica, Acobamba, Peru
{ruggerths.delacruz,roberto.chuquilin}@unh.edu.pe
[2] Escuela de Ingeniería Agroindustrial, Universidad Nacional Federico Villarreal, San Miguel, Peru
amiranda@unfv.edu.pe
[3] Universidad Católica de Trujillo Benedicto XVI, Trujillo, Peru
amiranda@uct.edu.pe
[4] Facultad de Ingeniería en Industrias Alimentarias, Universidad Nacional del Centro del Perú, Huancayo, Peru

Abstract. The effect of initial quinoa (*Chenopodium quinoa* Willd) seed water content and pressure on the yield and acceptability of the expanded quinoa was evaluated. A 2^2 factorial design was used, with a water content of 15 and 20% and expansion pressure of 140 and 160 psi. Sensory evaluation was performed using a 7-point hedonic scale with 32 untrained panelists. The study variables did not have a significant effect on yield, but their interaction affected the acceptability of the expanded quinoa. Through the Tukey test, treatment 3 (water content 15% and pressure 160 psi) was chosen; for having the highest acceptability (5.43 points) and its chemical composition was determined (per 100 g of sample): moisture (9.9 g), ash (1.9 g), protein (14 g), fat (5.3 g), fiber (2.05 g), and carbohydrates (68.9 g); total energy 379.3 kcal and digestibility 94.4 g/100 g. Expanded quinoa has a high digestibility and is especially recommended for infant feeding and as a breakfast cereal.

Keywords: Insufflated · Popped · Starch · Gluten-free diet

1 Introduction

Quinoa (*Chenopodium quinoa* Willd.), a C3 dicotyledonous species from the *Amaranthaceae* family, produces achene fruits (chenopod grains) with 1.5 to 4 mm rounded seed [1]. It can grow in various agro-ecological zones (coasts, valleys, highlands, salt flats, and subtropics) and is frost, salinity, and drought-tolerant [2].

The dried quinoa seed is consumed as a cereal grain and has gained importance worldwide due to its high nutritional value. The balance of essential amino acids, fatty

Y. Iano et al. (Eds.): BTSym 2021, SIST 207, pp. 535–542, 2023.
https://doi.org/10.1007/978-3-031-04435-9_56

acids, micronutrients, vitamins, and antioxidants is considered to be of high quality compared to common cereals [3]. As a gluten-free food grain with a low glycemic index, quinoa is a viable alternative for special diets and industries [4].

Expanded snacks are cereal grains that have undergone sudden changes in temperature and pressure, causing an expansion to occur and the initial size to increase. The principle of this process is the explosive vaporization of water inside the grain, conveniently combined with the physical effects of the food pressure and temperature [5].

Different studies indicate that protein and starch from expanded or insufflated cereals show improved digestibility. Based on these observations, we decided to evaluate the effects of the expansion process on the quinoa seed sensory and nutritional characteristics, as well as report the initial humidity and pressure parameters to apply this technology.

2 Materials and Methodology

2.1 Materials

30 kg of quinoa of Blanca variety were processed using a single-trigger expander barrel. The equipment characteristics are 1 kg capacity, 100 kg total weight, 0.70 × 0.45 × 0.74 m (long × width × height), 0–300 psi manometer, and an expansion time between 8 and 10 min.

2.2 Experimental Design

The experiment followed a 2^2 full factorial design and three repeats. The two factors were the seed initial water content (15 and 20%), and the expansion process pressure (140 and 160 psi). the treatments were as follows: T1: 15% water content and 140 psi; T2: 20% water content and 160 psi; T3: 15% water content and 160 psi; T4: 20% water content and 140 psi.

2.3 Quinoa Seed Expansion Process

Reception. The quinoa was received and examined. The seed water content was analyzed, to know the needed amount of water required to reach the required water content during the process.

Selection. This operation consisted in evaluating the seed quality. Seeds with no mechanical, insect, or microorganism damage were hand-selected. Seeds were also selected considering a constant diameter to obtain uniform cooking in the expander barrel.

Conditioning. Water was added to the seeds so that the final water content be 15 or 20% in the expander barrel.

Expansion. Once the required pressure inside the expander barrel is reached, its lid is opened quickly, causing the seeds to come out explosively and expand due to the sudden vaporization of internal humidity and the pressure drop (140 and 160 psi) resulting in the seeds expanding.

Sifting. Expanded seeds were sieved to eliminate shells, and broken, burned, or smaller expanded seeds. The resulting expanded seeds displayed a uniform and selected size.

Packaging. Expanded seeds were packed in 500 g 1 mm thick polyethylene bags. The bags were sealed for storage using an electric sealer.

2.4 Yield

The expansion process yield was obtained following this equation:

$$\eta = \frac{input}{output} \times 100 \tag{1}$$

Where:

- Input: Total quinoa weight before starting with the expansion process.
- Output: Total expanded quinoa weight after the expansion process.

2.5 Sensory Evaluation

32 untrained panelists realized a hedonic test on the final product to evaluate its general acceptability, using an evaluation sheet. A 7 point hedonic scale was applied, with values between "I extremely dislike the product" (1 point) to " I extremely like the product" (7 points) [6].

2.6 Chemical Composition

Protein, ash, fat, fiber [7], carbohydrates [8], and water contents were determined.

2.7 *In Vitro* Digestibility Assay

In vitro digestibility was determined following the multi enzymatic, *in vitro* pepsin, and hydrochloric acid digestions [9].

2.8 Statistical Analysis

Results were analyzed with Minitab v.17 (trial version). An analysis of variance (ANOVA) was applied to the 2^2 full factorial designs. Significant ANOVA results were further processed using the Tukey test.

3 Results and Discussion

Results show there is no effect of water content or pressure variations, or the interaction of these two factors, on the expansion process yield ($p > 0.05$). The acceptability demonstrates no effect of the water content or pressure, but a significant effect of their interaction ($p > 0.05$) (Table 1).

Figure 1 shows that higher water content and pressure generate higher yields, however, this observation is not significant. T1, T2, T3, and T4 had yields of 77.30%, 80.48%, 77.57%, and 77.75% respectively.

Figure 2 shows that water content and pressure are associated with lower acceptability of the final product; however, only the water content and pressure interaction were significant ($p > 0.05$).

During the expansion of the quinoa seed, its surface area and solid phase volume ratio increase, and a modification of its starch (crystallinity) occurs, thus improving the yield [10]. The expansion or insufflation of cereals is an operation that is carried out

Table 1. Analysis of variance.

Factor	Yield	Acceptability
	p	
Initial water content (A)	0.240	0.700
Pressure (B)	0.285	0.887
A × B interaction	0.368	0.038*

* Significance at 0.05.

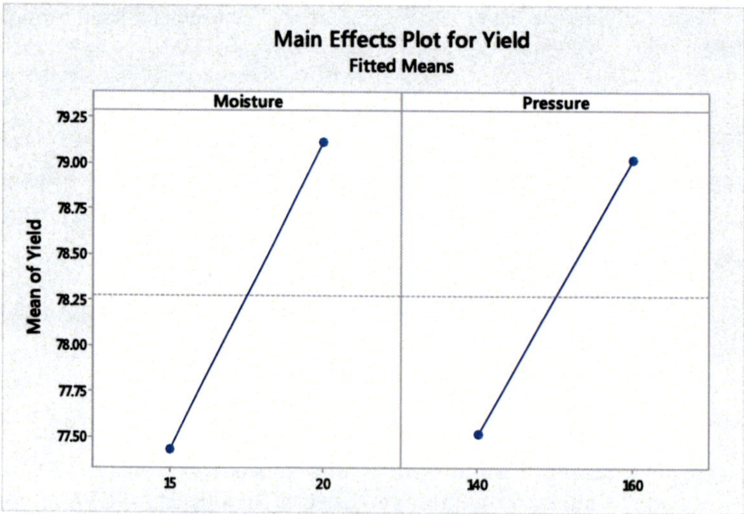

Fig. 1. Main effects plot for yield of expanded quinoa.

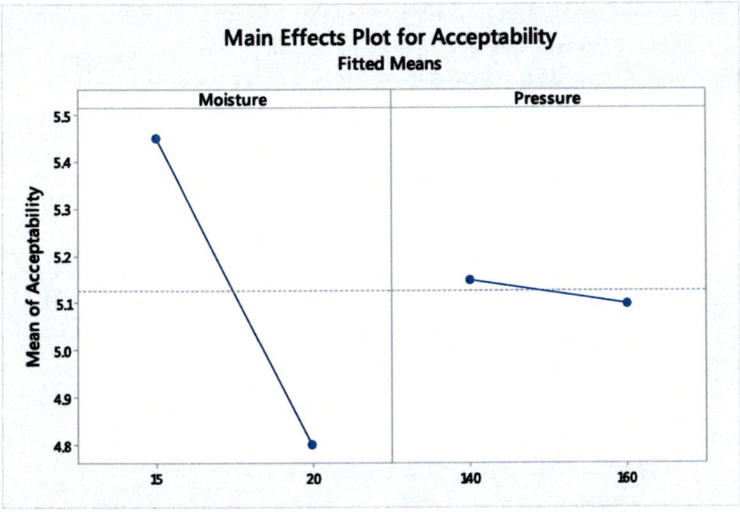

Fig. 2. Main effects plot for acceptability of expanded quinoa.

employing an expander barrel, which suddenly applies heat in such a way that the water vaporizes inside the grain, reaching high internal pressures. The outer tissue breaks and the grain expands, forming a foamy structure [10–12]. When the internal pressure of the blowing chamber reaches approximately 200 psi (about 9 to 12 min), the lid is opened, and the sudden pressure drop causes the blowing of the grain and its explosive discharge into a collecting chamber [13]. As the interaction of both variables is evident, it shows an interaction of the grain water content and pressure.

Based on yield, acceptability, and Tukey test (Table 2) results, the third treatment (T3) was chosen, for its higher acceptability (5.43 points) and for being significantly different from all other treatments.

Table 2. Tukey's test for acceptability.

Comparison	t	p
T1–T2	−0.10	1.00
T1–T3	2.77	0.03*
T1–T4	0.79	0.86
T2–T3	2.86	0.03*
T2–T4	0.89	0.81
T3–T4	−1.98	0.20
T1–T2	−0.10	1.00

* Significance at 0.05.

T3 proximate chemical composition (Table 3) indicates high carbohydrates concentration (68.9%); cereals are mainly composed of starch, the principal component in the human diet [14]. Although quinoa is classified as a pseudocereal, the presence of carbohydrates (starch) is noteworthy. Protein appears to be the second most common component in the expanded quinoa (14.0%). Protein content decreases during the expansion process, due to the process itself [11].

Table 3. Chemical composition and digestibility of T3.

Analysis	Result (per 100 g)
Water content	9.9 g
Ash	1.9 g
Fat	5.3 g
Protein (f = 6.25)	14.0 g
Carbohydrates	68.9g
Total energy	379.3 kcal
In vitro protein digestibility	94.4 g

Temperatures around 100 °C are likely to induce protein denaturation; when reacting with reducing sugars, the Maillard reaction may occur [15]. Fat is the third most common component, with 5.3%. Recent studies have shown quinoa has a high-quality oil content, is rich in vitamin E, and can be used as an ingredient in gluten-free products to improve their vitamin E content, and therefore, their general nutritional quality [16].

The water content of the expanded quinoa (9.9%) allows a longer shelf life of the product, but makes it more hygroscopic, therefore requiring a container that has a water vapor barrier. These results are comparable to those found for expanded quinoa of the Blanca (9.33%), Huancayo (7.28%), and Rosada (7.65%) varieties [17]. It should be noted that these quinoas were expanded at 30% humidity and 140 psi pressure.

The ash content (1.9%) was higher than observed in other studies: ash content in expanded quinoa was found 1.61% in Blanca, 1.73% in Huancayo, and 1.56% in Rosada varieties [17]. The ash contains minerals, and the raw material used from this region (Huancavelica) would probably contain more minerals since no theoretical support has been found to explain the influence of the expansion process on the mineral content in the final product.

Expanded quinoa digestibility was 94.4 g/100 g, indicating 94.4% of the protein being used in vitro. Under the same expansion process conditions, the expanded quinoa obtained in this study shows higher digestibility than from other varieties in the Puno area. Using the same enzymatic method, results were as follows: Pasankalla (81.83%), Koncalla (80.80%), and Ayara (72.88%) [5]. During the expansion process, the digestibility of these seeds improves. Moreover, an increase in protein content was observed, due to the loss of water content [10–12]. Food texturing is a combined process of physical, pressure, temperature, and humidity effects leading to different textures, desirable flavors, and improve digestibility [10–12].

4 Conclusions

We determined that initial water content of 15% and an expansion pressure of 160 psi would provide higher acceptability for the expanded quinoa (*Chenopodium quinoa* Willd.). Under these conditions, the chemical composition of the final product was water content (9.9%), ash (1.9%), fat (5.3%), protein (14.0%), carbohydrates (68.9%), total energy 379.3 kcal, and *in vitro* protein digestibility (94.4 g/100 g). The higher digestibility obtained during this process and the conservation of its chemical composition make expanded quinoa ideal for breakfast for children and adults.

References

1. Garcia, M., Condori, B., Del, C.C.: Agroecological and agronomic cultural practices of quinoa in South America. Quinoa Improv. Sustain. Prod. 25–46 (2015)
2. Xiu-shi, Y., Pei-you, Q., Hui-min, G., Gui-xing, R.: Quinoa industry development in China. Cienc e Investig Agrar Rev Latinoam ciencias la Agric **46**, 208–219 (2019)
3. James, L.E.A.: Quinoa (Chenopodium quinoa Willd.): composition, chemistry, nutritional, and functional properties. Adv. Food Nutr. Res. **58**, 1–31 (2009)
4. Gordillo-Bastidas, E., Díaz-Rizzolo, D.A., Roura, E., et al.: Quinoa (Chenopodium quinoa Willd), from nutritional value to potential health benefits: an integrative review. J. Nutr. Food Sci. **6**, 10–4172 (2016)
5. Yana Quispe, S.: Caracterización y determinación de la digestibilidad proteica de quinua insuflada en tres variedades (Chenopodium quinoa Willd). Universidad Nacional del Altiplano, Puno (2015)
6. da Silva, A.N., dos Santos Navarro, R.D.C., Ferreira, M.A.M., et al.: Performance of hedonic scales in sensory acceptability of strawberry yogurt. Food Qual. Prefer. **30**, 9–21 (2013)
7. AOAC: Official methods of analysis, 21st edn. Association of Official Analitical Chemists, Washington (2019)
8. Reyes-García, M., Gómez-Sánchez Prieto, I., Espinoza-Barrientos, C.: Tabla peruanas de composición de alimentos, 10th edn. Instituto Nacional de Salud, Lima (2017)
9. Mertz, E.T., Hassen, M.M., Cairns-Whittern, C., Kirleis, A.W., Tu, L., Axtell, J.D.: Pepsin digestibility of proteins in sorghum and other major cereals. Proc. Natl. Acad. Sci. **81**, 1–2 (1984)
10. Cabrera-Ramírez, A.H., Castro-Campos, F.G., Gaytán-Martínez, M., Morales-Sánchez, E.: Relationship between the corneous and floury endosperm content and the popped sorghum quality. J. Cereal Sci. **95**, 102999 (2020)
11. Mishra, G., Joshi, D.C., Panda, B.K.: Popping and puffing of cereal grains: a review. J grain Process. storage **1**, 34–46 (2014)
12. Vázquez-Carrillo, M.G., Santiago-Ramos, D., de Dios, F.-Cárdenas. J.: Kernel properties and popping potential of Chapalote, a Mexican ancient native maize. J. Cereal Sci. **86**, 69–76 (2019)
13. Kokini, J.L., Ho, C.-T., Karwe, M.: Food extrusion science and technology. Dry Technol. **11**, 417–418 (1993)
14. Espinoza Silva, C., Quispe Solano, M.Á.: Tecnología de Cereales y Leguminosas. Universidad Nacional del Centro del Perú, Huancayo (2011)
15. Rodier, L.C., Hartel, R.W.: Characterizing Maillard reaction kinetics and rheological changes in white chocolate over extended heating. J. Food Sci. **86**(6), 2553–2568 (2021)

16. Benito-Román, O., Rodríguez-Perrino, M., Sanz, M.T., Melgosa, R., Beltrán, S.: Supercritical carbon dioxide extraction of quinoa oil: Study of the influence of process parameters on the extraction yield and oil quality. J Supercrit Fluids **139**, 62–71 (2018)
17. Chagua Lazo, G.S., Palomino Villazan, L.L.: Estudio comparativo de la capacidad antioxidante y compuestos fenólicos en quinua (Chenopodium quinoa) expandida de tres variedades provenientes del departamento de Junín. Universidad Nacional del Centro del Perú, Huancayo (2014)

Iconic-Verbal Strategies to Develop Reading Comprehension in Children of 4 Years Old

Pascual Ayamamani[1](✉), Liliana Valenzuela[2], Luisa Flores[3], and Diana Quispe[4]

[1] Departamento de Ecoturismo y Humanidades, Universidad Nacional Intercultural de Quillabamba, Quillabamba, La Convención, Peru
pascual.ayamamani@uniq.edu.pe
[2] Institución Educativa Inicial N° 06 "Nuestra Señora del Carmen", Abancay, Apurímac, Peru
[3] Institución Educativa Particular N° "Virgen de La Estrella", Abancay, Apurímac, Peru
[4] Facultad de Ingeniería, Universidad Nacional Micaela Bastidas de Apurímac, Abancay, Apurímac, Peru

Abstract. The study area is a context where students present severe difficulties in reading comprehension; the objective is to demonstrate the level of contribution of the iconic-verbal reading strategy in the development of reading comprehension levels in communication. We used a quantitative and applied method; 24 children of four years old participated, established through non-probabilistic sampling; pre-experimental design of a single group, whose period of experimentation or fieldwork lasted three months. In measuring reading comprehension level, the learning assessment scale was applied, and the results before treatment show that 100% of children are at the "beginning" level of reading comprehension. After the application of the iconic-verbal reading strategy, progress was made towards the "in progress" (45.8%) and "expected achievement" (54.2%) levels of reading comprehension in the area of communication. In conclusion, the application of the iconic-verbal reading strategy contributes significantly to the development of reading comprehension.

Keywords: Early childhood education · Teaching strategy · Reading habit

1 Introduction

The results of the Programme for International Student Assessment (PISA) placed Peru in 64th place out of 77 participating countries in 2018 [1] in terms of reading, and, according to the 2018 Sample Assessment, 56.4% of students in the second grade of primary school are at the "in-process" level in reading. These results show the poor development of reading comprehension skills of children, especially in less favored social sectors of Peru. In this regard, [2] reveals that students from the high socioeconomic level obtain better performance, not only because they belong to that stratum, but also because the educational institution where they participate benefits them (advantageous educational offer), while students from the low socioeconomic level, in addition to

Y. Iano et al. (Eds.): BTSym 2021, SIST 207, pp. 543–550, 2023.
https://doi.org/10.1007/978-3-031-04435-9_57

their social stratum, participate in a precarious educational institution (disadvantageous educational offer).

[3] states that 51.5% of students in the second grade of primary school are at the process level, so the students partially achieved the expected learning for the 3rd cycle (they still have difficulties). In other words, students do not manage to improve their initially precarious performance, even with the time spent in school, especially when the beginning of learning to read is difficult for most children and they even tend to perceive learning to read as unsuccessful, a process in which significant individual differences are generated among those who learn to read [4].

In this context, reading and reading comprehension are being taken with importance by world organizations and national institutions; therefore, according to [1], reading comprehension is one of the co-communicative skills that require particular attention since there is an international consensus to raise the level of reading comprehension.

In the kindergarten that was the object of this study, the absence of effective pedagogical strategies that allow the development of reading comprehension in children was evident. For these reasons, it is necessary and of vital importance to address the study of the development of reading comprehension in initial training processes, where every reader must use functional strategies, understood as the set of structured actions that put into practice, and take advantage of the conditions and the available resources. It could be said that these strategies have in common with the procedures, for their usefulness to regulate the activity of people, to the extent that their application allows to select, persist, or abandon specific actions to achieve the objective.

Consequently, achieving the management and mastery of communicative competencies that allow students to obtain written, oral, and audiovisual information, as well as understand messages and assimilate them, a competence that is essential for the personal and social development of children and implies having demonstrated the contribution of the strategy of iconic-verbal readings in the development of the levels of reading comprehension.

Consider that "if the reading strategy is procedures and the procedures are teaching content, then we must teach strategies for the comprehension of texts" [5], especially when, according to [6], "the first necessary but not sufficient condition to achieve reading comprehension is accurate and fluent reading", considering that most students manifestly demand explicit teaching or application of specific reading comprehension strategies, whose requirement is based on the limited knowledge and precariousness in the mastery of reading skills.

To foment awaken children's clear perceptions of the content and meaning of iconic-verbal texts, it is necessary developing levels of reading comprehension.

2 Methodology

2.1 Focus

The study is quantitative, concerning the type of applied research. The level of research used is experimental since it allows to foresee the degree of improvement of reading comprehension levels under certain conditions or characteristics of the strategy of iconic-verbal readings.

2.2 Unit of Analysis

The study population consisted of children of 4 years old, who were studying in the Kindergarten N° 155, of the Bancapata Village Center, of the district of Tamburco, Apurímac region - Peru. The sample consists of 24 boys and girls, in whom the incorporation of iconic-verbal texts has experimented with a strategy to develop the levels of reading comprehension for three months (March–May 2018).

2.3 Techniques

The technique used was a systematic observation, using an "observation sheet" on the situation before and after the treatment, on the levels of reading comprehension in the area of communication in 4-year-old children. Verification of the facts was made, which allowed the children to act naturally and spontaneously, facilitating the collection of reliable information, based on the "scale of evaluation of learning in early education" considering a scale: (A) expected achievement, (B) in process and (C) at the beginning.

3 Results

3.1 Development of Reading Comprehension Levels

Figure 1 shows the notorious contribution of the iconic-verbal reading strategy in improving the reading comprehension level. On a rating scale of 1–3; more remarkable progress is seen in the literal and critical levels, whose rise is of 1.4 ranks in each case (from 1.1 and 1.2 to 2.5 and 2.6), going from the scale "in beginning" to the scale "in process" and "expected achievement", respectively. That is followed by the critical level in a range of 1.3 of rising, between before and after the treatment.

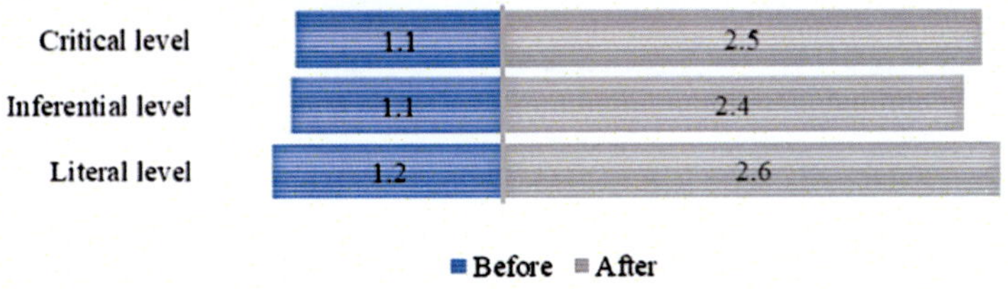

Fig. 1. Reading comprehension level in children, before and after treatment, on a rating scale of 1 = Beginning, 2 = In process, and 3 = Expected achievement.

3.2 Improved Reading Comprehension at the Literal Level

The results disaggregated by indicators of reading comprehension at the literal level (Fig. 2) show greater progress in the ability to "locate pictures and words in texts" and "say the content of the text read", in a range of 1.5 and 1.4 respectively, whose increase in reading comprehension at the literal level goes from the scale "at the beginning" (1.1 and

1.2) to "expected achievement" (2.6) between before and after the treatment, followed by the indicators that express the progress of 1.3 range, evidencing the achievement of learning in the programmed time; while the indicator with the smallest increase is the ability to "represent some fact that he/she likes most in the text", which goes from 1.0 to 2.1 (range of increase of 1.1), being located in the scale "in process", suggesting that he/she still presents difficulties and requires reinforcement.

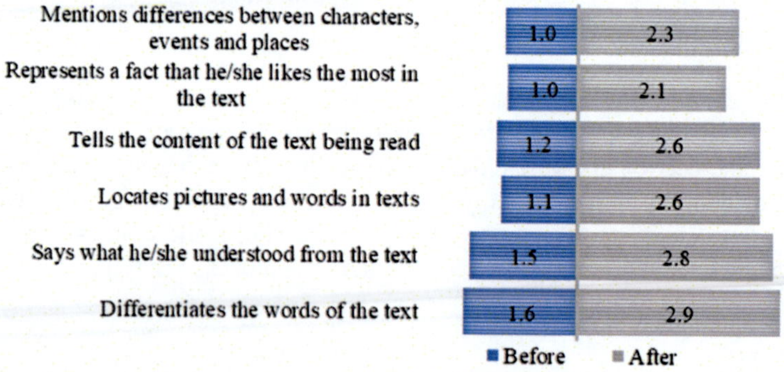

Fig. 2. Reading comprehension at the literal level, before and after treatment, on a rating scale of 1 = Beginning, 2 = In process, and 3 = Expected achievement.

3.3 Development of Reading Comprehension Levels

The disaggregated analysis of the improvement of reading comprehension at the inferential level (Fig. 3), corroborates the positive effect of the iconic-verbal reading strategy; to the extent that, the progress of 1.3 scores is shown in the indicator "tells what the text is about" (1.5) and "interprets texts from gestures and the message of the text" (1.0), moving up from the "at the beginning" level to the "in progress" and "expected achievement" levels, between before and after the treatment; and 1.2 of progress in the indicators "mentions the characteristics of the characters and places" and "expresses assumptions about the text".

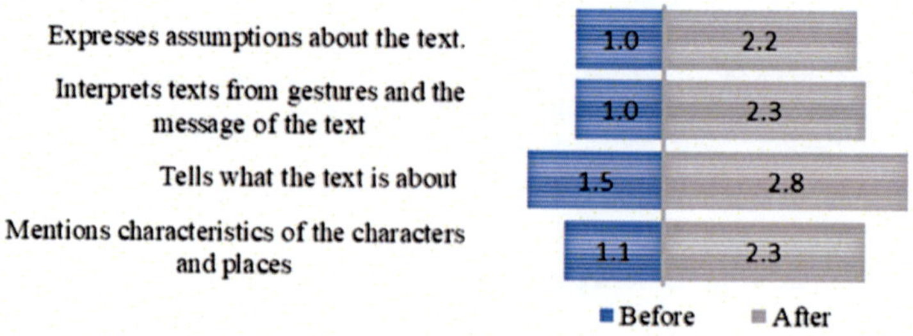

Fig. 3. Reading comprehension at the inferential level, before and after treatment, on a rating scale of 1 = Beginning, 2 = In process, and 3 = Expected achievement.

3.4 Development of Reading Comprehension Levels

The results (Fig. 4) show that reading comprehension at the critical level shows a significant increase in the ability to "say what he likes or dislikes about the text", by 1.5 points, between before (1.3) and after (2.8) the incorporation of the icon-verbal reading strategy; followed by the ability to "explain cause-effect relations of the text" (by 1.4 points) and "give an opinion about the reason for liking or disliking the text" (an increase of 1.3 to 1.4 points) and "to give an opinion on the reason for liking or disliking the text" (an increase of 1.3); the improvement in reading comprehension at a critical level, in all cases, from the "beginning" level to the "in-process" level and the "expected achievement" level.

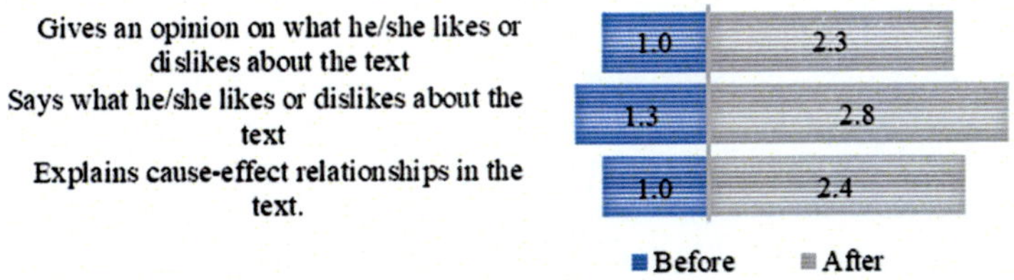

Fig. 4. Reading comprehension at the critical level, before and after treatment, on a rating scale of 1 = Beginning, 2 = In process, and 3 = Expected achievement.

A value $p = 0.000010 < 0.05$ was found, through the non-parametric Wilcoxon test; which evidences that the application of the iconic-verbal reading strategy contributes significantly to the development of reading comprehension levels in the area of communication, with a confidence level of 95%; specifying that the results of the statistical tests carried out for each level of reading comprehension, such as the literal level, inferential level, and critical level, reveal the same significance ($p = 0.000010 < 0.05$), and that the results of the statistical tests carried out for each level of reading comprehension, such as the literal level, inferential level, and critical level, reveal the same significance ($p = 0.000010 < 0.05$). $000010 < 0.05$) and confidence level (95%). These results are corroborated with the findings of Rivas (2019), whose study shows the p-value < 0.05, asserting "that there is a significant difference between the scores obtained by the students in the pre-test and post-test, with higher scores in the post-test" and that of [1] who based on the results of his study states that "the application of the program based on image readings improves the levels of reading comprehension in 5-year-old children", and shows T-test $= -13.3611 < 1.7459$ and p-value $= 0.0000$.

These revelations allow us to deduce and assert that the foundations for comprehending texts are established through the incorporation of appropriate strategies, such as the strategy of iconic-verbal readings, which allowed a substantial improvement in the reading comprehension levels of the children who were the object of the present study. It denotes "lights to the incidence (...) of the processes in the significant competence of the participants and, with it, to the intervention strategies" [7], in each of the reading comprehension levels; being these: a) literal level, basic phase of reading comprehension referred to the obtaining of ideas, information and meaning according to the text, which implies recognizing and remembering the facts as they appear expressed, b) inferential level, in this phase the reader goes beyond the text, establishing relationships between the parts of the text, this implies deducing information and conclusions that are not explicitly found in the text, and c) critical level, referred to the reader's evaluation and elaboration of judgments from the text and previous knowledge, as well as the elaboration of arguments.

The levels of reading comprehension constitute a thinking process that is generated progressively, starting from the basic phase of reading comprehension, which implies recognizing everything that appears explicitly in the text; followed by the process that is reformulated while reading; to then carry out a process of meta-cognition, which implies the formation of subjective judgments of one's own. That is, "achieving a good reading comprehension requires that the child has developed an adequate vocabulary and an appropriate ability to understand oral texts" [8], considering the phases of reading comprehension, a process that should start from the recognition of strengths and weaknesses of reading comprehension, to then enhance the strengths and mitigate the weaknesses; taking into account the strategy of iconic-verbal readings, since according to [9], children learn through the interpretation of images, making sense of them and trying to understand the contents found in the text, thus ensuring the development of reading comprehension at the literal, inferential and critical levels.

[10] consider the iconic representation is useful for the learning of students in the pre-operational stage and that of the concrete operations proposed by Piaget. "For that reason, the strategy of iconic-verbal readings, which consists of a text elaborated by an enunciator, containing a linguistic code and an image code ordered according to a structure, a point of view, and a purpose to address a statement, constitutes a great help to the reader in connecting with the text and developing reading comprehension. More specifically, the strategy of iconic-verbal readings implies the narration of texts with images, which usually represent characters, places, or a certain element of the narration, because after the narration the children appropriate the resource to narrate the story to the teacher" [11].

However, there is still no agreement on which are the skills that best contribute to the effective learning of written language and the development of reading comprehension skills, nor which are the most suitable teaching practices to facilitate it [12]. Furthermore, there are several conditioning factors in the student learning process, both from the perspective of demand (socioeconomic characteristics of students and their families), as well as from the educational supply (teaching and school characteristics).

4 Conclusions

It has been demonstrated that there are significant differences in the levels of reading comprehension in the area of communication in 4-year-old children, between before and after the treatment, where the p-value $= 0.000010 < 0.05$; therefore, it is possible to affirm that the application of the strategy of iconic-verbal readings contributes significantly in the development of the levels of reading comprehension in the area of communication in children of initial education, being able to be replicated in other educational environments of the same level.

When examining reading comprehension at the literal level in the area of communication in 4-year-old children, it was found that there are significant differences between before and after the treatment, so it is possible to assert, with a confidence level of 95%, that the strategy of iconic-verbal readings has a positive impact on the development of reading comprehension at the literal level in the area of communication in children in early education.

It has been found that the p-value = 0.000010 < 0.05, which shows that there are significant differences in reading comprehension at the inferential level in the area of communication in 4-year-old children; therefore, it is appropriate to argue that the application of the strategy of iconic-verbal readings contributes considerably to the development of reading comprehension at the inferential level in the area of communication in 4-year-old children, and can be replicated in other educational spaces at the initial level.

When evaluating reading comprehension at a critical level in the area of communication in 4-year-old children, it has been shown that there are considerable differences between before and after the treatment; therefore, it is appropriate to affirm, with a confidence level of 95%, that the application of the strategy of iconic-verbal readings contributes significantly to the improvement of reading comprehension at a critical level in children in early education, being advisable to incorporate it in the teaching processes at the early level.

References

1. Ulloa, A.: Programa de lectura de imágenes para mejorar los niveles de comprensión lectora en niños de cinco años de la Institución Educativa La Casa del Niño Florencia de Mora. Universidad Católica Los Ángeles Chimbote-Perú (2018). http://repositorio.uladech.edu.pe/handle/123456789/3042
2. Beltrán, B.A., Seinfeld, J.: La trampa educativa en el Perú: cuando la educación llega a muchos pero sirve a pocos. Repos la Univ del Pacífico - UP (2013). www.up.edu.pe
3. MINEDU:¿Cuánto aprenden nuestros estudiantes?. Resultados de la ECE 2016. Evaluación Censal de Estudiantes. Ministerio de Educación. p. 9 (2016). http://umc.minedu.gob.pe/wp-content/uploads/2017/04/DRE-Apurimac-2016-2.pdf
4. Mayor, M.Á., Zubiauz, B.: Desarrollo del lenguaje y preparación para la lectura en Educación Infantil. Padres y Maest Publicación la Fac Ciencias Humanas y Soc. 0(366):66 (2016). https://revistas.upcomillas.es/index.php/padresymaestros/article/view/6801
5. Solé, I.: Estrategias de Lectura. Barcelona, España (2000)
6. Fonseca, L., Pujals, M., Lasala, E., Lagomarsino, I., Graciela, M., Aldrey, A., et al.: Desarrollo de habilidades de comprensión lectora en niños de escuelas de distintos sectores socioeconómicos. Neuropsicol Latinoam. **6**(1), 41–50 (2014)
7. Valencia Leguizamón, M., Rodríguez León, D.: Interacción icono-verbal y recepción de libros álbum. Desafíos de la lectura en básica primaria. Sophia 15(1):48–59 (2019). https://revistas.ugca.edu.co/index.php/sophia/article/view/712
8. Junyent, A.: La comprensión lectora en los niños peruanos. Econ y Soc, pp. 32–38 (2016). https://www.cies.org.pe/sites/default/files/files/articulos/economiaysociedad/la_comprension_lectora_en_los_ninos_peruanos_a._junyent.pdf
9. Pinzas, J.: Guía de estrategias metacognitivas para desarrollar la comprensión lectora. Lima - Perú. (Documento técnico) (2006). https://www.researchgate.net/profile/Juana-Pinzas/publication/320007311_GUIA_DE_ESTRATEGIAS_METACOGNITIVAS_PARA_DESARROLLAR_LA_COMPRENSION_LECTORA/links/59c71e00a6fdccc7191ed997/GUIA-DE-ESTRATEGIAS-METACOGNITIVAS-PARA-DESARROLLAR-LA-COMPRENSION-LECTORA.pd

10. Retuerto, N.: Aplicación de técnicas ícono-verbales para mejorar la producción de textos literarios de los estudiantes del 3er grado de educación primaria de la IE N° 84173 "Quilca", Sihuas 2017. Universidad Nacional del Santa (2019). http://repositorio.uns.edu.pe/handle/UNS/3467
11. Silveyra, C.: Literatura para no lectores: la literatura y el nivel inicial. Homo Sapiens (2002)
12. Llamazares, M., Alonso, M.: Lectura compartida y estrategias de comprensión lectora en educación infantil. Rev. Iberoam Educ. 71:151–172 (2016). https://rieoei.org/RIE/article/view/9

Technological Resources for Teaching/Learning in Biomedical Areas in the COVID-19 Pandemic Context: Systematic Review

Tania Carola Padilla-Cáceres(✉), Luz Marina Caballero-Apaza, Vilma Mamani-Cori, and Sheyla Lenna Cervantes-Alagón

Universidad Nacional del Altiplano, Puno, Peru
tpadilla@unap.edu.pe

Abstract. The global health crisis caused by the Covid 19 pandemic has forced teaching and learning models to be rethought. The objective of this systematic review was to identify the technological resources used in virtual university education in biomedical careers in the context of the pandemic. It was carried out a literature search in the PubMed, Science-Direct, and Scielo databases for publications between 2020 and 2021, and 127 original articles were identified by the electronic search. Finally, 8 publications that met eligibility criteria were selected for the analysis. The results showed that various technological resources have been incorporated or consolidated during the virtual education of university careers in the biomedical area. They also showed acceptance, effectiveness, and, in some cases, satisfaction with their application. The question that remains is "Will effective training be guaranteed in the current scenario?" It is concluded that the biomedical programs incorporated different technological resources from those used in traditional education.

Keywords: Covid-19 · Teaching-learning · Educational technologies · University

1 Introduction

The SARS-CoV2 outbreak and the associated disease COVID-19 were reported for the first time in China in December 2019 [1, 2], being declared a pandemic and a global health emergency, abruptly and drastically changing the medical education [3, 4], forcing the adoption of contingency plans and programs in the study curriculum, incorporating virtual or distance/online education with smart work approaches, electronic learning platforms, among others [5, 6] in order to recover learning quickly [3, 7] by incorporating multiple technological platforms, including the use of video-based education, virtual rounds, virtual reality, 3D printed models, digital use of images on slides, virtual patient simulators, as well as telesimulation, transcending geospatially and temporally [8, 9], as clinical medical education required a change to contact without patients [10] facing many technological challenges focused in the efficiency of these delivery methods, interactivity and knowledge tests [11].

Y. Iano et al. (Eds.): BTSym 2021, SIST 207, pp. 551–560, 2023.
https://doi.org/10.1007/978-3-031-04435-9_58

In fact, the importance of technology became even more evident during the COVID-19 pandemic and the use of online platforms began to grow [12, 13], and the dissemination of its application in successful experiences will add to the growing body of evidence the usefulness of these tools in the acquisition of simple and complex clinical skills [14]. Therefore, the purpose of this systematic review is to identify and analyze the technological resources used in virtual university education in biomedical careers in the context of the COVID-19 pandemic.

2 Methodology

This systematic review complied with the Preferred Reporting Items for Systematic Reviews and Meta-Analyzes (PRISMA) statement [15]. Both the selection of the articles and the eligibility criteria as well as the qualitative analysis of the study were carried out by four reviewers. Data selection was carried out in electronic databases, evaluating the title and abstract of the manuscript. The full texts of all identified articles were also compiled to evaluate their eligibility for the analysis.

2.1 Eligibility Criteria

Those articles that used technological tools as a teaching/learning strategy and, in addition, have shown acceptability and satisfaction of the students of clinical courses in virtual university education in careers in the biomedical area in the context of the Covid-19 pandemic were considered.

Inclusion criteria were based on the PICOT questioning guidelines [16, 17]:

- P: Population - University careers in the biomedical area;
- I: Interventions – Use of technological tools as a didactic strategy in virtual university education;
- C: Comparator: Not applicable;
- O: Results- Acceptability and satisfaction with the technological tools used as a didactic strategy;
- T: Types of study: Cross-sectional/educational intervention - pre and post-test (applicative)/descriptive/observational).

The inclusion filters were cross-sectional and observational at the application level, on the use of technological tools as a didactic strategy in clinical courses in the biomedical area in virtual university education.

2.2 Exclusion Criteria

Other review articles, letters, lecture readings, editorials, personal opinions, and studies without abstracts were excluded. The search was limited to articles published from 2020 that have been applied and whose results were obtained during the COVID-19 pandemic.

2.3 Information Sources

A systematic electronic search for original articles was carried out in three databases that have been published in Spanish or English, in the PubMed and Sciencedirect databases the terms used were: "educational technologies" and "learning" and "University" and "Technological tools" and "teaching-learning" and "COVID-19"; while in Scielo we used the following keywords: Educational technologies, learning, University, technological tools, teaching-learning, and COVID-19.

In the three databases consulted, 127 studies related to technological tools, educational technologies, virtual university education, and the COVID-19 pandemic were identified and registered, of which 89 studies were withdrawn before starting the review because 11 studies presented duplication and the other 78 studies did not correspond to university clinical courses in the biomedical area.

Subsequently, the reviewers proceeded to review the 38 articles, coinciding that only 8 of them met the inclusion criteria of this study. The main reason for the exclusion was that the articles reflected a report of the teaching experience in the use of technological tools but not the student's experience, as well as that some of them did not show the results of the study (Fig. 1).

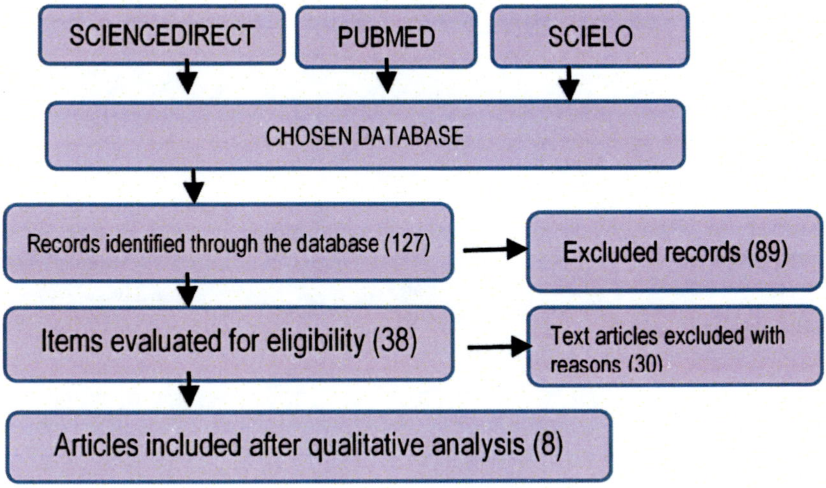

Fig. 1. Selection of articles via databases for systematic review according to PRISMA

2.4 Assessment of Bias Risk

The included studies received a risk of bias assessment for each reviewer following that described both in EPOC Resources for Reviewers and in the special data form of the Review Manager RevMan V 5.1 software package (The Nor-dic Cochrane Center, The Cochrane Collaboration, Copenhagen, Denmark) [18].

2.5 Selections of Studies

The study selection process was achieved in three stages:

1. Selection: In the first stage, all studies retrieved from the databases described considering the initial search criteria were submitted to a selection process by reading titles and abstracts;
2. Eligibility: In the second stage, the eligibility criteria were applied by reviewing the full text of the studies;
3. Inclusion: Finally, in the third stage, the studies were included to perform the qualitative synthesis.

It is important to mention that the following data were considered for each study: year of publication, country, research objectives, sample and technological tool as a teaching/learning strategy, acceptability, and satisfaction of the students.

3 Result

The eight selected articles that met the eligibility criteria were considered in this review (Table 1).

Table 1. Studies included in the systematic review.

Authors	Place of study /Year	Sample	Techniques/Strategies	Conclusions
Aguilar Gálvez, D., et al. [8]	Peru/2021	170 undergraduate students and 50 pediatric dentists	Virtual learning object (VLO)	VLO works in a similar way to asynchronous virtual teaching approaches
Furlan, R., et al. [19]	Italy/2021	15 medicine undergraduate students	VPS Virtual patient simulator for clinical diagnostic reasoning	By combining ITS and NLP technologies, it provides students with a reasoning tool for diagnosis, useful in restricted access

(*continued*)

Table 1. (*continued*)

Authors	Place of study /Year	Sample	Techniques/Strategies	Conclusions
Petrica, A., et al. [20]	Romania/2020	Emergency medicine professors, 69 students of fourth year that attended the Emergency medicine course	Knowledge assessment based on YouTube video analytics	The 360-degree video scenarios were effective in teaching Medical Emergencies
Drees, C., Ghebremedhin, E., Hansen, M. [21]	Germany/2020	213 preclinical students of medical education	Interactive E-learning software "Histologie für Mediziner"	The interactive elements of the software and educational design contributed to increase the essential factors for intrinsic motivation
Natsis, K, et al. [22]	Greece/2021	196 students of dentistry and medicine	Anatomy videos	The novel teaching modality (DEV series) can be used as educational material
Birrenbach, T, et al. [14]	Switzerland/2020	29 medicine students	Virtual reality simulation	Virtual reality simulation was as effective as traditional learning methods in training medical students, reporting benefits and user satisfaction

(*continued*)

Table 1. (*continued*)

Authors	Place of study /Year	Sample	Techniques/Strategies	Conclusions
Tan, S.H.X., et al. [23]	Singapore/2021	70 students of the Oral Health Therapy course	Home Simulation Model	The Simulation at Home model generates satisfaction and self-confidence in learning in students
Ramada R. Khasawneh.[24]	Jordan/2021	2263 students of the first 3 years of the medicine course	Online teaching	Students report that there is a gap and difficulty in understanding topics that require practice and visual guidance

3.1 Characteristics of the Articles Included After Qualitative Analysis

The investigations were carried out in different countries, Peru (1), Italy (1), Romania (1), Germany (1), Greece (1), Switzerland (1), Singapore (1), and Jordan (1), published between the years 2020 (3) and 2021 (5). Among the selected studies, one of them presented the pre-experimental type of educational intervention with before and after tests, 3 cross-sectional-descriptive studies, 4 experimental studies with a control group (pre and post-test). The sample size for included investigations ranged from 15 to 2263 university undergraduate participants. These studies evaluated different techniques and strategies such as the virtual learning object, the development of virtual patient simulators, knowledge evaluation through oral exams with the use of combined tools, interactive electronic learning software, videos according to the curriculum, and the application of educational virtual reality.

3.2 Effectiveness of Technological Resources for Teaching/Learning

After qualitative analysis, it was shown that the use of technological resources in the biomedical area is a novel and innovative teaching modality [21]. Petrica et al. [19] also in their descriptive study, perceived the interest of the students, and their acceptance of video technology being educational, accessible, and economical would also add value to the training. Furlan et al.[9], in their pre-experimental design, show that the use of a patient simulator will provide undergraduate medical students with a useful learning tool due to the restricted access to the clinic or laboratories during the health crisis. Experimental studies included in the systemic review, which are the highest percentage [8, 14, 21, 23], evidence the increase in knowledge after training regardless of the method used and indicates that it works similarly to the traditional teaching approaches [8, 24]. Drees et al. [20] note that interactive software scored significantly better compared to traditional learning programs. Experimental designs show that virtual reality simulation was as effective as traditional learning methods in training medical students, providing beneficial results [14].

4 Discussion

The results of this systematic review have shown the most widely used technological resources for teaching/learning in the areas of biomedical sciences in the context of the COVID-19 pandemic.

Biomedical programs consider in their study plans courses that require presence for interaction with patients, the use of laboratories, and the dissection of bodies, among others. Therefore, in this context of the pandemic and the need for social distancing, these activities have been suspended and resources for teaching/learning have been implemented, with an immediate migration to virtual learning [25].

Social networks (e.g., WhatsApp, Twitter, Instagram, and YouTube) began to be used more for communication with students, to teach seminars, and leave recorded sessions and videos that were easily accessible [26].

Although indeed, the online lectures of the theoretical courses did not have many drawbacks, the greatest difficulty was to implement those that required observations or interactions. The implementation of a flexible strategy was observed with the help of virtual learning objects (VLO) this tool helped to improve learning, leading the student to explore digital images, which are available to be reproduced at any moment thus consolidating the student's knowledge [8].

An educational video teaching modality was implemented in the online anatomy course. These videos were explained during the synchronous hours, allowing a fluid dialogue between the tutor and the student, and later used as self-learning material [21]. Students consider the experience and knowledge they gain from performing cadaver dissection are incomparable [23]. Some pilot educational projects have recorded and filmed scenarios using 360-degree video technology, which offers the possibility of a virtual experience, being able to explore from different angles, at their own pace [19].

The use of interactive software achieved a significant increase in student learning, allowing the sustained acquisition of knowledge Based on the cognitive theory of multimedia learning, which allows combining visual and audio information in working memory [20].

Simulators in biomedical programs have also been used by placing students in a context that imitates problems or situations similar to the reality that they may face in a way that favors the acquisition of skills prior to real contact with a patient [22]. Encouraging experiences were observed when using a virtual patient simulator based on natural language processing and an intelligent tutoring system for the clinical diagnostic process, training students in diagnostic reasoning [9].

Virtual reality (VR) has been another strategy that has been used by some biomedical programs [14], it uses computer systems to generate realistic pseudo-environments and that provides students with auditory, tactile, and visual sensations, giving them the possibility of a realistic interaction with the virtual environment [27], the students participating in this strategy were able to acquire simple and complex clinical skills [14], and there is evidence that virtual reality improves knowledge and skills more than traditional education or other types of digital education [28].

As it can be seen, biomedical programs have drastically changed the way they teach in this new context, redesigning their strategies for distance education. Students are not totally satisfied with online courses especially those that require practical understanding and visual orientation. The question remains: Will effective training be guaranteed in the current scenario for biomedical students who require clinical interaction? The obvious limitation was the few studies found that applied designs with statistical tests.

5 Conclusions

The findings of this review show that biomedical programs incorporated technological resources different from those used in traditional education and that some resources may remain permanently in current curricula. In some cases, biomedical students are not fully satisfied, especially where hands-on interaction is required. It is important to carry out other investigations emphasizing the effectiveness of the methodologies used.

References

1. Lu, R., Zhao, X., Li, J., et al.: Genomic characterisation and epidemiology of 2019 novel coronavirus: implications for virus origins and receptor binding. Lancet **395**, 565–574 (2020). https://doi.org/10.1016/S0140-6736(20)30251-8
2. Viner, R.M., Russell, S.J., Croker, H., et al.: School closure and management practices during coronavirus outbreaks including COVID-19: a rapid systematic review. Lancet Child Adolesc. Heal. **4**, 397–404 (2020). https://doi.org/10.1016/S2352-4642(20)30095-X
3. Okoye, K., Rodriguez-Tort, J.A., Escamilla, J., Hosseini, S.: Technology-mediated teaching and learning process: a conceptual study of educators' response amidst the Covid-19 pandemic. Educ. Inf. Technol. **26**(6), 7225–7257 (2021). https://doi.org/10.1007/s10639-021-10527-x
4. Bao, W.: COVID-19 and online teaching in higher education: a case study of peking university. Hum. Behav. Emerg. Technol. **2**, 113–115 (2020). https://doi.org/10.1002/hbe2.191
5. Gallenga, C.E., Agnifili, L., D'Aloisio, R., et al.: Virtual learning solutions in COVID-19 era: university italian ophthalmology department perspective. Eur. J. Ophthalmol. (2021). https://doi.org/10.1177/11206721211015576
6. Martinelli, S.M., Chen, F., Isaak, R.S., et al.: Educating anesthesiologists during the coronavirus disease 2019 pandemic and beyond. Anesth. Analg. **132**, 585–593 (2021). https://doi.org/10.1213/ANE.0000000000005333
7. Arezzo, A., Vignali, A., Ammirati, C.A., et al.: Is it possible to continue academic teaching in surgery during the COVID pandemic era? Minim. Invasive Ther. Allied Technol. 1–9 (2020). https://doi.org/10.1080/13645706.2020.1845210
8. Aguilar Gálvez, D., Noal, F.C., Arriola-Guillén, L.E., et al.: Virtual learning object for developing knowledge about the diagnosis and management of molar incisor hypomineralization. Int. J. Paediatr. Dent. (2021). https://doi.org/10.1111/ipd.12925

9. Furlan, R., Gatti, M., Menè, R., et al.: A natural language processing-based virtual patient simulator and intelligent tutoring system for the clinical diagnostic process: simulator development and case study. JMIR Med. Informatics **9**, e24073 (2021). https://doi.org/10.2196/24073
10. Wilson, T.E., Mulye, M., Akbar, S.: Opportunistic physiology: inserting physiology and pathophysiology content into virtually delivered clinical rotations. Adv. Physiol. Educ. **44**, 545–549 (2020). https://doi.org/10.1152/advan.00113.2020
11. Zaguia, A., Ameyed, D., Haddar, M., et al.: Cognitive IoT-based e-learning system: enabling context-aware remote schooling during the pandemic. J. Healthc. Eng. **2021**, 1–12 (2021). https://doi.org/10.1155/2021/7358874
12. Charitos, I.A., Del Prete, R., Inchingolo, F., et al.: What we have learned for the future about COVID-19 and healthcare management of it? Acta Biomed. 91:e2020126 (2020). https://doi.org/10.23750/abm.v91i4.10253
13. Favale, T., Soro, F., Trevisan, M., et al.: Campus traffic and e-learning during COVID-19 pandemic. Comput. Netw. **176**, 107290 (2020). https://doi.org/10.1016/j.comnet.2020.107290
14. Birrenbach, T., Zbinden, J., Papagiannakis, G., et al.: Effectiveness and utility of virtual reality simulation as an educational tool for safe performance of COVID-19 diagnostics: prospective, randomized pilot trial. JMIR Serious Games (2021). https://doi.org/10.2196/29586
15. Page, M.J., McKenzie, J.E., Bossuyt, P.M., et al.: The PRISMA 2020 statement: an updated guideline for reporting systematic reviews. BMJ **372**, 71 (2021). https://doi.org/10.1136/bmj.n71
16. Da Costa, C.M., De Mattos, C.A., Cuce, M.R.: Estrategia PICO para la construcción de la pregunta de investigación y la busqueda de evidencias. Rev. latino-am. Enferm. **15**, 1–4 (2007)
17. Riva, J.J., Malik, K.M.P., Burnie, S.J., et al.: What is your research question? An introduction to the PICOT format for clinicians. J. Can. Chiropr. Assoc. **56**, 167–171 (2012)
18. Cochrane Recursos de EPOC para revisores. https://epoc.cochrane.org/resources/epoc-resources-review-authors. Accessed 31 Oct 2021
19. Petrica, A., Lungeanu, D., Ciuta, A., et al.: Using 360-degree video for teaching emergency medicine during and beyond the COVID-19 pandemic. Ann. Med. **53**, 1520–1530 (2021). https://doi.org/10.1080/07853890.2021.1970219
20. Drees, C., Ghebremedhin, E., Hansen, M.: Development of an interactive e-learning software "histologie für mediziner" for medical histology courses and its overall impact on learning outcomes and motivation. GMS J. Med. Educ. **37**, 1–20 (2020). https://doi.org/10.3205/zma001328
21. Natsis, K., et al.: "Dissection educational videos" (DEVs) and their contribution in anatomy education: a students' perspective. Surg. Radiol. Anat. **44**(1), 33–40 (2021). https://doi.org/10.1007/s00276-021-02829-z
22. Tan, S.H.X., Ansari, A., Ali, N.M.I., Yap, A.U.: Simulation design and students' satisfaction with home-based simulation learning in oral health therapy. J. Dent. Educ. **85**, 847–855 (2021). https://doi.org/10.1002/jdd.12576
23. Khasawneh, R.R.: Anatomy education of medical students during the COVID 19 pandemic. Ent. J. Morphol. **39**, 1264–1269 (2021)
24. Aguilar-Gálvez, D., Noal, F.C., de Araujo, F.B., Arriola-Guillén, L.E.: Virtual learning object: an asynchronous solution for virtual learning in dentistry post COVID-19. J. Dent. Educ. **85**, 1123–1125 (2021). https://doi.org/10.1002/jdd.12439
25. Mian, A., Khan, S.: Medical education during pandemics: a UK perspective. BMC Med. **18**, 100 (2020). https://doi.org/10.1186/s12916-020-01577-y
26. Singh, K., Srivastav, S., Bhardwaj, A., Dixit, A., Misra, S.: Medical education during the COVID-19 pandemic: a single institution experience. Indian Pediatr. **57**(7), 678–679 (2020). https://doi.org/10.1007/s13312-020-1899-2

27. Babini, M.H., Kulish, V.V., Namazi, H.: Physiological state and learning ability of students in normal and virtual reality conditions: complexity-based analysis. J. Med. Internet Res. **22**, e17945 (2020). https://doi.org/10.2196/17945
28. Rourke, S.: How does virtual reality simulation compare to simulated practice in the acquisition of clinical psychomotor skills for pre-registration student nurses? A systematic review. Int. J. Nurs. Stud. **102**, 103466 (2020). https://doi.org/10.1016/j.ijnurstu.2019.103466

Current Situation of the Monitoring of Graduates and Labor Insertion in the Electrical Mechanical Engineering Career

Julio Fredy Chura Acero(✉), Omar Chayña Velasquez, Fredy Bernardo Coyla Apaza, Leonidas Vilca Callata, and Henry Shuta Lloclla

Ingeniería Mecánica Eléctrica, Universidad Nacional del Altiplano, Puno, Peru
{jchura,fredycoyla,lvilca,henryshuta}@unap.edu.pe

Abstract. The purposes of the studies carried out in this research carried out in the professional career of electrical mechanical engineering in coordination with the graduate monitoring area of the Universidad Nacional del Altiplano are i) to analyze the experiences of monitoring graduates and the strategies of labor insertion developed by the participating universities, with the intention of contributing to the incorporation of improvements in their institutional effectiveness processes, and ii) analyze the accreditation agreements, regarding observations of strengths and weaknesses on labor insertion and follow-up of graduates. The latter was specified through the review of the institutional accreditation agreements for the professional career of electrical mechanical engineering, accredited by SINEACE.

Keywords: Engineering · Graduate · Labor insertion · Monitoring

1 Introduction

Currently, universities must assume responsibility for the insertion of graduates in the labor market to improve their supply of education and training.

Fundamental economic conditions have radically changed and labor organization schemes demand the existence of a more solid link between training and professional skills. This requires cooperation in the area of university education and also indicates the role that universities should play as drivers of innovation and technological development. Now, universities need to ensure professional skills and services that do not end with obtaining a degree, but also take into account the need for continuous learning [1].

In this context, the monitoring of graduates is a matter of vital importance for universities because the evaluation of the professional and personal performance of graduates allows establishing indicators regarding the quality and efficiency of educational institutions.

Y. Iano et al. (Eds.): BTSym 2021, SIST 207, pp. 561–568, 2023.
https://doi.org/10.1007/978-3-031-04435-9_59

Continuous changes in the labor market indicate that the trend of modern society constantly blurs the boundaries between work, leisure time, education, and attendance. This has generated greater mobility and flexibility. Life and work career paths are no longer standardized, and, as a result, employability has become a key issue. Higher education, training, and the labor market have a vital need for rapprochement in this versatile scenario, especially in the area of insertion into the labor market of recent graduates. In recent years, higher education institutions have begun to focus on quality assurance to meet the needs of both their students and society concerning the labor market [2]. Consequently, the success of higher education institutions is measured primarily based on their students' performance regarding their employment status and social commitment.

In Peru, there is very limited information on the situation of university graduates, which allows analyzing what are the results of the university higher education and the quality of it. Currently, there is only an analysis of the occupational situation and period of permanence of the graduates of the electrical mechanical engineering career in the labor market of the Puno Region.

2 Referential Frame

This study deals with a currently underdeveloped topic, such as the situation in which graduates of a higher institution find themselves, specifying the terms that conceptualize the topic and delimiting the ideological framework in which its use is not only possible but also necessary. In this sense, it is convenient to disaggregate the three semantic elements it contains: the noun situation and the adjectives work and personal, which lead to consider a unique and concrete reality: the graduate, from different scientific perspectives.

This reality currently presents a poorly defined, extremely changing appearance, which does not allow it to be determined precisely based on a systemic and, at the same time, synchronous perspective; nevertheless, macrosocial trends seems to allow attempting to configure models that explain the behavior of society in the historical moment that is lived. Thus, the analysis carried out begins by drawing up an outline of economic and social trends in the world today, those that pose challenges that would have been unthinkable a few decades ago, although they have strongly influenced Peruvian society, requiring education to adapt to them.

Due to changes in the economy, the concept of work has changed [3]; with the postmodernity of globalization, they open job opportunities, but at the same time other job opportunities are canceled. And in this sense, education is also affected by the change of paradigms; Thus, knowledge is no longer deposited in colleges, institutes, or universities, but they are everywhere, accessible to anyone through various means that are not necessarily those of formal education.

The points considered contribute to the reinforcement of the idea of lifelong education [4], ending the conception of education in stages, taught in centers specially provided for it.

On the other hand, a very work-related education is required, but given this from the new considerations, where the content learned does not matter so much, but the quality of what has been done. In this sense, education will be highly competitive and institutions will be sufficiently and exclusively necessary licensed and authorized by law to teach it [5]. The aforementioned guidance poses a challenge for those receiving the education and for those who teach it, because it will be imperative to make education serves effectively for the formation of the person, before only the professional. Education will aim to teach people to find the information they need when they need it and where they are. This means that the work of the university must experience a change of purpose and orientation: from teaching knowledge "To learn to learn", and from disciplinary specialization to disciplinary integration, respectively. From the above, it follows that higher-level educational institutions will seek to obtain as a product graduates whose qualities are compatible with the new educational and labor paradigms, which is why the concept of a relationship between educational education and the graduate must broaden its significant content as a function of variables that exceed the limits of the academic.

Education, then, will form the person, within an axiological framework, certain attitudes, of substance and principle. The person at work will not only show how much they know and how many skills they have but how much it is worth, a value that the market will necessarily translate into economic terms.

University education conceived in this orientation goes from being a passive, discontinuous, abstract, disjointed, and unilinear teaching process, to an active, continuous, experiential, inclusive, and multiple learning process; the graduate must go from being an element outside the university to becoming part of the institution, with the capacity and presence to contribute to the increase in the quality of the service it offers.

But, at the same time, taking this course requires rethinking how the universities keep in contact with their graduates and, on everything, ensure their permanence in the institution, as it may be necessary to reject the role of the alumni associations, rather passive, and increasing the mechanisms and instruments that enable the relationship between the university and its graduates, perhaps considering these potential clients of the university [6].

In this sense, students, and also graduates, as users of the educational service, will acquire an important role in the configuration of the educational models, as they will play the role of catalysts of the quality of the educational service they receive [7].

The consideration of the different factors mentioned is encompassed at the same time for the concept of educational quality, for which, following deeming principles, mention is required of the effects of a process of implementation of decisions aimed at achieving total quality [8], translated in terms of excellence. The same author states that in this permanent search for quality, the conception of leadership is unavoidable both in professional training and in the training of the person [9]. The university graduates must become leaders, for which they must be trained in leadership processes and in the ability to communicate with others, it must be conducted in the deep knowledge of the human sense, in addition to being a specialist in the knowledge for which they were trained [10].

3 Methodology

Two information collection forms were prepared and applied, which were previously validated through consultation with experts.

- Form for Graduate Monitoring Programs or initiatives;
- Form for Labor Insertion Programs or initiatives.

The main focus of the information gathering was to characterize and analyze:

- The purposes and activities carried out;
- The age of the experiences;
- The responsible units;
- The state of formalization and the resources they have;
- The target audience it is aimed at;
- The evaluation mechanisms;
- The links with other university programs or with other units;
- The facilitators and obstacles in its implementation and development.

The situation of the follow-up of graduates and job placement seen from the accreditation of institutions and careers was carried out by reviewing the observations contained in https://bolsadetrabajo.unap.edu.pe/admin/reportes/reporte-egresados.

3.1 Definition of Variables and Indicators

The dependent variables are the labor situation of the graduate and the personal situation of the graduate.

The independent variables are academic training, work activity, the social importance of the career, economic importance, career choice, identification with the institution where he works, and satisfaction with work activity main who plays.

3.2 Material and Methods

The carried out study corresponds to a relational descriptive investigation, in which the method was applied statistically in its entirety. The population consisted of all the graduates of the EPIME (Professional Career of Electrical Mechanical Engineering), during the period 1995 to 2020, out of a total of 2,256 graduates. To select the sample, the EPIME graduates in the 2014–2020 period were considered (no academic degree, only graduate). For the present work, a pilot sample was taken in which the differentiated non-existence of the percentages of graduates working by career was observed. For the global study, a random sample of 509 graduates was used, this random sample of graduates was distributed during this period (Fig. 1).

Fig. 1. EPIME-UNAP graduate registration website

4 Result and Discussion

From the global study of EPIME graduates, period 2014–2020, it is found that graduates have the following characteristics (Table 1 and Fig. 2).

Table 1. Graduates per years

Year	Graduates
2014	60
2015	76
2016	74
2017	125
2018	78
2019	78
2020	18
Total	**509**

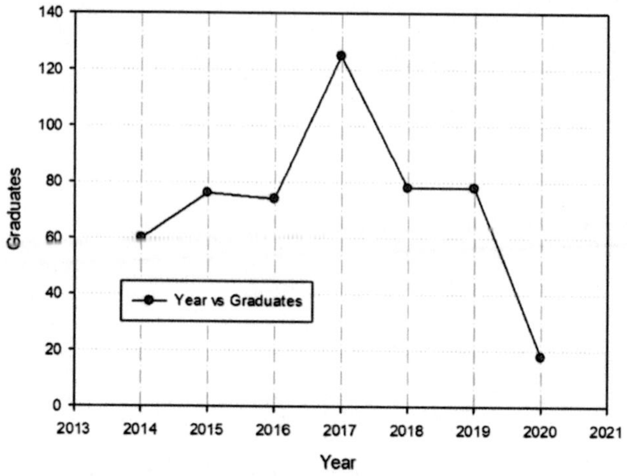

Fig. 2. Result of graduates per year

4.1 Main Activities Developed by the Programs

When consulting the main activities carried out by the Program, it is observed that they cover only some aspects of the proposed areas of action (Table 2 and Fig. 3); In other words, the expectations generated by the Program, inspired by the needs demanded of the institution today, are more complex than the activities carried out in practice.

Table 2. Main activities developed in the graduate follow-up programs

Year	Graduates
Alumni meetings	10
Dissemination of job opportunities	07
Maintenance of alumni website, social network (Facebook)	06
Training for graduates (strengthening skills through talks, workshops, conferences, seminars, update sessions, others)	05
Maintenance of databases of graduates	09
Dissemination and realization of cultural activities	08
Information gathering and validation of graduation profiles	06
Dissemination of continuing education programs or activities	07
Alumni Benefit Management	05
Labor market studies, employability	04
Survey application for Accreditation processes	03
Studies of interest in continuing education	01

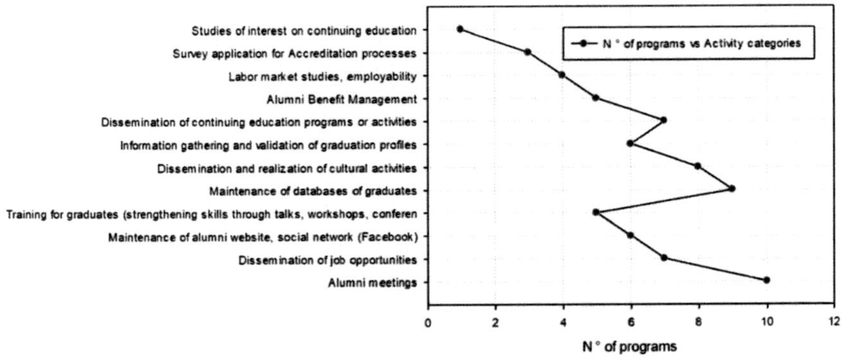

Fig. 3. Main activities developed by the programs

5 Conclusions

The following are the main conclusions that emerge from the analysis of the results presented here on the study of the status of programs to monitor graduates and support employment in higher education from the sample made up of the professional engineering career electrical mechanics.

Higher education institutions need to have systematic and reliable information on educational social needs, mainly concerning the professions offered by universities and their labor field. Hence, the evaluation of results and impact of the degrees of employability and labor insertion of the professionals of the universities is necessary for effective feedback of the curriculum and support decisions about the continuity and opening of careers according to the needs of today's society.

An important and essential tool to establish communication and the management of information related to the follow-up of graduates is that the institution has updated databases of its students and graduates.

It is appreciated that the efforts developed are still incipient and insufficient.

It has not been possible to verify the existence of systematic mechanisms for the follow-up of graduates; Two-thirds of the programs reviewed are recently created (6 years or less), and apparently, they arise from the need to ensure the quality of the training offered required in the career accreditation processes. On the other hand, the programs analyzed are initiatives that correspond rather to the enthusiasm and commitment of some managers and academics.

At present, given that the revised accreditation agreements indicate for the most part weaknesses, it is evident that few institutions considered in the study have formal and systematic processes that allow the use of information on graduates to make adjustments in the quality of professional training.

The process of monitoring graduates, job placement, and employability constitutes, then, an institutional responsibility that must be part of the quality assurance of professional training.

References

1. Wang, H.: Comparative study on the physics experimental teaching models between chinese and american universities. J. Anhui Agric. Univ. (Social Science Edition) **7**(1), 54–58 (2013)
2. Shi, C.: The rise of virtual reality: time machine or hallucinogen. J. Modern Commun. (J. Commun. Univ. China) **38**(6), 95–99 (2016)
3. Lu, Y., Yin, Z., Shao, Z.: Design of the dynamic traffic information prediction system based on multi-source information fusion. J. Autom. Instrum. **10**, 59–62 (2019)
4. Shi, L.: Rational cognition of the application of the virtual reality technology in education field. J. Dalian Educ. Univ. **33**(4), 37–38 (2017)
5. Huang, Y.: Research overview of the application of virtual reality technology in education. J. China Educ. Inform. **1**, 11–16 (2018)
6. Potkonjak, V., Gardner, M., Callaghan, V., et al.: Virtual laboratories for education in science, technology, and engineering: a review. J. Comp. Educ. **95**, 309–327 (2016)
7. Hu, A., Zhang, R., Yin, D., et al.: Image quality assessment using a SVD based structural projection. J. Sig. Process. Image Commun. **29**(3), 293–302 (2014)
8. Acero, J.F.C., Viamonte, W.R.L., Velasquez, O.C., Pari, E.R.P.: Virtual university education is an environment without losing its essence in the teaching and learning spaces. In: Proc 2020 IEEE Int Symp Accredit Eng Comput Educ ICACIT 2020, pp. 20–23 (2020). https://doi.org/10.1109/ICACIT50253.2020.9277672
9. Acero, J.F.C., Viamonte, W.R.L., Vilcapaza, W.M.S., Velasquez, O.C.: Matlab mobile as a support tool for the performance of students in engineering. In: Proc 2019 Int Symp Eng Accredit Educ ICACIT 2019 (2019). https://doi.org/10.1109/ICACIT46824.2019.9130340
10. Chura Acero, J.F., Beltrán Castañón, N.J., Limachi Viamonte, W.R., et al.: International Internship based on research collaboration to improve the educational management on engineering, pp. 1–5 (2021). https://doi.org/10.18687/laccei2021.1.1.379

Job Motivation and Mental Health Status of Care Person at the District Hospital La Esperanza During COVID-19

Miryam Lora-Loza[1(✉)], Marco Quispe Barra[2], Julia Alvarado-Alfaro[1], Juan Rodríguez-Vega[1], Davis Alberto mejía Pinedo[1], Rosa Estrella Pillman Infanson[1], and Maximo Amancio Moltalvo Atco[2]

[1] Universidad César Vallejo, Trujillo, Peru
mlora@ucv.edu.pe
[2] Universidad Nacional del Altiplano de Puno, Puno, Peru
mmontalvo@unap.edu.pe

Abstract. The objective of this study was to determine the relationship between job motivation and the mental health status of care personal of the District Hospital of Jerusalén - La Esperanza during COVID-19. The methodology used was descriptive correlational analysis with a hypothetical deductive method with a non-experimental quantitative approach and cross-sectional design. A total of 154 health care workers participated. Two types of surveys were considered; the first was for job motivation (JM) and the second for general mental health (PHQ-9) and its dimensions such as General Anxiety Disorder (GAD-7), Insomnia Severity (ISI), and Impact of Events Reviewed (IES-R). The Spearman correlation coefficient was used to determine the association between the two variables and their dimensions. In the results, we observed that most of the workers (56.5%) reflect a medium job motivation and a Mild level of depression ($Rho = 0.872$, $p < 0.000$). In addition, we observed that 41.6% of the workers reflect a medium job motivation and a mild general anxiety disorder ($Rho = 0557$, $p < 0.000$).

Keywords: Job motivation · Job performance · Spearman coefficient · Health · COVID-19

1 Introduction

At the beginning of the COVID-19 pandemic, no one imagined its impact on the relationship between job motivation and mental health status in healthcare workers. Just six months after the beginning of this pandemic, fears and panics were already verified not only from inside the healthcare centers but also from outside them, between 60 and 80% of healthcare workers were already affected in their mental health, but it was still not accepted as a medical problem or mental illness, the priority was the clinical care of the COVID-19 patient and not the mental health of healthcare workers [1].

In highly developed countries such as Japan, Canada, Australia, Russia, China, the United Kingdom, and Germany, support plans to care for the mental health of workers

Y. Iano et al. (Eds.): BTSym 2021, SIST 207, pp. 569–576, 2023.
https://doi.org/10.1007/978-3-031-04435-9_60

began to be activated, and the analysis of motivational aspects was prioritized to counteract the negative impact of COVID-19 on them. Psychoeducational programs for the management of learning elementary behaviors, e.g., teaching reflexes, impulses, character predispositions, and mental attitudes were quickly incorporated into motivational strategies and policies as essential elements of support for care personnel, regardless of the function performed [2]. In this way, they succeeded mainly in curbing the negative motivation to adapt to the new working conditions. The healthcare personnel opened up to cohesion and psychological stability processes, achieving that the states of anxiety, the severity of insomnia, the impact of events, which occurred inside and outside the workplace, were treated with absolute responsibility [3].

However, the capacity of these programs and even of the strategies and public policies implemented for this purpose presented significant limitations for their replication in the developing countries of Africa, Asia, Latin America, and the Caribbean, showing their impossible modeling as a process of treatment and care of mental health of the health worker in general. The hindering elements in these countries were the innate attitudes of individuals, intrinsic to their personality and professional development, which uniquely distinguished them from health workers in highly developed countries [4]. However, these motivational programs, strategies, and policies of attention to the mental health of the health worker, in the medium term, became very effective tools to work on adaptive behavior problems, where new job habits and customs became renewed products that encouraged or were part of homogeneous and undifferentiated behaviors. Following the logic of risk management in mental health in the ways of thinking, feeling, or acting, as a common denominator of their actions, they began to work on individual attitudes, a product of traditional methods and job style (before COVID-19). So much so that they soon became barrier elements, generating negative motivations at intrinsic and extrinsic levels. The extraordinary capacity of COVID-19 to mutate and spread aggressively could not be adequately counteracted [5].

There have not been organized or structured psychoeducational support programs for this case. Except in Argentina, Brazil, and Chile, where the problems of job motivation seek to improve the mental health of health workers, this is not the case in other countries because they are not considered non-priority issues, thus allowing the problem to advance, to report today, according to the WHO that it would be affecting more than 60% of health workers in developing countries around the world [6].

In Peru, due to the distinctive characteristics of the health care worker, the problem was ignored more than in any other part of the world. Furthermore, it does not consider the motivational elements that affect the worker's mental health within its human talent management policies, only taking into account the opinions of those whose true feelings are hidden behind the COVID-19 patient care needs. With them, work overload, failure to structure equitable schedules, long working hours, disorientation due to compliance with biosafety standards or care protocols, etc., become part of the problem that does not allow a labor motivation policy to be successful [7].

In other words, conservative attitudes have always been expressed regarding the need to intervene in the relationship between job motivation and the mental health status of health care personnel. Human talent management and administration policies are closed in structures of precarious rationality, not even allowing the possibility of identifying the role of a dimension of this relationship, affirming, conveniently, interests related to the welfare idea of attending the cases once the disease has been declared. Therefore, no level of predictive, preventive, and projective job is observed, only the promotion of a unilateral mentality that discards the relationship between motivation and mental health of the health worker as an area of intervention or jobs for labor policies and strategies [8].

This situation has placed the mental health of more than 70% of health personnel at high risk and the other 30% at moderate risk, exponentially exacerbating fears or panics in all health institutions in our country. The importance of the mental health of health personnel may not be valued. Still, it is necessary to recognize the preponderance of the negative rationalisms it produces or the changes it generates in behavior and job performance, even though there is no desire to acknowledge this problem in the discursive forms of scientific empirical logic [9].

In the District Hospital of Jerusalén - La Esperanza, it was found that more than 60% of the workers are affected in their mental health without knowing it. They act according to their professional criteria, leaving aside their feelings. They show neutral and acceptable attitudes. They move away from antagonisms and look for solidarity and cooperation, inside and outside the institution where they job. However, they manifest existential panics accompanied by an anxiety disorder, insomnia, and pessimism [10].

Consequently, the research aimed to determine the relationship between job motivation and the mental health status of the health personnel of the District Hospital of Jerusalén - La Esperanza, in times of COVID-19, 2021. Thus, this study provides new elements of analysis on job motivation and the mental health status of the health care personnel of the District Hospital, La Esperanza, during COVID-19.

2 Methodology

2.1 Design and Study Population

This study used descriptive correlational analysis with a hypothetical deductive method, a quantitative non-experimental approach, and a cross-sectional design. The results were described through simple and weighted frequencies obtained in the cross-sectional relationships, interpreting the correlations between variables and their dimensions. A population of 154 health care workers of both sexes of the District Hospital of Jerusalén - La Esperanza - Peru in the year 2021 [11] was considered. A non-probabilistic sampling by strategic allocation was applied to select the sample, assuming 100% of the population; the sample included workers of both sexes with more than one year of service in the institution.

2.2 Variables and Dimensions

Due to their nature, the study variables were qualitative measures on an ordinal scale and their dimensions. Likewise, two correlational variables were considered. The first variable was job motivation. The second variable was mental health status with its respective dimensions (General Anxiety Disorder (GAD-7), Insomnia Severity (ISI), and Impact of Events Reviewed (IES-R)).

2.3 Techniques and Data Instrument and Procedure

In this study, the person-to-person survey technique was used through a set of questions. The instruments used were of two types. The first instrument to measure the job motivation variable was a questionnaire of job motivation (JM) questions. The second instrument used was the general mental health questionnaire (Patient Health Questionnaire - PHQ-9) [12], which assesses depressive symptoms based on the occurrence of depressive episodes. This questionnaire consists of 9 items that respond to the Likert scale of 4 alternatives ranging from: never (0), rarely (1), occasionally (2), and frequently (3).

2.4 Statistical Analysis

Data [11] were transferred from a Microsoft Excel program to Package for the Social Sciences (SPSS) version 22 processing software in Spanish [13, 14]. Data were pre-coded with mutually exclusive numbers. Descriptive statistics (frequency tables) and inferential statistics were also used. Spearman's correlation coefficient was used to analyze and interpret the results. To determine if there is a significant relationship between the study variables, the statistical significance should present $p < 0.05$.

Spearman's correlation coefficient, ρ (Rho) [15], is a measure of the correlation between two random variables (both continuous and discrete). To calculate the coefficient, Rho (ρ) is expressed by the following formula:

$$\rho = 1 - \frac{6\sum d_i^2}{n(n^2 - 1)} \quad (1)$$

where, d_i is the difference between the ranks (x–y) and $n =$ is the number of elements in the ranking.

3 Results and Discussion

A total of 154 (100%) care workers of both sexes participated in this study. Table 1 shows the relationship between job motivation and level of depression. We observed that in times of the COVID-19 pandemic, 87 care workers (56.5%) reflected a middle level of job motivation and mild level of depression. However, 50 care workers (32.5%) reflected low job motivation and moderate depression, and 7 care workers (4.5%) represented high job motivation and a minimal level of depression. In addition, Table 1 also shows that the relationship between job motivation and the stories of depression of the care workers is significant ($p < 0.05$). Spearman's Rho coefficient (Rho = 0.872) indicates a moderate positive correlation.

Table 1. Relationship between job motivation and level of depression of general mental health (PHQ-9)

Level of depression (PHQ-9)	Job motivation								Spearman's correlation	
	High		Middle		Low		Total			
	N°	%	N°	%	N°	%	N°	%	Rho	*p-value
Minimal	7	4.5	1	0.6	1	0.6	9	5.8	0.872	0.000
Mild	0	0.0	87	56.5	0	0.0	87	56.5		
Moderate	0	0.0	8	5.2	50	32.5	58	37.7		
Total	7	4.5	96	62.3	51	33.1	154	100		

N° = Number of workers surveyed, % = percentage representation of N°.
*$p < 0.05$ (significant)

Table 2 shows the relationship between job motivation and the characteristic dimensions of mental health status (generalized anxiety disorder, insomnia severity, impact of the events reviewed). We observed that 64 care workers (41.6%) reflected a middle level of job motivation and a mild level of generalized anxiety disorder, 3 care workers (1.90%) reflected a low level of job motivation and a severe level of generalized anxiety disorder, and 6 care workers (3.9%) represented a high level of job motivation and a minimal/zero level of generalized anxiety disorder. The Spearman's Rho coefficient was 0.557, indicating a moderate positive correlation and a significant relationship ($p < 0.05$). Table 2 also shows the relationship between insomnia severity and job motivation, where 49 care workers (31.8%) present a middle level of job motivation and the subumbral level severity of insomnia, 4 care workers (2.6%) reflect a low level of job motivation and severity of insomnia and 6 care workers (3.9%) represent a high level of job motivation and normal level severity of insomnia. The Spearman's Rho coefficient was 0.385, indicating a moderate positive correlation and a significant relationship ($p < 0.05$). Finally, in Table 2 we observe the relationship between job motivation and the impact of events reviewed, where 44 workers (28.6%) represent a middle level of job motivation and a subclinical level impact events, 26 workers (16.9%) represent a low level of job motivation and mild level of impact events and 7 workers (4.5%) express a high level of job motivation and a subclinical level of impacts event. The spearman's Rho coefficient was 0.564, indicating a moderate positive correlation and a significant relationship ($p < 0.05$).

Table 2. Relationship between job motivation and the characteristic dimensions of mental health status (generalized anxiety disorder, severity of insomnia, impact of revised events)

		Job motivation								Spearman's correlation	
		High		Middle		Low		Total			
		N°	%	N°	%	N	%	N°	%	Rho	*p-value
General Anxiety Disorder (GAD-7)	Minimal	6	3.9	4	2.6	1	0.6	11	7.1	0.557	0.000
	Mild	0	0.0	64	41.6	6	3.9	70	45.5		
	Moderate	1	0.6	25	16.2	41	26.6	67	43.5		
	Severe	0	0.0	3	1.9	3	1.9	6	3.9		
	Total	7	4.5	96	62.3	51	33.1	154	100		
Severity of Insomnia (ISI)	Normal	6	3.9	0	0.0	5	3.2	11	7.1	0.385	0.000
	Subumbral	0	0.0	49	31.8	1	0.6	50	32.5		
	Moderate	1	0.6	41	26.6	41	26.6	83	53.9		
	Severe	0	0.0	6	3.9	4	2.6	10	6.5		
	Total	7	4.5	96	62.3	51	33.1	154	100		
Impact Event Discomfort (IES-R)	Subclinical	7	4.5	44	28.6	1	0.6	52	33.8	0.564	0.000
	Mild	0	0.0	41	26.6	26	16.9	67	43.5		
	Moderate	0	0.0	11	7.1	24	15.6	35	22.7		
	Total	7	4.5	62.3	96	51	33.1	154	100		

N° = Number of workers surveyed, % = percentage representation of N°.
*p < 0.05 (significant)

The results obtained in this study indicate that the majority (56.50%) of the healthcare workers of the District Hospital of Jerusalén - La Esperanza during COVID-19 show a medium level of job motivation behavior, a medium level of depression, and (41.6%) of workers in the dimension of generalized anxiety distortion showed a mild level.

Job motivation strategies ceased to have positive effects; on the contrary, they became the fuel for the expansion and consolidation of negative behaviors and attitudes reflecting unmet needs among healthcare workers [16].

According to WHO [1], healthcare workers had fewer mental health problems before the COVID-19 pandemic. During the pandemic, health care workers were very vulnerable because they were not clear about the real impact of COVID-19 [5]. The mental health status of the healthcare workers was improved by understanding their experiences, despite the limitations of operational capacity, which aimed to break down the exacerbated anxiety disorder, the severity of insomnia, and the impact of the events reviewed. This last component tried to generate psycho-labor barriers to reduce the replication of the negative experiences.

4 Conclusion

The results obtained in Table 1 of this study show a significant relationship ($p < 0.05$) and a moderate positive correlation with Rho = 0.672 between job motivation and mental health of care workers of the District Hospital of Jerusalén - La Esperanza. It can be concluded that most of the workers (56.5%) reflect a medium level between job motivation and depression during the COVID-19 pandemic.

Similarly, we conclude that the results obtained in Table 2 between the relationship between job motivation and the dimensions of PHQ-9 (General anxiety disorder, insomnia severity, and impact of the events reviewed). It is affirmed that there is a significant relationship ($p < 0.05$) and a moderate positive correlation with Rho = 0.557, Rho = 0.385 and Rho = 0.564 respectively. Based on the results obtained in Table 2, we conclude that most care workers (41.60%) during the COVID-19 pandemic reflected a medium level of job motivation and a mild level of generalized anxiety distortion.

References

1. World Health Organization: Coronavirus disease (COVID-19): weekly epidemiological update, 24 August 2020 (2020). https://apps.who.int/iris/handle/10665/334168?locale-attribute=de, Accessed 17 July 2021
2. Wang, Y., Zhao, X., Feng, Q., Liu, L., Yao, Y., Shi, J.: Psychological assistance during the coronavirus disease 2019 outbreak in China. J. Health Psychol. **25**(6), 733–737 (2020)
3. Kang, L., Li, Y., Hu, S., Chen, M., Yang, C., Yang, B.X., Liu, Z.: The mental health of medical workers in Wuhan, China dealing with the 2019 novel coronavirus. Lancet Psychiatry **7**(3), e14 (2020)
4. Lange, M., Joo, S., Couette, P.A., Le Bas, F., Humbert, X.: Impact on mental health of the COVID-19 outbreak among general practitioners during the sanitary lockdown period. Irish J. Med. Sci. **191**(1), 1–4 (2021)
5. Leira-Sanmartín, M., Madoz-Gúrpide, A., Ochoa-Mangado, E., Ibáñez, Á.: Psychological impact of COVID-19 pandemic and related variables: a cross-sectional study in a sample of workers in a spanish tertiary hospital. Int. J. Environ. Res. Pub. Health **18**(7), 3608 (2021)
6. Organización Panamericana de la Salud/ Organización Mundial de la Salud: Consideraciones psicosociales y de salud mental durante el brote de COVID-19 (2020). https://www.paho.org/es/documentos/consideraciones-psicosociales-salud-mental-durante-brote-covid-19, Accessed 15 Apr 2021
7. Gobierno de Perú: Coranavirus (COVID-19) en Perú (2021). https://www.gob.pe/coronavirus, Accessed 10 July 2021
8. de Pablo, G.S., et al.: Impact of coronavirus syndromes on physical and mental health of health care workers: systematic review and meta-analysis. J. Affect. Disord. **275**, 48–57 (2020)
9. Jiménez Calderón, A.C.: Satisfacción laboral y salud mental en los profesionales de salud de primera y segunda línea durante el inicio de la pandemia por SARS-COV-2 en el Hospital III Daniel Alcides Carrión–Essalud, Tacna–2020 (2020). https://repositorio.upt.edu.pe/handle/20.500.12969/1464, Accessed 10 July 2021
10. Ministerio de salud: Resolucion Ministerial RM_448-2020-MINSA (2020). https://cdn.www.gob.pe/uploads/document/file/903763/RM_448-2020-MINSA.pdf, Accessed 10 July 2021
11. Base de datos Work motivation and mental health status of the care staff of the District Hospital en Excel. https://docs.google.com/document/d/1ffOJbAvlz0VA67E8xqsqV3A6R0K7NEs7/edit#heading=h.gjdgxs, Accessed 10 July 2021

12. Kroenke, K., Spitzer, R.L., Williams, J.B.: The PHQ-9: validity of a brief depression severity measure. J. Gen. Intern. Med. **16**(9), 606–613 (2001)
13. Wagner, W.E., III.: Using IBM® SPSS® statistics for research methods and social science statistics. Sage (2019)
14. Reguant Alvarez, M., Vilà Baños, R., Torrado Fonseca, M.: La relación entre dos variables según la escala de medición con SPSS. REIRE Revista d'Innovació i Recerca en Educación **11**(2), 45–60 (2018)
15. Stuart, A.: Calculation of spearman's rho for ordered two-way classifications. Am. Stat. **17**(4), 23–24 (1963)
16. Raurell-Torredà, M., Martínez-Estalella, G., Frade-Mera, M.J., Rodríguez-Rey, L.C., de San Pío, E.R.: Reflexiones derivadas de la pandemia COVID-19. Enfermeria intensiva **31**(2), 90–93 (2020)

Quality and User Satisfaction of the Telemedicine Service in Cardiology During COVID-19

Miryam Lora-Loza[1](✉), Marco Quispe-Barra[2], Castro Uceda Maria[1], Juan Rodríguez-Vega[1], David Salinas-Mendoza[2], and Jenner Fabián-Ramírez[1]

[1] Universidad César Vallejo, Trujillo, Peru
mlora@ucv.edu.pe
[2] Universidad Nacional del Altiplano de Puno, Puno, Peru
{marcoquispe,dsalinas}@unap.edu.pe

Abstract. The objective of this study was to determine the relationship between the quality of telemedicine service in cardiology and user satisfaction during COVID-19 at the Victor Lazarte Echegaray Hospital - Trujillo, Peru, in 2021. The research methodology used is the hypothetical method deductive with a non-experimental quantitative approach and descriptive correlational analysis with cross-sectional design. For this study, a total of $n = 96$ patients (100%) were surveyed and received care in the telemedicine services in cardiology. Two types of survey instruments were used, the first SERVQHOS is to evaluate the quality of the telemedicine service, and the Second SERVQUAL is for user satisfaction and its dimensions such as reliability, responsiveness, security, empathy, and tangibility. We used the Spearman correlation coefficient to determine the association between telemedicine service quality versus user satisfaction and its dimensions. The results show a significant relationship of $p < 0.05$ between the two variables and a moderate positive correlation of $Rho = 0.525$. In addition, most of the patients surveyed (41.7%) during the COVID-19 pandemic indicated that the telemedicine service is regular and safety is acceptable.

Keywords: Telemedicine · Cardiology · SERVQUAL · SERVQHOS · COVID-19

1 Introduction

During the COVID-19 pandemic, people underwent significant changes in their lives [1]. In several countries, quarantine has been a mandatory measure to reduce person-to-person disease transmission. However, the COVID-19 pandemic has abruptly disrupted ambulatory medical care worldwide, especially in cardiology, causing anxiety in patients of all ages. Therefore, it created challenges in the medical personnel of health institutions to care for patients, maintaining a safe distance and on a larger scale. A new medical assistance service was designed in a fast, efficient, and safe way, obtained by a telephone,

Y. Iano et al. (Eds.): BTSym 2021, SIST 207, pp. 577–583, 2023.
https://doi.org/10.1007/978-3-031-04435-9_61

video, or broadband system, having as a universal theme the connection between the doctor and the patient [2].

Telemedicine is not a new technology. Its implementation was limited to selected patients. In addition, telemedicine lacks a physical examination in the consultation. However, despite the impossibility of physical diagnosis, telemedicine has been strengthening the health system to improve the quality of care for patients over time, achieving proper diagnoses for patients. It is worth mentioning that telemedicine will never replace traditional face-to-face visits. This induced change will continue in the future, and telemedicine consultations acquired a new prominence [3].

Research on telemedicine during the COVID-19 pandemic has provided information based on experiences [4]. These investigations [3, 4] developed a consensus document with different cardiology groups to guide cardiologists in facing the challenges caused by the new outpatient care approach to heart disease. Also, one of the aspects is to guarantee maximum efficiency of patient care. Another aspect they considered was examining and diagnosing the main most common features in patients with heart problems. Finally, they mentioned the future of consultations after the COVID-19 pandemic.

In EsSalud level III Hospitals, especially in the Víctor Lazarte Echegaray Hospital in Trujillo, considered as a reference hospital in the La Libertad Region, telemedicine care centers were implemented in cardiology, accepted by users by the health situation of COVID-19, under this perspective. The research question was posed as follows: Is there a relationship between the quality of telemedicine service in cardiology and the satisfaction of the user treated at the Víctor Lazarte Echegaray Hospital during the COVID-19 pandemic? Therefore, the study's objective was to determine the relationship between the quality of telemedicine service in cardiology and the satisfaction of the user treated at the Víctor Lazarte Echegaray Hospital during COVID-19.

2 Methodology

2.1 Design and Study Population

A descriptive correlational study was carried out with a cross-sectional design, and the hypothetical deductive method was used with a quantitative approach. In this study, a population of 624 patients treated in the cardiology services by telemedicine was considered during April 2021 at the Víctor Lazarte Echegaray Hospital of EsSalud - Trujillo, Peru [5].

In selecting the sample, a simple random non-probabilistic sampling was carried out for known populations, considering only 96 patients attended. The sample included older patients without psychological problems.

2.2 Variables and Dimensions

The study variables, by their nature, were qualitative measures on the ordinal scale, as well as their dimensions. Likewise, two correlational variables were considered. The first variable was the quality of the telemedicine service in cardiology, and the second variable was users' satisfaction with their respective dimensions (Reliability, Responsiveness, Security, Empathy, and Tangibility).

2.3 Techniques and Data Instrument and Procedure

The technique used to measure the variables and their dimensions was through patients' telephone surveys. The instruments used were of two types. The first adapted instrument was the SERVQHOS questionnaire [6] that evaluates the quality of telemedicine services in cardiology. The questionnaire consists of 19 items designed for the application. Items are scored on a Likert scale of 5 alternatives that range from: Much worse than expected (1), Worse than expected (2), It is as I expected (3), Better than expected (4), and Much better than expected (5).

The second adapted instrument was the SERVQUAL questionnaire [7, 8] used to evaluate user satisfaction with the telemedicine service in cardiology, which consists of 22 items distributed in 5 dimensions (Reliability with 5 items, Quality of response with 4 items, Safety with 4 items, Empathy with 5 items and Tangibility with 4 items).

2.4 Statistical Analysis

The data [5] collected by telephone from the patients through questionnaires were transferred from a Microsoft Excel program to a registry and statistical processing software Statistical Package for the Social Sciences (SPSS) Version 22 in Spanish [9, 10]. Data were previously coded with mutually exclusive numbers. In addition, descriptive statistics (frequency tables) and inferential statistics were used. We can analyze and interpret [11] the results using Spearman's correlation coefficient, given the variables' ordinal scale and a way to determine if there is a significant relationship among the variables under study; the statistical significance must present $p < 0.05$.

Spearman's correlation coefficient, ρ (rho) [12], is a measure of the correlation between two random variables (both continuous and discrete). The formula for calculating the Spearman rho correlation coefficient is as follows:

$$\rho = 1 - \frac{6\sum d_i^2}{n(n^2 - 1)} \quad (1)$$

Where d_i is the difference between the ranges (x–y) and n is the number of elements in the ranking.

3 Results and Discussion

This study included 96 (100%) patients surveyed during the COVID-19 pandemic. Table 1 shows the relationship between the quality of telemedicine service and user satisfaction, where we observe that 39 patients (40.6%) indicate that the quality of the telemedicine service is regular and user satisfaction is acceptable. However, 27 patients (28.1%) show that the user is satisfied with the good quality of telemedicine service in cardiology. Only 2 patients (2.1%) were affected by the poor quality of telemedicine service in cardiology. In addition, Table 1 shows the level of correlation between the variables of user satisfaction and the quality of telemedicine service, where the Spearman coefficient Rho = 0.488, in such a way that it is stated that there is a significant correlation with p-value = 0 < 0.05.

Table 1. Relationship between quality of telemedicine service in cardiology and user-satisfaction

User satisfaction	Quality of telemedicine service in cardiology								Spearman's correlation	
	Good		Regular		Bad		Total			
	N°	%	N°	%	N°	%	N°	%	Rho	*p-value
Satisfied	27	28.1	11	11.5	0	0.0	38	39.6	0.488	0.000
Acceptable	15	15.6	39	40.6	0	0.0	54	56.3		
Unsatisfied	0	0.0	2	2.1	2	2.1	4	4.2		
Total	42	43.8	52	54.2	2	2.1	96	100		

N° = Number of workers surveyed, % = percentage representation of N°
*$p < 0.05$ (significant)

Table 2 shows the relationship between the quality of telemedicine service and dimensions of user satisfaction (Reliability, Response Capacity, Safety, Empathy, and Tangibility). We observed that 33 patients (34.4%) indicate that the quality of telemedicine service is regular and reliability is acceptable. 26 patients (27.1%) indicate that the quality of telemedicine service is good and reliability is satisfactory. We also observe that the variables are significantly correlated $p < 0.05$ with a Spearman coefficient Rho = 0.357. Table 2 shows the relationship between telemedicine service quality and response capacity, where 35 patients (36.5%) indicate that telemedicine service quality is regular and response capacity is acceptable, 22 patients (22.9%) indicate that the quality of service is good and the response capacity is acceptable. The variables under study are significantly correlated $p < 0.05$ with a Spearman coefficient Rho = 0.325.

We also have the relationship between the quality of telemedicine service and safety, where 40 patients (41.7%) indicate that the quality of telemedicine service is regular and safety is acceptable. After 29 patients (30.20%) show that the quality of service telemedicine is good and security is satisfactory. The variables under study are significantly correlated $p < 0.05$ with a Spearman coefficient Rho = 0.525, indicating a moderate positive correlation. Furthermore, Table 2 presents the relationship between telemedicine service quality versus empathy, where 36 patients (37.5%) indicate that telemedicine service quality is regular and empathy is acceptable. 26 patients (27.1%) suggest that the quality of service is good and empathy is satisfactory. The variables under study are significantly correlated $p < 0.05$ with a Spearman coefficient Rho = 0.392. Finally, in Table 2, we observe the relationship between telemedicine service quality and tangibility, where 33 patients (34.4%) indicate that telemedicine service quality is regular and tangibility is acceptable. 21 patients indicate that the quality of telemedicine service is good and the tangibility is satisfactory. The variables under study are significantly correlated $p < 0.05$ with a Spearman coefficient Rho = 0.312.

The results obtained in this study during the COVID-19 pandemic indicated that most patients accepted a regular quality of telemedicine service in cardiology. Therefore, these results show that Victor Lazarte Echegaray Hospital adapted to this mandatory technological service due to the pandemic.

Table 2. Relationship between quality of telemedicine service in cardiology and dimensions of user satisfaction (reliability, response capacity, safety, empathy, and tangibility)

		Quality of telemedicine service in cardiology								Spearman's correlation	
		Good		Regular		Bad		Total			
		N°	%	N°	%	N°	%	N°	%	Rho	*p-value
Reliability	Satisfied	26	27.1	17	17.7	0	0.0	43	44.8	0.357	0.000
	Acceptable	16	16.7	33	34.4	0	0.0	49	51.0		
	Unsatisfied	0	0.0	2	2.1	2	2.1	4	4.2		
	Total	42	43.8	52	54.2	2	2.1	96	100		
Response capacity	Satisfied	20	20.8	13	13.5	0	0.0	33	34.4		
	Acceptable	22	22.9	35	36.5	0	0.0	57	59.4	0.325	0.001
	Unsatisfied	0	0.0	4	4.2	2	2.1	6	6.3		
	Total	42	43.8	52	54.2	2	2.1	96	100		
Safety	Satisfied	29	30.2	11	11.5	0	0.0	40	41.7		
	Acceptable	13	13.5	40	41.7	0	0.0	53	55.2	0.525	0.001
	Unsatisfied	0	0.0	1	1.0	2	2.1	3	3.1		
	Total	42	43.8	52	54.2	2	2.1	96	100		
Empathy	Satisfied	26	27.1	14	14.6	0	0.0	40	41.7		
	Acceptable	15	15.6	36	37.5	0	0.0	51	53.1	0.392	0.000
	Unsatisfied	1	1.1	2	2.1	2	2.1	5	5.2		
	Total	42	43.8	52	54.2	2	2.1	96	100		
Tangibility	Satisfied	21	21.9	14	14.6	0	0.0	35	36.5		
	Acceptable	20	20.8	33	34.4	0	0.0	53	55.2	0.312	0.002
	Unsatisfied	1	1.1	5	5.2	2	2.1	8	8.3		
	Total	42	43.8	52	54.2	2	2.1	96	100		

N° = Number of workers surveyed, % = percentage representation of N°
*$p < 0.05$ (significant)

Telemedicine was implemented for better continuous monitoring of patients with chronic diseases. Consequently, a study has been found in Colombia [13] where the telemedicine service during the COVID-19 pandemic was satisfactory for patients with early Rheumatoid Arthritis, with 90% satisfaction among surveyed patients.

On the other hand, patient satisfaction with the quality of telemedicine service adopted a communication centered with the patient, as an empathic and affectionate attitude, showing responsiveness to emotions, providing safety for better satisfaction of the telemedicine patient [14]. However, compared with the author Torres et al. [15], our research has a regular and high satisfaction expressed by users, obtaining a great capacity to respond to teleconsultations. In addition, the author proposes to expand teleconsultation services after the COVID-19 pandemic.

4 Conclusion

The results found in Table 1 of this study affirm that there is a moderate positive correlation (Rho = 0.488) between the quality of telemedicine service in cardiology and user satisfaction at the Victor Lazarte Echegaray Hospital, with a significant relationship $p < 0.05$. In addition, it can be concluded that most of the patients surveyed (40.6%) indicate that the quality of telemedicine service was regular and user satisfaction was acceptable during the COVID-19 pandemic.

In the same way, we conclude that the results in Table 2 show the quality of telemedicine service and dimensions of user satisfaction of the Victor Hospital patients Lazarte Echegaray. There is a significant relationship $p < 0.05$. They affirm a moderate positive correlation with Rho = 0.525, where most patients surveyed (41.7%) indicate that the quality of telemedicine service is regular and safety is acceptable during the COVID-19 pandemic.

References

1. Ministerio de Salud: Recomendaciones para evitar actividades con riesgo de contagion (2020). https://www.minsa.gob.pe/newsletter/2020/edicion-47/nota3/index.html. Accessed 13 May 2020
2. Cohen, E., Cohen, M.I.: COVID-19 will forever change the landscape of telemedicine. Curr. Opin. Cardiol. **36**(1), 110–115 (2021)
3. Barrios, V., et al.: Telemedicine consultation for the clinical cardiologists in the era of COVID-19: present and future. Consensus document of the Spanish Society of Cardiology. Revista Española de Cardiología (English Edition) **73**(11), 910–918 (2020)
4. Alexander, G.C., Tajanlangit, M., Heyward, J., Mansour, O., Qato, D.M., Stafford, R.S.: Use and content of primary care office-based vs telemedicine care visits during the COVID-19 pandemic in the US. JAMA Netw. Open **3**(10), e2021476–e2021476 (2020)
5. Base de datos validez y confiabilidad de instrumentos de recolección de datos. https://docs.google.com/document/d/1bbJCo4ObYHI729aFB_JnPQDzAmrdTRJ_/edit?rtpof=true. Accessed 12 Apr 2021
6. Henao Nieto, D.E., Giraldo Villa, A., Yepes Delgado, C.E.: Instrumentos para evaluar la calidad percibida por los usuarios en los servicios de salud. Revista gerencia y políticas de salud **17**(34), 69–80 (2018)
7. Numpaque-Pacabaque, A., Rocha-Buelvas, A.: Modelos SERVQUAL y SERVQHOS para la evaluación de calidad de los servicios de salud. Revista de la Facultad de Medicina **64**(4), 715–720 (2016)
8. Hoque, S.I., Karim, A.M., Hossen, M.R., Arjumand, D.: Evaluation of patients' satisfaction in telemedicine service quality: a case study on maizbhanderi foundation, fatikchari, bangladesh. Am. Econ. Soc. Rev. **8**(1), 1–10 (2021)
9. Wagner, W.E., III.: Using IBM® SPSS® statistics for research methods and social science statistics. Sage (2019)
10. Reguant Alvarez, M., Vilà Baños, R., Torrado Fonseca, M.: La relación entre dos variables según la escala de medición con SPSS. Revista d'Innovació i Recerca en Educación **11**(2), 45–60 (2018)

11. Schober, P., Boer, C., Schwarte, L.A.: Correlation coefficients: appropriate use and interpretation. Anesth. Analg. **126**(5), 1763–1768 (2018)
12. Shadmani, M., Marofi, S., Roknian, M.: Trend analysis in reference evapotranspiration using Mann-Kendall and Spearman's Rho tests in arid regions of Iran. Water Resour. Manag. **26**(1), 211–224 (2012)
13. Cortés, J., Rodríguez, G.M., Velandia, A.M., Cajas, L.J., Barón, R.A., Cajas, D.: Satisfacción del servicio de telemedicina en pacientes con artritis reumatoide. Revista Colombiana de Reumatología (2021)
14. Mason, A.N.: The most important telemedicine patient satisfaction dimension: patient-centered care. Telemedicine and e-Health (2021)
15. Torres-Domínguez, Y., Acón-Royo, D., De Toro-Salas, A., Cruz-Villalón, F., Rivera-Sequeiros, A.: Evaluación de la satisfacción del paciente en la teleconsulta durante la pandemia por COVID-19. Rev. Esp. Salud. Pública. **95**(16), 11 (2021)

Clinical Management for Hospitalization of Patients with COVID-19 and Resolution Capacity Functionality in I-4 Facilities - Contumazá-Cajamarca Health Network, 2021

Juan Rodríguez-Vega[1], Miryam Lora-Loza[1](✉), Marco A. Quispe Barra[2], Segundo Manuel Hoyos-Gallardo[1], André Torres Quispe[1], and Jhon Rengifo-castillo[3]

[1] Universidad César Vallejo, Trujillo, Peru
mlora@ucv.edu.pe
[2] Universidad Nacional del Altiplano de Puno, Puno, Peru
[3] Universidad Nacional de Trujillo, Trujillo, La Libertad, Peru
jrengifoc@unitru.edu.pe

Abstract. With the arrival of the COVID-19 pandemic, health systems were forced to adapt their intra- and extra-hospital processes through mechanisms of articulation, integration, and coordination between the organic units of public and private institutions, configuring a single national health system. In this sense, the objective of this study was to determine whether there is a relationship between clinical management for the hospitalization of patients with COVID-19 and resolution capacity functionality in a Health Network. A hypothetical deductive method was used with a descriptive-correlational analysis with a cross-sectional design. The sample consisted of 150 health care workers (100%). The selection was conducted in 3 level I-4 facilities of the Contumazá-Cajamarca Health Network, Peru, 2021. The dependent variable was resolution capacity functionality, and the independent variable was clinical management for the hospitalization of patients with COVID-19. The results suggest an average level both in clinical management and in resolution capacity functionality and the transport dimension in 45.3%, extending a significant relationship ($p < 0.05$) with Rho = 0.755 and Rho = 0.251, respectively.

Keywords: Clinical management · Hospitalization · Resolution capacity · COVID-19

1 Introduction

With the advent of COVID-19, health services management has revealed the most negative side of leadership. Many health systems in countries in Asia, Africa, Latin America, and the Caribbean are perceived as primitive organizations for not having an effective health response to the spread of COVID-19 infection; health facilities initially faced this

Y. Iano et al. (Eds.): BTSym 2021, SIST 207, pp. 584–591, 2023.
https://doi.org/10.1007/978-3-031-04435-9_62

health emergency with deficiencies in their response capacity functionality [1]. Health systems inevitably collapsed [2].

In countries such as Russia, China, Germany, Japan, and Canada, the health systems were high quality in clinical management for hospitalization of patients with COVID-19 and presented a good functionality of their resolution capacity to care for patients with COVID-19, they managed to adapt their intra- and extra-hospital processes to the demands of the current health situation [3], Their policies were adjusted to the criteria of equity and justice, assuming the challenges of containing the exaltation of the primary impulse caused by COVID-19, in such a way that processes such as patient referral and counter-referral were quickly rethought, making use of the most advanced medical technologies and intra- and extra-hospital care [4].

In countries such as Brazil, Chile, and Colombia, support mechanisms were installed, both psychological and organizational, to improve the clinical management of hospitalization of COVID-19 patients and resolution capacity functionality of health facilities, generating particular processes of primary interaction among health personnel and making these become cross-cutting work strategies, in such a way that by combining training and resilience efforts they were able to take control of these processes more successfully than other countries [5].

New strategies were created to confront the COVID-19 pandemic, allowing the construction of provisional hospitals with a high-resolution capacity in record time (10 days). The cooperation and accessibility scenarios for COVID-19 patients were transformed. New efforts were made to increase the number of health personnel, and labor policies were more flexible and permeable to the needs of clinical management for the hospitalization of COVID-19 patients. Better budgets were granted for equipment and maintenance of health services. Referral and counter-referral processes became more functional, practical, efficient, safe, and accessible [6].

In Peru, the disoriented action of the health personnel only generated disruptive behaviors of the clinical management for the hospitalization of the COVID-19 patient and the functionality of the resolution capacity of the institutions providing health services [7]. The managers of the Contumazá-Cajamarca Health Network had to reprogram the characteristics of clinical management for hospitalization and resolution capacity functionality, giving way to new ways of working, primarily referrals and counter-referrals of patients, following the experiences of primary health facilities in countries such as Italy, Spain, France, Hungary, and Belgium [8].

The problems described above are associated with a bureaucratization of clinical management processes in the hospitalization of patients with COVID-19. Still, it is not known whether they are related to resolution capacity functionality. From this perspective, the research objective was to determine whether there is a significant relationship between clinical management for the hospitalization of the patient with COVID-19 and resolution capacity functionality in the I-4 facilities of the Contumazá-Cajamarca Health Network, 2021. Thus, this study provides new elements of analysis for improving clinical management strategies during the hospitalization of the COVID-19 patient with resolution capacity functionality in the patient referral and counter-referral services.

2 Methodology

2.1 Design and Study Population

A descriptive correlational analysis study was carried out with a hypothetical, deductive method, quantitative approach, and cross-sectional design. The results were described through simple and weighted frequencies obtained in the cross-sectional relationships, interpreting the correlations and their dimensions. A population of 150 health care workers of both sexes (professionals, technicians, and administrative staff) of the Contumaza-Cajamarca health network in Peru in 2021 was considered [9]. To select the sample, a non-probabilistic sampling was applied for strategic interest, assuming 100% of the study population, including workers with at least one year's seniority in the institution, with no psychological problems.

2.2 Variables and Dimensions

The study variables by their nature were qualitative measures on the ordinal scale and their dimensions. Likewise, two correlational variables were considered, the first independent variable was the clinical management for COVID-19 hospitalization, and the second dependent variable was the functionality of resolution capacity with its respective dimensions (Functionality of the infrastructure, Registration of information, Flow of human capital, Flow of information, Transport, Supply and Functionality of reference/counter-reference).

2.3 Techniques and Data Instrument and Procedure

This study used the person-to-person survey technique through a set of questions. The instruments used were of two types. The first instrument was to measure the scale of evaluation of clinical management for hospitalization COVID-19; this questionnaire consists of 10 items that the general level of clinical management for hospitalization of patients.

The second instrument used was the EMFU-CARE multidimensional scale questionnaire, which measures the scale of the functionality of resolution capacity; this questionnaire consists of 28 items distributed in 7 dimensions (Functionality of the infrastructure with 4 items, Registration of information with 4 items, Flow of human capital with 4 items, Flow of details with 4 items, Transportation with 4 items, Its-ministry with 4 items and Functionality of reference/counter-reference with 4 items).

2.4 Statistical Analysis

The data [9] were transferred from a Microsoft Excel program to statistical processing software Statistical Package for the Social Sciences (SPSS) Version 22 in Spanish [10, 11]. The data were previously coded with mutually exclusive numbers. In addition, descriptive statistics (frequency tables) and inferential statistics were used. Spearman's correlation coefficient was used to analyze and interpret the results. To determine if there is a significant relationship between the variables in the study, the statistical significance must present $p < 0.05$.

3 Results and Discussion

A total of 150 (100%) health care workers of both sexes participated in this study. Table 1 shows the relationship between clinical management for COVID-19 patient hospitalization and functionality of resolution capacity during the COVID-19 pandemic, where we observed that 68 workers (45.3%) reflected regular clinical management and functionality of resolution capacity. However, 38 workers (25.3%) reflected bad clinical management and resolution capacity functionality, and only 21 workers (14%) reflected good clinical management and resolution capacity functionality. In addition, Table 1 shows that the relationship between clinical management for COVID-19 hospitalization and functionality of resolution capacity is significant ($p < 0.05$). The Spearman's Rho coefficient (Rho = 0.755) indicates a moderate positive correlation.

Table 1. Relationship between clinical management for patient hospitalization by covid-19 and resolution capacity functionality

Resolution capacity functionality	Clinical management for hospitalization								Spearman's correlation	
	Good		Regular		Bad		Total			
	N°	%	N°	%	N°	%	N°	%	Rho	*p-value
Good	21	14.0	1	0.7	0	0.0	22	14.7	0.755	0.000
Regular	2	1.3	68	45.3	7	4.7	77	51.3		
Bad	3	2.0	10	6.7	38	25.3	51	34.0		
Total	26	4.5	79	52.7	45	30.0	150	100		

N° = Number of workers surveyed, % = percentage representation of N°
*$p < 0.05$ (significant)

Table 2 shows the relationship between clinical management for COVID-19 hospitalization and the dimensions of the functionality of resolution capacity (Infrastructure functionality, Information recording, Human capital flow, Information flow, Transport, Supply, and Referral/counter-referral functionality). We observed that 66 workers (44.0%) reflect a regular level in clinical management and infrastructure functionality, only 3 workers (2%) reflect a bad level in clinical management and infrastructure functionality, and 12 workers (8%) reflect a good level in clinical management and good infrastructure functionality. The Spearman's Rho coefficient was 0.306, and there is a significant relationship ($p < 0.05$).

Next, we also observed the relationship between clinical management and information recording, where 49 workers (32.7%) reflect a regular level of clinical management and information recording, 24 workers (16%) reflect a bad level of clinical management and information recording, and only 15 workers (10%) reflect good clinical management and good information recording. The Spearman's Rho coefficient was 0.372, and there is a significant relationship ($p < 0.05$).

We also observed the relationship between clinical management and human capital flow, where 46 workers (30.7%) reflect a regular level of clinical management and human capital flow, 33 workers (22%) reflect a bad level of clinical management and human capital flow, and only 6 workers (4%) reflect good clinical management and good human capital flow. The Spearman's Rho coefficient was 0.415, and there is a significant relationship ($p < 0.05$).

Next, we observed the relationship between clinical management and information flow, where 40 workers (26.7%) reflect a regular level of clinical management and information flow, 22 workers (14.7%) reflect a bad level of clinical management and information flow. Only 13 workers (8.7%) reflect good clinical management and information flow. The Spearman's Rho coefficient was 0.288, and there is a significant relationship ($p < 0.05$).

In addition, we observe in Table 2 the relationship between clinical management and transport, where 68 workers (45.3%) reflect a regular level in clinical management and transport, 5 workers (3.3%) reflect a bad level in clinical management and transport, only 8 workers (5.3%) reflect good clinical management and good transport. The Spearman's Rho coefficient was 0.251, and there is a significant relationship ($p < 0.05$).

Next, we observed the relationship between clinical management and supply, where 39 workers (26%) reflect a regular level of clinical management and supply, 21 workers (14%) reflect a bad level of clinical management and supply, and only 10 workers (6.7%) reflect good clinical management and good supply. The Spearman's Rho coefficient was 0.261, and there is a significant relationship ($p < 0.05$).

Finally, we observe in Table 2 the relationship between clinical management and referral/counter-referral functionality, where 47 workers (31.3%) reflect a regular level of clinical management and referral/counter-referral functionality, 31 workers (20.7%) reflect a bad level of clinical management and referral/counter-referral functionality, only 12 workers (8%) reflect good clinical management and good referral/counter-referral functionality. The Spearman's Rho coefficient was 0.460, and there is a significant relationship ($p < 0.05$).

The results obtained in this study reflect a regular level in the majority of workers (45.3%) of the Contumaza-Cajamarca Health Network received from the relationship of clinical management for the hospitalization of the patient for COVID-19 and the functionality of resolution capacity in the I-4 facilities and its dimensions during the COVID-19 pandemic. Therefore, to the extent that health personnel perceives an improvement in resolution capacity functionality, this would improve the valuations related to clinical management for the hospitalization of the patient with COVID-19, so it is expected to achieve a better treatment of the patient with COVID-19 that contributes to its good clinical management.

About the relationship found in this research between clinical management for the hospitalization of the patient with COVID-19, as well as resolution capacity functionality of health facilities in the world, it is the most effective for corporate action in health institutions, whether it is the transformational evolution of the normative adaptability [12]. In that sense, it is known that evolutionary and transformational changes began long before the pandemic. Thus, COVID-19 sowed uncertainties, fears, and social and existential medical panics, dragging in its wake health institutions and all the economies of the world, both because of its great capacity to mutate, vary and expand in different ways, epidemiologically speaking, and at the individual level [7].

Table 2. Relationship between clinical management for hospitalization and dimensions of the functionality of the resolute capacity

		Clinical management for hospitalization								Spearman's correlation	
		Good		Regular		Bad		Total			
		N°	%	N°	%	N°	%	N°	%	Rho	*p-value
Infrastructure functionality	Good	12	8	3	2.0	1	0.7	16	10.7	0.306	0.000
	Regular	14	9.3	66	44.0	41	27.3	121	80.7		
	Bad	0	0.0	10	6.7	3	2.0	13	8.7		
	Total	26	17.3	79	52.7	45	30.0	150	100		
Information registration	Good	15	10.0	1	0.7	4	2.7	20	13.3		
	Regular	8	5.3	49	32.7	17	11.3	74	49.3	0.372	0.000
	Bad	3	2.0	29	19.3	24	16.0	56	37.30		
	Total	26	17.3	79	52.7	45	30.0	150	100		
Human capital flow	Good	6	4.0	4	2.7	1	0.7	11	7.3		
	Regular	15	10.0	46	30.7	11	7.3	72	48.0	0.415	0.000
	Bad	5	3.3	29	19.3	33	22.0	67	44.7		
	Total	26	17.3	79	52.7	45	30.0	150	100		
Information flow	Good	13	8.7	8	5.3	5	3.3	26	17.3		
	Regular	8	5.3	40	26.7	18	12.0	66	44.0	0.288	0.000
	Bad	5	3.3	31	20.7	22	14.7	58	38.7		
	Total	26	17.3	79	52.7	45	30.0	150	100		
Transportation	Good	8	5.3	4	2.7	1	0.7	13	8.7		
	Regular	17	11.3	68	45.3	39	26.0	124	82.7	0.251	0.000
	Bad	1	0.7	7	4.7	5	3.3	13	8.7		
	Total	7	4.5	96	52.7	45	30.0	150	100		
Supply	Good	10	6.7	11	7.3	1	0.7	22	14.7		
	Regular	10	6.7	39	26.0	23	15.3	72	48.0	0.261	0.000
	Bad	6	4.0	29	19.3	21	14.0	56	37.3		
	Total	26	17.3	79	52.7	45	30.0	150	100		
Referral/counter-reference functionality	Good	12	8.0	6	4.0	4	2.7	22	14.7		
	Regular	12	8.0	47	31.3	10	6.7	69	46.0	0.460	0.000
	Bad	2	1.3	26	17.3	31	20.7	59	39.3		
	Total	26	17.3	79	52.7	45	30.0	150	100		

N° = Number of workers surveyed, % = percentage representation of N°
*$p < 0.05$ (significant)

This evaluation of the processes inherent to the functionality of the infrastructure only occurs among healthcare personnel when changes are demanded and they are aware of what is happening in their institution. [13]. In this way, healthcare personnel confidently take on the reality of the new challenges imposed on them. Therefore, it is assumed that health personnel is developing skills and abilities to overcome institutional problems, demonstrating that those who supposedly have the responsibility to do and do not do it are those who precariously occupy power, both may be by internal corruption, or, by weakness or inability to live up to the demands imposed by the new reality in institutional work [14].

4 Conclusion

The results in Table 1 show that there is a significant relationship ($p < 0.05$) and a moderate positive correlation (Rho = 0.755) between clinical management and functional capacity. In addition, it can be concluded that most workers (45.3%) reflect a regular level between clinical management and functional capacity during the COVID-19 pandemic.

Similarly, we conclude that the results obtained in Table 2 between clinical management and the dimensions of the functionality of resolving capacity (Functionality of the infrastructure, Registration of information, Flow of human capital, Flow of information, Transport, Supply and Functionality of reference/counter-reference), the majority of the workers (44%) reflect a regular level in clinical management and transport. There is a significant relationship ($p < 0.05$) with Rho = 0.251.

References

1. The Business Standard: Healthcare workers in Chattogram dissatisfied over low-quality PPE (2020). https://www.tbsnews.net/coronavirus-chronicle/covid-19-bangladesh/healthcare-workers-chattogram-dissatisfied-over-low. Accessed 18 June 2021
2. Wölfel, R., et al.: Virological assessment of hospitalized patients with COVID-2019. Nature **581**(7809), 465–469 (2020)
3. World Health Organization: Coronavirus disease (COVID-19): situation report 166 (2020). https://apps.who.int/iris/handle/10665/333145?locale-attribute=pt&%20Accessed%205%20March%205%202020. Accessed 18 Mar 2021
4. World Health Organization: Report of the WHO-China joint mission on coronavirus disease 2019 (COVID-19) (2020). https://www.who.int/docs/default-source/coronaviruse/who-china-joint-mission-on-covid-19-final-report.pdf. Accessed 11 July 2021
5. Rede de Informação e Conhecimento: Coronavirus COVID-19 global cases by the center for systems science and engineering (CSSE) (2020)
6. Wang, Y., Wang, Y., Chen, Y., Qin, Q.: Unique epidemiological and clinical features of the emerging 2019 novel coronavirus pneumonia (COVID-19) implicate special control measures. J. Med. Virol. **92**(6), 568–576 (2020)
7. Sakib, N., Akter, T., Zohra, F., Bhuiyan, A.I., Mamun, M.A., Griffiths, M.D.: Fear of COVID-19 and depression: a comparative study among the general population and healthcare professionals during COVID-19 pandemic crisis in Bangladesh. Int. J. Ment. Health Addict , 1–17 (2021). https://doi.org/10.1007/s11469-020-00477-9
8. Pedersen, S.F., Ho, Y.C.: SARS-CoV-2: a storm is raging. J. Clin. Investig. **130**(5), 2202–2205 (2020)
9. Hoyos, S.: Base de datos y validación de datos en Excel (2020). https://docs.google.com/document/d/1fuQAIUDMBjh_u46XJGRpoVXtTchR04yj/edit?rtpof=true. Accessed 12 Aug 2021
10. Wagner, W.E., III.: Using IBM® SPSS® statistics for research methods and social science statistics. Sage (2019)
11. Reguant Alvarez, M., Vilà Baños, R., Torrado Fonseca, M.: La relación entre dos variables según la escala de medición con SPSS. REIRE. Revista d'Innovació i Recerca en Educación **11**(2), 45–60 (2018)

12. Marques, J.A.L., Gois, F.N.B., Xavier-Neto, J., Fong, S.J.: Prediction for decision support during the COVID-19 pandemic. In: Predictive Models for Decision Support in the COVID-19 Crisis, pp. 1–13. Springer, Cham (2021)
13. King, L.A., Hothi, S.S.: Remodelling elective hospital services in the COVID-19 era–designing the new normal. Future Healthcare J. **7**(3), e60 (2020)
14. Feinberg, M., Willer, R.: Moral reframing: A technique for effective and persuasive communication across political divides. Soc. Pers. Psychol. Compass **13**(12), e12501 (2019)

Motivation and Job Performance at the Regional Health Management of La Libertad During COVID-19

Marco A. Quispe Barra[1](✉), Miryam Lora-Loza[2], Yris Cedrón-León[2], Juan Rodríguez-Vega[2], Fanny Sofia Quispe Barra[1], and Vanesa Castillo-Castro[2]

[1] Universidad Nacional del Altiplano de Puno, Puno, Peru
qbarra_14@hotmail.com
[2] Universidad César Vallejo, Trujillo, Peru
mlora@ucv.edu.pe

Abstract. The impact of the COVID-19 pandemic directly affected labor dynamics, especially the motivation of health personnel. In this context, the objective of this study was to determine the behavior of the relationship between motivation and job performance of the staff of the Regional Health Management - La Libertad during COVID-19. We use the descriptive correlational analysis with a quantitative method and cross-sectional design in the methodology. For this study, a sample of 106 workers (100%) was identified in which two types of survey instruments were used. The first is for job motivation, and the second is for job performance and its dimensions, such as cognitive, affective, and psychomotor skills. The Spearman correlation coefficient was used to determine the relationship between job motivation versus job performance and its dimensions. The results affirm a significant relationship ($p < 0.05$) between the motivation and performance variables with their dimensions. Furthermore, it was found that there is a moderate positive correlation with Rho = 0.633. Most of the workers (N° = 52, 49.1%) during the COVID-19 pandemic presented a low-level behavior in motivation and the psychomotor capability of the job performance of the Regional Health Management - La Libertad.

Keywords: Motivation · Job performance · Spearman coefficient · Health · COVID-19

1 Introduction

During the pandemic, more than 80% of health sector workers worldwide were infected by COVID-19, with a high rate of mortality and associated morbidity in 1 out of 20 cases [1]. One of the first labor problems during the COVID-19 pandemic was the immense psychological burden on workers in the health sector; this caused erratic decision-making from the highest levels of government to the smallest health service systems, putting in its minimum expression, motivation, and job performance [2]. However, individual and

Y. Iano et al. (Eds.): BTSym 2021, SIST 207, pp. 592–597, 2023.
https://doi.org/10.1007/978-3-031-04435-9_63

collective interests were focused on searching for new motivational elements to provide psychological and emotional support to reduce the effects of the pandemic, especially fear and panic due to losing their quality of work comfort. Several people assumed that the impact of the pandemic would pass sooner rather than later, but it did not. Thus the negative emotions caused by underestimating the likelihood of contracting the disease and ignoring public health warnings were out of control.

As negative emotions increased, health personnel became more dependent on negative information about COVID-19. Especially due to the number of infected people within health institutions increased negative motivation and led health personnel to opt for individual behaviors, many of them not according to the circumstances. In addition, it generated strong emotional reactions; for example, in July 2020, 40% of health personnel had already been replaced, retired from their job, or would be requesting their permanent discharge because they are considered vulnerable and high-risk people [3].

Likewise, the labor frameworks do not correspond to the changes that the health sector and its contexts for evaluating job performance [4]. The central government economically and socially motivated health personnel to counteract their selfish behaviors and/or anguish that they presented before the impact of COVID-19 [5]. For this purpose, the behavior of the La Libertad Regional Health Management staff was more orderly, but it was not enough to keep up with the changes proposed by the MINSA. Generating scenarios of appointment and dismissal of staff, among others, caused more panic than COVID-19 itself since the sense of independence that it had achieved had been lost [6].

In the regional health management - La Libertad, practically 100% of the health personnel admit that their job performance was affected by a significant decrease in cognitive, affective, and psychomotor ability due to COVID-19. The study's objective was to determine the relationship between job motivation and job performance of the Regional Health Management - La Libertad workers during COVID-19. Based on the relevant literature and the proposed study method, the following hypothesis was established:

H1: Is there a relationship between the job motivation of the staff and the job performance in the regional health management - La Libertad?

2 Methodology

2.1 Design and Study Population

This study used descriptive correlational analysis with a quantitative method and cross-sectional design. The results were described through the simple and weighted frequencies obtained in the crossed quotients, explaining the correlations between the variables and their dimensions. A population of 400 workers of both sexes from the Regional Health Management - La Libertad, Peru in 2020 was considered. A simple random sampling was applied to select the samples. In this study, the sample consisted of 106 workers of both sexes with more than one year of service in the Regional Health Management - La Libertad.

2.2 Variables and Dimensions

The study variables by their nature were qualitative measures on the ordinal scale and their dimensions. Likewise, two correlational variables were considered. The first variable was job motivation and the second variable was job performance with its respective characteristic dimensions (cognitive ability, affective ability, and psychomotor ability).

2.3 Techniques and Data Instrument and Procedure

This study used the technique of person-to-person surveys through a set of questions. The instruments used were of two types. The first instrument to measure the level of motivation was a questionnaire of a set of motivational questions adapted from a study by Steers and Braunstein (1976). This questionnaire consists of 24 items that respond to a Likert scale of 5 alternatives ranging from: Totally disagree (1), Disagree (2), Neither agree nor disagree (3), Agree (4), and Totally agree (5) and their measurements respond to a category of three levels, among them, we have High (90–120 pts.), Medium (57–89 pts.) And Low (24–56 pts.). The second instrument used was the multidimensional job performance scale, a questionnaire adapted from TSU Dubraska, 2005. This questionnaire consists of 24 items divided into three dimensions which are: Cognitive Ability (10 questions), Affective Ability (8 questions), and Motor Ability (6 reagents). Each dimension responds to the Likert scale of 5 alternatives ranging from: Strongly Agree (5), Agree (4), Indifferent (3), Disagree (2), Strongly Disagree (1).

2.4 Statistical Analysis

The data [7] were transferred from a Microsoft Excel program to a statistical processing software Statistical Package for the Social Sciences (SPSS) Version 22 in Spanish [8, 9]. Data were previously coded with mutually exclusive numbers. We also use descriptive statistics (frequency tables) and inferential statistics to analyze and interpret the results, using the Spearman correlation coefficient, given the ordinal scale of the variables. It was established that there is a significant relationship between the variables under study when the statistical significance presents a $p < 0.05$.

Spearman's correlation coefficient, ρ (rho) [10], measures the correlation between two random variables that allows the value of ρ to be calculated. The data were ordered and replaced by their respective order. The coefficient rho (ρ) is expressed by the following formula:

$$\rho = 1 - \frac{6\sum d_i^2}{n(n^2 - 1)} \tag{1}$$

Where d_i is the difference between the ranges (x–y) and n is the number of elements in the ranking.

3 Results and Discussion

In this study, a total of 106 (100%) workers of both sexes participated in the survey. Table 1 shows the relationship between job motivation and job performance; where we observe that during the COVID-19 pandemic, 52 workers (49.1%) reflect a low level in relation to job motivation and job performance. However, only one worker (0.9%) gives a high level of the relationship, and 24 workers (22.6%) represent a medium level. Therefore, Table 1 also shows that the relationship between job performance and worker motivation is significant ($p < 0.05$). Spearman's Rho coefficient (Rho = 0.633) indicates that there is a moderate positive correlation.

Table 1. Relationship between job motivation and job performance

Job performance	Job motivation								Spearman's correlation	
	Low		Medium		High		Total			
	N°	%	N°	%	N°	%	N°	%	Rho	*p-value
Low	52	49.1	6	5.7	6	5.7	64	60.4	0.633	0.000
Medium	1	0.9	24	22.6	5	4.7	30	28.3		
High	1	0.9	10	9.4	1	0.9	12	11.3		
Total	54	50.9	40	37.7	12	11.3	106	100		

N° = Number of workers surveyed, % = percentage representation of N°
*$p < 0.05$ (significant)

Table 2 shows the relationship between job motivation and the characteristic dimensions of job performance. We observe that 43 workers (40.6%) present a low level in relation to cognitive ability and job motivation, 24 workers (22.6%) present a medium level, and the high level is represented by one worker (0.9%). Spearman's Rho coefficient was 0.544, indicating that there is a moderate positive correlation and there is a significant relationship ($p < 0.05$). Table 2 also shows the relationship between affective ability and motivation where 44 workers (41.5%) present a low level, 21 workers (19.8%) have a medium level, and two workers (1.9%) represent a high level of the relationship. Spearman's Rho coefficient was 0.568, indicating that there is a moderate positive correlation and there is a significant relationship ($p < 0.05$). Finally, we observe the relationship between psychomotor ability versus job motivation, where 52 workers (49.5%) represent a low level, 16 workers (15.1%) represent a medium level, and one worker (0.9%) represents a high level. Spearman's Rho coefficient was 0.603, indicating a moderate positive correlation and a significant relationship ($p < 0.05$). Based on these results, it can be observed that most of the workers have a low level of motivation concerning the dimensions of job performance (cognitive ability, affective ability, and psychomotor ability).

The results obtained in this study reflect the reality of the workers of the regional health management - La Libertad during COVID-19, showing behavior of most of the workers with a low level of job motivation and consequently a low level of job performance. Therefore, these results explain the failures in the workers' personal aspirations, interests, attitudes, and philosophy, even though they bear in mind the spirit of institutional work in the terms known before COVID-19 [1].

Table 2. Relationship between job motivation and the characteristic dimensions of job performance (cognitive, affective, and psychomotor ability)

Job performance		Job motivation								Spearman's correlation	
		Low		Medium		High		Total			
		N°	%	N°	%	N°	%	N°	%	Rho	*p-value
Cognitive ability	Low	43	40.6	4	3.8	5	4.7	52	49.1	0.544	0.000
	Medium	10	9.4	23	21.3	6	5.7	39	36.8		
	High	1	0.9	13	12.3	1	0.9	15	14.2		
	Total	54	50.9	40	37.7	12	11.3	106	100		
Affective ability	Low	44	41.5	4	3.8	5	4.7	53	50		
	Medium	9	8.5	21	19.8	5	4.7	35	33	0.568	0.000
	High	1	0.9	15	14.2	2	1.9	18	17		
	Total	54	50.9	40	37.7	12	11.3	106	100		
Psychomotor ability	Low	52	49.1	11	10.4	5	4.7	68	64.2		
	Medium	1	0.9	16	15.1	6	5.7	23	21.7	0.603	0.000
	High	1	0.9	13	12.3	1	0.9	15	14.2		
	Total	54	50.9	40	37.7	12	11.3	106	100		

N° = Number of workers surveyed, % = percentage representation of N°
*$p < 0.05$ (significant)

The impact of workers during the pandemic was one of dread, fear, and uncertainty in the health sector, and many health leaders had to face the problem, sometimes losing their motivation and their corporate and institutional work style [11].

The significant relationship between job motivation and the psychomotor ability of job performance shows the behavior of inability to detach from past job performance also serves to understand the current reality of their development, which were affected during the pandemic. In this way, performance evaluation processes are no longer positive, according to the logic followed by current actions: search for efficiency, efficiency, and effectiveness, and all this is due to the lack of factual support for new corporate and institutional actions [12].

4 Conclusion

The results obtained (Table 1) in this study affirm that there is a moderate positive correlation (Rho = 0.633) between job motivation and job performance of the workers of the regional health management - La Libertad, with a significant relationship ($p < 0.05$) in this way we affirm the hypothesis raised in this study. In addition, it can be concluded that most workers (No. = 52, 49.1%) present a low level between job motivation and job performance during the COVID-19 pandemic.

Similarly, we conclude that the results obtained in Table 2 show the relationship between job motivation and the dimensions of job performance (cognitive ability, affective ability, and psychomotor ability) of the workers of the regional health management - La Libertad. Where there is a significant relationship ($p < 0.05$) and they affirm a moderate positive correlation with Rho = 0.544, Rho = 0.568, and Rho = 0.603 respectively. Based on the results obtained in Table 2, we conclude that most of the workers during the COVID-19 pandemic presented a low level of motivation with respect to psychomotor ability with 49.1% (No. = 52), followed by effective ability with 41.5% (N° = 44), and, finally, the cognitive ability with 40.6% (N° = 43).

References

1. Wise, T., Zbozinek, T.D., Michelini, G., Hagan, C.C., Mobbs, D.: Changes in risk perception and self-reported protective behaviour during the first week of the COVID-19 pandemic in the United States. Royal Soc. Open Sci. **7**(9), 200742 (2020)
2. Lunn, P., Belton, C., Lavin, C., McGowan, F., Timmons, S., Robertson, D.: Using behavioural science to help fight the coronavirus (No. 656). ESRI working paper (2020)
3. Organización Internacional del Trabajo: El COVID-19 y el mundo del trabajo: Repercusiones y respuestas (2020). https://www.ilo.org/wcmsp5/groups/public/---dgreports/---dcomm/documents/briefingnote/wcms_739158.pdf. Accessed 18 Mar 2020
4. Stiff, C.: La teoría de juegos de la compra por pánico y cómo reducirla. La conversación (2020). http://theconversation.com/the-game-theory-of-panic-buyingand-how-to-reduce-it-134107. Accessed 18 Mar 2020
5. Cepal, N.U.: América Latina y el Caribe ante la pandemia del COVID-19: efectos económicos y sociales (2020)
6. Basol, M., Roozenbeek, J., van der Linden, S.: Good news about bad news: gamified inoculation boosts confidence and cognitive immunity against fake news. J. Cogn. **3**(1), 2 (2020)
7. Cedrón León, Y.Y.: Motivación y desempeño laboral en la Gerencia Regional de Salud La Libertad, en tiempos COVID-19 (2020). https://repositorio.ucv.edu.pe/handle/20.500.12692/48403. Accessed 5 Sept 2021
8. Wagner, W.E., III.: Using IBM® SPSS® statistics for research methods and social science statistics. Sage (2019)
9. Reguant Alvarez, M., Vilà Baños, R., Torrado Fonseca, M.: La relación entre dos variables según la escala de medición con SPSS. REIRE. Revista d'Innovació i Recerca en Educación **11**(2), 45–60 (2018)
10. Shadmani, M., Marofi, S., Roknian, M.: Trend analysis in reference evapotranspiration using Mann-Kendall and Spearman's Rho tests in arid regions of Iran. Water Resour. Manag. **26**(1), 211–224 (2012)
11. Brady, W.J., Crockett, M.J., Van Bavel, J.J.: The MAD model of moral contagion: the role of motivation, attention, and design in the spread of moralized content online. Perspect. Psychol. Sci. **15**(4), 978–1010 (2020)
12. Van Kleef, G.A., Oveis, C., Homan, A.C., van der Löwe, I., Keltner, D.: Power gets you high: the powerful are more inspired by themselves than by others. Social Psychol. Personality Sci. **6**(4), 472–480 (2015)

The Method of Logical Development of Regulatory Dependability Terminology

Nikolai I. Plotnikov[1(✉)], Gabriel Gomes de Oliveira[2], and Yuzo Iano[2]

[1] Scientific Research Project Civil Aviation Institute «AviaManager», Novosibirsk, Russian Federation
am@aviam.org

[2] School of Electrical and Computer Engineering, State University of Campinas, Campinas, Brazil
oliveiragomesgabriel@ieee.org, yuzo@unicamp.br

Abstract. This paper contains a brief historical overview of the evolution of the dependability theory. The problem of the terminology and long years of search for the definitions of dependability consists in the deficiency of the academic development of the subject matter in philological, philosophical, and logical terms. The evolution of the concept of dependability reflects the unsolved terminological problem in the dependability theory of technical objects. The problem of terminology largely consists of the ambiguous use and confusion of ontological terms. Deduction of such terms based on pseudo physical logic and introduction of the category of object's assignment is the main result of this paper in terms of the introduction of an alternative noncontroversial structure and content of dependability-related terms. The paper aims to examine the feasibility of applying the modern dependability theory to subsequent theoretical development and practical application of the concept of dependability of organizations, social groups, and individuals. The suggested approach is recommended to be used for the revision of the existing standards.

Keywords: Terminology · Dependability · Reliability · Assignment · Ontological terms · Property · State · Event

1 Introduction

Standard descriptions of dependability were developed for engineering, industrial products, machines, and devices. The research of dependability is associated with the theory of life safety and risks. This paper examines the problem of terminology as part of dependability standardization. Russian standards prioritized the dependability concept, where the regulatory descriptions, definitions, such as "the ability to perform the required (specified) functions, (an object's) ability to function", "to function as and when required", "functional dependability", "parametric dependability", "requirements specified in the documentation" are simply generalized by the category of assignment. Such

Y. Iano et al. (Eds.): BTSym 2021, SIST 207, pp. 598–608, 2023.
https://doi.org/10.1007/978-3-031-04435-9_64

descriptions are none other than an indication of the property of an object's assignment. For instance, an object's ability to move in space is a property of the assignment, not dependability. Thus, all the terminological searches in terms of dependability standardization demonstrate an unjustified reduction of the concept of assignment to the concept of dependability. The introduction of the category of assignment solves the problems of terminology in the dependability theory. The assignment is the property of an object defined by the natural origin or designed application. Dependability is a set of states as the measure of concordance with the assignment of an object. The paper aims to examine the feasibility of applying the modern dependability theory to subsequent theoretical developments and practical application of the concept of dependability of organizations, social groups, and individuals [1, 2].

2 Overview of Terminology Problem

2.1 Normative Base

Since the introduction of the term of industrial product dependability and the emergence of the first respective documents in the 1950s, there has been an ongoing discussion of the problem of dependability terminology, specificity as regards the specification of the fundamental term of "dependability".

The regulatory framework includes Russian and interstate standards: GOST 13377-67; GOST 13.337-75; GOST 27 002-83; GOST 27.002-89; GOST 27.002-2015; GOST R 51901.3-2007 (IEC 60300-2: 2004); GOST ISO 9000-2011; GOST R 27.002-2009. The concept of dependability is examined much wider in standards related to the terminology of quality, risk, safety: GOST15467-79; GOST R 51901-2002; GOST R 51897-2002; GOST R 51898-2002; GOST 51901.14-2005 (IEC 61025: 1990); GOST R 51901.12-2007; GOST R 53480-2009; GOST R ISO/IEC 31010-2011.

2.2 Semantic Descriptions of the Concept of Dependability

Semantic descriptions of the concept of dependability in the documents can be reduced to the following: "… property of the system to perform the task, … specified functions, … specified indicators within the given ranges, … parameters characterizing the ability to perform the required (specified) functions…" The same is true with the semantic descriptions in the IEC dependability standards: "the ability (of an object) to function as and when required".

The concept of *good state* of a technical device is the initial description that defines the content and scope of the concept of dependability. As machines can operate in a partially faulty state, the definition of a good/faulty state becomes blurry. The introduction of the term "failure" defines the impossibility to use an engineering product. That caused the definition of dependability through *reliability* and additionally: durability, maintainability, storability. Next, the concept of the *up state* was introduced. GOST R 27.002-2009 defines dependability through *availability*, while GOST 27.002-2015 [3] defines it through *ability*. Thus, the evolution of terminology from 1950 to this day shows roughly the following sequence of replacement of the concept of dependability with other generalized "explicative" concepts:

$$DEPENDABILITY \cong <(goodstate) \to (\text{reliability}) \to (\text{operability}) \to (\text{availability}) \to (\text{ability})> \quad (1)$$

A detailed account of the content and historical analysis of the regulatory documents on dependability terminology was presented in the papers [4, 5]. The introduction of standard [3] does not resolve the problem of standardization of the dependability terminology. Let us examine the basic definitions of this standard. Section 3 "Terms and definitions", Item 3.1 "Basic definitions": "3.1.5. dependability": The property of an object to maintain in time the ability to perform the required functions in specified modes and conditions of operation, maintenance, storage, and transportation. Note 2. Dependability is a composite property that, depending on the assignment and operating condition of the object, may include reliability, maintainability, restorability, durability, storability, availability or certain combinations of such properties". Further in the standard, the above properties are defined. In Item 3.2 "States", the following definitions are introduced: "3.2.1 good state: ..., 3.2.2 faulty state: ...". In other words, the above properties are also considered as states.

In Item 3.4 "Failures, defects, damage", the terms "failure" and "damage" are defined as events, while "defect" is defined as the object's non-compliance with the requirements specified in the documentation. The "causes of failure are provoked", while the "consequences of a failure are conditioned" in a simultaneous and joint total: phenomena, processes, events, and states. In Item 3.5.5, restoration is considered both as a process and an event. The standard is full of such confusions.

The concept of dependability, without any doubt, is abstract, i.e., is a category. That is the main reason for the terminology problem and unchanging need to define dependability through other concepts. To make matters worse, compound words are used, such as reliability, maintainability, durability, restorability, etc., which further complicates the solution of the problem. In logic, the definition of a concept that leads to a term is based on the content (signs), rather than the scope of such concept. If signs are successfully identified, the definition is considered to be strong, quantitative, suitable for object properties calculation. Otherwise, the definition comes down to a simple replacement with another concept or several concepts ("dependability is reliability", etc.). In other words, the definition of the term is descriptive, weak, and ill-suited for quantitative estimation. It is supposed, that identifying the signs of the content of the concept of dependability "directly" does not appear to be possible. The problem of the terminology and long years of search for the definitions of dependability consists in the deficiency of the academic development of the subject matter in philological, philosophical, and logical terms. Certainly, such research is to be conducted by experts in the appropriate fields of knowledge. Let us make our own contribution as regards the subject matter of this paper.

3 Research of the Dependability Terminology

3.1 The Philological Aspect

The philological examination of the concept of dependability is possible in the lexical and grammatical aspects. The word "dependability" is often searched for in technical literature with no tangible results. Normally, the definition is descriptive and repeats the content specified in technical documents. Encyclopedias, specialized and explanatory dictionaries enable defining the scope of a concept, but identifying the signs of the content, which is required for term derivation, proves to be unsuccessful.

In the lexical-grammatical aspect, the morphological structure of the word "dependability" consists of the prefix de-, root -pend- and suffix -ability. The lexical-grammatical class of the word dependability presents a suffix abstract noun that designates a quality (property) or state of an object, motivated by the adjective dependable with the meaning of an abstract sign. "Nouns with a suffix represented with the morpheme -ability with the meaning of "carrier of a sign" designate an abstract state [6, p. 164]; ... with the meaning of abstract sign, designate a state with an abstract meaning of a sign, property" [6, p. 177]. It has been theoretically established that nouns are motivated (governed) by adjectives in pairs: "dependable – dependability". Out of this context and the general linguistic definitions, it is impossible to identify, what exactly the noun "dependability" and the adjective "dependable" designate: a property, a state, or a sign of an object. Thus, attempted philological research does not bring us closer to the solution.

3.2 Analysis of Ontological Terms in Logic

Let us examine how terms, definitions, and reducibility of terms are established in logic. Abstract concepts (categories) are also called ontological terms (time, space, beginning, end, cause) that are later specified (defined) with logical terms. The theory of concepts, a sub-discipline of classical logic, has it that the scope and content of categories are the most difficult to define. Furthermore, the most abstract categories such as essence, entity, thing, quality, and others are not generalizable at all in terms of scope and their sign cannot be identified in terms of content.

First, the term (object) is specified. Then, the predicate (sign) is specified. The term and the predicate are correlated. "The predicate "red" (and the sign corresponding to the "redness") is one-place. The predicate "bigger" (and the size sign "bigger") is two-place. A predicate with the correlation ≥ 2 is n-place. One-place signs are called properties. N-place signs are called relationships [7, p. 61], i.e., statements with multiplace predicates. Reducing terms to simple ones is the most important problem of logical analysis of scientific knowledge. "Defining a term means establishing its meaning using other terms, whose meaning is already known" [7, p. 228]. Terms are subdivided into initial and derived ones.

Definition 1. The term t^1 is initial in relation to the term t^2, while t^2 is derivative in relation to t^1 if and only if t^1 is used in the creation (specification of the meaning) of t^2" [7, p. 62].

The creation (introduction) and definition of ontological terms is the most important part of the problem of this paper as regards the terms "property", "state", "event", "situation", "process".

The Categories of "Property" and "State". There are no definitions of the concepts of "property" and "state" in the dependability standards. Meanwhile, the difference between those concepts is not trivial. According to Aristotle, "*property* (hexis) is the manifestation of a certain activity by that which possesses and what it possesses; such an arrangement towards another, for example, health, is a certain property"..., "a transient property or *state* (pathos) is a property subject to possible changes; various manifestations of such properties and applications" [8, p. 244]. A property is a quality, a state is a quantity. Examples: (1) A body has the property of weight and can be in the states of movement and rest. (2) The property of water manifests itself in the states of liquid, solid, gaseous (vaporous), crystalline. It can be said that water has a "composite property" and manifests itself in the "sub-properties" of liquid, ice, vapor, snow. However, it is preferable to explain the states. Thus, in the existing standards, the definition of dependability as a *property* of an object is up to discussion and selection. In our opinion, the concept of dependability is a *state* of an object. That is substantiated below.

The Categories of "Event" and "Process". In the standard [9], the following definitions are introduced:

- 113-01-04 event: something that takes place, happens, occurs in a random point in space-time;
- 113-01-06 process: time sequence of correlated events.

These definitions do not satisfy the objectives of this paper, as they are descriptive, weak. In scientific literature, the definitions of the terms under consideration are often missing, are not reduceable to each other, or equated to each other. It would be sufficient to note that in the book "Situational management" [10] the concept of "situation" is not defined. According to Zinoviev, "If X is a statement, then $\downarrow$X is a term of event (or state). Events exist or do not exist in a certain given or any situation" [7, p. 166]. Clearly, the terms "event" and "state" are equalized to each other. It is explained that a situation is defined by specifying the following: a) spatial area, b) time, c) event or set of events, d) combination of (a, b, c) is expressed; or follows from the context. However, out of the above logical construct does not directly follow the definitions of the terms "event", "situation" and others. In our opinion, the definitions can be deduced only using the pseudo physical (pseudo logical) way or method.

Deduction of the Term through Pseudo Physical Logic. The pseudo physical logic (PL), a term proposed by Pospelov [10], of the correlation between time, space, causality, and their combinations allows deducing the definitions required for a subsequent explanation. Let $|O_i|$ be the term, the observed object. Let us represent the following definitions and examples of reality.

Definition 2. An object's transitions in mapping spaces are called *ascending* if directed towards higher dimensionality:

$$O_i : \langle s_i \rightarrow s_{i+1}, \ldots, \rightarrow s_{i+k} \rangle, i \in \overline{1, n} \quad (2)$$

where n is the number of an object's states.

Example: the moment of aircraft take off the runway and starts moving in a three-dimensions space.

Definition 3. An object's transitions in mapping spaces are called *descending* if directed towards lower dimensionality:

$$O_i : \langle s_i \rightarrow s_{i-1}, \ldots, \rightarrow s_{i-k} \rangle \quad (3)$$

where $k \in \overline{1, n-1}$ is the number of transitions.

Example: the moment of aircraft landing and transition to movement in a two-dimensional space.

Definition 4. An object's transitions in mapping spaces are called *symmetrical* if directed either way of the same (higher or lower) dimensionality:

$$O_i : \langle s_i \leftrightarrow s_{i\pm1}, \ldots, \leftrightarrow s_{i\pm k} \rangle \quad (4)$$

Example: a) moment of the takeoff run at the decision-making speed; b) the moment at the landing decision point height when the decision is made to land or go around.

Definition 5. Object mapping in ascending transitions is called the *involution* of object description data.

Definition 6. Object mapping in descending transitions is called the *convolution* of object description data.

Defining the terms "event", "situation" is only possible at the convergence of the temporal and spatial logic. A complete derivation of ontological terms is only possible at the convergence of the temporal, spatial, and causal logic, which requires additional research.

Definition 7. A representation of an object in transitions from space to space s_i in the moment in time t_i is called an *event*:

$$e_i : (t_i \langle s_i \leftrightarrow s_{i\pm1}, \ldots, \leftrightarrow s_{i\pm k} \rangle) \quad (5)$$

where e_i is the object mapping operator for the transition from a space to another space.

An event, set E, $e_i \in E$, is the association of the PL of relationships between the time and the space $e_i(t_i, s_i)$.

Definition 8. The set of associations of the PL of relationships between the time and the space is called *situation*.

In the above definitions and examples: a situation is a set (beginning of movement and stopping of an object, takeoffs and landings, and decisions making) of events in ascending, descending, and symmetrical transitions.

In standard [9], there is a concept of "time interval: part of the axis of time-limited by two moments", item 113-01-10. But there is no definition of the concept of "state". Let us introduce the following definition of state.

Definition 9. A *state* is a parameter, set of parameters of an object's properties within the observed time intervals.

In this definition, the concept of state is defined as the parameter of commensuration of indicators (combined and simultaneous). For instance, the speed of a vehicle is commensurated with two indicators: distance and time. Thus, in this part of the work, based on pseudo physical logic, the definitions of the concepts "situation", "state", "event", which are the most important for the purpose and objectives of this work, were derived.

4 Solving the Problem of the Dependability Terminology

4.1 The Structural Approach

It is proposed in this paper a structural approach to terminological research. The essence of such an approach consists of the following. If identifying the signs of the concept *content* is complicated, structuring the concept *scope* may be considered. The structuring is done using universal observation bases: time, space, groups, and their combinations: time-space, time-group, space-group [11]. It has been empirically established that the dependability of an object of any nature changes in time in steps, the so-called U-shaped profile: entry (below the norm), normalization, aging (below the norm). Subsequently, the dependability of any object can be reliably observed (measured, evaluated) according to the U profile. For that purpose, special terminology is required.

It is suggested a strict hierarchic structure of terms. An object has a name. The assignment (essence) of an object specifies its property. A property is observed (measured, evaluated) in states or parameters. States are formed by a set of indicators. Indicators are specified based on attributes, values. For the purpose of calculating an object's properties, an information unit (IU) is defined with n-placed elements:

$$\{1,\ \text{assignment} \subseteq 2,\ \text{property} \subseteq 3,\ \text{state(parameter)} \subseteq 4,\ \text{indicator} \subseteq 5,\ \text{attribute}\} \tag{6}$$

Definition 10. The IU is a single-point if the subset corresponding to the value is a single-point and cannot be divided into parts.

Definition 11. The precision is associated with the value.

Each element of the IU is observed (identified, defined) based on signs. An IU has a hierarchy, does not allow for ambiguous interpretation and arbitrary application of terms. The presented system of terms is a universal model, can be used for calculating the states of objects of any nature. The structure of terms is shown in a diagram in Fig. 1.

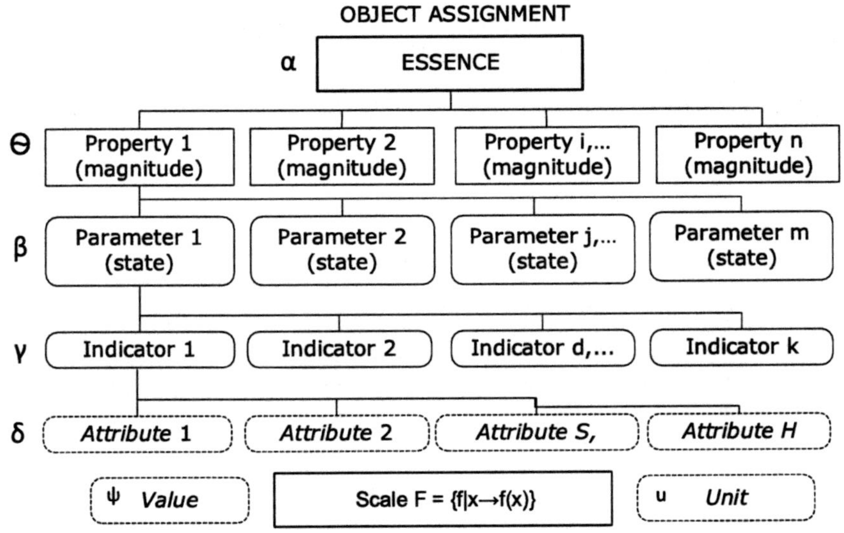

Fig. 1. Complex of computation terms

The structural approach also implies the search for and establishment of a more abstract (in relation to the researched one) umbrella term. In our opinion, the category of "assignment" is such a concept. Below is the substantiation.

4.2 The Notion of Assignment as Object Property

The concept of assignment is large in scope and is more abstract than the concept of dependability. Let us note that the quality standards (ISO) were developed under the assumption of the assignment being the compliance of an object's characteristic with the requirements [12]: "Quality: the degree of compliance of the sum and the intrinsic characteristics with the requirements". Russian standard [3] prioritized the dependability concept, where the regulatory descriptions, definitions, such as "the ability to perform the required (specified) functions, (an object's) ability to function", "to function as and when required", "functional dependability", "parametric dependability", "requirements specified in the documentation" are simply generalized by the category of assignment. Such descriptions are none other than an indication of the property of an object's assignment. For instance, an object's ability to move in space is a property of assignment, not dependability. If an automobile is a state of traffic jam and is not intended to also be a helicopter or an airplane, like in the movie Fantomas (1965), this is inherent in the project of its purpose - movement in two-dimensional space. Therefore, an absolutely reliable car will stand in a traffic jam "securely" and motionless.

Thus, all the terminological searches in terms of dependability standardization demonstrate an unjustified reduction of the concept of assignment of an object to the concept of dependability. The introduction of the category of assignment solves the problems of terminology in the dependability theory. It is proposed in this paper the following definition.

Definition 12. An assignment is the property of an object defined by the natural origin or designed application.

This definition is introduced for a simple object. The assignment of a complex object may be considered as a combination of property elements. If we adopt the proposed point of view that the concept of assignment is a property, then all the derived concepts are states. In the context of this paper, such states include dependability, safety (security), risk, efficiency, etc.

4.3 Derivation of Alternative Terminology of Dependability Theory

Given the above, let us present an example (model) of the development of an alternative dependability theory terminology. Only the basic definitions are suggested. This paper does not aim to complete the development of the standard.

Definition 13. Dependability is a set of states as the measure of concordance with the assignment of an object.

The alternative terminology definitions of dependability are given below (Table 1).

Table 1. Alternative dependability terminology

GOST 27.002-2015	Alternative definitions
3.2 states	States of dependability
3.2.1 good state: the state of an object, in which it complies with all the requirements specified in the respective documentation	Good state: a state that complies with the assignment of the object to allowable damage
3.2.2 faulty state: the state of an object, in which it does not comply with at least one of the requirements specified in the respective documentation	Faulty state: a state that does not comply with the assignment of the object due to defects
3.2.3 up state: the state of an object, in which it is able to perform the required function	Up state: a state that complies with the assignment of the object
3.2.4 down state: a state of an object, in which it is unable to perform at least one of the required functions due to reasons depending on it or due to preventive maintenance	Down state: a state that does not comply with the assignment of the object
3.4 failures, defects, damage	Events of disrupted dependability
3.4.1 failure: an event consisting of the disruption of an object's up state	Failure: an event of disruption of an object's up state
3.4.2 defect: each individual non-compliance on an object with the requirements specified in the documentation	Defect: an event of disruption of an object's good state

(*continued*)

Table 1. (*continued*)

GOST 27.002-2015	Alternative definitions
3.4.3 damage: an event consisting of the disruption of an object's good state under the condition of retained up state	Damage: an event of an object's disrupted good state under the condition of retained up state
3.5 maintenance, restoration, and repairs	Dependability restoration processes
3.5.2 maintenance: a set of organizational actions and technical operations aimed at maintaining the operability (good state) of an object and reducing the probability of its failures in the course of the assignment, storage, and transportation	Maintenance: the process aimed at maintaining the up and good state of an object
3.5.5 recovery: a process and event consisting of the transition of an object from the down state into the up state	Recovery: a process aimed at causing the transition of an object from the down state into the up state
3.5.9 repairs: a set of technical operations and organizational actions aimed at recovering the good or up state of an object and restoration of the operating life of the object or its components	Repairs: a process aimed at causing the transition of an object from the faulty state into the up state

5 Conclusion

In the work, the definitions of the concepts "situation", "state", "event", "assignment", "dependability" were derived. The example of terminology development does not provide for analogous use of terms like "good state". It is assumed that terms related to events and processes are to be carefully examined in terms of the philological and logical aspects in order to establish clear distinctions. Thus, among other things, instead of the term "defect" the term "breakdown" should probably be used. The evolution of the concept of dependability reflects the unsolved terminological problem in the dependability theory of technical objects. The problem of terminology largely consists of the ambiguous use and confusion of ontological terms. Deduction of such terms based on pseudo physical logic and introduction of the category of an object's assignment is the main result of this paper in example of the development of an alternative terminology in the dependability theory.

References

1. Plotnikov, N.I.: Methods of resource modeling of organizational objects. In: Kwasiborska, A., Skorupski, J., Yatskiv, I. (eds.) ATE 2020. LNITI, pp. 116–130. Springer, Cham (2021). https://doi.org/10.1007/978-3-030-70924-2_10
2. Plotnikov, N.I.: The bases of human-operator dependability theory. Dependability **2**, 94–97 (2015). [publication in Russian]
3. GOST 27.002-2015: Dependability in technics. Terms and definitions. Standartinform (eds), p. 28, Moscow (2016). [publication in Russian]

4. Netes, V.A., Tarasyev, Y.I., Shper, V.L.: Current issues of terminology standardization in dependability. Dependability **2**, 120–123 (2014). [publication in Russian]
5. Netes, V.A., Tarasyev, Y.I., Shper, V.L.: How we should define what "dependability" is. Dependability. **4**, 15–26 (2014). [publication in Russian]
6. Nauka (ed.): Russian Grammar, vol. 1, 2, p. 784, 710, Moscow (1980). [publication in Russian]
7. Zinoviev, A.A.: The Logic of Science. Mysl (ed.), p. 279, Moscow (1971). [publication in Russian]
8. Aristotle. Politics; Metaphysics; Analytics. Eksmo (eds) Moscow; Midgard (eds), Saint Petersburg, p. 960 (2008). [publication in Russian]
9. GOST IEC 60050-113-2015: International Electrotechnical Vocabulary. Part 113. Physics for electrotechnology. Standartinform (eds) Moscow (2016). [publication in Russian]
10. Pospelov, D.A.: Situational management. Nauka (ed.) p. 286, Moscow (1986). [publication in Russian]
11. Klir, G.: Architecture of systems problem solving. Radio i sviaz (ed.) p. 544, Moscow (1990). [publication in Russian]
12. GOST ISO 9000-2011: International standard. Quality management systems. Fundamentals and vocabulary. Standartinform (ed.) p. 43, Moscow (2012). [publication in Russian]

Regulatory Decision of Dependability Terminology

Nikolai I. Plotnikov[1] (✉), Gabriel Gomes de Oliveira[2], and Yuzo Iano[2]

[1] Scientific Research Project Civil, Aviation Institute «AviaManager», Novosibirsk, Russian Federation
am@aviam.org

[2] School of Electrical and Computer Engineering, State University of Campinas, Campinas, Brazil
oliveiragomesgabriel@ieee.org, yuzo@unicamp.br

Abstract. This paper presents the development regulatory decision of dependability theory and terminology. The development is a conceptual description of the automaton as the term structure of a fixed complexity that shows non-contradictory interrelations and clear dependability state transitions of an item. The description of the state structure of the automaton implies the subsequent development of a computing device for monitoring the dependability of items of any nature. The automation is described using terms and symbols from standards, as well as alternative definitions of dependability states. A review of the appropriate standards is to involve a new terminology. The operation of the dependability automaton reflects alternative transitions. Such research is to produce logical proof and substantiation of a set of coordinated, non-contradictory ontological terms: property, state, event, etc. The results will be used in technical standards for the purpose of construction and substantiation of special terms. The paper provides a theoretical and practical substantiation of applying individual provisions of the dependability theory of technology to develop the dependability theory of non-digital entities like representatives of civil aviation: pilots, flight crews.

Keywords: Terminology · Dependability · Reliability · Assignment · Regulation · Automaton

1 Introduction

In this paper, unlike in the standard [1], dependability is defined as a set of states, the measure of concordance with the purpose of an object. The purpose is defined as the property of an object attributed to the natural origin or designed application. Following such definitions, alternative definitions of dependability states have been developed. An observation of the dependability states of an item can be described with a common algorithm. The problem is resolved with the help of the automaton theory [2]. In accordance with the theory of automation, a definition of a dependability automaton is drawn up, which is specific for solving the problems of this work. The structure and algorithms of

Y. Iano et al. (Eds.): BTSym 2021, SIST 207, pp. 609–614, 2023.
https://doi.org/10.1007/978-3-031-04435-9_65

the automaton operation in processes, states, and events have been developed. In conclusion, the theoretical foundations and evidence of practical application in the field of civil aviation are presented.

2 Investigation Solutions in the Automaton Theory

In order to solve the problem, it is suggested to develop a dependability automaton (DA). The development is a conceptual description of the automaton as the term structure of a fixed complexity that shows non-contradictory interrelations and clear dependability state transitions of an item. The description of the state structure of a DA implies subsequent development of a computing device for monitoring the dependability of items of any nature.

2.1 Automaton Definition

Dependability automaton (DA) is designated as D (dependability) of an object as a deterministic, fully specified finite-state automaton. DA is defined by a set consisting of the following elements:

$$DA = \{X, S, Y, \delta, \lambda, s_0\} \tag{1}$$

where DA is dependability automaton;

X is the input alphabet of the automaton (set of input symbols):

$$X = \{x_1, \ldots, x_m\} \tag{2}$$

S is the automaton states:

$$S = \{s_0, \ldots, s_n\} \tag{3}$$

where s0 is the initial automaton state, sn is automaton n-state;

Y is the output alphabet of the automaton (set of output symbols):

$$Y = \{y_1, .., y_p\} \tag{4}$$

where δ is the specified indication of states at a set of input signals, the function of automaton transition from one state into another:

$$s_j = \delta_i(s_i, x_k) \tag{5}$$

where s_j is the subsequent state of the automaton, s_i is the current state of the automaton; x_k is the current input symbol;

λ is the specified indication of states at a set of output signals, the output function:

$$y_l = \lambda_i(s_i, x_k) \tag{6}$$

where y_l is the subsequent output symbol of the automaton, λ_i is i-state in a subset of output signals, s_i is the current state of the automaton, and x_k is the current input symbol.

The conditions are: sets X, S, Y are finite; the output symbol ($y_l \in Y$) depends on the input symbol ($x_k \in X$) and the current state of the automaton ($s_i \in S$); description entries of the automaton are defined at discrete instants in time. The deterministic automaton: a) from state s_i under the influence of signal x_k transitions into state s_i; at the output, y_h changes to y_l; b) for $(x_i, y_i) \in (X, Y)$, δ and λ are defined.

2.2 Structure of Automaton Dependability States

In the automation theory, the properties of items are examined in terms of being in states and transitioning between them. However, in the well-known scientific literature on the theory of automation, there are no basic descriptions and definitions of states, processes, events, their connections, and directions of transitions. In this work, a conceptual state diagram in the form of an elementary graph is developed. Dependability states change in terms of disruption and restoration of item purpose (Fig. 1).

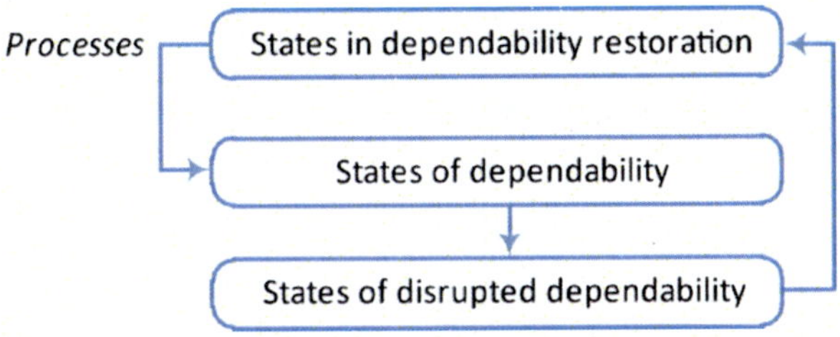

Fig. 1. Observation of dependability states

Such changes can be represented as a directed graph, whose nodes correspond to states, while the edges correspond to transitions between states. As the dependability restoration states are deterministic, they can be represented as processes, i.e. planned, consisting of activities, measures, procedures, operations. The states of disrupted dependability are random, therefore they can be considered as events. Thus, the property of an entity's purpose is observed when the states of dependability are observed as changes in events and processes (Table 1).

Table 1. States of the dependability automaton

Terms	States	Ω
Processes	Maintenance	(s_{eng})
	Repair	(s_{rep})
	Restoration	(s_{rest})
States	Up state	(s_{up})
	Perfect state	(s_{per})
	Imperfect state	(s_{imp})
	Down state	(s_{dw})
Events	Failure	(s_{fail})
	Defect	(s_{def})
	Degraded state	(s_{deg})

The automaton is described using terms and symbols of standards [1, 3, 4] and alternative definitions of states developed in this paper. A review of the appropriate standards is to involve a new terminology. For instance, the definition of the term "defect" clearly does not correspond to the technical sense. In standard [1], "defect" is defined as the non-compliance on an item with the requirements specified in the documentation. In standard [4], "defect" is defined as non-fulfillment of the requirement associated with the presumed or specified use. The basic states in this paper are set forth as follows the development of DA algorithms.

2.3 Development of DA Algorithms

The description of the DA operation consists of the translation of the standard terms into symbolic algorithms suitable for subsequent software development. Let us introduce the following symbols and construct the algorithms:

Dependability processes:

- $\overline{\mathrm{D}}$ is dependability;
- $\downarrow$D are dependability disruptions;
- $\uparrow$D are dependability restorations;
- $(s_j \rightarrow s_i)$ are the transitions from the current state into the subsequent state in dependability disruption events;
- $(s_j \leftarrow s_i)$ are the transitions from the current state into the subsequent state in dependability restoration processes;
- $\overleftarrow{\mathrm{S}} : (s_{rest} \subseteq s_{rep} \subseteq s_{eng})$ are subsets of dependability restoration states (processes);
- $\overrightarrow{\mathrm{S}} : (s_{fail} \subseteq s_{def} \subseteq s_{deg})$ are subsets of dependability disruption states (events);
- $(s_j \rightarrow s_i)|\overrightarrow{\mathrm{S}}$ are transitions amidst dependability disruption events;
- $(s_j \leftarrow s_i)|\overleftarrow{\mathrm{S}}$ are state transitions amidst dependability restoration events.

DA states are shown in the diagram (Fig. 2).

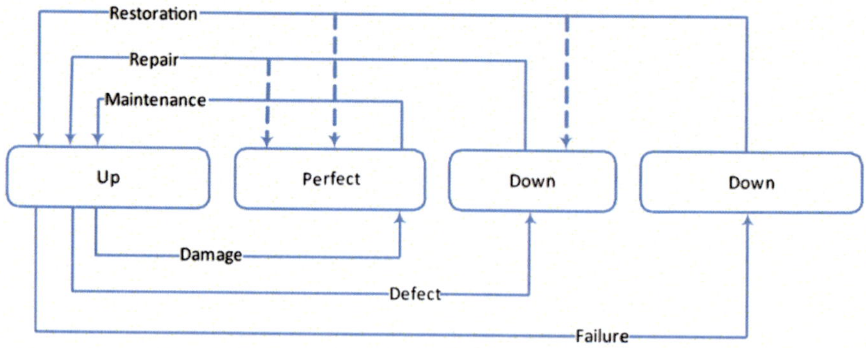

Fig. 2. Conceptual diagram of the dependability automaton

Dependability states:

- $\overline{\mathrm{D}} : s_{up}|\left(s_{up} \leftarrow s_{eng}\left(s_{per} \leftarrow s_{rep}(\leftarrow s_{rest})\right)\right)$ is the up state by maintenance, repairs, recovery condition;
- $\overline{\mathrm{D}} : s_{per}|\left(s_{per} \leftarrow s_{rep}(\leftarrow s_{rest})\right)$ is the good state by repairs, recovery condition;
- $\overline{\mathrm{D}} : s_{imp}|\left(s_{imp} \leftarrow s_{rest}\right)$ is the faulty state by recovery condition;
- $\overline{\mathrm{D}} : s_{up}|s_0$ is the down state.
- States in dependability disruption events:
- $\downarrow$ D: $s_{deg} \mid \ulcorner s_{eng}$is the damage by maintenance condition;
- $\downarrow$ D: $s_{def} \mid \ulcorner s_{rep}$is the defect by no-repairs condition;
- $\downarrow$ D: $s_{fail} \mid \ulcorner s_{rest}$is the failure by no-restoration condition.

States in dependability restoration processes:

- $\uparrow \mathrm{D} : s_{eng}|\left(s_{up} \leftarrow s_{per}\right)$ is the maintenance for a transition from the good state into the up state;
- $\uparrow \mathrm{D} : s_{rep}|\left(s_{up} \leftarrow s_{per}(\leftarrow s_{imp})\right)$ are the repairs for a transition from the faulty state into the up (good) state;
- $\uparrow \mathrm{D} : s_{rest}|\left(s_{up} \leftarrow (s_{per} \leftarrow \left(s_{imp} \leftarrow s_{dw}\right))\right)$ is the restoration for a transition from the down state into the up (faulty, good) state.

The operation of a DA, unlike the standards [1, 3, 4], reflects transitions and alternative transitions. Restoration is designed as a complete and partially incomplete process: a) transition from the down state into the up state; b) transition from the down state into the faulty state; c) transition from the down state into the good state. The states of DA summarize the resource hierarchy in terms of "restoration" $\subseteq$ "repairs" $\subseteq$ "maintenance". However, all technology dependability standards lack a substantiation of the term hierarchy.

3 Theoretical and Practical Implementation of DA

The findings contributed to the development of theoretical [5] and practical [6] dependability of organization, social groups, and individuals (Fig. 3).

The DA concept includes the development of the engineering design of an expert decision support system (ES) for the flight operation of an airline. The ES has functional modules, includes a knowledge base or ES shell, as well as named functional units: information assets transformation system (IATS); module for indicator data analysis and prediction of the states of pilot resources; solver or pilot resources management decision-making module. The DA is represented as the sum of pilot resources (SPR) consisting of three groups of properties of dependability: individual dependability resources (IDR), professional dependability resources (PDR), operational dependability resources (ODR). Such grouping is based upon the structural approach to defining standard terms consisting in the partition of abstract concepts using the example of the category of "dependability". Using that approach, a base of observation in time has been developed: IDR, the time of human evolution, PDR, the time of employment between ages 20 and 60, ODR, the

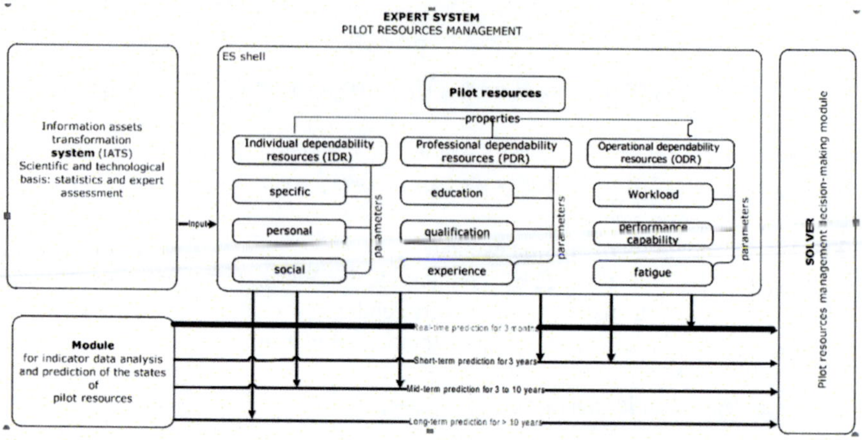

Fig. 3. Dependability automaton of an individual (commercial aviation pilot)

time from the duration of one flight up to a year. The new SPR structure allows defining various norms and limits, which improves flight efficiency and safety supervision [6].

4 Conclusion

Technical standards require prior philosophical, philological, and logical review. Such research is to produce logical proof and substantiation of a set of coordinated, non-contradictory ontological terms: property, state, event, etc. The results will be used in technical standards for the purpose of construction and substantiation of special terms. For instance, why the term "failure" is larger than the term "damage" in terms of scope and content in the physical and technical senses. The paper provides a theoretical and practical substantiation of applying individual provisions of the dependability theory of technology for the purpose of developing the dependability theory of non-digital entities.

References

1. GOST 27.002-2015: Dependability in technics. Terms and definitions. Standartinform (ed.) p. 28, Moscow (2016). [publication in Russian]
2. Vykhovanets, V.S.: Automation theory: a study guide for superior education establishments. Rio, P.G.U. (ed.) p. 124, Tiraspol, Russia (2001). [publication in Russian]
3. GOST IEC 60050-113-2015: International Electrotechnical Vocabulary. Part 113. Physics for electrotechnology. Standartinform (ed.) p. 62, Moscow (2016). [publication in Russian]
4. GOST ISO 9000-2011: International standard. Quality management systems. Fundamentals and vocabulary. Standartinform (ed.) p. 43, Moscow (2012). [publication in Russian]
5. Plotnikov, N.I.: Methods of resource modeling of organizational objects. In: Kwasiborska, A., Skorupski, J., Yatskiv, I. (eds.) ATE 2020. LNITI, pp. 116–130. Springer, Cham (2021). https://doi.org/10.1007/978-3-030-70924-2_10
6. Plotnikov, N.I.: Pilot resources. Reliability. Monograph. AviaManager (ed.) Novosibirsk (2013). [publication in Russian] load: http://aviam.org/images/sampledata/book/2015_prd.pdf

A Python Script to Generate a 3D Model of a Coaxial Cable

Daniel J. C. Pereira[1(✉)], Kenedy Marconi G. Santos[1], Douglas O. Campos[1], Polyane A. Santos[1], Lucas S. Ribeiro[1], Marcelo B. Perotoni[2], Tagleorge M. Silveira[3], Marcela S. Novo[4], and Willian F. S. Maia[5]

[1] Federal Institute of Education, Science, and Technology of Bahia, Vitória da Conquista, Brazil
{daniel.jonathas,douglascampos,lucasribeiro}@ieee.org, polyane@ifba.edu.br
[2] Federal University of ABC, Santo André, Brazil
[3] University of Aveiro, Aveiro, Portugal
tagleorge@ua.pt
[4] Federal University of Bahia, Salvador, Brazil
[5] Federal University of Minas Gerais, Belo Horizonte, Brazil

Abstract. This article presents the techniques used to create a parametric Python script to generate 3D models of braided coaxial cables. The script is designed to run on free and open-source FreeCAD software. FreeCAD has a Python console that allows the expansion of its functionality. The script was developed with the aim of using the generated 3D models to carry out simulations in electromagnetic simulation software. In the area of Telecommunications, for example, it is of fundamental importance to use coaxial cables that are as efficient as possible. Computer simulations allow the analysis of the behavior of coaxial cables, allowing the most efficient cables to be selected. For this, 3D models that can faithfully represent the electrical and mechanical parameters of cables are needed. However, given its constructive characteristics, modeling the cables in 3D is a difficult task. Thus, it is proposed that the developed script is an alternative to obtain 3D models of braided coaxial cables. The parameters used to generate the 3D model are presented, as well as the results of using the script to generate the 3D model of a coaxial cable, model RG58.

Keywords: Coaxial cable · Script · Python · FreeCAD · 3D modeling

1 Introduction

As communication technologies and devices are developed, the electromagnetic noise levels in the environment tend to increase. To guarantee the integrity of the transmitted signal, since the 20th century, coaxial transmission lines [1] are commonly used. Coaxial cables with braided shield generally do not have the

Y. Iano et al. (Eds.): BTSym 2021, SIST 207, pp. 615–622, 2023.
https://doi.org/10.1007/978-3-031-04435-9_66

same shielding efficiency compared to cables with solid shielding [2], however, they have the advantages of being more flexible and having lower production and installation costs. In computer simulations, it is necessary to have a good parameterization of the entire measured and simulated arrangement, since the physical parameters are directly responsible for the electromagnetic performance of cables and antennas. The parameterization of details with reduced dimensions, such as, for example, the holes of stranded coaxial cables, which are commonly of the order of 1 mm^2 in area, is not simple, since the dimensions of the arrangement to be simulated are usually of a few tens of centimeters to a few meters. So, the modeling and simulation of coaxial cables are complex, given that the aspect ratio is too large for an adequate mesh to be laid on the structure. In [3], for example, it was noted the need to use a realistic model that would faithfully represent the simulated coaxial cable geometries. In [1], the author uses software to create three-dimensional models of braided coaxial cables, however, according to the author, to simulate the electromagnetic behavior of the model used, a high level of robustness would be necessary.

Thus, this article aims to present the parametric three-dimensional modeling of a braided coaxial cable using the Python programming language, with the possibility of generating three-dimensional models only by varying the physical parameters of the cable. In addition, the example of using the script for RG-58 coaxial cable is shown.

2 The Script

The script was designed based on the representation of the coaxial cable obtained from [4], displayed in Fig. 1. It was designed in Python language and intended for use in FreeCAD, an open-source parametric 3D modeling software [5].

The parameters used in the cable shaping are:

- Number of segments (N_c);
- Number of filaments in each segment (N_f);
- Diameter of each filament (D_f);
- Diameter of inner conductor (D_c);
- Diameter of internal insulation (D_i);
- Diameter of shield (D_s);
- Diameter of outer insulation (D_e);
- Angle of each filament with respect to the longitudinal direction of the cable (α);
- Cable length (l).

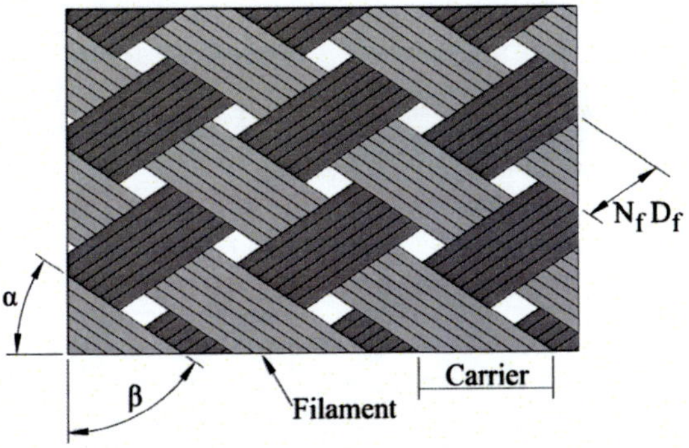

Fig. 1. Detail of the shield of a coaxial cable, adapted from [4]

From the basic parameters, parameters useful in cable modeling are calculated:

- Angle of each filament with respect to cable cross-section (β);
- Step of each filament (p);
- Angle of each segment (γ), illustrated in Fig. 2.

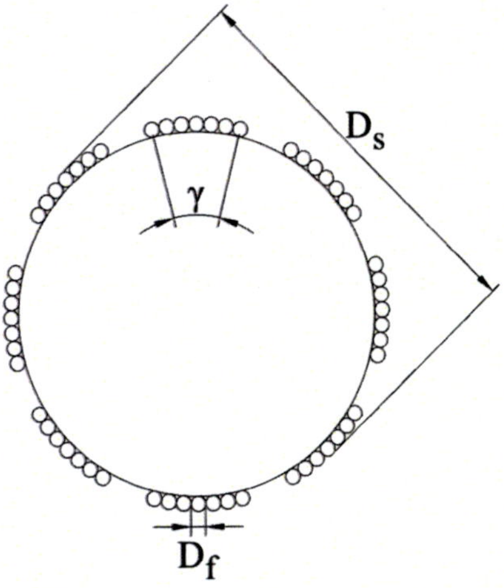

Fig. 2. Each segment angle detail

The angle of each filament in relation to the cable cross-section is given by (1).

$$\beta = 90 - \alpha \tag{1}$$

The step of each filament is given by (2).

$$p = \pi D_s \tan(\beta) \tag{2}$$

The angle of each segment is given by (3).

$$\gamma = 2(N_f - 1) \arctan\left(\frac{D_f}{D_s}\right) \tag{3}$$

A segment is defined in the script as an array of filaments. For creating a filament, in principle, a path represented as a helix is defined. The base of a filament is represented as a circle, which is then extruded through the path created initially. Two types of paths are defined: right-oriented and left-oriented.

To create the filament path, the parameters are inserted in the script according to the Filament Path Creation Code.

Filament Path Creation Code

```
FreeCAD.ActiveDocument.addObject("Part::Helix", "fil")
FreeCAD.ActiveDocument.fil.Pitch = filamentPitch
FreeCAD.ActiveDocument.fil.Height = cableLength
FreeCAD.ActiveDocument.fil.Radius = shieldRadius
```

The base of the filament is built from the Filament Base Creation Code.

Filament Base Creation Code

```
filPos = FreeCAD.Placement(FreeCAD.Vector(shieldRadius,
    0,0), FreeCAD.Rotation(0,0,0,1))
filBase = Draft.makeCircle(radius=filDiameter/2, placement=
    filPos, face=True, support=None)
```

To generate the filament, the base is then extruded along the created path, according to the Filament Creation Code.

Filament Creation Code

```
FreeCAD.ActiveDocument.addObject("Part::Sweep", "Filament")
FreeCAD.ActiveDocument.getObject("Filament").Sections = [
    FreeCAD.ActiveDocument.Circle, ]
FreeCAD.ActiveDocument.getObject("Filament").Spine = (
    FreeCAD.ActiveDocument.Filament, [])
```

The shield is defined as an array of segments. A right-oriented segment and a left-oriented segment are generated, and then an array of each type of segment is made, where the number of cable segments is divided equally between right-oriented segments and left-oriented segments. The following box presents the Code for Creating a Segment.

Code for Creating a Segment

```
Draft.makeArray(FreeCAD.ActiveDocument.getObject( "Filament"), FreeCAD.Vector(0,0,0), carrierAngle, numberFilament)
```

The inner conductor is defined as a cylinder, as shown in the Code for Creating the Internal Conductor. The inner insulation is defined from the Boolean operation of the difference between a cylinder of the diameter of the inner insulation and a cylinder of the diameter of the inner conductor, as shown in the Code for Creating Internal Isolation. External insulation is defined similarly.

Code for Creating the Internal Conductor

```
FreeCAD.ActiveDocument.addObject("Part::Cylinder", "conductor")
FreeCAD.ActiveDocument.getObject("conductor").Radius = conductorDiameter/2
FreeCAD.ActiveDocument.getObject("conductor").Height = cableLength
```

Code for Creating Internal Isolation

```
FreeCAD.ActiveDocument.addObject("Part::Cylinder", "insulationAux")
FreeCAD.ActiveDocument.getObject("insulationAux").Radius = insulationDiameter/2
FreeCAD.ActiveDocument.getObject("insulationAux").Height = cableLength

FreeCAD.ActiveDocument.addObject("Part::Cylinder", "conductorAux")
FreeCAD.ActiveDocument.getObject("conductorAux" ).Placement = FreeCAD.Placement(FreeCAD.Vector(0,0, -cableLength), FreeCAD.Rotation(FreeCAD.Vector(0,0,1), 0))
FreeCAD.ActiveDocument.getObject("conductorAux").Radius = conductorDiameter/2
FreeCAD.ActiveDocument.getObject("conductorAux").Height = cableLength

FreeCAD.ActiveDocument.addObject("Part::Cut", "CutInt")
FreeCAD.ActiveDocument.getObject("CutInt").Base = FreeCAD.ActiveDocument.insulationAux
FreeCAD.ActiveDocument.getObject("CutInt").Tool = FreeCAD.ActiveDocument.conductorAux
```

3 Practical Application

To demonstrate the use of the script, a 3D model of an RG-58 coaxial cable was generated, with the parameters described in Table 1, adapted from [4].

Table 1. RG-58 coaxial cable parameters

Parameter	Value
N_c	16
N_f	7
D_f	0.122 mm
D_c	0.900 mm
D_i	2.950 mm
D_s	3.100 mm
D_e	4.950 mm
α	26°

Figure 3 shows the shield produced by the script from the data in Table 1.

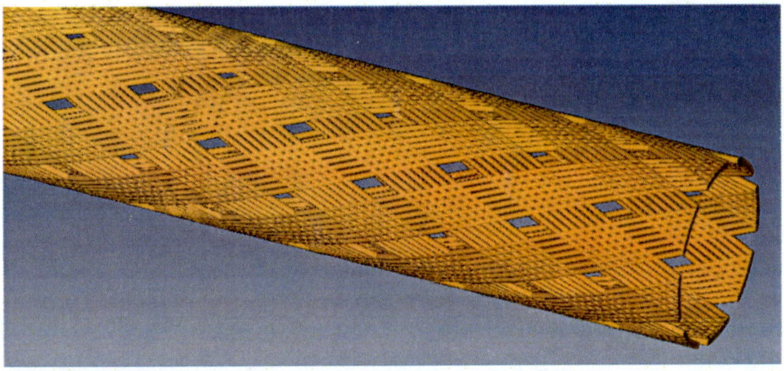

Fig. 3. RG58 cable shield generated from script execution in FreeCAD

Running the script with the data of Table 1 in FreeCAD, the result presented in Fig. 4 was obtained. The model generated in FreeCAD through the script can be exported in the most diverse formats, to be, later, imported into electromagnetic simulation software. Supported formats include STL, OBJ, DAE, STEP, among many others, which are easily imported into the main electromagnetic simulation software available on the market. The definitions of the materials are carried out directly in the software chosen to carry out the simulations so that the developed script only presents definitions of the format and dimensions of the cable.

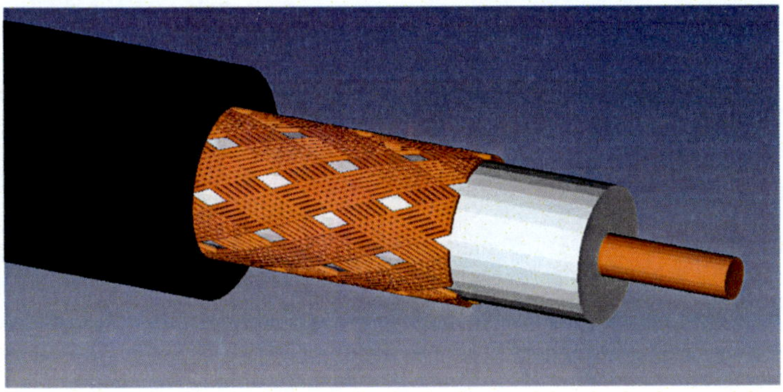

Fig. 4. RG58 3D model generated from script execution in FreeCAD

4 Conclusion

The script is useful to generate 3D models of braided coaxial cables, being one of the possible solutions in situations in which the realization of computer simulations using realistic models for coaxial cables proves necessary, as, for example, the situation presented in [3]. Also, since Python is a free programming language and simple to learn, the script is easy to maintain. The Python language is extremely powerful and versatile. It is cross-platform and can be expanded through numerous libraries. With Python, it is even possible to perform electromagnetic simulations, using the FDTD simulator [6]. FreeCAD is open-source software, which is related to the risks that it will be discontinued in the near future and the script will become unusable [7].

As suggestions for future work, we can list: theoretical validation of the model created from its evaluation in software for electromagnetic simulation, such as CST Microwave Studio, FEKO, and HFSS, and the creation of an interface easy-to-use graphics for FreeCAD, so you do not need prior knowledge to use the script.

References

1. Pornin, C., Vuong, T.P., Xavier, P., Angenieux, G.: Shielding effectiveness assessement of a coaxial cable design with a combination of numerical and analytical solutions. In: 2019 International Symposium on Electromagnetic Compatibility - EMC EUROPE, pp. 1061–1065. IEEE (2019). https://doi.org/10.1109/EMCEurope.2019.8872020
2. Santos-IFBA, K.M.G., et al.: Electric field radiated by the RG-58 coaxial cable and connectors versus cellflex coaxial cable and connectors with total shielding. In: 2019 SBMO/IEEE MTT-S International Microwave and Optoelectronics Conference (IMOC), pp. 1–3. IEEE (2019). https://doi.org/10.1109/IMOC43827.2019.9317632
3. Otin, R., Verpoorte, J., Schippers, H.: Finite element model for the computation of the transfer impedance of cable shields. IEEE Trans. Electromagn. Compat. **53**(4), 950–958 (2011). https://doi.org/10.1109/TEMC.2011.2146257
4. Xiao, P., Du, P.-A., Zhang, B.: An analytical method for radiated electromagnetic and shielding effectiveness of braided coaxial cable. IEEE Trans. Electromagn. Compat. **61**(1), 121–127 (2019). https://doi.org/10.1109/TEMC.2018.2814629
5. Jürgen, R.: FreeCAD: Your own 3D parametric modeler (2001)
6. Laporte, F.: Python 3D FDTD Simulator (2021)
7. Morgan, L., Finnegan, P.: Benefits and drawbacks of open source software: an exploratory study of secondary software firms. In: Feller, J., Fitzgerald, B., Scacchi, W., Sillitti, A. (eds.) OSS 2007. ITIFIP, vol. 234, pp. 307–312. Springer, Boston, MA (2007). https://doi.org/10.1007/978-0-387-72486-7_33

Analysis and Validation of a Computational Model for a Two Bipolar Drain Wire Cables

Douglas O. Campos[1(✉)], Kenedy M. G. Santos[1], Daniel J. C. Pereira[1], Polyane A. Santos[1], Lucas S. Ribeiro[1], and Tagleorge M. Silveira[2]

[1] Federal Institute of Education, Science, and Technology of Bahia, Vitória da Conquista, Brazil
{douglascampos,lucasribeiro}@ieee.org, polyane@ifba.edu.br
[2] University of Aveiro, Aveiro, Portugal
tagleorge@ua.pt

Abstract. The following article presents the results of an experiment carried out in the electromagnetic compatibility laboratory of SENAI CIMATEC history. Computer simulations of a twinaxial cable were designed with the electric field measurements. The CST Microwave Studio software was used for computational modeling of a twinaxial cable. The result of the simulations was studied according to the influences of different dielectric radii without frequency of electric field radiation, specifically when radiation present is given by the phenomenon of common-mode current. As a result, two resonant frequencies close to 300 MHz and 900 MHz, respectively, and for such values the radiated electric field has greater intensity, so there is a need to use another frequency different from the resonance frequencies obtained, thus cutting the problem and avoiding electromagnetic interference for the 50 cm length of the twinaxial cable used in an on-board electronic system. The practical measurements were carried out in a semi-anechoic chamber, submitting the twinaxial cable in the frequency range from 30 MHz to 1000 MHz.

Keywords: Electromagnetic compatibility · Twinaxial cable · Common-mode radiation · Computational modeling

1 Introduction

Currently, due to the high prices of fossil fuels, electric vehicles are proving increasingly promising as a solution to this problem, becoming more present in the world automobile scenario.

The considerable increase in embedded systems in vehicles, aircraft, and even electronics, as well as the increasing technological advances in electronic and telecommunications systems, show the importance of Electromagnetic Compatibility - EMC. There are several techniques and equipment that reduce the emission of unwanted electromagnetic waves. The use of shielded cables, for

Y. Iano et al. (Eds.): BTSym 2021, SIST 207, pp. 623–629, 2023.
https://doi.org/10.1007/978-3-031-04435-9_67

example, becomes fundamental in this context, as it is essential in the transmission/reception of signals, mobile telephony, TV, and radio AM/VHF - FM [1].

Specifically to electric vehicles one of the components of studies in which the EMC acts around the frequency inverter, which is responsible for controlling the rotation speed and torque of the electric motor. In general, inverters are immediately associated with a source of electromagnetic noise in which they may have the potential to cause disturbances in the functioning of components of the electronic modules of the onboard vehicle. In order to minimize these effects during the communication between the inverter and the onboard exchanges, shielded conductors 2 are used.

In 3, it is proposed the development of an analytical technique to evaluate the radiated electric field and the shielding efficiency of blind cables whose shielding is perforated. For the validation of the technique used, the entire array is compared with computer simulations using the CST software, the same used in this work.

In 4, it is stated that electromagnetic radiation from a cable is one of the most important factors for electromagnetic disturbances in frequency inverters. In the case of electric cars, such disturbances can affect the integrity of the vehicle if there is a leakage of an electromagnetic field interfering with the onboard electronics. Thus, the analysis of unwanted electromagnetic emission is made from calculations and simulations presented below.

Thus, this article aims to present a modeling of a biaxial cable, analyzing the data obtained from measurements performed in a semi-anechoic chamber and comparing practical and simulated results between 30 MHz and 1000 MHz.

2 Common-Mode Radiation

The presence of a dielectric between two conductors makes it possible to analyze the shielded conductor as a cylindrical capacitor. Figure 1 shows the cable under a ground plane and the parasitic distributed capacitance formed between the opening and ground plane. It is known that the reactance for a capacitive coupling can be modeled by Eq. (1). At high frequencies, this coupling provides a displacement current flow across the capacitance and returns the source through the conducting ground plane [5].

For low frequencies, the use and arrangement of the system that includes a shielded cable become trivial, as only the current in a differential mode acts on the conductor. For high frequencies, if the shielding is not efficient, it is possible that the capacitive coupling between the inner conductors of the biaxial cable and the ground plane close to the conductor may occur, and consequently, a common-mode current distribution, responsible for the unwanted radiation, arises.

$$X_c = \frac{1}{j\omega C} \tag{1}$$

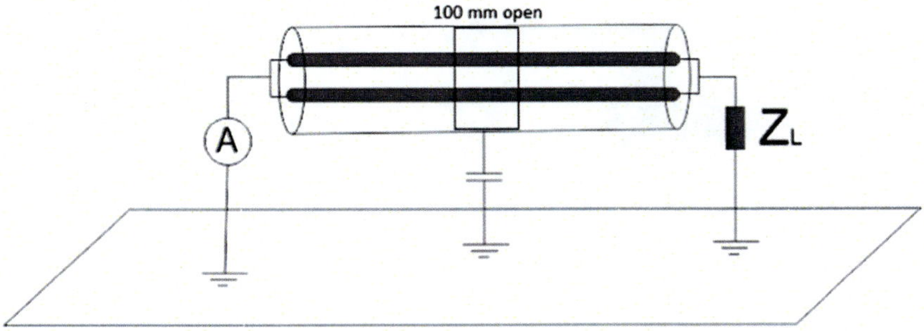

Fig. 1. Two bipolar Drain Wire cables (2 × 0.5 mm^2 with 100 mm aperture) with a capacitive coupling between the shield and the ground plane

Considering a shielded cable subjected to a sufficiently high-frequency signal, and analyzing the relationship between electromagnetic radiation and common and differential-mode current, it is observed that the differential-mode current flows from the source to the inner conductor, and returns to the source in the opposite direction, so as to nullify the resulting magnetic field. Thus, for a biaxial cable with an excellent shielding factor, there is only differential-mode current and no electromagnetic wave radiation. With small openings in the biaxial cable mesh and the capacitive coupling with the ground plane, a common mode current flows between the conductors and the ground.

Even considering its small amplitude current, compared to differential-mode, the common-mode can radiate and compromise the proper functioning of an electronic system. Figure 2 presents a circuit model for the proposed problem. In the existence of common-mode radiation, the cable can be modeled as a monopole or dipole antenna [6], and its electric field is calculated by the Eq. (2).

$$E = \frac{4\pi 10^{-7} f \cdot l \cdot I_{cm} \cdot sin(\theta)}{r} \tag{2}$$

where:

- E = Electric field;
- f = Frequency;
- I_{cm} = Common-mode current;
- l = Cable length;
- r = Distance between the reference point and the cable;
- θ = Angle between the reference point and an antenna.

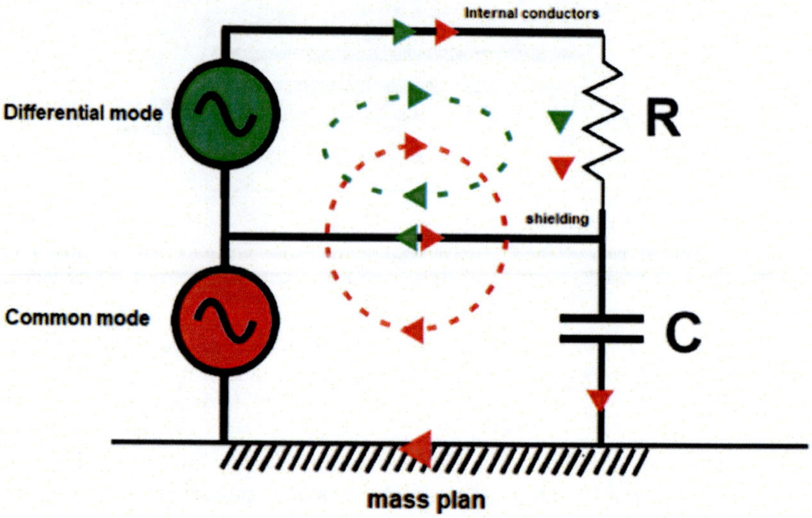

Fig. 2. Common and differential mode current

3 Methodology

Initially, electric field measurements were carried out from 30 MHz to 1000 MHz, in a Semi-Anechoic Shielded Chamber, with an arrangement composed of a biaxial cable on a ground plane. A metallic ground plane was used, whose function was to create a capacitive coupling in the system, providing the flow of common-mode current. Zinc material was used to make the arrangement. The cable used in this configuration had two inner conductors and an outer shield.

To facilitate data collection, a 100 mm opening was made in the cable shield, as suggested in [5]. Figure 3 shows the proposed arrangement for laboratory tests. The system was supplied with a constant power of 50 W, and the radiated electric field was measured at a distance of 50 cm from the cable using an electric field probe. Made as measured, in sequence, the CST Microwave software was used to model the arrangement used in the indications, shown in Fig. 4. The measurement point used in the simulation was also positioned at a distance of 50 cm. After collecting the measured and simulated data, a comparison was made using both curves to validate the proposed modeling.

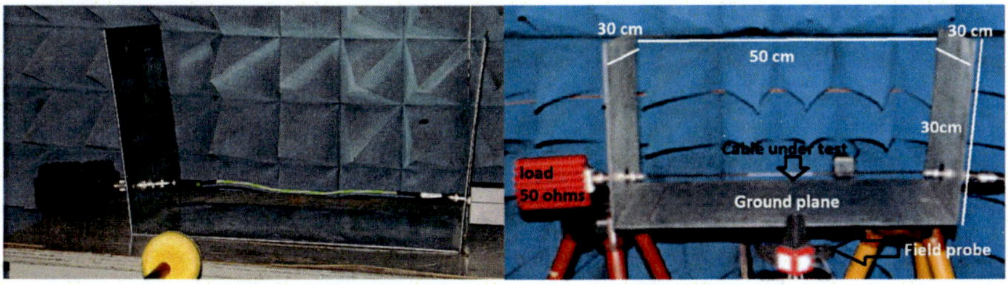

Fig. 3. Two bipolar drain wire cables Condumax above the ground plane with the aperture

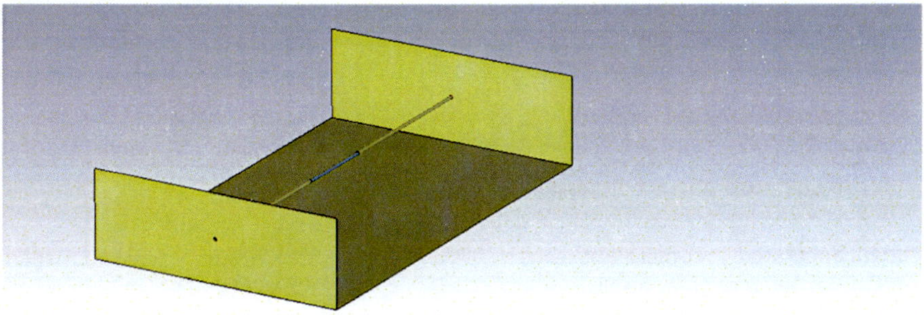

Fig. 4. Virtual model of two bipolar drain wire cables

4 Results

From the measured and simulated data, it is possible to observe a satisfactory degree of coherence between them, as shown in Fig. 5. However, there is a small frequency shift, something that is related to the imprecision in the assembly of the experimental arrangement. In other words, the experimental arrangement does not count as exact measurements of the simulated model, since precise cutting equipment was not used. In this arrangement, constant power was also used in the supply of this system.

From the experimental data, it is possible to observe that the evaluated frequency ranges have two resonances. At around 300 MHz and around 900 MHz, the measured electric field takes on considerably large values. Resonance frequencies indicate ranges in which the use of the cable is not recommended due to the high levels of emitted noise. It is possible to assess that it is prudent to use this type of cable with this length at frequencies other than the resonance frequencies, as high levels of an electric field can interfere with the operation of other electronic devices. Thus, electromagnetic compatibility would not be impaired. In the electromagnetic analysis of noise from a vehicle, for example, or from any other equipment, a simulation that respects the length of the cables is necessary, since this variable modifies the resonant frequency range.

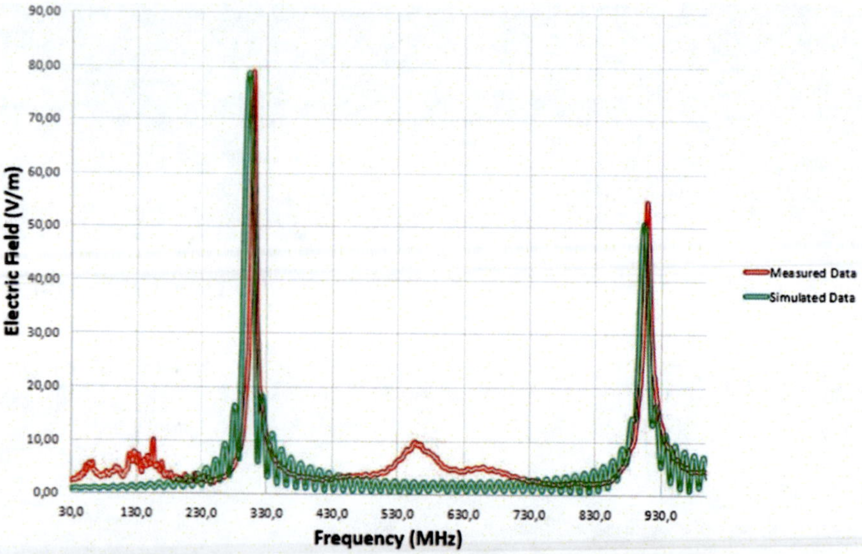

Fig. 5. Simulated and measured electric field - Cable with constant shield and small aperture of 100 mm (L = 50 cm)

5 Conclusion

This article presented a computer simulation using the CST Microwave software, where a good agreement was found between practical and simulated data in the frequency range between 30 MHz and 1000 MHz.

More specifically, Fig. 5 presented the results of the proposed modeling, resulting in two resonance frequencies close to 300 MHz and 900 MHz. They represent frequencies where the radiated field has greater intensities, having the potential to affect the operation of the electronics onboard cars. Electrical devices, having as a contour for this problem would be to use such a cable with 50 cm for a resonance frequency different from the one found.

In view of the electromagnetic simulations performed computationally, they allow a much lower cost than the practical tests carried out in electromagnetic compatibility laboratories. Thus, computational modeling becomes increasingly relevant, making errors in the development of functions that can present EMC problems, in which it is possible to identify the frequency range in which the radiation reaches critical levels, affecting the embedded devices present in electric vehicles.

Acknowledgment. The authors would like to thank SENAI-CIMATEC plant in Salvador-BA (Brazil) for the use of semianechoic chamber and devices. CAPES/CNPq and FAPESB for the financial support.

References

1. Santos, K., et al.: Evaluation of surface transfer impedance of coaxial cables. IEEE Lat. Am. Trans. **18**(03), 598–603 (2020). https://doi.org/10.1109/TLA.2020.9082732
2. Boonpirom, N.: The EMI improvement on active filter using common-mode noise circuits balanced method. In: 2015 18th International Conference on Electrical Machines and Systems (ICEMS), pp. 1325–1328. IEEE (2015)
3. Xiao, P., Du, P.-A., Zhang, B.: An analytical method for radiated electromagnetic and shielding effectiveness of braided coaxial cable. IEEE Trans. Electromagn. Compat. **61**(1), 121–127 (2019a). https://doi.org/10.1109/TEMC.2018.2814629
4. Shi, G., Zhang, Y., Liao, Y., Ying, X.: Characteristics of the electromagnetic radiation emission from coaxial cable of a type inverter. In: 2015 IEEE 6th International Symposium on Microwave, Antenna, Propagation, and EMC Technologies (MAPE), pp. 458–461. IEEE (2015)
5. Santos, K.M.G., Novo, M.S., Fontgalland, G., Perotoni, M.B.: A simple test for measuring the shielding effectiveness of coaxial cables and connectors. Int. J. Appl. Electromagn. Mech. **55**(1), 151–166 (2017). https://doi.org/10.3233/JAE-170032
6. Ott, H.W.: Electromagnetic Compatibility Engineering. John Wiley & Sons Inc., New York (2011)
7. Pornin, C., Vuong, T.P., Xavier, P., Angenieux G.: Shielding effectiveness assessment of a coaxial cable design with a combination of numerical and analytical solutions. In: 2019 International Symposium on Electromagnetic Compatibility - EMC EUROPE, pp. 1061–1065. IEEE (2019)
8. Xiao, P., Qiu, Y., Liu, Z., Bian, L., Li, G.: A method of analyzing the impact of the wiring parameters on the electromagnetic coupling to PCB inside electronic equipment. In: 2019 12th International Workshop on the Electromagnetic Compatibility of Integrated Circuits (EMC Compo), pp. 1–3. IEEE (2019b)
9. Edminister, J.A., Nahvi-Dekhordi, M.: Eletromagnetismo, 3rd edn. Bookman (2013)
10. Paul, C.R.: Introduction to Electromagnetic Compatibility. John Wiley & Sons Inc., New York (2006)
11. Andrade, C.L., da Silva, V.L., Novo, M.S., Santos, K.M.G.: Avaliação da impedância de transferã̂ncia de cabos coaxiais RG213 e RGC213 de fornecedores distintos. In: Anais do XXIII Simpósio Internacional de Engenharia Automotiva. Editora Edgard Blücher, São Paulo, pp. 720–729 (2015)

Author Index

A
Acero, Julio Fredy Chura, 561
Akabane, Ademar Takeo, 97
Akkari, Alessandra Cristina Santos, 334, 378, 428, 521
Alarcón, Roger, 494
Almeida, Alecssander Daniel de, 239
Altamirano, Segundo Francisco Segura, 443
Alvarado-Alfaro, Julia, 569
Alvarez, Alejandro, 123, 479
Alves, Gisele Letícia, 436
Anari, Shokofeh, 386
Apaza, Fredy Bernardo Coyla, 561
Aquino, Janet, 494
Araujo, Ana Caldas, 487
Arizaca-Cusicuna, Jorge L., 412, 420
Arteaga-Llacza, Pedro, 529, 535
Arthur, Rangel, 370
Atco, Maximo Amancio Moltalvo, 569
Ayala, Diego, 403
Ayamamani, Pascual, 543
Ayuba, Nuhu, 151
Azarias, Fábio Luis Fernandes, 115

B
Bacega, Paolo Rodrigo de Oliveira, 73
Bahmanyar, Danial, 56
Barra, Fanny Sofia Quispe, 592
Barra, Marco A. Quispe, 584, 592
Barra, Marco Quispe, 569
Bartolo Gotarate, Félix, 301
Behrens, Frank Herman, 143
Bendechache, Malika, 386
Benites, Cristian, 412
Bernaola, Merly Liliana Yataco, 443
Bianchini, David, 163, 172
Bonello, Daniel Katz, 198
Boñón, Cesia, 469
Brancalhone, João, 35
Bravo, Jessie, 494
Buratto, Davi F., 143

C
Caballero-Apaza, Luz Marina, 551
Cabrera, Lucia Isabel Chamán, 308, 314
Callata, Leonidas Vilca, 561
Campos, Douglas O., 615, 623
Carbajal, Ricardo, 479
Cardenas, Diana Mercedes Castro, 443
Cardeñoso, Franklin, 412
Carrión-Barco, Gilberto, 286
Carvalho, Bruno Campos Simoni de, 73
Carvalho, Marcius F. H. de, 349
Castillo, Pedro Felipe Arce, 436
Castillo-Castro, Vanesa, 592
Castro Cárdenas, Diana Mercedes, 249
Castro, Consuelo Ivonne Del Castillo, 308
Cedrón-León, Yris, 592
Cervantes-Alagón, Sheyla Lenna, 551
Cherre, Joseph, 123
Chuma, Euclides Lourenço, 73, 229, 269
Chuquilín-Goicochea, Roberto, 451, 529, 535
Colomeu, Talita Cristina, 269
Cruz, Gabriel, 412

D
da Silva, Lara Poliny Nogueira, 342
Davila-Andrade, Henry L., 420

Y. Iano et al. (Eds.): BTSym 2021, SIST 207, pp. 631–633, 2023.
https://doi.org/10.1007/978-3-031-04435-9

de Andrade, Alexandre Acácio, 458
de Andrade, Fábio Ramos Dias, 342
de Freitas Gomes Hernandez, Marli, 88
de Gea Neves, Marina, 269
de Godoy Ricci, Marlucy, 428
De La Cruz-Marcos, Ruggerths, 451, 529, 535
de Mello Filho, Luiz Vicente Figueira, 276, 294
de Mendonça, Bárbara Santiago, 106
de Moraes Gomes Rosa, Maria Thereza, 428, 436
de Oliveira, Gabriel Gomes, 1, 16, 25, 35, 56, 163, 172, 198, 213, 229, 257, 276, 386, 598, 609
de Souza Lemos Farali, Larissa, 458
Del Castillo Castro, Consuelo Ivonne, 314
Demanboro, Antonio Carlos, 163, 172
Díaz, Silvia Juana Saavedra, 308, 314
do Carmo, Lucas Antônio Gomes, 44
dos Santos, Talison Alvarenga, 502
Durini, Daniel, 403

F
Fabián-Ramírez, Jenner, 577
Ferreira, Denise Helena Lombardo, 44, 188
Flores, Christian, 123, 133
Flores, Luisa, 543
Flores, Patricia Viveros, 487
Frazão, Danilo, 363
Fuentes, Denny John, 286

G
Galvani, Italo, 294
Germán, Nilton, 494
Gonçalves, Caroline Luiz, 334
González Cusi, Bryan Martin, 322
Guimarães, Daniela Helena Pelegrine, 428, 436

H
Hoyos-Gallardo, Segundo Manuel, 584
Huamán-Tovar, William, 451

I
Iano, Yuzo, 1, 16, 25, 35, 73, 163, 172, 198, 213, 229, 239, 257, 276, 349, 598, 609
Infanson, Rosa Estrella Pillman, 569
Ito, Julia Yumi, 378
Izario, Daniel, 35
Izario, Karine, 35

J
Jácobo-Domínguez, Verónica Náthaly, 469
Jafarzadeh Ghoushchi, Saeid, 386
Junior, Luiz Ariovaldo Fabri, 88

K
Kolecha, Higor de Paula, 349

L
Laraia, Lilian Regis, 276
Li, Fiorella, 494
Libório, Luis Augusto, 370
Lima, Leonardo Bruscagini de, 239
Lloclla, Henry Shuta, 561
Lopes, Gabriela Cantarelli, 151
Lopez, Joshi, 479
Lora-Loza, Miryam, 569, 577, 584, 592

M
Maia, Willian F. S., 615
Mamani, Raul, 412
Mamani-Cori, Vilma, 551
Manhiça, Jemis Dievas José, 97
Maquen-Niño, Gisella Luisa Elena, 286
Maria, Castro Uceda, 577
Medina, Fiorella Medina, 487
Mendes, Jakeline Pertile, 188
Minango, Juan, 229
Minango, Pablo, 229
Miranda-Chávez, Helí, 529, 535
Miranda-Jara, Angélica, 529, 535
Monga, Himanshu, 1
Mota, Lia Toledo Moreira, 115
Motta, Gustavo Bertozzi, 239
Murray, Victor, 479

N
Negrete, Juan Carlos Minango, 25, 257
Negrete, Pablo David Minango, 25, 257
Neto, Adolfo Blengini, 349
Neto, Umberto Bonello, 198
Neves, João Emmanuel D.'Alkmin, 88
Nobre, Augusto Gonçalves, 342
Noel, Julien, 123
Noritomi, Pedro Y., 239
Novo, Marcela S., 615

O
Oliveira, Gabriel Gomes de, 73, 239, 349

P
Padilla-Cáceres, Tania Carola, 551
Paes, Vanessa Marques, 521
Pajares, Luis Alberto Chunga, 308, 314
Palanivel, G., 1
Parhizkar, Mahboub, 386
Passos, Gabriela Vieira, 428
Pedro, Paulo Sérgio Martins, 88
Peralta, Joaquin, 133
Pereira, Daniel J. C., 615, 623
Pereira, Júlio César, 370
Pereira, Luciana, 334
Perotoni, Marcelo B., 615
Pinedo, Davis Alberto mejía, 569

Plotnikov, Nikolai I., 598, 609
Podeleski, Fabiana Silva, 349

Q
Querido, Amanda Faria, 436
Quispe, André Torres, 584
Quispe, Diana, 543
Quispe, Reencarnacion, 412
Quispe-Barra, Marco, 577
Quispe-Boza, Noemi, 535
Quispe-Solano, Miguel Ángel, 529, 535

R
Rajinikanth, V., 56
Rangel-Magdaleno, Jose, 403
Ranjbarzadeh, Ramin, 386
Razmjooy, Navid, 56
Rengifo-castillo, Jhon, 584
Ribeiro, Anderson Orzari, 502
Ribeiro, Lucas S., 615, 623
Rodrigues, Lásara Fabrícia, 106
Rodriguez, Herbert Lazo, 487
Rodríguez-Vega, Juan, 569, 577, 584, 592
Roque-Velásquez, Betty Yolanda, 469

S
Salcedo, Victor, 412
Salinas-Mendoza, David, 577
Sánchez-Onofre, Arcadio, 451
Santos, Kenedy Marconi G., 615, 623
Santos, Polyane A., 615, 623
Saturno, Miguel, 479
Schwarz, Matias Oliveira, 239
Segura Altamirano, Segundo Francisco, 249
Siegle, Alex Restani, 213
Silva, Edgard, 363
Silva, Maria Luisa Damas, 44
Silveira, Franciane Freitas, 378, 458, 502, 521
Silveira, Tagleorge M., 615, 623
Solano Dávila, Olga Lidia, 301, 322
Solano-Cavero, Jessica Karin, 286
Soubache, I. D., 1
Sugahara, Cibele Roberta, 44, 188

T
Takahira, Ricardo, 276
Tataei Sarshar, Nazanin, 386
Thiagarajan, Y., 1, 16
Torres, Henry, 412

U
Utus-Ramires, Domingo, 529

V
Valdivia, Carlos, 494
Valenzuela, Liliana, 543
Vaz, Gabriel Caumo, 1, 16, 25, 35, 73, 163, 172, 198, 213, 229, 239, 257, 276, 349
Velasquez, Omar Chayña, 561
Villaça, Gabriel Matsumoto, 239
Vitulas Quille, Yasmani, 512

Y
Yataco Bernaola, Merly Liliana, 249

Z
Zancanela, Débora Akemi Ide, 44
Zuloaga-Cachay, José Fortunato, 286